Lecture Notes in Networks and Systems

Volume 111

The series "Lecture Notes in Networks and Systems" publishes the latest developments in Networks and Systems—quickly, informally and with high quality. Original research reported in proceedings and post-proceedings represents the core of LNNS.

Volumes published in LNNS embrace all aspects and subfields of, as well as new challenges in, Networks and Systems.

The series contains proceedings and edited volumes in systems and networks, spanning the areas of Cyber-Physical Systems, Autonomous Systems, Sensor Networks, Control Systems, Energy Systems, Automotive Systems, Biological Systems, Vehicular Networking and Connected Vehicles, Aerospace Systems, Automation, Manufacturing, Smart Grids, Nonlinear Systems, Power Systems, Robotics, Social Systems, Economic Systems and other. Of particular value to both the contributors and the readership are the short publication timeframe and the world-wide distribution and exposure which enable both a wide and rapid dissemination of research output.

The series covers the theory, applications, and perspectives on the state of the art and future developments relevant to systems and networks, decision making, control, complex processes and related areas, as embedded in the fields of interdisciplinary and applied sciences, engineering, computer science, physics, economics, social, and life sciences, as well as the paradigms and methodologies behind them.

**** Indexing: The books of this series are submitted to ISI Proceedings, SCOPUS, Google Scholar and Springerlink ****

More information about this series at http://www.springer.com/series/15179

Tatiana Kolmykova · Ekaterina V. Kharchenko
Editors

Digital Future Economic Growth, Social Adaptation, and Technological Perspectives

Springer

Editors
Tatiana Kolmykova
Russian Federation
South-West State University
Kursk, Russia

Ekaterina V. Kharchenko
South-West State University
Kursk, Russia

ISSN 2367-3370 ISSN 2367-3389 (electronic)
Lecture Notes in Networks and Systems
ISBN 978-3-030-39796-8 ISBN 978-3-030-39797-5 (eBook)
https://doi.org/10.1007/978-3-030-39797-5

This Springer imprint is published by the registered company Springer Nature Switzerland AG
The registered company address is: Gewerbestrasse 11, 6330 Cham, Switzerland

Introduction: Regarding the Future of the Digital Economy

The global and wide interest to new opportunities that appear based on digital technologies has been stimulating—in recent five years—the institutionalization of the practices of their application in all spheres and sectors of economic activities. The growing trust to widely accessible and the leading digital technology, of which the most well-known and perspective is AI, predetermined the directions of technological progress and the path along which socioeconomic development will take place in the future 5 years. There's no doubt that humanity will see the digital future.

The modern stage of development of the global economy is a start of a new wave of the economic cycle. Similarly to the Age of Discovery (formation of the common global economy) in the seventeenth century, complication of the systems of knowledge and information (end of the age of encyclopedias and segmentation of knowledge) in the eighteenth century, change of the spheres of influence in the global economy (establishment of colonial relations) in the nineteenth century, and social unification (overcoming gender and other differences) and invention of new forms of energy (nuclear energy) in the twentieth century, at present—in the twenty-first century—the socioeconomic reality is changes, and its new shades appear.

The deep character of the current digital transformations predetermines the increased risks of their implementation and the necessity for risk management. The multi-aspect character of the risks that emerged in the conditions of the digital economy predetermined four basic directions of scientific research, which are of very high theoretical and practical significance.

First direction: growth of the digital economy. After the 2008 global financial crisis, the world expects acceleration of economic growth, and by 2020 these expectations have reached their peak. A decade of stagnation and economic growth that balanced at the zero level, unstable inflation, and unemployment are to be replaced by quick growth of GDP, deflation, and revival of the labor market. However, despite the common expectations, digital technologies have not yet accelerated economic growth. The reasons of this contradiction are not obvious and require targeted comprehensive elaboration.

However, the essence and methodological approach to determining the competitiveness of economy changed drastically. It is based on digital technologies, the activity, and effectiveness of application of which largely determine not only companies' success in the sectorial markets but also the possibility of entering and presence at most markets. Also, the problem of imbalance of the global economic system became very important. Developed countries are leaders in the global digital transformations, which aggravated the position of developing countries. That's why it is necessary to search for the mechanisms of accelerating their economic growth.

Second direction: social adaptation to the digital economy. Application of digital technologies in professional and entrepreneurial activities and in the practice of households forms a new—digital—type of thinking. It makes a modern human expand the limits of world perception and communications. There's an unprecedented breakthrough in linguistics—together with digital communications between humans there is machine learning—which envisages communications between humans and machines. Modern human has to find his place in the new world, and the task of science is to help him with that.

Third direction: technological perspectives of the digital economy. There's a classification of the spheres of economy according to the character and level of complexity and modernity of the applied technologies. High technologies are most interesting and valuable, as they allow manufacturing complex goods—e.g., optical, pharmaceutical, and technical products.

Around the world, leading technologies are implemented, intellectual companies are created, and "smart" regions are formed. This allows achieving an unprecedented scale of production, efficiency, and effectiveness—but also causes high technological risks. Autonomous nature of robots causes the risk of their insufficient control and losses of human control over AI. Quick growth of the markets of digital technologies will sooner or later be replaced by a downward phase of the economic cycle—which will lead to a crisis. Thus, it is necessary to search for alternative directions of technological development for diversification of technological progress and provision of its anti-crisis effect.

Fourth direction: ecological costs of the digital economy. This direction is derived from the previous three directions. The problems of climate changes have been in the focus of humanity's attention for a lot of decades, and, though they were reflected in the adopted global goals in the sphere of sustainable development and the activities of international organizations, they remain unsolved and they continue growing. One cannot ignore the coming ecological crisis, as this could set back the achieved progress and put into doubt the perspective of supporting the usual way of living.

All these directions are studied in the volume "Digital Future: Economic Growth, Social Adaptation, and Technological Perspectives". The volumes include the most interesting and original works of the fundamental and applied character as a result of the multidisciplinary international conference "Climate Changes and Economy of the Future: Global Transformation", which was held on November 13–14 in Pskov State University (Russia).

The book contains four parts. The part I is devoted to Growth and Development of the Digital Economy. It studies the issues of acceleration of economic growth, management of the global digital competitiveness of economy, and provision of balance of economic systems. The part II considers the Social Consequences of the Digital Future and Perspectives of Adaptation to Them. It also considers the issues of the role of human in a new socioeconomic context, formation of a favorable legal field, and formation of digital communications.

The part III contains chapters on Technological Perspectives and Future Horizons of Development of the Digital Economy, in which the authors try to determine the future vectors of development of the digital economy and technological risks and threats to its stability and sustainability. The part IV dwells on Ecological Costs of the Digital Future, the Problem of Climate Changes, and its solution with the help of economic, social, and legal mechanisms.

Contents

Part I
Growth and Development of the Digital Economy: A View into the Future

Chapter 1
About the Place and Role of Risk Management in the Adoption of Management Decisions at the Enterprise

Yaroslav S. Potashnik, Viktor P. Kuznetsov, Ekaterina P. Garina, Elena V. Romanovskaya and Natalia S. Andryashina

Abstract Objective of work: Present the nature and characteristics of approaches to the place and role of risk management in the design and implementation of enterprise management decisions, as well as the effectiveness of this approach. The degree of scientific elaboration of the problem. Theoretical, methodological and methodological issues of risk management are reflected in the writings of many specialists, including: V. A. Akimova, A. G. Badalova, K. V. Baldin, T, Barton, S. W. Valditsev, M. B. Gracheva, G. B. Kleiner, N. Taleb, M. G. Laposta, F. H. Nayta, V. L. Tambovtsev, T. Flynn, B. Hedges, D. Hubbard and others. Research Methodology: methods such as system approach, logical analysis, and synthesis, abstraction, sociological survey, method of expert assessments were applied. Results of the research. Fragmented and inclusive approaches have been identified. It was noted that in a fragmented approach, risk management was seen as a complementary, discrete process. Due to a number of drawbacks, the use of a fragmented approach can lead to significant losses for enterprises. Within the framework of an inclusive approach, risk management is the main process, along with others, of developing and implementing management decisions. In doing so, risk managers are involved as deeply in decision-making as possible, starting with the first stage of decision-making. Conclusion: Identified approaches differ in the place and role of risk management in the design and implementation of enterprise management decisions. A fragmented

Y. S. Potashnik (✉) · V. P. Kuznetsov · E. P. Garina · E. V. Romanovskaya · N. S. Andryashina
Minin Nizhny Novgorod State Pedagogical University, Nizhny Novgorod, Russia
e-mail: yaroslav.sandy@mail.ru

V. P. Kuznetsov
e-mail: kuzneczov-vp@mail.ru

E. P. Garina
e-mail: e.p.garina@mail.ru

E. V. Romanovskaya
e-mail: alenarom@list.ru

N. S. Andryashina
e-mail: natali_andr@bk.ru

T. Kolmykova and E. V. Kharchenko (eds.), *Digital Future Economic Growth, Social Adaptation, and Technological Perspectives*, Lecture Notes in Networks and Systems 111, https://doi.org/10.1007/978-3-030-39797-5_1

approach is traditional for many businesses, requiring less attention from top management and implementation costs. At the same time, it is less effective. An inclusive approach is more costly, but significantly improves the quality and effectiveness of management decisions.

Keywords Management decision · Risk management · Efficiency · Enterprise

JEL Code L6 · M5

1.1 Introduction

Modern enterprises make most management decisions under conditions of uncertainty. The success of such solutions depends to a large extent on the extent to which risks have been taken into account and effectively addressed. Due to the high importance, scientists and practitioners proposed numerous methods that allow identifying, evaluate and implement a response to risks. The legal and regulatory framework for enterprise risk management has been developed. The works of specialists are regularly published, expanding ideas about the nature and management of risks, including such authors as B. Hedges, N. Taleba, D. Hubbard and others. However, until now there are, in our view, some methodological issues that are not fully reflected in the publications. In particular, approaches that define the place and role of risk management in the design and implementation of management decisions, and characteristics of these approaches and the effectiveness of their application need to be clarified. This article is devoted to the study of this issue.

1.2 Theoretical Basis of Research

The Management Decision is a choice that a manager must make in order to fulfil the responsibilities of the post [1]. Typically, the manager chooses from pre-defined options to achieve the purpose for which a management decision is made. As a rule, selected and further implemented actions, which, in the opinion of the manager, will allow to achieve the most desired goals with the least cost and negative consequences.

Management decisions in enterprises can be made in conditions of certainty, risk and uncertainty [2]. Certainty is called conditions when it is possible to predict with high accuracy the outcome of a given course of action, i.e., there is a very high probability that there will be one pre-calculated outcome. For example, if an enterprise places surplus money in a bank deposit of a very sustainable bank, at the end of the deposit, an enterprise with a very high probability will return its money with a pre-known interest. There are few decisions taken with certainty in business practice. They are often elements of larger, more comprehensive solutions.

Risk conditions are mentioned when the outcome of the action option cannot be accurately predicted because different outcomes are possible, but it is possible to determine the likelihood of its occurrence through mathematical processing of statistical data. For example, in conditions of risk, decisions are usually made on insurance of the property of the enterprise, the frequency of repair and replacement of equipment, etc. [3].

Uncertainty is when the outcome of the course of action may be multiple and there is no mathematical calculation of the probability of each of them at the decision-making stage. In such circumstances, for example, it is usually necessary to decide whether to start producing and selling a new product for the enterprise or to enter a new market.

As practice has shown, in view of rapid and often unpredictable changes within and outside enterprises, the need for management decisions within a very short time frame, most management decisions are now made by enterprises under conditions of uncertainty [4]. In order to make rational and correct choices in these circumstances, decision makers need to know what risks are associated with a given course of action, i.e., random events which, if any, may affect the outcome. It is necessary to imagine what is and what the magnitude of the impact is and what can be done to influence the likelihood and impact of risks to the achievement of goals. The search and processing of this information is carried out within the framework of activities that are commonly identified as risk management. The main stages of risk management as a process are presented in [5].

Depending on the novelty and scope of the management decision, its relevance to the enterprise, risk management may be performed by the manager (his team), the decision maker or the enterprise's risk management specialists. They may also involve external consultants. The first two approaches have traditionally been preferred because they are simpler in an organization, less costly, to better manage risks and to develop internal competencies in risk management [6].

1.3 Methodology of Research

During the study the following methods were used: Systemic approach, logical analysis and synthesis, abstraction, sociological survey, method of expert assessments.

1.4 Analysis of the Study Results

Research into the literary sources and activities of Nizhny Novgorod enterprises enabled the authors to identify two main approaches to the role and place of risk management in the design and implementation of management decisions: fragmentation and inclusion.

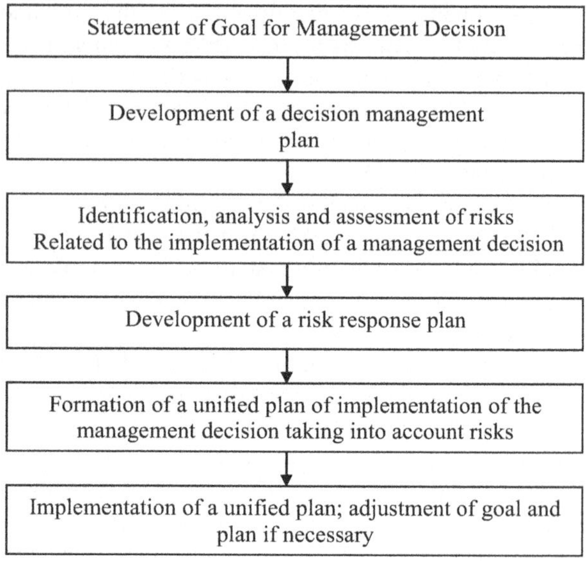

Fig. 1.1 Place of risk management in the development and implementation of a management solution according to a fragmented approach

In the framework of the fragmented approach, risk management is considered as an auxiliary process carried out discretely, i.e. only at certain stages of the development and implementation of management decisions. The place and content of risk management within the framework of the fragmented approach can be enlarged in the form of a diagram (Fig. 1.1).

As part of a fragmented approach, the goal and implementation plan for the management solution are developed without taking into account or minimizing risks. Consideration of the risks begins with the third phase and involves the identification, analysis and assessment of the risks associated with the management decision. If the risk review indicates an unacceptably low probability of achieving the goal and implementation of the decision in the specified parameters, the goal and/or implementation plan of the decision shall be adjusted. Further, a risk response plan is developed, which becomes part of a single plan for the implementation of the management decision. A single plan is then implemented with an adjustment if necessary.

A fragmented concept of risk management corresponds to a functional approach to enterprise management and, as the study conducted by the authors showed, is currently quite common. In our view, however, it has a number of shortcomings, including [7]:

(1) As management goals and implementation plans are formulated prior to a risk study, objectives are often initially set and plans are developed with an unacceptably low probability of success; hence, there is a general need to re-evaluate management goals and plans with corresponding resource losses.
(2) Risk management professionals are not involved in decision-making from the outset and are seen by other actors as outsiders to the process, impeding rapid

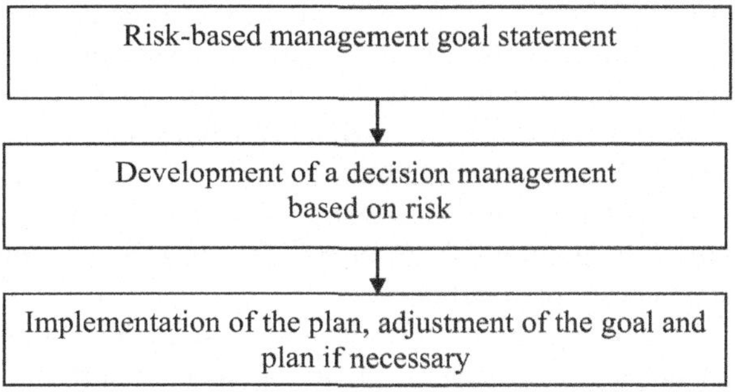

Fig. 1.2 The place of risk management in the development and implementation of management decisions in accordance with an inclusive approach

decision-making and implementation, thus providing a basis for conflict and productivity reduction.

An inclusive approach can, in our view, significantly minimize the shortcomings described above. The place and content of risk management within the framework of an inclusive approach is presented in Fig. 1.2.

In an inclusive approach, possible risks have already been taken into account in the formulation of goals and development of the management solution implementation plan. The underlying purpose and plan of the assumption (e.g. concerning the level of consumer demand, different types of costs, cost and affordability of funds, staff and other) are tested for validity and determined the respective distribution of probabilities [8]. Risks are identified, analyzed and evaluated. Determines whether the alternative formulation of goals and plans is realistic. The ones that meet requirements most in terms of achieving the desired and feasibility are selected.

In terms of implementation of the administrative decision specifying the measures to respond to risks [3], sequence, time and cost of implementation, and responsible employees of the company (Fig. 1.2).

Inclusive concept corresponds to the project approach to enterprise management. As part of an inclusive concept, risk management specialists participate in the development and implementation of management decisions from the very beginning and

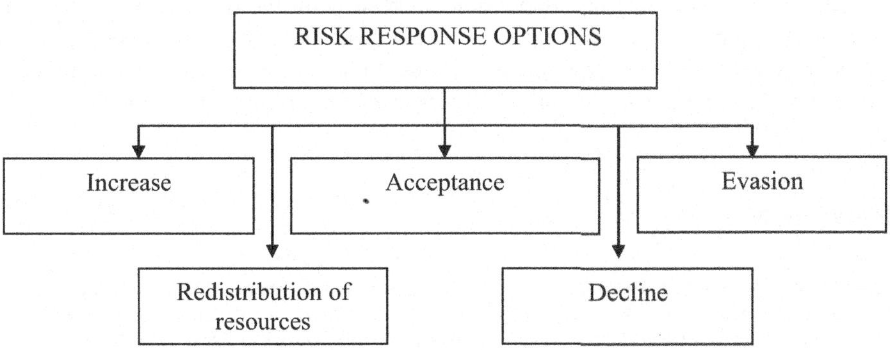

Fig. 1.3 Risk response options

are one of the main participants in these processes. Risk management specialists contribute to a culture of enterprise risk management, are responsible for developing and implementing an effective risk management methodology in different areas of organizational decision-making [9]. Help to test assumptions, identify risks, analyze and evaluate them (quantitative and/or qualitative). Develop and participate in the implementation of risk response measures. Together with direct implementers, continuously monitor risks throughout the implementation period and, where necessary, adjust past goals and plans.

According to the research conducted by the authors [10, 11], the practical use of an inclusive approach to risk management requires higher skills from risk managers, deep awareness and involvement in all spheres and business processes of enterprises, support of their activities by the management of enterprises. Enterprises often have to bear additional costs, for example, involving more competent staff, improving the skills of employees in the area of risk or providing technical equipment for risk management. According to research, the volume of additional investments is on average 5–10% of the cost of risk management in the framework of the fragmented approach. However, by improving the quality and effectiveness of management decisions, these investments become profitable investments. The cost-effectiveness of risk management based on an inclusive approach is estimated by the authors to be 20–25% higher than the cost-benefit values of the fragmented approach.

1.5 Conclusions

Two main approaches to the place and role of risk management in the development and implementation of management decisions are identified: fragmentation and inclusion. As part of a fragmented approach, risk management is seen as an auxiliary discrete process. Due to a number of drawbacks, the use of a fragmented approach can lead to significant losses for enterprises. Within the framework of an inclusive approach, risk management is the main process, along with others, of developing and implementing management decisions. In doing so, risk managers are involved as deeply in decision-making as possible, starting with the first stage of decision-making. An inclusive approach requires higher costs and considerable support from the management of the enterprise. At the same time, it allows you to get significant gains by improving the quality and effectiveness of management decisions.

References

1. Mescon M, Albert M, Hedouuri F (2006) The basis of management. M. Williams (eds), p 672
2. Garina EP, Garin AP (2017) The use of modern methods of quality management to improve the efficiency of production of complex product. Vestnik NGIEI 11(78):111–119
3. Garina EP, Garin AP, Vinogradov SA, Batsyna YV (2019) Establishment of a risk management system for enterprise project practices. Financ Econom 3:459–463

4. Špilevskaya EV, Medvedeva OB (2016) Issues of licensing practice and valuation. Estim Contract Work Construct 10:14–19
5. Knight FX (2003) Risk, uncertainty and profit. Translated from the English. M. Case (eds), p 360
6. Risk management standards. Federation of European Risk Management Associations, 2008, p 16
7. Garina EP, Garin AP, Kuznetsov VP, Popkova EG, Potashnik YS (2018) Comparison of approaches to development of industrial production in the context of the development of a complex product. Adv Intell Syst Comput 622:422–431. https://doi.org/10.1007/978-3-319-75383-6_54
8. Kuznetsov V, Kornilov D, Kolmykova T, Garina E, Garin A (2017) A creative model of modern company management on the basis of semantic technologies. Commun Comput Inf Sci
9. Smirnova ZV, Kochnova K (2019) A training of employees of service enterprises using information technology. Vestnik of Minin University 2019. The United Nations 7, no. 7, no. 1, p 1
10. Antonyan LV (2019) The role and place of risk management in company management. www.bkg.ru/library/articles/?ELEMENT_ID=3299
11. Taleb N (2016) Black swan. Under the sign of unpredictability. M. CoLibri, p 736

Chapter 2
Assessment of Restructuring Efficiency of Industrial Enterprises

Anastasia D. Efremova, Victor P. Kuznetsov, Ekaterina P. Garina, Elena V. Romanovskaya and Natalia S. Andryashina

Abstract Introduction. In this article the authors propose to evaluate the results of restructuring of industrial enterprises by means of correlation of qualitative and quantitative indicators of activity of the company. Materials and methods. In the article several different methods of assessment of restructuring efficiency of enterprises are considered. Basically, the work of scientists is based on the assessment of sustainability. But since the restructuring process is directly accompanied by changes in the organizational structure, the authors decided to add an assessment of the effectiveness of the organizational of management structure. The following methods were used: observations, comparisons, abstraction, statistical methods, analysis, synthesis, generalization. Results. The decision on efficiency of restructuring process is proposed to be made based on a calculation of the following indicators and their comparison with normative values: indicators of enterprise stability in the market, economic indicators of enterprise activity, production and technological stability and efficiency of management system. Conclusion. The article proposes an algorithm for assessing the efficiency of restructuring on the basis of sustainability, economic indicators and indicators of efficiency of the management system.

Keywords Business activity · Organizational structure · Restructuring · Enterprise sustainability · Efficiency

A. D. Efremova (✉) · V. P. Kuznetsov · E. P. Garina · E. V. Romanovskaya · N. S. Andryashina
Minin Nizhny Novgorod State Pedagogical University, Nizhny Novgorod, Russia
e-mail: asya_efremova@bk.ru

V. P. Kuznetsov
e-mail: kuzneczov-vp@mail.ru

E. P. Garina
e-mail: e.p.garina@mail.ru

E. V. Romanovskaya
e-mail: alenarom@list.ru

N. S. Andryashina
e-mail: natali_andr@bk.ru

T. Kolmykova and E. V. Kharchenko (eds.), *Digital Future Economic Growth, Social Adaptation, and Technological Perspectives*, Lecture Notes in Networks and Systems 111, https://doi.org/10.1007/978-3-030-39797-5_2

JEL Code L6 · M5

2.1 Introduction

A logic of modern economic development leads to the fact that the economic aspect begins to play an increasing role in the management of a large enterprise. On the basis of the above objectives, the efficiency of a disaggregation process can be assessed by means of indicators and methods of assessing the economic situation of the enterprise, in which the main estimated indicators are: business performance and profitability indicators.

Scientists work in assessing the effectiveness of reorganization activities is based on sustainability assessment [1–3]. Despite the fact that scientific literature offers a quantitative assessment of the performance of enterprises, the results of these studies are not correlated with the analysis of the effectiveness of a new organizational structure.

2.2 Theoretical Basis of the Study

Restructuring of an industrial enterprise affects its organizational structure. Organizational structure of enterprise management is an ordered set of services that manage its activities, relationships and subordination.

Issues affecting organizational structures and management of the organization began to be dealt with in the early twentieth century. F. Taylor, M. Weber, H. Emerson and many other classics of economic science [4, 5] saw enterprise management as a separate system. This issue remains relevant in our time. Over time, the economy has changed and processes have changed along with its laws of functioning. Enterprise management system was not an exception. Modern authors, Russian and foreign, who summarized the old theories and introduced into them a new current can be attributed L. I. Abalkina, I. Ansoff, S. N. Bobylev, R. L. Daft, P. F. Drucker, A. A. Dinkin, M. Mescona, V. L. Quint and others. Despite a fairly large amount of literature, which describes the theory of building efficient enterprises, the possibility of free application in practice is characterized by not all provisions of these theories. In order to manage an infrastructure of the enterprise, i.e. to form the optimal set of its elements, it is necessary to formalize processes of assessment and management, as well as develop methods and models for operational management in case of various changes in the market infrastructure.

The decision on the efficiency of the restructuring process is proposed based on the calculation of the following indicators and their comparison with regulatory values (Fig. 2.1).

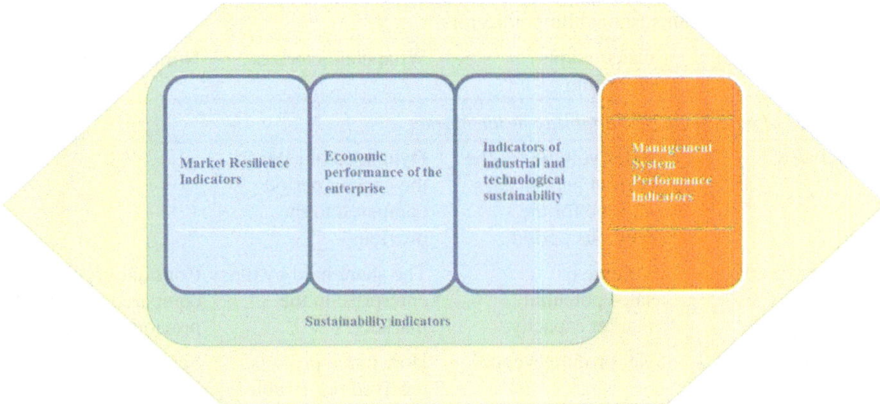

Fig. 2.1 Restructuring performance indicators

Conditionally assessment indicators of restructuring efficiency can be divided into two groups: indicators of sustainability and efficiency of the management system [1]. Sustainability indicators are divided by the authors into three subgroups: stability of the enterprise in the market, economic performance of the enterprise and production—technological stability (Table 2.1).

Restructuring process involves an inevitable change in the organizational structure of an enterprise [6]. Therefore, when deciding on the conduct and effectiveness of a disaggregation procedure, it becomes necessary to evaluate the effectiveness of the organizational structure of management.

The main approach was made by Z. M. Rybalkina and V. G. Kulikova consisting in assessing the manageability of the organization by the degree of goals achievement, which is determining the ratio between the number of completed and set goals.

According to Table 2.2, enterprises can be divided into 4 types according to the level of manageability.

1. Highly manageable. At this level of management is characterized by the achievement of intended goals.
2. Normative controlled. At this level of management, the enterprise fully and in time achieves the set goals and practically does not deviate from them.
3. Low-controlled. In this case, the enterprise is characterized by constant failures in operation. Achievement of the set goals occurs either with a large deviation in time or does not occur at all.
4. Uncontrollable. With this level of manageability, the enterprise cannot achieve its objectives, and there are large losses of resources.

This approach shows only qualitative characteristics of assessment of efficiency of enterprise management. The authors propose to supplement this approach with quantitative indicators.

The analysis of scientific works allowed to identify the main coefficients used in the evaluation of the management system for its effectiveness [7, 8]:

Table 2.1 Business unit sustainability indicators

Indicator	Calculation algorithm	What characterizes	Normative value
Indicators of enterprise sustainability in the market			
Sales Index	Sales volume for the current period/sales volume for the previous period	Dynamics of sales in the current period compared to the previous	>1
Company market share	Volume of sales/potential market capacity	The share held by the enterprise in the market	Constant growth compared to the previous period
Profitability of sales	Net profit/Revenue	How much profit is received per 1 ruble of revenue	Set by industry average
Economic performance of the enterprise			
1. Business performance indicators			
Asset turnover ratio	Net proceeds from sales of products/Average annual value of assets	Turnover rate of total capital of the enterprise	Set by the company
Working capital turnover ratio	Net proceeds from sales of products/Average annual value of current assets	Turnover rate of working capital	Set by the company
Receivables turnover ratio	Net proceeds from sales of products/Average annual amount of receivables	Rate of repayment of receivables	Set by the company
2. Performance indicators of the enterprise			
Return on equity	Net profit/Equity	Efficiency of using the organization's own capital	Set by industry average
Return on assets	Profit from ordinary activities before tax/Balance sheet currency	Level of profit generated by all assets of the enterprise that are in use according to a balance sheet	Set by industry average
Profitability of sales	Net profit/Revenue	How much profit is received per 1 ruble of revenue	Set by industry average

(continued)

Table 2.1 (continued)

Indicator	Calculation algorithm	What characterizes	Normative value
Indicators of production and technological sustainability			
Production index	Volume of production for the current period/volume of production for the previous period	Dynamics of production in the current period compared to the previous	>1
Power utilization factor	Actual output of products/maximum output of products	Percentage of production capacity utilization at the enterprise	Set by the company
Fund return	Production/average annual value of fixed assets	Production of finished products per 1 ruble of fixed assets	Set by the company
Proportion of defective products	Volume of defective products/total output	The share of defective products in total production	Set by the company
Costs per ruble of commercial products	Total cost of production and sale of products/value of produced products in current prices	The share of costs in the price of products	Set by the company

1. Manageability coefficient, which characterizes the ratio of the number of actual personnel subordinated to the head to the same normative value [9]. The calculation is made according to the formula 2.1:

$$\mathrm{Cm} = \frac{1}{z} \times \left(\sum_{1}^{m} \frac{H\phi}{H\text{н}} \right) \tag{2.1}$$

z is the number of levels of control;

m is the number of management personnel at the appropriate level;

Hf and Hn—number of actual and normative value of personnel under the supervision of a manager.

Standard Cm = 0, 5—1.

2. Coefficient of employee labor efficiency of management apparatus Czy is calculated according to formula 2.2:

Table 2.2 Characteristics of manageability levels of enterprises [10]

Indicator	Level of management			
	High managed	Normally managed	Hard to manage	Uncontrollable
Nature of goals achievement	Systematic		Probabilistic	
Completeness of goals achievement	Achievement and maintenance of planned and overplanned goals	Maximizing the full achievement of goals	Partial achievement of goals or full, but with overexpenditure of resources	Goals are not achieved
Timeliness achievement of goals	Early or on time	On time	Violation of deadlines	With significant violation of the deadlines
Utilization of resources	Savings	Within the norm	Overexpenditure	Significant losses
Effect	More than normal	Within the norm	Less than normal	No effect
Manageability level	(more than 70%)	Average (50… 69.9%)	Low (20… 49.9%)	Critical (up to 20%)

$$\mathrm{Czy} = \frac{\mathrm{Zc.c.y.} * \mathrm{Zp.p.y.}}{\mathrm{Zc.p.y.} * \mathrm{Zp.c.y.}} - 1 \tag{2.2}$$

Zc c.y.—total amount of costs spent on performance of managerial duties in the current year;

Zc p.y.—total amount of costs spent on performance of management duties in the previous year;

Zp—the total amount of costs spent on production and sale of products for the year.

3. Coefficient of economic efficiency of management activity Ce shows how many rubles of profit P (income) falls on one employee from the administrative apparatus and is calculated according to the formula 2.3:

$$\mathrm{Ce} = \frac{P}{Sm} \tag{2.3}$$

Evaluation of the effectiveness of the form of management is based on achievement of established benchmarks in a specific situation and on an assessment of the effectiveness of an organization. An impact of the enterprise can be analyzed on: achievement of proposed benchmarks; adequacy of the purpose of a current management situation; strengthening the position in the external environment.

Methodology of study

The following methods were used in the article: observations, comparisons, abstraction, statistical methods, analysis, synthesis, generalization.

Analysis of the results

Since the main change in a process of disaggregation is restructuring of the organizational structure, all estimates should be calculated to the effectiveness of organizational structure.

When correlating qualitative and quantitative indicators, the following table is obtained for assessing the manageability of the enterprise (Table 2.3).

According to the table, the final conclusion about the efficiency of restructuring is made, based on the assessment of the level of manageability of a business unit and indicators of its sustainability.

Table 2.3 Indicators of assessment efficiency of restructuring in terms of manageability of organizational structure

Indicator	Management level			
	High managed	Normally managed	Hard to manage	Uncontrollable
Quality indicators				
Nature of goal achievement	Systematic		Probabilistic	
Completeness of goals achievement	Achievement of planned and overplanned goals	Maximizing the full achievement of goals	Partial achievement of goals	Goals are not achieved
Timeliness achievement of goals	Early or on time	On time	Violation of deadlines	With a significant violation deadlines
Quantitative indicators				
Labor efficiency, Clc	Savings compared to prior period	Within the previous period	Overexpenditure over prior period	Significant losses compared to prior period
Economic efficiency of the management apparatus, Ce	More than the previous period	Within the previous period	Less than previous period	No effect
Manageability level, Cm	(more than 70%)	Average (50… 69.9%)	Low (20… 49.9%)	Critical (up to 20%)
Indicators of enterprise sustainability (Table 2.1)	Above the normative value	Within the regulatory value	Small deviations from the standard value	Significant deviations from the standard value

The method of evaluating the effectiveness of the results of restructuring on the basis of diversification of enterprises includes 16 steps, each of which consistently calculates 16 indicators, which in the future are given to the integral indicator (II). To do this, it is proposed to add all the indicators, setting for each of them a certain standard value according to the formula 2.4.

$$\text{IP} = \sum \Delta(\text{Is, Tg, Rp, Ca, Ct., Ctr, Rre, Ip, Kpu, Fo, B, Z, Czy, TC, Cmf}), \quad (2.4)$$

where

$\Delta = \frac{i\phi}{iH}$—the ratio of the actual indicator to the accepted standard value

Is—sales volume index (standard value ≥ 1)

Tg—growth rate of the company's market share (normative value ≥ 1)

Rp—profitability of sales (standard value is taken by industry average)

Ca—asset turnover ratio (standard value is set by the company)

Ct—turnover ratio of working capital (standard value is set by the company)

Ctr—turnover ratio of accounts receivable (standard value is set by the company)

Rre—return on equity (regulatory value is taken by the industry average)

Ra—profitability of assets (standard value is taken by industry average)

Ip—index of production (standard value ≥ 1)

Kpu—Power Utilization (Standard $= 1$)

Fo—capital return (normative value is set by the company)

B—the ratio of defective products (the standard value is set by the company)

Z—costs per ruble of commodity products (the regulatory value is set by the company)

Czy—labour efficiency ratio (standard value < 1)

TC—rate of growth of economic efficiency of the control apparatus (normative value ≥ 1)

Cmf—manageability factor (standard value ≥ 0.7)

The integral indicator includes 16 evaluation criteria, and the normative value of each of which the author compared with unit. The method of expert assessments was used to find the threshold value of the indicator. Which revealed that when the integral indicator reaches the mark of 10 points and below should be considered possibility of structural changes of the enterprise.

2.3 Conclusion

As a result, it is proposed to evaluate the efficiency of restructuring enterprises on the basis of the assessment of the overall strategic stability of the enterprise. Specifically, evaluate using the following indicators: stability of the enterprise in the market, economic indicators of the enterprise, production and technological stability and efficiency of the management system.

Sharing these indicators will provide a comprehensive assessment of the effectiveness of the measures, by examining various aspects of undergoing certain changes.

References

1. Garina EP, Kuznetsov VP, Romanovskaya EV, Andryashina NS, Efremova AD (2018) Research and generalization of design practice of industrial product development (by the example of the domestic automotive industry). Qual-Access Success 19(S2):135–140
2. Kuznetsov VP, Garina EP, Romanovskaya EV, Kuznetsova SN, Andryashina NS (2018) Organizational design and rationalization of production systems of a machine-building enterprise (by the example of the contract assembly workshop). Espacios 39(1)
3. Mizikovsky IE, Druzhilovskaya TY, Druzhilovskaya ES, Garina EP, Romanovskaya EV (2018) Accounting for costs and expenses: problems of theory and practice. Adv Intell Syst Comput 622:152–162
4. Quint VL (2012) Strategic management and economics in the global emerging market. M.: Business Atlas, 627 p
5. Mescon M, Albert M (2017) Franklin Hedoury Foundations of Management Williams Publishing House, 672 p
6. Sidorova E Yu, Stepanov AS (2015) Functional model of increase and evaluation of management efficiency at production organization in conditions of competition, Bulletin Saratov university New Vol Economy. Management Law. vol 15, issue 3, p 298
7. Bobyleva AZ, Sazhina MA, Bobylev SN, Peganova OM, Kashirova AV, Zhavoronkova EN, Tsarenko AS, Lviv OA (2011) Management in the conditions of instability of the financial and economic system: strategy and tools. In: Bobyleva AZ (ed) M.: Publishing House of Moscow University, 224 p
8. Salakhieva MF, Semenova OV (2012) Evaluation of efficiency of restructuring of industrial enterprise. Econ Anal: Theory Pract 13, https://cyberleninka.ru/article/n/otsenka-effektivnosti-restrukturizatsii-promyshlennogo-predpriyatiya (Date of access: 24.10.2018)
9. Markova SM, Narkoziev AK (2018) Production training as a component of professional training of future workers. Vestnik of Minin University, vol 6. 1 p
10. Gerasimov BI, Konovalova TM, Spiridonov SP, Sapetna NI (2008) Complex economic analysis of financial and economic activity of the organization, Publishing House of Tambov State Technical University, 142 p
11. Garina EP, Garin AP, Vinogradova SA, Batsyna Ya V (2019) Management of investment programs at the stage of their implementation. Financ Econ 3, 464–467

Chapter 3
Business Model of Contract Productions

Svetlana N. Kuznetsova, Victor P. Kuznetsov, Ekaterina P. Garina, Elena V. Romanovskaya and Alexander P. Garin

Abstract Purpose of the work: to form a modern approach to organize production, which involves a multi-level distributed supply chain, which includes specialized contract services and production. In the conditions of products transformation into services, the profit zone of product companies goes to the area of data collection generated by consumers and technologies of their processing. At the same time, product companies strive for maximum flexibility in the organization of production and supply chains. Degree of scientific development of the problem. Questions about contract production have been developed in the works of scientists, as: Aniin V. M., Baldin K. V., Barker A., Blank I. A., Valinurova L. S., Gokhberg L.M., Egorov A. Y., Zelensky Yu. B., Ivanov G. I., Ihonina L. M., Igoshin N. V., Kleyton M., Pilipenko A., Prigozhin A. I., Twiss B., etc. The methodology for studying the content of discussions consists of: development of conceptual apparatus, statistical (quantitative) methods, forecasting methods, estimation methods of state policy in the field of digital economy. The results of the study: the further evolution of the contract manufacturing business involves the development of cross-sectoral and cross-cutting competencies among contract producers and the application of successful experience gained in some industries, to solve similar tasks in others. Already today, world leaders are creating consortia, inviting leading contract companies to meet the challenges needed for future competitiveness. This study and rating of Russian contract productions summarize the results of analytical work in order to identify features

S. N. Kuznetsova (✉) · V. P. Kuznetsov · E. P. Garina · E. V. Romanovskaya · A. P. Garin
Minin Nizhny Novgorod State Pedagogical University, Nizhny Novgorod, Russia
e-mail: dens@52.ru

V. P. Kuznetsov
e-mail: kuzneczov-vp@mail.ru

E. P. Garina
e-mail: e.p.garina@mail.ru

E. V. Romanovskaya
e-mail: alenarom@list.ru

A. P. Garin
e-mail: rp_nn@mail.ru

T. Kolmykova and E. V. Kharchenko (eds.), *Digital Future Economic Growth, Social Adaptation, and Technological Perspectives*, Lecture Notes in Networks and Systems 111, https://doi.org/10.1007/978-3-030-39797-5_3

and regularities of development of contract manufacturing companies. Conclusion. In this article, change management between the client and the contractor was considered as another aspect of high-performance standards to ensure high quality, safety and the efficiency of the final product.

Keywords Contract · Customized products · Technology · Process · Business model

JEL Code L6 · M5

3.1 Introduction

Since the beginning of 2010, the global industry has seen a downward trend in growth due to the saturation of commodity markets and the exhaustion of productivity gains within the existing technological way. A similar trend is observed in Russia. On the background of a slowdown in economic growth, which in 2000 was primarily associated with the favorable situation in world commodity and energy markets, Russian industry is in great need of both updating fixed assets and introducing modern technologies and models for organizing industrial production necessary to ensure the competitiveness of manufactured products. In conditions of the "fourth industrial revolution" proclaimed by the industrial leaders, and the world industry is transforming at great speed [1]. Global industrial giants are implementing industrial models aimed at shifting from traditional mass conveyor production to flexible, reconfigurable production systems that allow to produce customized products. Digitalization of both production processes and products opens up huge opportunities to increase productivity, reduce time and financial costs, and individualize products (Markova 2019). Production companies operating in a competitive market today face a number of challenges of their competitiveness:

- Reduction of product life cycle;
- Increasing importance of service component in consumer and industrial products;
- Necessity of accelerated introduction of advanced production technologies and creation of new consumer qualities of products;
- Need for constant optimization of technological, production, logistics and management processes.

According to the established definition in business terminology, contract production is the production of products at facilities of an independent manufacturer, which ensures full compliance with the technological cycle and quality control of finished products according to customer requirements.

This understanding of contract production as outsourcing exclusively of production functions became widespread in 1970–1980, when companies-manufacturers of final products began to look for ways to reduce costs by placing orders on external capacities for the manufacture of various parts and components, which previously they had to produce themselves, according to their specifications and drawings.

The development of this model, despite the obvious economic benefits, led to a further complication of communication with a growing number of suppliers. More processes and products a company outsourced, then more the overhead costs associated with supply chain management increased. This was particularly affected in situations related to the development of new product models [2].

3.2 Methodology

In 1990, this model began to be replaced by another one—multi-level model. This meant that instead of contacting many manufacturers from raw materials to complex products, the OEM (Original Equipment Manufacturer) sought to work directly only with those companies that collected complex aggregates and systems, they are usually large (Tier 1 suppliers).

These companies, in turn, worked with Tier 2 suppliers, i.e. manufacturers of any nodes and assemblies. Then came the third-tier suppliers (Tier 3), manufacturers of individual parts, materials, and components. Saturation of markets and intensification of competition forces producers to shorten the life cycle of products, and to implement innovative solutions faster. At the same time, the cost of mass production should be strictly controlled in order to ensure the competitive price of the final product. Therefore, large end product manufacturers since 1990 have focused on research and development, new product development, marketing, and service, while retaining final assembly and only those production operations that determine the competitiveness of products and whose transfer to other companies has too high risks. As a result, tier 1 suppliers have played a significant role in the processes related to development and production of units and systems.

If at the beginning of outsourcing their task was only to produce according to the customer's documentation in compliance with their terms, price, and quality of products, today they have to provide scientific development, the introduction of new technologies, to offer effective technological solutions as part of a strategy for the development of the final product. Their share in the value of the final product is increasing and this trend will soon continue, as is the trend of deepening specialization in value chains [3].

To date, a number of terms have been established in supply chain management, including key terms [4]:

OEM (Original Equipment Manufacturer)—the term in different sources is interpreted differently. The most relevant interpretation in accordance with understanding of this term by manufacturers and contract service companies is as follows: OEM is a company whose business model is built around product development, marketing and sales. A significant part of production processes is outsourced.

At the same time, a company, as a rule, itself creates and accumulates intellectual property and reserves the right to design, technological, and other developments.

CM (Contract Manufacturer) is a manufacturing company that produces products under contract for other companies under the trademark of the customer. OEMs are customers for CM-companies.

ODM (Original Design Manufacturer) is a contract manufacturer that produces products on the order of an OEM company, but at the same time performs partially or fully develop the product and own the intellectual property. Such companies tend to specialize in a small number of certain types of products.

EMS Provider (Electronic Manufacturing Service Provider) is a contract manufacturer in the field of electronics, which not only produces products commissioned by OEM-companies but also offers a wide range of additional services (design, development of new products, supply chain management, testing, logistics, etc.). In the field of electronics, large suppliers of the 1st level, as a rule, have already switched to the model of production service.

The EMS model described above is one of the examples of open high-tech contract manufacturing services. Open high-tech contract production service—a set of specialized services for design, development, engineering, production and maintenance of customized products of any technological complexity and any volume at the capacity of an independent manufacturer, which provides a complete solution of the task, full compliance with the technological cycle and strict control product quality according to customer requirements. These are primarily services with high added value. The business model of open high-tech contract production service is connected with provision of services for the production of a batch of products of any technological complexity from small to large series. The main source of income of a contract producer is provision of services, not the production of a particular type of product. The value-added structure of the service is dominated by the intellectual component rather than directly by production services. Key customers of these services are typically large and medium-sized enterprises that are unable to deliver certain products within the specified time frame, or in the case of unprofitability or inefficiency of their production by the customer.

Key principles of open high-tech contract production service [5]:

- Deep specialization in specific research, engineering or production competencies;
- Strict compliance with the quality of products and service, at a level much higher than the average market;
- Development and production of customised products of any technological complexity precisely in time (concept just in time);
- Flexible, transparent pricing, with the customer's understanding of the price depending on the complexity of tasks. As a result, the cost of rendered services will always be lower than that of vertically integrated companies;
- High speed and flexibility of development of new technological processes and products;
- Reducing the cost of engineering and production processes. Minimization of technological and managerial costs;
- Reputation and trust—the most important asset of the contract company.

3.3 Results

In the framework of this study, we studied and analyzed the practice of functioning and development of foreign contract manufacturing using the example of countries: USA, UK, Netherlands, Switzerland, Belgium, China, Taiwan, Republic of Korea. Russian contract engineering and manufacturing companies are also analyzed. In general, based on the results of the analysis of foreign and Russian contract technology and production service companies, it is possible to conclude that there are a number of features in business models [3].

Contract companies are highly qualified in end-to-end technologies and processes across multiple industries. At the same time, solutions developed to solve problems in one industry can be successfully applied to solve similar problems in other industries. Conditionally, two approaches to formation of technological competencies that make up the strategic competitive advantages of companies can be distinguished:

- Expansion of the range of mastered competencies by creating new divisions or acquisitions of other companies;
- Deepening of specialization in the range of already mastered competencies by investing in own innovative developments of applied technological processes.

Making an order on the principle of "just in time" with high-quality standards.

The ability to withstand tight time limits when executing an order is a prerequisite for the successful operation of a contract company in the market. Foreign companies pay great attention to this, most of them take various measures to optimize time costs at all stages of the life cycle of the order. Contract companies constantly maintain communication with the customer during the execution of the order. Many companies have tools to track the status of an order online, which indicates a high degree of transparency and manageability and allows their customers to control deadlines in any a moment in time. An example of further development of this approach is a fully automated order processing system.

The critical competitive advantage of contract manufacturing companies is the ability to tightly control production costs and ensure minimum overhead costs. Minimization of costs, along with deep specialization and technological competence of contract manufacturers, is a key factor in the refusal of food companies to maintain their own production capacity in favor of working with contract companies.

While for a large vertically integrated product company, the overhead rate for production (workshop) costs maybe 100–200% or more, as a rule, does not exceed 15–20%. This is achieved through three factors:

- Maximum utilization of production facilities and distribution of costs for their maintenance among many customers;
- Use of flexible advanced production technologies;
- An effective horizontal management system that allows you to make decisions with minimal time and financial costs.

Contract companies seek to diversify their clients' portfolio. This is done both to maximize the load of production lines and reduce costs, and to reduce dependence on one or more specific customers.

The presence of multiple clients allows for a contract company to also gain interdisciplinary experience, which contributes to the development of its competencies and the expansion of the range of solutions and services offered.

Availability of non-productive services up to full specialization in services.

In addition to providing production services, contract companies can provide services in the field of technology consulting and project management.

For example, the American company Nytec specializes in technology consulting in electronics and virtual reality, offering consulting and management services at any stage of creating new product: from concept to design, testing, production, and delivery [6]. Most foreign contract manufacturers provide development, design and engineering services, considering these high-margin services as the main source of profit.

Often these competencies, involving deep expertise in development and engineering, as well as their own unique developments and technologies, are the main source of a competitive advantage contract manufacturer.

Implementation of this model is possible due to a company having the necessary competencies in the field of optimization of production and technological processes, integration and robotization of production systems.

Some contract companies that provide production services operate on the principle of fabless, that is, do not have their own production capacity at all. These companies provide services in product development and engineering, project management and supply chain management. At the same time, production takes place at the capacity of other contract manufacturers.

Like customer product companies, contract manufacturers form a stable cooperative network of contract companies (cloud of cooperation) that meet the criteria formulated on the basis of customer requirements. Major contract manufacturers constantly develop their network of suppliers and carry out activities aimed at supporting their competences.

As a result, contract manufacturers build flexible production systems based on advanced production technologies in order to provide the quality required by the customer with strict control of cost and timing and to provide flexibility and ability to produce small series of customized products with minimal costs.

Contract companies apply digital design and modeling, product lifecycle management automation, CNC machine tools and machining centers, 3D printing equipment, digitalization and robotization of production [7]. At the same time, many companies invest heavily in the development of their own advanced production technologies and equipment.

Foreign contract companies, especially those operating in the global market, use an advanced intellectual property protection system to minimize the risk of leakage or disclosure.

This is achieved by technical means of information protection (data exchange protocols, special rooms, information security systems, etc.) and by means of a well-thought-out protection policy information (including the control of access of employees to information, reliable destruction of technological and design documentation of the customer immediately after execution of the order and other measures). The legally contracted manufacturer guarantees the safety and non-disclosure of the intellectual property of the customer at the conclusion of the contract.

In order to avoid conflicts with customers, including those related to the use and copying of the customer's intellectual property by unscrupulous companies, contract manufacturers often do not create products under their own brand. Cases of separation of contract type business from the product business are not uncommon. In such cases, the parent product company seeks to separate a contract business into a separate business unit until the sale to other owners.

Another alternative is the development of its own brand for a contract manufacturer is the output of products under his own brand in market segments, which are not represented by companies—customers of a contract manufacturer.

Flexibility in scale of activity and proximity to the client Foreign contract manufacturers are most often ready to accept an order for the development and production of a single product or prototype, as well as capable of mass production. At the same time, they usually implement a strategy of maximum proximity to the client, which involves readiness to perform non-standard orders and provision of services that allow to increase efficiency of the client's business.

An example of this strategy is the creation or purchase of factories by contract manufacturers in countries of strategic interest for large OEM and ODM customers. However, the deployment of production in other countries is usually carried out within the framework of long-term cooperation with several large customers. For example, Foxconn created production sites in India for several major customers, such as Xiaomi and Apple (Egorova et al. 2017).

As their specialization expands or deepens, contract companies maintain a horizontal management structure while maintaining low management costs.

Effective contract business is possible only in the presence of an effective system of quality management of products and services. Contract companies accredit internal production processes in accordance with the requirements of the international certification system. Most often, companies receive a quality management system certificate ISO 9001, as well as specific industry certificates and private accreditations of large corporate customers [8].

3.4 Conclusion

A modern decentralized model of the organization of the industry allows to increase the efficiency of production by deepening the specialization of contract companies. The final manufacturer focuses on key competencies (typically research and development, production of individual assemblies and systems, assembly of final products, marketing, and service). Everything else is outsourced to contract companies, which allows the end producer to get rid of the significant costs of creating and maintaining excess production capacity [9]. With the development and deepening of specialization, contract companies integrate more closely with the business of their customers, offering solutions to improve product quality and optimize production processes. This leads to the fact that end manufacturers consciously abandon an increasing number of production operations, up to a complete abandonment of their production capacities, in favor of working with contract manufacturers. This allows end producers to focus their resources on research and development and consumer work without participating in the "technological race" of production efficiency. Successful by world standards, contract technology, engineering, and manufacturing companies are highly specialized on specific technological competencies services for solving problems related to the development, engineering, mastering of production and serial production of complex high-tech products. Setting the customer task (a product company or an integrator in the supply chain) can be absolutely anything up to the development of products or technology from scratch. It is the high intellectual component and deep knowledge of the customer's business that is the basis of forming the profit of effective contract business.

The total output of products on the basis of contract production amounted to 61,125.94 million rubles, and the volume of tax revenues amounted to 42,096.01 million rubles [10].

References

1. Markova SM, Narkoziev AK (2019) Methods of research of the content of professional education. Vestnik of Minin University, vol 7, no 1(26), p 2
2. Pavlov II (2015) System of control of state and municipal procurements in the framework of the Federal Law of April 5, 2013 N 44-FL "On contract system in the sphere of procurement of goods, works, services to ensure state and municipal needs", Taxes, no 1, pp 27–30
3. Potrashkov S (2015) The process of coordination of the contract with the sole supplier under the Law on the contract system. Adm Law 1:35–39
4. Garina E, Kuznetsov V, Yashin S, et al (2017) Management of industrial enterprise in crisis with the use of in company reserves. In: Overcoming uncertainty of institutional environment as a tool of global crisis management: conference on overcoming uncertainty of institutional environment as a tool of global crisis management location, Athens, Greece Date: APR, Book Series: Contributions to economics, pp 549–555
5. Kuznetsov VP, Garina EP, Romanovskaya EV, Kuznetsova SN, Andryashina NS (2018) Organizational design and rationalization of production systems of a machine-building enterprise (by the example of the contract assembly workshop). Espacios 39(1), Article No 25
6. Parshina AA, Shpilevskaya EV, Garina EP (2018) Improving the competitiveness of domestic production through the implementation of management concepts of complex product development. Humanit Socio Econ Sci 3(100):113–118
7. Romanovskaya EV, Kuznetsov VP, Agafonov VP, Andrashina NS, Artemyeva MV (2018) Design and reengineering of business processes: textbook. University of Minin, Nizhny Novgorod, 200 pp
8. Khalapin AA, Alekseev VM (2016) Theoretical foundations of public procurement. Symbol Sci 9–1:167–169
9. Garina EP, Garin AP, Vinogradova SA, Batsyna YaV (2019) Management of investment programs at the stage of their implementation. Financ Econ 3:464–467

10. Kuznetsova SN, Romanovskaya EV, Artemyeva MV, Andryashina NS, Egorova AO (2018) Advantages of residents of industrial parks (by the example of AVTOVAZ). Adv Intell Syst Comput 622:502–509. https://doi.org/10.1007/978-3-319-75383-6_64

Chapter 4
Strategy of Development of Regional Infrastructure for Creating Innovative Production Complexes in the Digital Economy

Maria Kuzmina, Oksana Avdeyuk, Catherine Kuzmina, Irina Tarasova and Alevtina Rayushkina

Abstract One of the directions of strategic development of the Russian economy in the conditions of digitization is transformations in regions by creation and development of innovative production complexes through improvement of the regional infrastructure. Dynamics of increase of production, economic, and social indicators are largely determined by the level of development of the infrastructure in regions with innovative production complexes. Possessing the production and innovative potential, each region has to have such regional infrastructure that would allow ensuring growth of the indicators of socio-economic development and would become the basis for the digital economy.

Keywords Infrastructure · Regional infrastructure · Production complex

4.1 Introduction

In the modern conditions of the digital economy, new strategies of development are formed—they are connected to transformation in its structure and the necessity for transition to hi-tech and innovative path of production. Serious changes at the macroeconomic level could be performed through regional transformations due to more

M. Kuzmina (✉) · O. Avdeyuk · C. Kuzmina · I. Tarasova · A. Rayushkina
Volgograd State Technical University, Volgograd, Russia
e-mail: mi_kuzmina@mail.ru

O. Avdeyuk
e-mail: oxal2@mail.ru

C. Kuzmina
e-mail: katerina993@yandex.ru

I. Tarasova
e-mail: tarasova.irina.aleks@gmail.com

A. Rayushkina
e-mail: ap@vstu.ru

T. Kolmykova and E. V. Kharchenko (eds.), *Digital Future Economic Growth, Social Adaptation, and Technological Perspectives*, Lecture Notes in Networks and Systems 111, https://doi.org/10.1007/978-3-030-39797-5_4

flexible adaptation to the conditions of the external environment. That's why one of the directions of strategic development of the Russian economy in the conditions of digitization is transformations in regions by creation and development of innovative production complexes through improvement of the regional infrastructure. Dynamics of increase of production, economic, and social indicators are largely determined by the level of development of infrastructure in regions with innovative production complexes.

In the existing conditions, each region has to have such regional infrastructure that would allow ensuring growth of the indicators of socio-economic development and would become the basis for the digital economy.

4.2 Regional Infrastructure in the Digital Economy

Creation and development of innovative production complexes in a region, which allow forming and raising the competitive advantages of the territorial entity and independent companies, are one of the ways of solving this problem. The most important factor of creation and development of innovative production complexes is usage of digital technologies [1–5].

It is obvious that infrastructure largely determines the dynamics of creation, growth, and development of the innovative production complexes. Regional infrastructure is very important, in view of large territory of Russia and unequal distribution of labor, natural, material, and financial resources and digital technologies.

Therefore, regional infrastructure is primarily the characteristics of general conditions of creation of innovative production complexes and effective development of material production in the conditions of the digital economy. It helps solving scientific and technical problems, raising the level of information and digital provision, and optimizing the production ties on the region's territory and beyond it. Eventually, sustainable growth of production of hi-tech products is achieved, development of production relations takes place, and digital technologies are implemented.

Thus, regional infrastructure forms the basis for functioning of innovative production complexes. It consists of the territorial and sectorial interconnected system of organizations, types of activities, and material and technical objects that ensure the needs for innovative production complexes for resources, products, works, services, and digital technologies, and creation of the factors for effective reproduction functioning. However, in the existing conditions of development and modernization, proper attention is not paid to regional infrastructure. That's why its creation and improvement is often spontaneous and rarely depends on the needs of created innovative production complexes.

4.3 Strategy If Development of Regional Infrastructure

For the purpose of elimination of the problem aspects in the functioning and modernization of regional infrastructure, it is necessary to develop a strategy of its development. Let us determine the conditions that influence the formation of the strategy of development of regional infrastructure for creation and development of innovative production complexes:

(1) readiness of production complexes for implementing the strategy of their development;
(2) territorial and inter-territorial opportunities for creation of innovative production complexes;
(3) external business environment of innovative production complexes;
(4) mission and tactical tasks of production complexes in the region's development;
(5) elaboration and resource provision of the companies' actions in innovative production complexes;
(6) availability of information and digital technologies in the regional infrastructure.

In view of a complex mechanism of implementing the strategy of development of regional infrastructure for creation of innovative production complexes, let us show the process of its formation (Fig. 4.1). It is possible to distinguish five stages in the process of development of the strategy of development of regional infrastructure for innovative production complexes in the conditions of the digital economy.

The first stage of development of the strategy is preliminary—it will be a start of preparation for construction of new and development of the existing regional infrastructural elements. Collection and analysis of data will allow determining the production and economic potential of the region, as well as its investment and innovative capacity. New elements of the infrastructure should ensure servicing of the growing number of the technological and production spheres of economy, guarantee the conditions for creation and effective functioning of innovative companies, stimulate the improvement of production and economic types of activities and relations, and support continuous dynamic growth and development of the region.

The second stage of development of the strategy envisages creation of new and modernization of the existing production infrastructure. This stage envisages determination of conditions and elements of creation of new objects and reconstruction (modernization) of the infrastructural elements, according to the innovative terms of production and application of digital technologies.

The third stage envisages solving the tasks of transport elements of the infrastructure. This stage envisages the design of new and modernization of the existing transport and magistral infrastructure—i.e., construction of main lines, lines of communications, transport networks, and data transfer networks.

The fourth stage is design and creation of the digital infrastructure, which will allow increasing the production and material resources, thus flexibly adapting the innovative production complex to the modern requirements of doing business. The digital infrastructure implements the most leading and effective models of business.

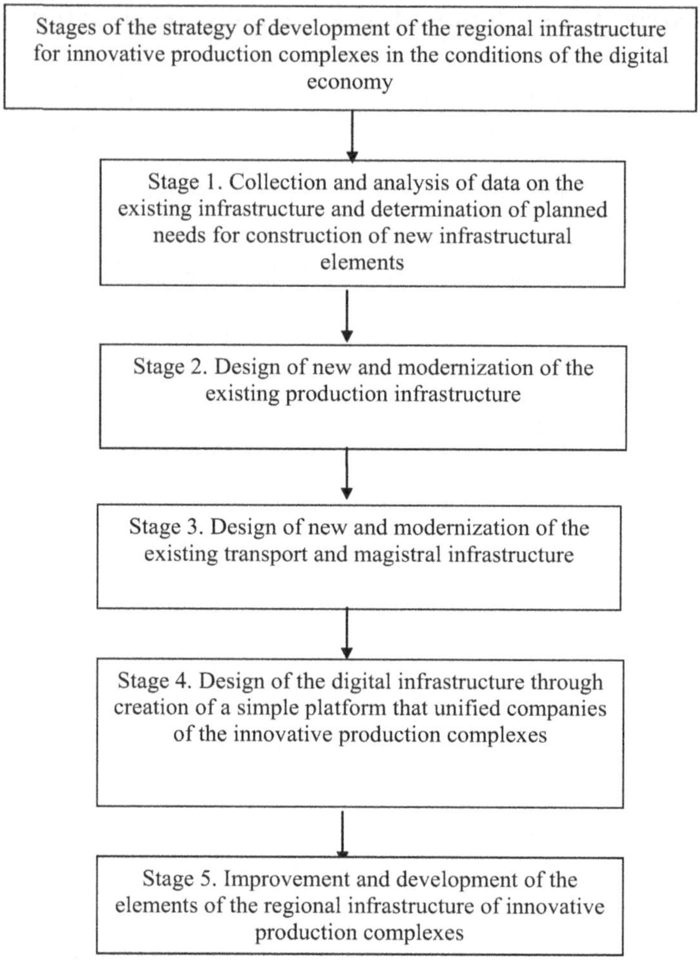

Fig. 4.1 Stages of development of the strategy of development of regional infrastructure for innovative production complexes in the conditions of the digital economy

The fifth stage includes preparation and implementation of development of the elements of the regional infrastructure. This stage is a complex program that includes the measures on improvement of the regional infrastructure within development of the innovative industrial complex. The indicators of effectiveness of development of infrastructure in the context of each element have to reach maximum indicator that are envisaged by the program of strategic development of the innovative industrial complex.

Development of the strategy of development of regional infrastructure for creation and development of innovative production complexes is based on four components: resource, organizational, innovative, and digital.

The resource component of the process of planning and development of the strategy of development of infrastructure of the innovative industrial complex includes resource management, formation of movement of resources, effective usage of the resources, and optimal resource capacity of production. The resulting indicators of effective regional infrastructural provision could be such indicators as profitability of production, renewal of production funds, full usage of production capacity, investment provision, and growth of indicators of profit.

The innovative component of the process of development the strategy of development of the infrastructure industrial complex ensures creation of the innovative component and stimulates the innovative activity. The resulting indicators of effective innovative regional infrastructural provision could be such indicators as intensity of technologies exchange, scientific and technical provision, ratio of production innovative products, and presence of fixed and soft assets.

The organizational component is based on unification and interaction of resources and the organizational component. As a result of this interaction, the strongest factors are used for further development and improvement of the regional infrastructure.

The digital component contains digital technologies of the innovative production complex, which could be presented by implementation of the technologies of digitization in the structure of management and the production process. For the purpose of controlling the parameters of the technological cycle, it is possible to use sensors and wireless technologies of data transfer, as well as digital automatized and robotized systems of production management.

For implementing the strategy of development of regional infrastructure it is necessary to select and determine the level of execution. Figure 4.2 shows the directions of the strategies of development of regional infrastructure that require a complex program solution. Implementation of the strategy of development is shown structurally in the form of three blocks of planning and implementation depending on the level of execution.

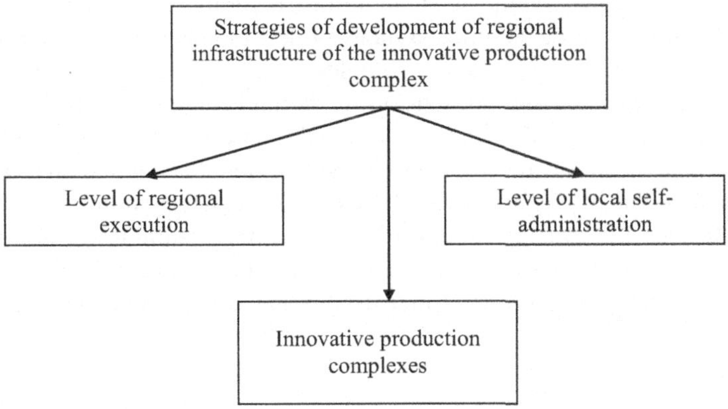

Fig. 4.2 Strategy of development of regional infrastructure

The first block of strategies of development envisages implementation at the level of regional execution. This block envisages the following directions of development:

(1) Creation of new and development of the existing innovative production complexes, so called "growth points", and the objects of the production, transport and digital infrastructure;
(2) Creation of favorable economic and innovative conditions of functioning and development of innovations in the region based on development of the legislative and normative basis;
(3) Participation in inter-regional programs on joint creation of regional infrastructural objects.

The second block of the strategies of development envisages implementation by the local administration bodies. The following directions of strategic development are considered:

(1) liquidation of disproportions in development of productions and infrastructure, which implement the local reproduction cycles;
(2) stimulating the development of the innovative production complexes;
(3) complex improvement of the local legislative and normative basis for stimulating the activities of the innovative production complexes, which are of top-priority for this territory, and creation of favorable conditions for attracting investments for active development of infrastructural objects.

The third block of strategies of development includes the component of innovative production complexes. The following directions of strategic development are studied here:

(1) creation of the system of certification of innovative products;
(2) acceleration of creation of the innovative technological and production basis;
(3) creation and production of competitive innovative products.

The problems of development of the infrastructural complex should be solved within development of the common strategy of development, in view of priorities of creation of the innovative production complexes and improvement of the indicators of region's activities in the conditions of the digital economy. For this, the strategy of development of regional infrastructure for the purpose of creation and effective interaction with the innovative production complex should be based on usage of new digital technologies of resource consumption and exchanges in industry and infrastructure. Besides, the strategy should stimulate the dominating usage of innovations during selection of top-priority directions of development of production complexes and coordination of the interests of industrial production participants.

4.4 Conclusions

Thus, in the economic and innovative activities of a region the infrastructure is a complex, multi-level, and hierarchical system that ensures the process of creation and development of innovative production complexes and stimulates the implementation of the usage of digital technologies. Regional infrastructure forms the conditions for implementing the process of industrial production and development of the digital economy, providing the necessary resources, and creates preconditions for development of industrial complexes, which are involved in this process. This stimulates the production complexes for obtaining additional revenues by means of systemic transformations and usage of the results of scientific, technological, digital, and innovative potentials.

References

1. Burkaltseva DD, Tsohla SY, Guk OA, Borovskaia LV, Bondar AP (2016) Basic forms of institutional changes of economy in Russia. Int J Appl Bus Econ Res 14
2. Fomin NY, Dyrdonova AN, Andreeva ES, Girfanova EY (2017) Clusterbased model of socioeconomic and innovative development of regions. In: Conference proceedings "Political sciences, law, finance, economics and tourism, vol 3, pp 771–778
3. Albiter LM (2011) Managing the development of the production infrastructure of industrial complexes: theory and methodology: monograph. SGASU
4. Shevchenko SA (2018) Influence of technological progress on organization's activities. In: Shevchenko SA, Kuzmina EV, Kuzmina MI (eds) Financial economics, no 7, 1, 114–117
5. Shevchenko SA, Kuzmina EV, Trunina VF, Kuzmina MI (2019) Corporation: conceptual approaches to the essence of the notion. Econ Entrep 13(2)(103):348–351

Chapter 5
Contemporary Issues of Effective Land Desertification Control in the South of Russia

Alla A. Murasheva, Pavel V. Klyushin, Svetlana V. Savinova, Viktor M. Stolyarov and Pavel P. Lepekhin

Abstract The paper presents the analysis of the agricultural exploitation of land resources of Russia at present, points out the problems in agriculture and identifies the main desertification control activities. This work is aimed at developing a number of guidelines for the control of degradation of agricultural lands in the south of Russia as a result of desertification. Research methodology consists of the analysis of modern condition of farm land and definition of threats of degradation of land due to desertification on the basis of a wide variety of archived and current statistical sources on the use of available land of the Russian Federation. As a result, the authors of this research revealed consistent patterns and trends in the quantitative condition of farm land, developed several operational guidelines contributing to the slowdown of desertification of the most valuable productive lands.

Keywords South of russia · Analysis · Available land · Issues · Agriculture · Desertification control

JEL Q240 renewable resources and conservation: land

A. A. Murasheva (✉) · P. V. Klyushin · S. V. Savinova · V. M. Stolyarov · P. P. Lepekhin
State University of Land Use Planning, Moscow, Russia
e-mail: amur2@nln.ru

P. V. Klyushin
e-mail: klyushinpv@gmail.com

S. V. Savinova
e-mail: savinova2010@gmail.com

V. M. Stolyarov
e-mail: vms88@inbox.ru

P. P. Lepekhin
e-mail: shampolamo@gmail.com

T. Kolmykova and E. V. Kharchenko (eds.), *Digital Future Economic Growth, Social Adaptation, and Technological Perspectives*, Lecture Notes in Networks and Systems 111, https://doi.org/10.1007/978-3-030-39797-5_5

5.1 Introduction

Desertification is a process that results in the loss of solid vegetation mantle by the natural ecosystem; this vegetation cannot self-regenerate in future. Self-regeneration is only possible once in a while—in case of elimination of a constant human-caused pressure. Active and often unreasonable human economic activities (deforestation, excessive exploitation of pastures, irrational utilization of water resources in case of irrigation of lands etc.) in arid regions, occupying about 30% of the land area, posed a real threat of disruption of the ecological balance, turning lands into barren and dangerous for neighboring regions that have not yet been affected by the desertification process. According to UNEP, each year 6 million hectares of productive, irrigated lands turn into deserts globally. In general, 40 million hectares of irrigated lands suffered as a result of the desertification process. Territories with the highest degree of desertification amount to: in Asia—about 19%, in Africa—23%, in Australia—45% and in South America—about 10% of the total area of land. The countries of the African continent are especially affected. In particular, in Mauritania, due to the progressive drought, 250 thousand km^2 turned out to be threatened by intense desertification. The Sahara desert is extended to the south at an average speed of 6 km/year, hence, out of 200 thousand hectares of dry-land suitable for agricultural work, only 50 thousand hectares remain.

By the early third millenium, desertification started threatening to 3.6 billion hectares of arid lands. This amounts to 70% of potentially productive arid lands, or ¼ of the total area of land surface; what is more, this data excluded the area of natural deserts. This process affects about 1/6 of global population. Desertification may occur in different climatic conditions, but it is especially rapid in hot, arid areas. The intensity and the scale of degradation of land are increasing in many regions of the world. 20% of arable land, 30% of forest land and 10% of ranchland are subject to degradation. According to some estimates, the life of 1.5 billion people, i.e. one fourth of global population, directly depends on the lands that are subject to degradation. The consequences of degradation of land are decline in fertility, population migration, food insecurity, destruction of natural resources and ecosystems, as well as biodiversity loss due to changes in the habitat of various species of plants and animals, including changes at the genetic level.

Russia is located in eastern Europe and northern Asia, occupying about $^1/3$ of the territory of Eurasia and $^1/9$ of land. As of January 1, 2018, according to Rosreestr, the total area of available land of the Russian Federation was 1712.5 million hectares, of them agricultural lands occupied 383.2 million hectares [1]. As at year 2019, Russian Federation includes 85 subjects: 22 Repubics, 9 Krais, 46 Oblasts, 4 autonomous districts and 1 autonomous region; Moscow, St. Petersburg and Sevastopol are cities of federal significance—individual subjects of the Russian Federation. In total, there are about 157 thousand settlements in the country [2, 3].

5.2 Methodology

The total area of farm land as part of agricultural lands, according to the data provided by Rosreestr, as of 01.01.2018 [1] is 197.8 million hectares (197,785 thousand hectares), including the total area of arable land—116.2 million hectares (58.8%), pastures—57.3 million hectares (28.9%), hayland—18.7 million hectares (9.5%), layland—4.3 million hectares (2.2%), perennial plantings—1.2 million hectares (0.6%). The area of nonfarm land in the structure of agricultural lands in 2017 was 185.4 million hectares (Fig. 5.1).

Only slight variations could be observed in the changes of areas of farm land as part of agricultural lands for the period from 2010 till 2018; their total area and structure didn't change compared to the year 2011 [1, 4, 5]. At the same time, according to the analysis of changes in the area of agricultural lands for the last five years, during the period from 2010 till 2014 there was a reduction of the total area of agricultural lands in the Russian Federation by 13.5 million hectares, and in 2016 there was an increase by 1.5 million hectares as a result of annexation of Crimea by Russia (Table 5.1) [6].

The annual reduction of the area of lands of this category could be observed during the period from 2010 till 2013 [1, 4, 5]. However, according to the Federal Service for State Registration, Cadastre and Cartography for 2013, the total area of agricultural lands has increased by 0.4 million hectares compared to the previous year in connection with the change of designated purpose to agricultural lands 1022

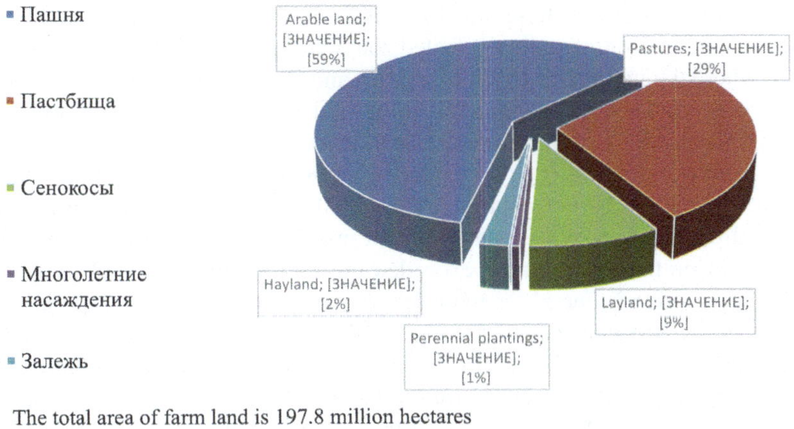

Пашня	Arable land
Пастбища	Pastures
Сенокосы	Hayland
Многолетние насаждения	Perennial plantings
Залежь	Layland

Fig. 5.1 The structure of farm land as part of agricultural lands in the Russian Federation as of 01.01.2018

Table 5.1 The structure and changes in the total area of agricultural lands (as of January 1, then-year), million hectares

Indicators	2010	2011	2012	2013	2014	2015	2016	2017	2018
The total area of agricultural lands	400.0	393.4	389.0	386.1	386.5	385.5	383.7	383.6	383.2
Including farm land	196.1	196.1	196.3	196.2	196.2	196.2	197.7	197.7	197.8
Including:									
Arable land	115.3	115.1	115.1	115.1	115.1	115.1	116.3	116.2	116.2
Layland	4.2	4.4	4.4	4.4	4.4	4.3	4.3	4.4	4.3
Perennial plantings	1.2	1.2	1.2	1.2	1.2	1.2	1.2	1.2	1.2
Hayland	18.6	18.6	18.6	18.6	18.6	18.7	18.7	18.7	18.7
Pastures	56.8	56.8	56.9	56.9	56.9	56.9	57.2	57.2	57.3

thousand hectares of land of other categories, including 930.1 thousand hectares of reserve lands, the main area of which is included in nonfarm land of agricultural lands within the territory of Krasnoyarsk Krai (850.6 thousand hectares). Since 2016, the total area of agricultural lands have stabilized at the level of 383.2 million hectares, the total area of farm land have stabilized at the level of 197.8 million hectares.

At present, environmentally neglected zones cover about 15% of territory of Russia [7], where the main production capacities and the most productive farm land are concentrated, and the most important thing is that more than 60% of the population lives here. Although in the mid-1990s, there was a short-term stabilization of environmental conditions (due to decline in industrial and agricultural production), ecological situation did not improve. After 2000, given the industrial growth in the context of degradation of state environment protection system, the volume of negative impact on the environment exceeded the level of the year 1987. At present, ecological situation is one of the main reasons for the deterioration of public health, decrease in life expectancy, and rise in mortality.

As early as in the 50–70s of the 20th century, the state of the majority of lands became environmentally unsatisfactory in the main industrial and agricultural areas of Russia. In 1989–2000, the process of the loss of fertility of soils accelerated due to erosion, violations in utilization of land, reduction in the volumes of fertilizer application, chemical and radiation pollution of environment. In Russia, about 40% of agricultural lands are subject to wind erosion, and about 18% of agricultural lands are subject to water erosion. In 35 subjects of the Russian Federation, desertification occurs (especially in the Volga region, Pre-Caucasian region, Dagestan, Voronezh Oblast, Orenburg Oblast, Omsk Oblast, Transbaikal).

Of special danger is the degradation of land (desertification), decrease in biological productivity in the arid belt of Russia [8], where these processes are aggravated by drought. According to the State Reports of the Federal Agency for the Cadastre of Real Properties (Rosreestr) [1, 4, 5], the total area of the territory that is subject to desertification is more than 50 million hectares. Desertification became especially extensive in the Caspian Sea region: in the Republic of Kalmykia (4.4 million hectares) and the Republic of Dagestan (2.4 million hectares), in the Chechen Republic (1.5 million hectares), Stavropol Krai (2.1 million hectares), Astrakhan Oblast (4.4 million hectares), Volgograd Oblast (11.4 million hectares), Rostov Oblast (10.1 million hectares), Saratov Oblast (2.2 million hectares), as well as in the Altai Territory (Kulunda Steppe)—(6.1 million hectares), the Republic of Tuva (even lands—2.6 million hectares). Desertification problem becomes especially acute in Krasnodar Krai, in the south of Saratov Oblast, in Orenburg Oblast, in Omsk Oblast and the Republic of Buryatia. This being said, the reasons for desertification are by 87% stipulated by the irrational utilization of natural resources, and only 13% are stipulated by environmental conditions.

All these negative processes caused the insufficient productivity of irrigated cultivated lands and the low stability of crop production. Therefore, in the 1980–90s, a significant part of degraded irrigated lands was decommissioned and transferred to the category of rain-fed lands or lands that were withdrawn from agriculture. One of the main reasons for the deterioration of the ameliorative condition of irrigated cultivated lands lies in the low feasibility of design solutions for the construction of most irrigation systems due to the absence or poor quality of forecasting. Over the past thirty years, since the construction of the main melioration systems, the distribution of production forces in agriculture, the structure of its production in individual regions, and the sum of business goals and incentives have changed significantly (Table 5.2).

At the same time, it is necessary to point out that desertification and other degradation processes resulted in the fact that the largest amount of unused arable land is in the Northwest Federal District—42% of the total area of arable land, and in the Far Eastern Federal District—37%. More than 20% of arable land is not used in the Siberian Federal District (24%) and the Central Federal District (23%). The condition of unused arable land is as follows: 7.8 million hectares or 37% are bushy and wooded, 1.8 million hectares or 8.5% are subject to erosion, and 600 thousand hectares or 3% are boggy and waterlogged. Hence, almost a fifth of arable land is degraded.

At present, works are underway in Russia for the creation of a regional environmental scheme of desertification control as a basis for the complex economic development of any region. The UN is already implementing the Project addressing the desertification problems, making provision for the preservation of landscape and ecological balance in various parts of the world that are subject to the development of desertification processes, with due account for social and economic problems.

North Caucasus economic region is one of the biggest regions of Russia. This region includes the Republic of Adygeya, the Republic of Dagestan, the Republic of Ingushetia, the Kabardino-Balkarian Republic, the Karachay-Cherkess Republic, the Republic of North Ossetia-Alania, the Chechen Republic, as well as Krasnodar Krai and Stavropol Krai and Rostov Oblast. The area of the region is 355.1 thousand km^2

Table 5.2 Areas agricultural lands of Russia that are subject to degradation processes [9]

Natural agricultural areas	Overmoistened	Eroded	Wind-eroded	Saline and saline alkaline
	% of the total area of farm land			
Northern economic region	29.8	5.4	0.3	0.8
Northwest economic region	26.7	6.3		
Central economic region	20.4	15.6	0.1	
Volga-Vyatka economic region	14.3	24.8		
Central Black Earth economic region	4.5	26.4	2.6	1.9
Volga economic region	3.3	26.6	12.7	37.4
North Caucasus economic region	8.5	27.3	22.6	21.1
Ural economic region	6.5	25.9	7.2	14.8
West Siberian economic region	20.3	6.6	12.9	35.1
East Siberian	7.8	9.8	14.3	3.8
Far Eastern	36.5	7	0.8	4.3
Russian Federation	11.8	19.1	9.8	18.1

and occupies 2% of the country. This region is located in the far south of European part of the Russian Federation and extends from the Lower Don to the Greater Caucasus Mountain Range. The region is surrounded by three seas: the Black, Azov and Caspian. North Caucasus economic region is one of the most densely populated regions of the country. More than 18 million people live within its territory. The average density of population is 50 people per square km. The population of the North Caucasus differs from that of other regions of the country by its multinationality (Table 5.3).

5.3 Results

The resources of the most important granary of the European part of Russia have been used thoughtlessly for many years. In pursuit of increasing yields, land improvement was not carried out in the required amount: a huge load on arable land resulted in the reduction of natural fertility of soils, their massive erosion. In addition, an immense amount of fertilizers and toxic chemicals accumulated in the soils and water bodies of the district. Contaminated waters from Ukraine and other states fall into the Sea of Azov and the Black Sea, too. Therefore, the region's ecosystems are in danger.

On the farming lands that are classified as agricultural lands, as well as included in the land redistribution fund, in the Russian Federation, 17.8% of the total area of farm land is subject to water erosion, of which 12.1% of arable land, and wind erosion—8.4% and 5.3% respectively. 2.4% of the total area of farm land are subject

Table 5.3 Territory and population by the subjects of the North Caucasus economic region of the Russian Federation as of 01.01.2019

Subjects of the Russian Federation	Area		Population			
	km^2	As a percentage of figure for the entire region	Size of population	As a percentage of figure for the entire region	As a percentage of rural population	Density of population, people/km^2
The Republic of Adygeya	7792	2.05	454762	2.02	53.0	58.4
The Republic of Dagestan	50270	13.20	3085738	13.69	55.2	61.4
The Republic of Ingushetia	3685	0.97	497393	2.21	40.0	137.0
The Kabardino-Balkarian Republic	12470	3.27	864455	3.84	47.9	69.5
The Karachay-Cherkess Republic	14277	3.75	465563	2.07	57.2	32.6
Krasnodar Krai	75485	19.66	5648254	25.08	45.6	74.8
The Republic of Crimea	26081	6.85	1911816	8.49	41.0	73.3
Rostov Oblast	100967	26.51	4202337	18.65	31.7	41.6
The Republic of North Ossetia-Alania	7987	2.10	699312	3.10	35.8	87.6
Stavropol Krai	66160	17.25	2795103	12.41	414.0	42.3
Sevastopol	1080	0.28	443211	1.97	7.0	513.2
The Chechen Republic	15647	4.11	1456951	6.47	65.6	93.1
North Caucasus economic region	380821	100	22524895	100		59.14
Russian Federation	17125191	100	146781095	–	25.8*	8.56

*Note** As at January 1, 2014

to cumulative impact of water and wind erosion. In addition, overmoistened and waterlogged lands account for 12.3%, and saline and alkaline soils account for 20.1% of the total area of farm land. In Russia, 24.3% of lands are subject to water erosion, and 40.2% of lands are subject to wind erosion.

The North Caucasian Federal District plays a critical part in the provision of the population of the country with food supplies—almost 8% of domestic agricultural products are produced here; this is the 5th place, giving way to the regions of Central Russia, the Volga region, the South and Siberia. In 2016, agricultural producers of the North Caucasus produced about 9% of the total volume of milk, 10.9% of the total volume of grain and 7% of the total volume of cattle and poultry meat, 15%

of the total volume of vegetables and 13% of the total volume of fruits. In addition to a significant contribution to the grain harvest, the regions of the North Caucasus with account of the climate pattern can become a center of import substitution for vegetables, fruits, grapes and their derivative products [10–12].

The North Caucasian Federal District is located in the south of European part of Russia, in the central and eastern parts of the North Caucasus. The total area of the district is 170.7 thousand km^2 and it is comprised of 6 republics and Stavropol Krai. The share of agriculture in the gross product of the district is about 15% (about 4% for Russia). This disctrict accounts for 45% of gathered vine crop in the Russian Federation, more than 10% of grain, fruits, berries and vegetables, as well as more than 5% of sugar beets. The area of farm land in the district is 11354.1 thousand hectares or 84.1% of the total area of agricultural lands. At the same time, almost half of it (49.8%) accounts for Stavropol Krai and 28.4% accounts for the Republic of Dagestan. The five republics account for only 2476.4 thousand hectares, or 21.8%, that is, the significative entities of the region in terms of agricultural production are the above-mentioned Krai and the Republic [10–12].

The farms of the North Caucasian Federal District keep 10.6% of cattle stock and 38.2% of sheep and goat stock out of the total amount of stock kept in of the farms of the Russian Federation. The North Caucasian Federal District accounts for 7.4% of milk and 43.8% of wool produced in the Russian Federation. In addition, the district develops pig farming, poultry farming and bee farming. All of the above bears evidence of high potential of this region, and, at the same time, there are many inner problems that impede its implementation. This district has all the necessary components for the high performance of the agroindustrial complex. At the same time, given the availability of necessary means of production, the North Caucasian Federal District is distinguished by low quality of life of population and high unemployment with a share of rural population ranging from 50 to 65%. In this regard, the Decree of the Government of the Russian Federation on the development of the south of Russia was adopted. High-quality grain production is guaranteed in Stavropol Krai, the steppe regions of Dagestan, and the Chechen Republic. Maize cultivation is expedient in the foothills of the Kabardino-Balkarian Republic, the Republic of North Ossetia-Alania and the Republic of Dagestan, as well as on irrigated lands of Stavropol Krai (Table 5.4).

Climatic conditions and the historically developed high level of development of the agroindustrial complex (primarily cattle breeding, sheep breeding, vine culture, horticulture, olericulture, and the food industry) make the agricultural sector of the economy one of the key sectors for the development of the North Caucasian Federal District. This being said, the region as well as the entire country is characterized by a number of endemic problems impeding the accelerated development of agriculture (Table 5.1). The region has experienced a sharp increase in the rate of degradation of land and desertification processes. Thus, at present, only within plain territory of Dagestan, desertification covers more than 2.5 million hectares of agricultural and forest lands, while in the northern part of Dagestan about 40% of lands are subject to severe desertification, and 8% of lands are subject to extremely severe desertification; 70 thousand hectares became open sands.

Table 5.4 Characteristics of agricultural lands of the subjects of the North Caucasian Federal District of the Russian Federation as of 01.01.2019

Subjects of the Russian Federation	Agricultural lands, thousand hectares	Total area of farm land		Area of arable land		
		Thousand hectares	As a percentage of figure for the entire district	Thousand hectares	As a percentage of the total area of farm land	Per citizen
Russian Federation	386135.8	196163.3	–	115121.0	58.7	0.79
The Karachay-Cherkess Republic	817.2	603.1	5.3	145.3	24.1	0.31
Stavropol Krai	6108.6	5657.1	49.8	3931.2	69.5	1.40
The Kabardino-Balkarian Republic	711.8	627.6	5.5	284.5	45.3	0.33
The Republic of North Ossetia-Alania	374.4	315.2	2.8	184.2	58.4	0.26
The Republic of Ingushetia	150.7	140.1	1.2	81.5	58.2	0.17
The Chechen Republic of Ichkeria	994.8	790.4	7.0	285.6	36.1	0.20
The Republic of Dagestan	4345.8	3220.6	28.4	467.5	14.5	0.16
North Caucasian Federal District	13503.3	11354.1	100	5379.8	46.6	0.55

In Stavropol Krai, 115 thousand hectares of saline lands were found; 21 thousand hectares are subject to desertification. The area of pastures reduced by 7.8 thousand hectares, and the area of farm land decreased by 42.2 thousand hectares. According to all available information, the area of solonetzic soils and alkaline soils here is about 1.5 million hectares, including 750 thousand hectares of arable land, which are common in chernozem soils (396 thousand hectares) and chestnut soils, and the mean score of human-caused salinization for landscapes of Stavropol Krai is 1.52. At present, 1670 thousand hectares are subject to deflation, which amounts to 30% of the total area of farm land. There are 84% of erosion threatening lands in the Krai together with rockiness—bogginess, salinization, alkali soils and solonetzic complexes, and waterlogging across all the landscapes of Krai have a degree of degradation from low to average, then the other three—from medium to high. Hence, the highest man-caused impact was registered from water erosion, as well as a joint manifestation of water and wind erosion, and deflation.

In the majority of cases, not one but several negative processes develop within the territory of the Kabardino-Balkarian Republic: flooding, water and wind erosion (deflation), salinization, increased area of overcompacted and stony lands etc. Thus, the most catastrophic condition (5 points—catastrophic) is observed within the territory of three districts—Leskenskiy, Mayskiy and Elbrusskiy, which is 133.8 thousand hectares or 18.84% of the total area of agricultural lands. Besides, in three administrative divisions—Baksan district and Baksan and Prokhladny urban districts—a total score of 4 points (very high) of degradation processes is observed, which is another 82.5 thousand hectares of agricultural lands or 11.62%, that is, catastrophic and very high total score of degradation occupy 30.46% of the total area of agricultural lands, which creates great problems for the experts of agricultural production.

In Dagestan, as a result of intensive development of natural resources, including disturbed hydrological regime of soils (phreatic decline), there is a degradation of relict (rare but formerly widespread) and oak forests in the Samur delta, as well as in the Kizilyurt and Kumtorkali districts on an area of 20 thousand hectares. In Dagestan, ecological disaster zones have already been identified: Kizlyar pastures, blacklands, Agrakhana creek, Tarkitau mountain—natural monument, Kaitag forests. Almost the entire territory of the plain Dagestan is subject to the influence of wind erosion. Out of 7 million hectares of blacklands and Kizlyar pastures, 1.5 million hectares accounts for the territory of the Republic of Dagestan. Out of this area, nearly 550 thousand hectares are subject to desertification, 130 thousand hectares of them are subject to severe desertification. Aerospace surveys showed that the movement of sands from neighboring regions was directed to the republic (Sukhokumsk, Terekli-Mekteb). According to experts, the area of open sands increases annually in the Caspian Sea Region by 40–50 thousand hectares [6, 11, 12].

Favorable environmental conditions of the North Caucasian Federal District have stipulated top-priority agricultural development for the crop production, including specializing in vine culture, horticulture, olericulture, and cattle breeding with the development of dairy and meat areas, as well as wool-and-meat sheep breeding, goat breeding. The economy of agricultural organizations improved somewhat, the activity of agroindustrial formations, as well as the work on social development of rural areas became more intensive. That said, the problems associated with the achievement of progressive economic development of the agroindustrial complex of the region still remain unsolved. Among them, it is necessary to highlight and urgently solve the following: technical and technological inferiority of agriculture of the North Caucasian Federal District due to insufficient income of agricultural commodity producers for the implementation of modernization and transition to the innovative development; restricted access of agricultural commodity producers to the market in the context of its imperfect infrastructure, increasing monopolization of trade networks, poor development of cooperation in the area of production and sales of agricultural products; slow rate of social development of rural areas, drop in rural employment in case of poor development of alternative types of activity, low social rating of agricultural labor, inadequate resource provision at all levels of funding.

5.4 Conclusion

The following activities aimed at preventing desertification and soil erosion must be performed within the territory of the southern regions of the Russian Federation:

- management of the long-term environmental monitoring of desertification processes and erosion, including stationary observations of hydrometeorological, agrochemical, biological, hydrological stations, posts, platforms etc., periodic surveys of the territory including aerospace, cartographic etc.;
- optimization of the use of land resources, optimization of the structure of farm land, specialization profile of farms, normalized utilization of pastures;
- amelioration, a system of measures, including safeguarding foresting, combating erosion of soils, amelioration of alkaline soils;
- expansion of water resources, including regulation of land runoff, search for and production of fresh groundwater, protection of surface water and groundwater from pollution
- adaptive landscape land use, ensuring high and sustainable productivity;
- vegetative reclamation of pastures;
- research studies, fundamental research of desertification factors;
- the use of social and economic mechanisms for desertification control, state supervision over nature consuming use, provision of incentives to environmental production works;
- increased role of interregional and international cooperation in desertification and soil erosion control.

In addition, in the future, for the North Caucasus economic region, including for the North Caucasian Federal District, there is a need to develop agroindustrial, recreational, port and trade complexes with all sectors contributing to their growth. Economic success will be achieved with social stability in the region, elimination of interethnic tension. That said, without strengthening the economy and improving the quality of life it is difficult to expect stability in this region. In the development of the regional economic and social policy in this difficult area, it is necessary to consider national religious traditions, values and priorities, many of which prevent mountain people from understanding the standards of European industrial society in accordance with which Russia is generally developed.

In connection with the above, it would be effective to develop a methodology for a comparative assessment of the economic feasibility of measures to prevent degradation of land and its restoration depending on the degree of damage and to conduct on its basis an assessment of the expediency of such restoration. In this regard, the improvement of the feasible decision-making system to justify both the methods for prevention of degradation and restoration of degraded lands that are subject to secondary salinization and to justify government spending for these purposes, is a highly urgent problem for science and practice. The importance of this problem was noted as far back as in the Federal Target Program "Conservation and restoration of fertility of soils of agricultural lands and cultivated lands as the national heritage of

Russia in 2006–2010 and until 2013", approved by the Decree of the Government of the Russian Federation No. 99 of February 20, 2006.

References

1. Report on the status and use of agricultural land (2019) Federal State Budgetary Institution "Rosinformagroteh", Moscow
2. Khlystun V, Murasheva A, Emelyanova T, Klyushin P (2017) Methodological foundations of the development of the real estate market. State university of land use planning, Moscow
3. Khlystun VN (2016) "Efficient use of land as a factor in the success of import substitution", Economics of agricultural and processing enterprises, No. 1, pp 58–61
4. State (national) report on the state and use of the lands of the Russian Federation in 2005–2006, and the Land Fund of the Russian Federation as of January 1, 2010–2013, http://rosreestr.ru/ (Accessed 4 Sept 2019)
5. Report on the status and use of agricultural land (2014) Federal State Budgetary Institution "Rosinformagroteh", Moscow
6. Kashtanov AN, Lisetskiy VN, Shvebe GI (2009) Fundamentals of landscape ecological farming. Kolos, Moscow
7. State report "On the State and Environmental Protection of the Russian Federation in 2017" (2018), Ministry of Natural Resources of Russia; NPP "Cadastre", Moscow
8. The scientific basis for preventing the degradation of soils (lands) of agricultural lands in Russia and the formation of systems for the reproduction of their fertility in adaptive landscape agriculture: Vol 1. Theoretical and methodological foundations for preventing the degradation of soils (lands) of agricultural lands (2013) Soil Institute of them. V.V. Dokuchaev Russian Agricultural Academy, Moscow
9. Methods for assessing the degree of degradation of agricultural land: scientific. Edition/Federal State Budgetary Institution Research Institute "Rainbow" (2015), IP Vorobiov O.M., Kolomna
10. Klyushin PV, Musaev MR, Savinova, SV, Avaev R (2015) Rational use of agricultural land in the North Caucasus Federal District and the Republic of Dagestan, Land management, cadastre and land monitoring, vol 2, pp 23–31
11. Klyushin PV, Musaev MR, Savinova SV, Shalov TB (2016) Rational use of agricultural land in the North Caucasus Federal District, Moscow
12. Klyushin PV, Musaev MR, Shapovalov DA, Savinova S (2016) Ecology of land use of agricultural land in the North Caucasus Federal District. South of Russia: Ecology, Development 11(2):132–142

Chapter 6
Business That Brings Pleasure: Fundamental Principles and Conceptual Attitudes for the Management of a Modern Enterprise

O. V. Konina

Abstract This research is aimed at identifying the fundamental principles and conceptual attitudes for the management of a modern enterprise in the interests of taking pleasure in doing business. An original hypothesis has been made according to which many modern entrepreneurs do not take pleasure in doing business due to the immature research and methodology basis and the lack of targeted management measures, preventing them from getting happy. In order to test this hypothesis, in June-July 2019, a survey of 5000 owners of individual, small and medium enterprises who independently manage their businesses (in the regions of the Southern Federal District of the Russian Federation) was conducted in Russia. The results of research verified the hypothesis and showed that doing business that brings pleasure is a challenging task, which is not always possible to implement without a clear guideline to follow. Using the example of enterprises of the Southern Federal District of the Russian Federation in 2019, it has been substantiated that entrepreneurs are interested in taking pleasure in doing business, but they cannot achieve this in the majority of cases (57.32%). Taking pleasure in doing business determines the level of happiness of entrepreneurs and is an essential condition for building pleasure economy. The author has formed the research and methodology basis of the concept of "business that brings pleasure": conceptual attitudes and fundamental principles for following this concept have been defined. The developed model of management of a modern enterprise in accordance with the concept of "business that brings pleasure" is intended to serve as a guidemark for its implementation by modern entrepreneurs.

Keywords Business that brings pleasure · Doing business · Management of an enterprise · Pleasure economy

JEL codes: J28 · L26 · M21

O. V. Konina (✉)
Volgograd State Technical University, Volgograd, Russia
e-mail: koninaov@mail.ru

T. Kolmykova and E. V. Kharchenko (eds.), *Digital Future Economic Growth, Social Adaptation, and Technological Perspectives*, Lecture Notes in Networks and Systems 111, https://doi.org/10.1007/978-3-030-39797-5_6

6.1 Introduction

A key priority for the development of modern economic systems is the formation of pleasure economy, in which the highest quality of life would be achieved. In it, every person can unlock their potential and experience joy from meeting their needs. It should be pointed out that pleasure economy is relying on the ideas of capitalism and is a model of social market-based economy. Therefore, unlike other models focused on the solution of the optimization problem of satisfaction of universal needs (from communism, foe example), pleasure economy takes into account the lack of availability of resources, and its target function is related not to the full, but the maximum possible satisfaction of needs of society.

Pleasure economy is formed when Pareto optimum is established: the level of happiness in society, by which fuller satisfaction of needs of some social groups results in lower satisfaction of needs of other social groups. Another important feature of pleasure economy is the fact that satisfaction of needs in it is not a goal in itself, but the instrument for taking pleasure. Therefore, the range of academic problems of pleasure economy as a concept of modern economic theory consists in the goal to ensure (if absent) or maximize (if present) pleasure taken while meeting needs in society.

This problem is manifested differently—we shall examine them in accordance with Maslow's hierarchy of needs. The needs for essential products and services (physiological needs, safety), which are at the lower levels of the hierarchy, are considered in the World Happiness Report, and also include social support, freedom, lack of corruption, generosity, level of GDP per capita, and healthy life expectancy [8].

The needs of the middle levels (belonging, respect) are more complex and under current conditions of economy management they are often characterized by artificial nature—they are formed by means of marketing (advertising) by manufacturers and sellers of products and services, based not on the objective need for them, but on fashion and on beliefs in their benefits and advantages. After the purchase, consumers often do not get pleasure or are even disappointed, because the products and services they purchased are not as useful and beneficial as they had expected. This manifestation of the problem of taking pleasure while meeting needs is studied within the scope of the concept of Consumer Society at the interface of philosophical, social, and economic sciences.

And finally, the needs of the highest level (self-fulfillment: cognition, aesthetic needs, disclosure of one's abilities) are the most complex and are realized in the education market and the labor market. Getting pleasure from finding a job by profession, achieving good performance and building a career is studied within the framework of the Concept of higher-performance incentives. At the same time, taking pleasure in doing business is the least studied manifestation of the problem solved in pleasure economy and a gap in its concept. Hence, the relevant area of studies on pleasure economy is the elaboration of research and methodology basis for taking pleasure in doing business.

This research is aimed at identifying the fundamental principles and conceptual attitudes for the management of a modern enterprise in the interests of taking pleasure in doing business.

6.2 Materials and Method

In the existing academic and economic literature, the problem of taking pleasure in doing business is considered through the prism of state enterprise management. An example is the writings of Battersby [2], Broome [5], Cikic et al. [6], Dirks [7], Horton and Sloan [9], Manning and Bungay [13], Morozova et al. [14], Ostrovskaya et al. [16], Konina et al. [11], Konina and Nanetadze [10], Popkova et al. [18], Sergi and Scanlon [19], Tekin and Konina [21].

In turn, the problem of taking pleasure in management of an enterprise is studied from he perspectives of a hired manager in corporate entrepreneurship. In this case, he realizes his needs in the implementation of labor (not entrepreneurial) potential. Examples of publications on this topic are the writings of Albert [1], Belhabib et al. [3], Boukid [4], Malaquias and Hwang [12], Nagano [15], Parham [17], Strizhova and Evdokimenko [20].

Thus, in consequence of a review of existing researches and publications on a given topic, it was found that it is multi-aspect. Although many aspects of this topic are fairly well elaborated in the publications of various authors, the aspect of improving the practice of managing their own (individual, small and medium) enterprise to take (or maximize) pleasure in this process is still understudied and needs further studies, which this paper actually deals with.

An original hypothesis H0 has been made in this research, according to which many modern entrepreneurs do not take pleasure in doing business due to the immature research and methodology basis and the lack of targeted management measures, preventing them from getting happy. This hypothesis is tested on the example of Russia in 2019 as a country with a significant share of individual, small and medium businesses in the structure of entrepreneurship (61.54%)—according to Rosstat (2019), the are totally 4,561.7 thousand enterprises in Russia of which 238.3 thousand pcs are small business entities and 2,568.8 thousand pcs are individual business entities. Russia is also the preferred object for the study, since it occupies the middle position (68th of 147th) in the global ranking according to the level of happiness [8].

In order to test this hypothesis, in June-July 2019, a survey of 5,000 owners of individual, small and medium enterprises who independently manage their businesses (in the regions of the Southern Federal District of the Russian Federation) was conducted in Russia. Respondents were asked to answer the following questions: "Do you take pleasure in doing business? (yes/no)", "Does the fact of taking pleasure in doing business influence your inner feeling of happiness? (yes/no)", "How happy are you? (%)", "How high are the following barriers on the way to taking your pleasure in doing business? (please assess in points from 1 to 10) "high workload (a lot of things to do), recurring losses, multitasking, misbehavior of employees, lack of interest in business specialization”.”

6.3 Results

The results of this survey are shown in Figs. 6.1, 6.2.

As can be seen from Fig. 6.1, 42.68% of entrepreneurs take pleasure in doing business. 89.14% of respondents pointed out that taking pleasure in doing business has an impact on their inner feeling of happiness. 59.47% of entrepreneurs assessed their own level of happiness at 59.47%. Hence, taking pleasure in doing business is

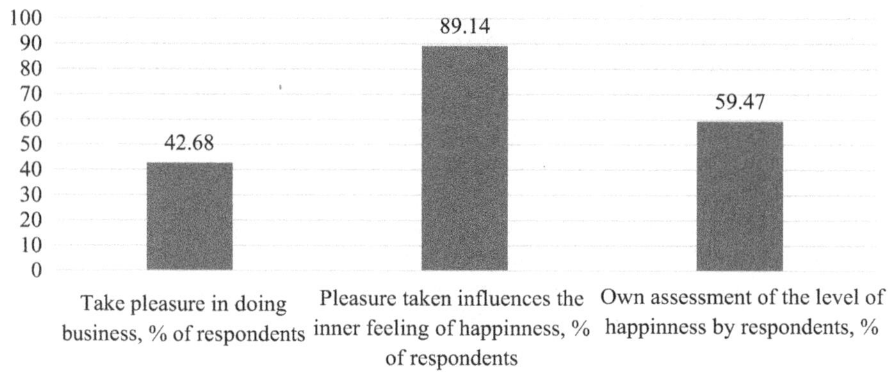

Fig. 6.1 Pleasure in doing business and its influence on the level of happiness of entrepreneurs. *Source* calculated and plotted by the author

Fig. 6.2 Barriers on the way to doing business that brings pleasure. *Source* calculated and plotted by the author

very important in the system of needs of an entrepreneur. Taking pleasure in doing business determines the level of happiness of entrepreneurs. There are barriers on the way to doing business that brings pleasure that are shown in Fig. 6.2.

As is shown in Fig. 6.2, the most serious barrier on the way to doing business that brings pleasure consists in recurring losses (9.12 points). There are also such significant barriers as high workload (7.86 points), multitasking (4.68 points), misbehavior of employees (2.19 points) and lack of interest in business specialization (22.19 points).

In order to overcome the identified barriers during the management of a modern enterprise, it is suggested that one should rely on the following conceptual attitudes:

- recognition of the possibility in principle for taking pleasure in doing business;
- pursuance of taking pleasure in doing business;
- treatment of doing business as a source of pleasure and a method of becoming happy.

The following fundamental principles for the management of a modern enterprise are proposed for doing business that brings pleasure:

- principle of delegation of powers and responsibility: abandoning independent execution of all organizational, managerial and other functions by an entrepreneur and the selection of regular or temporary assistants;
- risk insurance principle: creation of reserves or acquisition of insurance services to increase the entrepreneur's tolerance to additional costs and losses;
- teambuilding principle: selection of personnel in such a way that they can form a coordinated team subordinate to the manager (abandoning saving on human resources).

It is also recommended to apply a creative approach to doing business, which implies active introduction of innovations and the pursuance of increasing the interest of even a business that is initially unattractive to the entrepreneur and launched exclusively for the purpose of realization of profit. The proposed conceptual attitudes and principles were used as a basis for the development of a model of the management of a modern enterprise in accordance with the concept "business that brings pleasure" (Fig. 6.3).

As can be seen from Fig. 6.3, the model of the management of a modern enterprise in accordance with the concept "business that brings pleasure" ensures such benefits for entrepreneurs as taking pleasure in doing business, increase in effectiveness of entrepreneurship (happy entrepreneur will be more successful), the fullest satisfaction of needs (according to Maslow's hierarchy of needs) and "balanced happiness" (taking pleasure in satisfaction of needs at all levels), as well as the increase in the level of happiness. Besides, this model creates benefits for the economic system: the increase in the level of happiness (and improvement of the quality of life of population) in society and the formation of pleasure economy: competitive (territory that is attractive in terms of residency and doing business), efficient (costs for business bring happiness) and stable (happy society has advanced capability of withstanding disturbing factors and crises).

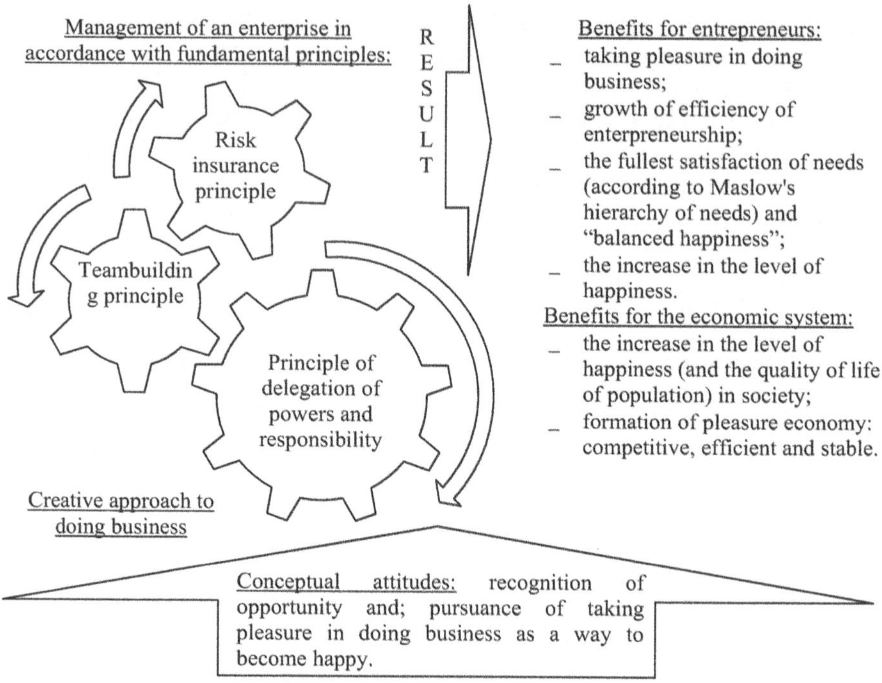

Fig. 6.3 Model of the management of a modern enterprise in accordance with the concept "business that brings pleasure". *Source* developed and compiled by the author

6.4 Conclusion

The results of research verified the hypothesis H0 and showed that doing business that brings pleasure is a challenging task, which is not always possible to implement without a clear guideline to follow. Using the example of enterprises of the Southern Federal District of the Russian Federation in 2019, it has been substantiated that entrepreneurs are interested in taking pleasure in doing business, but they cannot achieve this in the majority of cases (57.32%). Taking pleasure in doing business determines the level of happiness of entrepreneurs and is an essential condition for building pleasure economy.

In the course of research, the research and methodology basis of the concept of "business that brings pleasure" has been formed: conceptual attitudes and fundamental principles for following this concept have been defined. The developed model of management of a modern enterprise in accordance with the concept of "business that brings pleasure" is intended to serve as a guidemark for its implementation by modern entrepreneurs.

References

1. Albert DP (2017) Mixing business with pleasure preface. Int J Appl Geospatial Res 8(2):V–VI
2. Battersby ME (2015) Business and pleasure: attending trade shows is a valuable part of business strategy-with tax benefits. Spec Fabr Rev 2015(1):28–39
3. Belhabib D, Campredon P, Lazar N, Kane EA, Pauly D (2016) Best for pleasure, not for business: Evaluating recreational marine fisheries in West Africa using unconventional sources of data. Palgrave Communications 2(1):15–50
4. Boukid F (2017) 16th European Young Cereal Scientists and Technologists Workshop: Business and Pleasure in Thessaloniki, Greece. Cereal Foods World 62(4):186–187
5. Broome B (2018) Mixing business and pleasure. Arch Rec 206(4):88–92
6. Cikic J, Jovanovic T, Nedeljkovic M (2018) Business and/or pleasure - Gender (In)equalities in rural tourism in Vojvodina. J Agric Sci Technol 20(7):1341–1352
7. Dirks D (2012) Business and pleasure. Confrontation 112(1):132–145
8. Helliwell JF, Layard R, Sachs JD (2019) World Happiness Report – 2019. https://worldhappiness.report/ed/2018/. Accessed 2 Sept 2019)
9. Horton S, Sloan D (2016) Accessibility for business and pleasure. Interactions 23(1):80–84
10. Konina OV, Nanetadze ED (2019) Managing a modern university: the role of business processes in the structure of establishments of higher professional education. Adv Intell Syst Comput 726(1):1053–1059
11. Konina OV, Natenadze ED, Simonov AB (2014) Entrepreneurial culture and management style: overview and formation mechanism. World Appl Sci J 30(9):1109–1112
12. Malaquias RF, Hwang Y (2017) Mixing business and pleasure: empirical implications for trust in mobile banking. J Electron Commer Res 18(3):212–224
13. Manning E, Bungay V (2017) Business before pleasure': the golden rule of sex work, payment schedules and gendered experiences of violence. Cult, Health Sex 19(3):338–351
14. Morozova IA, Popkova EG, Litvinova TN (2019) Sustainable development of global entrepreneurship: infrastructure and perspectives. Int Entrep Manag J 15(2):589–597
15. Nagano H (2008) Pleasures in the management of a venture business born from a university. Zair Kankyo/Corros Eng 57(10):435–440
16. Ostrovskaya VN, Tyurina YG, Konina OV, Pupynina EG, Natsubidze AS (2017) 'Perspectives of elimination of "institutional gaps" in foreign economic activities of subjects of SME within the global crisis management. Contrib Econ 9783319606958:213–218
17. Parham S (2018) From the agora to the modern marketplace food markets as landscapes of business and pleasure. In: Routledge handbook of landscape and food. A franchise: Routledge Handbooks, Oxford (England), Routledge, pp 445–461
18. Popkova EG, Morozova IA, Litvinova TN (2017) New challenges for human capital from the positions of its infrastructural role in the system of entrepreneurship. In: Human capital: perspectives, challenges and future directions. Nova Science Publishers, Inc, New York (United States), pp 257–275
19. Sergi BS, Scanlon CC (eds) (2019) Entrepreneurship and development in the 21st century. Emerald Publishing, Bingley, UK
20. Strizhova EA, Evdokimenko AS (2018) Business leaders in finance - the key measures of motivation. Organizatsionnaya psikologiya 8(4):24–38
21. Tekin AV, Konina OV (2019) The role of information and communication technologies in the process of strategic management of entrepreneurial structures activities: the budget and financial aspect. Adv Intell Syst Comput 726(1):269–278

Chapter 7
Self-administration and Development of Local Communities of North Caucasus: By the Example of Kabardino-Balkar Republic and Karachay-Cherkess Republic

Aslan M. Chechenov, Murat Z. Shogenov, Gyuldzhan K. Azamatova, Zarema A. Atabieva and Natalia N. Reshetnikova

Abstract *Purpose*: this paper is aimed at studying the problem of formation of local self-government. Besides, the paper is focused on the issue of strategic control of development of municipal districts, which includes mechanisms of state administration, local self-government, traditional self-organization of local community in the territory of North Caucasus as exemplified by the two Republics—Kabardino-Balkaria and Karachay-Cherkessia. *Design/methodology/approach*: The tools of systemic, structural functional, sociocultural, and institutional approaches were used in the process of substantiation of theoretical provisions, data analysis, conclusions and recommendations that allowed analyzing and taking the changes of local communities in the North Caucasus as a whole. Modern government institutions in the territory of the North Caucasus (based on the data from Kabardino-Balkaria and Karachay-Cherkessia) are analyzed as the studied material. The analysis of this paper was based on the results of field studies that were performed in 2014–2018 within the scope of the North Caucasian Complex Expedition from 2014 till 2018. A review of the literature on the selected topic has shown that the role and significance of

A. M. Chechenov (✉) · M. Z. Shogenov · G. K. Azamatova · Z. A. Atabieva
Kabardino-Balkaria State University, Nalchik, Russia
e mail: a.chechenov@mail.ru

M. Z. Shogenov
e-mail: shogen@inbox.ru

G. K. Azamatova
e-mail: azagulya2007@mail.ru

Z. A. Atabieva
e-mail: atab_zara@mail.ru

N. N. Reshetnikova
Don State University, Rostov-on-Don, Russia
e-mail: nata.dstu@yandex.ru

Rostov State University of Economics, Rostov-on-Don, Russia

T. Kolmykova and E. V. Kharchenko (eds.), *Digital Future Economic Growth, Social Adaptation, and Technological Perspectives*, Lecture Notes in Networks and Systems 111, https://doi.org/10.1007/978-3-030-39797-5_7

self-government and the development of local communities of the North Caucasus are discussed in many papers of such scholars as Chechenov et al. [1], Kumykov et al. [2], Magomedov et al. [3], Smagina et al. [4], Popkova et al. [5], Azamatova et al. [6], Lokova et al. [7], Krasnokutskiy et al. [8], Reshetnikova et al. [9], Zmiyak et al. [10], Reshetnikova and Magomedov [11]. *Findings*: The authors have identified modern institutional self-governments at the local level, dynamic pattern of transformation of relations between the state and society, as well as problems and prospects of improvement of local self-government in the context of development strategies of rural settlements. *Originality/value*: The novelty of the approach consists in the substantiation of the systemic concept of local communities as specific territories that have a certain commonality of cultural, historical and state development.

Keywords Local community · Local self-government · Development · North caucasus

7.1 Introduction

In addition to public administration which acts as external within the scope of activities of local communities, there is local (municipal) administration. This type of self-administration is an organizational mechanism which allows monitoring matters of local significance in reliance upon the local code, with due consideration of interests of the local community.

It is worth noting that responsibilities of the state have not yet been fully established at the local level. If we analyze the law about local government, we can find many contradictory points in it.

It has been noted in academic literature that during the crisis, the state dealt with problems that were related to local issues in a certain territory, in an alienated manner. This circumstance led to the fact that territories such as state farms and collective farms were abandoned, and the social structure in the village was severely affected. The situation hasn't changed at present: the powers of local authorities are only documented in black and white, while in fact they cannot exercise their powers, because they have neither tangible nor organizational resources for that.

Another point of view, which is reflected to a lesser extent in the academic literature, is the great opportunities available of local authorities.

Judging from this perspective, existing mechanisms of self-government in the territory of the North Caucasus act as the only effective mechanism in the context of present-day reality [12].

The self-government concept, by definition, could not emerge out of nothing, let alone losing its relevance today. To understand the current situation of this concept, one should know that the idea of a social contract appeared as early as at the origins of Greek philosophy, and more specifically in the worldview of sophists and epicureans. Later, during the Middle Ages, new time, this concept acquired various modifications, which always attracted the attention of forward-minded individuals and still maintain their popularity up to this day.

A European vision of this question is covered in papers by G. Hobbs, J. Locke, Charles-Louis de Montesquieu and Jean-Jacques Rousseau, who wrote about the state and its essence within the framework of the theory of social contract. The essence of this theory was that people could no longer be at feud and decided to create a single state, transferring part of their freedoms to it. That is, based on the social contract—an agreement accepted by all, the people exchanged their freedoms for a state guarantee of security and protection against external enemies in case of an attack [13]. Discussion of problems related to the development of local self-government and relations that established between the people and the state, the problems of the legitimacy of power and justice constitute a huge part of the political thought existing in the West. Contractual theories are supported by such scientists as John Rawls, Ronald Dvorkin, James Buchanan, David Gauthier, Thomas Scanlon [14, pp. 130–142].

And if we take a look at the real state of things existing today, the local government strategies can have various combinations. One of such situations is the situation at which state mechanisms are concentrated in the center, while self-organization mechanisms are prevalent in the territory of distant regions—high mountain regions.

So, what combinations of management types exist today? In order to answer this question, we have to analyze the historical block dealing with the formation of local self-government, and touch upon the data obtained in the course of field studies.

7.2 Materials and Method

In order to assess the efficiency of local self-government strategies, we shall turn to the results of the studies that were obtained as a result of the North Caucasian Complex Expedition. This research was conducted for 4 years (2014–2018) in the territory of the entire North Caucasus. The measurements were taken in settlements of Karachay-Cherkessia and Kabardino-Balkaria. Data collection methods: questionnaire, interviewing and statistical data processing.

When we give consideration to the two Republics: Kabardino-Balkaria and Karachay-Cherkessia, we should point out their typical features for the North Caucasian federal district in the sociopolitical and economic system.

But on the other hand, these territories act as twins in terms of the ethnic composition of the title peoples, who represent the elite public authorities in this territory. In the North Caucasus, these two republics only preserved their ethnicity within the framework of administrative and territorial arrangement which can be observed in these territories since the Soviet era. The unification of ethnic groups within the framework of administrative arrangements occurs due to small size of territory and historical ties between the mountain and plain districts (Table 7.1). At the same time, the republics present different cases of the relationship between the state and the local community. For example, distinctive features are most pronounced within the

Table 7.1 Characteristics of the title ethnic groups of Kabardino-Balkaria and Karachay-Cherkessia

Name	Language	The share of ethnic groups in the number of population of republic (%)[2]			Main form of traditional farming
		1989	2002	2010	
The Kabardians	Adygei group (Caucasian family)	49	55	57	Hill-and-plain agriculture
The Circassians		9.7	11.3	12	
The Balkars	Turkic group (Altai family)	9.6	12	13	Hill animal breeding
The Karachays		31	38.5	41	

framework of greater concentration of power. Kabardino-Balkaria is characterized by greater concentration of power, which is expressed in the control of state (republican) authorities over local authorities. And in the territory of Karachay-Cherkessia, greater autonomy on the part of the local community as against republican authorities can be observed [15, pp. 89–94].

7.3 Results

Modern government institutions at the local level are undergoing profound changes. This is primarily due to the reassessment of the role and the significance of the local level in the state. Local community includes the layers of the Soviet period, legal innovations and the modern period of centralized management.

Soviet period. In the Soviet era, local communities were the element of the machinery of government. At local levels, there were also three branches of government that functioned at the state level: Soviet, party and economical.

On the face of it, the major part was granted to the Soviet power represented by village councils. This structure was headed by the chairman. He and the secretary of the council who carried out all the bureaucratic work, held positions paid from the state budget, i.e. were full-time officials. The remaining deputies of the village council combined their activities with their primary employment.

Party power in the village played a relatively minor role (compared to that of the district and the city and at higher government levels). The secretary of the village party organization was the main assistant to the collective farm chairman in resolving personnel issues.

The collective farm chairman or the director of the state farm had more power in the village than the party secretary. It was them, and not the secretaries of rural party organizations or the chairmen of village councils, that the stake was placed on. Given significant state subsidies, collective and state farms maintained the social sphere and gave jobs to most villagers (in fact, there were hardly ever any issues associated with obtaining employment or receiving salaries). There was no sufficient culture medium for conflict resolution. The role of traditional institutions in the settlement of conflicts was reduced (the ubiquitous state depreciated traditional institutions). Traditional institutions of the intra-village organization (congregations, council of elders, religious organizations et al.) went into decline during the Soviet period. However, there were still some niches where the state could not get. These included the adat which regulated the use of land in the mountains or could help in resolving intrafamilial conflicts.

Touching upon the period of liberal reforms of the 1990s, we should note that the institutions of councils formally functioned until the 90s. The party did not have much power either, and the collective farm chairman did not have the absolute power to make decisions. Traditional institutions were the most effective, but regulated only parts of the relationship inside a rural settlement. These democratic changes became the basis of mythopoetics [16, pp. 3–22]. This may include such statements as "local self-government is the foundation of Russian statehood", "local government is the most democratic form of government." In 1991, the Law of the RSFSR No. 1550-1 of 06.07.91 "About local self-government in the Russian Soviet Federated Socialistic Republic" was adopted, the provisions of which established the principle of competence of local self-governing authorities. In accordance with this document, elected bodies were formed, which managed their own resources, including the local budget. This local self-government body was answerable not to the people, but to the superior bodies. In 1993, the activities of the Soviets subsided, and the heads of administrations were granted authoritative powers.

Thus, towards mid 1990s, three types of institutions were present in he villages:

(1) Administration—an official local self-governing authority which is controlled by superior public authorities.
(2) Collective farms—former collective and state farms, which became known as joint stock companies.
(3) Traditional institutions—governing bodies that have been formalized by the state, for example, the Council of Elders, and unformalized institutions—the joint use of public lands.

When we give consideration to the time period since the beginning of the 2000s, one cannot but touch upon municipal reform. In 2003, the Federation Council approved Federal Law No. 131-FZ "On general principles of organization of local self-government in the Russian Federation". This document was an attempt by the state to take control of the local level.[1] During this period, there was an accrescence of power both at the federal and regional levels. Within the framework of

[1]It should be noted that there have been several attempts like that in the history of the Russian state. The most destructive of them are as follows: (1) Stolypin's reform of the early XX century, aimed at the destruction of the conservative communal pattern in the village and the development of elements of free competition, the revitalization of entrepreneurship; (2) Collectivization of the 1920–30s, which was primarily aimed at organizing institutions that promote public administration and control at the local level; (3) Khrushchev's reforms, which resulted in the liquidation of the so-called unpromising rural settlements (mostly small and remote), the massive relocation of highlanders to the plain, etc.

municipal reform, local self-government obtained functions related with social and economic development. Subsidies were directed to activities within the framework of administrative system and on the operation of economic projects. Federal Law on local self-government was to have dramatically changed the principles of local self-government: heavier regulation of local self-governing authorities, unification of approaches used within the scope of this process, arrangement of competencies implemented by local self-governing authorities.

Local self-governing authorities are formed according to various principles. In some regions, the head is selected from among the deputies, but in some placed the head of administration is elected directly. In many rural settlements of Kabardino-Balkaria, the head is appointed upon the recommendation of the district administration. The head is in charge of the local administration and representative body. Elections are held in Dagestan, during which heads of municipal districts are appointed. 90% of the heads of administrations are elected in such a manner. The heads of municipal districts in the Republic of Daghestan, as well as and in Kabardino-Balkaria, are also the heads of local administration and head the representative public authority. Mixed model of local self-government is in common practice in Karachay-Cherkessia—direct elections exist here within the framework of municipalities of the first level.

Table 7.2 shows key villages of Republics under consideration with population of different ethnicities, different quantitative composition, economic and living environment, located at different distances from administrative centers.

Staryy Cherek, Islamey and *Aushiger* are rural settlements in Kabardino-Balkaria. These villages are located in piedmont region with productive soils. These territories are located close to main transport routes, which has a beneficial effect on the development of turnover of commodities and the movement of the population in the republic as a whole.

Staryy Cherek is a village that belongs to the most economically active settlements in the republic. Population growth has been observed in recent years. The economy is dominated by agriculture and gardening. Horticulture in hothouses and greenhouses gained ground. Products are sold in outdoor agricultural markets located in the village along the federal highway. The land is distributed by the administration of municipal district. There is a shortage of pastures. The main pastures are located beyond the river Cherek. The maximum amount of developed land was in the 1970–1980s. At present, the areas under pastures are decreased, and the share of arable land leased by private owners and various enterprises is higher than during the period of maximum development. The lack of a convenient bridge and periodical floods complicate cattle grazing and damage pastures.[2] The proximity to the district center counterbalances the lack of work opportunities in the village. Part of the population is employed in the capital of the Republic. The municipal infrastructure (water supply, gas supply, removal of refuse) is relatively well developed, but there are not enough opportunities for leisure activities. Management in the village is represented by both formal and

[2]Commentary interview in the course of filling out a questionnaire with a resident of Old Cherek on 19.05.2014.

Table 7.2 Characteristics of key villages

Name	Main ethnic groups	Number of population	Proximity to administrative centers	Special aspects of economic and living environment
Staryy Cherek	The Kabardians	6500	Close to the district center and relatively close to Nalchik	Piedmont and low-mountain territories with limited land potential
Islamey		11,500		
Aushiger		4850		
Krasnyy Kurgan	The Karachays	3200	Proximity to the district center and Kislovodsk	
Kardonikskaya	The Russians, the Karachays	7800	Relative proximity to the district center	
Zhankhoteko	The Balkars (60%) and the Kabardians (40%)	1350	Far removed from administrative centers	Low mountains with the lack of arable land and pastures
Khasan'ya	The Balkars	10,800	Nalchik's suburb	
Zhankhoteko	The Circassians	1200	Relative proximity to the district center	
Nizhnyaya Teberda	The Karachays	1200		Medium-altitude mountains with the lack of agricultural lands
Elbrus	The Balkars	5370	Far removed from administrative centers	High mountains with the lack of agricultural lands and high potential in terms of pastures

informal institutions. There are strong family-lineal ties regulating the access to resources and local self-government. Other formal social institutions regulating the interior life of the settlement, have preserved as well (council of elders—council of veterans of labor and war, female council, village community council).[3]

Islamey is a Kabardian village which is the largest of the sample, with more than 11 thousand people. Population growth has been observed in recent years. It is located near the district center Baksan and relatively close to the capital of the Republic—Nalchik. The land potential is limited, but proximity to the district center

[3]Commentary interview in the course of filling out a questionnaire with a resident of Old Cherek on 18.02.2015.

counterbalances the lack of work opportunities in the village for the people. Favorable soil and climatic conditions contribute to home gardening, but the development of intensive gardening on the lands in the vicinity of the village rented by large agricultural firms has led to a decrease in demand for products from households, and, accordingly, household incomes. There is an acute shortage of land for residential construction. In order to solve this problem, scarce agricultural land is allocated, stirring up disputes between the village administration and local entrepreneurs who need land. There are several housing and public utilities problems, the most pressing of which are water supply and removal of refuse. The problem of water supply is especially critical during the period of irrigation and is accompanied by conflicts between the upper and lower parts of the village. Local self-government council is formally present and is represented by people who are held in respect in the village—headmaster, doctors, veterans, entrepreneurs etc. The proximity to the agglomeration of the district center and the inclusion of the population in its economy and institutional environment gradually eliminate the importance of traditional rural institutions.

Aushiger is a large Kabardian settlement which is located in the Cherek district of Kabardino-Balkaria. A slight population growth has been observed in recent years. Agriculture, especially horticulture, is traditionally developed in this district, but the land potential is limited. A specific resource is thermal springs which are popular with tourists and are rented by actors of the republican level, yet of rural descent. The state is present in its formal institutions, mainly controlling local resources. Traditional social institutions, such as the Council of Elders, are there too, but they are rather more representative. The actual administration in the village is represented by informal leadership, reputable villagers included in the networks at the republican level. The cohesion of these networks makes it possible to solve issues relevant to the village and mobilize the population for welfare activities whenever necessary.

Krasnyy Kurgan—this populated locality has a good geographic location: The Kislovodsk—Karachayevsk—Cherkessk highway is located nearby. It is mainly inhabited by the Karachays. Compared to other populated localities, one may observe the growth of population size (currently 4100 people, and in 1990–3120 people) here. The territory of the settlement has a stud farm, sewing shops and a number of tourist attractions. Favorable territorial location and the availability of jobs attracts the population to stay in this territory. There are many lands in the territory of Krasnyy Kurgan.

Kardonikskaya cossack village is located in the mid-mountain region of Karachay-Cherkessia. There are a large number of pastures in this area. The Karachays are dominant among the population of this territory. During the Soviet period, this cossack village had a well-developed industrial sector, which fell to decay at present.

The village of *Zhankhoteko* is located far away from district centers. This settlement is located at the junction of the Balkarian and Kabardian natural areas. The ethnic composition of the population is diverse: the Balkars (58%) and the Kabardians (40%). The number of population in the village has decreased in recent years. The former administrative affiliation of the village changed in the early 1990s, having changed from the Kabardian Baksan district to the Balkarian Elbrus district. Before perestroika, the main position was held by the chairman of the collective farm, a Kabardian, and the second position was held by the chairman of the village council—a Balkar. A Balkar is currently the head of local administration. Work on household plots allows people to live most comfortably in this territory. Sometimes there are interruptions of the supply of water during irrigation in the village. Due to the lack of land, the areas of such plots have been decreased twofold in recent times.

Khasan'ya is one of big villages located in Kabardino-Balkaria (in 2005, it was included in Nalchik urban district). The population size has increased in recent years. A high level of unemployment has been registered in this village, yet low standard of living has not been registered here. The territorial nearness to Nalchik reduces the deficit of various services and goods in this settlement. Executive authorities are represented by the head of administration of Nalchik urban district. The local council of elders includes the most respected rural people from among both the elder and the younger generation. Due to population growth, a key problem in this populated locality is the shortage of housing and land for construction, which caused a number of conflict situations accompanied by self-seizure of land. At the same time, the objective need for land for construction is accompanied by speculations aimed at transfer of the ownership to land with a view to selling it in the following.

Khumara is a settlement in Karachay-Cherkessia. The location of its territory on the border of low and middle mountains affects the low amount of agricultural land, which in turn causes out migration. Due to the lack of earnings potential, people are forced to leave this territory. For example, during the Soviet period, people were employed in the mines, and now there is no such possibility. Today, Khumara resembles the ancestral nest of the urban Circassians, where the younger generation comes to visit their relatives.

Nizhnyaya Teberda is a village, the territory of which is located near the transit route to Dombay. There is not a huge amount of land here, which forces the population of this territory to search for income sources. But the business interests of the district and republican levels are specifically due to favorable territorial location of the settlement.

Elbrus is located in the Baksan River Valley. This territory is suitable for the development of tourism, which allowed changing the image of this settlement, making it an urbanized settlement. Agriculture is highly important to residents of this territory. Most often, local agricultural products are used for the tourism industry in this territory. However, the land potential is severely limited by the lands of national park. The deficit of agricultural land is mainly due to the large number of pastures and hayfields. The tourist attraction of the district impacts its costs. The growing flow of tourists coupled with obsolete municipal infrastructure represented mainly by the Soviet period, put significant pressure on the ecology of the village. Development is also hindered by the fear of major investment and infrastructure projects, as a result of which the local population fears losing their income sources. Management in the village is represented by both formal and informal institutions. A high level of social control ensures local security which is necessary to maintain the flow of tourists and relevant revenues. Internal conflicts are resolved not publicly, but mainly through the use of traditional institutions.

The analysis of the key characteristics of these villages allows us to conclude that modern transformations at the local level have led to significant transformations in the institutional structure. The domination of state organizational mechanisms is gradually changing towards the development of local institutional forms. Three types of institutions are distinguished in the current institutional structure:

- traditional institutions which were elaborated many centuries ago. Some of them are formalized while others are still informal;
- institutions of the Soviet heritage which even after significant reforms of the post-Soviet period still maintain their influence;
- modern institutions and rules that are implemented in connection with new political and economic conditions.

7.4 Conclusion

Based on the analysis performed, currently local self-government is a set of state, traditional and hybrid institutions of various historical layers.

Today, we cannot observe the substitution of state institutions for local self-government institutions; on the contrary, the state seeks to return to the local level, which is due to economic matters. Local self-governing authorities (village administration) have no full powers. Local self-governing authorities is a hybrid of local self-government and the state. Judging from its formal features, local self-government have sweeping powers, while in practice local administrative institutions depend on the state. Such hybrid administration is demonstrated within the scope of activities of district administrations: in fact, they are local self-government, and they in actual practice depend on national social policy at the local level. In our opinion, many issues of developing social support and economic support for the rural population from administrative-state institutions must gradually change over to the actual local self-government.

Acknowledgements The work has been made within the framework of the basic part of State Order No. 0669-2018-0007.

References

1. Chechenov A, Gaunova I, Kushkhova K (2014) The impact of national development programs on the institutions at the local level. From the perception of local conflicts to the use of development opportunities: collection of research papers. Issue 2/under the editorship of Yan Kioler, Alexey Gunia, Murat Shogenov. Berlin-Nalchik: Kabardino-Balkaria State University, p 211
2. Kumykov AM, Shogenov MZ, Chemaev NA, Shogenova FZ, Reshetnikova NN (2017) Informal patterns of civil society and social stability at a local level. Espacios T 38(57):28

3. Magomedov MG, Khubiev BB, Atabieva ZA, Kushkova AF, Ugnich EA (2017) The concept of social human rights in the conditions of the destruction of the social sphere. Espacios T 38(54):31
4. Smagina NN, Magomedov MG, Buklanov DA (2017) Sustainable competitive advantage of the international business tourism on the regional level, Overcoming uncertainty of institutional environment as a tool of global crisis management, Popkova EG (ed), pp 541–548
5. Popkova EG, Bogoviz AV, Lobova SV, Romanova TF (2018) The essence of the processes of economic growth of socio-economic systems. Stud Syst Decis Control 135:123–130
6. Azamatova GK, Shidov AK, Vindizheva AO, Lyuev AH (2018) Social aspects of change of economic behavior of the Russian youth. J Organ Behav Res 3(1):307–316
7. Lokova MY, Khanova MN, Azamatova GK, Vindizheva AO, Reshetnikova NN (2019) Social consequences of the impact of information technologies on the values of modern youth in the conditions of the global. In: Future of the global financial system: downfall or harmony (lecture notes in networks and systems - LNNS), vol 57. Springer Nature Switzerland AG, pp 176–182
8. Krasnokutskiy PA, Ugnich EA, Taranov PM (2019) Modern scientific and methodological approaches to assessing the competitiveness of the region: application problems. June 2019. IOP Conf Ser Earth Environ Sci 272(3):032179
9. Reshetnikova N, Magomedov M, Buklanov D, Zakharchenko E (2019) The international business cooperation and its influence on enterprise financial security under globalization. In: Popkova E (eds) The future of the global financial system: downfall or harmony. ISC 2018. Lecture notes in networks and systems, vol 57. Springer, Cham, pp 294–308
10. Zmiyak SS, Ugnich EA, Krasnokutskiy PA (2019) Generation and commercialization of knowledge in the innovational ecosystem of regional university in the conditions of information economy establishment in Russia. In: Advances in intelligent systems and computing, vol 726. Springer, Berlin, pp 23–31
11. Reshetnikova NN, Magomedov MG (2020) Influence strategic competitive advantage international business cooperation in the frame of financial crisis. In: Solovev D (eds) Smart technologies and innovations in design for control of technological processes and objects: economy and production. FarEastCon 2018. Smart innovation, systems and technologies, vol 138. Springer, Cham, pp 399–408
12. Adat (2003) Caucasian cultural district. Traditions and modernity. International research institute of the peoples of Caucasus. Moscow-Tbilisi
13. Nozick R (2008) Anarchy, state, and Utopia/ Robert Nozick; translated from English by Pinsker, B. under the editorship of Kuznetsov Y, Kuriaev A, IRISEN, p 424
14. Chalyy VA (2011) Significance of theories of social contract for the development of political philosophy in Russia – Bulletin of the Immanuel Kant Baltic Federal University, 2011, No 12. pp 130–142
15. Gunia AN, Tenov TZ, Shogenov MZ, Chechenov AM (2015) The impact of various strategies on the character if conflict settlement and development of North Caucasian republics. Nauchnaya Mysl' Kavkaza. – Rostov-on-Don, 2015 3(83):89–94
16. Tishkov V (2001) Local management in a complex society: Russian experience. From the book "Local management of multiethnic communities in the CIS countries". Moscow, Aviaizdat, pp 3–22
17. Dahrendorf RG (2002) Der moderne soziale Konflikt: Essay zur Politik der Freiheit. Translated from German, p 311

Chapter 8
Modernization of Management Mechanisms of a Non-profit Organization Endowment

E. V. Markina, O. A. Polyakova, A. S. Lozhechko and L. A. Sugrobova

Abstract The rapidly developing non-profit sector of the economy of the Russian Federation in recent years has faced a shortage of financial resources that was caused primarily by insufficient financing of the social field. Non-profit organizations got a new tool for attracting donations in 2007 with the beginning of the formation of endowments. However, the general volatility of the economy and the weak development of charity in Russia currently contribute to the formation of specific risks for non-profit organizations. It is possible to eliminate these risks both by transforming the best foreign experience in this area, taking into account Russian specifics, and by applying strategic management methods specific to the corporate sector. This refers to the allocation of stimulating and stabilization parts within large endowments. This will allow smoothing out the effect of general economic cycles on the profitability of funds and on their ability to fulfil their obligations in the long-term outlook. Identification of key risks of the functioning of endowment will allow you to create a special risks map. This tool will serve as the basis for a risk-based approach in the strategic management of endowment. In addition, the specificity of any endowment requires the implementation to the general management strategy section devoted to the marketing. Obviously, standard marketing tools intended primarily for-profit organizations must undergo a certain transformation, due to other goals of non-profit organizations functioning, introduction of new tools and methods for attracting donations, as well as the possibilities of the digital economy. Such tools include crowdfunding and fundraising, the use of which at present in Russian practice seems to be insufficient. Taking into account that the largest number of endowments in modern Russia was created in the field of education, and their profitability remains at a rather

E. V. Markina (✉) · O. A. Polyakova · A. S. Lozhechko · L. A. Sugrobova
Financial University under the Government of the Russian Federation, Moscow, Russia
e-mail: EVMarkina@fa.ru

O. A. Polyakova
e-mail: OPolyakova@fa.ru

A. S. Lozhechko
e-mail: ASLozhechko@fa.ru

L. A. Sugrobova
e-mail: lgubzheva@gmail.com

T. Kolmykova and E. V. Kharchenko (eds.), *Digital Future Economic Growth, Social Adaptation, and Technological Perspectives*, Lecture Notes in Networks and Systems 111, https://doi.org/10.1007/978-3-030-39797-5_8

low level, we consider that the proposals set forth in the article will allow solving key problems in this field.

Keywords Endowment · Non-profit organization · Strategic management · Crowdfunding · Fundraising

8.1 Introduction

A non-profit organization endowment should be understood as a set of assets donated to a non-profit organization in order of their further investment and receipt of additional income to finance the needs of a non-profit organization. At the same time, a specific feature of the endowment is the inability to spend on the indicated goals of the fund itself—only management income is used. It should be noted that there is no single definition of the endowment in foreign scientific literature. In scientific publications and reference books, as a rule, the following substantial features of funds are put on top:

- fund assets are a set of cash or property received as a gift or inheritance [6];
- these assets are subject to mandatory investment in order to obtain a positive financial result [13, 15];
- this financial result is subject to specific use, while the endowment itself is not spent [3].
- The Russian practice does not have significant differences from the global one—the authors mainly make a correction for the specifics of legislation [9, 12].

The formation of an endowment allows a non-profit organization to provide relative independence from one-time donations. The receipt of income from the management of an endowment depends primarily on the activity of the management company itself. In addition, endowments, based on best international practices, are able to provide maximum transparency in attracting and using donations that is especially important for the charity field [8].

8.2 Problem Statement

Today in the world there are 104 endowments with a volume of more than 1 billion dollars. Most of them operate in the United States and have a long and rich history of formation. And, although the history of endowments is usually counted from the Roman emperor Marcus Aurelius, who created one endowed chair to representatives of the four main schools of philosophy of his time (about 176 A.D.) [10], the modern history of university endowments began with arrangement of such payments in 1502 at Oxford and Cambridge.

The total amount of endowments researched and acting as members of NACUBO (National Association of College and University Business Officers) is estimated at

Table 8.1 List of the largest endowments of universities in the United States of America and Canada

No.	University	Endowment, billion USD (2019)
1	Harvard University	39.2
2	The University of Texas System	30.8
3	Yale University	30.3
4	Stanford University	27.7
5	University of Toronto	2.0
6	McGill University	1.3
7	University of Alberta	1.0
8	Queen's University	0.8

Source Authors

617 billion US dollars. On average, endowments amounted to a significant value—10% of the operational budget of educational institutions in 2018.

According to official data provided by the institutions participating in the research, about $9.9 billion of new donations in favor of endowments were received in 2018. Thus, according to NACUBO, in 2018, university and college expenses of more than $21 billion were funded from endowments, of which 49% was used to pay scholarships to students.

The amounts of university endowments in the USA and Canada as of 2019 are shown in Table 8.1.

In pre-Soviet Russia, in comparison with foreign practice, endowments appeared not so long ago—simultaneously with the widespread dissemination of charity ideas in the Russian Empire at the end of the 19th century [7].

So, the industrialist A. V. Buryshkin in 1882 made a significant donation in favor of a hospital for chronic patients. At that, more than half of the amount of the donation (RUB 240,000) was, in the modern sense, an endowment. The fund itself was not spent directly, and the interest earned from managing it was used for alimentation of patients.

Also, it is worth noticing the story of the merchant H. S. Ledentsov [16], who, according to his will, left about RUB 2 million to the "Society of Friends of Humanity" to support inventions and scientific discoveries. Separately, the will stipulated the possibility of using only the interest received from inviolable capital and the possibility of using only 10% of the funds to pay for the organizational activity of the fund itself.

In the twentieth century, under the domination of state ownership and a high level of centralization of financial resources in the State budget of the USSR, the development of endowments in the USSR stopped and continued only in the 21st century along with market transformations. It seems possible to solve separate problems appeared during the functioning of endowments in Russia, using the best foreign experience in this area.

8.3 Research Questions

From the point of view of the theory of finance, an endowment is a stock form of organization of financial resources of a non-profit organization. This form has common features of funds: isolation, legal regulation and target nature. At the same time, in comparison with other funds, the endowment also acquires a number of specific features. Among them, there are distinguished the absence of the need to distribute the share of net assets, the presence, as a rule, of a large number of sub-accounts (related to particular direction of donations), higher shares of tax-free donations (for donors) [17]. Donors often make large donations to pre-defined goals [6].

In world practice it is customary to distinguish three main approaches to the role of endowments in university financing: Anglo-Saxon (with a strong emphasis on endowments, the value of which can reach significant amounts), European (the role of endowments is one order less, other sources of financing play a dominant role), Asian (the role of endowments is minimal) [1]. In Russia, the role and quantity (see Fig. 8.1) of endowments continues to grow annually.

At the beginning of 2019, 156 operating funds were registered in Russia, that own at least 186 endowments. Among them, the majority operates in the field of education—61%, in the field of social assistance 14% of endowments operate. In the field of culture and art—13%, and in the field of sports—4%. The median return of endowments in Russia is about 10.8%, the average—is 10.4%.

The largest endowments in Russia are listed in Fig. 8.2.

However, at the moment endowments do not play a significant role in supporting education, science and other fields of social life of the society. Less than 10 endowments in Russia have the volume of accumulated cash over RUB 1 billion [12].

The purpose of forming endowments is to increase the value of its constituent assets using various financial instruments and methods, or to keep it unchanged in case of negative external effect. The purpose of using the endowment is additional financing of the current needs of a non-profit organization, mainly due to income from endowment management.

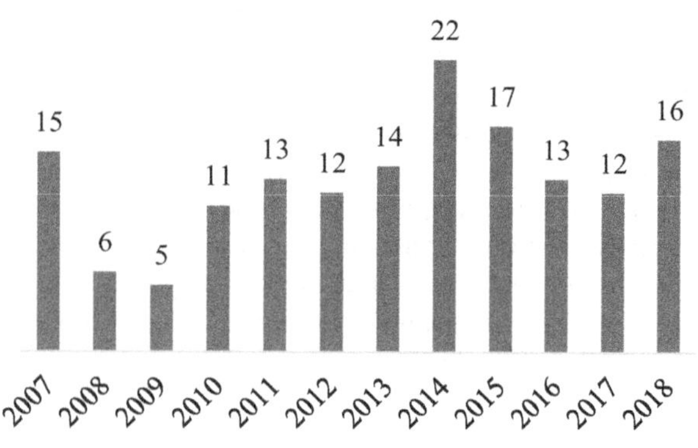

Fig. 8.1 The quantity of established endowments, 2007–2018. *Source* Authors

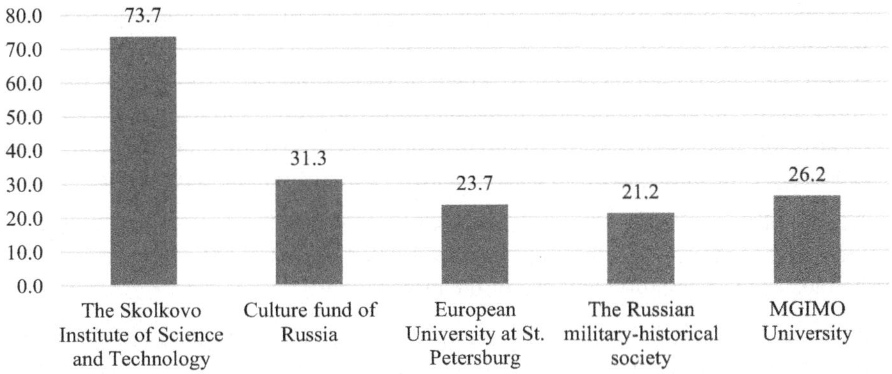

Fig. 8.2 The largest endowments in Russia, millions of dollars. *Source* Authors

Achieving these goals runs across a number of restrictions that do not allow endowments to take their rightful place among the sources of social institutions financing:

- the absence of a systematic approach to assessment and analysis of risks of endowments functioning;
- low level of transparency of endowments functioning;
- low financial stability of endowments;
- the absence at the majority of endowments in the Russian Federation of a methodology for determining the maximum amount of spending the funds received from fund management;
- the absence of promotion strategy (marketing strategy) of the endowment among potential donors.

8.4 Purpose of the Study

The main goal of the study is to determine the conceptual foundation of endowments functioning; identify the main risks associated with the attraction and use of endowments, and approaches to their elimination; to develop a general concept of a strategy for promoting an endowment among potential donors.

8.5 Research Methods

Based on the above features of the endowment, it seems justified to distinguish two of its functions:

- stimulating (its implementation is aimed at additional financing of the priority directions of the non-profit organization activity at the expense of income from

managing the funds of the endowment in conditions of impossibility to spend the fund itself).
- stabilizing (its implementation is aimed at increasing the endowment, that will allow, firstly, increasing income from managing this fund, and secondly, will allow creating an additional reserve within the endowment to implement the stimulating function of the endowment in case of a lack of management income).

The implementation of these functions requires the development of a specific conceptual base for its functioning. This conceptual base should be based on specific principles of the formation and use of the endowment:

- the principle of stability of goals (setting clear and unambiguous goals for the functioning of the endowment, that should be of a strategic nature and unchanged);
- the principle of dynamism (quick change in the tactical tasks of the endowment functioning, taking into account the influence of macroeconomic factors while the strategic goal remains unchanged);
- the principle of financial stability (the ability of a nonprofit organization to uninterruptedly finance separate expenses by means of endowment in the long term) [2, 19];
- the principle of long-term (functioning of the endowment at least within the period of existence of the non-profit organization itself);
- principle of openness (maximum informational transparency of the process of formation and use of endowment);
- principle of social responsibility (functioning of the endowment should be accountable, ethical, carried out taking into account the interests of all interested parties, be systematic and significant).

Risks of endowments functioning are directly related to the violation of these principles and affect the activity of a particular participant in the formation and use of the fund. For example, the most critical risks for nonprofit organizations are the risks of worsening macroeconomic situation, reduction of donations, change in the preferences of donors or reduction of their quantity [18], as well as loss of financial stability.

To eliminate the effect of this significant risk, it is necessary to manage the stimulating and stabilization parts of the endowment using various financial instruments and methods. It appears that the management of the stimulating part should be based on an active strategy of funds investment, and the stabilization management should be based on a passive strategy. On the one hand, the ongoing active investment strategy will allow both financing current expenses and increasing the endowment itself. On the other hand, a "safety cushion" will be created, the income from the management of which in the short term will be used solely at the lack of income from the management of the stimulating part.

To reduce the impact of these risks, it is also possible to develop a methodology for determining the optimal amount of funds expenditure from income received from the management of the endowment.

Analysis of the existing in world practice methods of determining the optimal amount of funds expenditure allows us to distinguish five main groups of methods [5]:

- simple methods (methods of using all the income received and the method of annual decision-making on the level of expenses) [14];
- methods based on the value of assets (establishing a specific rate of expenditure and the moving average method);
- methods based on inflation;
- α-β method [11];
- hybrid methods (as an example, the author cites the so-called Stanford rule and Yale rule).

Based on the α-β method, it seems possible to develop a regulatory and target method for determining the maximum scope of expenses:

The generalized calculation formula is presented below.

$$E_{lim} = A_a * N_a + A_\beta * N_\beta$$

where E_{lim} is the maximum scope of expenses realized by means of income from the management of the endowment;

A_a is the market value of the assets that make up the stimulating part of the endowment;

N_a is the norm of expenses realized by means of income from the management of the stimulating part of the endowment;

A_β is the market value of the assets that make up the stabilizing part of the endowment;

N_β is the norm of expenses realized by means of income from the management of the stabilizing part of the endowment.

In order to ensure the safeguard of the endowment, it seems justified to establish the norm of expenditure N_α—at a level not higher than the average rate of return of the stabilizing part of the endowment. The expenditure rate of N_β, due to the extreme nature of the use of the corresponding income, will be determined on the basis of the actual need at the lack of other funds. In this case, it is possible to use the method of using all the income received.

The risk of reducing donations can be eliminated through the development and active implementation of an endowment promotion strategy (marketing strategy). In the nonprofit sector, it is preferable to talk about social marketing. Its ultimate goal is to improve the well-being and prosperity of a man and society, to achieve a given social effect. When developing a social marketing strategy, it is necessary to take into account the one-sided nature of charity, the need for a more thorough individual approach to the target audience [4, 20].

The marketing strategy for promoting an endowment is a part of the marketing strategy of a non-profit organization. At that, it should be accepted and implemented as an independent document. The endowment promotion strategy should include the use of such methods as SWOT-analysis, SNW-analysis, PEST-analysis, VMOST-analysis, GAP analysis. In addition, the marketing strategy should contain information on groups of potential donors, persons responsible for organizing and conducting marketing events, instruments and methods of interaction with potential donors.

In the digital economy, various online platforms and social networks play an important role in social marketing. In these conditions, such instruments as fundraising and crowdfunding appeared and today are more actively used.

8.6 Findings

As mentioned above, an important role in the management of endowments is played by the systematization of the risks of their functioning. This can be realized by constructing a risk matrix according to the most common probability-damage scheme. In this matrix, the probability of occurrence of risk is placed vertically (extremely unlikely, unlikely, possibly, probably, very likely), and horizontally—the level of damage from its implementation (insignificant-low-medium-significant-catastrophic).

Taking into account the specifics of the functioning and promotion of the endowment, as well as the specifics of the work of the nonprofit sector, this risk matrix will be primarily filled in on the basis of the conclusions obtained during the development and updating of the marketing strategy for promoting the endowment.

The publication of this marketing strategy will allow increasing the openness and transparency of the mechanisms of the endowment operation. We consider it necessary to publish the following information:

- legal status and relations of a non-profit organization with a management company and donors;
- goals and objectives of the functioning of the endowment;
- domestic policy and rules for the formation, management and use of the endowment;
- degree of independence in making investment and other management decisions of the organization directly managing the endowment;
- annual audit results;
- events to reduce the risks of the endowment functioning;
- self-assessment results of a non-profit organization and a managing organization.

In addition, all processes related to attracting donations from donors should be completely open and transparent. This is actually for crowdfunding and fundraising events, a significant part of which is implemented on electronic platforms.

Thus, the measures given in this article to ensure transparency and openness of endowments functioning and increase their financial stability will contribute to the development of the institution of endowments in Russia, the elimination of a significant part of the existing problems.

8.7 Conclusion

Taking into account the world experience in the formation and use of endowments, it is worth recognizing that this institution can serve as a starting point for the breakthrough development of social organizations in the non-profit field in the Russian Federation.

The limiting factors in this case are problems that have a rather wide range of questions as their source. The key, in our opinion, are the problems of financial stability and transparency of the endowments functioning.

Not enough developed culture of charity in Russia, in our opinion, has one of its reasons also the distrust of potential donors to the institution of charity. Trust increasing is possible, first of all, by increasing the openness of endowments, publishing their goals and objectives, internal policies, mechanisms for attracting and endowments expenditure, conducting open events with widespread use of crowdfunding and fundraising mechanisms, strict adherence to the principles proposed in this article and conceptual foundations of the endowments functioning.

At that timely identification, analysis and prevention of risks arising in the framework of the formation and use of endowments, including those on the basis of a risk matrix, will not only reduce the loss of non-profit organizations, but also demonstrate to donors the extreme interest of the non-profit organization itself in the existence and operation of the endowment.

References

1. Balatsky EV (2017) The market of university endowments in Russia: development trends and opportunities. Sci Gov Sci 1(23):149–178 (in Rus.)
2. Borodina NM (2009) Management of financial stability of the enterprise in the conditions of financial crisis. Arch Sci Art 12–1:184–186 [in Rus.]
3. Dimmock SG (2012) Background risk and university endowment funds. Rev Econ Stat 94(3):789–799
4. Dolnicar S, Lazarevski K (2009) Marketing in non-profit organizations: an international perspective. Fac Commer Pap 26(3):275–291 (in Rus.)
5. Dyachkova EA (2016) Revenue management of university endowment. Educ Stud 1:225–244 (in Rus.)
6. Irvin R (2010) Definition and management of endowment. In: Anheier H, Toepler S (eds) Book: international encyclopedia of civil society, 1st edn. Springer, pp 632–636
7. Komissarova E (2012) Endowment of non-profit organizations in Russia: history and modern situation. Mod Law Mag 7:11–15 (in Rus.)
8. Kostyan VV, Khizhnyak VD (2017) Analysis of the practice of creating endowments in Russia. In: Yu F, Elzesser LA, Pavlov (eds) HNRI National development 2017 conference, St-Petersburg, Russia, pp 53–55 (in Rus.)
9. Lagutina OV (2017) Fundraising activities in the Russian challenge from donations virtuous persons of public education to endowments. Proc Southwest State Univ Hist Law 1(22):128–135 (in Rus.)
10. Lynch JP (1972) Aristotle's school; a study of a Greek educational institution. University of California Press 19–207:213–216

11. Mehrling P (2005) A robust spending rule: the α-β approach. Forum Future High Educ 67–72
12. Revunov AA, Rodionova SV (2015) Diversification of financing of higher education institutions: educational fundraising. Eurasian Sci Community Ltd Liabil Co 5–1(10):118–122 (in Rus.)
13. Rosen H, Sappington A (2019) The impact of endowment shocks on payouts. J High Educ 90(5):690–716. https://doi.org/10.1080/00221546.2019.1621115
14. Sedlacek VO, Jarvis WF (2010) Endowment spending: bulding a stronger police framework. Commonfund White Paper
15. Sherlock MF, Gravelle JG, Crandall-Hollick ML, Stupak JM (2015) College and university endowments: overview and tax policy options. Congressional Research Service, Washington, DC
16. Shnol SE (2010) Heroes, villains, conformists of domestic science. Librokom, Moscow, Russia, p 720 (in Rus.)
17. Smith D, Shawky H (2011) Endowment and Foundation Funds. book: institutional money management: an inside look at strategies, players, and practices, 1st edn. Publisher, Wiley, pp 295–308
18. Subanova OS (2015) Economic risk in the activity of the endowments. System analysis in design and management. In: 2015 XIX international scientific and practical conference (pp 308–312) (in Rus.)
19. Tyan NG, Kholevitskaia AA (2017) The essence of financial risks and their relationship with financial stability. Success Mod Sci Educ 2:87–90 (in Rus.)
20. Tinyakova VI, Sheryanova VP (2014) Implementation and promotion of non-commercial projects: foreign experience and Russian reality. Vestn Saratov State Socio-Econ Univ 5(54):103–107 (in Rus.)

Chapter 9
Moving Towards "Digit": Via the Management of Economic Security to Ensure Sustainable Development

Evgeniya K. Karpunina, Irina M. Podmolodina, Olga I. Averina, Igor A. Kuznetsov and Svetlana V. Solonina

Abstract The pace of scientific and technological progress and the intensive use of information and communication technologies in daily life, production activities of enterprises and in the process of public administration indicate the transition of economic systems of most countries of the world to the digital stage of their development. In such conditions, the concept of sustainable development adopted by the international community, aimed at ensuring a balance of economic, social and environmental components, is undergoing a number of changes. The article proves that in the digital economy there is a decrease in the stability of the economic system due to the high degree of fluctuations of digital artifacts. Their application leads to permanent changes in the system of production and consumption at the micro level and projects transformations at the macro level. The authors argue about the emergence of a whole range of new risks that reduce the economic security of the country, and attempt to identify and systematize them. The article reveals the content of the conceptual model of risk-oriented management of economic security proposed by the authors to achieve the goals of sustainable development. In addition, the authors propose a method of step-by-step implementation of the risk- oriented economic security management model to achieve sustainable development goals, and

E. K. Karpunina (✉) · I. A. Kuznetsov
Tambov State University Named After G.R. Derzhavin, Tambov, Russia
e-mail: egenkak@mail.ru

I. A. Kuznetsov
e-mail: igkuz73@mail.ru

I. M. Podmolodina
Voronezh State University of Engineering Technology, Voronezh, Russia
e-mail: podmin@mail.ru

O. I. Averina
National Research N.P. Ogarev, Mordovia State University, Saransk, Republic of Mordovia, Russia
e-mail: oiaverina@mail.ru

S. V. Solonina
Kuban State Technological University, Krasnodar, Russia
e-mail: Svevic@mail.ru

T. Kolmykova and E. V. Kharchenko (eds.), *Digital Future Economic Growth, Social Adaptation, and Technological Perspectives*, Lecture Notes in Networks and Systems 111, https://doi.org/10.1007/978-3-030-39797-5_9

demonstrate the mechanism of its application on the example of a specific economic security risk.

Keywords Digital economy · Economic security · Threats · Risks · Sustainable development · Economic security management

JEL Classification Codes O11 · O33 · P40

9.1 Introduction

At a certain stage of development, when scientific and technological progress largely began to determine the nature of economic development, society faced a number of contradictions between real life and life in harmony with nature, between real development and the environment, between the interests of modern and future generations, between rich and poor countries and people. During this period, the concept of sustainable development was created, aimed at achieving a balance in economic, social and environmental aspects [30].

In the conditions of digitalization the economies of the world meet new requirements, conditions and mechanisms of management. This is reinforced by the action of information technology factors with varying degrees that affect the stability of the economic system not always, in a positive aspect.

The emergence of failures and deviations in the normal functioning of economic systems requires a detailed study of the risks and threats to economic security that the digital environment carries. In addition, the need to develop tools to respond to these threats and their rapid leveling in order to achieve greater sustainability is more urgent than ever.

9.2 Methodology

The scientific hypothesis of the study is the assumption that in the digital economy, the stability of the economic system becomes more vulnerable and is subject to fluctuations from digital artifacts (their application leads to permanent changes in production and consumption at the micro level and projected transformations at the macro level).

At the same time, digital artifacts can have both positive and negative impact on the sustainable development of the economy. In case of negative impact, mechanisms to reduce the economic security of the state are launched. As a result, the functioning of the entire economic system is threatened. Risk-oriented management of economic security in the context of sustainable development components can be a way out of this situation.

The study is based on the classical understanding of sustainable economic development, which is reflected in the Brundtland report: «development that meets the needs of the present without compromising the ability of future generations to meet their own needs» [30].

The need to develop a concept of sustainable development (that is, a model of civilization development, proceeding from the need to maintain a balance between solving social, economic problems and preserving the environment) was due to the threat of environmental catastrophe from scientific and technological progress. The authors of the concept and the researchers who later formed its scientific basis systematized the main components of the stability of the economic system [4, 17].

Such authors as Drucker [8], Machlup [19], Umesao [27] in their studies point to the features of the development of the economic system on the digital vector. Some scientists argue that the useful power of society is manifested in the form of knowledge and technology and acts as a means of supporting sustainable development [7].

However, in modern conditions, with the ever-increasing use of computer infrastructure and constantly changing environmental conditions, more and more of the recorded information becomes digital. Digital artifacts can also have a negative impact on the stability of the economic system and exacerbate the problems of economic security.

The methodology of economic security management in the digital environment remains insufficiently developed, some attempts are made to form a threat leveling tools [12, 22, 25]. The authors of the study are trying to adapt the theoretical research of risk management [15, 3, 6] to risk management of digitalization and sustainability of the economic system on the basis of their qualitative analysis, systematization of the types of risks, identify their sources, their effects, scenario analysis of dangerous events, studying the consequences of their occurrence, the development of techniques of neutralization and formation of risk management strategies.

9.3 Results

Interpretation of the concept of sustainable development allows us to identify the following subsystems, reflecting the relationship in the system «person -society-nature»: economic (is based on the theory of the maximum flow of total income that can be produced, provided at least the preservation of the total capital with which this income is produced [1]; social (focused on people and ensuring the stability of social systems with a fair division of benefits and the preservation of social capital); environmental (aimed at ensuring the integrity of the ecosystem and biodiversity conservation); institutional (acting as an integrator and forming the basis for the sustainable functioning of the economic system). Since the subjects of sustainable development use different forms of capital in the process of their activities, the economic subsystem becomes a priority in the concept of sustainable development, and the social, environmental and institutional subsystems are slave [21].

The principle of sustainable development is based on the law of conservation of power (flow of energy) [16] and its projection is the principle of "preserving the development of living systems" by Vernadsky [28] and Bauer [2]. This principle is ensured by the undiminished efficiency of using the useful power of society in interaction with the world environment [18].

Supporters implemented under the auspices of the United Nations concept of sustainable development of society and economy slant in the direction of humanization of social-economic life of society, to ensure effective control over the efficiency of natural-resource potential of land in the interests of the entire population of the planet, respect for the rights and freedoms of citizens, social protection of the population in the ways of a more equal distribution of income and capital etc. in Spite of this, the summit "Rio + 10" in 2002 Johannesburg noted the deepening of the global environmental crisis as a consequence of the wrong course of development of the world economy. The past decade since Rio 92 has shown that humanity reacts to global changes in environmental quality not as a whole system, but as a set of different interests of States that protect their national interests. Scenario forecast for the coming decades shows that the trend of destruction of the biosphere, degradation of nature, deterioration of living conditions of people will continue and even increase [23].

In the process of evolution, the concept of sustainable development meets another concept that determines the current state of society and the economy-the concept of digital development, leading to the formation of the digital economy.

Schmidt and Cohen described what is happening with the term "digital age" and linked it to the active development and implementation of information and communication technologies (ICT), covering a variety of hardware and software related processes [24]. In the study of digital transformations scientists focus on transforming the way people interact with each other and with their natural environment [5], on changing the nature of relationships between companies, on the growing importance of digital technologies in providing new opportunities for the future.

The ongoing changes have led to rapid growth of the ICT sector and related sectors (in European countries alone, the value added of the ICT sector amounted to 632 billion euros in 2017 due to the introduction of digital technologies by European enterprises [9], the level of penetration of information and communication technologies in most countries of the world (in 2019—4.1 billion Internet users, 333.8 million registered domain names [26], 3.5 billion social media users worldwide [29], intensive development of e-Commerce (this year the sector generated 3.45 trillion dollars in sales [26], the scale of automation and robotics (the total market volume of industrial robots in 2017 amounted to $ 48 billion dollars, including software [20], caused significant changes in the system of production and consumption at the micro level, which projected transformational processes at the macro level of the economic system.

However, the opinions of researchers are divided when it comes to the impact of digital processes on the socio-economic system and its sustainable development in the coordinates "person-society-nature".

The results of the comparison of the determinants of the concept of sustainable development and the concepts of digital development of different countries are presented in Table 9.1.

Of the currently developed and approved digital development strategies of the world's leading countries (Australia, Austria, Czech Republic, Estonia, Canada, USA, UK, Japan, China, Hungary, Sweden, Switzerland, Norway, France, Germany, Netherlands and so on) only a few reflect the issues of social, environmental and institutional improvement (primarily in Japan, USA, China). Basically, everything is aimed at achieving the goals of economic and technological leadership.

In the digital economy, the sustainability of the economic system is becoming increasingly vulnerable and is subject to fluctuations from digital artifacts. At the same time, digital artifacts can have both positive and negative impact on the state of sustainable economic development.

The more a country becomes involved in digital processes, the more likely it becomes to face new challenges and uncertainties.

Factors and conditions that give uncertainty to the digital environment are: high speed of transformation of unique knowledge into a public good, a short period of preservation of its competitive advantage [11]; formation of competitive advantage not due to the possession of knowledge, but due to the ability to quickly create this unique knowledge; development of social networks and platforms that determine social dynamics on a global scale; formation of network relations between companies that expand the zone of risk and uncertainty and create new conditions for a competitive environment; virtualization of economic activities, in which the status of goods acquires a "trademark"; instant global movement, achieved through high-speed exchange of data, information and knowledge; the cross-cutting nature of development, that is, the penetration of digital technologies in all spheres of society [13].

The emergence of risks generated by the economic system and external fluctuations in the process of digital change requires special research attention. Let us present the author's vision of the problem in the form of a conceptual model (Fig. 9.1).

The analysis of risks of economic security as a result of digitalization of economy allows them to systematize on the following types:

- system risks (risks arising in different segments of the economic system and having a systemic nature, for example, the lack of its own element base for digital development, dependence on the resources of digital technologies of other States, the risk of "digital inequality") [22];
- structural (the use of digital artifacts leads to permanent changes in production and consumption at the micro level, that is, it has "structural effects"; this, in turn, can lead to permanent changes at the structural and institutional level and therefore occur at the macro level (for example, significant changes in the labor market and rising unemployment)) [14];
- corporate (arising in the process of digital restructuring of individual enterprises, changing business models, adaptation to new conditions; leakage of corporate

Table 9.1 Comparison of determinants of sustainable and digital development

Determinant	The concept of sustainable development	The concept of the digital development
Economic determinant	– Financial resources and mechanisms for their rational use – Transfer of environmentally friendly technologies – Cooperation and development of productive capacity – International cooperation to enhance sustainable development and related domestic policies	– Digitalization and interconnection of products, value chains and business models (Germany) – Increasing competitiveness as a production base for high value-added products (Germany) – Global free exchange of information (USA) – Trust and security online (USA) – Access (infrastructure development) and digital skills (USA) – Innovation and new technologies (USA) – Full ICT-independence (China) – Acceleration of value creation by robots and information and analytical technologies based on CPS (Japan)
Social determinant	– Combating poverty – Dynamics of demographic processes and sustainable development – The development of education, literacy, training programs – Protection and improvement of human health – Ensuring sustainable development of places of mass residence of people	– Growth in the number of jobs (Germany) – The balance between work and life through flexible work models (Germany) – Acquisition of digital skills by employees (Germany) – Implementation of migration strategies within the framework of intersectoral labour movement (Germany) – Ensuring personal information security (China) – Forecasting the needs and welfare of citizens to ensure uninterrupted experience and quality service (China) – Increase the quality of life of the population (China) – Improving the comfort of use and manufacturability of household items (China) – Providing everyone with a safe and secure comfortable and healthy life, providing everyone with the opportunity to realize their desired lifestyle (Japan) – Addressing complex social issues (population growth, sustainable development, population ageing) (Japan) – Building social consensus through the exchange of national vision among all stakeholders (Japan) – Definition of individual happiness of mankind (Japan)

(continued)

Table 9.1 (continued)

Determinant	The concept of sustainable development	The concept of the digital development
Environmental determinant	– Protection of water quality and provision of drinking water – Protection of oceans, all types of seas and coastal areas integrated approach to planning and management of land –resources – Management of fragile ecosystems: deserts and arid zones, mountainous areas – Sustainable agriculture and rural development – Atmosphere protection – Solid waste recycling and sewage issues, toxic chemicals, hazardous waste – Processing and disposal of radioactive waste – Combating desertification – Biodiversity conservation – Environmentally friendly biotechnology	– Creating an "ecosphere " conducive to digitalization (China) – Development of Autonomous vehicles to reduce the harmful effects on the environment (China) – Improving the quality of the environment (China) – Increasing the power of the individual in the economy (Japan)

(continued)

Table 9.1 (continued)

Determinant	The concept of sustainable development	The concept of the digital development
Institutional determinant	– The integration of environmental interests and development principles in the decision-making process – Science and sustainable development – International legislative instruments and mechanisms – Provision and exchange of information for strategic decision-making – Strengthening of major population groups	– Consumer protection in the implementation of electronic Commerce (China) – The creation of conditions for the rule of law (China) – Creation of adaptation mechanisms to optimize population size, income level and education level (China) – Building institutions that reflect society's view of citizen-satisfying digital services and clusters to further integrate and bridge digital and economic divides (USA) – Creating new values by integrating cyberspace and physical space (CPS) with the rapid development of ICT as a background (Japan)

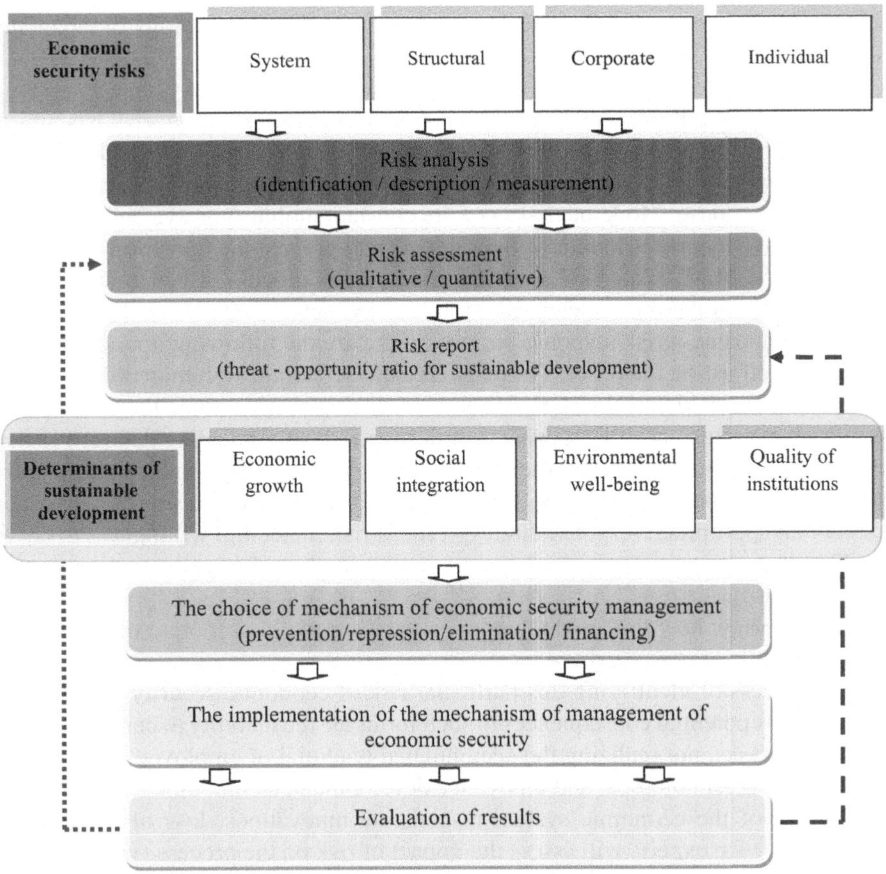

Fig. 9.1 Conceptual model of risk-oriented management of economic security to achieve sustainable development goals (created by the authors)

information, cyber espionage, failure of information systems, low level of digital skills of personnel);

- individual (inherent in individuals and associated with changes in the mechanism of formation of needs, consumption patterns, the emergence of new requirements for competencies and skills, manipulation of personal data, leakage of personal data, cybercrime, as well as many psychological problems associated with the excessive use of information and communication technologies [12].

The conceptual model of risk-oriented economic security management for achieving sustainable development goals involves the identification of sources of economic security risk, objects of its impact, the analysis of scenarios of dangerous events at all stages of assessment, research and evaluation of the consequences of their occurrence, the development of methods of neutralization, drawing up a plan for economic security management [6]. We describe in more detail the methodology of the model implementation (Table 9.2).

The methodology given in the table requires an explanation in terms of working with the map of differentiated risks of economic security. So if the risk has a high degree of impact and a high probability of occurrence, and does not meet the sustainable development goals, it falls into the category of "critical risks" and requires an immediate response. Moderate risks require constant monitoring, systematic control and corrective measures to reduce their impact and prevent the transition to the critical category. Minor risks will be monitored, and the process of economic security management in relation to them will not be carried out.

When choosing a risk response strategy, there are the following options: [1] risk prevention-adjusting the process of managing the economic system in order to eliminate the threat of risk; [2] risk repression—reducing to acceptable limits the probability or consequences of an event with a high risk of a negative nature; [4] risk elimination—bringing the identified and assessed risks to an acceptable level in order to assess the impact of risks, concentration and distribution of resources, development of an appropriate response strategy [10, 5] risk financing: willingness to cover possible losses (for example, by forming special funds and reserves to compensate for losses) [6].

We will show how this model works on the example of corporate risk-low level of digital skills of personnel.

The process of identifying this particular risk of economic security allows us to determine its potential consequences in such forms as: the inability to perform certain production tasks, not realizing the accumulated potential of employees, the decline in the pace of technological and innovative development, technological dependence, low ability of the economic system to generate innovations, loss of competitive advantages. Five experts will assess the impact of risk on the process (values from 0 to 100: 70, 80, 30, 90, 65 respectively) and the probability of its occurrence (from 0 to 1: 0.9, 0.8, 0.9, 0.7, 0.6 accordingly). The risk value will be: expert 1–63, expert 2–64, expert 3–27, expert 4–63, expert 5–39. Our risk enters the risk register with an average value of 51.2.

We will differentiate the risk depending on the degree of its impact on the objects of influence (on the process of implementation of innovative tasks, the speed of response to management teams, the process of interaction with contractors and consumers, the quality of goods and services produced, the competitiveness and efficiency of the economic system). It is assigned an average risk exposure value with a score of 3 (from the range 0 to 5). According to the results of the matrix of probability and consequences, our risk falls into the zone of critical risks. In addition, there is a high level of correlation of this risk with the possibilities of sustainable development of the economic system (low level of digital skills of personnel is one of the key factors of sustainability, since it is a key characteristic of labor as a factor of production, determining the possibility of achieving economic growth, as well as the quality of social integration in the digital environment depends on it). This means that the low level of digital skills of staff is a risk that requires a non-urgent impact on the part of

Table 9.2 Methodology of implementation of the conceptual model of risk-oriented management of economic security to achieve sustainable development goals

Action	Stage 1	Stage 2	Stage 3	Stage 4	Stage 5	Result
The compilation of a risk register	Identification of each specific risk and identification of its potential consequences	Selection of experts	Assessment of the risk impact on the process (from 0 to 100)	Estimation of probability of occurrence of risk (from 0 to 1)	Risk assessment (multiplication of the degree of risk influence by the probability of risk occurrence)	Register of risks with their average values
Differentiation of risks depending on the degree of their impact on the objects of impact, probability and consequences of their manifestation, correlation with the possibilities of sustainable development	Analysis of the impact of risks on the objects of influence (from 0 to 5)	Drawing up a matrix of probability and consequences	Identification of areas with high/medium/low risk exposure and varying probability of occurrence	Identification of critical, moderate, minor risks	Correlation of identified risks with sustainable development opportunities (economic growth/social integration/environmental well-being/quality of institutions), risk selection	The map of differentiated risks
Economic security management	Planning of response to economic security risks	Choice of response strategy in the context of each differentiated type of risk	Development of measures for prevention, repression, elimination, risk financing	Distribution of responsibility for the implementation of activities	Evaluation of results	The performance report of the economic security management

management. When choosing a response strategy, risk prevention (depending on the level of digital development of the economic system) or its repression is possible.

Thus, the proposed model allows us to assess the scale of threats to the economic system and take effective response measures.

9.4 Conclusions

The study allowed to substantiate the scientific hypothesis about the decrease in the stability of the economic system in the digital economy due to the high degree of fluctuations on the part of digital artifacts, which have a different nature and intensity of the impact on the economy and society.

The authors revealed that the concept of sustainable development is continued in the currently implemented concepts of digital development (they contain economic, social, environmental and institutional components), however, the priority in both concepts is given to the economic determinant.

Systematization of risks of economic security (system, structural, corporate, individual) is made, and also action of mechanisms of decrease in economic security of the state is characterized. The advantages of risk-oriented management of economic security in order to ensure sustainable development are substantiated.

References

1. Aitzhanova D, Omarov A (2016) Mechanisms of transition to sustainable development of Kazakhstan's economy in the context of globalization. Stat, Account Audit 4(63):72–78
2. Bauer E (1935) Theoretical biology. VIEM, Moscow
3. Bernstajn P (2000) Protiv bogov: Ukroshchenie riska, ZAO. Olimp-Biznes, Moskva
4. Bobylev S, Khodjaev A (1997) Economics of nature management. TEIS, Moscow
5. Boyd D, Crawford K (2012) Critical questions for big data: provocations for a cultural, technological, and scholarly phenomenon. Inf, Commun, Soc 15(5):662–679
6. Chernenky A (2016) Application of risk-oriented approach in the construction of quality management. Int Sci Res J 8(50):92–96
7. Dobrynin A, Chernykh K, Kupriyanovsky V (2016) Digital economy-various ways to effective application. Int J Open Inf Technol 4(1):4–11
8. Drucker P (1969) The age of discontinuity. HarperCollins, New York
9. European commission (2018) Digital economy and society index for 2018, https://ec.europa.eu/digital-single-market/en/news/digital-economy-and-society-index-2018-report (accessed 14 Dec 2018)
10. Gorulev D (2018) Economic security in the digital economy. Tech Technol Probl Serv 1(43):77–84
11. Gribanov Y (2019) Factors and conditions of digital transformation of socio-economic systems. Bull Altai Acad Econ Law 2(2):253–259
12. Karpunina E, Gorbunova O, Moiseev S, Cheremisina T (2019) Resistance is not hopeless...regarding the policy of countering information threats of economic security. In: Vision 2020:

Education excellence and management of innovations through sustainable economic competitive advantage proceedings of the 33nd international business information management association (IBIMA) in Granada. Spain, pp 2679–2687
13. Karpunina E, Gorchev Y, Stromova M (2017) Virtual space of economic entities: the myth of the digital economy or the object of project management? Socio-Econ Phenom Process 12(6):112–118
14. Karpunina E, Konovalova M, Shurchkova J, Isaeva E, Abalakin A (2020) Economic security of businesses as the determinant of digital transformation strategy. In: Popkova E, Sergi B (eds) Digital economy: complexity and variety versus rationality. Springer, Cham, pp 251–261
15. Knight F (2003) Risk, uncertainty and profit. Delo, Moscow
16. Kuznetsov P (1973) Universal language for formal description of physical laws. Moscow state University, Moscow
17. Kuznetsov O, Kuznetsov P, Bolshakov B (2000) System nature-society-person: sustainable development. Noosphere, Moscow
18. Kuznetsov O, Bolshakov B (2014) Noospheric laws and model of global sustainable development. Stand Living Popul Reg Russ 1(191):90–97
19. Machlup F (1962) The production and distribution of knowledge in the United States. Prinston University Press, Princeton
20. McKinsey Global Institute (2017) A future that works: automation, employment and productivity, https://docviewer.yandex.ru/view/316254058/ (accessed 13 Aug 2019)
21. Myakisheva E (2015) The principle of sustainable development as the basic principle of strategic management of the territory of advanced socio-economic development. In: Innovations in the management of regional and sectoral development proceedings of the all-Russian scientific and practical conference with international participation, pp 164–168
22. Popov E, Semyachkov K (2018) Problems of economic security of digital society in the conditions of globalization. Econ Reg 4:1088–1101
23. Rossinskaya M, Bugaeva M (2010) Problems of realization of the concept of sustainable development at the regional level. Eng Bull Don, http://www.ivdon.ru/ (accessed 24 Dec 2018)
24. Schmidt E, Cohen J (2013) The new digital age. Reshaping the future of people, nations and business, Knopf, New York
25. Senchagov V (2005) Russian economic security. General course: the textbook, Business, Moscow
26. Statista (2019) Digital Market Outlook, https://www.statista.com/outlook/digital-markets (accessed 2 Sept 2019)
27. Umesao T (1991) Theory of information industry-the dawn of the coming era of ectodermal industry. Chuo Koronsha, Tokyo
28. Vernadsky V (1988) Philosophical thoughts of the naturalist. Nauka, Moscow
29. Web-canape (2019) Vsya statistika interneta na 2019 god – v mire i v Rossii, https://www.callcentrehelper.com/consumers-expect-brand-communications-136541.htm (accessed 1 June 2019)
30. World Commission on Environment and Development (1987) Our common future. Oxford University Press, Oxford

Chapter 10
The Way to Perfection and Safety: About Motives of Business Development in the Digital Environment

Evgeniya K. Karpunina, Yuri A. Salikov, Oksana N. Suslyakova, Olga N. Gorbunova and Roza A. Avdeeva

Abstract Digitalization processes affect all participants of economic relations-consumers, producers, the state. They change the forms and content of their interaction. The research objects of this article are the ongoing changes in the ways of doing business and competition of business structures, approaches to interaction with consumers and ways to meet their needs. The authors show the changing roles of the main economic agents in the digital economy. Special emphasis is placed on the study of the behavior of business structures, whose boundaries are becoming less rigid, and their activities are becoming more transparent. The article points to the emergence of alternative ways of selecting resources by business structures and changing requirements for them. New forms of formal and informal business cooperation are reflected, their advantages and disadvantages are revealed. The authors prove that only the continuous search for new effective tools to meet consumer needs and deliver pleasure to a person (on the basis of various forms of communication with the consumer) is able to ensure the success of the business structure and its competitive advantages in the digital environment. The attention of the authors is

E. K. Karpunina (✉) · O. N. Gorbunova
Tambov State University Named After G.R. Derzhavin, Tambov, Russia
e-mail: egenkak@mail.ru

O. N. Gorbunova
e-mail: oon26@mail.ru

Y. A. Salikov
Voronezh State University of Engineering Technology, Voronezh, Russia
e-mail: saural@rambler.ru

O. N. Suslyakova
Financial University Under the Government of the Russian Federation, Kaluga Branch, Kaluga, Russia
e-mail: finans-11@mail.ru

R. A. Avdeeva
Kuban State Technological University, Krasnodar, Russia
e-mail: arozita@mail.ru

T. Kolmykova and E. V. Kharchenko (eds.), *Digital Future Economic Growth, Social Adaptation, and Technological Perspectives*, Lecture Notes in Networks and Systems 111, https://doi.org/10.1007/978-3-030-39797-5_10

focused on the study of threats to economic security as a side effect of digitalization. They increase vulnerability of the business structure and bring it real economic damage, jeopardize the whole meaning of its existence.

Keywords Digital economy · Economic security · Economic development · Threats · Businesses · Business activity

JEL Classification Codes E71 · L21 · L26 · O11 · O14 · O33 · P4

10.1 Introduction

High-tech development is replacing industrial development, a wide range of digital technologies is replacing traditional ones, inter-country borders are disappearing and globalization interactions are expanding. All these processes affect the transformation of the value chain in the digital environment, and, therefore, change the role and place of business structures and their motives for development in the new environment.

The business structure should be in direct contact with the consumer and focus on the permanent change of his tastes and desires, maintain long-term relationships with him. This will allow it to remain sustainable and cope with both expected and unexpected situations, form competitive advantages, achieve business goals and open new opportunities.

The risks of the digital economy are currently poorly understood, they form an ever-changing structure of aspects of management of the expanded value chain in conjunction with a number of ethical, legal, regulatory issues, issues of corporate social responsibility.

The choice of the research topic, its goals and objectives, structure are due to the complexity, discussion, relevance and relative undevelopment of the problems described above.

10.2 Methodology

The authors of this study will have to answer the following questions and find theoretical and methodological explanations for their solutions.

1. In the digital environment, there is a qualitative shift in the path companies do business and compete, the way they interact with consumers and the method they meet their needs.
2. Consumer pleasure becomes a key motive of business activity in the digital environment, which stimulates business structures to improve, create new tools to meet consumer needs, build digital communication channels. This forms their new competitive advantages.

3. A side effect of digitalization of business activity is the emergence of new challenges and threats that can reduce the level of economic security of the business structure and cause real economic, social and psychological damage. This problem requires the development of adequate threat level response measures.

The theoretical basis of the study is:

- fundamental works of researchers of digital development [1, 2], as well as scientific research of modern authors [3, 4];
- scientific publications devoted to the problems of business development in the conditions of digitalization [5];
- theoretical studies of behavioral Economics that determine consumer behavior [6, 7];
- studies determining individual areas of scientific knowledge in the field of economic security [8–10].

In fact, attempts at theoretical analysis of business activities in the digital environment are few, and methodological approaches to the study of this problem have not yet been defined.

Research methods: dialectical method of cognition, system-functional method, method of comparative analysis, as well as scientific abstraction, induction and deduction, grouping, classification, economic and statistical methods of information processing, expert evaluation method.

10.3 Results

The environment of functioning of modern business structures can be characterized by the following technological trends: sensory systems and mobility; augmented man; postclassical computing and communications; digital ecosystems; advanced artificial intelligence and Analytics [11].

The intensive development of the process of creating 3D sensor cameras, augmented reality cloud platforms, drones for light cargo delivery, flying Autonomous transport, fully Autonomous driving make the movement and digitization of any objects automated with minimal human involvement [3].

So, according to Bloomberg at the end of December 2018, Sony Corporation is increasing the production of 3D cameras in connection with the interest of smartphone manufacturers, in particular, Apple. Technologies of creation of biochips, personification, development of biotechnologies, creation of systems of the augmented intelligence, emotional artificial intelligence, immersive working spaces are directed on improvement of cognitive and physical abilities of the person, and also preservation of his health, intelligence and force.

Postclassical computing and communications are carried out through changes in traditional computer architectures. The changes have resulted in faster processors, greater RAM chip capacity, and increased performance that help accelerate human-to-human, human-to-device, device-to-device interaction.

According to Gartner's definition, digital ecosystems are web connections between businesses, people, and things sharing a digital platform. Such ecosystems have evolved as digitalization has transformed business processes. Enhanced Analytics is the Autonomous or semi-Autonomous exploration of data or content using sophisticated tools that go beyond traditional business concepts.

The development of artificial intelligence technologies (the latest artificial intelligence, boundary Analytics, explicable artificial intelligence and so on) makes the collection, processing and analysis of data more perfect.

In these circumstances, the activities of value creation companies go beyond the usual framework, which is becoming less clear [12].

Firstly, there is a change in the roles of the main economic agents.

In the digital economy, the role functions of many economic agents are changing: the consumer becomes the top of the entire production chain, not only in terms of demand formation, the source of needs. Now the consumer is increasingly cooperating with computers, collecting information and presenting it in the form of new needs and desires. The consumer actively joins in process of creation of products, makes the rationalization contribution to their development, acts at the same time as the supplier. Traditional consumer-producer relations take the form of cooperation (and sometimes are aggravated in the direction of self-creation of value by the consumer). The boundaries of business structures are becoming less rigid (especially in connection with the spread of the phenomenon of open innovation), and their activities are becoming more transparent [13]. The term "prosumer" appeared as a reflection of this trend, it combines the concepts of "produce" and "consumer" [14]. In these conditions, the state becomes the main defender of the rights and interests of the consumer, stimulating producers to meet consumer demands (Fig. 10.1).

Secondly, there are alternative ways of selection of resources by business structures and requirements to them change.

Selection of the factor of production "labor" is important. As in the industrial economy, personnel qualifications are taken into account, but the requirements for digital, analytical and high-tech skills are now coming to the fore [15]. The use of new technologies in business requires professionalism, technological efficiency, ease and speed of finding the necessary information, the use of new communication channels with the possibility of being in the access zone. There are new opportunities for modeling scenarios of situation development (forecasting), future analysis, resource optimization, scenario playback, risk forecasting. The requirements for a significant increase in the speed of response of specialists to the request of the counterparty are formed.

The company's focus is shifting from the employees it employs to the consumers. In addition, altruistic volunteers often provide digital services without payment. This leads to the replacement of the system of hiring professional workers with voluntary work, or work on the basis of people's contracts for the provision of certain services [12]. For example, in European countries, residents often install solar panels or small

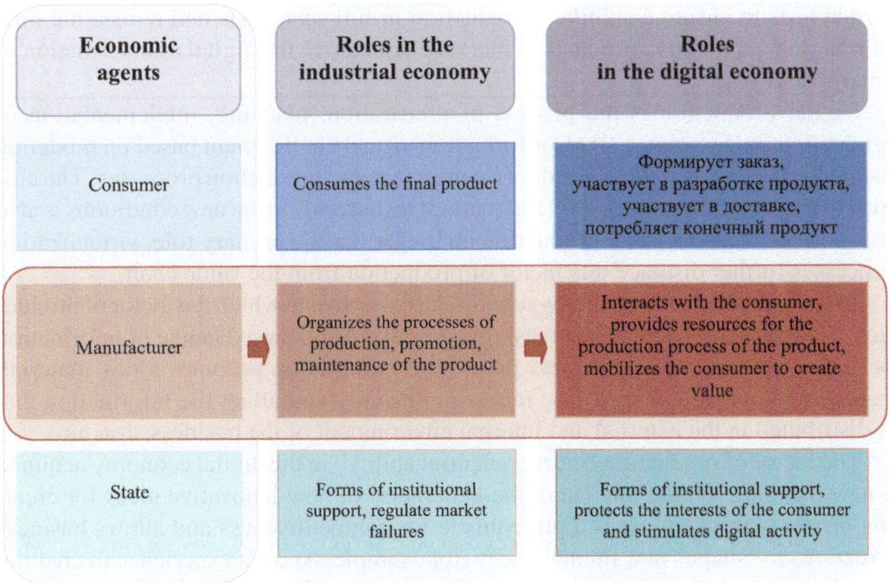

Fig. 10.1 Changing the role of economic agents in the digital economy

combined heat stations on roofs, and thus they assume the role of electricity producers and distribute excess electricity to the public grid.

There are new forms of attracting specialists (such as freelancing) and proactively concluded service contracts. According to CB Insights, in the United States, one in three employees have already joined the ranks of freelancers, by 2020 their share will grow to 40% of the total number of employees. 51% of top managers surveyed by CB Insights expect to use the services of temporary executors within the next 3–5 years [15].

The features and characteristics of the production teams of the era of transition to the digital economy are changing: a prerequisite is the staff's professional development, retraining and training on a permanent basis, changing the mentality of employees and corporate culture. Development of adaptive abilities of workers and reduction of resistance to constantly occurring changes, ensuring integration into these changes on the basis of generation of new ideas becomes a requirement of time [16]. Teamwork becomes the main method of organizing the work of the production team, and project management—the most effective method of management.

The requirements for the capital factor of production are also changing. In the conditions of industrialization, the production capabilities of the machines and equipment used determined the main efficiency of production. Fundamentally new high-tech approaches in product design based on a multi-level matrix of targets and resource constraints, digital automation platforms and systems of intelligent assistants designed for the development of digital twins (Digital Twin) products and production, the development of virtual stands and polygons, the implementation of virtual tests to ensure a significant reduction in full-scale tests and reduce the time of bringing competitive products to market characterize the digital transformation of enterprises [17].

Digital production is the process of preparation, planning, implementation of production processes in a single information virtual environment based on modeling using digital counterparts of products, equipment and production processes. The attitude to the factor of production "land (natural resources)" in the new conditions is also transformed. This factor of production no longer plays a primary role, virtualization processes further displace this factor of production from the value chain.

At the same time, the economy remains an economy, in which the factor of production "information" becomes increasingly dominant, and the exchange of information becomes the basis of all business processes: information becomes a raw material, new technologies tend to diffuse more significantly and affect the information that is distributed in the external and internal environment of the business structure.

The factor of production "entrepreneurial ability" in the digital economy acquires a new meaning: on the one hand, the generation of new innovative ideas for creating products and services is a prerequisite for competitiveness and allows business structures to "shape" new human needs (for example, Apple's experience in creating the IPhone), and on the other hand, entrepreneurial ability enables the integration of vertical and horizontal value chains (from product design and material supply to the manufacturing process itself, including logistics and service) [18]. At the same time, entrepreneurial solutions can be worked out using augmented reality technologies.

Thirdly, formal and informal business cooperation is taking on new forms.

In the Deloitte study, collaboration is recognized as one of the "pillars of increasing the sustainability" of the business structure in the digital environment. Collaboration-the ability to develop trusting relationships with value chain partners and other strategic entities [19].

In the digital economy, collaboration is increasingly taking the form of networking. A clear advantage of the network interaction of business structures is that they are always in the process of updating, because their main elements are characterized by mobility and allow optimal adjustment in accordance with changes in the market, technology and other environmental factors. The results-based adaptation process allows managers to expand the choice of modern ways to achieve the goals of the organization in a changing environment.

In networked organizations, there is often inter-organizational interdependence of resources instead of intra-firm allocation of resources, so business structures included in a single network constantly strive to improve relationships to strengthen common competitive advantages. At the same time, the human factor, social relations, are considered as the main resource and competitive advantage of the business structure. Networks are formed and maintained on the basis of informal relations based on trust that are reinforced through business relationships. Trust gradually develops as a result of growing knowledge of mutual interests [20].

Fourthly, the continuous search for new effective tools to meet consumer needs and deliver pleasure to the person determine the success of the business structure and its competitive advantages.

Long-term relationships with consumers require business structures to constantly communicate with them. In the industrial economy, companies divided consumers into "active" and "inactive", inactive customers fell out of their target audience. In a digital environment, businesses use the art of communication to reactivate inactive customers. Facebook, Instagram, means of telephone and mobile communication, Internet channels, CRM-systems, mobile applications, readiness tracking systems and many other things are used by companies for this purpose. Companies have not only free customers, but also those who regularly pay for information about the company's activities and change the product line through effective communication [21].

Personalization that increases response, conversion, and revenue is at the heart of this communication [22].

However, marketers have studied personalization technologies and come to the conclusion that it is impossible to set up rules for each individual user, try to predict every aspect of the customer's actions and create a highly personalized service experience. A 2018 consumer report from Accenture also found that consumers increasingly feel the digital experience has fallen short of expectations. They are more likely to shop with a brand that treats them personally. The main problem is that most approaches to personalization are based on assumptions and guesses without attracting the very customers the company wants.

Of course, personalization concerns individual users, reflects their behavior, interests, needs and desires, interaction models. It is conversations and dialogue with customers that act as a connecting link that ensures the trust of customers. Business structures can meet new customer expectations, "create continuous, two-way digital conversations that resemble customer conversations with a salesperson or service partner, conversations that treat them as individuals rather than a set of users" [22].

According to the Accenture survey, 75% of consumers believe they will find value in creating and managing a "needs profile," making interactions more personal and improving their brand experience. These need profiles are created through continuous communication, reflect true preferences, and go beyond what the consumer buys or consumes. They reflect the details of the needs and explain why customers make certain choices. These live profiles also evolve with the consumer through their interaction with the brand of the business structure.

Gartner pointed to the following possible stages of interaction between the business structure and the consumer about the creation and consumption of a new digital product (Hype cycle): "launch"—the start-up stage, which promises a breakthrough and a solution to many consumer problems, "peak of high expectations"—excessive enthusiasm and inadequate expectations of the product lead to failures in practical application, "the abyss of disappointment"—the excitement subsides because the views do not correspond to the results obtained, "the slope of enlightenment"—bad experience forces to reconsider ways of application of a product and to start from its real possibilities, "plateau of productivity"—the product receives recognition and evolves, and its advantages become generally recognized [11].

Therefore, the motive for the business structure in the process of creating a new product will be to achieve a "plateau of productivity", that is, maximum satisfaction of human needs and providing him with pleasure from the consumption of the product.

Fifthly, threats and risks to the economic security of a business structure, which can cause real economic, social and psychological damage, become a side effect of the digitalization of business activities.

The global CGMA study showed that reputational risks and taking measures to eliminate the consequences for the value of the company as a result of adverse events deserve increasing attention [23].

The leaking of information, crashing of computer systems, kibershpionazh, as well as risks associated with the partnership—the risk of default (when the partner does not deliver a product or service, the product or service does not meet the required quality, the effective coordination of operations is impossible, which leads to the failure to achieve goals, the planned joint innovation does not work) and the risk arising from the relationship (the partner uses the results of intellectual activity of the business structure, non-compliance of remuneration to profit participation, cultural risks, risks of regulatory non-compliance) can become adverse events in the digital environment [24].

The achievement of sustainable competitiveness in the digital age by business structures depends on an adequate understanding and consideration in practice of the possibilities of the digital environment and the challenges and threats to business and its stakeholders, including total control over the state of all business processes regardless of space and time.

New forms of financing (crowdfunding), participation of local companies in global supply chains, access to international markets with products arise due to continuous on-line possession of complete information about the business structure. These forms of activity are also subject to information threats to economic security, which can reduce the vulnerability of business structures.

The dangers associated with the loss of employee independence over themselves, the growth of social inequality due to uncontrolled penetration into their private lives, new criminality, information security gaps are new threats to the digital environment. All these threats can bring real economic damage to the business structure, and in some cases-cause bankruptcy and withdrawal from the market.

The need to develop adequate response measures to the level of threats is actualized in the context of reducing the economic security of business structures. In this context, a risk locator should be created that can anticipate problems and look at things from a different angle. This will allow the business structure to form an early warning system and use new opportunities.

The establishment of an internal rapid response system can prevent an incident from escalating into a crisis or natural disaster, ensure the correct placement of people and the proper procedures for the prompt restoration of normal activities.

10.4 Conclusions/Recommendations

Thus, the study proves that the digital environment creates conditions for a qualitative shift in the ways of doing business and competition of companies, approaches to interaction with consumers and ways to meet their needs. It is shown that the roles of the main economic agents change, alternative ways of selection of resources by business structures and the nature of changing requirements to them are determined. It is argued that formal and informal business cooperation acquires new forms.

It is proved that the success of the business structure and its competitive advantages are determined by the continuous search for new effective tools to meet consumer needs and deliver pleasure to a person. At the same time, long-term relationships with consumers require constant communication of business structures with them, as a result of which their new competitive advantages are formed.

Identified threats to economic security, which increase the vulnerability of the business structure. These threats can bring real economic damage, and in some cases-cause bankruptcy and withdrawal from the market.

References

1. Drucker P (1969) The age of discontinuity. HarperCollins, New York
2. Castels M (2000) Information age: economy, society and culture: transl. with English under the scientific the editorship of O. I. Shkaratan, HSE, Moscow
3. Tabernaculov A, Kaufmann Y (2019) The blockchain in practice. Alpina Publisher, Moscow
4. Sukharev O (2012) Information economy, transaction costs and development. J Econ Theory 1:80–93
5. Adams M, Bennett S (2018) Corporate governance in the digital economy: the critical importance of information management. J Gov Guide 70(10):38–54
6. Arieli D (2010) Predictable irrationality: the hidden forces that determine our decisions. Mann, Ivanov and Ferber, Moscow
7. Tyler R, Sunstake K (2018) The Architecture of choice: how to improve our decisions about health, well-being, and happiness. Mann, Ivanov and Ferber, Moscow
8. Popov E, Semyachkov K (2018) Problems of economic security of digital society in the conditions of globalization. Econ Reg 4:1088–1101
9. Karpunina E, Konovalova M, Shurchkova J, Isaeva E, Abalakın A (2020) Economic security of businesses as the determinant of digital transformation strategy. In: Popkova E, Sergi B (eds) Digital economy: complexity and variety vs. rationality. Springer, Cham, pp 251–261
10. Kolosov A (2010) Economic security and economic systems. RAGS, Moscow
11. Gartner (2019) Gartner top 10 strategic technology trends for 2019. Available at: https://www.gartner.com/smarterwithgartner/gartner-top-10-strategic-technology-trends-for-2019/. Accessed 10 Oct 2019
12. CIMA and CGMA (2016) Rethinking the value chain: ethics, risk, governance at every link. Available at: https://www.cfin.ru/management/finance/valman/Ethics-risk-and-governance.shtml. Accessed 15 Oct 2019
13. Richter K, Pakhomova N (2018) Digital economy as innovation of the XXI century: challenges and chances for sustainable development. Prob Mod Econ 2(66):22–31
14. CGMA and Transparency International (2014) The role of transparency in the value chain: anti-bribery legislation. Available at: https://www.cgma.org/content/dam/cgma/resources/downloadabledocuments/ethics-risk-and-governance.pdf. Accessed 03 Sept 2019

15. Karpunina E, Isaeva E, Moiseev S, Gorbunova O (2019) Challenges of the digital economy: how does human capital determine the economic security of the state? Drucker's Bulletin, No 4
16. Mordashov A (2018) How industry 4.0 is changing the control. Available at: https://hbr-russia.ru/liderstvo/lidery/a24981. Accessed 02 Apr 2019
17. Amelin S, Shchetinina I (2018) Organization of production in the digital economy. Organ Prod 26(4):7–18
18. Kovalchuk U, Stepnov I (2017) Digital economy: transformation of industrial enterprises. Innov Manage 11:32–43
19. Marchese K, O'Dwyer J (2014) From risk to resilience—using analytics and visualisation to reduce supply chain vulnerability. Deloitte Review, issue 14
20. Hagedoorn J (1993) Understanding the rationale of strategic technology partnering: interorganizational modes of cooperation and sectoral differences. Strateg Manag J 14:371–385
21. Tynerblain (2009) The conversation economy. Available at: http://tynerblain.com/blog/2009/09/01/the-conversation-economy/. Accessed 27 October 2018
22. Data and Marketing Association (2018) Why brands must start listening in the conversation economy. Available at: https://dma.org.uk/article/why-brands-must-start-listening-in-the-conversation-economy. Accessed 14 Feb 2019
23. CGMA (2013) From insight to impact unlocking opportunities in big data. Available at: https://www.cgma.org/resources/reports/downloadabledocuments/from-insight-to-impact-executive-summary.pdf. Accessed 13 Aug 2019
24. Das TK, Teng BS (2001) Trust, control, and risk in strategic alliances: an integrated framework. Organ Stud 22(2):251–283

Chapter 11
Creation of the Territory of the Advancing Socio-Economic Development as a Way to Diversify the Economy of a Single-Industry City

Dmitry D. Talalaev, Elena A. Yakovleva, Elena V. Korolyuk, Irina V. Kosorukova and Alexander S. Astakhin

Abstract The relevance of the research topic is due to the existence of the problem of low level of socio-economic development of a significant number of Russian single-industry towns. This problem was formed in the process of transition from a planned to a market economy. It was caused by the bankruptcy of previously active and dynamically developing city-forming enterprises in the transition period. This led to a decrease in the level and quality of life of the population of single-industry towns, completely dependent on these enterprises. Currently, a special research interest is the search for new effective tools for the development of single-industry towns and improve their socio-economic development. In the scientific work the definition of the concept of "single-industry city" is given, its criteria and characteristics are highlighted. The necessity of state support in the development of single-industry towns is substantiated. It is proved that one of the main ways to assist in the diversification of the economy of single-industry towns is the creation of territories of advanced socio-economic development in them, which are characterized by special conditions and have their own characteristics. In the article the experience of creation

D. D. Talalaev (✉)
Tambov State University Named After G.R. Derzhavin, Tambov, Russia
e-mail: talalaev.dimasik@mail.ru

E. A. Yakovleva
Voronezh State Forestry Engineering University Named After G.F. Morozov, Voronezh, Russia
e-mail: elena-12-27@mail.ru

E. V. Korolyuk
Kuban State University (Branch in Tihoretsk), Tihoretsk, Russia
e-mail: lenkor@bk.ru

I. V. Kosorukova
Moscow University for Industry and Finance "Synergy", Moscow, Russia
e-mail: ikosorukova@synergy.ru

A. S. Astakhin
Belgorod State National Research University, Belgorod, Russia
e-mail: aastaxin@yandex.ru

T. Kolmykova and E. V. Kharchenko (eds.), *Digital Future Economic Growth, Social Adaptation, and Technological Perspectives*, Lecture Notes in Networks and Systems 111, https://doi.org/10.1007/978-3-030-39797-5_11

of the territory of the advancing social and economic development "Kotovsk" was studied, problems of its functioning and directions of its development are defined.

Keywords Single-industry town · City-forming enterprise · Territory of advanced social and economic development

JEL Classification Codes K2 · L5 · R1 · R5

11.1 Introduction

In scientific sources, the term "single-industry city" is used, which is formed by combining the concepts of "single-industry" and "city" [1].

This term has been widely used in the Russian Federation relatively recently: in 1990, due to the need to determine the list of municipalities whose socio-economic development is directly dependent on the economic condition of the city-forming enterprises. The Russian specificity lies in the fact that at the end of the 20th century there were a large number of city-forming enterprises. They were created and initially functioned in a planned economy. In the transition to a market system of management, they were uncompetitive. This fact caused crisis situations in many Russian cities where they were located.

Thus, on the map of Russia began to appear municipalities with huge problems of social and economic development, the solution of which became impossible without the intervention of the Federal center. Therefore, today the state must solve the important task of finding and justifying the ways of socio-economic development of single-industry towns. This requires the creation of a theoretical and methodological basis for the long-term development of single-industry towns.

11.2 Methodology

The scientific hypothesis of the study consists in the assumption that the creation of territories of advanced socio-economic development in single-industry towns can serve as a factor for improving the socio-economic situation in such settlements. This will be accompanied by an increase in the gross regional product, the creation of new jobs, an increase in average wages, as well as a decrease in the outflow of qualified personnel.

In the scientific literature, there are many points of view regarding the concept of "single-industry city". According to the Russian researcher Granberg "a single-industry city is a city on the territory of which there is only one city-forming enterprise" [2]. Lipsits defines "monocity as a municipal formation on the basis of the city-forming enterprise" [3]. Neshadin gives the following definition: "a single-industry city is a city whose economic base is one large production and several technologically related enterprises" [4].

Gureeva said "mono settlement (town, village) where there is a close relationship between the functioning of the major (backbone) of the enterprise and socio-economic aspects of life of the people and the market mechanisms of the growth can significantly affect the fate of the settlement" [5].

Western theorists consider the concept of "single-industry city" from the position of the city-forming principle: here they highlight its important feature-industry binding. Therefore, a single-industry city arises as a result of the development of a particular industry [6, 7].

According to Linda Carlson, "a single-industry city is a city that is owned by a company". In her research, she considered the vector of development of single-industry settlements in the northwestern United States (near the city of Seattle) in order to identify the relationship between the main employer in the settlement and its population. According to the author, all multi-disciplinary entities are subjected to processes of socio-economic depression, as a result of which the presence of the main employer and the population in the settlement is reduced [8].

Hardy green called it a "one-company town". Attention in his works he paid to successful examples of management of the company's own single-industry town. To a greater extent, the companies, according to the author, had an operational approach to the population of single-industry towns in order to obtain the maximum possible profit. Some settlements the author cites as a positive example of the correct attitude of the employer to the population. For example, in the city of Corning in New York state, the reason for the high standard of living of the population was the direct interest of the company in its development [9].

As you can see, the basis of all the above definitions of the term "single-industry town" is a city-forming enterprise, whose economic development has a huge impact on the life of a single-industry town.

The methods of comparison, classification, grouping, and economic analysis were used.

11.3 Results

Consider the specifics of the development of single-industry towns in the Russian business practice.

Today in the Russian Federation there is 319 single-industry towns with about 15 million inhabitants [10]. The largest single-profile municipality is the Russian city of Togliatti, whose population is 702.7 thousand people (as of January 1, 2019). The next largest single-industry city is Novokuznetsk with a population of 554.9 thousand people. The city of Naberezhnye Chelny closes the three largest single-industry towns in Russia, its population is currently 531.0 thousand people.

Given the total population of single-industry towns and the fact that the population of some of them exceeds the population of many administrative centers of the subjects of the Russian Federation, it is obvious the enormous impact of the results of their economic development on the economy of the state as a whole. It is worth noting that the most sparsely populated single-industry town in the Russian Federation is the village of Bering Chukotka Autonomous Okrug, which as of 1 January 2019 has a total population of 816 people.

The institutional framework regulating the activities of Russian single-industry towns is the Decree of the RF Government from July 29, 2014 No. 1398-R "On the List of monoprofile municipal formations of the Russian Federation (single-industry towns)". This document reflects the classification of single-industry towns by risk factors of the socio-economic situation on the basis of the so-called "traffic light" principle. According to this principle, there is a division of all single-profile municipalities in Russia into three zones: red-the most risky zone; yellow-the zone of medium risks; green is a zone with relatively prosperous socio-economic development [10].

The Russian legislation defines specific criteria for identifying the type of single-industry city. In accordance with the proposed government methodology, at the first stage, single-industry towns are checked for belonging to the red zone, that is, the most risky. The implementation of at least two criteria of the methodology automatically allows to attribute a single-industry city to the red zone (Table 11.1).

If a single-industry city does not fall into the range of criteria of the red zone, then the possibility of its attribution to the green zone is revealed. The criteria for classifying a single-industry city as a green zone are also presented in Table 11.1. If the characteristics of a single-industry city do not allow it to be attributed to the green zone, it automatically falls into the yellow zone.

Currently, the largest number of Russian single-industry towns are assigned to the yellow zone (154 single-industry municipalities), 94 single-industry municipalities have features characteristic of the red zone. Only 71 single-industry municipalities in Russia meet the criteria of the green zone (Fig. 11.1).

Note that in the period 2017–2019, 6 single-industry towns moved from the red zone to the yellow zone. This is evidence of a smooth process of stabilization of the socio-economic situation in these settlements.

The Ministry of economic development of the Russian Federation is the supervising Federal Executive body for monitoring the socio-economic situation of single-profile municipalities. According to the resolution of the Government of the Russian Federation "about criteria of reference of municipalities of the Russian Federation to single-industry towns and categories of single-industry municipalities of the Russian Federation depending on risks of deterioration of their social and economic situation" the Federal security service of the Russian Federation carries out monitoring of public opinion in such settlements [11]. The results of social surveys are one of the criteria for determining the types of single-industry towns.

Socio-economic development of mono-profile municipalities in the Russian Federation is characterized by the presence of various problems. First of all, these problems are caused by the crisis situation of the city-forming enterprises [12]. In such a situation, conditions are created for the growth of unemployment, a decrease in the level of income of the population of single-industry towns, a reduction in local

Table 11.1 Criteria for singling out the types of single-industry towns depending on the socio-economic situation

Critical situation	The criteria for inclusion in the "red zone"	The criteria for inclusion in the "green zone"	The criteria for inclusion in the "yellow zone"
Finding the city-forming enterprise in a state of bankruptcy	Yes	No	All other settlements
Intentions of the city-forming enterprise on implementation of reduction of workers, in %	10% or more	3% or more	
Unfavorable assessment of specialization of the industry market of the city-forming enterprise	Yes	No	
Assessment of the level of registered unemployment	Exceeds 2 times the average national level	Does not exceed the national average	
Assessment of the socio-economic situation in the single-industry town by population	Negative	Positive	

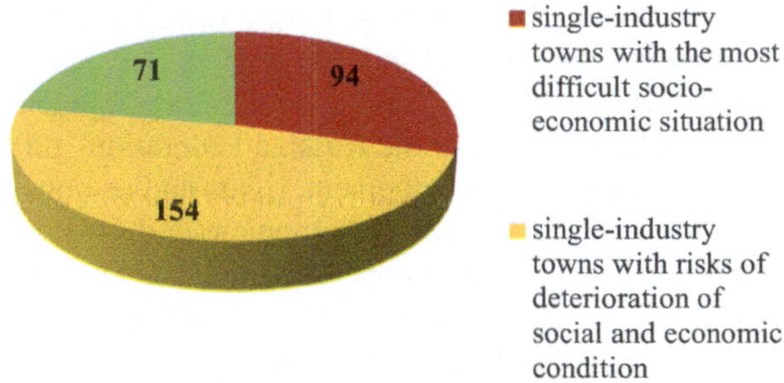

Fig. 11.1 The ratio of different types of single-industry towns in Russia depending on the socio-economic situation in 2019

budget revenues. In addition, the population is beginning to decline as a result of migration.

The government of the Russian Federation is taking measures to support single-profile municipalities. So, now the non-profit organization "Fund of development of single-industry towns" is created and functions [13].

To implement the main objectives of this Fund, budget funding in 2019 in the amount of 4.7 billion rubles is provided. In 2016 the Presidium of the presidential Council for strategic development and priority projects approved the passport of the priority program "Integrated development of single-industry towns" [14].

Reducing the dependence of single-industry municipalities on the economic condition of the city-forming enterprises is a key goal of the program. According to the passport of the program, 230 thousand new jobs (not related to the activities of city-forming enterprises) were to be created by the end of 2018. In fact, the value of this indicator reached 406.5 thousand jobs [15].

However, temporary jobs were also taken into account in the calculation of this indicator. The volume of investments in fixed assets, including extra-budgetary, by the end of 2018 was expected to reach 170 billion rubles. In fact, it was 1.8 trillion rubles, which is more than 10 times the plan. However, the calculation took into account all sources of funds, including loans and own funds of the population for shared construction.

In this regard, the accounts chamber of the Russian Federation criticized the state program for the development of single-industry towns. According to the accounts chamber of the Russian Federation, measures for state support of single-industry towns have not solved the main task of improving the well-being of their population [15].

The Ministry of economic development of the Russian Federation agreed with the conclusions of the accounts chamber. Thus, a new draft program of integrated development of single-industry towns is currently being developed.

One of the main objectives of the program "Integrated development of single-industry towns" is the creation of territories of advanced socio-economic development in single-industry towns in order to improve the socio-economic situation in single-industry municipalities by diversifying the economy and reducing the dependence of single-industry towns on the state of the city-forming enterprise.

Federal law "About territories of advancing socio-economic development in the Russian Federation" [16] and Government Decree from "On peculiarities of creation of territories of priority socio-economic development in the territories monoprofile municipal formations of the Russian Federation" (2015) establish rules for the creation, management, development, rights and responsibilities of residents of areas of advancing socio-economic development in these towns.

Article 2 of Federal law specifies that "the territory of advancing socio-economic development is part of the territory of the RF subject, including closed administrative-territorial education with special legal regime of entrepreneurial and other activities to create favorable conditions for attracting investments, ensuring accelerated socio-economic development and create favorable conditions for the viability of the population" [16].

The peculiarity of the territory of advanced socio-economic development is that the state provides its residents with various preferences, for example, tax incentives, customs procedures of the free customs zone, simplified connection to infrastructure facilities, and more.

On the territory of the Tambov region there are two single industry towns: the working village of Znamenka and the city of Kotovsk (currently both belong to the green zone). The RF Government decree "About establishment of territories of priority socio-economic development "Kotovsk" awarded the city the status of the territory of advancing socio-economic development [17].

In accordance with the tripartite agreement between the Ministry of economic development of the Russian Federation, the administration of the Tambov region and the administration of the city of Kotovsk within 10 years, the number of jobs created by residents of the territory of advanced socio-economic development will be 2,237 jobs, the number of residents of this zone will be 15 companies, the volume of investments made by these residents will be 10,745 million rubles [14].

The status of a resident of the territory of advanced socio-economic development "Kotovsk" implies the receipt of the following tax benefits until the end of 2027:

- reduction of income tax by 100% for the first 5 years and up to 13% for the next 5 years;
- land tax reduction by 100%;
- reduction of corporate property tax by 100%;
- reduction of insurance premiums to state extra-budgetary funds to 7.6%.

The condition of state support is to obtain the status of a resident of the territory by the investor, as well as the conclusion of an agreement on the implementation of activities in the territory of advanced socio-economic development "Kotovsk".

The criteria for the competitive selection of investment projects for the provision of state support to investors are:

- Minimum investment—15 million rubles.
- During the first year after the inclusion of a legal entity in the register of residents of the territory of advanced social and economic development, the minimum amount of capital investment carried out within the framework of the investment project implemented by the said resident in respect of the relevant economic activities is 2.5 million rubles.
- The minimum number of new permanent jobs created as a result of the implementation of the investment project by a resident of the territory in respect of the relevant economic activities is 10 units.
- Compliance with economic activities of the investor in the implementation of the investment project to the list of economic activities included in classes of all-Russian classifier of types of economic activity in which the special legal regime of entrepreneurial activity on the territory of advancing socio-economic development "Kotovsk".

A legal entity registered in the city of Kotovsk that intends to implement an investment project and acquire the status of a resident of the territory of advanced socio-economic development "Kotovsk" for the conclusion of the agreement shall submit an application to the administration of the city of Kotovsk in accordance with the legislation [18].

The application shall be accompanied by a passport of the investment project, a business plan, copies of constituent documents, a certificate of the absence of the applicant's debt on taxes, fees and other mandatory payments to the budgets of the budget system of the Russian Federation and state extra-budgetary funds, an expert opinion on the investment project (business plan) and more.

1. For one and a half years of functioning of the territory of the advancing social and economic development "Kotovsk" there were the following problem situations:
2. The first residents appeared only at the end of 2018.
3. As of April 1, 2019, within the territory of the advanced socio-economic development of Kotovk, only 11 jobs were created (with a planned indicator of 290 jobs), the volume of investments amounted to 8.3 million rubles, while it was planned to 1 919 thousand rubles. Of course, such indicators do not have a positive impact on the economy of the city. However, it is worth noting that the number of residents amounted to 6 companies, while it was planned to attract 5 companies for this period. All 6 companies have become residents in the month of December 2018–March 2019. According to the plan to attract residents to the territory of advanced socio-economic development "Kotovsk" until the end of 2019, the status of resident of this territory will receive 4 more companies.
4. In accordance with the Decree of the government of the Russian Federation, residents of this territory of advanced social and economic development can open production according to 26 permitted economic activities [18].
5. Article 427 of the Tax Code of the Russian Federation defines the procedure for granting reduced tariff insurance premiums, which are applied to a resident of the territory of advanced social and economic development, who has received such status not later than within three years from the date of creation of the relevant territory [19].
6. The tax code of the Russian Federation defines that residents of the territory of advanced social and economic development can take full advantage of the preferential tax regime, if they are on the General tax system. In this regard, small and medium-sized businesses that are on the simplified tax system in Russia and wish to expand production and become residents of such a territory, unfortunately, do not have preferences in full [19].

11.4 Conclusions

Directions of development of territories of advanced social and economic development in the Russian single-industry towns are: introduction of changes in the legislative acts regulating functioning of territories of advanced social and economic development.

For the purpose of more active development of the territory of the advancing social and economic development "Kotovsk" and diversification of economy of the city, it is expedient to cancel restrictions on the all-Russian classifier of types of economic activity. You should consider doing the investor any activity permitted by the legislation of the Russian Federation, in addition to classified in accordance with the national classification of economic activities in any of the following types: hunting, trapping and shooting of wild animals, manufacture of tobacco products manufacture of alcoholic beverages, wholesale trade and retail, in addition to trade in goods of own production activities financial and insurance activities for the organization and conduct of gambling and betting, organization and holding of lotteries, etc.

It is necessary to extend benefits on insurance contributions to state extra-budgetary funds regardless of the moment of obtaining the status of a resident of the territory of advanced social and economic development.

The Federal Law "About territories of advanced social and economic development in the Russian Federation" should be extended to individual entrepreneurs.

It is necessary to introduce benefits for residents of the territories of advanced social and economic development, which are on the simplified tax system, by reducing the applied interest rates.

Thus, at the moment it is too early to draw conclusions about the effectiveness of the territories of advanced socio-economic development. The creation of such zones in the Russian Federation began relatively recently, in practice, there are a large number of problems associated with their functioning. At the moment, both at the Federal and regional level, consider different ways to solve these problems, so the improvement of the situation in single-industry towns is possible over time. There is no doubt that the functioning of the territories of advanced social and economic development will have a positive impact on the stability of the economies of single-industry towns, attract investment, increase tax payments to budgets of all levels.

References

1. Kryukova A, Bochkovskaya A (2018) Single-industry town—the territory of advanced development. In: Education and science in modern realities of the international scientific and practical conference, Cheboksary, Russia, pp 122–127
2. Granberg A (2001) Fundamentals of regional economy: textbook for universities. HSE, Moscow
3. Lipsits I (2000) Mono-profile cities and city-forming enterprises. Review report, vol 1, Chronicler, Moscow
4. Neshadin A (2016) Single-industry towns of Russia. http://dom.viperson.ru/wind.php?ID=631953&soch=1. Accessed 12 Oct 2019
5. Gureeva M (2011) Economic problems of single-industry towns in the conditions of the world financial crisis, vol 2. Vestnik Rossiyskogo Novogo Universiteta, pp 37–45
6. Leadbeater D (2004) Mining towns and the new Hinterlend crisis, Canadian Dimension, vol 5. Dimension Publications Inc., pp 41–44
7. Dale B (2002) An institutional approach to local restructuring—the case of four Norwegian mining towns. Eur Urban Reg Stud 9(1):5–20
8. Carlson L (2003) Company towns of the Pacific Northwest. University of Washington Press, Seattle
9. Green H (2010) The company town: the industrial Edens and Satanic Mills that shaped the American economy. Basic Books, NY, 248 pp

10. The Decree of the Russian Federation Government (2019) The List of monoprofile municipal formations of the Russian Federation (single-industry towns), No. 1398-R from July 29, 2014 (as amended on March 18, 2019). Sobranie Legislation of the Russian Federation, No 31, Article 4448
11. Resolution of the Government of the Russian Federation (2014) About the criteria for classifying municipalities of the Russian Federation to single-industry categories of single-industry municipalities of the Russian Federation depending on the risks of deterioration of their socio-economic situation (July 29, 2014, No 709)", Collection of legislation of the Russian Federation, No 31, Article 4425
12. Karpunina E, Konovalova M, Shurchkova J, Isaeva E, Abalakin A (2020) Economic security of businesses as determinants of digital transformation strategy. In: Digital economy: complexity and variety vs. rationality. Lecture notes in networks and systems, 251–260. Springer
13. Resolution of the Federation Council of the Federal Assembly of the Russian Federation (2017) About measures taken by the Government of the Russian Federation to implement the priority program "Integrated development of single-industry towns (March 22, 2017, No 59-SF)". Legislative Assembly of the Russian Federation, No 13, Article 1792
14. Ria (2018) Tambov authorities have issued a regulatory framework for the activities of Kotovsk as territory of special social and economic development. https://ria.ru/20180807/1526110027.html. Accessed 20 June 2019
15. Vestifinance (2018) The accounts chamber criticized the state program of development of single-industry towns. https://www.vestifinance.ru/articles/118216. Accessed 20 June 2019
16. Federal law of the Russian Federation (2014) About territories of advanced social and economic development in the Russian Federation (29.12.2014, No 473-FZ, as amended on December 27, 2018). Collection of legislation of the Russian Federation, No 1, Article 26 (Part I)
17. Resolution Of the government of the Russian Federation (2017) About the creation of the territory of advanced socio-economic development "Kotovsk" (December 22, 2017, No 1607). Collection of Legislation of the Russian Federation, No 1, Article 350 (Part II)
18. Resolution of the Administration of the Tambov Region (2018) About approval of the procedure for concluding an agreement on the implementation of activities in the territory of advanced socio-economic development "Kotovsk" (August 6, 2018, No 795). http://tamlife.ru/files/phpkqIZNQ.pdf. Accessed 20 July 2019
19. Federal Law of the Russian Federation (2000) About approval of the Tax code of the Russian Federation, part 2 (August 5, 2000, No 117-FZ) as amended on 31.05.2018. Collection of Legislation of the Russian Federation, No 32, Article 3340

Chapter 12
The Problems of Administrative Liability for the Unfair Usage of Digital Platforms

Alexander Yu. Sokolov, Kirill S. Krotov, Bela Yu. Dzhamirze and Vera N. Batova

Abstract *Purpose*: The purpose of this research is determination of most significant features of the problem, mentioned above and the development of approaches to its solution. Taking into account the newness of this topic in the Russian legal space, the most productive research method in considered work is comparative method. It allowed to construct analogy with law enforcement practice of some foreign countries, which already created the necessary system of prohibitions and liability. *Design/methodology/approach*: During the research authors have collected the full information on the legal status of a subject, the abusing of its dominance and also have analysed the existing approaches to the implication of this status to the institutes of the digital economy. The search of information was conducted among scientific literature. It let to collect 20 sources, which helped to formulate the understanding of the abuse of dominance by using of digital platforms. Besides scientific literature, authors have analysed the law enforcement practice of foreign countries and some popular bills, discussed by Russian theoretics and practices of administrative and antitrust law. Basic research methods are formal legal (during the analysis of terms "dominance", "digital platforms") and comparative (during the comparison of status of digital platforms within the context of antitrust legislation in different countries. *Findings:* The result of the research is the author's recommendations, oriented to the development of the system of prohibitions for dominant subjects, existing in Russian legislation, and also reformation of administrative liability for delicts in considered sphere. *Originality/value*: The intensive development of digital economy and corresponding technologies raises the issue on the necessity of the development of legal

A. Yu. Sokolov (✉) · K. S. Krotov · V. N. Batova
Saratov State Law Academy, Saratov, Russia
e-mail: aysockolov@mail.ru

K. S. Krotov
e-mail: kirillkrotov@mail.ru

V. N. Batova
e-mail: batova.v.n@yandex.ru

B. Yu. Dzhamirze
Maykop State Technological University, Maykop, Russia
e-mail: bella901@mail.ru

T. Kolmykova and E. V. Kharchenko (eds.), *Digital Future Economic Growth, Social Adaptation, and Technological Perspectives*, Lecture Notes in Networks and Systems 111, https://doi.org/10.1007/978-3-030-39797-5_12

rules, which no longer respond to contemporary challenges. The realization of goods and services through the Internet is supported by government and convenient for customers, but such kind of trading is complex both for legislators and lawyers. One of the most discussable issues in this sphere is the issue on legal status of a subject, abusing dominance, and also the necessity of the adoption of new legal rules, regulating the deterrent force of a state to companies, reached the status of a dominant (or close to it) on a relevant market, by using of technological supremacy.

Keywords Digital economy · Abuse of dominance · Digital platform · Administrative liability · Network effects · Unfair competition

JEL-Codes K130 · K210 · K240

12.1 Introduction

One of the key attributes of the market economy is its permanent development, based on free competition of economic entities. The selecting through the economic competition unavoidably sets the trend to monopolization on relevant markets. The expression of such a trend is the concentration of market power in hands of the few group of economic entities, oriented to the most profitable of trading conditions for themselves. This circumstance obtains the new level of actuality in connection with the implementation of technological solutions. Particularly, it is development of digital economy, known as the "economic activity, whose key factor of production is the digital data, the processing of big data and using results of their analysis in comparison with traditional forms of economy allow significantly increase the efficiency of different kinds of production, technologies, equipment, keeping, realization, delivery of goods and services".

The optimal way to the data operating is the special software, which was some secondary technologies, but became the self-sufficient asset and even the mean of capital concentration. The using of such kind of asset, beside the main focus to simplification of the exchange of goods and services, could be executed unfairly, in purpose of the repression of competitors, their suppression and restrictions of market entry. In this context such categories of administrative law, as "abuse of dominance" and "unfair competition" need the review and renovation in purpose of the adoption to new conditions.

12.2 Materials and Method

The necessity of the adoption of legal system to contemporary market conditions, full of digital technologies, was reviewed by Belikova [1, 2], Puzyrevskiy [3], Yakovleva [4], Haucap and Heimeshoff [5], Egham [6], Dotsenko and Ivanov [7], Vartaev et al. [8], Richter and Slowinski [9].

12.3 Results

The most clear example of new challenges of antitrust regulation in global practice is the series of investigations in regards to international information company Google. These investigations were executed in different countries and were anyhow connected with anticompetitive actions regarding to competitors [10–12]. First of all, mobile operation system Android was created as opened, generally accessible system. At the same time, intense competition with Apple corporation, which has bring to the factual capturing of significant part of the global market of mobile operation systems, made further development of Android as free distributed product non-profitable [5]. This is why Google has been factually limiting its competitors actions on the contract level.

The first landmark case within this context in the Russian legal practice became the case of Federal Antimonopoly Service of the Russian Federation (FAS RF) vs. Google, initiated upon the application of the Russian company Yandex. The essense of complaint is the restriction of competition by mean of the prohibition by Google for pre-installation of all of Yandex services to mobile devices Fly, Explay and Prestigio. Taking into account the product, general for Yandex and Google (software and relevant services), such prohibition factually restricted Google competitors (particularly, Yandex) to promote their goods and services, because software has the unity with digital technic, because the one is the user content of another.

There is an issue of the level of such unity. Each customer, which buys computer or other digital device, could chose a software, which will provide the connection between hardware possibilities and customer needs. This point was especially actual in the early twentieth century, when customers have usually bought a personal computer, and then have installed an operation system (Windows, for example). At the same time, the development of digital technic market and relevant technologies, set the trend to the maximal easiness and usability of digital products. The significant achievement of this trend is the manufacturer's ambition to give an off the shelf solution, which works out of the box and does not need additional knowledge or technical support.

Additionally, the struggle for convenience has formulated the market of operation systems and technological solutions, whose so convenient for customers, that competition on such market comes to nothing in some moment, until the moment of the origin of principally new solutions—innovations. In this reason the supporting and further acceleration of innovative development are the key trends in antitrust regulation of digital economy [7].

With that, together with governmental supporting of innovative development of digital field, the issue of public restraining regulation becomes more complex. Though the necessity of restraining governmental influence to the economy in purpose of the prevention of abuse of economic rights seems undoubted and almost generally received, the origin of new products in digital sphere makes the issue on the freedom of its administration discussable. For example, the discussed above case FAS RF vs. Google contains the issue on the unity of hardware and software, which could be or could not be dictating by a manufacturer of the last one. The development of digital space has factually bring to situation, when exactly software became a principal in the struggle for a customer. This is why such monopolists, like Google could unilaterally influence to global market of innovations and technological solutions. European and American legal practice cover this circumstance by the legal category Standard Essential Patents, which obligates owners of patents, having key value for one or another field, to afford the relevant rights to interested persons on justified, reasonable and nondiscriminatory conditions under threat of legal liability [9].

Within the framework of the case of FAS RF vs. Google this obligation was violated, what showed the shortage of special legal regulation in Russia. The violation had multilateral character and was oriented to the maximal restraining of competition. As representative of Yandex Yu. Vecher noted, "first of all we meet with situation, when Google did not allow to manufacturers to install our searching system on the homepage of a device. Later we have seen restraints for other applications – for example, for "Yandex. Browser". And now we have the total prohibition for pre-installation of all of our services to devices of some manufacturers. As we know, in 2014 Google has signed the new contracts as minimum with three manufacturers of Android-devices, ex-partners of Yandex – Fly, Prestigio and Explay. The contracts have the prohibition for pre-installation of any our services, including services, which have never compete with Google products. For example, "Taxi", "Market" or "Subway" [8]".

Remarkably, that at the begin of the proceeding inn 2015 FAS RF has qualified Google actions according to the Article 14 of the Federal Law "On the Protection of the Competition" ("Unfair Competition") [13], then qualification was changed to the Article 10 ("Abuse of Dominance") [14]. As the result, Google was fined 438.1 millions roubles (approximately $6.6 millions) by antitrust agency for the abuse of dominance [15].

According to the conditions of the Federal Law dated on 26 July 2006 № 135-FL "On the Protection of the Competition" (Competition Law), unfair competition is defined as any actions of economic entities (group of entities), focused on the getting advantages during entrepreneur activity, bending Russian legislation, business customs, requirements of fairness, reasonableness and justice to other economic entities—competitors and caused or could caused damage their commercial goodwill.

The prohibition, established by the same law, which covers the abuse of dominance defines this violation as acts (non-acts) of economic dominant, which bring or could bring to denial, restraint, removal of the competition and (or) infringement of interests of other subjects (economic entities) in the sphere of entrepreneur activity or indefinite range of customers.

The term "dominance" itself is regulated by the Article 5 of the Competition Law: this is the condition of economic entity (group of entities) or few of economic entities (group of entities) on the market of a certain good, which gives to such economic entity (group of entities) the possibility to have a paramount impact to the general conditions of commodity circulation on the relevant market, and (or) to take out of market other economic entities, and (or) inhibit access to this product market for other economic entities. Along with this definition, the Law established the series of conditions, variable depend on the type of market and other circumstances. One of the attributes among such circumstances, is a certain share of a relevant market [2]. The presence of such attribute means, that certainly quantitative meaning as market share, is determinative, what oftentimes received the argued criticism [4]. For example, the using of such criterion makes a lot of objective difficulties in the sphere of digital commerce, because digital platforms are close to the dominance in the sphere of trade and services and now are also spreading to the sector of industry [16]. In connection with this, it is natural, that the European practice embarked on the course of the understanding of dominance as the "market power" [1], as the level of the influence of an economic entity, which allows him or her to restraint competition unilaterally and to operate independing on competitors and customers [3]. This approach seems more flexible and perspective, taking into account even legally statued multivariance of situations and circumstances, what influence to necessary market share. On the one hand, it makes frames for an executor of law, but on the other hand, it degrades the efficiency of the using of legal prohibition, by creating of a prospect, when the real economic situation is not covered by actual legal rules. In this connection, we suggest the abolishment of the legal requirements to certain market share on a relevant market.

The necessity of the adoption of legal antitrust legislation to new digital reality translates into the series of amendments, which actively discussed on the moment of the preparation of this article—so called "Fifth Antitrust Package" [17].

One of the suggested reforms is the suggestion on the establishment of terms "digital platforms" and "network effects". The first of these—"infrastructure, placed in the Internet, used for the organization and providing of cooperation between sellers and buyers". The networks effects understanded as the ability of an economic entity, acting in Internet, to influence to market conditions (for example, to prices of goods), by using of information (big database). With that, a digital platform, which took no less, than 35% of a market of "substitute services" and has reached the total revenue no less, than 400 million of roubles (approximately $6 million) per year will be recognized as a dominant.

We have to note, that the Ministry of the Economic Development of the Russian Federation has estimate the bill negatively, by pointing a series of directions of legal regulation, whose need further elaboration [18]. Particularly, it concerns the absence of the term of "substitute services" and broad definition of "digital platforms".

12.4 Conclusion

Taking into account the conditions of actual legislation, whose regulate the status of a dominant subject together with new trends in digital trading, we could make conclusion on the necessity of reformation of legislation in purpose of the adoption of its conditions to real economic relations. First of all, there is a need to correct the rules, whose regulate the status of dominance by the abolishment of the depending on share market, what have to make positive effect to the flexibility of antitrust regulation, what is necessary for the regulation of digital economy. Moreover, existing global and Russian legal practice point at the necessity of special prohibition to the using of dominance in purpose of the restraint of competition on digital market. Such acts could be expressed in the tying of anticompetitive conditions on cooperation for competitors, competition cessation and other. As the most multilateral and efficient remedy of administrative coercion, the administrative liability in this sphere also needs special regulation by means of legal detalization of the corpus of administrative violation, established by the Article 14.31 of the Russian Code of Administrative Violations—"The Abuse of Dominance on the Goods Market".

References

1. Belikova KM (2006) Abuse of a dominant position in the legislation, doctrine and law enforcement practice of the European Union and the common market of South America-Argentina, Brazil and Peru. Law and Policy 79(7):38–52
2. Belikova KM (2013) Limits of permitted market dominance and control over mergers and acquisitions of large parties of property in BRICS on the example of Russia, India and South Africa. Lawyer 20:21–26
3. Puzyrevskiy SA (2013) Violation of the antitrust laws through the abuse of a dominant position. Econ Law 5:27–49
4. Yakovleva AA (2013) Dominance and abuse of the commodity market in European Union law. Compet Law 3:25–28
5. Haucap J, Heimeshoff U (2014) Google, Facebook, Amazon, eBay: is the Internet driving competition or market monopolization? Int Econ Econ Policy 11(1–2):49–61. Available at: https://doi.org/10.1007/s10368-013-0247-6. Data accessed 18 Aug 2019
6. Egham UK (2015) Gartner says smartphone sales surpassed one billion units in 2014. Available at: https://www.gartner.com/en/newsroom/press-releases/2015-03-03-gartner-says-smartphone-sales-surpassed-one-billion-units-in-2014. Data accessed 18 Aug 2019
7. Dotsenko AV, Ivanov AYu (2016) Antitrust, digital platforms and innovation: Google's case and developing approaches to protecting competition in the digital environment. Law 2:31–45
8. Vartaev RS, Garaev RZ, Kovalenko AI (2016) Abuse of dominance of digital platforms (Google case in Russia). Mod Compet 10(5(59)):89–141
9. Richter H, Slowinski PR (2019) The data sharing economy: on the emergence of new intermediaries. Int Rev Intellect Prop Compet Law 50(1):4–29
10. Antitrust: commission probes allegations of antitrust violations by Google. Available at: http://europa.eu/rapid/press-release_IP-10-1624_en.htm. Data accessed 18 Aug 2019
11. Google agrees to change its business practices to resolve FTC competition concerns in the markets for devices like smart phones, games and tablets, and in online search. Available at: https://ftc.gov/news-events/press-releases/2013/01/google-agrees-change-its-business-practices-resolve-ftc. Data accessed 18 Aug 2019
12. Statement of the Department of Justice's Antitrust Division on Its Decision to Close Its Investigations of Google Inc.'s Acquisition of Motorola Mobility Holdings Inc. and the Acquisitions of Certain Patents by Apple Inc., Microsoft Corp. a…. Available at: http://justice.gov/opa/pr/statement-department-justice-s-antitrust-division-its-decision-close-its-investigations. Data accessed 18 Aug 2019
13. FAS Russia opened a case against Google. Available at: https://fas.gov.ru/news/6113. Data accessed 18 Aug 2019
14. The decision and order in case No. 1-14-21/00-11-15 5, 2015. Available at: https://br.fas.gov.ru/ca/upravlenie-regulirovaniya-svyazi-i-informatsionnyh-tehnologiy/ad-54066-15/. Data accessed 18 Aug 2019

15. The FAS media: FAS appointed a Google penalty in the amount of 438 million rubles. Available at: https://fas.gov.ru/publications/866. Data accessed 18 Aug 2019
16. Report of the Federal Antimonopoly Service on the state of competition in the Russian Federation for 2017. Available at: https://fas.gov.ru/documents/658027. Data accessed 18 Aug 2019
17. The FAS media: FAS presented the fifth Antimonopoly package. Available at: https://fas.gov.ru/publications/14947. Data accessed 18 Aug 2019
18. Gazette No. 173 (2897) (0810) technology and media (2018) The fifth package got an fail, 5 Oct

Chapter 13
The Choice of Strategic Priorities for the Development of the Organization: A Methodological Approach

Sergey G. Emelyanov, Yulia V. Vertakova, Olga A. Kryzhanovskaya and Olga Y. Nepochatykh

Abstract The development of an organization in an unstable environment implies the need for a competent choice of strategic development priorities. A variety of alternative options make choice difficult. The use of conventional scientific traditional methods of choosing strategic alternatives is a prerequisite, nevertheless, the development and application in practice of a unified methodology for choosing strategic priorities will allow a more effective management decision regarding the strategic development of the organization in a shorter time frame, which will provide a competitive advantage in the future. The purpose of the scientific work is to develop scientific and practical recommendations on the development of methods for choosing strategic priorities for the development of the organization. The authors substantiated the methodology for choosing strategic priorities for the development of the organization, based on the dual use of the matrix tools of strategic management (building the matrix "industry attractiveness - competitiveness"), expert assessment methods (Delphi method), multi-criteria selection method (Churchman-Ackoff) taking into account the action of the "Pareto principle". The proposed methodology for the selection of strategic priorities of the organization was tested on the example of a specific company and proved its universal character.

Keywords Strategic priorities · Strategic management · Matrix analysis · Multi-criteria selection · Methodological approach

Keywords C51 · L25 · O1

S. G. Emelyanov (✉) · Y. V. Vertakova · O. A. Kryzhanovskaya · O. Y. Nepochatykh
Southwest State University, Kursk, Russia
e-mail: rector@swsu.ru

Y. V. Vertakova
e-mail: vertakova7@yandex.ru

O. A. Kryzhanovskaya
e-mail: morozikolya2008@yandex.ru

O. Y. Nepochatykh
e-mail: holga1306@yandex.ru

T. Kolmykova and E. V. Kharchenko (eds.), *Digital Future Economic Growth, Social Adaptation, and Technological Perspectives*, Lecture Notes in Networks and Systems 111, https://doi.org/10.1007/978-3-030-39797-5_13

13.1 Introduction

In conditions of external uncertainty and limited resources, you usually have to choose priorities from several options.

Our study allows us to argue that there are a sufficiently large number of matrix methods and methodological approaches to choosing alternatives, choosing organization development strategies, however, at present, there is no unified methodology for choosing strategic priorities for the development of an organization. We justified the methodology for choosing strategic priorities for the development of the organization, based on the dual use of the matrix tools of strategic management (building the matrix "industry attractiveness - competitiveness"), expert assessment methods (Delphi method), multi-criteria selection method (Churchman-Ackoff) taking into account the action of the "principle Pareto".

13.2 Materials and Method

The problem of choosing strategic priorities for the development of the organization is relevant today. Despite the fact that numerous attempts have been made to unify and parameterize the procedure for choosing strategic priorities (alternatives, strategies, directions, etc.), the development of a unified organization has not yet been developed. Features of the strategic management of organizations were studied by such foreign scientists as Vissem [13], Doyle [2], Karloff and Karlof [3], Kotler [6], Thompson and Strickland [10], etc. Moreover, the emphasis on the choice of strategic priorities was made in the works of such researchers as Bowman and David [1], Kleiner [4], Vertakova et al. [11, 12], Kryzhanovskaya [7], Kosinova et al. [5] and others.

Despite the significant contribution of foreign and Russian scientists, the issues of improving the methodological approach to the selection of strategic priorities for the development of the organization remain not fully explored and require further study.

Currently, a number of rules have been developed for choosing strategic priorities for the development of the organization. To concretize these rules, it is necessary to determine a number of factors that are considered when choosing priorities, to formulate criteria for each of them, compliance with which is mandatory.

It should be noted the existence of a variety of tools for selecting priorities. Quite often, in practice, such methods are implemented as the "preference analysis method", "analogy method", "scenario method", "goal tree method", etc. [12].

The "Matrix of Priorities" is another tool that allows you to focus on the most priority and potentially significant opportunities for "breakthroughs" when deploying a strategy.

Matrix tools can be used both independently and in some combination, which is determined by the goal and the task. The most famous and effective among them are the Boston Advisory Group (BCG) matrix, the Abel matrix, the Shell/PDM model, the General Electric–McKinsey matrix, the ADL/LS model, the Ansoff matrix, the Thompson-Strickland matrix.

Formation and selection of a strategy using analytical and procedural methods for its development is most rational to carry out on the basis of a systematic approach by the method of hierarchy analysis (MAI), Polygon of alternatives, Churchman-Ackoff and other methods of multi-criteria selection.

13.3 Results

The essence of the author's method is that in the case of the McKinsey matrix, the Delphi method allows you to determine indicators of the level of product competitiveness and segment attractiveness for business.

During the study, experts are provided with all information available about the markets or business entities.

During the first round, experts are tasked with assessing the attractiveness of the market being studied and the competitiveness of the product being studied on a scale of 1 to 10, where the obtained values are interpreted as follows: from 0–3 points: low, from 4–7 points: average, from 8–10 points: high. Experts should definitely get acquainted with the results and conclusions of analysts, after which another round is held.

Depending on what final assessment the product received on competitiveness and the market on attractiveness, its position in the matrix changes (Table 13.1).

The strategic lines of the organization's behavior, the implementation of which is necessary on the basis of the occupied quadrants of the matrix for choosing the strategic priorities of the organization's development, are related to the description of the elements of the McKinsey matrix.

Based on Churchman-Ackoff methods, as well as taking into account the "Pareto rule", a multi-criteria selection of strategic priorities for the development of the organization was carried out.

Thus, the author's methodology is based not only on the results of applying matrix tools when conducting strategic analysis, choosing strategic directions (McKinsey matrix), but also taking into account and coordinating the opinions of experts using

Table 13.1 Matrix selection of strategic priorities for the development of the organization

		Product competitiveness		
		0–3 points	4–7 points	8–10 points
Market appeal	8–10 points	1	2	3
	4–7 points	4	5	6
	0–3 points	7	8	9

the Delphi method, as well as multi-criteria selection by the Churchman-Ackoff method, taking into account the implementation of "Pareto principle".

On the example of Motors-Kursk CJSC (the main activity is trade in motor vehicles; additional activities are maintenance and repair of motor vehicles (car insurance services together with insurance companies, car body repair and painting, car repair and maintenance of all brands, exchange of a used car for a new one (Trade-IN program); trade in automotive parts, components and accessories) the practical aspect of the application of the improved tools for choosing strategic priorities for the development of the organization.

We conducted an analysis of the industry—the automotive industry of the Russian Federation, identified the main trends, problems and development priorities taking into account the strategic development priorities reflected in the developed Strategy for the development of the automotive industry of the Russian Federation until 2025 [9].

Next, a matrix strategic analysis of the development of Motors-Kursk CJSC was carried out. To analyze the external and internal environment, qualitative and quantitative (based on the opinions of the expert group) SWOT, PEST and SNW analysis were built.

The following conclusions can be drawn from the analysis: the company has many advantages that can be promoted in the industry, these advantages help the company to cope with the crisis, but there are also disadvantages that, on the contrary, slow down profit growth, attract new customers, and a weak enterprise strategy does not allow achieve specific goals of the organization. The enterprise needs to introduce innovations, which will help increase turnover, attract new customers and reduce costs.

Using the BCG matrix for CJSC Motors-Kursk, we identified the most promising and weakest products. To select the strategic development priorities of CJSC Motors-Kursk, we tested the developed methodology for selecting strategic priorities, described in the second section of the study.

To determine the indicators of the level of competitiveness of the product and the attractiveness of the segment for business, an expert assessment was carried out according to the Delphi method. The study was conducted on the example of the company Motors-Kursk. 10 people took part in the examination. The examination was carried out in writing, in the form of mailing profiles. The following question was asked in the questionnaires: "Rate the attractiveness of the market being studied and the competitiveness of the organization being studied on a scale of 1 to 10".

The examination was carried out in 2 rounds.

1. Determination of the attractiveness of the studied market. During the second round, experts revised their grades.

After the second round, the examination was completed, as expert opinions were agreed upon (coefficient of variation = 13.85%). The data obtained during two rounds are presented in Table 13.2.

2. Determining the competitiveness of an organization. After the second round, the examination was completed, as expert opinions were agreed upon (coefficient

Table 13.2 General results of the examination of tours

Type	Interval	Average rating	Dispersion	Variation (%)
1	7.0–8.75	7.9	2.10	18.34
2	7.5–8.75	8.2	1.29	13.85

Table 13.3 The General results of the examination of tours

Type	Interval	Average rating	Dispersion	Variation (%)
1	1–2	1.6	0.93	60.38
2	2–3	2.4	0.27	21.52

of variation = 21.52%). The data obtained during two rounds are presented in Table 13.3.

In expressing their opinions, the experts were guided, among other things, by the following data: the growth of the car sales market in Russia is expected in the next 5 years in physical terms from 6 to 9% per year (depending on the macroeconomic situation in the country), which will entail the expansion of markets service, which also operates the investigated company.

In the period from 2013 to 2018, the number of cars, per 1000 population of the region, increased by 28.3%. At the same time, there is a steady positive trend towards a further increase in this indicator. Based on expert assessments, the enterprise under study occupies the following position in the matrix:

Market attractiveness: 8.2 points;

Organization competitiveness: 2.4 points, i.e. the enterprise under study is in the 1st quadrant (Fig. 13.1).

The first quadrant (high segment attractiveness—low business competitiveness in the segment) has the following basic strategic directions of development:

Entering the market or expanding current activities in the market is impractical without a pronounced competitive advantage.

Given the current existence in the segment, it is recommended to consider the possibility of creating competitive advantages, while not trying to occupy a large market share only at the expense of the current product range.

		Product competitiveness		
		0-3 points	4-7 points	8-10 points
Market appeal	8-10 points	1	2	3
	4-7 points	4	5	6
	0-3 points	7	8	9

Fig. 13.1 Matrix for the selection of strategic development priorities of CJSC Motors-Kursk

As follows from the above, in order to propose activities, the company is faced with the task of developing a strategy to increase competitiveness.

The following strategic priorities presented in Table 13.4 can be considered as alternatives.

To select one of the proposed alternatives, the company's management formulated the following criteria for assessing strategic priority, presented in Table 13.5 [8].

Next, the criteria for assessing strategic priority by importance were sorted and assigned to them on a scale of 1 to 11. At the first stage of the analysis, we compare the estimates using the Churchman-Ackoff method. The total values are presented in Table 13.5.

Table 13.4 Strategic development priorities of CJSC Motors-Kursk

Strategic priorities	Description
Strategic priority No. 1	The introduction of public services for cars that meet the needs of users in the implementation of transportation, without personal ownership of vehicles, which avoids the cost of ownership of the vehicle (payment of transport tax, costs of maintenance, repair and others) For example, "car-sharing" (car sharing) is a type of short-term car rental. The benefits of car sharing include the ability to rent per minute, returning the car to almost anywhere within a certain area (usually inside the city), free parking near stations and airports, low cost of travel. Prospects for the development of the Russian market suggest an increase in the share of cars used by the "car sharing" model to 10% (more than 200 thousand units) by 2025
Strategic priority No. 2	The provision of the "e-hailing" service, providing the opportunity to order a car through a mobile device or computer (Uber, Yandex taxi and others). Such services have several advantages for ordering a taxi, such as: ease of ordering and payment, fixed price, transparency of pricing, the ability to track the route via GPS and others Global trends dictate the growing popularity of such services - on the horizon until 2030, according to an aggressive scenario, the share of passenger cars can reach 9% of total sales by 2030 (10 million public cars, 115 million cars in total in 2030) The conservative scenario assumes that a total of 126 million cars will be sold in 2030 and only 3 million of them will be purchased for sharing
Strategic priority No. 3	The introduction of Mobility as a service technology is a solution that provides for the planning and implementation of real-time multimedia trips from point A to point B using various modes of transport and allowing you to choose the optimal route according to many criteria, including user preferences and financial capabilities. It is predicted that by 2025 the global market for MaaS will be 1 trillion. $12, while the Russian market is tentatively estimated at $58 billion (5.8% of the total market) and 50 million. 13 users

Compiled by the authors taking into account the main trends in the development of the industry: The development strategy of the automotive industry of the Russian Federation for the period until 2025 [Electronic resource] / Official website of the Government of the Russian Federation // Access mode http://static.government.ru/media/files/EVXNIplqvhAfF2Ik5t6l6kWrEIH8fc9v.pdf

Table 13.5 Assessment of strategic priorities and their utility according to the Churchman-Ackoff method

No.	Criteria	V'_i	Strategic priority no. 1	Strategic priority no. 2	Strategic priority no. 3
1	Expected profit	0.17	0.85	1.36	0.51
2	Risk level	0.15	0.45	1.05	0.9
3	Payback period	0.13	0.52	0.91	0.91
4	Level profitability	0.12	0.6	0.72	0.96
5	Volume of investments	0.10	0.4	0.6	0.7
6	Market condition	0.08	0.56	0.48	0.56
7	Technical support	0.07	0.28	0.35	0.49
8	Project support complexity	0.06	0.12	0.3	0.42
9	Availability of labor resources for project implementation	0.05	0.5	0.3	0.5
10	Perspective direction	0.04	0.12	0.24	0.32
11	Other project investments	0.03	0.21	0.12	0.12
12	Utility		0.50	0.76	0.67

According to the results of applying the Churchman-Ackoff method, strategic priority No. 2 should be chosen. Nevertheless, the well-known Pareto law, the so-called "Pareto principle" or "80/20 principle", which states that 20% of the effort contributes to 80% of the result, we used as a basic setting in the analysis of the criteria for choosing strategic priorities of the organization: choosing the right minimum important actions, you can quickly get a significant part of the planned full result, while further improvements are ineffective and may be unjustified (Figs. 13.2, 13.3, and 13.4).

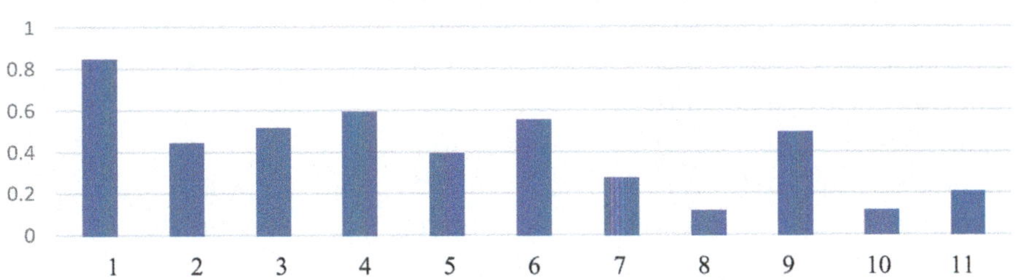

Fig. 13.2 Assessment of criteria 1 of the strategy

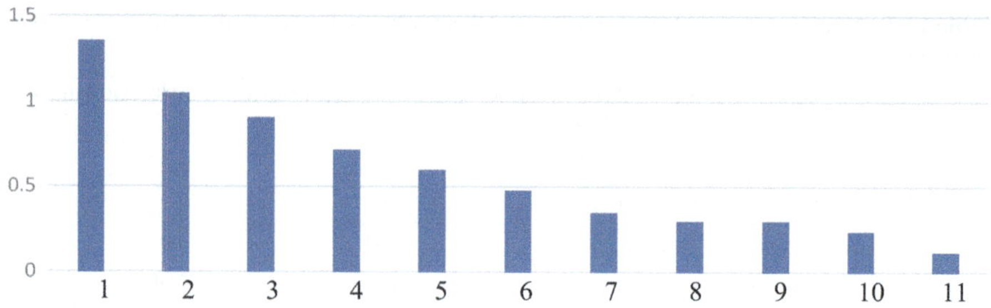

Fig. 13.3 Evaluation of criteria 2 of the strategy

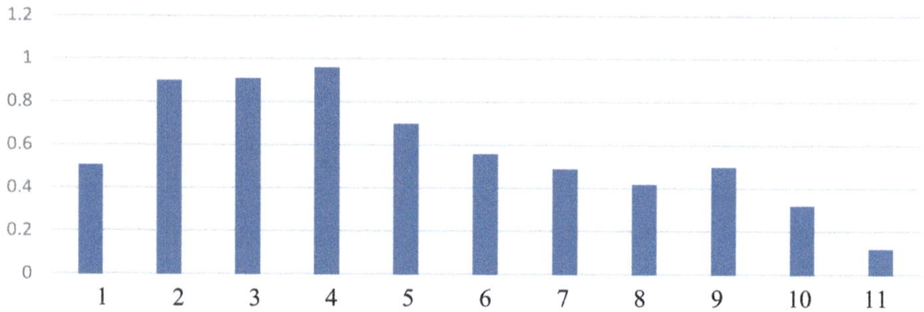

Fig. 13.4 Assessment of criteria 3 of the strategy

The final conclusion about the choice of strategic development priorities for the organization Motors-Kursk CJSC under study was made on the basis of a similar utility calculation procedure (Table 13.6).

Based on the study, we can conclude that it is most advisable to choose strategic priority No. 2, the provision of e-hailing services, i.e. the ability to provide cars to order through a mobile device or computer.

Table 13.6 Assessment and selection of strategic priorities of ZAO "Motors-Kursk"

Criteria	V_i'	Strategic priority no. 1	Strategic priority no. 2	Strategic priority no. 3
Expected profit	0.17	*0.85*	*1.36*	0.51
Risk level	0.15	0.45	*1.05*	*0.9*
Payback period	0.13	0.52	*0.91*	*0.91*
Level profitability	0.12	0.6	0.72	*0.96*
Utility		0.1445	0.507	0.393057

13.4 Conclusion

Application of the author's methodology for choosing strategic priorities for the development of the organization will allow the selection of a strategic line of behavior to achieve strategic priorities for the development of the organization.

It should be noted that this technique can be adapted to various management objects (including state and municipal enterprises). In this case, based on the dual use of the matrix tools of strategic management, it becomes possible to choose the most optimal strategy option, taking into account the specific characteristics of organizations.

References

1. Bowman C, David C (1987) Strategic management: corporate strategy and business policy. Palgrave Macmillan, Basingstoke, United Kingdom
2. Doyle P (1999) Management: strategy and tactics. Peter, St. Petersburg
3. Karloff B, Karlof B (1989) Business strategy: a guide to concepts and models. Macmillan Reference Books, London and Basingstoke, p 450
4. Kleiner G (1998) Strategic decision-making mechanisms and strategic planning at enterprises. Econ Issues 9:46–66
5. Kosinova N, Tolstel M, Sazonov S, Vaysbeyn K (2016) Development of methodological approach to enterprise's financial strategy based on comprehensive evaluation of its strategic potential. Eur Res Stud J 19(2):21–33
6. Kotler P (1998) Marketing management. Prentice Hall, Upper Saddle River, NJ
7. Kryzhanovskaya O (2019) The cholistic approach to the choice of strategic priorities of organization development. J News Southwest State Univ. Series: Economics. Sociology. Management. 9(2(31)):92–105
8. Kryzhanovskaya O, Sevriukova L, Plotnikov V, Vertakova Y (2016) The prospects for the implementation of industrial policy in countries in the medium term. In: Proceedings of the 27th international business information management association conference—innovation management and education excellence vision 2020: from regional development sustainability to global economic growth, IBIMA 2016, pp 845–856
9. The Government of the Russian Federation (2019) The development strategy of the automotive industry of the Russian Federation for the period until 2025. Available at: http://static.government.ru/media/files/EVXNIplqvhAfF2Ik5t6l6kWrEIH8fc9v.pdf. Accessed 12 Aug 2019
10. Thompson AA, Strickland AJ (1990) Strategic management. BPI-Irwin, Homewood
11. Vertakova Y, Klevtsova M, Babich T (2016) Identification of the new research areas and development of the existing ones by methods of morphological analysis and synthesis. J Econ Ann-XXI 157(3–4):4–7
12. Vertakova Y, Kurbanov A, Plotnikov V (2017) Approaches to the implementation of strategic innovation and investment projects for development industry of the territory. In: Proceedings of the 30th international business information management association conference, IBIMA 2017—vision 2020: sustainable economic development, innovation management, and global growth 2017, January, pp 1851–1858
13. Vissem H (2000) Strategic management and entrepreneurship opportunities for future prosperity. Finpress, New York, p 186

Chapter 14
Aspects of Regulating Individual's Bankruptcy Under the Laws of Foreign Countries

Gennady G. Egorov, Irina B. Oreshkina, Tatyana K. Krasilnikova and Larisa E. Strilets

Abstract *Purpose*: The article is a study of the peculiarities of regulating the legal status of individuals bankruptcy based on the norms in force in various international systems of countries with more developed regulators of this procedure. *Design/methodology/approach*: The methodological basis of the study is a complex of dialectical, historical and legal, systemic, structural, structural and functional methods that determine how to implement international legal achievements in national legal systems. *Findings*: Carrying out a review of legal norms valid in different countries of the world, the authors determine the need to develop, with the help of international experience, new approaches to ensure legislative regulation of personal bankruptcy procedures, taking into account the interests of all participants, while examining the experience of international legal regulation. The main features and shortcomings of such procedures were identified based on the specifics of a number of international entities at the state level. The main purpose of using this approach is to take into account achievements in the field of research in different countries of the world. Moreover, in recent years, especially abroad, there has been a steady increase in the category of court cases related specifically to the bankruptcy of individuals, which confirms the high degree of relevance of the topic under study. *Originality/value*: the lack of financial stability in the countries of the world during the ongoing economic crisis will always cause financial ties aimed at optimizing all relations, including the violation of financial obligations, which entails the need for legal mechanisms to stabilize the financial situation in the country.

G. G. Egorov (✉) · I. B. Oreshkina · T. K. Krasilnikova · L. E. Strilets
Volgograd State University (Volzhsky Branch), Volzhsky, Russia
e-mail: egorov@vgi.volsu.ru

I. B. Oreshkina
e-mail: orechkina-ib@mail.ru

T. K. Krasilnikova
e-mail: kun_krasilnikova@mail.ru

L. E. Strilets
e-mail: urexspert@yandex.ru

T. Kolmykova and E. V. Kharchenko (eds.), *Digital Future Economic Growth, Social Adaptation, and Technological Perspectives*, Lecture Notes in Networks and Systems 111, https://doi.org/10.1007/978-3-030-39797-5_14

Keywords Personal bankruptcy · Legal reception · Legal forms of drawing law · Ensuring the interests of the debtor · National financial security

JEL Code K 100

14.1 Introduction

In recent years, we can note a steady growth in relations that have caused individuals to have debt obligations. Such legal relations can also serve, in case of violation of the refund, the reason for recognizing the participant as an insolvent person, whose status is further confirmed through recognition of it as bankrupt and application, subsequently, of the relevant legal procedures to it. The primary reasons for the emergence of such relations are often a variety of consumer loans, the increasing popularity of which is observed due to imposed credit cards, the spread of digital systems and the general availability of this type of lending. The current situation usually entails adverse consequences entailing both a change in the social status of the person and an increase in debt obligations of any nature. The logical result of this phenomenon, as a rule, is the positive growth dynamics of insolvent individuals, which generates an avalanche of negative social consequences. These include the loss of a job, as well as a variety of diseases, family disputes, which only intensifies the "stagnation" of international financial systems.

All this determined the need for a detailed review of the experience of bankruptcy in countries around the world.

14.2 Methodology

The methodological basis for the analysis of the financial manifestation of the peculiarities of the regulation of recognition procedures for individuals who are not able to repay loan obligations and lack of ability to regulate the protection of the interests of all participants in the relationship, as well as difficulties in determining the credit base during the personal bankruptcy procedure, combine both scientific and practical problems of using personal bankruptcy in order to protect the interests of all participants and are based on historical and legal, systemic kturnyh and structural and functional methods.

14.3 Results

The legislation of foreign countries is quite different as the history and approaches in this area. At the same time, it is characterized by the presence of all kinds of differences existing between legal systems (first of all, the Anglo-American and continental legal systems), which have a number of significant features that affect the final goals of both civil law and its institution itself—personal bankruptcy (bankruptcy of individuals). The norms valid in the legal systems of different States related to the studied aspect of bankruptcy have both a number of similarities and a set of significant features that allow them to be distinguished. First of all, it is a certain focus on the liquidation or rehabilitation of the debtor (the so-called procreditor and debtor rules). This allows us to draw a conclusion about the basic tasks facing the law-making entity in the analysis of social prerequisites for the regulation of bankruptcy legislation. These tasks, despite the relatively universal nature of the existing bankruptcy rules, but at the same time and not aimed at rigid adaptation to a specific legal system, allow us to conclude that it is necessary to build a legislative framework governing all aspects of personal bankruptcy, taking into account the involvement of international experience, allowing you to choose those solutions that will most fully and reasonably eliminate legal conflicts in the studied area.

Bankruptcy of individuals has been used for quite a long time and with more or less success in various foreign countries. The study of foreign experience is relevant at the present stage of the bankruptcy procedure of individuals for a number of reasons. First, this mechanism of release of citizens from debt is used in Russia only since October 1, 2015. Despite the fact that the idea of declaring bankrupt not only a business entity, but also of an ordinary citizen has been known for a long time, there is no stable practice in the country to resolve cases of this category. Secondly, there are a number of problems in applying the provisions of Chapter X of the bankruptcy Law that require resolution, which is also possible through the use of foreign experience [11].

The characteristic of the norms in Chapter X of the bankruptcy Law, devoted to the bankruptcy of citizens who are not individual entrepreneurs, shows that foreign legal experience in the field of consumer activity entailing bankruptcy of a person was taken into account when adopting legislative acts. So in particular in the USA (Chapter 13 of the bankruptcy Code of 1978), in Germany (the bankruptcy Law of 1994), in Sweden (the bankruptcy Law and the debt relief Act adopted in addition to it), etc. [4].

V. G. Golubtsov, conducting a comparative legal detailed analysis of a number of functional features of bankruptcy proceedings in Russia and the United States, argues that the use of foreign experience to eliminate legal conflicts in the considered indicators should significantly positively optimize law enforcement practice in the Russian Federation [2].

In the United States, the problem of growth of outstanding debt obligations in the field of consumer lending is becoming more urgent every year. In order to optimize this situation, the act on prevention of abuse in bankruptcy and protection of consumer rights (Bankruptcy Abuse prevention And Consumer Protection Act) was adopted, which entered into force on October 17, 2005 [9].

In the United States, the right to bankruptcy is one of the constitutional rights, as it is aimed at protecting the economic interests of citizens as a weak party in relations with legal entities and the government [5].

The US law, as in a number of other countries, aims to regulate the status of a natural person who is a debtor. American law allows the legal procedure of assistance to persons in financial difficulties, while initially focusing on the introduction of alternative forms of changes in debt obligations. However, as representatives of public organizations have repeatedly noted, the situation does not reduce its relevance [7]. Most citizens of the country have a lot of loans, allowing through the opening of the new to close the obligations of the old, creating an endless credit debt. Thus, only lending rates change. Credit institutions that require repayment of debt, and do not seek to get all the debt. It is more profitable to claim at the moment only interest on the obligation and reduce the probability of repayment of the debt by the borrower, who is offered new loans with more favorable rates. And only if the creditor fixes the "absolute" insolvency of the debtor the procedure of personal bankruptcy begins to be used in practice [3].

The debtor can take advantage of the competitive process. The transfer of part of the existing property may allow the citizen to satisfy the "appetites" of creditors and allow him to keep his future income.

However, in addition to the first option, it is also possible to draw up a debt management plan, which will allow the debtor to keep all his property, but at the same time donate part of his future income [1].

A similar separation is seen as positive. We once again note that the issue of restoring solvency should occur only at the initiative or with the consent of the debtor. Otherwise, it will only delay the already lengthy bankruptcy proceedings.

We would like to note that there are also specialized courts in the United States that hear bankruptcy cases. At the present stage, it is difficult to give an unambiguous answer to the question of the need to create such vessels in Russia. Considering the fact that in each arbitration court, special "bankrupt" compositions of judges have been created and are functioning to resolve bankruptcy cases. The mechanism of bankruptcy of citizens in the United States is more debugged and is used everywhere. We can observe that although in our country more than 680 thousand people are subject to the bankruptcy law (according to the National Bureau of Credit Histories located at https://ria.ru/economy/20170228/1488904516.html), only a few want to use their right. Most citizens do not have a clue about the existence and characteristics of these procedures, and many simply do not have the financial ability to finance bankruptcy. Therefore, the creation of specialized courts that have proven their worth in the United States is seen only in the future.

In Germany, at Konkursordnung, which was previously in effect until December 31, 1994, there were no special rules applicable for the bankruptcy of individuals. In the process of reforming the institution of personal bankruptcy at the legislative level, a special procedure was formulated to declare a consumer insolvent, that is, bankrupt. The grounds, procedure for introducing and the course of the consumer insolvency proceeding are regulated by §304–314 part 9 Insolvenzodnung.

The course of the procedure depends on who submitted the application. If the application is filed by creditors and not by the debtor, then in this case the requirements for the attempt of an out-of-court or judicial agreement do not apply (paragraph 3 §306 Insolvenzodnung). If an application to the court was filed by one debtor or debtor along with creditors, then the accelerated proceedings are opposed by the mutual resolution of the problems of the insolvent debtor and his creditors [10].

Under German law, it is not enough for the debtor to have only signs of bankruptcy for the right to appeal to the court with a corresponding application. The debtor is charged with pre-trial settlement of the dispute. Moreover, the obligation is not formal, the debtor must take real steps to resolve the dispute.

Bankruptcy of individuals is already quite long and with more or less success is applied in various foreign countries. The study of foreign experience is relevant at the present stage of development of the personal bankruptcy for a number of reasons. Firstly, this mechanism of exempting citizens from debt has been used in Russia only since October 1, 2015. Despite the fact that the very idea of declaring bankrupt not only a business entity, but also an ordinary citizen has long been known, there is no and could not be a stable practice in resolving cases in this category. Secondly, there are a number of problems in applying the provisions of Chapter X of the Bankruptcy Law, which require their resolution, which is also possible through the use of foreign experience [11].

"If the attempt to settle the debts was unsuccessful, the debtor has the right to apply to the court for the introduction of a consumer insolvency procedure over his property. In a statement, the debtor is required to explain why the out-of-court debt settlement agreement was unsuccessful according to the sentence 1 paragraph 1 § 305 Insolvenzodnung" [10].

One of the features of German bankruptcy law is that after the completion of the bankruptcy procedure, creditors may present their claims to the debtor that were not satisfied, from the point of view of the law, in the implementation of the bankruptcy procedure.

Subsequently, instead of the right to unlimited collection of the remaining debt, the legislative acts of the German legal system introduced a priority to ensure interests, primarily in general civil circulation, introducing, by analogy with American law, the possibility of freshstart for the debtor.

Such mechanisms for the pre-trial settlement of disagreements between the debtor and creditors are not enough in the domestic bankruptcy law. Conditions must be created when both parties will be interested in using pre-trial mechanisms. These procedures should not be strictly regulated by the state, but conditions must be created when negotiations are inevitable and the refusal to pre-trial settlement of the conflict should be justified.

Consider the situation with bankruptcy of individuals in Asian countries. So, for example, according to the current norms of law adopted in the Democratic People's Republic of Korea, an individual is allowed to restore his solvency out of court. These mechanisms for the settlement of debt obligations, in essence, are a form of financial restructuring of credit obligations accumulated by an individual, and are applied only if there is a financial debt of the person to organizations.

The Credit Counselling and Recovery Service was established in 2002 with the organizational and financial support of the Government of the Republic of Korea. It is engaged in the restructuring of financial debts.

The conditions for applying debt restructuring, in addition to the nature of the debt, is also the amount of the debt. The following indicators are determined: for citizens who are not employed in entrepreneurial fields of activity, the allowable financial debt should not exceed the figure of 1.5 billion won (approximately about 60 million Russian rubles). Individual entrepreneurs are allowed a higher level of possible debt—about 3 billion won (about 120 million Russian rubles). An additional means of analyzing a person's guilt and financial solvency is the approved period of analysis of "delays" in repaying debts; according to accepted standards, they cannot exceed a period of three months. At the same time, participation in the program, which allows reducing the level of debts, is possible only if the debtor has a confirmed, and at the same time, constant source of income, which will ensure his living and financial needs of family members. An additional indicator is the mandatory presence of a total balance, after payment of all social needs, in the amount of not less than 150% of the living wage approved in the country.

If the debtor has no income at the time of the appeal, then he is granted a delay of up to 2 years [6].

This mechanism is very effective, and citizens are interested in its application. One of its advantages is that it is not public. Individuals do not want to publicly cover their financial problems and seek to solve them out of court.

The legislation of Japan, from the point of view of a number of authors, lags behind a number of countries and is more classical, based on the national values of the country. Legally, it does not contain, in relation to citizens, the term "bankruptcy". If it becomes necessary to repay the debt obligations of an individual, the legislator provides for their redemption by the state and the further application of the "debt elimination" procedure. The legal consequences of such a procedure for citizens are a change in its social status, which entails an indefinite violation of the rights of a person. In particular: debt obligations are forgiven to a citizen-debtor, but at the same time, there is a lifelong ban on taking managerial positions in companies, work as a lawyer, notary, work as a person with financial obligations, this also affects accounting activities, guardians of minors. In this case, the situation where the debtor at the time of the censure of rights was already a ward deserves special attention, the law allows the return of such incompetent citizens to guardianship carried out by the country's authorities [8].

Declaring a citizen bankrupt in Japan is associated with a number of significant restrictions. This follows from the fact that the debtor's guilt in his bankruptcy is presumed. A citizen who cannot cope with his financial situation is not worthy to engage in certain activities. In fact, the root causes of bankruptcy are not seen in difficult life circumstances, but in personality traits.

14.4 Conclusions/Recommendations

To summarize, in Russia a number of scientists insist on the need to use a simplified model for regulating the personal bankruptcy. At the same time, the legislator's readiness in the future to toughen the consequences of using a bankruptcy procedure by a citizen, for example, as it was implemented in Japan, deserves special attention. It seems to many authors that the idea of tightening the consequences of declaring a citizen bankrupt takes place. But, of course, legislative bodies should use it differentially There is no arguing about the need for legal regulation of the consequences of declaring an individual bankrupt, especially for entities that, at the time of bankruptcy proceedings did not possess property that could be levied and could not satisfy debt obligations to half of their creditors.

In such cases, legal debt relief is carried out. A person does not make efforts to restore his solvency and cannot even "donate" part of his property to creditors. Obviously, the right does not completely blame the debtor, which is why he found himself in such a deplorable financial condition. It is necessary to lay the foundation for a citizen to be more responsible in planning his budget. In modern Russia, there is a need to eliminate the consequences of a protracted economic crisis. It is believed that in the future mechanisms will be developed to prevent the personal bankruptcy.

Foreign experience will be incredibly useful for the development of the mechanism of personal bankruptcy, for its optimization, resolution of conflicts and can be used to eliminate problems that arise when applying the norms of Chapter X of the Bankruptcy Law.

References

1. Badahova LR (2010) Nekotorye aspekty bankrotstva grazhdan v SShA. Obshchestvo i pravo 4(31):283
2. Golubcov VG (2016) Zarubezhnye i nacionalnye standarty regulirovaniya instituta nesostoyatel'nosti (bankrotstva) grazhdanina. Russ Juridical J 5:13
3. Egorov GG, Melihov VM (2019) Problemy normativno-pravovoj reglamentacii i pravoprimenitelnaya praktika priznaniya fizicheskogo lica bankrotom v Rossijskoj Federacii. Vestnik Saratovskoj gosudarstvennoj yuridicheskoj akademii 2(127):148–158
4. Ershova IV (2016) Bankrotstvo hozyajstvuyushchih subektov: textbook for bachelors. In: Ershova IV, Enkova EE. M.: Prospect, p 280
5. Kirillovyh AA (2015) Bankrotstvo fizicheskih lic: novacii zakonodatelstva o nesostoyatel'nosti. Zakonodatelstvo i ekonomika 6:11
6. Pirogova ES, ZHukova YuD (2015) Osobennosti nesostoyatelnosti (bankrotstva) fizicheskih lic po zakonodatelstvu Respubliki Koreya. Predprinimatelskoe pravo. Appendix « Pravo i Biznes » 4:47
7. Tarakanov VV, Inshakova AO, Dolinskaya VV (2019) Information society, digital economy and law. In: Popkova EG (ed) Ubiquitous computing and the Internet of Things: Prerequisites for the development of ICT/studies in computational intelligence, vol 826. Springer Science+Business Media, Cham, pp 3–15
8. Frolov I.V. (2016) Bankrotstvo grazhdanina: problemy vvedeniya i modeli pravovogo regulirovaniya. Zakony Rossii: opyt, analiz, praktika 2:95–102

9. Chuprova AYU (2007) Otvetstvennost fizicheskih lic za narushenie anglijskogo zakonodatel'stva o bankrotstve. Mezhdunarodnoe publichnoe i chastnoe pravo 1:115
10. Shishmareva TP (2015) Institut nesostoyatelnosti v Rossii i Germanii. M., p. 392
11. Strilec LE (2016) Nekotorye voprosy pravovogo regulirovaniya bankrotstva fizicheskogo lica (razd. 1, gl. 11.). In: Mironova YuI, Egorova GG (eds) Aktualnye problemy yuridicheskoj teorii i praktiki (Actual problems of legal theory and practice: Collective monograph). Volgograd Scientific Publishing House, Volgograd, pp 123–128

Chapter 15
Protection from Digital Cartels at Electronic Auctions

Aleksandr Bortenev, Marina Kozlova and Olga Sergacheva

Abstract *Purpose*: The purpose of the article is to determine the main legal remedies from "digital cartels" at electronic auctions, therefore, the specific features of cartel collusions in electronic auctions for state and municipal procurement as well as a mechanism to counter them are examined. *Design/methodology/approach*: the authors use dialectical, as well as comparative-law research methods. *Findings*: The article highlights that "digital cartels" as part of a larger-scale phenomenon of harmonizing market strategies threaten the competition, detects common situations of cartel collusion occurring at electronic auctions, the main approaches of government authorities on their detection and legal remedies against them, examines international experience in this area. The approaches of the antimonopoly authorities of Great Britain, Italy, and Russia are studied. Anonymity, complicated detection and usage of a digital environment for collusion requires new cartel-combating legal means. Such means may include toughening administrative and criminal liability, harmonizing antimonopoly law and law on the protection of information confidentiality. Antimonopoly authorities are transforming their operation techniques by using the latest technologies, means, and standards to detect and combat digital cartels. It is common to use software aimed at detecting cartels by digital footprints and a combination of indirect evidence of collusion. *Uniqueness/value*: the authors have revealed specific features of digital cartels, including anonymity, complicated detection, and usage of the digital environment for collusion. Also, they have determined legal means to suppress cartel collusions, as well as new standards of proving, including both direct and indirect evidence. It has been found that digital cartels are just one of the elements of anticompetitive behavior since digital means allow business agents

A. Bortenev (✉) · O. Sergacheva
Volgograd Institute of Management, The Russian Presidential Academy of National Economy and State Administration, Volgograd, Russia
e-mail: bortenev-ai@vlrg.ranepa.ru

O. Sergacheva
e-mail: sergacheva-oa@vlrg.ranepa.ru

M. Kozlova
Volgograd State University, Volgograd, Russia
e-mail: kozlova@volsu.ru

T. Kolmykova and E. V. Kharchenko (eds.), *Digital Future Economic Growth, Social Adaptation, and Technological Perspectives*, Lecture Notes in Networks and Systems 111, https://doi.org/10.1007/978-3-030-39797-5_15

to align not only their behavior at the electronic auction but also coordinate a market strategy in general.

Keywords Competition · Electronic auction · Bid rigging · Digital cartels · Restricted competition · Protected competition

JEL Classification Code K 210

15.1 Introduction

Electronic auctions for state and municipal procurement have become the most widespread in the Russian Federation. Here the law provides for a whole range of electronic procedures. This is fully justified since ensuring state and municipal needs as an extremely important area for the state economy should be in close attention of the legislator. Only if the legal regulation of procurement procedures is sufficiently high-quality, it becomes possible to conclude an effective state or municipal contract [10].

The electronic auction can be defined as the procedure for entering into a contract effected through the software on the Internet when the bidders interact through the exchange of electronic documents. Auctions through the Internet in terms of publicity and accessibility are fully consistent with the basic idea of the auction as a way to select a contracting party that most meets the auctioneer's interests. However, new opportunities associated with electronic auctions are not always used by bidders on the terms of good faith. Thus, the number of purely electronic ways to restrict competition increases. First and foremost, we are talking about the so-called digital cartels, that is, bid-rigging at electronic auction to reduce or increase the price, grant the win of a predetermined bidder. This phenomenon tends to spread in the long-term and therefore requires effective legal remedies.

15.2 Materials and Methods

The issue of cartel collusions at electronic auctions is addressed in the works of Russian academicians, such as O. A. Sergacheva [10] as well as foreign researchers, such as A. Ezrachi, M. E. Stucke [5], C. Frediani [7], D. J. Lynch [9], V. Šmejkal [11]. At the same time, the law does not have effective means to combat digital cartels at disposal, therefore the research of this area is still relevant. Given the development of digital technology and a significantly increasing number of an electronic auction, this issue is discussed in the materials of judicial and antimonopoly authorities. For example, Judgment in Case No. A40-188927 / 17 as of March 14, 2018, Conclusion on the Facts of the Case No. 1-11-166 / 00-22-17 as of April 10, 2018, Clarification No. 3 of the Presidium of the Russian Federal Antimonopoly Service dated February 17, 2016, as well as statistics Global Cartel Enforcement [8], CMA Launches Digital

Tool to Combat Bid-Rigging [2], Statistics of Processing Complaints on Tendering by the Central Office of the Russian Federal Antimonopoly Service [12]. The article uses the comparative-law method, description, and interpretation; theoretical methods of formal and dialectical logic; statistical data analysis.

15.3 Findings

In international practice, cartel collusion in e-commerce is an issue that has been dealt with for many years. As noted in the international mass media, "nevertheless, the development of artificial intelligence and powerful algorithms help to set up stronger cartels that can support higher prices thanks to consumers and in spite of traditional security regimes" [9]. Obviously, in addition to advantages like simplification and expedition of procurement procedures, lower prices for goods due to increased competition caused by expansion of the market and growing number of potential participants, electronic auctions also lead to negative consequences such as abuse of new digital algorithms to monopolize the market and establish digital cartels that manipulate prices during electronic procurement. As experts emphasize, "cartel agreements are now concluded not in separate rooms filled with cigar smoke, but in fundamentally new anonymous forms using digital algorithms" [7].

Cartels in most legal systems present an illegal agreement per se, i.e. regardless of whether it was effective and whether it caused any damage to bidders [5]. The data provided by Allen and Overy in their annual report of 2017 on global cartel control is of interest. The report notes that "it is predicted that violations of competition law in the digital area will become more substantial. The growing number of digital cartels disclosed in the EU is also the outcome of focusing on electronic commerce and regulating the operation of digital platforms" [8].

Innovative methods to combat cartels (including when conducting electronic auction) have been introduced recently in the United Kingdom. In 2017, the Competition and Markets Authority of the United Kingdom launched a special social media campaign designed to increase legal literacy in competition law (one of the sides of antimonopoly compliance) and collect information on likely cartels. Informants are offered a reward of up to £ 100,000 for information on the cartel. The second direction is the test launch of a digital tool "Screening for Cartels", which collects and processes data on ongoing electronic auction and detects collision. This free software uses algorithms to detect unusual patterns of customer behavior and pricing, which may evidence that there has been bid-rigging [2].

Another key area that is intensively discussed in international academic literature is the potentially harmful impact of new digital technologies on competition and the market in general and on the effectiveness of electronic auction in particular. This is the so-called "algorithmic price parallelism" [11]. This technology appears like a market control strategy that is not prohibited by working competition law of the EU or the USA except for the rare cases of the so-called abuse of collective dominant position in EU law [11]. However, with the introduction of modern pricing algorithms, the same type of parallel or sequential behavior can be disastrous for price competition in online markets. PC algorithms make it possible to process such a volume of data that the underlying assumptions which have served a basis of protection of competition until now are no longer working. Thus, competition as a phenomenon will fall under threat, and, therefore, all mechanisms to maintain it, for example, electronic auction, will become meaningless. The Russian antimonopoly laws also contain requirements aimed at preventing competition to be restricted during auctions, including the electronic ones.

Initially, electronic procedures for mandatory auctions have been introduced to ensure the availability of information, transparency of all procedures, and provide sufficient competition at auctions. However, electronic auctions have created new opportunities for violation of competition rules.

The most common violations of arranging and tendering procedures under Art. 18.1 of Federal Law No. 135-ФЗ "On the Protection of Competition" dated July 26, 2006 (hereinafter also—Law on Protection of Competition) are the establishment of illegal, unadministered requirements for bidders by the auctioneers, bids; restriction of access to participation in auctions by electronic platforms, also by establishing excessive requirements for documents submitted in registration stage, and a complicated registration procedure; unreasonable non-admission of applicants to participation [12].

At the same time, proving unacceptable agreements, in particular, cartel agreements, and concerted actions at the auction is the most complex task for antimonopoly authorities. Bid rigging is an agreement between business agents on the terms of winning at auction. These agreements are prohibited in Clause 2, Part 1, Article 11 of the Law on Protection of Competition.

Proving cartel collusion is often quite complicated, however, the use of new technologies in tendering opens up new opportunities for proving a cartel's existence. So, for example, both closed and open data of the digital environment, such as e-mail correspondence, WhatsApp, Skype, Facebook data, Dropbox, etc. can be investigated, but delete correspondence with quickness and ease is additional trouble. Along with that, the necessary information is difficult to get under legal prohibitions and restrictions on communications, if bids are submitted from a mobile device.

Bid rigging can be proved by both direct and a combination of indirect evidence.

In particular, the following facts and circumstances may be considered as indirect evidence of a cartel agreement at electronic auction: using the same IP address (account) by bidders when submitting bids and participating in electronic auction; issuing certificates of electronic digital signatures for the same physical entity [1].

For example, two legal entities were adjudged offenders of the law on the protection of competition relating to the conclusion of an oral cartel agreement. The antimonopoly authority came to the conclusion that these organizations have made legally significant acts, such as submitting bids, offering and signing contracts from one IP address, jointly using the infrastructure and equipment belonging to one of them. Also, there was a matching of the accounts where bid files were made up and altered, which indicates the use and exchange of bid files and the coordination of bids' preparation for auctions. The commercial (arbitrazh) court disallowed the claim to deem illegal and denounce the resolution of the antimonopoly authority on bringing to administrative liability [4]. The judgment has entered into force.

Another specific circumstance serving as evidence of cartel collusion is the exploitation of auction robots. Auction robot is a tool available to bidders in their account on an electronic platform, which allows automating the participation in tendering. The user sets the robot's parameters in his account: the step of the auction robot's operation, as well as the minimum (best) bid on the contract price that can be made by the auction robot during the auction. After activation, the bidder agrees to use the robot during the auction and confirms that the bids on the contract price submitted by the auction robot and recorded in the system are the bids submitted on behalf of this bidder. Thus, the independent utilization of robots by bidders does not contradict the tendering rules and the current legislation. However, settings of auction robots agreed by several persons to determine the winner or fix prices at the auction will point at cartel collusion. Besides, this poses the question: who will be held liable if the collusion is reached without the participation of a person—program user, developer, or another person.

Russian Federal Antimonopoly Service had examined 14 electronic auctions with the participation of two legal entities engaged in medical supplies. Among other facts, it was found that the auction robots of both organizations were configured to submit minimum price offers; bids and price offers were filed, as well as electronic auction robots were designed by bidders using a single infrastructure (IP-addresses); properties of files with the first parts of bids matched; powers of attorney for certificates and digital signature keys of both legal entities were issued to one person. The Commission of Russian Federal Antimonopoly Service came to the conclusion on determining the acts of legal entities under Cl. 2, Part 1, Art. 11 of the Law on the Protection of Competition, which resulted in the conclusion and participation of business agents-competitors in the agreement (cartel), which in turn led (could lead) to fixing the prices at auctions under consideration [3].

Artificial intelligence capabilities lead to the emergence of stronger cartels in e-commerce, new threats to competition, and discreditation of the very tendering idea. Digital cartels using developing technologies present a new challenge to the antimonopoly regulation system. Cartel collusions are a priori prohibited, but, nevertheless, it is predicted that their number will rapidly grow, especially in the digital area. Antimonopoly authorities have to apply adequate measures to detect and restrict the cartel's activities, in particular, the research of digital media data.

The digital cartels pose several new challenges to the law. Thus, cartel agreements at electronic auctions can be reached without the direct participation of a person, and several legal issues, for example, related to the party liability, arise. Legislation on the protection of communication secrets in some cases prevents the antimonopoly authorities from getting the necessary information, for example, in the event of submitting a bid from a mobile device but not from a desktop computer. Also, criminal liability in the event of collusion of business agents and customers at the auction is not provided. Therefore, it is required to harmonize various directions of the current legislation and toughen liability to prevent cartel collusion in any area, especially the digital one.

15.4 Conclusion

The anonymity, complicated detection and use of a digital environment for collusion should be called the specific features of digital cartels. Legal remedies are applied to suppress cartel collusions, but the greatest troubles arise when detecting digital cartels. A variety of cartel detection methods can be applied, such as encouraging the provision of information on the digital cartels; the use of software aimed at detecting cartels by digital footprints. Evidence relating to digital cartels can be both direct and indirect, namely utilization of the same IP address to perform legally significant acts at the auction by different parties, agreed settings of auction robots, etc. The antimonopoly authority can control any electronic auction held for state and municipal needs using specialized software designed to prevent cartels.

At the same time, it should be taken into account that digital cartels can be just one of the elements of anticompetitive behavior because digital tools allow business agents to align not only their behavior at the electronic auction but also coordinate a market strategy in general. It seems that such a strategy is the main challenge of the digital age to competition law.

Thus, the procedure for electronic auction can be infringed, competition can be restricted to the same extent (if not to a greater one) like at traditionally held auction with help of various means available exclusively in the digital environment. Besides, the antimonopoly authorities seek and find tools to overcome and prevent such offenses.

References

1. Clarification No. 3 of the Presidium of the Federal Antimonopoly Service of Russia “Proving Unacceptable Agreements (including Cartels) and Concerted Actions in Commodity Markets, Including Tendering” as of February 17, 2016. https://fas.gov.ru/documents/562597. Accessed 20 June 2019
2. CMA launches digital tool to fight bid-rigging (2017). https://www.gov.uk/government/news/cma-launches-digital-tool-to-fight-bid-rigging. Accessed 20 June 2019
3. Conclusion on the Facts in the Case No. 1-11-166 / 00-22-17 by Commission of the Federal Antimonopoly Service of the Russian Federation as of April 10, 2018. https://solutions.fas.gov.ru/ca/upravlenie-po-borbe-s-kartelyami/ats-25131-18. Accessed 10 May 2019
4. Judgment in Case No. A40-188927/17 as of March 14, 2018. https://kad.arbitr.ru/Card/65769a33-0502-41c7-9ed6-9ad71a5f6438. Accessed 25 June 2019
5. Ezrachi A, Stucke ME (2017) Artificial intelligence and collusion: when computers inhibit competition. https://illinoislawreview.org/wp-content/uploads/2017/10/Ezrachi-Stucke.pdf. Accessed 23 June 2019
6. Federal Law No. 135-ФЗ dated July 26, 2006 “On Protection of Competition” Collection of the RF legislation, 2006, No. 31 (Part 1), Cl. 3434
7. Frediani C (2016) Cartelli Digitali: Dimmi Chi Sei e Ti Faro ‘Il Prezzo Piu’ Alto. La Stampa Technologia, December 30. https://www.lastampa.it/2016/12/30/tecnologia/cartelli-digitali-dimmi-chi-sei-e-ti-far-il-prezzo-pi-alto-jfIrwELIVEbwCEf6fXn24I/pagina.html. Accessed 20 June 2019

8. Global Cartel Enforcement (2017) Full-year cartel report of Allen and Overy. http://www.allenovery.com/SiteCollectionDocuments/2017%20Full%20Year%20Cartel%20Report.pdf. Accessed 23 June 2019
9. Lynch DJ (2017) Policing the digital cartels. Financial Times, January 8. https://www.ft.com/content/9de9fb80-cd23-11e6-864f-20dcb35cede2. Accessed 14 May 2019
10. Sergacheva OA (2017) Revising the legal regulation of essential conditions of the state (municipal) contract in the legislation of the Russian Federation. Pravovaya Paradigma 16(1):132–138
11. Šmejkal V (2017) Cartels by robots: current antimonopoly law in search of an answer. Intereulaweast 4(2):1–18
12. Statistics of Processing Complaints on Auctions by the Central Office of the Federal Antimonopoly Service of Russia (2017) https://fas.gov.ru/documents/602704. Accessed 15 July 2019

Chapter 16
Integration of the Evolutionary and Genetic Approach and the Spatial Approach to the Region's Sustainable Development Assessment

Alexander V. Plyakin, Elena A. Orekhova, Alina V. Konyashova and Sergey M. Aparin

Abstract *Purpose*: Urgency of solving the scientific problem of regional economy connected with the assessment of the productive capacity and economic activity defining the region's sustainable development. *Design/methodology/approach*: It is extremely important to evaluate the state of productive capacity and economic activity of the regional social and economic system basing on synthesis of the evolutionary and genetic approach, as well as the spatial approach providing objective identity, classification, visualization and assessment of factor components for the region's sustainable development. The article gives the assessment results for factor components of the municipal districts' productive capacity and economic activity; the districts' spatial variability was also studied. The research result is represented by the methodology for assessing the sustainable development of municipal districts. *Findings*: The article demonstrates the effectiveness of using themed digital maps providing the visual side of spatial predicted patterns in distributing indexes of productive capacity and economic activity in municipal districts. The key characteristic of the region's sustainable development assessment is its interdisciplinary nature provided by integration of the evolutionary and genetic, as well as spatial approaches based on contemporary GIS technologies. *Originality/value*: The results' scientific value includes the theoretical and methodological explanation for method tools used for reasoning the decisions made in the area of managing the region's sustainable social and economic development. According to the authors, the spatial approach based on the spatial analysis methods will enable to bring up serious thought on the meaning of the region's sustainable social and economic development opening up certain perspectives to improve methods and means for regional and municipal management, include spatial analysis methods into the process of making management decisions.

A. V. Plyakin (✉) · E. A. Orekhova · A. V. Konyashova · S. M. Aparin
Volzhsky Institute of Economics, Pedagogy and Law, Volzhsky, Russia

A. V. Konyashova
e-mail: konyashova@list.ru

T. Kolmykova and E. V. Kharchenko (eds.), *Digital Future Economic Growth, Social Adaptation, and Technological Perspectives*, Lecture Notes in Networks and Systems 111, https://doi.org/10.1007/978-3-030-39797-5_16

Keywords Sustainable development · Region · Productive capacity · Economic activity · Evolutionary and genetic theory of production factors · Spatial analysis · GIS

JEL Classification C31 · Q01 · R12

16.1 Methods

The region's sustainable development (STB) is characterized by the state of productive capacity (PED) and economic activity (EA) of municipal districts on its territory. For assessing sustainable development we need indicators that can be formed on basis of the evolutionary and genetic theory of production factors [1–3]. According to the theory, the six basic production factors—labor (A), engineering and technology (T), natural resources (M), institutional (Ins), organizational (O) and informational (Inf)—determine the endogenous "core" of the social and economic system development. Based on this knowledge, the productive capacity potential and social and economic system activity may be represented in six-dimensional form characterized by the set of complex sustainable development indicators. This suggested approach allows to take into account the six types of the region's sustainability: sustainability of labor force (STB_A), sustainability of production (STB_T), sustainability of natural environment (STB_M), sustainability of institutions (STB_{INS}), sustainability of organizations (STB_O); sustainability of informational environment (STB_{INF}). Each type of the region's sustainable development (STB_i) can be described with the function of productive capacity potential (PED_i) and economic activity (EA_i) development, i.e. $STB_i = F(PED_i; EA_i)$.

The methodological approach to creating the model of the region's sustainable development suggested by the authors enabled to form a new system of aggregated indicators characterizing conditions, resources and factors of sustainable development for municipal districts. The value of conditional municipal gross product (Q) was taken as basic indicator to assess the region's productive capacity potential, and the number of enterprises and organizations showing concentration of economic activity (O) in the region was taken for assessing the region's economic activity.

Basing on estimate statistical indicator—the value of conditional municipal gross product (Q)—we can calculate and assess the dynamics of relative indicators characterizing labor potential $PED_A = F(Q/A; A/Q)$; engineering and technology potential $PED_T = F(Q/T; T/Q)$; natural resources potential $PED_T = F(Q/M; M/Q)$; institutional potential $PED_{INS} = F(Q/Ins; Ins/Q)$; organizational potential $PED_O = F(Q/O; O/Q)$ informational potential $PED_{INF} = F(Q/Inf)$. Further, taking into account the basic estimate statistical indicator—the number of organizations and enterprises (O)—we can calculate the relative indicators of economic activity: labor activity $EA_A = F(O/A; A/O)$; production activity $EA_T = F(O/T; T/O)$; activity of using natural resources $EA_M = F(O/M; M/O)$; institutional activity (lawmaking, law violation, etc.) $EA_{INS} = F(O/Ins; Ins/O)$; organizational activity (i.e. activity of existing organizations) $EA_O = F(O/O_u; O_u/O)$; informational activity (i.e. need of organizations in informational and communicational technologies and services) $EA_{INF} = F(O/Inf; Inf/O)$.

To compare the results of the calculated six factor indicators of sustainable development taken according to different scales, you need to find the relative non-dimensional index M_i for each municipal district, which shows the degree of approximation of the absolute sustainability indicator Q_i to the maximum Q^{max} and minimum Q^{min} indicators [4]. The received normalized index values for productive capacity, economic activity and sustainable development of municipal districts are classified within the range from 0 to 1. Thus, the sustainable development of municipal districts may be characterized by the value of these six aggregated indexes: labor sustainability index (M_1); production sustainability index (M_2); natural environment sustainability index (M_3); institutional environment sustainability index (M_4); organizational environment sustainability index (M_5); informational environment sustainability index (M_6). Integral indicator of municipal districts' sustainable development can be calculated as a simple average of the values of these six aggregated indexes $M_1, \ldots, M_6$.

The indicators of productive capacity and economic activity in the region are the initial data for assessing sustainable development in municipal districts on the territory of Volgograd region [5–8] (Table 16.1).

Twelve aggregated indexes K_i of productive capacity and economic activity factor components were calculated, together with their integral indexes as a simple average of aggregated indexes [9, 10].

To assess the spatial variability and asymmetry of factor components of productive capacity and economic activity in municipal districts we used the following statistical characteristics: average value, middle line, mean square deviation, asymmetry ratio, degree of excess, variation coefficient.

To determine the relative uniformity zones (clusters) of productive capacity and economic activity in nearby municipal districts, the verification of local variations of productive capacity and economic potential aggregated indexes was performed using Voronoi clustering method. The technological platform for the study was the ARCGIS system with its spatial analysis tools [11, 12].

16.2 Introduction

Methodological principles for assessing sustainable development of a region and municipal districts based on various index systems can be found in a number of published works [13–17]. However, the authors of this study suggest their own conception, according to which sustainable development of the regional social and economic system is a function of at least two of its system characteristics, one of which is the region's productive capacity potential, and the other is its economic activity, i.e. STB = F (PED; EA).

According to the authors, the productive capacity and economic activity are characterized by their spatial variability, assessment of which enables to conclude existence of sustainable development of municipal social and economic systems on the

Table 16.1 Initial indicators for assessing the sustainable development of municipal districts

Factor component	Productive capacity indicators	Economic activity indicators
Labor	Labor capacity per employee, thous. RUR	Labor capacity of organizations, number of workers per organization
Production	Effectiveness of using basic funds, thous. RUR	Value of relative municipal product, thous. RUR; Fixed investment amount, thous. RUR
Natural resources	Crops productivity, thous. RUR per 1 ha of agricultural land	Amount of polluting substances emitted into atmosphere, tons; Amount of water used from natural sources, thous. of cubic m; Areas of basic crops, thous. ha
Institutional	Number of registered crimes per 100,000 people in municipal districts; loss of working hours caused by production accidents, days	Amount of organizational loans, thous. RUR; Debit debts of organizations, thous. RUR
Organizational	Effectiveness of organizations, thous. RUR per organization	Relative share of profit-making organizations, per cent; Organizational budget income of municipal entities, mln. RUR per management organization
Informational	Effectiveness of using informational resources, thous. RUR per each ruble spent on information and communication technologies (ICT)	Expenditures on ICT, thous. RUR per organization; Number of workers using ICT per organization, persons

regional territory. The region's sustainable social and economic development is the result of balanced inter-municipal relations, on the one hand, and external unbalanced interaction of the region with its macro community, on the other hand. Imbalance of inter-municipal relations destroys unity of the regional social and economic system, causes leveling of opportunities for regional development in external environment, and becomes a cause of the region's social and economic development decrease. An obvious condition for the region's sustainable development is its compliance with the general law of distribution of the municipal productive capacity and economic activity spatial variability that displays dynamic balance of inter-municipal relations. On the contrary, spatial variability of productive capacity and economic activity different from the general law of distribution is the result of presence of one or several external dominating factors breaking balance within regional channels of material, labor, informational and other resources in the region. The latter can be characterized by existence of serious spatial asymmetry of the municipal productive capacity and economic activity. However, you may not see any spatial interference of productive capacity and economic activity areas in municipal districts.

Use of spatial approach together with GIS involvement additionally enables to visualize the ranging results of municipal districts according to the level of their sustainable development and introduce the revealed spatial regularities on the regional territory in a form of a themed index map [9].

Considering all data mentioned above, the purpose of the research is to implement the region's productive capacity and economic activity index system into ARCGIS, which will enable to perform ranging of municipal districts according to their level of sustainable development, find out spatial trends.

16.3 Results

GIS-created index maps visualize the spatial distribution of productive capacity and economic activity of municipal districts on the territory of Volgograd region in 2016 (Fig. 16.1).

Serious spatial variability of productive capacity in municipal districts was detected, which explains existence of processes cardinally differentiating it on the regional territory (Table 16.2).

Variability of productive capacity in municipal districts of Volgograd region is changeable enough, having three main axes of rather high level of productive capacity in the region's north, south and center. The greatest spatial variability and asymmetry of productive capacity in municipal districts exist in its labor, production, organizational and informational components (variability ratio over 1.0).

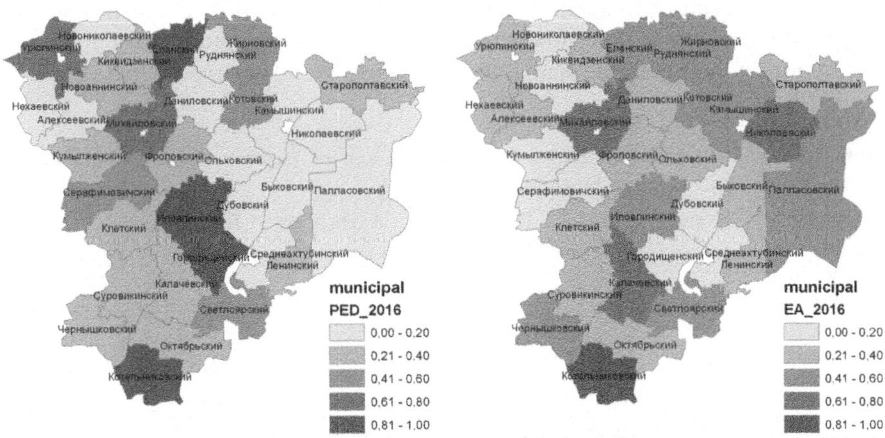

Integral index of productive capacity (K_{PED}) Integral index of economic activity (K_{EA})

Fig. 16.1 Integral indexes of productive capacity and economic activity in municipal districts of Volgograd region

Table 16.2 Variability of productive capacity factor components in municipal districts of Volgograd region

Productive capacity factor components PED	Average	Middle	RMS Deviation	Asym. Coef.	Excess Coef.	Var. Coef.
PED_A	0.19	0.01	0.28	1.51	4.54	1.53
PED_T	0.23	0.11	0.30	1.40	3.86	1.30
PED_M	0.45	0.45	0.25	0.56	2.87	0.55
PED_{INS}	0.61	0.65	0.25	−0.71	2.92	0.41
PED_O	0.20	0.02	0.30	1.40	3.73	1.50
PED_{INF}	0.16	0.02	0.253	1.87	6.07	1.56
PED	**0.35**	**0.24**	**0.28**	**1.03**	**2.92**	**0.80**

Variability of economic activity in municipal districts is also a serious variable with significant spatial asymmetry represented by one axis of economic activity in the region's south (Table 16.3).

The greatest spatial variability and asymmetry of economic activity in municipal districts exists for the production and informational components (variability ratio 1.0 and over). The state of economic activity in municipal districts should be described as more balanced within the region, compared to their productive capacity conditions.

Spatial analysis of local variability using Voronoi maps enabled to find out clustering characteristics of municipal districts on the region's territory (Fig. 16.2).

The found contrast in spatial distribution of municipal district clusters according to the indicators of their productive capacity and economic activity proves the assumption of their possible spatial incompliance. We may suppose that economic

Table 16.3 Statistical characteristics of economic activity factor component variability in municipal districts of Volgograd region

Economic activity factor components (EA)	Average	Middle	RMS Deviation	Asym. Coef.	Excess Coef.	Var. Coef.
EA_A	0.38	0.36	0.25	0.54	2.48	0.66
EA_T	0.11	0.02	0.2	3.02	13.43	1.82
EA_M	0.38	0.33	0.3	0.74	2.37	0.79
EA_{INS}	0.79	0.94	0.29	−1.58	4.18	0.37
EA_O	0.39	0.38	0.21	0.34	3.44	0.54
EA_{INF}	0.24	0.16	0.23	1.81	5.95	0.96
EA	**0.35**	**0.37**	**0.22**	**0.67**	**3.59**	**0.63**

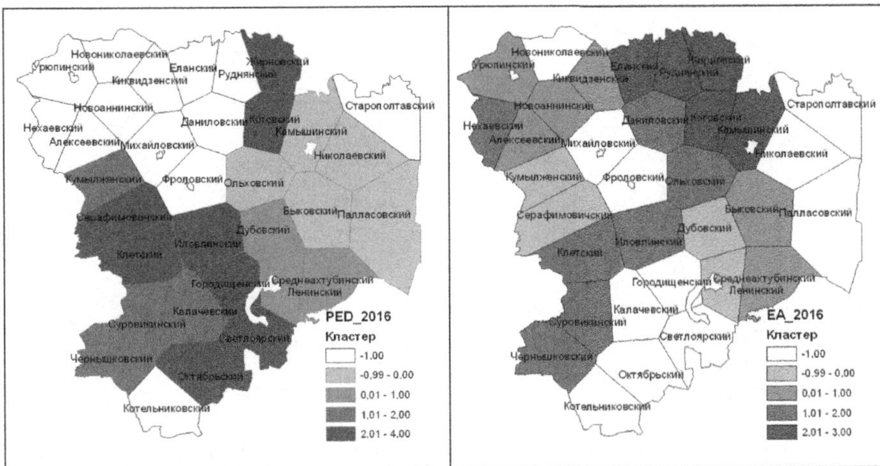

Clusters of municipal districts according to their indicators of productive capacity	Clusters of municipal districts according to their indicators of economic activity

Fig. 16.2 Local spikes of productive capacity and economic activity values in municipal districts of Volgograd region (received using Voronoi clustering statistical method)

activity in municipal entities compensates lack of their productive capacity providing the region's sustainable development increase.

The index map created in GIS characterizes the specifics of spatial distribution of sustainable development indicator values in municipal districts of Volgograd region in 2016 (Fig. 16.3).

As the result of analysis performed in the study, we could find out that maximum spatial variability and asymmetry of sustainable development indicators in municipal districts are typical for the labor, production, organizational and informational components of sustainable development (Table 16.4).

Performed integral assessment of municipal districts' sustainable development allowed to find those characterized by the highest values of their sustainability factor indicators (Table 16.5).

The state of sustainable development in municipal districts can be evaluated according to the 5-grade scale using the following value ranges of integral grade (I):

(1) very low level of sustainable development ($0 <= I < 0.21$);
(2) sustainable development below average ($0.21 <= I < 0.4$);
(3) middle level of sustainable development ($0.41 <= I < 0.6$);
(4) sustainable development above average ($0.61 <= I < 0.8$);
(5) high level of sustainable development ($0.81 <= I < 1$).

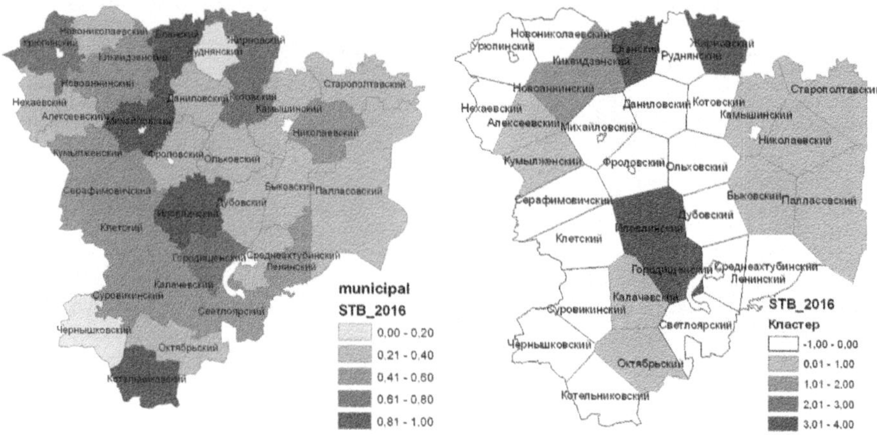

Integral index of sustainable development (K_{STB})

Clusters of sustainable development in municipal districts

Fig. 16.3 Sustainable development assessment for municipal districts of Volgograd region

Table 16.4 Statistical characteristics of variability of sustainable development factor components (STB) in municipal districts of Volgograd region

Sustainable development factor components (STB)	Average	Middle	RMS Deviation	Asym. Coef.	Excess Coef.	Var. Coef.
STB_A	0.33	0.28	0.23	0.79	2.49	0.84
STB_T	0.26	0.31	0.12	0.96	2.53	1.19
STB_M	0.6	0.23	0.65	−0.52	2.65	0.38
STB_{INS}	0.73	0.20	0.76	−1.67	6.43	0.27
STB_O	0.44	0.25	0.42	0.25	2.55	0.56
STB_{INF}	0.25	0.27	0.10	1.28	3.43	1.08
STB	**0.48**	**0.24**	**0.42**	**0.65**	**2.89**	**0.5**

The level of sustainable development in municipal districts of Volgograd region was evaluated as below average (I = 0.33). However, municipal districts of rather high level of sustainable development were detected in the region's southern, northern and central parts. The spatial analysis of sustainable development local variability using Voronoi's statistical methods partly supports the result. We can assume that the state of sustainable development in municipal districts of Volgograd region is provided by the labor, natural resources, institutional and organizational factors determining their competitive advantage in the region.

Table 16.5 Integral assessment of sustainable development in municipal districts of Volgograd region

Name of sustainable development factor		Compliance with general distribution of productive capacity indicators	Compliance with general distribution of economic activity indicators	Total grade
		Yes = 1, No = 0	Yes = 1, No = 0	
1	Labor	0	1	0.5
2	Production	0	0	0
3	Natural resources	1	0	0.5
4	Institutional	1	0	0.5
5	Organizational	0	1	0.5
6	Informational	0	0	0
	Total grade	0.33	0.33	
	Integral grade			0.33

16.4 Conclusion

The performed study brought up the following results:

- methodology of assessing sustainable development of municipal districts was formulated enabling to differentiate municipalities in GIS according to the value of selected sustainable development indicators for its six components: labor, production, natural resources, institutional, organizational and informational;
- it was discovered that compliance with the general law of spatial variability distribution of productive capacity and economic activity factor components in municipal districts represents the state of sustainable development in the region due to the dynamic balance of intermunicipal relations;
- detecting clusters of municipal districts similar to each other in their sustainable development indicator values enables to determine potential readiness of municipal districts to the active intermunicipal cooperation and sustainable economic development;
- the methodology of sustainable development of the region was suggested based on measuring the variability of the region's productive capacity and economic activity factor components using bar chart analysis. The level of sustainable development can be assessed according to the 5-grade interval sustainability scale;
- the effectiveness of creating themed GIS maps providing high-quality imaging of spatial regularities in distribution of productive capacity, economic activity and sustainable development of municipal districts based on example of Volgograd region.

In conclusion we would like to emphasize that integration of evolutionary and genetic approach together with spatial approach to assessing the sustainable development of municipal districts discovers additional opportunities in solving the tasks of managing sustainable social and economic development of the region.

References

1. Inshakov OV (2003) "Development nucleus" in the light of the new factors of production theory, Economics of Contemporary Russia, no 1, pp 11–26
2. Inshakov OV (2004) Potential of the evolutionary approach in an economics of contemporary Russia, Economics of Contemporary Russia, no 4, pp 42–52
3. Inshakov OV (2001) Factors and functions of human being: achieving of a new measure, VolSU Publ., Volgograd
4. Orekhova EA, Plyakin AV (2017) Safety structure of social and economic development of municipal entities: spatial analysis practice, Regional Economy: South of Russia, 2(16):112–124
5. Statistical Review (2017) City and municipal districts of Volgograd region, 2016, Territorial body of Federal statistical service in Volgograd region, Volgogradstat, Volgograd
6. Statistical Review (2017) Indicators of informational society of Volgograd region in 2016, Territorial body of Federal statistical service in Volgograd region, Volgogradstat, Volgograd
7. Statistical Review (2017) Basic organizational funds of Volgograd region, Territorial body of Federal statistical service in Volgograd region, Volgogradstat, Volgograd
8. Statistical Review (2017) Labor and employment in Volgograd region, Territorial body of Federal statistical service in Volgograd region, Volgogradstat, Volgograd
9. Plyakin AV, Orekhova EA (2018) Spatial display of economic activity in municipal districts. In: Proceedings of the international scientific conference "Competitive, sustainable and secure development of the regional economy: response to global challenges", Advances in Economics, Business and Management Research, vol 39, pp 578–582
10. Plyakin AV, Orekhova EA (2017) Indicators of sustainable development of municipal entities: approaches to construction and practice, Regional Economy: South of Russia, 4(18):106–113
11. ArcGIS Desktop. Creation and publication of maps, data and analytics (2018). Available at: http://desktop.arcgis.com. Accessed 5 Aug 2019
12. ArcGIS for Desktop: examing local variations (2018). Available at: http://desktop.arcgis.com/en/arcmap/latest/extensions/geostatistical-analyst/examining-local-variation.htm. Accessed 5 Aug 2019
13. Vazhenina TM (2014) Methods of economic assessment of sustainable development in provincial municipal entity, Corporate management and innovative economic development of North. Journal of Science and Research Center for Corporate Law, Management and Venture Investment and Syktyvkar State University, no 3, pp 11–19
14. Kazanskaya AN et al (2015) Assessment of social, ecological and economic state of territories as a tool of municipalities' sustainable development, Online Journal Technologies of Technogenic Safety, 1(59):216–224
15. Korabeynikov IN, Yegorychev SA (2014) Methodology of assessing regional trends of sustainable development in municipal entities, News of Orenburg State Agricultural University, 4(48):210–214

16. Ogorodnikov SS (2018) Use of cluster analysis for assessing social and economic state of municipal entities, Sustainable Development of Science and Education, 2:51–55
17. Khisamutdinov IA (2013) Social, ecological and economic assessment of the region in frames of transfer to sustainable development, Regional Economy: Theory and Practice, 4:9–15

Chapter 17
Strategic Audit as an Integral Part of External State (Municipal) Financial Monitoring in the Modern Context

Aleksandr Mordvintsev, Dmitrii Polyakov, Sergei Sazonov and Aleksei Dmitriev

Abstract *Purpose*: the article justifies the necessity and expediency of introducing a strategic audit into the activities of the control and accounting bodies as one of the directions to improve the system of external state (municipal) financial monitoring. *Methodology*: research methods include structural and statistical analysis, synthesis, and economic forecasting. *Findings*: The article explores the changes that have occurred in recent years in approaches to the management of economic relations at government and municipal level in Russia. The Development Strategy for the Accounts Chamber of the Russian Federation until 2024 has been discussed and priority areas of development have been revealed. One of them is the development of a methodology and the further implementation of strategic audit tools. The article also deals with theoretical approaches to defining "strategy", "strategic audit" and presents the practical experience using strategic audit elements. It has been found out that strategic audit is a new area of financial monitoring with its distinctive features. The history of a strategic audit in public administration is closely connected to strategic planning. The strategic audit is aimed at control (monitoring) of activities related to the accomplishment of national and industry-specific goals of the socio-economic development of territories specified in strategic planning documents. *Rationale*: The content and methodology of strategic audit both at the federal, regional and municipal level are emerging now. In the framework of this article, the authors investigate some areas that haven't been studied before and predict further development of strategic audit tools.

A. Mordvintsev (✉) · D. Polyakov
Volgograd Control and Accounts Chamber, Volgograd, Russia
e-mail: mai121263@gmail.com

D. Polyakov
e-mail: akos-volg@yandex.ru

S. Sazonov
Volgograd State Technical University, Volgograd, Russia
e-mail: sazonovsp@mail.ru

A. Dmitriev
Local Agency of the Federal Service of State Statistics for Volgograd Region, Volgograd, Russia
e-mail: lesha_dmitriev@outlook.com

T. Kolmykova and E. V. Kharchenko (eds.), *Digital Future Economic Growth, Social Adaptation, and Technological Perspectives*, Lecture Notes in Networks and Systems 111, https://doi.org/10.1007/978-3-030-39797-5_17

Keywords Strategic planning · Strategic audit · Integrated monitoring · Development strategy · Control and accounting bodies · External financial monitoring

JEL Classification Codes G28 · H50 · H72

17.1 Introduction

As the economic situation in Russia gets complicated, state authorities and local self-government bodies (within the scope of their powers) faced with the need to take effective measures to change approaches to the management of economic relations in the public and municipal sectors.

The State Duma adopted on June 20, 2014 Federal Law No. 172-ФЗ "On Strategic Planning in the Russian Federation" (hereinafter referred to as "Law No. 172-ФЗ"), which establishes the legal framework for strategic planning in Russia, coordination of state and municipal strategic management and budgetary policies, powers of state authorities (federal and regional), local self-government bodies and the procedure for their cooperation with public, academic and other institutions in strategic planning [1].

It follows from the definition provided for by law that strategic planning is the activity of its participants aimed at the goal-setting, planning, forecasting and programming socio-economic development of the Russian Federation both at the federal, regional, and local levels.

State authorities of all levels and local self-government bodies have started an introduction of strategic planning and further coordination of activities under strategic planning documents.

Today, almost every constituent entity of the Russian Federation as well as many municipalities have developed their strategies for social and economic development. And every year the number of development strategies and plans is increasing.

17.2 Methodology

Research methods include structural and statistical analysis of the working legislation, regulatory documents, strategic development documents of Volgograd, a report on the activities of the Volgograd Control and Accounts Chamber for 2018, as well as synthesis of information accessed from open sources to build a full picture of particular applications of strategic audit and its impact on more effective accomplishment of goals and measures of strategic development.

17.3 Findings

Within the implementation of this federal law, Volgograd developed a Strategy for the socio-economic development of the city until 2030 in 2017 (approved on January 25, 2017, under No. 53/1539, hereinafter referred to as Strategy 2030). This is a basic strategic planning document that determines priorities, goals, and objectives socio-economic development of Volgograd and municipal administration until 2030.

The mission of Strategy 2030 is to provide every assistance in the social and economic development of the city by consolidating the local community interests and uniting territorial resources to create comfortable living conditions for citizens. The mission outlined goals for execution of projects on the development of socially-oriented non-profit organizations (that carry out activities in the sports education, art, outreach activities, charity, etc.), establishment of favorable conditions for the performance of business projects, urban planning, tourism, improvement of transport-logistic system, etc. [2].

Strategy 2030 sets out 4 main areas of Volgograd development, highlighting priority goals in each of them… (Table 17.1).

Besides, Strategy 2030 provides for target indicators and actual figures for each area.

The President of the Russian Federation issued Decree No. 204 "On National Goals and Strategic Objectives of the Development of the Russian Federation until 2024" on May 7, 2018 (hereinafter referred to as Decree No. 204).

Decree No. 204 aimed at achieving breakthrough science-technology and socio-economic development of the Russian Federation, increasing country's population, raising the living standards, providing comfortable living conditions, as well as opportunities for self-fulfillment and showing the talents of each person entrusted Government of the Russian Federation with ensuring the national goals to be achieved [4].

Under Decree No. 204, the development of the Russian economy since 2018 has been formed by national goals and national projects to achieve them until 2024.

We should note that any amendments due to issuance of the Decree No. 204 haven't been introduced into Strategy 2030. A similar situation can take place in other municipalities, as well as in the constituent entities.

Along with that, the authors of the article think that the adoption of Decree No. 204 surely requires an adjustment of regional and municipal development strategies that have been approved before its entry into force, since they cannot be implemented in a departure from national goals and projects stipulated by the Decree.

The target parameters of regional and municipal development strategic areas should be aligned with the principal state development strategy and national projects.

The described changes in approaches to the management of economic relations (the widespread use of strategic planning and the implementation of national projects) have aroused a need to improve the system of financial monitoring of the activities of economic entities in the public sector.

Table 17.1 Volgograd strategic development areas

Area	Goals
Human capital development	Improving the demographic situation in Volgograd The development of a single educational space Cultural development The development of physical education and mass sports Encouraging active youth
Development of an innovative economy	Development of the manufacturing sector Development of the transport-logistic complex Development of domestic and inbound tourism Consumer market and service sector development Creating a favorable investment climate Establishing a favorable business environment
Improving the quality of the urban environment	Urban landscaping of Volgograd territory Ensuring environmental security Ensuring transport mobility of the population Providing affordable and comfort housing Rehabilitation of unused territories
Development of local self-government	Development of civil society Developing a system for the provision of services Development of a single information space Ensuring the safety of city residents Making and promoting the image of Volgograd

Source Developed by authors using the materials [3]

The quality of life depends on the strategy of various bodies of state and municipal level, therefore, a well-established monitoring (audit) mechanism and timely adjustment of strategic goals (if necessary) is the key to the successful development of municipalities, regions, and the country in general.

Law No. 172-ФЗ attribute the control and accounting authorities, that are vested with the powers to monitor and supervise the execution of strategic planning documents affirmed (approved) by state authorities and local self-government bodies, to participants in strategic planning.

Since it's necessary to assess the prospects for achievement of the target development indicators, their feasibility, and justification, as well as to form feedback in the course of strategic documents' execution, the strategic audit has become especially important. The strategic audit tools are being actively implemented in the activity of the control and accounting bodies.

Pursuant to the federal law No. 41-ФЗ as of April 05, 2013 "On the Accounts Chamber of the Russian Federation", a strategic audit is used to assess feasibility, risks, and progress when achieving the goals outlined in strategic documents.

This fact is distinguishing strategic audits from other types intended primarily to assess the current situation.

According to the projections of the Chairman of the Accounts Chamber of the Russian Federation Aleksei Kudrin, the share of strategic audit in the activities of control and accounting bodies over the next 6 years will exceed 50% of the total number of measures kept. In his words, the goal of control and accounting bodies should be an alternative assessment of strategic projects' implementation [5].

Thus, the descriptive and verifying activity of the control and accounting bodies should shift towards predicting the possible progress in the medium and long term.

Due to specific nature and preventive orientation, the strategic audit should become one of the tools to enhance the management of socio-economic processes and the responsibility of key participants through expanded monitoring of the progress of strategic goals and state program goals and expanded assessment of the reasons of achievement or failure to achieve some indicators.

Development Strategy of the Accounts Chamber of the Russian Federation for 2018–2024 provides for the intensive introduction of strategic audit tools to assess programs and projects being accomplished by state authorities and local self-government bodies. Within this strategy, the strategic audit has deemed a part of a combined audit consisting of financial, performance, and strategic audits… (Table 17.2).

The main purpose of the activities of the control and accounting bodies of Russia within the new development strategy is to broaden the opportunities to fulfill national goals by upgrading the system of state and municipal supervision. One of the indicators of the goal achieved is the share of the strategic audit in the total number of monitoring activities carried out by the control and accounting bodies.

The high-priority objective following the goal set is shaping a strategic vision among the heads of state bodies, also using the methodological development of a strategic audit.

Nowadays, the strategic audit is not widespread in the practice of control and accounting bodies due to the unavailable methodological base. Some regional and municipal control and accounting bodies start introducing elements of a strategic audit into their activities but relying on the already developed techniques of supervision over the implementation of programs and budget execution.

Table 17.2 Description of audit types

Type	Goal
Financial audit	Monitoring legality and efficiency of resource spending, including auditing accounting statements, as well as the legality of municipal funds and property utilization
Performance audit	Auditing the efficiency and effectiveness of municipal resources utilization, which allows identifying and fixing systemic problems in the activities of the audited bodies
Strategic audit	Auditing the match between funds spent and strategic goals achieved, supervision and adjustment of development programs to raise their efficiency for the solution of current problems

Source Developed by authors using the materials [6]

As an example, let's address the Accounts Chamber of the city of Tomsk, which partially uses the strategic audit tools in monitoring implementation of the Development Strategy of Tomsk until 2020. The main goals of the strategic audit are to assess:

- the adequacy of the goals and objectives of municipal programs;
- consistency of project goals and objectives with regulatory documents, including strategies and programs for socio-economic development.

The practice of holding strategic audit measures has shown that a substantial part of municipal programs is low-quality. The project objectives are mostly formal. Not being supported by need-evidencing facts, municipal programs looked like expenditure items without the justification of goals and objectives. Audit opinions indicate that financing of about 50% of ongoing projects is half of the actual need for capital, which significantly increases the risks of their non-fulfillment. Based on these opinions, regional and municipal control and accounting bodies concluded that there is a need to upgrade the financing system of regional and municipal programs, as well as to refine the monitoring mechanism of their implementation [7].

The work plan of the Volgograd Control and Accounts Chamber for 2019 also provides for measures to assess the consistency of projects accomplished by the municipal bodies with their socio-economic development strategy of Volgograd. According to the financial monitoring standard for the expert assessment of municipal programs, projects are assessed by the following criteria:

- consistency of the measures provided for in the program with municipally-important issues;
- compliance of the project materials with legislative and other regulatory documents;
- the accordance of the project with a municipal strategy for socio-economic development;
- the integrity of project goals and objectives, their concordance with the activities held within the project;
- justification of state/state/another financing amount;

- sufficiency of activities planned to achieve the ultimate goal of the program and resources for implementation.

We should note that the Volgograd Control and Accounts Chamber had previously applied the strategic audit tools for external municipal financial monitoring. As a result of the activities held in 2018, Volgograd Control and Accounts Chamber submitted 173 proposals on improvement and adjustment of draft regulatory documents. 144 proposals were taken into consideration, 26 of them concerned adjustments to departmental target and municipal programs focused on the fulfillment of the strategically important goals in development areas of Strategy 2030.

When estimating the budget execution of Volgograd, the Control and Accounts Chamber stressed the risks of non-fulfillment of some municipal and departmental target programs implemented under Strategy 2030. In this regard, the city administration timely had been adjusting planned events, including that related to alteration of funding volumes for the effective implementation of programs, following the chamber's proposals.

As a part of the preliminary supervision over the establishment of the municipal budget, Volgograd City Council carried out an expert assessment of the draft decision of the Volgograd City Council "On the budget of Volgograd 2019 and the planning period 2020-2021". Experts assessed the first stage of implementation of Strategy 2030 and found out that strategy should be adjusted in terms of some indicators due to overfulfillment or underfulfillment. Thus, experts have revealed the following:

- figures projected for 2019, in particular, industrial production growth, are insufficient for the achievement of strategic goals;
- part of the target indicators set forth in the program documents are not balanced;
- in the projection, there are no target indicators of strategic goals' progress that determine the satisfaction of the population with the quality of strategic program implementation [8].

In addition, the Chamber also inspected the conformance of funding volumes declared in the passports of the municipal programs and uncovered some discrepancies in the funding volumes for 3 municipal programs within strategic development areas.

On the back of reporting audit findings for 2017, the Chamber found out that performance indicators haven't been achieved within the program "Landscaping". It had been also revealed that the actual funding of the subprogram on the support of young families under the municipal program "Housing" is almost twofold lower than the planned amount [8].

The outcomes of the activities held and proposals developed by the Volgograd Control and Accounts Chamber are intended to adjust program documents and improve the conditions for the fulfillment of strategic goals indicated in Strategy 2030, as well as bring its indicators in line with national goals and indicators of the Volgograd region development strategy.

17.4 Conclusion and Recommendations

The analysis shows that introduction and application of strategic audit in the activities of the control and accounting bodies is directed towards achievement of strategic goals and more effective implementation of program activities and refinement of the management system.

Summing up, we can state that strategic audit is becoming an integral part of expert assessment of strategic goals' progress.

The application of financial and performance audit tools is focused on the assessment of the current situation and taking certain measures, as a rule, in the short term. A strategic audit allows looking into the future to some extent and developing an action plan that can help to achieve target indicators in the least cost-intensive and the most effective way. In combination with a financial and performance audit, a strategic audit will provide the opportunity to provide the most accurate and comprehensive forecasts of strategic goals in the short, medium and long term.

High-qualitative projection, in turn, will make it possible to timely take the necessary preventive measures to adjust managerial decisions and actions that contribute to more effective implementation of programs and projects within the target development strategy.

References

1. On Strategic Planning in the Russian Federation: [Federal Law of the Russian Federation: adopted by the State Duma of the Federal Assembly of the Russian Federation on June 20, 2014, under No. 172-ФЗ: most recent]. Accessed from the computer-assisted legal research system ConsultantPlus, 2019
2. The Strategy for Socio-economic Development of Volgograd until 2030 [Electronic resource]—URL: http://www.volgadmin.ru, free. Accessed 25 Apr 2019
3. Concerning the Budget of Volgograd for 2019 and the Target Period 2020–2021 [Electronic resource]—URL: http://www.volgadmin.ru, free. Accessed 25 Apr 2019
4. Presidential Decree No. 204 as of May 7, 2018 On national goals and strategic goals of the development of the Russian Federation Until 2024. Accessed from the computer-assisted legal research system ConsultantPlus, 2019
5. Aleksei Kudrin: over six years, the share of strategic audit in the activities of the Accounts Chamber will exceed 50 percent [Electronic resource]—URL: http://www.ach.gov.ru/press_center/news/35245, free. Accessed 1 May 2019
6. Development Strategy of the Accounts Chamber of the Russian Federation for 2018–2024 [Electronic resource]—URL: http://www.ach.gov.ru/about/document/strategy-2018-2024.php, free. Accessed 16 Apr 2019
7. The place and role of strategic audit in the activities of the Tomsk Accounts Chamber, the practice [Electronic resource]—URL: http://spgt.tomsk.ru, free. Accessed 27 Apr 2019
8. Report on the activities of the Volgograd Control and Accounts Chamber for 2018 [Electronic resource]. URL: http://www.kspvolg.ru, free. Accessed 27 Apr 2019

Chapter 18
Assessment of the Impact of the Policy of Import Substitution on the Competitiveness of Enterprises

Sergey A. Konovalenko, Egor A. Lopatin, Mikhail N. Trofimov and Genadiy S. Shkabin

Abstract *Purpose*: The purpose of the research is to assess the impact the strategy of import substitution on the competitiveness of economic entities in the context of the sanction policy of foreign partners against Russian enterprises. *Design/methodology/approach*: Impact assessment is carried out on the basis of the assessment tools generally accepted in the system of economic analysis. Analyzing modern methodological approaches the authors substantiate methods and the system of indicators used to assess the effectiveness of the import substitution policy and its impact on competitiveness of enterprises (evidence from the Russian company OOO ZVK BERVEL, Ryazan [OOO ZVK BERVEL—the high-strength fastener plant in Rybnoye (Ryazan Region, Russia), a limited liability company under the laws of the Russian Federation]). *Findings*: Assessing the effectiveness of import substitution policy at the enterprise level the authors calculate the dependence on imported production, show the positive trend in the most indicators of import substitution, but reveal both positive and negative impact of import substitution on the competitiveness of the enterprise. They make prediction research of the import substitution program and give recommendations to increase the effectiveness of the import substitution policy for the enterprise. *Originality/value*: The results of the research will be useful for assessment of the competitiveness of an enterprise before and after the implementation of the import substitution program. The main ways to increase the effectiveness of the import substitution policy at the enterprise level are proposed.

S. A. Konovalenko (✉) · E. A. Lopatin · M. N. Trofimov
Kikot Moscow University of the Ministry of Internal Affairs of Russia (Ryazan Branch), Ryazan, Russia
e-mail: sergey_marsel@mail.ru

E. A. Lopatin
e-mail: 138075@yandex.ru

M. N. Trofimov
e-mail: trofimov_fanat1k@mail.ru

G. S. Shkabin
Kikot Moscow University of the Ministry of Internal Affairs of Russia, Moscow, Russia
e-mail: uprzn@ya.ru

T. Kolmykova and E. V. Kharchenko (eds.), *Digital Future Economic Growth, Social Adaptation, and Technological Perspectives*, Lecture Notes in Networks and Systems 111, https://doi.org/10.1007/978-3-030-39797-5_18

Keywords Export · Import · Import substitution · Competitiveness · Principles and methodology of economic analysis

JEL Classification F21 · R1 · L53

18.1 Introduction

Studying the effectiveness of import substitution is impossible without statistical analysis of the export-import relationship in Russia's total turnover [6]. According to statistics, Russia's exports reached a value of $452.1 billion in 2018. Traditionally, the basis of Russia's exports was fuel and energy products, whose share in the total turnover of export goods exceeded 63.7%. Metals held the leading positions in the export structure with the specific indicator 9.9%. The products of the chemical industry occupied the third position with the specific indicator 6.1% in 2018. Exports of machinery and equipment increased insignificantly (by 2.7%), which may indicate a small share of this category of goods in the total volume of world technologies. Food products, which showed a growth of 20.2% in value terms and 19.2% in physical terms, completed the list of the most important export commodities.

According to the statistical analysis conducted by the Ministry of Economic Development of the Russian Federation, in 2018 Russia's imports reached the value of $240.5 billion and grew by 5.1% compared with 2017.

The largest share in the commodity composition of imports accrued to machinery and equipment, which reached 47.3% in 2018. The share of chemical products in the commodity structure of imports was 18.3% in 2018, while the import content of food products and raw materials for their production in 2018 was 12.4%. This is the lowest value over the last 10 years, which indicates a nationwide increase of effectiveness of import substitution policy.

As a result of the study of macroeconomic indicators in the field of foreign economic activity some conclusions were made:

First, there is a technological dependency of Russian companies from developed nations in the structure of imports, evidenced by the total share of machinery and equipment, which remains at 47% of imports.

Second, the import substitution has entered a new phase of growth and is growing in intensity, as evidenced by the reduction of import content in goods strategically important for the Russian economy in the total turnover.

After studying the scientific approaches, it can be concluded that import substitution is above all a non-market retaliatory measure of a comprehensive action (legal, political, social and institutional) aimed at replacing foreign goods (works, services) with domestic developments.

The import substitution in Russia is based on a system of regulatory acts [5]. In order to increase the competitiveness of domestic products on the international market, Federal Law of the Russian Federation of December 2016, 19 No. 415-FZ

"Concerning the federal budget for 2017 and the scheduling period of 2018 and 2019" provides the following measures of state support for import substitution:

- granting subsidies to organizations for loan interest payment in 2019 (8.9 billion rubles);
- granting preferences to domestic companies through the partial reimbursement of the costs of research work on innovative projects in 2019 (5.6 billion rubles);
- granting subsidies to Russian producers for the reimbursement of the costs of commissioning work in 2019 (1.2 billion rubles).

In accordance with the Order of the Government of the Russian Federation of August 2015, 04 No. 785 "Concerning the Governmental Commission on Import Substitution", types of strategic goods were designated and priority production for import substitution was established. The priority economy sectors were:

- ingredients and machines for the food industry;
- heavy engineering industry;
- machine-tool industry;
- chemical industry;
- pharmaceutical industry;
- conventional weapons industry, civil aircraft [1].

Measures of state support for the production of certain types were enshrined in the legal acts:

- in the Order of the Government of the Russian Federation of August 31, 2017 No. 1876-r (support for agricultural engineering industry);
- in the Order of the Government of the Russian Federation of August 31, 2017 No. 1877-r (support for automotive industry products);
- in the Order of the Government of the Russian Federation of August 31, 2017 No. 1878-r (support for railway engineering products).

The system measures of import substitution policy in the general economic policy of Russia are based on the following aspects:

1. Stimulating exports until 2030.
2. Setting-up state owned companies in accordance with the import substitution policy.
3. Adoption legal regulations "concerning import ban of some products disrupting economic security of the state and originating from foreign countries" [8].

The strengths and weaknesses of the impact of the policy of import substitution on the Russia's economy are presented in the Table 18.1.

Table 18.1 Assessment of the impact of the policy of import substitution on the economy and competitiveness of the state

Strengths	Weaknesses
Strengthening the industrial base of domestic producers	Possible deterioration of the quality and competitiveness of domestic products on the world market
Increase in production of industries with state support	Reduction of turnover and violation of economic relations with traditional partners
The ability to support strategic industries [12]	Risks of suing domestic producers by the WTO and other international organizations
Reducing the capital outflow. Improving the financial condition of companies involved in import substitution	Closing access to international capital. Outflow of skilled personnel

Source Compiled by the authors

18.2 Methodology

The methodological basis of the research is the techniques, procedures and methods generally accepted and widely used in the study of economic processes in economic analysis and statistics.

The following modern methodological approaches to assessment of the impact of the policy of import substitution in the real economy are highlighted:

1. The assessment of the potential of import substitution.
2. The assessment of the effectiveness of import substitution expenditures.
3. The coefficient method for assessing the effectiveness of import substitution [10].

Recently, methodology of Matveeva L.G., Chernova O.A. and Klimuk V. gains currency which allows assessing the effectiveness of import substitution at the level of an individual enterprise and involves the calculation of ten coefficient indicators [7]. The advantages of this method of calculating are in the possibilities of:

1. Calculations using the tools of economic analysis.
2. Assessing the effectiveness of import substitution policy at the enterprise level.
3. Comparing the results with the industry average values and indicators of another company.

The assessment of the impact of the policy of import substitution on the competitiveness of an enterprise should be based on the following principles:

1. Assessment should be taken first of all at the micro-level.
2. Tools of economic analysis and mathematical statistics should be used.
3. The results of the analysis should be focused on calculating the economic benefit (increase of competitiveness) [4].

Using data of OOO ZVK BERVEL, Rybnoye, Russia, the high-strength fastener plant in the Ryazan Region (Russia), the assessment of the impact of the policy of import substitution on the competitiveness of enterprises is conducted. Primary data are given for the calculation of 100 production unit (ELAX abrasive polishing wheels) (Table 18.2).

Additional data are given in the Table 18.3.

Table 18.2 Basic design data

No	Component	Years (total cost, in rubles)			Purchase under the import substitution program in 2018
		2016	2017	2018	
1	Composite material «Organit»	505,000	525,000	542,000	Imported[a]
2	Fibreboard	205,500	211,000	221,400	Domestic
3	Polymer material «Bimax -10»	242,000	244,000	267,000	Imported[a]
4	Composite material «Kriminit»	120,500	135,800	140,400	Domestic
5	Composite material «Denfill»	185,600	188,000	192,000	Imported[a]
6	Composite material «Well-Pex Type A»	33,500	35,400	36,400	Imported[a]
7	Composite material «Well-Prep Refill»	6,800	7,500	8,400	Imported[a]
8	Steel	1,100	1,150	1,200	Domestic
9	Composite material «Alumina trihydrate»	22,100	24,300	27,500	Imported
10	Composite material «Clearfil Core»	21,430	26,500	28,800	Imported
11	Composite material «Prisma»	407,500	425,000	444,000	Domestic
12	Composite material «Doxer»	69,800	87,500	92,000	Imported
13	Composite material «Кегг»	31,450	33,250	37,200	Imported[a]
14	Roamed polyurethane	173,000	175,200	182,000	Imported
15	Resins	15,000	15,000	15,200	Domestic
16	Interconnect материалы	102,000	101,500	103,500	Domestic
17	Epoxide материалы	54,200	55,250	57,000	Imported
	Total	2,196,480	2,291,350	2,394,000	

[a]Imported material replaced by domestically produced materials

Source Compiled by the authors

Table 18.3 Assessment of the effectiveness of the policy of import substitution in OOO ZVK BERVEL in 2016–2018

Indicator	Method of calculating	Years			Percentage ratio of 2018 to 2016
		2016	2017	2018	
Coefficient of import dependence (physical)	Share of imported parts required for the production, in the total number of parts 0.647 0.647 0.647 100.0	0.647	0.647	0.647	100.0
Coefficient of import dependency of production (value)	Ratio of total cost of imported parts to the total cost of all parts setting cost-accounting value	0.602	0.617	0.633	105.1
Actual coefficient of import coverage (physical)	Ratio of the number of domestic parts used for the production to the number of imported parts	0.545	0.545	0.545	100.0
Actual coefficient of import coverage (value)	Ratio of the cost of domestic parts used in the manufacturing and the cost of imported parts	0.633	0.634	0.630	99.5
Potential coefficient of import coverage (physical)	Ratio of the number of domestic parts to be replaced in the process of manufacturing to the number of imported parts	2.4	2.4	2.4	100.0
Potential coefficient of import coverage (value)	Ratio of the cost of domestic parts to be replaced in the process of manufacturing to the cost of imported parts	2.65	3.01	5.18	195.5
Coefficient of import substitution (physical) – before the program starts – after its implementation	Ratio of the number of domestic parts used in the manufacturing process to the total number of parts	– –	– –	0.353 0.706	– –

(continued)

Table 18.3 (continued)

Indicator	Method of calculating	Years			Percentage ratio of 2018 to 2016
		2016	2017	2018	
Coefficient of import substitution (value) – before the program starts – after its implementation	Ratio of the cost of domestic parts used in the manufacturing process to the total cost of parts	– –	– –	0.387 0.838	– –

Source Compiled by the authors

According to the calculations, the dependence of OOO ZVK BERVEL on imported production at the end of 2018 is 64.7%. The positive trend in most indicators of import substitution after the implementation of the investment project is demonstrated. So, the potential coefficient of import coverage (value) increased from 2.65 to 5.18 (almost doubled), and the coefficient of import dependency of production (value) increased from 0.602 to 0.633 (by 5.1%) and all the evidence goes to prove a decrease in dependence from imported component parts.

The interim result of the investment project (implementation of the ELAX abrasive polishing wheels) was the increase of the coefficient of import substitution from 0.353 to 0.706 in physical terms and from 0.387 to 0.838 in value terms.

The main ways to increase the effectiveness of the import substitution policy for OOO ZVK BERVEL should be:

- design and implementation of alternative technologies;
- design and implementation of modern and safe technologies for the use of industrial waste.

One of the most important aspects in the system of assessing the impact of import substitution on competitiveness is a prediction research of the proposals made for the implementation of the import substitution program of OOO ZVK BERVEL [11].

The method of data extrapolation "time series" is used to conduct the prediction research [3].

$$y_t \equiv a + bt$$

where y_t is a graduation; t is a serial number; a and b are parameters of the primary line;

$$\left[\sum y \equiv a \times n + b \times \sum t_t \quad \left[\sum yt \equiv a \times \sum t + b \times \sum t^2\right.\right.$$

The inferential calculations are presented in the Table 18.4.

$a = \frac{464,828}{3} = 154,942.7 \quad b = \frac{170,586}{2} \equiv 85,293$

$y2016 \equiv a + bt = 154{,}942.7 + 85{,}293\ (-1) = 69{,}649.7$ thousand rubles

Table 18.4 Analytical graduation of time series

Years	Revenue, in thousand rubles	t	t^2	yt	$\overline{yt}$
2016	54,415	−1	1	−54,415	69,649.7
2017	185,412	0	0	0	154,942.7
2018	225,001	1	1	225,001	240,235.7
Total	464,828	0	2	170,586	464,828.1

Source Compiled by the authors

$y2017 \equiv a + bt = 154{,}942.7 + 85{,}293\ (0) = 15{,}492.7$ thousand rubles
$y2018 \equiv a + bt = 154{,}942.7 + 85{,}293\ (+1) = 240{,}235.7$ thousand rubles
$y2019 \equiv a + bt = 154{,}942.7 + 85{,}293\ (2) = 325{,}528.7$ тыс. руб.
Prediction data for 2019 are presented in the Fig. 18.1.

As is seen in the Table 18.4 and Fig. 18.1 the revenue of the organization in 2019 will be 325,528.7 thousand rubles on condition that the trends of past periods (sequential replacement replacement of imported components, implementation of modern and safe technologies for the use of industrial waste) will persist.

Competitiveness is determined by such general performance indicators as cost per production unit; profitability of an innovative product; revenue or net profit of the company; payback period. A number of specialists [2] recommended to assess competitiveness with the use of such indicators as: NPV (net present value) or IRR (internal rate of return of the project).

The comparative analysis of the impact of the import substitution program on the overall competitiveness is presented in the Table 18.5.

According to the Table 18.5, the profitability of the ELAX products significantly exceeds the profitability of Airon, the global manufacturer from the USA (in particular by 122.8% before the import substitution project and by 166.7% after the introduction of the import substitution program) relative to 18.4%. The payback period for 1 ruble of investments in the ELAKS project is potentially 164 days (before the import-exchange program) and 137.2 days (after the introduction of the import substitution program). The comparable rate of the foreign competitor is more than

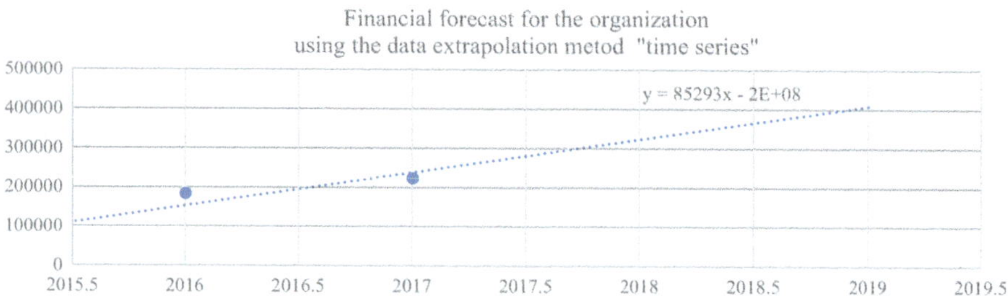

Fig. 18.1 Predictive study of the revenue changes without the implementation of the import substitution program proposed to OOO ZVK BERVEL *Source* Compiled by the authors

Table 18.5 Analysis of competitiveness evidence from OOO ZVK BERVEL taking into consideration the prediction values for 2019

Cost items/Competitiveness indicators	ELAX products (Russia), in rubles, before the import substitution program	ELAX products (Russia), in rubles, after the import substitution program	AIRON products (the USA), in rubles, at the exchange rate of the Bank of Russia[a]
Total cost, in rubles	61,715,000	122,052,521.2	8,025,478,000
Average per unit price, in rubles	1,550	1,298	1,900
Potential production output per year, in production units	88,710	250,792.52	5,000,000
Unit costs, in rubles	695.69	486.66	1,605.1
Potential revenue	137,500,000	325,528,700	9,500,000,000
Production profitability, %	122.8	166.7	18.4
Payback period, in days	164.27	137.22	309.19

[a]According to the financial statements of companies for 2018 and forecast 2019
Source Compiled by the authors

309 days. However, the realization of competitive advantages (potential production output) is currently impossible due to lack of funding and state support for enterprise development [9].

18.3 Results

The process of import substitution has both positive and negative impact on the competitiveness of enterprises. On the one hand, the implementation of import substitution programs at the federal and regional levels causes a decrease in the competitiveness and quality of domestic products because of limited competition in the market. On the other hand, import substitution is becoming an important element in the long-term trend of economic growth and can significantly improve the financial condition of domestic producers and ensure the introduction of new technologies into the production process creating prerequisites for the competitive recovery in the domestic economy in the future. Assessment of the impact of import substitution on competitiveness, in our opinion, should be conducted at the level of economic entities, and the assessment of this impact should be based on the principles and methodology of economic analysis.

18.4 Conclusion

The following conclusions were made as a result of the study.

First, the assessment of the impact of import substitution on competitiveness is based on the principles and methodology of economic analysis, but this impact cannot be unambiguously assessed. The import substitution initiates replacement of more expensive foreign components with domestic components, which generally reduces the cost of production and causes a dramatic improvement in the financial and economic indicators of the competitiveness of organizations. However, product degeneration and growth of unnecessary losses in general can lead to the reputational loss and market volume reduction in the future.

Second, the import substitution in a single case (for example, in case of OOO ZVK BERVEL) indicates an increase in the organization's competitiveness, which is evidenced by the positive dynamics of financial indicators. However, it is difficult to assess the true and fair view of the effectiveness of this process, because it is not clear, how changes (replacement with domestic components) affected the quality of the product and, as a result, the technical and production indicators of the organization's competitiveness.

Third, the study verified a hypothesis, that for the rational import substitution and competitiveness achievement it is necessary to eliminate the lack of funding, as well as an active investment promotion in the high-tech sector of the economy.

References

1. Chernova OA, Klimuk VV (2017) Rational import substitution in industry: assessment tools. Bull Tomsk State Univ Econ 38:43–51
2. Chernova OA, Klimuk VV (2016) Rational import substitution as an imperative for the implementation of a new model of development of the Russian economy. Bull Samara State Univ Econ 38:32–37
3. Czerwinski EA (2015) Import substitution in the Republic of Belarus: methods of analysis and directions of improvement. Belarusian science, Minsk, p 198. ISBN 978-985-08-1924-6
4. Fałat-kilijańska I, Luty P (2019) How private equity funds boost the competitiveness of companies (based on Polish cases). J Fin Stud Res, Vol 2019, Article ID 630543. https://doi.org/10.5171/2019.630543
5. Konovalenko SA, Kornilovich RA (2017) Principles of formation of system of innovative development of the enterprise. Econ Entrepreneurship 9–2(86-2):525–527
6. Lisov A (2017) There is a silver lining. Lawyer, no. 40, p. 3
7. Matveeva LG, Chernova OA, Klimuk VV (2015) Evaluation of the effectiveness of import substitution policy in industry: methodological tools. Bull Far Eastern Federal Univ Econ Manage 3(75):3–14
8. Podlupa VA (2017) Analysis and evaluation of potentials of private commercial organizations. The Auditor 7:53–57
9. Perrotta C (2014) Innovation in technology-enhanced assessment in the UK and the USA: Future scenarios and critical considerations. Technology, 23. https://doi.org/10.1080/1475939x.2013.838453
10. Semenov MV (2017) Import substitution. Ind Acc Taxation 4:52–60

11. Takács Ildikó-Csilla (2018) Strategic innovation management through organizational change. J Innov Bus Best Pract, vol 2018, Article ID 863719. https://doi.org/10.5171/2018.863719
12. Trofimov MN (2016) Indicative method of evaluation of investment activity of business structures. In: Kadyrgulov IR (ed) In the modern society: science, technology, education, 2016 proceedings of the Russian scientific conference with international participation, in 4 volumes, pp 239–244

Chapter 19
Motives of Entrepreneurial Activity in Regions with Limited Potential for Self-Development

Zhanna A. Zakharova, Vitaly V. Kislinsky and Zoya I. Bakhur

Abstract The article emphasizes the relevance of the motivation of entrepreneurial activity to achieve a wide range of motives, including the pleasure of doing business in regions with limited potential for self-development. The aim of the work is to determine the existing motives for entrepreneurial activity in regions with limited potential for self-development, and identify contradictions in the development of entrepreneurial activity of the studied regions. In the course of the work, the authors used general scientific methods: comparisons and analogies, analysis and synthesis, abstract-logical method in describing the qualitative institutional environment that contributes to the expansion of the motives of economic activity of capital. The study determined that the existing regions with limited capacity of self-development of the institutional environment of entrepreneurial activity is not conducive to the expansion of motives of economic activities of capital (gaining new and retaining existing market, increase of sales growth rate, obtaining satisfied the rate of profit, etc.). Also in the work it is revealed that one of the main motives of entrepreneurial activity in the studied regions is the self-preservation of the enterprise (survival in the long term), which with adequate use of various sources (for example, budget, a certain "debt economy") contributes to the capacity of self-development of regions. The results obtained can be used by public authorities, both at the Federal and regional levels, as an information base for the development and implementation of the necessary measures of state support for entrepreneurial activity in order to create a quality institutional environment for its implementation in regions with limited potential for self-development, as well as by specialists studying this problem. The subject of further exploratory research may be the identification of underutilized sources of

Z. A. Zakharova (✉) · V. V. Kislinsky · Z. I. Bakhur
Vladimir Branch of Russian Presidential Academy of National Economy and Public Administration, Vladimir, Russia
e-mail: zjane77@mail.ru

V. V. Kislinsky
e-mail: acidman74@mail.ru

Z. I. Bakhur
e-mail: baxurz@mail.ru

T. Kolmykova and E. V. Kharchenko (eds.), *Digital Future Economic Growth, Social Adaptation, and Technological Perspectives*, Lecture Notes in Networks and Systems 111, https://doi.org/10.1007/978-3-030-39797-5_19

capacity building for self-development of regions and the development of tools to create conditions for its provision.

Keywords Motives of entrepreneurial activity · Institutional environment · Region · Self-development of the region

JEL Classification Codes O 18 · R 58

19.1 Introduction

The topic of motivation of entrepreneurial activity has always been of interest to a large number of researchers, since the motives of the entrepreneur is the driving force of any economic system (organization, municipality, region and country as a whole). Due to the fact that one of the requirements of the market economy is the long-term orientation of private capital (long-term contracts, long-term projects), the most important should be the motives aimed at achieving strategic targets. This aspect is particularly relevant at the level of the regional economy. In view of the fact that even in some regions of the center of Russia there are conditions for entrepreneurial activity that do not contribute to the motivation of business to expanded reproduction. In this connection, entrepreneurial activity and the ability of the regions to ensure sustainable socio-economic development in the long term are reduced.

19.2 Methodology

The methodological basis of the research is general scientific research methods (analysis and synthesis, deduction and induction, normative and positive methods, etc.), quantitative econometric methods (summary and grouping of statistical data, statistical indices).

Also, according to the methodology proposed by Zh.A. Zakharova and V.V. Bogatyrev [1], the main indicators and criteria of self-development of the regions of Central Russia were calculated: Vladimir, Ivanovo, Kaluga, Kostroma, Yaroslavl, Moscow, Ryazan regions and the city of Moscow. Based on the analysis of criteria according to official statistics for 2017, the studied regions are differentiated into two types-self-developing and regions with limited potential for self-development. The self-developing regions include: the city of Moscow, Moscow and Kaluga region. Further, based on statistical data, some indicators of business development in two types of regions were considered: self-developing and with limited potential for self-development.

19.3 Results

As is known from economic theory, one of the main motives of an entrepreneur is to make a profit. However, approaches to the study of entrepreneurial motivation have evolved significantly. Thus, the classic of economic theory K. Marx considered profit maximization as the main motive of entrepreneurial activity [2].

Modern researchers emphasize that the motive of obtaining the maximum rate of profit of the current period can not provide sustainable development in the long term, that is, create conditions for self-development. Thus, in the "Institutional economy" edited by academician D.S. Lvov, it is noted that companies motivated not to maximize, but to a satisfied rate of profit demonstrate higher viability and adaptability to solving long-term problems [3]. However, the motives of entrepreneurial activity are influenced by external globalization processes, namely the trends of "digitalization of the economy" occurring all over the world. And in these conditions there is a significant transformation and expansion of the spectrum of motives of entrepreneurial activity [4]. Namely, to achieve strategic goals that take into account the necessary processes (globalization, digitalization, etc.), the most important are the motives associated not with maximizing profits, but with the conquest or retention of market share, with obtaining a satisfactory rate of profit, with an increase in sales growth, survival in the long term, the disclosure of potential opportunities for scientific and technological progress; reducing the cost of production; improving the quality of products and services, and, of course, getting pleasure from doing business.

At the same time Y. Schumpeter, drawing attention to the fact that one of the main motives of entrepreneurial activity is a constant desire, and the ability to innovate stressed that the continuous reproduction of entrepreneurs "new combinations" of factors of production will self-develop and the economic system itself [5]. That is, the creation of conditions for the constant introduction of innovations in the process of economic activity creates prerequisites for its self-development.

However, the specified range of motives of business activity, including receiving pleasure from housekeeping reflects a stereotype of the Western mentality and is based on the situation developing in stable economic systems of the Western countries [6]. In fact, the expansion of entrepreneurial motives depends on the institutional environment in which private capital operates. The main characteristics of a quality institutional environment that contributes to the expansion of business motives include: a steady level of demand; availability of capacious sales market and channels of promotion of goods; availability of resources, which implies developed markets for goods, services, works, labor, money, information; developed market infrastructure, which includes trade, road transport infrastructure, communications, organization of monetary circulation, insurance, etc.; trust, formed by the state economic policy; healthy, normal competition, minimizing monopolies; information provision of entrepreneurs about the state of the market, prospects for its development, intentions of public authorities, etc. [7]; a stable tax system, the necessary guarantees from the state; stability and stable rules of the game in the markets; a stable currency, etc.

There is no doubt that the main role in creating a high-quality institutional environment in the regions, contributing to the motivation of business to the target settings, which are the main condition for ensuring the self-development of the region, is assigned to public authorities, both at the Federal and regional levels. And according to a number of researchers, in a market economy there are regions, territories that can develop themselves and territories that have limited potential for self-development [8–12]. According to the authors, regions with limited potential for self-development are territories that in a modern market economy cannot independently respond to changes in environmental factors (mainly public demand) due to internal sources alone, which does not create conditions for ensuring a sustainable state in the long term. At the same time, it should be noted that the regions objectively have different heterogeneous institutional environment for entrepreneurial activity (Table 19.1).

Thus, in the conditions of regions with limited potential for self-development, the main characteristics of the institutional environment include: low capacity of the local market; compressed demand due to the low level of wages and incomes of the population; limited access to various kinds of investment resources; insufficient infrastructure provision, which hinders the development of integration both interregional and international; etc.

Thus, one of the most significant indicators in the development of entrepreneurial activity in the region is the "critical mass" of entrepreneurs. However, its value is not optimal for 2017 in many of the studied regions (Table 19.2).

As is known in theory and practice in a market economy, the normal indicator by which one can judge the "critical mass" of entrepreneurs is the ratio of the population of the territory to the number of enterprises and organizations 10–15 to 1 [13]. That is, the optimal indicator is when one organization accounts for 10–15 people of the population. However, according to Table 19.2, it should be noted that the above criterion is achieved only in the city of Moscow, and in the regions studied by us it is 2–3 times lower than the optimal value, which indicates an insufficient mass of entrepreneurs in the regions of Central Russia. The lowest values of the indicator in Vladimir (42.3 people per 1 organization) and Kostroma region (39.0 people per 1 organization). And in the Yaroslavl region, the indicator is better than in such self-developing regions as the Moscow and Kaluga regions. The deterioration of the value of this indicator in all the studied regions over the past year indicates significant limitations in the development of entrepreneurial activity in the regions.

In addition to the indicator of the critical mass of enterprises and organizations, it is necessary to consider the development of entrepreneurial activity of small businesses. At the same time, as noted by a significant number of researchers, under certain conditions, the development of small business provides a high level of self-organization of the economically active population and, accordingly, self-employment, which can also be considered as one of the characteristics of self-development of the region [14, 15]. And the leading indicator here can be the number of individual entrepreneurs and small business organizations per 10,000 people of the population of the region, with an optimal value of more than 200 units.

Table 19.1 The main characteristics of different types of regions (self-developing and with limited potential for self-development)

Regions	Characteristic
Self-developing	• Significant market capacity • High share (more than 40% in the structure of the regional economy) of self-developing activities • Optimal set of economic activities • A high degree of openness of the region's economy (the share of foreign trade turnover of the region in GRP is more than 30%) • High share of profitable enterprises (at least 80%) • High innovative potential • Higher quality of life, income, RFP, by the share of paid services in the structure of household expenditure, compared with the average for the country, which prepares the individual interest of the population in living and employment in the region
With limited potential for self-development	• Narrowness of the local market • Not significant differentiation of economic activities, which makes the economy of the region less resistant to various fluctuations in the conjuncture, both domestic and foreign markets • Not a significant degree of openness of the region's economy (the share of foreign trade turnover of the region in GRP is about or less than 30%) • High share of unprofitable enterprises (30% and above) • Low indicators of quality of life of the population on incomes, on salary, on an indicator of a share of paid services in structure of expenses of households in comparison with averages on the country • Low share of investments in fixed assets (less than 25%); - etc.

Thus, according to Table 19.3 in 2017 in the city of Moscow, the number of individual entrepreneurs and small business organizations per 10,000 people of the population is 2.2 times higher than the normative value. However, such self-developing regions as the Moscow and Kaluga regions are represented by a small number of small enterprises. Yaroslavl (226 units), Ivanovo (204 units) have a high value. In addition, during the period from 2012 to 2016, it should be noted a significant increase in the number of individual entrepreneurs and small business organizations in all the studied regions. However, in 2017 in the Vladimir, Ivanovo, Kaluga and Ryazan

Table 19.2 Indicator of critical mass of entrepreneurs in the regions of Central Russia

Region	2005	2010	2015	2016	2017
Vladimir region	*53.3*	*40.0*	*39.1*	*40.1*	*42.3*
Ivanovo region	37.8	32.8	28.9	30.0	32.0
Kaluga region	36.7	37.2	35.2	36.3	37.9
Kostroma region	40.5	38.3	36.5	37.6	39.0
Nizhny Novgorod region	39.0	36.1	31.0	30.0	34.1
Ryazan region	37.3	34.2	33.3	34.6	37.6
Yaroslavl region	31.9	28.2	27.1	26.9	27.9
Moscow region	34.7	31.6	28.1	30.1	31.3
Moscow city	9.0	10.0	10.5	12.2	12.5

Calculated by the authors on: Regions of Russia. Socio-economic indicators. 2018: stat. sat./Rosstat. Moscow, 2018, pp. 39, 48, 554, 555

Table 19.3 Number of individual entrepreneurs and small business organizations per 10,000 people in the region, units

Region	2012	2013	2014[a]	2015	2016[b]	2017[c]
Vladimir region	125	134	131	143	146	141
Ivanovo region	173	189	198	178	212	204
Kaluga region	125	129	135	134	154	144
Kostroma region	127	131	102	143	148	153
Nizhny Novgorod region	107	118	109	127	158	190
Ryazan region	129	136	139	151	182	167
Yaroslavl region	212	202	172	173	226	226
Moscow region	106	108	107	117	164	169
Moscow city	201	200	201	239	417	445

[a]Compiled by: Regions of Russia. Socio-economic indicators. 2016: stat. sat./Rosstat. Moscow, 2016. P. 1326
[b]Calculated by the authors: Regions of Russia. The main characteristics of the subjects of the Russian Federation -2017 URL: http://www.gks.ru/bgd/regl/b17_14s/Main.htm (accessed 05.06.2019)
[c]Compiled by: Regions of Russia. Socio-economic indicators. 2018: stat. sat./Rosstat. M., 2018, pp. 572–573

regions there is a decrease in this indicator, which, once again, emphasizes the presence of restrictions (deterioration of the General economic situation, restriction of access to various resources, etc.), constraining the implementation of business activities in the regions. For example, in the Ivanovo and Yaroslavl region's economy is dominated of small forms of managing (the number of individual entrepreneurs and small businesses per 10,000 population is 204 and 226 units, respectively), which corresponds to the optimum value. However, in conditions of limited access to financial resources and investment, low sales opportunities, infrastructure constraints, it can not provide self-movement of the region. In addition, instability from external destructive influences is too great in such economic systems [16].

An important point of motivation of entrepreneurial activity (the desire for innovation, expansion and retention of the sales market, increase in sales growth) is the presence of entrepreneurs' own capital. Here it is necessary to look at the financial results of enterprises, the share of unprofitable enterprises (Tables 19.4, 19.5, and 19.6).

Thus, according to Table 19.4 in such regions as Vladimir, Kostroma, Ryazan, Moscow regions for the period from 2016 to 2017, the financial results of organizations decreased, and in the Ivanovo region, although there is an increase in these results, but they are very insignificant compared to other regions (about 1.6 billion

Table 19.4 Financial results of enterprises and organizations (net financial result, profit minus loss), million rubles

Region	2013	2014	2015	2016	2017
Vladimir region	22,055	18,951	31,117	36,601	32,668
Ivanovo region	−2670	−1632	−2797	−200	1594
Kaluga region	12,908	−15,761	−6887	24,113	37,477
Kostroma region	8272	6544	8799	13,606	6315
Nizhny Novgorod region	80,427	32,135	90,207	144,131	143,309
Ryazan region	17,888	18,716	22,908	33,412	31,164
Yaroslavl region	8080	−2198	4903	22,044	37,605
Moscow region	208,655	88,579	79,333	579,971	400,345
Moscow city	2,397,441	1,182,043	2,371,081	4,649,271	2,633,786

Compiled by: Regions of Russia. Socio-economic indicators. 2018: stat. sat./Rosstat. Moscow, 2018. P. 1044

Table 19.5 Balanced financial result (profit minus loss) of small enterprises, million roubles

Region	2013	2014	2015	2016
Vladimir region	12,685	10,050	11,890	10,420
Ivanovo region	4917	6451	7707	8317
Kaluga region	9779	5574	7937	4685
Kostroma region	4946	5388	5454	1789
Nizhny Novgorod region	25,871	−8805	37,164	35,935
Ryazan region	10,414	10,392	7723	5473
Yaroslavl region	72,865	91,139	44,114	34,210
Moscow region	10,446	9140	9672	30,882
Moscow city	333,322	154,243	256,820	690,986

Compiled from the collections: "Small and medium-sized businesses in Russia. 2017" and "Small and medium entrepreneurship in Russia. 2015," URL: http://www.gks.ru/wps/wcm/connect/rosstat_main/rosstat/ru/statistics/publications/catalog/ (accessed 17.05.2019)

Table 19.6 Share of unprofitable enterprises, % of the total number of enterprises

Region	2013	2014	2015	2016	2017
Vladimir region	30.2	33.0	34.5	30.3	31.4
Ivanovo region	33.2	37.3	40.3	40.5	39.1
Kaluga region	31.0	34.8	34.8	31.4	30.8
Kostroma region	37.0	35.9	35.6	31.8	42.9
Nizhny Novgorod region	28.2	31.2	31.6	31.1	28.5
Ryazan region	27.3	25.0	24.6	28.0	24.8
Yaroslavl region	40.1	38.4	36.6	32.8	33.8
Moscow region	30.5	30.8	30.2	25.4	30.2
Moscow city	31.5	35.4	34.3	29.5	33.9

Compiled by: Regions of Russia. Socio-economic indicators. 2018: stat. sat./Rosstat. Moscow, 2018. P. 1054

rubles). Also, in such regions as Vladimir, Kaluga, Kostroma, Ryazan and a number of other regions, there is a decrease in the financial results of activities and small businesses (Table 19.5).

These points, as well as the increase in the share of unprofitable enterprises in the regional economy (Table 19.6) suggest that regional enterprises have limited opportunities to invest at their own expense, and a third of enterprises do not have this opportunity (to develop at the expense of their own resources, namely to invest in renewal and innovation) at all. In this regard, we can say that these trends do not motivate entrepreneurs to constantly introduce innovations, which does not create conditions for self-development of the region. In fact, a significant part of the business one of the main motives for entrepreneurial activity is survival (for a number of years, some enterprises in various economic activities are unprofitable and at the same time remain afloat).

Thus, according to Table 19.7, a high proportion of unprofitable enterprises for 2017 is observed in such regions as Kostroma (42.9%), Ivanovo (39.1%), Yaroslavl (33.8%) regions and the city of Moscow (33.9%).

In confirmation of the fact that the lack of own funds constrains the innovative activity of enterprises, in those regions where there is the highest share of unprofitable enterprises and insignificant financial results of organizations (Kostroma and Ivanovo region), according to 2017, there is the lowest proportion of organizations engaged in various innovations among the considered regions (Kostroma 2.8%, Ivanovo-4.2%) (Table 19.7).

In addition, it should be noted about some contradictions in the development of entrepreneurial activity of the regions studied by us. Thus, foreign experience shows that the region's economy can successfully self-develop when the share of unprofitable enterprises does not exceed 20% of the total number of registered enterprises [17]. At the same time, according to Rosstat, the average share of unprofitable enterprises in Russia as a whole in 2017 is a little more than 30%. At the same time, it is possible to allocate self-developing regions (Moscow, Kaluga region in which

Table 19.7 The share of organizations engaged in various kinds of innovation, % of the total

Region	2013	2014	2015	2016	2017
Vladimir region	10.7	12.6	11.2	10.4	9.0
Ivanovo region	8.4	6.3	4.4	3.2	4.2
Kaluga region	10.9	9.7	10.9	8.5	9.0
Kostroma region	7.0	6.0	8.2	8.6	2.8
Nizhny Novgorod region	15.4	14.3	13.5	12.8	11.1
Ryazan region	11.4	13.1	12.7	12.3	12.1
Yaroslavl region	11.0	10.3	8.7	7.1	8.3
Moscow region	8.4	8.7	8.0	8.5	8.9
Moscow city	18.3	18.8	19.7	16.1	14.3

Compiled by: Regions of Russia. Socio-economic indicators. 2018: stat. sat./Rosstat. Moscow, 2018. P. 946

indicators of "critical mass" of businessmen aren't reached) and (Moscow, with high specific weight of the organizations which are carrying out innovations) in which the share of unprofitable enterprises makes more than 30%. Thus, this contradiction is that the high proportion of unprofitable enterprises does not mean the complete absence of conditions for self-development of the region.

In addition, from the point of view of the market economy (neoclassical economic theory), the main purpose of a commercial enterprise is to make a profit. Taking into account this circumstance it is possible to reveal the contradiction between the theory and practice of economic activity, consisting that taking into account aspiration of the unprofitable (inefficient) enterprise to self-preservation (survival) the region with a high share of unprofitable enterprises, but more qualitative institutional environment, can show some conditions for self-development. At the same time, the question arises, at the expense of what sources this economic activity is supported?

There is no doubt that one of the sources of support for the economic activity of the city of Moscow, as shown by modern research [18], is the possibility of significant investment at the expense of budgetary resources (27.8% in the structure of investments in fixed capital).

Also, as such a source in the city of Moscow, Moscow and Kaluga region is the growth of debt of enterprises (legal entities) on loan obligations.

Thus, according to the data, the debt on loans granted by credit institutions to legal entities for the period from 2013 to 2017 increased in all the studied regions, except for the Ivanovo and Kostroma regions. At the same time, in self-developing regions, the growth and volume of this debt per organization is very significant (Table 19.8).

Thus, according to Table 19.8, the indicator of debt on loans granted by credit institutions to legal entities per 1 organization for 2017 has a high value in the self-developing regions of Moscow (9.04 million rubles), Moscow region (5.88 million rubles), Kaluga (4.8 million rubles). At the same time, in the Yaroslavl region, where the level of innovative activity of enterprises is close to the self-developing regions (Kaluga and Moscow regions), the debt per 1 organization amounted to 3.5 million.

Table 19.8 Debt on loans in rubles provided by credit institutions to legal entities per 1 organization (in million rubles)

Region	2013	2014	2015	2016	2017
Vladimir region	2.19	2.49	2.46	2.3	2.58
Ivanovo region	1.3	1.19	1.01	1.03	1.1
Kaluga region	3.4	4.19	4.37	4.73	4.8
Kostroma region	2.09	2.04	1.8	1.62	1.6
Nizhny Novgorod region	3.3	3.65	3.3	3.0	3.53
Ryazan region	2.5	2.64	2.5	2.5	2.9
Yaroslavl region	2.78	2.81	2.7	3.13	3.5
Moscow region	3.65	4.52	4.46	5.31	5.88
Moscow city	5.7	6.88	6.68	8.4	9.04

Calculated by the author on: Regions of Russia. Socio-economic indicators. 2018 Stat. sat./Rosstat. Moscow, 2018, pp. 1024, 554, 555

This is significantly higher than in other regions with insufficient potential for self-development, especially in Ivanovo (1.1 million rubles) and Kostroma (1.6 million rubles) regions. In addition, in addition to debt on loans, the sources of support for economic activity in the regions studied by us are wage debts, as well as debt to suppliers and to the budget, and possibly other sources. For example, accounts payable per 1 organization in self-developing regions as of 2017 amounted to 17.2 million in Moscow. in the Moscow and Kaluga regions about 11.5 million rubles, which is from 2.5 to 9 times higher than the value of this indicator in regions with insufficient potential for self-development (Vladimir—4.3 million rubles, Ivanovo—1.8 million rubles, Kostroma—2.2 million rubles, Yaroslavl—4.5 million rubles).

19.4 Conclusions

Thus, the data show that one of the main motives for entrepreneurial activity in regions with limited potential for self-development is the self-preservation of the enterprise and, accordingly, survival in the long term. However, the desire unprofitable (inefficient) businesses to self-preservation (survival) even in regions with insufficient "critical mass" of entrepreneurs and a high share of unprofitable enterprises (Moscow and Kaluga region), but higher-quality institutional environment can enhance the capacity of self-development in this region due to the formation of a "debt economy" (growth of debts: loans to suppliers; wages before the budget), and possibly other sources;

In such regions as Ivanovo, Kostroma, Yaroslavl, Vladimir and Ryazan regions, the existing institutional environment, namely, unacceptable for these regions (for example, the "economy of small forms" of the Yaroslavl and Ivanovo regions) conditions of access of entrepreneurs to credit resources, limited opportunities of equity capital, low market capacity and limited sales opportunities, compressed demand due to low incomes, insufficient infrastructure development, etc., they do not motivate entrepreneurial activity to achieve strategic targets (the desire for innovation, expansion and retention of the sales market, increase sales growth rates, etc.), which does not create conditions for self-development of these regions.

All of the above requires an adequate policy of state regulation, and possibly other approaches to its implementation (integration approach) (Borodin and Erokhin [19], aimed at creating a quality institutional environment that promotes the expansion of motives for entrepreneurial activity, including pleasure in regions with limited potential for self-development.

References

1. Zakharova ZhA, Bogatyrev VV (2018) Interconnection between the categories of region's self-development and population's living standards. Adv Intell Syst Comput 622:440–447
2. Marx K (1988) Capital. Criticism of political economy, vol 1. The process of capital production. Politizdat, Moscow, 891 pp
3. Lvov DS (2001) Institutional economics: textbook (Series "Higher education"). INFRA-M, Moscow, 318 pp
4. Klutina SV (2005) Monopoly and local monopoly as its type: background, methodology, theory and practice. Ivanovo State University, Ivanovo, 285 pp
5. Schumpeter JA (1982) The theory of economic development. Progress, Moscow, 455 pp
6. Tuyakbaeva AE, Polutova MA (2012) Entrepreneurial motivation as a special case of labor motivation. Bull Chita State Univ 6:70–75
7. Babaev BD, and others (2017) Regional reproduction: fundamental questions of theory and practice (on materials of regions of the upper Volga region). In: Babaev BD, Babaev DB, Borovkova NV, Yezerskaya SG, Zaitseva NE, Nikolaeva EE, Popova TK, Sergeeva ME; under the General ed. Ivanovo State University, Ivanovo, 308 pp
8. Zakharova ZA (2017) On approaches to the sources of self-development of the region. Multilevel Soc Reprod. Questions Theor Pract 12(28):107–114
9. Bukhvald EM (2018) Regional development policy in modern Russia: priorities, institutions, tools: Scientific work. IE RAS, Moscow, 250 pp
10. Tatarkin AI, Tatarkin DA (2010) Self-developing territorial economic systems: dialectics of formation and functioning. Bull Chelyabinsk State Univ 3(184); Economy 24:5–12
11. Sharp Jeff S, Agnitsch Kerry, Flora Vern Ryan And Jan (2000) Social infrastructure and community economic development strategies: the case of self-development and industrial recruitment in rural Iowa. J Rural Stud 18:405–417
12. Flora C, Flora JL, Green GP, Schmidt FE (1991) Economic development through local self-development strategies. Agric Hum Values 8(3):19–24
13. Babaev BD (2011) Tandem: the market and the state. Theor Econ 4:55–61
14. Buchvald EM (2016) Legal and institutional problems of state policy of development and support of small and medium-sized enterprises. Finance Theor Pract 3:6–18
15. Rudenko M (2018) Development and realization of economic potential of the population as a basis of sustainable development of regions. Federalism 2:79–96
16. Avtonomov AM (2005) Disclosure of the category of "self-development of the region" in modern economic literature. Bull Volga, series 9, vol 4, part 2, pp 46–51
17. Buvaltseva VI (2002) The Essence and content of the process of self-development all-Russian scientific conference "Science and education". Collection of scientific works. Publishing House of Chengtu, Novosibirsk, pp 25–29

18. Kuznetsova OV (2016) Differences in the attractiveness of Russian regions for domestic and foreign investors. Voprosy ekonomiki 4:86–102
19. Borodin VA, Erokhin AV (2015) Regional integration as a factor of increasing the potential of self-development of the territory. Reg Econ Theor Pract 47:2–12

Chapter 20
Institutionalization of Land Property Titles' Market as a Factor of Ensuring Economic Competitiveness

Liliya Oganesyan and Elena Fedyunina

Abstract The institutionalization of the land property titles' market is presented in the article as a system of state measures on the enhancement of the institutional market regulators aimed at the efficient use of land resources. Based on the monitoring of the performance of local markets in 2018–2019 in three economic zones of the Volgograd region with different land quality (high-quality, medium-quality, low-quality), we have revealed and ranked price and institutional factors by the level of importance. Three groups of issues justifying the necessity of policy on further institutionalization of the land property titles' market have been identified. The first issue is related to the controversial mechanism for disposal of titles to share in land under institutional restrictions. Joint ownership in agricultural land contributes to the development of informal relations, the manifestation of opportunistic behavior, and the growth of transaction costs. The second issue arises from the first one and demonstrates the conditions of local markets' performance. After the expert assessment, we ranked the factors restraining the development of market relations by the level of importance: high transaction costs of land circulation, registration of land ownership title, opportunistic behavior and contracting. The third issue is caused by information asymmetry of agricultural lands in information systems and almost restrains the solution of the above issues. The solution of the issues revealed is systemic, therefore it requires the execution of a set of measures to enhance the market and institutional mechanisms for the performance of land property titles' market, which will provide conditions for the development of competitive relations in the agricultural economy.

Keywords Land property titles' market · Dual market model · Market institutionalization · Transaction costs · Specification of property titles · Institutional regulators

JEL Classification Code D 23 · Q 15

L. Oganesyan (✉) · E. Fedyunina
Volgograd State Agrarian University, Volgograd, Russia
e-mail: oganesian@bk.ru

E. Fedyunina
e-mail: elena.fedyunina@mail.ru

T. Kolmykova and E. V. Kharchenko (eds.), *Digital Future Economic Growth, Social Adaptation, and Technological Perspectives*, Lecture Notes in Networks and Systems 111, https://doi.org/10.1007/978-3-030-39797-5_20

20.1 Introduction

In Russia, the establishment of institutional conditions for the performance of the land property titles' market is a long-term and controversial task due to particular stages of land reform. As a result, the object of transactions on the domestic market of land property titles is not only a full or limited set of powers but also a share in the land, disposal of which is associated with higher transaction costs and an institutional trap. According to the concept of land reform, the function of the land share institution was to conserve national wealth in the hands of rural citizens and prevent its monopolization, respectively the restrictions were introduced into the mechanism for land share disposal. Ultimately, a favorable situation has been created for the development of a market of limited property titles to agricultural land, which is determined by the simplified and available procedure for limited titles under minimal costs. Thus, institutional elements of the mechanism for the performance of land property titles' market have become prevailing in the course of exchanging full and limited titles.

20.2 Materials and Methods

The research methodology of land property titles' market is based on the basic principles of microeconomics and economic theory of property titles described in studies of leading foreign and domestic scientists: Alchian and Demsetz [1]; Coase [2]; North [3]; Peyovich [4], pp. 117–118; Furubotn and Pejovich [5]; Kapelyushnikov [6]; Tambovtsev [7].

Market analysis is built on the methodological approaches of domestic researchers [8–12], the author's concept [13, 14] and on the following provisions:

- the performance of land property titles' market is determined by a combination of formal rules, informal requirements and a mechanism for monitoring of compliance with contracts;
- the structural transformation of property stipulates changes in the system of economic incentives and the behavior of economic agents;
- the effectiveness of the mechanism for the exercise of property titles is determined by transaction costs.

20.3 Findings

The results of monitoring local markets' performance conducted in three zones of the Volgograd region that are characterized by fertility and quality of agricultural lands, the level of market localization (regional center, district center, rural settlement), allow us to highlight three groups of issues reflecting controversial phenomena during the performance of land property titles' market under institutional constraints.

The first group of issues is connected with an ambiguous mechanism for the exercise of titles to a land share. The Institute of Land Share was introduced into the Russian economy at the first stage of land reform and privatization of agricultural land (1990–1992) to conserve agricultural land and restrain their monopolization. Having completed its mission, later land share became a restraining factor to the development of land relations. Under the land legislation of Russia, owners of land shares have a preemptive right of land property utilization, but at the same time – limited ownership and disposal rights. However, full-fledged disposal of the property requires a transformation of the land share into a land plot, thereby taking it out of the total land area, which is associated with high transaction costs. Besides, this procedure is possible provided that the area of land share is higher than the minimum established by law. In practice, this principle leads to an unexplainable situation when the owner cannot exercise his titles to sale and rent of a land share. This problem has become topical and has gained importance at the governmental level. According to statistical data, joint ownership is prevailing in the Russian ownership structure of agricultural land. For example, Rosreestr reports that the most common forms of households engaged in the production of agricultural products used land shares for farms (89%), business partnerships and companies (51.7%), and producers' cooperatives (45.2%) [15]. A similar situation is unfolded in zones of the region under study, where the property of citizens ranges within 60–80%, while the majority of land share owners are dissatisfied with the mechanism for the exercise of their titles (Fig. 20.1).

The expert assessment showed that 45.8% of executives and 31.2% of agricultural staff recognize the inconsistency of the mechanism for exercise of the land share and its negative impact on the development of market relations. The assessment of experts in terms of land quality is different. The impact of this factor is higher in areas with more fertile agricultural lands (37.9%) than in areas with medium-quality lands (24.7%) and the low-quality ones (32.1%) (Table 20.1).

The specifics of exercising titles to a land share in the Russian agrarian economy is that the basic elements of the scope of joint ownership titles are subject to restrictions which, on the one hand, blocks the disposal right, and, on the other hand, ensure the rights of possession and use at farms and private households intended for agricultural activities to be exercised. Thus, the titles of the land share owner are limited and blurred in modern conditions, which necessitates a revision of institutional norms and rules for the exercise of agricultural land joint ownership.

The second group of issues exhibits the dominant action of institutional factors on the performance of the land ownership titles' market mechanism: high transaction costs and, the opportunism risks, institutional regulators and restrictions (regulations and rules, tax rate), the factor of trust, the weak position of the state on the market, and unavailable bidders' selection criteria. Transaction costs, the risks of opportunistic behavior and contracting in the medium and long term are of the highest importance in the structure of transaction costs. For example, the potential lessee bears landmarking costs to speed up the procedure for state and municipal lands' lease transactions.

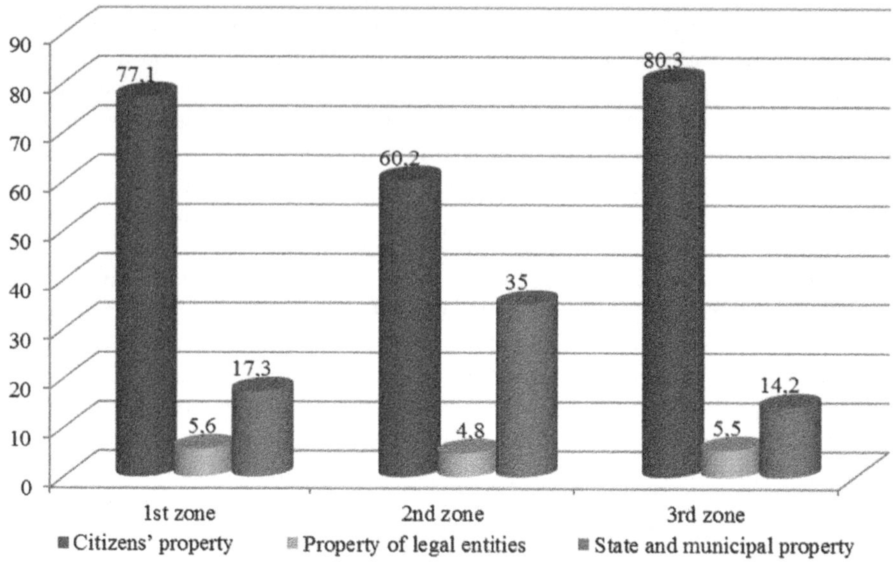

Fig. 20.1 Distribution of agricultural lands by ownership forms in three zones of the Volgograd region for 2018, (percent) *Note* The first zine is fertile lands (Uryupinsky and Elansky districts), the second zone is medium-quality lands (Olkhovsky district), the third zone is the low-quality lands (Kletsky district). *Source* Developed and drawn up by the authors

But another bidder may win while landmarking costs are not reimbursed. Thus, the potential lessee bears unreasonable transaction costs of land plot registration to effect a transaction. The research shows that high transaction costs are explained by inefficient institutional regulations, asymmetry of information on the market and a low level of knowledge of legal regulations and rules by ordinary market participants.

Risks of opportunistic behavior most often occur on the market for economic management titles in relations with citizens-land owners and in auctions for leasing state and municipal lands. In the first situation, the risks are associated with the overdue payment of rent by landowners, and in the second one—with unavailable bidder selection criteria, which contributes to fraud, extortion by unconscientious bides who are not engaged into agricultural activities and do not live in this area. The goal of extortion is that fraudsters extort reimbursement for rejection of further tendering.

A suspension of the sale of state and municipal lands was introduced in the zones under study, which contributes to the development of the market for economic titles and the containment of the market for full-fledged ownership titles. As for the significance of the tax rate as a regulator, most experts believe that the rate is small and acceptable. The position of local authorities in the management and control of land resources is weak in all zones (Table 20.1). The results of the monitoring show that the return on land ownership use and motivation to the continuity of agricultural activities are declining under high transaction costs, low prices of agricultural products and

Table 20.1 Factors of negative impact on the development of the market of ownership titles to agricultural land in local zones, percent of respondents

Factors	Zones under study		
	1st zone High-quality lands	2nd zone Medium-quality lands	3rd zone Low-quality lands
Low prices for agricultural products	55.2	43.8	57.1
Monopolization of land by a small group of owners	44.8	30.1	32.1
The plot to be purchased is in joint ownership	37.9	24.7	32.1
High prices for agriculturally used lands	32.8	24.7	17.9
High costs in the land circulation system	31	34.2	35.7
The inflated sales price of agricultural land out of state and municipal lands' system	29.3	32.9	20.2
Poor legal and economic awareness of ordinary land market participants	22.4	35.6	31
Local government support is unavailable	22.4	24.7	26.2
Rental land use is a prevailing form	20.7	30.1	10.7
Land plot pledge issues	19	32.9	11.9
Working land tax	17.2	31.5	33.3
Land profiteering and transfer of agricultural land to other categories	17.2	20.5	26.2
Preemptive right to purchase the land by local authorities	17.2	26	15.5
High costs of land ownership registration	15.5	38.4	23.8
The necessity for informal relations when effecting land sale/purchase transactions	6.9	19.2	9.5

Source Developed and made up by the authors on the back of local markets monitoring in 2018–2019

existing institutional restrictions, which leads to unstable employment and outflow of young people from rural areas.

The third group of issues reflects the occurrence of information asymmetry due to the changing methodology of accounting for agricultural land in the region and the consequent provision of false information. The study shows that real estate data on land types are not updated and their area in the land category structure is not monitored in the source documents of the unified state register. As a result, double accounting of areas of land categories and land types takes place. In this situation, it is necessary to revise the accounting methodology, meet the principles of land management documentation and develop an electronic base of archival data on land management documentation.

20.4 Conclusion

The studies carried out by us makes it possible to outline the following directions of institutionalization of the land ownership titles market that ensure the efficient use of land resources and contribute to the enhancement of the regional economy's competitiveness:

1. Improving the mechanism for the exercise of titles to a land share for effective use of property titles to agricultural land:

   - Expanding the title to the land share disposal aimed at simplifying the procedure for conclusion of lease relations for owners who cannot convert their land share into the land plot due to the area less than the minimum established by law;
   - Amending land legislation that prohibits the subdivision of inherited shares through reimbursement to other heirs.

2. Laying favorable institutional conditions for the performance of land ownership titles' market:

   - Making amendments to land legislation that provides for the differentiation of rental rates depending on the types of agricultural land, since current rental rates for all types of land (arable land, pastures, hayfields) are equal, and productivity and profitability indicators are different;
   - Establishing specialized logistics centers based on public-private partnerships, which will allow manufacturers to sell products at more competitive prices and increase their motivation to the expansion of agricultural production;
   - Introducing requirements to bidders in lease state and municipal agricultural lands to reduce opportunism risks. It is proposed to introduce the criteria such as a developed business plan for the use of the land plot and organization of agricultural production, the practice of agricultural production and compulsory residence in this area for a land lease term;

- It is necessary to fix the expenses related to the purchase of agricultural land (land shares) from owners to exercise the preemptive right under applicable law when making up regional and local budgets. The return on the lease of municipal land is accumulated in local budgets and can be used for socio-economic development of rural areas;
- It is necessary to supplement the land legislation and provide for reimbursement of landmarking costs to the bidder who was the first to rent this land plot in the event of a failed bid at the expense of budget funds, or the winning bidder.

3. Upgrading land accounting and performing land management and cadastral works:

   - The Federal Agency for Land Management and Cartography endowed with the design of topographic and agricultural maps, the arrangement and provision of land management, accounting of land and surveillance of their use in Russia should be established to reinforce control over the primary accounting of agricultural and other lands and obtain data in the land cadastral documentation;
   - The data of the state data fund received in the course of privatization is to be converted into electronic form and entered in the unified state real estate register to save and digitally process the primary documentation on the privatization of agricultural land;
   - Land management maps of agricultural lands and land categories, district land management charts, maps (plans) of land management objects and other land management documentation for municipal entities of the first and second level are to be drawn up.

Putting into practice the above measures in terms of directions can help protect ownership titles to agricultural lands; reduce transaction costs, prevent the opportunistic behavior, speculative transactions, monopolization; lay more favorable conditions for the efficient use of land resources, the development of competitive and contractual relations in the agrarian sector of the Russian economy.

Institutionalization of the market will make it possible to expand the employment sector and attract young specialists to the countryside that will ensure in turn competitive and sustainable development of the agrarian economy.

Acknowledgements The study was conducted under the support of the Russian Foundation for Basic Research (RFBR) and Administration of Volgograd Region in the framework of a scientific project No. 18-410-342001, p_мк "Monitoring the performance of local agricultural land markets".

References

1. Alchian AA, Demsetz H (1973) The property titles' paradigm. J Econ Hist 33:16–27
2. Coase RG (2000) The nature of the firm. In: Halperin VM (ed) Milestones in economic thought. Theory of the firm in 3, vol V. School of Economics, St. Petersburg
3. North D (2004) Performance of the economy in time. Otechestvennye Zapiski 6(21)
4. Pejovich S (1996) The market for institutions versus the strong hand of the state: the case of eastern Europe. Dallago B, Mittone L (ed) Economic Institutions, Markets, and Competition. Edward Elgar, Cheltenham, pp 111–126
5. Furubotn EG, Pejovich S (1972) Property titles and economic theory: a survey of recent literature. J Econ Lit 10(4):1137–1162
6. Kapelyushnikov RI (1990) Economic theory of property titles (methodology, basic concepts, scope of problems). Kuznetsova VN (ed). Institute of Economics of USSR Academy of Sciences, Moscow
7. Tambovtsev VL (2015) Titles, forms, and regimes of property. Econ Sci Modern Russia 1(68):7–18
8. Anoshina YuF (2011) Economic disposal of land property in the agricultural sector. STU, Oryol
9. Inshakov OV, Kharebava RP (2001) Land reform: basic principles and execution mechanisms. Publishing House of Volgograd State University, Volgograd
10. Inshakov OV (2003) "The core of development" in the context of the new theory of production factors. Econ Sci Modern Russia 1:11–25
11. Kresnikova NI (2003) Mechanisms of the land market performance. Land Bull 3:10–15
12. Makarov AN (2007) Disposal of land property: systemic and institutional-evolutionary approaches. Publishing House of Kazan State University, Kazan
13. Fedyunina E, Oganesyan L, Vorob'yov A (2019) Institutional aspects of the development of local markets for agricultural land titles. In: The second international scientific conference on 'competitive, sustainable and safe development of the regional economy' (CSSDRE 2019). Economics, business and management research, vol 83. Atlantis Press, pp 402–406. https://doi.org/10.2991/cssdre-19-19
14. Oganesyan LO, Fedyunina EN (2019) Monitoring the performance of local agricultural-land markets: research methods and results international. In: Proceedings of the international scientific conference "Far East Con" (ISCFEC 2018). Advances in economics, business and management research, vol 47. Atlantis Press, pp 1244–1249. https://doi.org/10.2991/iscfec-18-18
15. Zakharova JYu et al (2017) State (national) report on the land state and utilization in the Russian Federation in 2017, Abramchenko VV (ed). https://rosreestr.ru

Chapter 21
Methodology for Assessing the State and Development of Russian Regional Markets for Additive Products

Roman Yu. Skokov

Abstract *Purpose*: To develop a methodology for assessing the state and development of regional markets for additive products in order to obtain timely and quality information for decision-making on the management of the situation in the field of their consumption and circulation. *Design/methodology/approach*: Interdisciplinary synthesis of economic knowledge with economic sociology, psychology, addictology and drug addiction played an important role in the study. This highlighted the need for an institutional approach. Methods of expert assessment, analysis, synthesis, analogies, comparisons, monographic, indicative, statistical were used. *Findings*: For Russia, with its variety of climatic-geographical, national-religious, social and economic features of the regions of government and society in the markets of addictive goods at the regional and local levels is an actual field of economic research. The competitive model of functioning of the main Russian markets of additive goods does not allow to focus on social tasks in state regulation and makes it difficult to apply the experience of the relevant Nordic model of local policy. Russia is building a regulatory model based on economic interests. In countries close to the consumption stereotype, the state monopoly model with the leading role of public health and social policy authorities effectively functions at the national level. Projects are being implemented at the local level to bring together government authorities, businesses, young people, parents and the media. And these efforts focus on regulating the availability of additive products to consumers. It is recommended to build a model of local regulation in Russia similar to the Finnish PAKKA. *Originality/value*: In order to assess the state and development of the alcohol situation in the regions, a methodology has been developed that includes social and economic blocks of the system of indicators, the information base for which is the data of UISIS, The Federal Service for Alcohol Market Regulation, legal systems, VTsIOM and experts.

Keywords Additive products · Alcoholic beverages · Digital addiction · Government regulation · Regional markets · Local politics · Methodology · Evaluation

---

R. Yu. Skokov (✉)
Volgograd State Agrarian University, Volgograd, Russia
e-mail: rskokov@mail.ru

T. Kolmykova and E. V. Kharchenko (eds.), *Digital Future Economic Growth, Social Adaptation, and Technological Perspectives*, Lecture Notes in Networks and Systems 111, https://doi.org/10.1007/978-3-030-39797-5_21

JEL Code H75 · L83 · R13 · E14

21.1 Introduction

According to the research, 45.9 million Adults consume alcoholic beverages (VTsIOM), 37.6 million people—Tobacco products (Ministry of Health), about 29 million consume energy-tonics (Romir), 7.3 million use drugs (FSKN) [1]. The functioning of these markets has a far-reaching impact on the state, society, business and consumers. On the one hand, in 2016 the Russian budget received 466.9 billion rubles excise taxes on tobacco, 164.2 billion rubles excise taxes on strong alcoholic beverages, and 148.3 billion rubles excise taxes on beer. On the other hand, the high mortality from the causes of etiology related to alcohol—56,300 people (Rosstat), drugs—90,000 people (FSKN) and smoking—330,000 to 400,000 people (Experts). The number of crimes committed by persons under the influence of alcohol was 440,200 in 2016 (Rosstat). The share of intoxicated crimes in the total number of crimes increased from 8.7% in 2007–2011 to 17.8% in 2012–2016. The level of use of legal production facilities, for example, for alcoholic beverages was less than 30%, while the share of the de-formalization was 70%. The share of illegal cigarettes in 2018 reached 8.4% of the total Russian tobacco market against 4.5% in 2017.

The main markets for additive goods are alcoholic beverages, tobacco products, electronic cigarettes, energy drinks, drugs, gambling. The information revolution has led to a wide range of potentially addictive goods: Mobile phones, video games and digital games, social networks, etc. In 2018, gaming disorder is included by the World Health Organization in the 11th Revision of the International Classification of Diseases.

Information dependencies can also lead to physiological disturbances, negative social and economic effects that have yet to be assessed and which are increasingly needed to be regulated in society. Now society and the state are looking for tools to regulate information dependencies (digital or information drug markets).

In 2019, Italian government ministers have drafted a law aimed at sending smartphone-dependent teenagers to rehabilitation centers at public expense. This pathology is called "nomophobia" (so-called "no-mobile-phone phobia"—an anxiety over not having access to social networks or messaging apps). Courses for teachers will be organized in Italian schools for the prevention and elimination of children's telephone addiction. An information campaign is also being prepared for parents, who often underestimate the harm caused to their children by the uncontrolled use of smartphones.

Australia, Belgium, Canada, France, Malaysia, Nigeria, France, Uzbekistan, Uganda and the United Kingdom have developed recommendations on the use of mobile devices by children, including in educational institutions [2]. France has adopted a law banning all types of mobile phones, tablets and smart watches in schools. In Belgium and the UK, the ban on the use of mobile phones is decided on a school-by-school basis. Since 2019, mobile phones have been banned in schools of Ontario (Canada) and New South Wales (Australia). In Germany, according to sociological surveys, 86% of respondents were in favour of such a ban. There are various restrictions in the cantons of Switzerland.

The public and authorities are actively discussing the ban on mobile phones in Russian schools. Regional and municipal education authorities are encouraged to consider restricting the use of mobile devices in schools to students, teachers and parents [2].

Until science has developed special tools to regulate information addictions [3, 4], it is necessary to borrow them from the experience of alcohol, tobacco, drug and gaming addiction regulation.

Assessing the actions of public authorities and society in the markets for additive goods at the regional and local levels is a relatively new but growing field of economic research. Restrictions imposed by the national authorities of the Russian Federation prevent the adoption of independent measures by the authorities of the constituent entities of the Russian Federation and local authorities. The WHO Committee of Experts on problems relating to the use of alcohol recommends that strategies (e.g. anti-alcohol policies) and options be adapted and implemented at the national and subnational levels, taking into account the specific cultural and legal contexts as well as the local characteristics of alcohol problems [5]. This approach is relevant for Russia because of the diversity of climatic-geographical, national-religious, social-economic peculiarities in the regions. Monitoring of the level of consumption of additive goods and its consequences in the subjects should provide timely and high-quality information, to make prompt decisions on the management of the situation in the field of their turnover and to assess their effectiveness.

21.2 Materials and Method

Theoretical-methodological basis of the given research are works of the Russian and foreign scientists in the field of the theory of the specific markets of additive goods, consumer behavior, state regulation of their social and economic development, estimation of efficiency of public administration.

Interdisciplinary synthesis of economic knowledge with economic sociology, psychology, addictology and drug addiction played an important role in the study. This highlighted the need for an institutional approach.

Methods of expert assessment, analysis, synthesis, analogies, comparisons, monographic, indicative, historical-logical and statistical were used.

21.3 Results

The priority task of the state policy to reduce the consumption of addictive goods and their negative consequences in the subjects is the creation of an appropriate monitoring system. There is a concept of state policy, adopted in 2009, aimed at reducing the extent of alcohol abuse and alcohol prevention among the population of the Russian Federation for the period up to 2020. Among its priorities is the establishment of a system for monitoring and analysing the dynamics of indicators of alcohol abuse [6]. However, over the past ten years, such a system has not actually been created in Russia.

The system of monitoring the drug situation in Russia operates at the Federal [7] and regional levels [8, 9]. Institute of economics, the Ural branch of Russian Academy of Sciences has developed a theoretical and methodological approach to the creation of a system for monitoring the drug situation in Russia and its regions [10].

There are no systems in place to monitor consumption of tobacco products, energy beverages and gambling in Russia.

The approaches to assessing the social and economic consequences of alcohol consumption in Russia, as a rule, are epidemiological in nature (A. Soloviev, P. Krasnova, S. Vinnik, A. Avagyan, O. Beketova, D. Halturina and A. Korotayev, A. Podlazov, S. Aksyutina and N. Ovsyankina).

There are not many methods of integral comparative assessment of the alcohol situation in the constituent entities of the Russian Federation. For example, they are presented in the study of B. E. Gornyi and A. M. Kalinina [11] within the framework of the Federal Project "Sober Russia" and the All-Russia State Television and Radio Broadcasting Company [12].

IA "Rating" made a rating of regions of the Russian Federation on the absence of harmful habits, which is based on six indicators characterizing the prevalence of alcoholism, drug addiction and smoking among the population. These include alcohol and tobacco consumption, the number of drunken and drug-induced crimes, alcohol poisoning deaths and drug-related offences [13].

Russian Public Opinion Research Center (VTsIOM), as part of a study on healthy lifestyles, assesses the population's exposure to harmful habits of smoking and drinking. Among adults, the share and frequency of alcohol and cigarette consumption is estimated [14], as well as a passion for gambling for money [15]. The information obtained cannot be used for regional assessments, as it's based on all-Russian data. It is advisable to extend these studies to constituent entities of the Russian Federation and municipal entities.

Structural changes in the alcohol policy of the European Union, initiated in 1990 [16–19] have led to changes in state regulation in the Nordic countries. International free trade agreements and the unification of excise taxes on alcoholic beverages have reduced the possibilities of traditional measures at the national level and led to a shift of alcohol policy to the municipal level. A study by Holmila and Warpenius on the effectiveness of alcohol policy recommends the Finnish experience for local communities—The PAKKA Project [20]. In 2004, PAKKA focused on local alcohol markets. The project has now moved beyond it to represent a model of local alcohol, tobacco and gambling policies [21, 22]. Emphasis is placed on the social and commercial accessibility of alcohol, tobacco and slot machines to intoxicated consumers and minors. The model for harm reduction in society brings together key actors—public authorities, economic operators, young people, parents, and the media. The Finnish PAKKA project is led by the Ministry of Health and Social Policy.

Regular quantitative information to assess the achievement of goals and the success of the PAKKA operating model is derived from: trial shopping; restaurant reviews; school health checkups for alcohol, tobacco and gambling; police statistics (e.g., night-time violence statistics in public places); statistics of control (e.g., revoked licenses due to violations). Different indicators are used for schools by types of additive goods. Such as tobacco (daily smoking, retail sale of cigarettes to minors, retail purchase of electronic cigarettes by minors, purchase of electronic cigarettes by minors through mediation, etc.), snuff (daily consumption, purchase of snuff through minors, etc.), gambling (monthly games for money, etc.), alcoholic beverages (drinking alcohol on a weekly basis, drinking at least once a month, retail purchase of alcohol by minors, alcohol by minors, etc.), alcoholic beverages (drinking alcohol on a weekly basis, drinking at least once a month, retail purchases of alcohol by minors, alcohol is acquired through the mediation of minors, alcohol is obtained by minors from the family circle, etc.). The Ministry of Health of the Russian Federation needs to make such Nordic practice systematic in the constituent entities and municipalities of Russia.

The Nordic model of local policy on addictive goods and an assessment of the effectiveness of its implementation allow to focus on the social dimension—health. Economic issues are brought up to the level of state monopolies, which are preserved in the alcohol retail sector (Finland, Iceland, Norway, Sweden) [17]. Their practice is based on the leading role in the development of policies in the field of addictive goods by public health and social protection authorities. In Norway, the Ministry of Health and Social Affairs acts on behalf of the State monopoly (Vinmonopolet) [23]. Vinmonopolet is obliged to pay an annual fee to the state budget, which is used to carry out preventive measures in the field of health care and social sphere.

In wine countries (Greece, France, Portugal, Italy, Spain, etc.), where the southern European consumer culture is formed, the economic problems of the alcohol industry is an important agricultural and political issue. As a rule, there is a market competitive model with some licensing restrictions. Alcohol policy is aimed at the development of small and medium-sized grape farmers, protection of local producers, improvement of product quality and promotion of products in foreign markets.

Gambling monopoly and gambling can only be conducted in Finland by Veikkaus Oy [21, 22]. In Finland, the Parliament and the Government make key decisions regarding gambling policy. The gambling prevention unit is owned by the Ministry of Social Affairs and Health, which has been entrusted to the National Institute of Health and Social Welfare (THL). The Ministry of Social Affairs and Health is responsible for managing the activities of Veikkaus Oy. Municipalities have a responsibility to promote municipal health and to provide services related to prevention and reduction of gambling harm.

In Russia, the Ministry of Health has in fact been left to participate in the development of healthy lifestyle programmes with an extremely limited budget, which is not supported by economic and administrative levers. Markets for alcoholic beverages, tobacco products and energy beverages are competitive. The main regulators of issues in this area are economic and agrarian agencies and committees. The relevant agencies and law enforcement agencies (e.g. Federal Service for Alcohol Market Regulation, Drug control) are mainly focused on countering the shadow supply of addictive goods.

Domestic methods of assessment of the state and development of regional markets of additive goods lack indicators of availability of additive goods, which are the most important in Nordic practice. Along with the indicator of population per retail outlet characterizing physical accessibility, it is proposed to observe the economic accessibility by means of monitoring. Namely, by correlating the consumer price of vodka with the price of beef and the ability of the population with different income levels to buy legal products. However, it should be taken into account that the availability of legal products is important so that the population does not go into the de-formalization. At the same time, the level of accessibility should not lead to an increase in the consumption of additive goods. In this case, it is necessary to search for the optimum In general, consumer spending by households on additive goods should be reduced. The quality of products in the constituent entities of the Russian Federation is also the most important indicator to be included in the monitoring. VTsIOM surveys of consumers of additive goods should be conducted regularly by region.

Such indicators of assessment of the state and development of markets as effective excise tax rate, level of the de-formalization, profitability of expenses of organizations, share of profitable/loss-making organizations, level of use of production capacities, level of remuneration of employees of the branch are important from the point of view of the economy of the state and business. However, they are secondary in relation to social indicators, which are presented in the assessment methods discussed above. This approach prevails in scientific research in sociology, psychology, and medicine [11, 24–26]. In Nordic and other countries close to northern consumption, the state monopoly model of the addictive markets allows to focus society on social objectives and to leave the economic ones within the monopoly management system, on behalf of which the state health and social protection agencies act. In domestic practice, the economic effects of addictive markets remain an important part of public and private budgets, so economic indicators should be evaluated. In order to maintain the priority of social indicators, it is proposed to divide the indicators into social and economic blocks.

The economic block includes the PAKKA model for suspended and revoked licenses. From the point of view of the economic interests of the state, it is important to reconcile effective and efficient excise tax rates. For business, profitability of costs, the share of profitable organizations, remuneration of industry workers are important.

In order to assess the state and development of the alcohol situation in the Russian regions, a system of indicators has been built, broken down into social and economic blocks—Table 21.1.

Table 21.1 Indicators of rating assessment of alcohol situation in Russian regions

Name of blocks and indicators	Symbols	Data source	Methodists
Social	S		
Sale of AP* per capita for adults per year, litre of absolute alcohol	S_1	UISIS ***	"Sober Russia", IA Rating
Sale of strong AP per capita for adults per year, litre of absolute alcohol	S_2	UISIS	B. E. Gornyi and A. M. Kalinina
The number of deaths from accidental AP poisonings per 100,000 populace, 1,000 people	S_3	UISIS	"Sober Russia", B. E. Gornyi and A. M. Kalinina, IA Rating
The number of crimes committed by persons under the influence of alcohol per 100,000 populace, units per year	S_4	UISIS	"Sober Russia", IA Rating
Availability of legal AP to the population of the Russian Federation in terms of average per capita income, %	S_5	UISIS	R. Skokov
Ratio of consumer price of vodka with 40% alcohol content and higher to beef price (1 l/1 kg)	S_6	UISIS	R. Skokov
Population per one AP retail outlet, 1,000 people/1ro	S_7	UISIS, Reg **	PAKKA
Appropriate quality of domestic and imported AP from the number of samples taken is established, %	S_8	UISIS	R. Skokov
Consumer expenditures of households on AP, % of total consumer spending	S_9	UISIS	R. Skokov
Number of patients diagnosed with alcoholism and alcoholic psychosis for the first time, per 100,000 populace	S_{10}	UISIS	B. E. Gornyi and A. M. Kalinina

(continued)

Table 21.1 (continued)

Name of blocks and indicators	Symbols	Data source	Methodists
Number of patients diagnosed with alcoholism and alcoholic psychosis registered in medical institutions, per 100,000 populace	S_{11}	UISIS	"Sober Russia"
Proportion of adult population not using AP, %	S_{12}	VTsIOM	R. Skokov
Ban on sale of AP within 24 h, hours	S_{13}	Consultant, Garant	"Sober Russia"
Ban on sale of AP during the year, days	S_{14}	Consultant, Garant	"Sober Russia"
Economic	E		
Detected violations as a result of inspections to detect the illegal production and trafficking of ethyl alcohol and AP per 100,000 populace	E_1	UISIS	"Sober Russia"
Ratio of the number of suspended licenses to the total number of valid licenses, %	E_2	Reg	PAKKA, R. Skokov
Ratio of the number of revoked licenses to the total number of valid licenses, %	E_3	Reg	PAKKA, R. Skokov
Ratio of effective excise tax rate to valid excise tax rate, %	E_4	Experts	R. Skokov
De-formalization level, %	E_5	Experts	S. V. Aksyutina, N. M. Ovsyankina
Profitability of organizations' costs, %	E_6	UISIS	R. Skokov
Share of profitable/loss-making organizations, %	E_7	UISIS	R. Skokov
Share of use of production capacities, %	E_8	UISIS	S. V. Aksyutina, N. M. Ovsyankina
Level of remuneration of the industry's employees, rubles	E_9	UISIS	R. Skokov

Note *AP—alcoholic products; **Reg—Federal Service for Alcohol Market Regulation; *** UISIS—The Unified Interdepartmental Statistical Information System (UISIS)

The social and economic blocks should be assessed separately. The final score on blocks for each constituent entity of the Russian Federation is calculated by summing up the places occupied by this region for each of the indicators. Regions are ranked from the best (1st place) to the most problematic (85th place). The lower the score, the better the region is and the higher its position.

To test the methodology requires disclosure of information on the subjects of the Russian Federation from Rosstat, the Federal Service for Alcohol Market Regulation and VTsIOM, which is required to calculate the indicators of S_5, S_6, S_7, S_8, S_{12}, E_8, E_{16}, E_{17}. For indicators E4, E5 it is necessary to define the responsible agency for monitoring [1].

21.4 Conclusion

Predominantly competitive model of functioning of the main Russian markets of additive goods does not allow to concentrate in regional regulation and assessment of its efficiency on the social component and imposes difficulties in applying the experience of adequate Nordic model of local policy. Russia continues to build a regulatory model based on economic interests. In the meantime, the state and society have significant negative effects. In countries close to the consumption stereotype, the state monopoly model with the leading role of public health and social policy authorities effectively functions at the national level. Models that bring together public authorities, economic operators, young people, parents and the media are implemented at the local level. Efforts are focused on regulating the social and commercial availability of addictive goods to intoxicated consumers and minors. It is recommended to build a model of regulation similar to Finnish PAKKA in Russia. The approaches to assessing the social and economic consequences of alcohol consumption in Russia, as a rule, are epidemiological in nature. In order to assess the state and development of the alcohol situation in the regions, a methodology has been developed that includes social and economic blocks of the system of indicators, the information base for which is the data of UISIS, The Federal Service for Alcohol Market Regulation, legal systems Consultant/Guarant, VTsIOM and experts. The methodology requires disclosure of information on the constituent entities of the Russian Federation from the responsible authorities.

Acknowledgements The reported study was funded by RFBR and the Volgograd Region Administration according to the research project no. 19-410-340006, r_a.

References

1. Skokov RY (2019) Model for institutional regulation of addictive markets/Model mechanizma instituzionalnogo regulirovanyia rinkov addictivnyh tovarov. Rus J Econ Theory 2:279–292

2. Methodological recommendations on the use of mobile communication devices in general education institutions (approved by the Federal Service for Supervision of Consumer Rights Protection and Human Welfare and the Federal Service for Supervision of Education and Science on August 14, 2019 No MP 2.4.0150-19/01-230/13-01) (2019)/Metodicheskie rekomendacii ob ispolsovanii ustroistv mobilnoi svyasi v obsheobrazovatelnih organizaciyah (utverjdeno Federalnoyi slujboi po nadzoru v sphere zachity prav potrebitelei I blagopoluchyia cheloveka I Federalnoi po nadzoru v sphere obrazovanyia I nauki ot 14 avgusta 2019g. No MP 2.4.0150-19/01-230/13-01) (2019). http://www.consultant.ru/document/cons_doc_LAW_331786/. Accessed 30 Aug 2019
3. Skokov RY. Economic theory of addictive markets/Ekonomicheskaya teoriya rinkov addictivnyh tovarov. Volgograd State University, Volgograd
4. Örnberg J, Ólafsdóttir H (2009) How to sell alcohol? Nordic alcohol monopolies in a changing epoch. Nordic Stud Alcohol Drugs 25:129–153
5. WHO expert committee on problems related to alcohol consumption. Second report (2006), Geneva from 10–13 October 2006 (2006), Женева, 10–13 октября 2006 г., P.53
6. Order of the Government of the Russian Federation of 30 December 2009, No. 2128-p "On the concept for the implementation of state policy to reduce the scale of alcohol abuse and prevent alcoholism among the population of the Russian Federation until 2020" (2009)/Rasporyashenie Pravitelstva RF ot 30.12.2009 No 2128-p "O Konzepzii realisazii gosudarstvennoi politiki po snisheniyu masshtabov zlouputrebleniyia alkogolnoi produkziei I profilactike alkogolisma sredi naselenia rossiiskoi Federazii na period do 2020 goda" (2009). http://www.consultant.ru/document/cons_doc_LAW_96236/b88ce90c1402afee8a43afd0f24d507e67cc2a83/. Accessed 30 Aug 2019
7. Resolution of the Government of the Russian Federation of 20.06.2011 No. 485 "On approval of the Regulation on the state system of monitoring of the drug situation in the Russian Federation" (2011)/Postanovlenie Pravitelstva RF ot 20.06.2011 N 485 "Ob utverjdenii Polojenyia o gosudarstvennoi sisteme monitoringa narkosituazii v Rossiiskoi Federazii (2011). http://www.consultant.ru/document/cons_doc_LAW_115450/b6b640b95b329c237f7326862d9e82be99357b42/. Accessed 30 Aug 2019
8. Methodology and procedure for monitoring, as well as criteria for assessing the development of the drug situation in the Russian Federation and its constituent entities (third revision)/Metodika i poryadok osushestvlenyia monitoringa, a takje kriterii ozenki razvitya narkoindustrii v Rossiiskoi Federazii i eie subektah (tretyi peresmotr) (approved by the minutes of the SAC (State Anti-Drug Committee) meeting dated 15.02.2017 No. 32) (ред. от 11.12.2017) (2017). https://base.garant.ru/70809092/. Accessed 30 Aug 2019
9. Regulations on monitoring and evaluation of the development of the drug situation in the Volgograd region (2019)/Polojenyie ob osushestvlenii monitoringa I ozenke razvitiyia narkosituazii na territorii Volgogradskoi oblasti (2019). http://www.volgograd.ru/gubernator/koord/antinark/monitoring-polojenie.php. Accessed 30 Aug 2019
10. Tatarkin AI, Kuklin AA, Klevakin AN (2014) Practical role of science in conditions of modern evolution of social-economics relations in the sphere of drug trafficking. Drug Control 3(36):3–9
11. Gornyi BE, Kalinina AM (2016) Integral estimation of the alcohol situation at the regional level. Rus J Prevent Med Pub Health/"Profilakticheskaya meditsina" 19(3):34–40
12. REGIONS sobriety rating—2018 (2018)/Reiting trezvosti REGIONOV—2018 (2018). http://www.trezvros.ru/calendar/702. Accessed 30 Aug 2019
13. Which regions have less bad habits (2018)/V kakih regionah menshe vsego vrednyh privichek (2018). http://riarating.ru/infografika/20181211/630113813.html. Accessed 28 Aug 2019
14. A healthy lifestyle: monitoring (2019)/Zdorovyi obraz jisny: monitoring (2019). https://wciom.ru/index.php?id=236&uid=9713. Accessed 30 May 2019
15. Gambling zones in Russia: pro and con (2014)/Igornyie zony v Rossii: za i protiv (2014). https://wciom.ru/index.php?id=236&uid=9713. Accessed 16 July 2014
16. Babor T et al (2003) Alcohol: no ordinary commodity. Research and Public Policy, Oxford Medical Publications, Oxford University Press

17. Österberg E(2002) Alkoholipolitiikan murros 1990-luvulla. In: Heikkilä M, Kautto M (ed) Suomalaisten hyvinvointi, Helsinki, STAKES, Gummerus Kirjapaino Oy, pp. 230–249
18. Treno A, Holder H (2004) Prevention at the local level. In: Heather N, Stockwell T (eds) The essential handbook of treatment and prevention of alcohol problems. Wiley, Chichester, West Sussex
19. Wallin E, Norström T, Andréasson S (2003) Alcohol prevention targeting licensed premises: a study of effects on violence. J Stud Alcohol 64:270–277
20. Holmila M, Warpenius K (2012) Community-based prevention of alcohol-related injuries: possibilities and experiences. Int J Alcohol Drug Res 1(1):27–39
21. Terveyden Ja Hyvinvoinnin Laitos (2019). http://www.thl.fi/fi/web/alkoholi-tupakka-ja-riippuvuudet/ehkaisevapaihdetyo/ehkaisevan-paihdetyon-menetelmat/verkko-pakka-ehkaisevaanpaihdetyohon. Accessed 30 августа 2019
22. Terveyden Ja Hyvinvoinnin Laitos (2019). https://thl.fi/fi/web/alkoholi-tupakka-ja-riippuvuudet/rahapelit. Accessed 30 августа 2019
23. Normative acts of foreign states, (2017)/Normativnyie akty inostrannyh gosudarstv (2017). http://www.ved.gov.ru/rus_export/partners_search/torg_exp/?action=showproduct&id=4195. Accessed 6 May 2017
24. Artyukhov IP, Gorny BE, Mazharov VF (2011) Regional, age and gender specific features of mortality from external causes of working age population of the Krasnoyarsk region (2005–2009). Far East Med J 2:99–101
25. Kladov SY, Konobeyevskaya IN, Karpov RS (2010) Alcohol and premature death rates of the population of Tomsk Region 9(1):126–129
26. Sakharov A (2012) Organizational model of alcohol situation monitoring in a separate subject of the Russian Federation/Organizazionnayia model monitoringa alkogolnoi situazii v otdelnom subekte Rossiiskoi Federazii. Sib Her Psychiatry Addict Psychiatry 1(70):80–82

Chapter 22
Analysis of Clustering Processes in the Regions' Innovative Sustainable Development Formation

Irina B. Fomina, Tatiana V. Kushnarenko, Alexey N. Tabakov and Vadim V. Jakovenko

Abstract

Purpose: The purpose of this work is to study the state of clustering processes and to analyze their impact on the dynamics of innovative development in the region in terms of increasing its stability under conditions of development nonlinearity. *Design/methodology/approach*: The paper analyzes the organizational structure of the economy, its diversity and complexity influence the regional development sustainability. The relevance of the problem is due to the fact that in the conditions of development and transition to the digital economy, the regional enterprises' involvement in the complex cluster structures' activities increases. The paper is based on evolutionary-synergetic methods used for sustainable development factors' analysis; the authors analyze the system complexity and economic entities' interaction organization coordination on the conditions of maintaining the equilibrium and transition to a state of criticality and crisis. The history of regional innovation clusters is briefly described. It is emphasized that innovation clusters, which unite large, medium and small producers, provide conditions for modernization and moving away from the raw material economy to the progressive development of the regions and the economic system of the country as a whole. *Findings*: The reasons are revealed, on the one hand, determining the need for clustering in a number of industries and spheres of economic activity, the activation of these processes in the last decade in the Russian regions, and, on the other hand, slowing down the processes of clustering, which negatively affects the innovative activity of economic entities and prevents the creation of

I. B. Fomina (✉) · A. N. Tabakov · V. V. Jakovenko
Russian Presidential Academy of National Economy and Public Administration, Moscow, Russia
e-mail: fominamail@mail.ru

A. N. Tabakov
e-mail: ale-tabakov@yandex.ru

V. V. Jakovenko
e-mail: vino10@mail.ru

T. V. Kushnarenko
Don State Technical University, Rostov-on-Don, Russia
e-mail: tvkushnarenko@mail.ru

T. Kolmykova and E. V. Kharchenko (eds.), *Digital Future Economic Growth, Social Adaptation, and Technological Perspectives*, Lecture Notes in Networks and Systems 111, https://doi.org/10.1007/978-3-030-39797-5_22

a platform for sustainable development formation in the regions. *Originality/value*: The conclusions drawn in the work can be used in the development of cluster policy in the framework of determining the directions of industrial and investment policy of the regions.

Keywords Economic system stability · Systems theory · Cluster approach · Innovation cluster · Innovative cluster ecosystems · System entropy · Synergetic effect · Self-organization processes

JEL-Code F63 · O31 · D59 · D89

22.1 Introduction

The existence of an economic system is possible provided that it has mechanisms that ensure sustainability. The main regularities of long-term economic development include: cyclical trajectory, expressed in the alternation of waves, the presence of gaps between the cycle phases, in the changes in the growth factors' structure and technological structure. The nature of the economic system development in this case largely depends on the structure of relations, mechanisms of coordination of economic entities' interaction, its institutions. This explains the existence of different national models of well-being, reproduction mechanisms for the parameters of population life level and quality, different levels of state regulation of economic and social spheres, distribution of the national product and financial resources.

In terms of development, the dynamism and complexity of economic processes increases. The economy exhibits the properties of developing systems of the highest complexity. The assumption used in scientific research that the evolutionary development of economic systems represents a transition from one equilibrium state to another should be recognized as erroneous.

Increasing the adequacy of the description of the processes occurring in the economic system is possible with the use of approaches and methodological research tools based on the evolutionary-synergetic paradigm. In accordance with the concepts of the theory of systems, a system is understood as a union of elements, as a result of which there is a new quality inherent in this union. The manifestation of a new property means that new ordered structural relations have arisen.

22.2 Materials and Method

The analysis is based on the evolutionary theory of socio-economic systems' development at the meso-level. The study of clustering processes was carried out on the basis of the methodology of complex socio-economic systems' self-organization considered through the prism of economic synergetics.

22.3 Results

The elements' functioning o ensures the preservation of the system, provided that the elements take into account and ensure the achievement of the system purpose. Ensuring the stability of the system requires compliance with a number of conditions: protection from external negative impacts, the ability to restore the development potential, preservation of elements and improvement of system connections.

According to the entropy approach, a system can be between two opposite states. According to the law of entropy change, going beyond the formed entropy interval is impossible without significant structural transformations of the economic system [1].

Mechanisms to ensure the stability of the system are associated with the processes of its self-organization. There are three types of mechanisms: passive, active and combined. Accordingly, the first are implemented in the form of absorption or damping of disturbances, the second are focused on the formation of new structures, connections, functions, ways to eliminate the negative impact of disturbances on the system, the third combine elements of the previous two types.

Economic development is linked to the improvement of technology, the rapid growth of information and the quality of knowledge. In the economic system, there is a system of hierarchies, different levels of which include complex elements (transnational corporations, holdings, integrated corporate structures), which have a large number of interrelated and interdependent economic entities.

Diversity reflects the level of the system complexity, heterogeneity of elements, differences in the forms of relations and connections. Under the conditions of environmental parameters' constancy, the decrease in diversity increases the stability of the system. The processes of self-organization in complex systems are associated with the phenomenon of self-organized criticality, which manifests itself in the fact that in the process of development, even in the absence of external influences equilibrium in the system is periodically disturbed [2].

The presence of equilibrium and stability indicators can serve as signs of stagnation period, followed by a phase of transition of the system into an unstable state, when even minor impacts can initiate the crisis development and the associated significant changes in the structure and intra-system relations. The unstable state of the system (bifurcation point) is characterized by the presence of many different directions of development, while the choice of one or another option has a probabilistic nature [3].

In conditions of instability, the transition to a new stage of sustainable development requires the timely formation of new effective institutions, the development of a set of checks and balances to smooth out existing contradictions and create new forms of economic actors' coordination. New forms of economic relations provide convergence of goals and increase the efficiency of economic actors and the system as a whole as a result of the synergistic effect associated with reducing the costs of scaling up production, innovation, use of partners' experience and know-how.

The evolutionary development of the economy and institutions is one of the conditions for sustainable and crisis-free development. Regional innovation cluster structures, in modern terminology—"cluster ecosystems", generating innovations, ensure the evolution of economic relations. This makes it possible to determine the probability of economic cycles' various phases occurrence and to form mechanisms for sustainable economic development. At the same time, the development of digital information technologies creates favorable conditions for improving the network structure on the terms of mutually beneficial partnership for all cluster members.

The cluster nature of economic relations correlates with the solution of the problem of the regional economic system stability [4]. The emergence and further development of clusters in the Russian economy can be considered a natural phenomenon. In the USSR, in the early 1960s, a General scheme of productive forces distribution was developed for the first time in the world practice. Here, the prospects for the country development within 15 years were presented [5].

As part of the General scheme, an important place was occupied by plans for territorial development in the sectoral context in the form of Schemes for the main industries' development. The development Schemes determined the technical and economic parameters, location, timing of construction and commissioning of new production facilities for each industry, taking into account the calculation of the need for the products in the relevant industry and the requirements of the optimality of their creation total cost.

The formation of clusters in Russia is associated with targeted transformations of the planned type of economic management in the second half of the 1980s. The results of the planned economy improvement were the provision of financial and economic independence to enterprises, the elimination of the vertical scheme of planning, reporting, stimulation, quality control of production processes and products inherent in the bodies of sectoral and territorial administration. As a result, not only the vertical links "Ministry—General Directorate—Association—Enterprise" were destroyed, but also the horizontal interaction between economic entities and regions, which was established in the 1960s, was lost.

In the first half of the 1990s, the destruction of all types of economic ties in Russia was brought to an absolute limit. The country's leadership lost interest in the issues of territorial and sectoral planning. At the same time, many research institutes that served as industry think tanks almost completely ceased their work. The triumph of anarchy and total barter in the economy against the background of sane financial and tax legislation absence led to the completion of destructive processes and the complete breakdown of the economic system [6].

It should be noted that during this period the idea that the market would settle everything without the state intervention was actively promoted. Macroeconomic regulators such as monetary policy, currency regulation, the stock market and antimonopoly regulation replaced the mechanisms of economic planning. These mechanisms weren't able to fully cover all economic governance issues. The consequence of this was a decrease in investment activity, almost no innovation or their implementation, rising unemployment, deteriorating quality of life, increased criminalization and large-scale growth of the shadow economy. It became clear that without the activation of the role of the state in the economy, it will be difficult to create conditions for its progressive movement.

At the very end of the 1990s and in the first decade of the twenty-first century, attempts were made to form new economic relations. This period is characterized by the development and adoption of a set of legislative acts in the tax, budget, financial, banking, insurance, labor, corporate and other spheres of activity.

However, the efforts made were not enough to improve the efficiency of the economy. The uneven development of the regions increased, a large number of depressed areas appeared with a very small number of donor regions.

A study by the HSE Institute for statistical research and economic knowledge found that the per capita GRP in ten regions was higher than the national average, with only a quarter of the country's population living there in total. Other ten outsider regions had a 40% below average GRP, with about 6% of the country's population living there [7].

Thus, the destruction of existing economic ties between individual economic actors at the micro level, between individual industries, as well as between regions at the meso level led to the imbalance of the entire economic system and became the main obstacle to economic growth.

During the last quarter of the twentieth century, and especially in the first two decades of the twenty-first century, there have been global changes in the economic system. The formation of a new organizational order takes into account the non-linearity of the innovation process. The global economy is dominated by mobile networking, internal reconfiguration, which leads to synergistic effects.

Special attention of scientists and specialists is paid to clustering processes in Russia. The researchers analyzed the current state of the processes of formation and implementation of cluster initiatives, which made it possible to draw a number of important conclusions about their nature, pace of development, impact on the level of competitiveness of regions and mega-regions. An example of this kind of research is the analysis of cluster ecosystems' formation in the Southern Federal Okrug [8].

The cluster approach can be considered as the most effective form that appeared in the process of evolutionary development, creation and implementation of innovations. Thus, in M. Russell's study [9] it is shown that the evolution of innovations' creation process passed the following stages:

- innovations carried out by manufacturers;
- end-user driven innovation;
- strategic innovations initiated by the state;
- open innovation driven by a large number of participants;
- network collaborative innovations, the implementation of which can be carried out only on the basis of collaboration, understood as the highest form of interactive cooperation of a large number of participants.

N. V. Smorodinskaya's approach, who proposes to study clusters in three dimensions: as a model of industrial agglomeration, as a class of economic projects, as a variety of ecosystems, is truly interesting [10].

The viability of innovation ecosystems in the form of innovation clusters can be maintained under a number of conditions.

1. Innovation ecosystems must have some freedom to implement market mechanisms with long-term stable legislation in order to create and maintain their viability. The immutability of legislative norms will allow the participants of the innovation cluster to work in a calm environment, without financial and organizational complications.
2. Each of the innovation ecosystem participants must at all times maintain its status as a legally independent person. This will allow each participant to ensure legal viability, to be protected from external influences emanating from management structures, or from other more powerful and large participants of the collaboration.
3. Participants of innovation ecosystems should have a simple free access to the procedure of state registration of innovations and comply with the rules of copyright. This will allow the innovations' authors freely and without loss of economic benefits to interact with other actors in the process, accelerate the development and implementation of innovations.

Currently, the scientific literature studies the problems of formation of not separate innovation clusters, usually solving problems within one or more related industries, but at the level of clusters united in cluster systems or innovation ecosystems [11].

Many researchers include a large number of participants into the concept of "innovation ecosystems", like business structures, universities, research centers, government officials, professional communities, individual information environments [12–14].

Innovative cluster ecosystems, uniting large, medium and small producers, provide conditions for modernization and moving away from the raw material economy to the progressive development of the regions and the economic system of the country as a whole.

Innovative ecosystems, for example, those in the agro-industrial complex allow to modernize, reduce costs, improve the quality and competitiveness of products, meet the needs of the population in high-quality and at the same time affordable food [15].

Cluster ecosystems are structural elements in which the processes of formation and distribution of added value, capitalization of population's income are improved in market conditions.

The development of innovation clusters allows to create a hierarchy structure that supports the optimal level of diversity, allowing, in turn, to assess the possibility of crisis phenomena and ensure sustainable development of regions and the country as a whole.

22.4 Conclusions

The evolution of the economic system is manifested in the alternation of phases of the cycle: rise, stagnation, crisis. The state of the economic system is determined by the presence of ordered and effective structural relations between its elements and levels (mega-, macro-, meso-, microeconomics). The crisis period is characterized by the bifurcation behavior of the economic system, a distinctive feature of which is the possibility of transition to different development options. The passage of the crisis (bifurcation point) should be carried out consciously and be subordinated to the implementation of the goals of the economic system's progressive sustainable development.

Ignoring the interests of any group of system participants will inevitably lead to inefficient subsequent development of the ecosystem. Thus, the predominance of the interests of the Federal center to the detriment of regions and enterprises' interests can unbalance the delicate balance in which subsequent attempts to stabilize the system will not be able to have the proper impact, and the mechanisms used will be ineffective and useless.

In this situation, one of the possible positive factors influencing the processes of stabilization of the system may be the mechanism of cluster ecosystems' formation.

The prevailing innovative component in the formation of cluster relations can become a self-organizing force capable of having a stabilizing effect on the economy both regionally and nationally on the basis of the evolution of clusters.

Against the background of periodically emerging economic crises caused by the laws of cyclical development of the world economy in accordance with the theory of economic development phases in the medium-term cycles of Juglar (10–12 years), the creation of sustainable innovative cluster ecosystems will play a stabilizing role for individual industries, regions, macro-regions and the country as a whole.

References

1. Tabakov AN (2012) Theoretical foundations of the study of sustainability and self-organization of economic systems: monograph/A. N. Tabakov—Volgograd: Publishing house of the Volgograd branch of FSBEI RANEPA, 2012, 172 p
2. Bak P (2013) How nature works theory of self-organized criticality [Text] = How nature works. The science of self-organized criticality/Per Buck; entry Art. G. G. Malinetsky; transfer from English [V. E. Podobeda, A. V. Kondratiev, O. L. Shevchuk]—Moscow: URSS: LIBROCOM, 269 p
3. Prigozhin I (2000) End of certainty. Time, chaos and new laws of nature [Text]/Prigozhin I—Izhevsk: Research Center "Regular chaotic dynamics", 208 p
4. Evstigneeva LP, Evstigneev RN (2012) Economics as a synergetic system [Text]/Evstigneeva LP, Evstigneev RN—Ed. 2nd—Moscow: URSS: [LIBROCOM], 266 p
5. General methodology for the development of a General scheme for the development and deployment of the productive forces of the USSR for 1971–1980—M.: Economics (1966). Source: https://institutiones.com/general/85-2008-06-12-12-49-44.html

6. Volynsky AI, Gareev TR, Gareev VE, et al (2018) Mesoeconomics: status and prospects [Text]: [monograph]. In: Mayevsky VI, Kirdina-Chandler SG, Deryabinoy MA (eds) Institute of Economics of the Russian Academy of Sciences—Moscow: IE RAS, 314 p
7. Zubarevich, NV (2009) Regional development and regional policy over a decade of economic growth. J New Econ Assoc **1–2**(1-2):161–174
8. Inshakov O, Inshakova E, Beloglazova S. Clustering as a way to ensure the competitiveness of Russian regions: comprehension of current state and development prospects. In: Proceedings of the international scientific conference «Competitive, sustainable and secure development of the regional economy: response to global challenges» (CSSDRE 2018). https://www.atlantis-press.com/proceedings/cssdre-18/25896384
9. Russell MG et al. Transforming innovation ecosystems through shared vision and network orchestration/Conference paper, July 2011
10. Smorodinskaya NV (2014) Territorial innovation clusters: world landmarks and Russian realities. In: XIV April international scientific conference on the problems of economic and social development. Book 3./Ans. ed. Yasin E. G.—M.: Publishing House of the Higher School of Economics, pp 389–402
11. Ketels CHM (2011) Clusters and competitiveness: Porter's contribution, Chap. 10. In: Huggins R, Izushi H (eds) Competition, competitive advantage, and clusters: the ideas of Michael Porter, pp 173–192. Oxford University Press
12. Bramwell A et al. Growing innovation ecosystems: university-industry knowledge transfer end regional economic development in Canada. University of Toronto. Final Report. May 15, 2012
13. Gloor PA (2006) Swarm creativity. Competitive advantage through collaborative innovation networks. Oxford University Press, New York
14. Mercan B, Goktas D (2011) Components of innovation ecosystems. Int Res J Financ Econ 76:102–112
15. Fomina IB, Kushnarenko TV. The clustered nature of the sustainable development of the region (using the example of the South of Russia). In: Advancer in economics, business and management research (AEBMR), vol 39. «Competitive, sustainable and secure development of the regional economy: response to global challenges» (CSSDRE 2018), May 2018, pp 764–767. https://www.atlantis-press.com/proceedings/cssdre-18/

Chapter 23
On the Peculiarities of Translation of the Names of Administrative and Economic Institutions in German and Russian in the Context of Integration and Differentiation Processes

Svetlana V. Yakovleva, Olga A. Obratneva, Elena A. Smirnova and Irina I. Khrapchenkova

Abstract The article presents a comparative analysis of the translation of the names of administrative and economic institutions in German and Russian. On the basis of the analysis, a thesaurus of Russian and German names of institutions and their correlation with a certain sphere of activity is compiled. The study allows to reveal the correspondence/discrepancy of administrative and economic institutions of Germany and Russia to their spheres of activity, the tendency of mutual language integration in the context of transformational and globalization processes in all spheres of society around the world.

Keywords Cross-cultural communication · National identity · Administrative and economic system · Theory of translation · Translation equivalence · Culture-specific vocabulary · Integration · Differentiation · Language development

JEL Classification Codes H10 · R

23.1 Introduction

Modern society is a complex system of specially organized human activity. The largest spheres of human society are economic, political, social and spiritual. Each

S. V. Yakovleva (✉) · O. A. Obratneva · E. A. Smirnova · I. I. Khrapchenkova
Pskov State University, Pskov, Russia
e-mail: atevsa@inbox.ru

O. A. Obratneva
e-mail: olga.obr@gmail.com

E. A. Smirnova
e-mail: aspirpskgu@rambler.ru

I. I. Khrapchenkova
e-mail: lehrdeutsch@gmail.com

T. Kolmykova and E. V. Kharchenko (eds.), *Digital Future Economic Growth, Social Adaptation, and Technological Perspectives*, Lecture Notes in Networks and Systems 111, https://doi.org/10.1007/978-3-030-39797-5_23

of them is a certain system with its own type of organization and management, its own laws of development and traditions. A special place among the above-mentioned spheres is taken by the economic sphere, whose purposes are production, exchange, distribution, consumption of goods and services.

Due to universal internationalization, growth of digitalization in various spheres of social life and lives of individuals, to increase of informatization of living space, expansion of international contacts in various fields of human activity, including the economic sphere, there is a need to study features of national mentalities and national linguistic cultures, processes of language differentiation and integration, as well as discussion of cross-cultural interaction questions, which require optimal translation technology with the purpose of adequate language transcoding and achieving understanding and effective solution of job tasks.

The geographical position of Pskov, its traditional, historically established cross-border cooperation, as well as close cross-cultural relations with partners in Germany stimulates the necessity of an intermediary language formation not only in the cultural but also in the economic sphere.

The aim of the study is to determine the peculiarities of translation of the names of administrative and economic institutions in Russia and Germany, to compare their functioning systems in the context of integration and differentiation processes.

Objectives: to determine the scheme of communal services in Russia, on the basis of Pskov, and in Germany, for example, on the basis of Gera and Munich; to compare the system of communal services in Pskov with the system of communal services of Gera and Munich; to compare the names of administrative and economic institutions in the German and the Russian languages.

23.2 Materials and Method

The methods used in the study are method of descriptive analysis (observation, generalization, interpretation, and classification), method of comparative analysis, semantic method, method of linguistic modeling, method of structural analysis, method of component analysis; methods of cognitive analysis are also applied.

At the moment, the topic of translation of administrative and economic vocabulary requires further research in the linguistic literature, and there is not only a need to translate the names of administrative and economic institutions, but also to study the structural features of these organizations.

When comparing the administrative and economic systems of the two countries: Russia and Germany, it is worth mentioning that similar concepts have different content, although many of the reforms carried out by Russian government to optimize the work of utilities, based on the European experience in the system of functioning of the enterprises in this sphere.

The theory of translation considers different ways of transition from units of one language to units of another to express already stated thoughts. The concept of "method" Sdobnikov V. V. defines as "a psychological operation that implements any action, or the basic rule of achieving the goal, which reflects the objectively existing laws of reality" [1]. He distinguishes the interlinear method of translation (sign or direct translation) and transformational translation (semantic or indirect translation).

The interlinear method of translation assumes that in the mind of the translator, after receiving the sign of the target language emerges a foreign language equivalent. When using the transformational method of translation denotation is identified, that means that a process of recognition of the semasiological connections of the sign of the target language takes place. After determining the denotation, the translator begins to search for a lexical unit of the translated language or a speech formula that is able to mark the identified denotation or situation.

The interlinear method of translation includes such techniques as transcription, transliteration and replication. The transformational method involves the use of lexical and grammatical transformations during the process of the translation of the sentence containing culture-specific vocabulary.

Translation of culture-specific vocabulary often requires the addition of explanatory words for the correct and accurate expression of the meaning of the statement. This technique is used due to differences in the systems of norms of the two languages and can have its own varieties: generalization, modulation or semantic development, compensation, explication or descriptive translation, omission technique, i.e. "refusal to transfer semantically redundant words whose meanings are irrelevant or easily restored in the context in translation" [2].

F. A. Arkhipov discerns the following lexico-semantic transformation [3]: adding words, phrases and sentences; the omission of lexical items; lexical deployment, i.e. the replacement of a word with a phrase; lexical convergence, i.e., replacing a phrase with a word; shift, i.e. replacement of one superordinate concept with another; transcription, i.e. a phonetic imitation of the word and transliteration, i.e., a alphabetic imitation. A. L. Burlak presents the classification that includes seven types of transformations: borrowing [4], replication, literal translation, transposition, modulation, equivalence and adaptation.

23.3 Results

In this study, a thesaurus of administrative vocabulary was formulated, their correspondences in German and Russian were analyzed, equivalents to concepts that do not have correspondences and not recorded in the dictionary, the peculiarities of translation of such vocabulary were investigated, schemes of the functioning of communal services were made.

The names of administrative and economic institutions belong to the socio-political field of translation studies, to the sphere of communal services. The system of functioning of communal services in Russia and Germany is very similar.

Communal services in Russia and Germany are engaged in providing consumers with different types of energy: electricity, gas, heat, water. Municipal services of both countries run sanitary and technical activities, which include cleaning territories, sanitary cleaning of households, drainage systems. Transport enterprises of both countries belong to administrative and economic institutions. The system of communal services in Russia and Germany includes land tenure services, buildings and networks, repair and maintenance organizations that serve the housing stock, cemeteries, crematoria and organizations that provide funeral services. It is also worth mentioning that there are more state-owned companies in Russia than in Germany, where private ownership prevails. In this regard, the structure of housing and communal services in Germany is wider than in Russia. Considering this situation, it was elucidated that this fact is also reflected in the difference between organizational and legal forms of organizations compared.

In Russia, due to the optimization process, the bodies of administrative and economic institutions tend to merge, while in Germany, the organizations that provide communal services to people have a greater number.

It is also important to note that in Germany exist such organizations whose analogues in Russia are not so widespread, for example, the use of wind power. At the moment, this sphere of power engineering is only in its first stage but has a high potential. Russia has a smaller number of companies whose activities are connected with solar power. Nowadays, solar power is one of the perspective directions of renewable power engineering.

In the course of study, 50 terms related to the sphere of housing and communal services were recorded. Translation of these terms was carried out in different ways: literal translation, replication, descriptive translation, etc.

Literal translation, or word-for-word translation, refers to the transition from the target language to the translation language, which leads to the creation of a correct and idiomatic text. The equivalence of messages is ultimately based on the identity of situations, which allows us to assert that the language of translation contains some characteristics of reality that are not present in the target language [2]. Among the names of administrative and economic institutions studied, eight have equivalents in Russian. This refers to terms such as die Stadtwerke (городские коммунальные службы), die Kraftwerke (электростанции), der Straßenbau (дорожное хозяйство), die Gasversorgung (газоснабжение), die Abfallverwertung (утилизация отходов), das Heizkraftwerk (ТЭЦ), der Verkehrsbetrieb (транспортное предприятие), das Straßenbauamt (дорожно-строительное управление).

Replication is a borrowing of a special kind, when a particular syntagma is borrowed from a foreign language and the elements that compose it are literally translated. Thus, it comes to either replication of the expression with the use of syntactic structures of the translation language, with the implementation of some new expressive elements into it, or replication of the structure with the implementation of new structures into the language. Ten terms have been translated using replication die Umwelt-Akademie e.V (НКО «Академия окружающей среды»), Solardach München-Riem GmbH (ООО «Солнечная энергия Мюнхен-Рим»), Tivoli-Kraftwerk (Электростанция Тиволи), Kraftwerk Bäckermühle (Электростанция Бекермюле), Münchner U-Bahnbewachungsgesellschaft mbH (MUG) (Метрополитен Мюнхен), Münchner

Linien GmbH & Co. KG (ML) (Сети Мюнхен и Ко), Energie Südbayern GmbH (ESB) (ООО «Южнобаварское энергоснабжение»), Ostthüringer Wasser und Abwasser GmbH (OTWA) (ООО «Водоснабжение и система канализации Восточной Тюрингии»), Bayerngas (ООО «Газ Баварии»), Gasversorgung Germering GmbH (ООО «Газоснабжение Гермеринг»).

When naming public utilities, owners often add their first and/or last name. In this case replication is used, and the proper name is transcribed or translated. The study found seven similar terms: Straßen u. Tiefbau Reichardt GmbH (ООО «Наземное и подземное дорожное хозяйство Рэйхардт»), Grieneisen Bestattungen GmbH (ООО «Ритуальные услуги Гринейзен»), Abendruh Bestattungen Volker Uhlig (Похоронное агенство Абендру Фолькер Илихь), Bestattungen Hanrieder (Агенство ритуальных услуг Ханридер), Müller Willibald Straßenbau GmbH (ООО «Дорожное хозяйство Мюллер Виллибальд»), Birmann & Haum Containervermietung GmbH (ООО «Аренда утилизационных контейнеров Бирманн и Хаум»), Entsorgungen Zwigart e.K (ИП Компания по утилизации Цвигарт).

Description differs from replication on the point that in descriptive methods of transfer of the new formations the invariant of translation becomes mainly the meaning of a foreign language unit regardless of the nature of its relations with the external structure of the word, while in replication the invariant of translation is the form of the lexical unit of the original language (though not sound or graphical form, as in transcribing or transliteration, but lexical or lexico-morphological), the content remains as if "behind bracket". Descriptive translation consists of conveying the meaning of a German word by means of a more or less common explanation. Descriptive translation with usage of transcription has been applied to thirteen names of the following organizations: die INFRASERV Dienstleistungen Gera GmbH (ООО «Обслуживание транспортного парка ИНФРАЗЕРФ»), Zweckverband Wasser/Abwasser Mittleres Elstertal (Целевое объединение «Водоснабжение и канализация Эльстерталь»), M-Wasser (Водоснабжение Мюнхена М-Вассер), M-Strom (Коммунальная служба по энергоснабжению города Мюнхен), SWM Bayernwind (Коммунальные службы города Мюнхен по ветроэнергетике Байервинд), Aqua Komm GmbH (ООО Аква Комм и Ко»), GeraNetz GmbH (ООО «Электрогазоснабжение Геранетц»), Alpha Entsorgung & Service GmbH (ООО «Альфа Утилизация и Сервис»).

When translating the names such methods of translation as hyperalization were also used: Versorgungs GmbH (ООО «Коммунальные службы по энергоснабжению города Мюнхен»), Entsorgungen Zwigart e.K (ИП Утилизация Цвигарт), Birman & Haum Containervermietung GmbH, as well as concretization: Abfallwirtschaftbetrieb Munchen (AWM) (Предприятие по утилизации г. Мюнхен), die Hessischen Energie AG (АО «Энергоснабжение Гессен»).

Most frequently used methods of translation were Equivalent translation (16%), replication (20%), replication and transliteration of proper names (37%) and less often descriptive translation (10%), hyperalization (9%) and specification (8%).

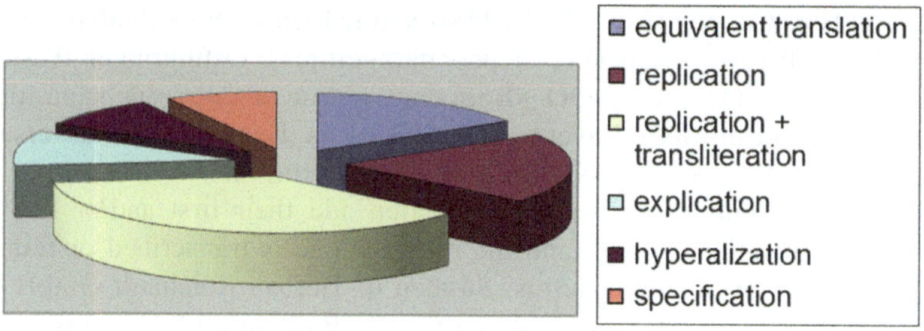

Fig. 23.1 Translation methods of the administrative and economic institutions' names in German

As an illustration of the methods of translation used in the process of thesaurus compiling, the following diagram can be shown (Fig. 23.1).

23.4 Conclusions

1. The structure of administrative and economic institutions in Russia (Pskov) and Germany (Munich) has much in common. Significant differences between companies providing housing and communal services are included in the organizational and legal form. In Russia, due to the optimization process, the bodies of administrative and economic institutions tend to merge, while in Germany, the organizations that provide communal services to have a greater number. In Germany, exist such organizations whose analogues in Russia are not so widespread (wind, solar power).
2. In the theory of translation, different ways of transition from units of one language to the units of another are considered, among which the most widespread are the following: borrowing, replication, word-for-word translation, a combination of transliteration and descriptive translation, transposition, modulation, equivalence, adaptation. The most frequent methods of translation of the names of administrative and economic institutions are the following: replication, adding of qualifiers, transliteration of proper names and less often descriptive translation, hyperalization, specification.
3. Difficulties in translation of administrative and economic vocabulary are connected in some cases with the lack of correspondence in the target language and the problem of adequacy and accuracy of their translation. The maximum possible (linguistic) degree of preservation of the contents of the original during the translation process, the semantic proximity to the original in varying degrees, achieving of mutual understanding and successful cooperation within the framework of cross-cultural communication is possible only on condition of studying and excellent command of translation methods, the formation of which help compiling of thesaurus and analysis of the administrative vocabulary of two countries (Germany and Russia).

4. The results of the study can be used in further lexicological and terminological studies, in training courses in linguistics, cognitive linguistics, sociolinguistics. Compiled thesaurus in the field of housing and communal services can be used in lexicographic practice in the compiling of different specialized dictionaries.

References

1. Komissarov VN (2002) Linguistic translation study in Russia: study book. ETS, p 258
2. Haugen E (2002) Language contact, The new in linguistic. Progress 6:61–80
3. Arkhipov AF (2008) Written translation from German into Russian. KDU, p 336
4. Melikhov NV (2000) Textbook of economical translation (German). Part 1. Study book of German. NVI Thesaurus, pp 69–72

Part II
Social Consequences of the Digital Future and Perspectives of Adaptation to Them

Chapter 24
Imperatives of Formation of the Information Society in the Context of the Modern Global Challenges

Yuriy I. Sigidov, Evgeniy V. Skubriy, Liubov N. Orlova, Nina V. Tsuglaeva and Saglar S. Ashtaeva

Abstract *Purpose*: The purpose of this work is to consider the current changes in society and the global economy under the influence of development of information technologies and their wide implementation into various spheres of public life. Information technologies contain potential opportunities and cause certain threats, which have to be neutralized before their emergence. The existing threats are of the global character and contain potential risks. Appearance of such risks will influence the work of the whole complexes and spheres of national economy, as global digital technologies establish information ties, along which information is transferred to all participants of this process. A lot of countries try to take into account and, if possible, prevent these risks, but the emerging frequent failures in the work of financial markets, hacks of international banking systems and internal information resources prove low effectiveness of the existing systems of protection of information systems. *Methodology*: The authors use the methods of systematization and logical analysis of text; for processing and evaluation of statistical material, the authors use the methods of graphic and table analysis of data, horizontal and vertical analysis, generalization, and generation of ideas. *Results*: The modern information society has a lot of opportunities and global perspectives for the countries that actively implement

Y. I. Sigidov (✉)
Kuban State Agrarian University, Krasnodar, Russian Federation
e-mail: tbu_kuban@mail.ru

E. V. Skubriy
Academy of Civil Protection EMERCOM of Russia, Khimki, Russian Federation
e-mail: skubriy@mail.ru

L. N. Orlova
Financial University Under the Government of the Russian Federation, Moscow, Russian Federation
e-mail: Lyuba_orl@mail.ru

N. V. Tsuglaeva · S. S. Ashtaeva
Kalmyk State University Named After B.B. Gorodovikov, Elista, Russian Federation
e-mail: palestinka-30@mail.ru

S. S. Ashtaeva
e-mail: ashtaeva.s@mail.ru

T. Kolmykova and E. V. Kharchenko (eds.), *Digital Future Economic Growth, Social Adaptation, and Technological Perspectives*, Lecture Notes in Networks and Systems 111, https://doi.org/10.1007/978-3-030-39797-5_24

the information and communication technologies; however, there are risks and lack of protection of the national information infrastructure. The authors determine the main tendencies of development of informatization in all spheres of public life, prove reduction of the level of differentiation of the subjects of the Russian Federation as to the indicator of information development, and study dynamics and statistics of application of the tools of protection of the information infrastructure by organizations and population. *Recommendations*: The studied main directions of supporting the level of information protection and security in Internet in the conditions of frequent cyber attacks on the critical information infrastructural objects will allow for protection of national interests in the cyber space, without cyber fraud and cyber terrorism.

Keywords Information society · Cyber attacks · Cyber space · Information inequality

JEL Code O13 · O32 · Q13 · Q16 · P52

24.1 Introduction

The modern tendencies of globalization and integration of the global economic ties between countries are predetermined by development of the information and communication technologies, which allow—in the real time—for remote operations, development of partnership relations, and establishment of telecommunication connections. The state vector at formation of the information society in Russia, adopted in 2017, is aimed at creation of a modern public mode based on large-scale implementation of the modern technologies of remote access and control over a lot of processes, according to the international principles of creation of the information society.

According to the first official document in this sphere—the Strategy of development of information society in the Russian Federation—the main direction of dissemination of the information and communication technologies in society is intensive application—by all social groups and market structures—of information and digital technologies. For this purpose, the state set the task of providing equal access to Internet for all users from various parts Russia. Another direction of dissemination of the ICT in society is large-scale application of mobile communications even in remote parts of Russia. Implementation of the Big Data technologies into systems management at various levels of authorities and business allows for significant reduction of time required for analysis and evaluation of data arrays and for receipt and processing new information online, with further generation of final solutions for future development.

The results of implementation of the Strategy of development of information society in Russia proved that population eagerly accepts new technologies and digital solutions (Fig. 24.1).

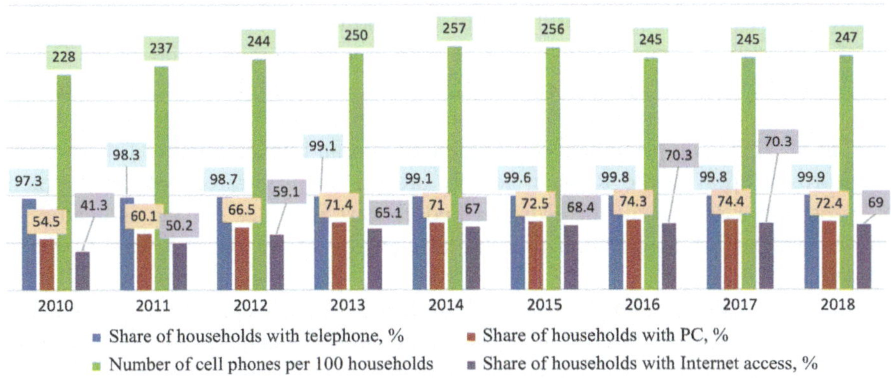

Fig. 24.1 Dynamics of development of information society in Russia in 2010–2018

Assessment of the dynamics of development of the information society determines the main imperatives of its formation. These could be growth of the number of mobile communications users, growth of the number of PC's in households with Internet access, and growth of Internet users. Figure 24.1 shows dynamics of these users in 2010–2018; growth of almost all criteria shows full-scale implementation of the information and communication technologies into everyday life of the Russian population.

The current processes are accompanied by certain risks, which are caused by internal and external cyber attacks. In the conditions of global informatization, technologies and politics unite in their influence on public interests. The telecommunication channels of information transfer become a powerful weapon for generating the political mood of the direction of future development in the conditions of aggravating geopolitical oppositions of the leading countries of the world. As a matter of fact, they are in a state of information war, distorting the real situations and supporting cyber espionage. A lot of terrorist organizations and crime groups take responsibility for terrorist attacks via Internet and perform provocations for popularization of their activities. Unfortunately, technologies cannot differentiate the quality of transferred information, so in this case they stimulate the dissemination of negative tendencies and support tense social and political situation in certain countries. Counties that actively implement new technologies and develop and improve internal production will be able to reach competitive advantage and monopolistic position in the world market of goods, services, and technologies.

24.2 Materials and Method

The theoretical and applies issues of informatization of social mode are studied in the works Fedotova et al. [1, 2], Sukhodolov et al. [3], Asaliev [4], Golubev et al. [5], Fedotova et al. [6], Sibirskaya and Shestaeva [7], Kusnetsov et al. [8], and

Popova [9]. The foundations of the normative and legal regulation of the processes of informatization in the public sector of economy are studied in the works Orlova [10], Plotnikov et al. [11], Kayl et al. [17], Glubokov and Skubriy [12], Kravets et al. [13], Kovazhenkov et al. [14], Skubriy and Vitushkina [15], Romanova et al. [16], Kayl et al. [17]. The principles of implementation and development of the process of informatization in the Russian society are set in the Strategy of development of information society in the Russian Federation. Statistics for the indicators of informatization are available at the official web-site of the Federal State Statistics Service.

However, despite the large number of publications on adjacent topics, the issues of information security of the critical objects of the information infrastructure are poorly studied. In the process of studying the main threats to the information society, the main principles of co-existence of states, companies, and people in the cyber space were formulated. Emergence of new types of cyber weapons and cyber technologies for unlawful influence on information resources requires thorough research of the opportunities of the modern cyber space. For this, the authors here use the methods of graphic presentation of information, trend analysis, comparison, analogy, and systematization.

24.3 Results

Formation of the global information society is based on the network interaction between government bodies, manufacturers of goods and services, and citizens of various countries—i.e., large network of communication between its participants. There appears a danger that a participant could use it for unlawful purposes, which will influence the economic, social, and political well-being of other users. Thus, it is necessary to strengthen the information security. In the conditions of the information society, information is a dominating feature of the modern times—it determines the state of society, which it influences. The global network Internet is not a fully independent resource. Internet is integrated into social, political, economic, and demographic spheres of society, and it reacts to all changes that take place in a country and at the international arena. This is a powerful channel of information transfer, and its quality will depend on the owner and participants of this process.

The Russian official document "Doctrine of information security" treats this notion as a state of protection of person, society, and state from internal and external information threats, which ensures realization of constitutional rights and liberties of human and citizen, decent quality of life and living standards, sovereignty, territorial integrity, sustainable socio-economic development of the Russian Federation, and security of the state. This state should be supported by technical means, which will create certain barriers against unsanctioned influences on the information environment. These are protection of personal data; protection of banking, financial personal, and operating accounts of the customers of financial and credit institutes; protection

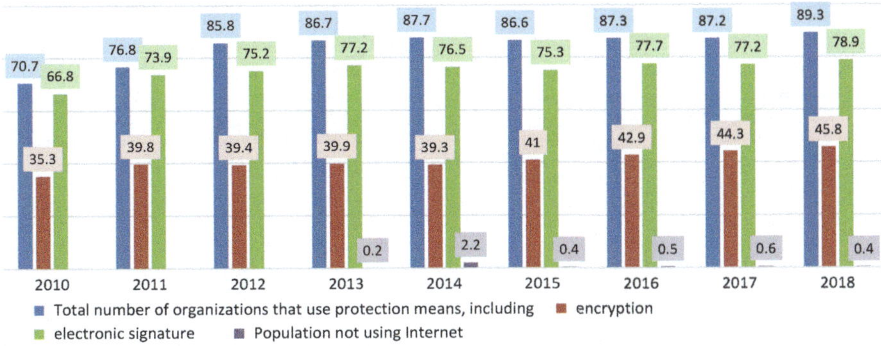

Fig. 24.2 Dynamics of the indicators of information security of the Russian Federation, %

against hacks and unlawful actions; protection from negative and provocative information. For this purpose, a lot of organizations purchase special means and software for protecting their information platforms, people try not to use Internet for personal purposes, and governments create specialized structures for protection of the national interests in the cyber space. The official statistics of Russia includes only the indicators on protection of information and population's usage of Internet into monitoring of information security. Statistics on the options of protection of internal information are shown in Fig. 24.2.

Figure 24.2 shows positive dynamics of the indicator in 2010–2018. Thus, the number of organizations that use encryption in their information channels and electronic signature grew. As for the population, the indicator has not changed much—but in 2018 there still were people who did not use Internet for protecting personal data.

Apart from the risks of loss of information and personal data, information technologies bring threats of so called "digital inequality". This notion is such state of infrastructure that does not have equal opportunities and conditions for usage of information technologies for work, and there is not possibility for online consultations of medical employees. Thus, the information society is differentiated not only as to geographical, political, and social principle, but also as to the information and telecommunication principle. This criterion characterizes the indicator "Level of differentiation of the subjects of the Russian Federation as to integral indicator of information development" (Fig. 24.3).

The statistics show that differentiation of the subjects of the Russian Federation as to the information indicator decreases, which substantiates the fact of increase of accessibility of information technologies on the whole territory of Russia.

The annual rating of regions of the Russian Federation as to the level of digital development shows that over the recent years the same regions are among top regions of Russia. Top 10 for 2017 and 2018 is shown in Fig. 24.4. At the global scale, this indicator is very high, as digital inequality between developed countries and developing countries of Africa and Asia is much higher. It should be noted that new notions "info-rich" and "info-poor" appear, and opposition between countries of the

Fig. 24.3 Dynamics of the indicator of the level of differentiation of subjects of the Russian Federation as to information development, %

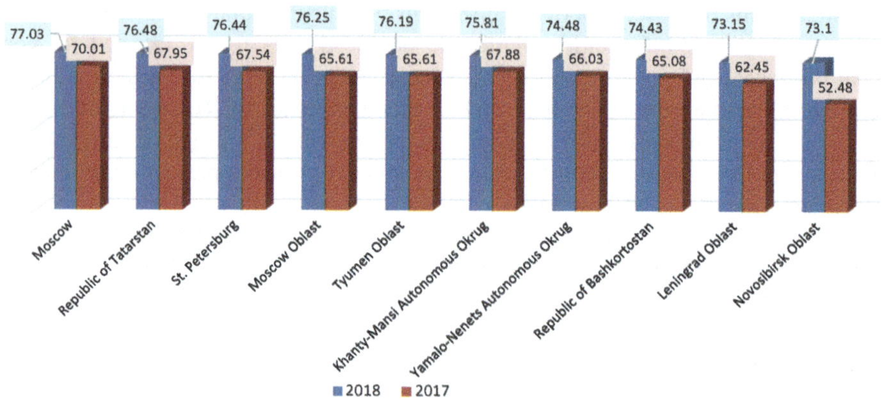

Fig. 24.4 Top 10 regions as to the indicator of digital development, %

West and Africa increases. The geo-political situation aggravates, and the "digital gap" grows. The processes of digitization ensure remote access and reduce the time required for a lot of operations, but cause social differentiation of society.

The group of UN government experts has been emphasizing the fact that the modern global information space is becoming an arena for political, economic, and military battles between the leading countries of the world. The UN General Assembly defines national responsibility of states for supporting information security. The UN supported Russia's initiative for creation of the Cyber Code—collection of 13 principles of peaceful cooperation in Internet. Besides, 2019 was the UN commission for development of recommendations for implementing the principles of peaceful usage of the information and telecommunication technologies in inter-government and national interaction. The main principles of cyber security and cyber responsibility of countries should consist in the following:

- observation of the UN Code, restraining from usage of information technologies for interference with national policy of other countries;
- protection of own critical information infrastructural objects and rights for them,
- supporting security of goods and services, aimed at ICT and channels of supply, for avoiding their application by other countries for their monopolistic advantages;
- protection of personal rights in the information environment;
- managing the Internet space on equal terms with other countries;
- restraining from unjustified accusations of other states regarding damaging information activity;
- restraining from activities that damage the groups of quick reaction CERT/CSIRT and from purposeful damage to critically important objects of infrastructure,
- exchanging information on the vulnerability factors in the sphere of ICT, restraining from dissemination of these factors;
- cooperation with private organizations and civil society in the sphere of application of the norms of responsible behavior in the information space in view of their potential role;
- development of measures for building trust, aimed at increase of predictability and reduction of probability of risk of emergence of conflicts;
- developing UN's role in preparation of international legal norms in the sphere of information security and peaceful regulation of international conflicts in the sphere of ICT.

A special danger comes from cyber attacks on the critical objects of information infrastructure. For example, according to the statistics of the National coordination center for computer incidents, there were 4.3 billion cyber attacks on the critical information infrastructure of Russia in 2018. As the practice shows, the main goals of cyber attacks were objects of the credit and financial sphere—38% (accounts of legal entities and individuals and budget accounts); services of government bodies—35% (Fig. 24.5).

The large number of cyber attacks on the objects of critical infrastructure proves attractiveness of the information environment and its accessibility for unlawful

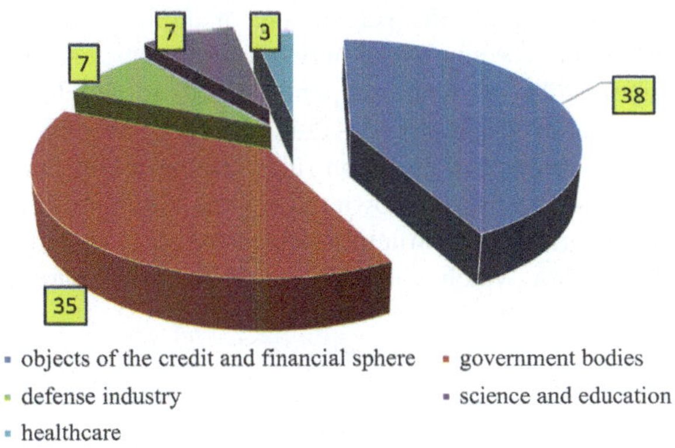

Fig. 24.5 The main critical objects of information infrastructure—targets of cyber attacks, %

actions. We see that cyber frauds and cyber terrorists attack everyday life of people and strive to destroy the functional ability of the government system.

In the dynamically forming information society, not only ICT but also technologies of malicious attacks develop, and new types of information weapons appear. These technologies are used as an additional, auxiliary, weapon for formation of a certain public opinion not only within a country but also at the international arena. The main task of such influence and cyber attacks is blocking and destroying the infrastructural information objects that service the main spheres of public life within a country.

Information becomes a global resource, which could be purchased, supplied, and assessed. Information opposition between countries is based on manipulation of the existing information, under the influence of which a party achieves the necessary actions and decisions. The level of state's influence in the information space depends on quality of information—the higher its exclusive character, the more influence it has on the opposite party. Countries spend a lot of money for development of new information technologies and means of protection from cyber attacks and develop and adopt the normative and legal documents for waging information wars. Full transition of all system of law and order and the existing public mode to the information platform could lead to loss of control over this system—which proves low level of protection of technologically developed countries in the cyber space.

24.4 Conclusion

It is possible to conclude that, firstly, information becomes an independent resource, possession of which allows dictating requirements in the international political space. Secondly, global informatization stimulates maximum cooperation of the national economic systems, up to erasure of political borders and economic limitations. Thirdly, "diffusion of power" takes place—the political sphere has new representatives from the spheres of education, healthcare, social sphere, and industry. Fourthly, the information society actually works online, which shows its speed and convenience of work. Fifthly, the very nature of public power changes, international relations develop according to the network principles, and control over information, knowledge, public opinion, and public moods grows. ICT is a resource that is generated, accumulated, and stored—thus becoming an importing management tool. Sixthly, due to absence of common rules and norms of behavior in the information space, each country forms its own rules—but it increases the existing contradictions. That's why supranational structures have to develop and adopt international norms and rules of behavior, which would be mandatory for all participants of the information space.

References

1. Fedotova GV, Lomakin NI, Tkachenko DD, Gontar AA (2018) Peculiarities of digital transformation of the system of bank's economic security. In: Popkova EG (ed) The future of the global financial system: downfall or harmony: materials of conference, Limassol, Cyprus, April 13–14, 2018. Springer Nature Switzerland AG, Cham (Switzerland), 2019, pp 1104–1112 (Book ser. Lecture notes in networks and systems; vol 57)
2. Fedotova GV, Chugumbaev RR, Chugumbaeva NN, Sukhinin AV, Kuzmina EV (2018) Increase of economic security of internet systems of credit organizations. In: The future of the global financial system: downfall or harmony, vol 57. Springer Nature Switzerland AG, pp 922–931 (2018)
3. Sukhodolov AP, Popkova EG, Kuzlaeva IM (2018) Internet economy: existence from the point of view of micro-economic aspect. Stud Comput Intell 714:11–21
4. Asaliev AM (2018) Formation of professional competencies of employees for the needs of the digital economy. Bull Plekhanov Russ Univ Econ 6(102):67–76
5. Golubev SS, Dyundik EP, Skubriy EV (2018) Methods and models of solving the classification tasks during innovative forecasting of technological development with the usage of intellectual analysis of "Big Data". Bull Mosc City Pedagog University Ser: Econ 4(18):8–16
6. Fedotova GV, Ilyasov RK, Gontar AA, Ksenda VM (2019) The strategy of provision of tax security of the state in the conditions of information economy [E-source]. In: Gashenko IV, Zima YuS, Davidyan AV (eds) Optimization of the taxation system: preconditions, tendencies and perspectives. Springer, pp 217–228 (2019). https://www.springer.com/us/book/9783030015138 (Ser. Studies in systems, decision and control)
7. Sibirskaya EV, Shestaeva KA (2016) The contents of innovative in the Russian Economy. In: Lula P, Pojer T (eds) Knowledge-economy-society. Contemporary aspects of economic transformation. Cracov University of Economics, Krakow, Poland, pp 27–37
8. Kuznetsov SY, Tereliansky PV, Shuvaev AV, Natsubize AS, Vasilyev IA (2016) Analysis of innovate solutions based on combinatorial approaches. ARPN J Eng Appl Sci 11(17):10222–10230
9. Popova L, Litvinova T, Ioda E, Suleimanova L, Chirkina M (2015) Perspectives of the growth of economic security by clustering of small innovational enterprises. Eur Res Stud J 18(Special Issue):163–172
10. Orlova LN (2016) The main principles and approaches to managing the sustainable innovative development of economic systems. Management in Russia and abroad, No 3, pp 3–9
11. Plotnokov V, Fedotova GV, Popkova EG, Kastyrina AA (2015) Harmonization of strategic planning indicators of territories' socioeconomic growth. Reg Sect Econ Stud 15-2(July–December):105–114
12. Glubokov MV, Skubriy EV (2018) Specifying the existing methods of evaluation of effectiveness of procurement of products for state needs, with the usage of the mathematical methods of technical and economic analysis. Econ Entrep 3(92):1116–1121
13. Kravets AG, Gurtjakov AS, Darmanian AP (2013) Enterprise intellectual capital management by social learning environment implementation. World Appl Sci J 23(7):956–964
14. Kovazhenkov MA, Fedotova GV, Kurbanov TK, Uchurova EO, Tserenova BI (2019) Verification of state programs of geographically-distributed economic systems. In: Popkova EG (ed) The future of the global financial system: downfall or harmony: materials of conference, Limassol, Cyprus, April 13–14, 2018. Springer Nature Switzerland AG, Cham (Switzerland) (2019), pp 1043–1053 (Book ser. Lecture notes in networks and systems, vol 57)
15. Skubriy EV, Vitushkina MG (2019) Interpretation of the results of analysis of the parameters of internal instability of shipbuilding companies. Bull Mosc City Pedagog University Ser: Econ 2(20):8–21
16. Romanova TF, Andreeva OV, Meliksetyan SN, Otrishko MO (2017) Increasing of cost efficiency as a trend of public law entities' activity intensification in a public administration sector. Eur Res Stud J 20(1):155–161

17. Kayl II, Epinina VS, Bakhracheva YS, Velikanov VV, Korobova SI (2017) Effectiveness and efficiency of public management of socio-economic processes at the city level // Overcoming Uncertainty of Institutional Environment as a Tool of Global Crisis Management, pp 185–190

Chapter 25
Innovative Social and Dominant Development of Region and Sustainable Functioning of the Social Complex

Yury A. Tsypkin, Roman A. Kamaev, Stepan V. Orlov, Inessa S. Feklistova and Hanna S. Pakulina

Abstract The purpose of the research is to develop the conceptual foundations of innovative social and dominant development of a region and sustainable functioning of the social complex. For substantiating the recommendations on formation of social and dominant innovative economy of regions and sustainable functioning of the social complex the authors use the systemic approach. The authors prove that in the conditions of the digital economy social development is the dominant idea of economic priorities of a state. The authors also specify the essence of managing the innovative social and dominant development of a region. It is proved that the dominant idea of regional management is social goals, not economic interests. The factors of formation and development of social and dominant innovative economy of a region are substantiated. The practical importance of the research consists in the possibility to use the obtained conclusions and results during improvement of the legislative basis and preparation of the federal and regional programs of development of economy with orientation at the social and innovative spheres. The authors determine the main tendencies of formation and development of socially-oriented innovative economy of the region and consider the main directions of social and dominant development of innovative economy of Russia's regions. The measures for formation of

Y. A. Tsypkin (✉) · S. V. Orlov
State University of Land Use Planning, Moscow, Russia
e-mail: tsypkin@valnet.ru

S. V. Orlov
e-mail: orlov@duma.mos.ru

R. A. Kamaev
Moscow University of Finance and Law, Moscow, Russia
e-mail: RKamaev@fadm.gov.ru

I. S. Feklistova
Central Russian Institute of Management, Branch of RANEPA (Russian Presidential Academy of National Economy and Public Administration), Orel, Russia
e-mail: Inessa_pakulina@mail.ru

H. S. Pakulina
Kharkiv National University of Civil Engineering and Architecture, Kharkiv, Ukraine
e-mail: alevtina.pakulina@gmail.com

T. Kolmykova and E. V. Kharchenko (eds.), *Digital Future Economic Growth, Social Adaptation, and Technological Perspectives*, Lecture Notes in Networks and Systems 111, https://doi.org/10.1007/978-3-030-39797-5_25

social and dominant innovative economy of regions and sustainable functioning of the social complex are substantiated: development of long-term indicative plans for formation social and dominant development of regions, in which social goals become the dominant idea; transition to the perspective budgeting, which ensures formation of social and dominant innovative economy; implementation of new forms of public expertise and public control over the process of implementation of strategic plans and programs of development of socially-oriented innovative economy of regions; increase of openness and public accountability of the executive authorities of regions by online posting of the results of the activities on transformation of the economy into effective social and dominant innovative economy.

Keywords Social and dominant development · Innovations · Social complex · Effectiveness · Region

JEL Code I38 · O35 · R58

25.1 Introduction

In the conditions of the digital economy, social development is the dominant idea of economic priorities of the state. Speaking of social and dominant development, it should be emphasized that its dominant idea is social goals.

Achievement of the high level of development of economy and institutional infrastructure is possible only in case of orientation at the fullest and effective implementation of creative potential and creative abilities of each individual, based on formation of new social innovative factors and conditions of life activities of regions' population. Practical implementation of this strategic goal dictates the necessity for organization and implementation of a socially-oriented innovative maneuver in the national and regional economy, which would ensure expanded reproduction of labor potential and material goods, service and socio-economic relations.

Importance of the topic of the performed research is high also due to the fact that transformation of Russia's regions' economy requires new mechanisms of management, which are adequate to strategic goals of the country's socio-economic development in the conditions of the digital economy. The issues of the level and quality of life are on the foreground of the national, economic, and social policy. That's why Russia's government's main priority is specific person and region's community—as the main object and subject of management of the region's innovative social and dominant development.

The purpose of the research is to develop the conceptual foundations of innovative social and dominant development of region and sustainable functioning of the social complex. The following tasks are to be achieved for this: specifying the essence of managing the region's innovative social and dominant development; substantiating the factors of formation and development of social and dominant innovative economy of the region; determining the main tendencies of formation and development of socially-oriented innovative economy of the region in the modern conditions; considering the main directions of social and dominant development of innovative economy of regions; developing the measures for formation of social and dominant innovative economy of regions and sustainable functioning of the social complex.

Methodology. The authors use the analytical, economic and statistical, economic and mathematical, and other methods of research. For substantiating the recommendations on formation of social and dominant innovative economy of regions and sustainable functioning of the social complex the authors use the systemic approach.

Results. In the context of the performed research, the essential characteristics of the system of management of social and dominant development of the innovative economy of the region consist in integration into one whole of socio-economic, innovative and technological, and investment processes with the dominance of social goals. In the conditions of harmonious combination of the above processes, development of the region's economy acquires sustainable character, which confirms to the strategic goal of a social state—provision of citizens' well-being and social protection.

Sustainability of development of the state and the socio-economic system is ensured by rational implementation of personal, public, and state priorities and interests, as well as systemic balance in all spheres of life activities.

Transformation of the regions' economy in the social and dominant innovative economy requires implementation of a social maneuver that would stimulate effective and real socialization of capital. As of now, capital is not socialized—instead, it is individualized by a group of persons. That's why Russia is peculiar for a large gap between the rich and the poor.

Socialized capital is socially-oriented and distributed between a large number of people. The more people in society possess capital, the more chances for its increase they have. That's why socialized capital is not capital for the sake of capital, but capital for the sake of human, who could increase it for himself and for the whole society.

Integrity of the economic and social goals of regional development consists in the fact that region's social development depends on its economic level of development [1]. Therefore, solution of certain social problems is determined by region's economic opportunities; in its turn, social factors influence the effective development of the regional economy [2–4].

The core of economy of the social and dominant innovative type of development is the innovative and scientific and technical activities, which ensure receipt of new knowledge and their further transformation into science-drive products and innovations in the activities of the social complex of Russia and its regions [5]. Based on this provision, we think that effective interaction of the system of education and companies of the leading spheres of region's economy is possible—it could lead to sustainable development of Russia and regions by means of creating advantages based on progressive growth of the scientific and technical and intellectual potential.

In the course of the research, we determined the main tendencies of formation and development of the region's social and dominant innovative economy in the modern conditions. Firstly, improvement of the forms and methods of social partnership of the state, business, and society with the dominance of social goals. In the social and dominant innovative economy, the state, public, and personal interests are not opposed to each other—instead, they are well-balanced [2–4].

We think that it is expedient for public authorities and business structures to start a dialog, as increase of society's well-being is possible only with the help of common efforts of the government and private organizations based on well-balanced policy that takes into account the interests of employers, employees, investors, local communities, and public organizations. With this approach, flexible financial and economic mechanisms and methods, as well as methods of self-management replace strict administrative mechanisms of management, which stimulates effective struggle against corruption.

Secondly, development of socially responsible business and provision of its participation in social programs of the regions and the state on the whole. It is necessary to emphasize the importance of development of social responsibility of business, which becomes an element of development of the region's socio-economic system, being a source of training of personnel and the basis of formation of "middle class" and decrease of social tension in regions [6].

Thirdly, changes in the region's community's social life, its spiritual development and moral priorities. Complication of production means, implementation of new technologies, and increase of the role of the subjective factor in the region's development requires high personal qualities and professional skills.

The socio-cultural sphere, which ensures harmonious development of the region's community, stimulates the formation of personality, which becomes an active subject of the social and dominant innovative economy.

One of the main directions of social and dominant development of innovative economy of Russia's regions is creation of the scientific and methodological basis of formation and social and dominant development of innovative economy at the regional level based on the unified information network, which includes: multi-level online data base on intellectual property, sectorial automatized data bases on scientific and innovative works, and state register of research works.

It is necessary to created conditions in which business well treat innovations not as an obligation before the state but as a model of behavior, which is critically important for the company's development, increase of effectiveness, and gaining leadership positions in the markets of goods and services in the conditions of the digital economy [7].

The most important direction of social and dominant development of economy is changing the priorities of innovative activities in regions from innovations in the technical and economic sphere to innovations in the social complex: remote education, informatization of social protection entities, development of a network of centers of internet access, implementation of the programs for development of speaking skills of population and increase of computer literacy, stimulation of creation of non-profit organizations that involve the population into the innovative processes and disseminate information on innovative activities.

Managerial activities for solving the strategic tasks of increase of population's well-being and development of social functions of the economy and social activity of society ensure sustainable growth of effectiveness of production and open social reserves and sources of the region's progress, increasing its competitiveness in the conditions of the digital economy.

The performed analysis of the innovative activities of large and medium companies showed that there are a lot of scholars and entrepreneurs in this sphere, as well as required elements of the innovative infrastructure [8]. However, the innovative processes do not have proper influence on the development of the Russian economic system. The level of Russia's companies' innovative activity is still behind the indicators of the European countries. The level of intensity of technological and non-technological innovations constitutes is lower by two times as compared to European countries. Also, the Russian economy is peculiar for relatively low intensity of expenditures for innovations as compared to sales volumes—2% (Sweden—5.5%, Germany—4.6%). Return from these expenditures is also low—growth of their volumes is not accompanies by increase of the share of innovative products in the total volume of sales (about 5% in 1995–2018).

The performed analysis showed very low growth of the number of companies and organizations of small and medium entrepreneurship that conduct innovative R&D. Small and medium entrepreneurship, which is one of the most important forms of implementation of the modern stage of the technological progress and which is very susceptible to technical innovations (as compared to large corporations), has weak influence on the innovative processes in Russia's regions.

The performed research allowed determining measures for formation of social and dominant innovative economy of regions. Firstly, development of long-term plans of formation social and dominant development of regions, in which social goals become the main priority and the dominant idea. Unlike the directive planning, the indictors are not of the mandatory character for companies and organizations of the social complex, production, and private sector.

Despite the offers of scholars and practitioners, indicative planning has not yet been implemented—which reduces the effectiveness of state regulation of the processes of socially-oriented innovative development of regions' economy.

Secondly, transition to perspective budgeting, which ensures formation of social and dominant innovative economy in the region, is expedient. It is also necessary to forget budget support and to develop the system of effective support for development of innovations based on the credit and commercial support and provision of economic subsidies, with development of the schemes and sources of return of allocated resources, as well as attraction of foreign investments and assets of non-budget funds. We offer the following scheme of evaluation of effectiveness of the programs of innovative development (Fig. 25.1).

Thirdly, it is necessary to implement new forms of public expertise and public control over the process of implementation of strategic plans and programs of development of socially-oriented innovative economy of regions. This will allow for a transfer from the principle of responsibility of executive authorities for development of plans and programs of formation of the socially-oriented innovative economy to the principle of responsibility for its effective development. Personal responsibility of the officials for action or inaction during implementation of the strategic tasks

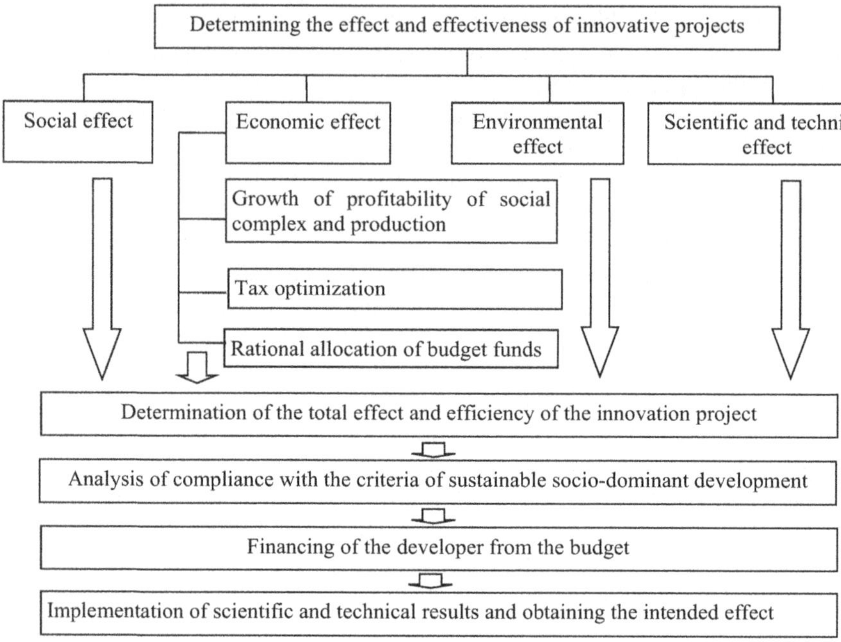

Fig. 25.1 Scheme of formation and evaluation of the effectiveness of projects of socio-dominant innovative development of the region

will stimulate the increase of effectiveness of managing the social and dominant development of the region.

Fourthly, the tasks of increase of openness and public accountability of the region's executive authorities by online posting of the results of the activities for transformation of economy into effective and social and dominant innovative economy become important. Creation of a system of constant monitoring and evaluation of quality of execution of state functions according to the strategic goals and tasks will allow for quick identification of excessive functions and administrative barriers and increase of effectiveness of interaction of the state, business, and local community for formation and development of socially-oriented innovative economy of the region.

Region's social complex is a totality of the types of economic activities that are aimed at satisfying the population's needs that include the following: systems of education and healthcare, culture and housing and utilities infrastructure, transport services for population, trade and restaurant sphere, tourism and recreational activities. Region's social complex is peculiar for common territorial placement, close interconnection of its elements, and dependence on the processes of international integration [2–4]. Insufficient development of one or several types of economic activities could lead to slowdown of development of the social complex on the whole or its destruction.

The scientific novelty of this definition of the region's social complex consists in the fact that the complex's components are types of economic activities, according to the Classification of the types of economic activities, not the spheres. Also, we'd like to emphasize the influence on the state and development of the region's social complex performed by the processes of international integration.

An important scientific problem is provision of sustainable development of the social complex. The main purpose of region's social development—and, therefore, planning of the social complex—is increase of population's living standards and formation of human capital on the whole as the main production force. It could be noted that it is a rather general goal, and due to this it cannot be modeled—though it has to be taken into account in the process of planning. Regional plans regarding the increase of population's living standards should be supplemented by the efforts of the production groups of the social complex, communities, and districts. Also, the plans of social development must have the indicators that would stimulate the decrease of the existing differentiations.

A lot of factors, which are interconnected ambiguously, influence the development of the social complex. This influence could be direct and indirect. The functioning of the social complex should be considered as an inseparable part of the whole socio-economic system of regions, which has 30% of the fixed assets and 30% of employees. It should be noted that it also has the highest intellectual potential.

This shows a high role of the social complex in region's life. Its value for the economy could be considered in two aspects: 1st—this segment of the economy determines the level of development of the regional economy. That's why without a flexible and effective system of financing its full functioning is impossible. In the long-tern, development of the social complex should have significant changes, which might include growth of investments, which will improve its infrastructural potential; further diversification of services of the social complex; activation of development of the services market; increase of the educational, professional, and qualification level of personnel potential—which, in its turn, will lead to increase of production effectiveness.

Region's social complex is a part of the social sphere. Social sphere is a part of the economic system which functional purpose is provision of services of the social character. In this case, the notion "sphere" should be treated as a totality of the types of economic activities with clear commonness of functions, special purpose in public division of labor, and a range of characteristics that distinguish these types of economic activities from material production—common stages of production and consumption, direct contact with consumer, orientation at individual consumer or society on the whole, and local character of consumption due to impossibility of transportation and accumulation of services.

Provision of sustainable development of the social sphere is an important scientific problem. The main purpose of social development—and, therefore, planning of the social sphere—is increase of population's living standards and formation of human capital on the whole as the main production force. This is a rather common goal, which cannot be modeled—though it should be taken into account in the process of planning.

It is necessary to improve accessibility to all elements of the infrastructure of the social complex and the corresponding services for each citizen. Solving this task, the social sphere creates preconditions for comprehensive development of a person. Planning and forecasting the achievement of social equality, it is possible to solve the tasks of development of personality and formation of the main productive force of public development. Increase of the living standards, formation of healthy way of life, and reduction of social differentiation allow a person to express himself through the main goal of social development and through the hierarchy—"tree of goals". The process of specification of goals from the highest level to the lowest level reminds of the process of growing of a tree. This method—building a "tree of goals"—should be used each time when the goal is large and complex, so it could be directly connected to the means of its achievement. However, such formulation of the program of social development envisages the necessity for detailed determination of goals for different levels of organization of society and each group, according to their role in the process of public production.

The modern culture and reduction provide people with large opportunities in realization of personal strivings during selection of a profession: determination of the character of leisure and personal life. An important role here belongs to the social sphere, which could be considered as a phenomenon of life activities of a modern society and socialization of a person. The social sphere envisages close interconnection between human and culture, science, and art—for satisfying its needs in the harmony with the interests of a people and the whole civilized world and subjecting the actions and desires to the moral and spiritual requirements of the cultural civilized sphere.

Participating in socio-cultural processes and creating new cultural phenomena, as well as developing—based on this—the internal world, human should get acquainted with the existing world of culture, which was created by the labor and mind of previous generations. Social sphere is not a static state of a certain system of material and moral values but a dynamic process, which is constantly developing and in which a human create and consumes values, with his self-development.

Social sphere returns scientific potential to society. We think that this is the only social institute due to which scientific knowledge develops and get disseminated. Achievements of science in the world are stimulated by the development of the social sphere. This influence is indirect. Knowledge that is transferred through education is the most important condition of social development on the whole. Social sphere intensifies interconnection between the foundations of spiritual life and is a relatively integrated phenomenon of public and economic processes, being implemented in public and economic processes of consciousness, moral culture, and everyday views of people. This sphere will be participating in socio-economic and cultural processes. A real possibility of formation of a personality is the criterion of development of the modern system of social purpose.

At present, Russia's economy faces complex tasks, which require implementation of new technologies of services and improvement of the structure of the social complex, as well as search for the paths of increase of effectiveness of its functioning, optimal managerial decisions, and active social policy. We think that accelerated economic growth and overcoming of poverty are the most important strategic tasks of the national economy. The most important threats to the country's security—especially, during the financial crisis—were manifested in the social complex. For the purpose of creation of an effective system of economic security in the social sphere, there's a necessity for a new approach to monitoring of quantitative indicators in the social complex with further qualitative evaluation.

Active social policy should be based on the state's combined usage of the mechanisms and methods of different types, aimed at accelerating the adaptation of able-bodied citizens to the market system, increase of guaranteed social protection for disabled citizens who cannot make their living in a competitive market environment without government's support; expansion of investments into "human capital"—health, education, training of personnel, education of youth, culture, science, and ecology.

Innovations in the social complex are aimed at solving the current social problems. A large part of innovations in the social complex have non-profit character and envisage large return period, with influence on large groups of people. These specific features determine the necessity for the systemic approach to implementation of innovations in the social complex.

It is necessary to note a high scientific level of the Russian studies on the issues of state management of socio-economic development. However, there is no unified scientific paradigm of usage of the systemic approach in state regulation and formation of the strategy of socio-economic development, and there are no technologies of selection of strategies and priorities of development or methods of evaluation of effectiveness of achievement of goals and the tools of their implementation [9]. The scientific and practical significance of these problems predetermined selection of the topic of this research.

The role of the social sphere in society will be growing [10]. In the conditions of globalization and integration of the Russian economy into the European and world market, the issue of acceleration of development of the social sphere, which requires creation of the corresponding technical, structural, organizational, and managerial preconditions, becomes very important. The world experience shows that without creation of a modern market of high-quality services, developed transport and communications, financial and credit sphere, complex of business services, educational sphere, and healthcare a country cannot become a part of the global economy.

25.2 Conclusions/Recommendations

1. The authors have specified the essence of managing the innovative social and dominant development of region and proved that the dominant idea of regional management is social goals, not economic interests.
2. The authors have substantiated the factors of formation and development of the social and dominant innovative economy of the region: unity of the economic and social goals of regional development with the social dominance; creation

of a system of continuous high-quality education and training of personnel and specialists; provision of effective interaction of the social complex and companies of the leading spheres of the region's economy; development of the scientific and technical and intellectual potential of the region and expansion of the sphere of the activities of innovators and inventors.

3. The authors have determined the main tendencies of formation and development of socially-oriented innovative economy of the region in the modern conditions: dominant development of the region's social complex; improvement of the forms and methods of social partnership of the state, business, and society; development of social responsibility of business; changes in the cultural life of region's society, its spiritual development, and value priorities.
4. The authors have considered the main directions of social and dominant development of innovative economy of regions of Russia: creation of the scientific and methodological basis of formation and social and dominant development of the innovative economy at the regional level; change of the priorities of innovative activities in regions from innovations in the technical and economic sphere to innovations in the social complex, provision of dominance of the social goals of innovations; increase of the volumes of investments into development of the innovative activities of the social complex; development of companies and organizations of small and medium entrepreneurship of the social complex and production, which deal with R&D.
5. The authors have offered measures for formation of the social and dominant innovative economy of regions and sustainable functioning of the social complex: development of long-term indicative plans of formation of social and dominant development of regions, in which social goals become the dominant idea; transition to perspective budgeting that ensures formation of the social and dominant innovative economy; implementation of new forms of public expertise and public control over the process of implementation of strategic plans and programs of development of socially-oriented innovative economy of regions; increase of openness and public accountability of executive authorities of regions by online posting of the results of the activities on transformation of economy into highly-effective social and dominant innovative economy.

The practical significance of the research consists in the possibility of application of the obtained conclusions and results during improvement of the legislative basis and preparation of the federal and regional programs of development of economy with orientation at the social and innovative spheres.

References

1. Pakulina AA, Pakulin SL (2016) Upravlinnia stalym rozvytkom suchasnoho pidpryiemstva (Management of sustainable development of a modern enterprise). Traektoriâ Nauki (Path Sci) (Electronic Journal) 3(8). http://pathofscience.org/index.php/ps/article/view/50. Accessed 30 Aug 2019

2. Feklistova IS, Pakulin SL, Pakulina AA, Pakulina HS (2019) Sustainable development of the economic base of cities in modern conditions. In: Proceedings of the scientific achievements of Europe: IX international scientific conference, United Kingdom, Edinburgh, Mar 9, 2019. SI Universum, London, pp 23–32
3. Feklistova IS, Pakulin SL, Pakulina AA, Pakulina HS (2019) Improvement of public-private partnership as a factor of sustainable development of territories. In: Proceedings of the Universum View 12: international scientific-practical conference, Ukraine, Lviv, January 5, 2019. LLC "Nylan-LTD", Vinnitsa, pp 115–125
4. Feklistova IS, Pakulina HS, Pakulin SL, Pakulina AA (2019) Land management as a determining factor of sustainable development of the city. In: Proceedings of IX international scientific conference "Scientific achievements of Europe", Edinburgh, Mar 9, 2019. SI Universum, London, pp 14–22
5. Tsypkin YA, Feklistova IS, Pakulin SL, Pakulina AA (2017) Sustainable development of the agricultural sector of the regional economy. In: Proceedings of II international scientific conference "Innovations in science and technology", Oct 31, 2017. SI Universum, London, pp 25–34
6. Feklistova IS, Pakulin SL, Pakulina AA, Pakulina HS (2018) Resource support of sustainable socio-economic regional development. In: Proceedings of the theoretical achievements for practice: XXVIII international scientific conference, USA, Morrisville, Sept 22, 2018. Lulu Press, Morrisville, pp 29–35
7. Pakulin SL, Feklistova IS, Pakulina AA, Pakulina HS (2019) Strengthening public-private partnership as a factor of sustainable regional development. In: Proceedings of the Universum View 12: international scientific-practical conference, Ukraine, Lviv, January 5, 2019. LLC "Nylan-LTD", Vinnitsa, pp 33–37
8. Pakulina HS, Pakulin SL, Feklistova IS, Pakulina AA (2019) Sustainable development of the regional social complex. In: Proceedings of the Universum View 13: international scientific-practical conference, Ukraine, Rivne, January 19, 2019. LLC "Nylan-LTD", Vinnitsa, pp 11–17
9. Tsypkin Y, Feklistova I (2019) Assessing the efficiency of management and land use in the agrarian sector of municipalities. In: IOP conference series: earth and environmental science, vol 274 012089. IOP Publishing. https://iopscience.iop.org/article/10.1088/1755-1315/274/1/012089/pdf. Accessed 30 Aug 2019
10. Pakulin SL, Tsypkin YA, Feklistova IS (2018) The implementation of the strategy of sustainable development of the social sphere. In: Proceedings of the Innovations in modern world: VII international scientific conference, United Kingdom, London, June 1, 2018. SI Universum, London, pp 18–24

Chapter 26
Increase of Quality of Provision of Social and Information Services to Population in the Conditions of Development of the Digital Economy

Yury A. Tsypkin, Roman A. Kamaev, Stepan V. Orlov, Alevtyna A. Pakulina and Ludmila L. Kalinichenko

Abstract The purpose of the research is to develop and to substantiate the measures for increasing the quality of provision of social and information services to population in the conditions of development of the digital economy. For substantiating the recommendations for increase of quality of provision of social and information services to population in the conditions of development of the digital economy the authors use the systemic approach. The authors substantiate the interconnection between the market of social and information services, determinants of their development, and specifics of interaction. Also, the authors determine the conditions of information interaction as a factor of increase of quality of provision of social and information services to population within the systemic approach in the conditions of development of the digital economy. The authors develop a model of provision of quality of social and information services to population in the conditions of development of the digital economy based on consideration of the influence of the factor of information interaction. The determined determinants of development, specifics, conditions, and mechanisms of information interaction in the market of social and information services allowed for development of a scientific and methodological tools that ensure the increase of quality of provision of social and information services to population in

Y. A. Tsypkin (✉) · S. V. Orlov
State University of Land Use Planning, Moscow, Russia
e-mail: tsypkin@valnet.ru

S. V. Orlov
e-mail: orlov@duma.mos.ru

R. A. Kamaev
Moscow University of Finance and Law, Moscow, Russia
e-mail: RKamaev@fadm.gov.ru

A. A. Pakulina · L. L. Kalinichenko
Kharkiv National University of Civil Engineering and Architecture, Kharkiv, Ukraine
e-mail: alevtina.pakulina@gmail.com

L. L. Kalinichenko
e-mail: dkll1905@gmail.com

T. Kolmykova and E. V. Kharchenko (eds.), *Digital Future Economic Growth, Social Adaptation, and Technological Perspectives*, Lecture Notes in Networks and Systems 111, https://doi.org/10.1007/978-3-030-39797-5_26

the conditions of the digital economy. The indicators, developed from the position of consideration of information interaction, the principles of openness and accessibility of information, consideration of standards and indicators, characterize the most significant criteria of evaluation of quality of provision of social and information services (satisfaction with quality of services and quality of received service), the mechanisms and tools of expert evaluation, the results of obtained statistical information and information via telecommunication channels, and the level of integration and automatization of information systems that are implemented in organizations of social services.

Keywords Quality of provision of services · Social and information services · Systemic approach · Efficiency · Population

JEL Code I38 · M38 · O35 · O38 · R58

26.1 Introduction

The scientific problem of provision of quality in the modern conditions is the most important factor in increase of population's living standards. Quality, being a very complex and multi-aspect category, is present in all spheres of material production and public relations and ensures competitiveness of products and services. Russian and foreign scientific studies offer multi-vector approaches to defining the notion "quality", which do not contradict each other, allowing looking at this problem from different positions. Various aspects of this notion are distinguished from the philosophical, social, economic, legal, statistical, information, technical, production, mathematical, and consumer points of view.

The social aspect of quality is manifested through attitude of certain social groups or the whole population towards the studied object (e.g., attitude of target consumers to the corresponding service). At present, the information component in the social sphere interacts with educational services; services connected to culture and leisure; healthcare; social provision of citizens; housing and utilities infrastructure; physical culture and sports, etc. [1, 2]. Accordingly, in the social sphere, the processes of provision of services' quality are implemented not only by means of material realization but are strengthened from the positions of the information content [3]. Quality of all types of social services and ideology of socio-economic behavior of human is influenced by informatization, which actively influences the process of redistribution of material, labor, and time resources in society [4]. There is a range of preconditions in this direction which allow treating social and information services as a separate type.

Active development of the technical means of transfer, processing, and provision of information (including wireless technologies) and changes in traditions and customs of public and production relations set before the state the task of preventing the underrun in this sphere. The audience of the Russian Internet constituted 62 million in 2018. In Russia, there are two cell phone numbers per capita.

A complex and multi-aspect process of increasing the quality of social and information services that are provided to population requires solving an important scientific task—application of the systemic approach to increase of quality of provision of social and information services to population of Russia and consideration of the influence of the factor of information interaction. The importance and complex factor of the studied phenomena are the basis for selecting the topic of the research.

The purpose of the research is to develop the measures for increasing the quality of provision of social and information services to population in the conditions of development of the digital economy. The following tasks were solved by the authors: substantiating the interconnection between the market of social and information services, determinants of their development, and specifics of interaction; determining the conditions of information interaction as a factor of increase of quality of provision of social and information services to population within the systemic approach in the conditions of development of the digital economy; developing a model of provision of quality social and information services to population based on the influence of the factor of information interaction in the conditions of development of the digital economy.

Methodology. The authors use various methods, including analytical, economic and statistical, economic and mathematical, etc. For substantiating the recommendations for increase of quality of provision of social and information services to population in the conditions of development of the digital economy the authors use the systemic approach.

Results. Formation of the post-industrial society in Russia is impossible without an effective structural transformation of the Russian economy for development of the social sphere. Development of the social economy sets the foundations of stability of society, reduces social tension, and stimulates confidence in the future. The key denominator of growth and globalization of economy is the sector of social services, which consumption ensures permanently growing needs of a human and transition from the industrial to post-industrial stage of the society's development.

The object of a social service is human's physical state, educational and cultural level, and soft assets and values, which ensure the increase of human's quality of life. The object of the market of social and information services is human's needs for changing his qualitative physical state (certain psycho-biological spheres) in everyday life, educational and cultural level, and fixed (soft) asses and values, which participate in the process of interaction between performer and consumer of the services and are based on the processes of information exchange of knowledge.

Belonging to the market of social and information services in the conditions of development of the digital economy is determined by the following determinants:

- in the conditions of volatile prices, information flows dominate over non-material flows in the social and information services;

- in the conditions of differentiation of market offers, social and information services are aimed at creation of conditions that stimulate satisfaction of human needs for various consumer groups;
- social and information services are susceptible to changes of the external environment and dynamics of the market situation;
- social and information services, provided to population, have to conform to such characteristics of quality as functional purpose, economy, safety of services, safety for the external environment, culture of services, etc.;
- social and information services should be treated as certain tools of state regulation, which are aimed at provision of solution to such key social problems as unemployment, reduction of living standards, migration of population, etc.

The market of social and information services in the conditions of development of the digital economy is a very complex management system, within which, in the conditions of the change of consumer preferences, the focus is moved to increase of quality of the process of services provision. A social and information service is a process of receipt and provision (to user) of data on satisfaction of everyday, medical, psychological, pedagogical, legal, and urgent social services, as well as information support for employment and increase of communicative potential of handicapped recipients, which ensures effective usage of the limited resources of the society based on integration of information interaction of all participants of this process.

The social aspect of quality is manifested in attitude of certain social groups (or the whole society) to the studied object (e.g., attitude of target consumers to the corresponding service). In the conditions of development of the digital economy, the information components, which is distinguished in the structure of the objects of the Russia's social sphere, interacts with the following main services:

- healthcare (it is necessary to distinguish the activities of healthcare establishments, their specific location, state, and equipment. At this level, personnel structure is considered);
- educational services (the main stages of education include preschool, primary, secondary, vocational, higher, and post-graduate education);
- services that are connected to culture and leisure (the objects include libraries, art galleries, exhibitions, parks, museums, theatres, various concert organizations, zoos, monuments, cultural sites, etc.);
- social provision of citizens (various types of specialized establishments, which provide help for various groups of population);
- housing and utilities infrastructure (information reflects the state of the housing fund, level of living conditions, organization and production activities of various companies and services that provide population with water, natural gas, heat, etc.) [5];
- physical culture and sport (the information component reflects the network of the existing and perspective sports premises, their location, personnel structure, and the number of people who work out).

The system of social and information services is considered and analyzed based on distinguishing such structural components that would allow for detailed determination of the mechanisms of optimization of this system of social services. The main components of this system in the conditions of development of the digital economy are as follows:

- various objects of social and information services, which include people and groups of people that have needs for the social services and certain interests in solving life problems;
- subjects of social and information services are personnel of various social organizations, as well as consumers of this service (clients);
- established standards and norms of provision of social and information services;
- financial, property, managerial, and personnel resources of organizations that provide social and information services;
- received result of effectiveness and quality of provided social and information services.

In the conditions of development of the digital economy, the market of social and information services is a very complex system, within which, in the conditions of the changes of consumer preferences, the emphasis is moved to increase of quality of the process of provision of service [6, 7]. The main factors that influence the quality of provision of social services in Russia are presence and state of the documents based on which the establishment that provide such services function; location of the establishment; personnel structure of the organization; presence of specialized equipment; state of information, which includes the list and characteristics of services, standards, and requirements of social services, interconnection between quality, condition of provision, and cost of services, communication means for effective communication with services' consumers, and obligations during execution of services; presence of external and internal departments of control over the organization's activities.

Intensive growth of the value of information in the conditions of development of the digital economy is accompanied by increase of the role of education, formation of new social standards of quality, creation of "e-governments" based on effective feedback, etc. [8]. However, the influence of information interaction on quality of provision of social and information services to population has not been thoroughly studied by Russian scholars. The developed scheme of interaction subjects of the system of social and information services shows that the need for social and information services appears due to necessity for protecting the population from social risks, which acquire the form of loss of labor income. The key task of management within provision of social and information service is reduction of the volume of state expenditures, improvement of the quality of the system of provision of public services to individuals and organizations, and increase of trust to public authorities from private sector and society. The state uses regulators to define the main rights and obligations of the subjects of interaction in the system of quality of provision of social and information services to population. It is necessary to take into account the fact that information determination of the process of information interaction during provision of a service largely depends on the volume and quality of information.

This dependence has to be evaluated according to the level of computerization of the process of provision of services and availability of resources that are used for storing, processing, and supply of information (individual services, consultation and educational services, etc.). In the conditions of the developed information society, this task is solved by using the existing possibilities of e-government, web-sites of municipal entities, organization of Internet receiving offices, online votes, population surveys, provision of consultation services, etc.

Information interaction within provision of quality of provision of social and information services to population should be treated as a dynamic process of exchange of information and documents (including in the electronic form) between subjects (consumer-supplier), aimed at increasing the quality of the activities of performer for provision of a service. Implementation of the principles and mechanisms of information interaction for increasing the quality of provision of social and information services in the conditions of development of the digital economy requires distinguishing the top-priority types of services for the types of economic activities in view of their social importance, evaluation of the volume of consumed market of these services, and necessity for improving the system of state management. This should ensure increase of quality and competitiveness of business and labor resources and increase the rates of socio-economic development of Russia by supporting the effective managerial decisions in the public and private sectors of economy.

The contents and requirements to increase of quality of provision of social and information services are connected to a large intellectual component, increase of new knowledge, constantly growing role of application of the achievements of technological progress, high demand—in the labor market—for professionalism and moral characteristics of personnel, wide specter of types and forms of services, combination of budget and market conditions of economic activities, constant change and expansion of demands, requirements, and preferences of consumers, and complexity of evaluating the quality of the provided social and information services. From the positions of the systemic approach and consideration of the influence of formation of interaction, we developed a model of provision of quality of provision of social and information services to population, which is based on totality of basic and applied theories and the basic approaches (systemic and functional; structural; communicative; integrative). The theoretical component is presented by the totality of methods, approaches, and means of provision, which are applied for the systemic evaluation of the level of provision of a social and information service.

The main principles that ensure the increase of effectiveness of information interaction of the process of provision of social and information services to population in the conditions of development of the digital economy are as follows:

- openness, transparency, free access to receipt, transfer, production, and dissemination of information (limitation of access is possible only according to the law);
- confidentiality and security of information (protection of privacy and provision of national security);

- state and public control over collection and storing of data and organization of access to information (feedback principle);
- topicality, completeness, authenticity, continuity, and timeliness of information;
- adaptation of information interaction based on usage of unified forms, reliability of software and hardware.

The vector of information interaction is aimed at increasing the quality and effectiveness of the process of provision of social and information services to Russian population in the conditions of development of the digital economy and it stimulates the solution of the following tasks:

- increase of effectiveness of usage of all resources of the socio-economic system;
- stimulation of entrepreneurial activity and increase of the level of state's social responsibility before society;
- resource provision, development of the market, and increase of quality of provision of social and information services;
- growth of labor efficiency in various types of economic activities of the Russian economy;
- increase of effectiveness of the budget financing of the social sphere and overcoming of the stagnation processes in the economy.

Improvement of the system of quality management is implemented through improvement of the processes that ensure the required quality of provision of a social and information service. That's why managing the quality of provision of these services required a quantitative assessment of interaction between the processes, which allow determining the level of influence of certain processes on the key processes of the system of quality management—which eventually influences the achievement of the main goals of provision of social and information services. The level of quality of provision of social and information services to population in the conditions of development of the digital economy, in view of information interaction, should be evaluated with the following group of indicators:

- completeness of information on a service that is posted on the official web-site (measured in points);
- level of topicality of information on a service that is posted on the official web-site (measured in points);
- presence and level of accessibility of means of feedback on the official web-site with recipients of services (measured in points);
- level of accessibility (for handicapped persons) to the organization's information products (measured in points);
- share of obtained services which quality could be evaluated with the help of the telecommunication channels (%);
- share of people who are satisfied with quality of services along the telecommunication channels;
- share of customers who are satisfied with attentiveness and politeness of the organization's employees on telecommunication channels;

- number of justified complaints of the service's recipient on quality of services that are provided by the organization on telecommunication channels;
- level of competence of the organization's employees;
- level of automatization of the process of provision of the service;
- level of automatization of interaction between the subjects of the process of provision of a social and information service;
- level of integration into the information space;
- level of standardization based on modularity, universal character, and flexibility of the architecture;
- level of correspondence of the unified functional model of e-government.

The formed indicators, developed from the position of accounting of information interaction and the principles of openness and accessibility of information, consideration of the standards and indicators characterize the most significant criteria of assessment of quality of provision of social and information services (satisfaction with quality of services and quality of obtained service), mechanisms and tools of expert evaluation, results of obtained statistical information and information along telecommunication channels, and level of integration and automatization of information systems, which are implemented in organization of social services.

26.2 Conclusions/Recommendations

1. In the conditions of development of the digital economy, the essential determinants of provision of services are as follows: aiming at satisfaction of constantly growing needs of people; non-material obtained useful result; direct or indirect interaction between consumer and service provider; as a rule, domination of actions over material embodied objects. Constantly growing consumption of services is predetermined by increase of the level of material provision of people; growth of living standards; active urbanization, which forms new types of services; changes in demography (increase of the number of elderly people, who require specific services); various technological changes, which raise quality and diversity of service. At the same time, there appears a contradiction between different economic interests of various market subjects and strategic (current) interests of society's development. This contradiction is manifested in various target landmarks of the market economy and social sphere of Russia. As increase of quality of provision of social services influences the society's development, the level of their development could be considered a significant indicator of the level of social servicing of Russians in the conditions of development of the digital economy. In its turn, increase of quality of life depends on the level of implementation of technological progress, development of information and telecommunication technologies in the market economy, constant change of the normative and legal information, and active computerization in all spheres of public life.

2. The research has been performed for the purpose of solving a scientific task of development of scientific and methodological tools that would ensure the increase of quality of provision of social and information services to population of Russia through consideration of the influence of the factor of information interaction; the research is based on regularities and principles of economics, application of the systemic approach to the studied phenomena and processes in the sphere of increase of quality of social and information services, and usage of scientific works of Russian and foreign scholars and existing normative documents and laws.
3. The objective tendencies of development of the world market of information technologies are exponential increase of the volumes of data, dissemination of various cloud computations, usage of mobile devices, and development of technologies of interaction in social networks and corporate environment [6, 7]. Based on analysis of various approaches to categories "market of information services" and "market of social services", the following authors' definitions are offered:
   - social and information service is a process of receipt and provision (to user) of data on satisfaction of everyday, medical, psychological, pedagogical, legal, and urgent social services, as well as information support for employment and increase of communicative potential of handicapped recipients, which ensures effective usage of the limited resources of the society based on integration of information interaction of all participants of this process;
   - market of social and information services is a system of organizational and economic, legal, and information and technological relations on satisfaction of human need for changing his qualitative physical state (or certain psychobiological spheres) in everyday life, increase of the educational and cultural level, and management of fixed (soft) assets and values by interaction between performer and consumer, based on the processes of exchange of information knowledge.
4. The dynamic market of social and information services in Russia contains the information, normative and legal, technical, technological, organizational, everyday, social, medical, pedagogical, labor, and psychological components and includes certain interconnected sectors [9]. Belonging to the market of social and information services in the conditions of development of the digital economy is determined by the following determinants:
   - in the conditions of volatile prices, information flows dominate over nonmaterial flows in the social and information services;
   - in the conditions of differentiation of market offers, social and information services are aimed at creation of conditions that stimulate satisfaction of human needs for various consumer groups;
   - social and information services are susceptible to changes of the external environment and dynamics of the market situation;

- social and information services, provided to population, have to conform to such characteristics of quality as functional purpose, economy, safety of services, safety for the external environment, culture of services, etc.;
- social and information services should be treated as certain tools of state regulation, which are aimed at provision of solution to such key social problems as unemployment, reduction of living standards, migration of population, etc.

5. During selection of the indicators of quality and provision of efficiency of the processes of provision of a social and information service it is necessary to form a totality of quantitative and qualitative criteria of the assessment processes [10]. For this purpose we developed an algorithm that allows establishing the priorities of the distinguished indicators/factors of evaluation of quality of social and information services and forming an integral criterion based on the indicators of information interaction. For evaluating the quality of provision of social and information services to population in the conditions of development of the digital economy the following criteria are offered: cost of the process and necessary resources for providing a service; dependence of service's quality on other processes; completeness and timeliness of provision of service: level of solving the material and financial problems of the service recipient as a result of information interaction with service provider; non-material efficiency/effectiveness of provision of service to the customer as a result of information interaction with service provider. The offered algorithm and mechanism of assigning points to the criteria of quality and indicators/factors reflect the significance of each of the studied elements and ensure the usage of the analytical approach to correct selection of the calculation results through usage of the formalized algorithm.
6. The performed research showed that from the positions of assessment of the criteria of quality of provision of social and information services to population, the most significant are as follows: completeness and timeliness of the service; level of solving material or financial problems of the service recipient as a result of information interaction with the service performer; non-material efficiency (effectiveness) of providing a service to the customer as a result of information interaction with the service performer. The calculations that were performed according to the developed algorithm show that the following factors are most important in the multi-criterial evaluation of the quality of a social and information service based on consideration of the indicators of information interaction: share of customers who are satisfied with attentiveness and politeness of the organization's employees on telecommunication channels; number of justified complaints of the service's recipient on quality of services that are provided by the organization on telecommunication channels; share of obtained services which quality could be evaluated with the help of the telecommunication channels; level of correspondence of the unified functional model of e-government; level of competence of the organization's employees. The offered model uses the combined approach of the theory of qualimetry and usage of the methods of analysis of hierarchies and ensures a complex and multi-criterial assessment

of quality of a social and information service based on the indicators of information interaction through consideration of the results of the export and formal approaches.

Practical implications. The distinguished determinants of development and specifics, conditions, and mechanism of information interaction in the market of social and information services allowed developing scientific and methodological tools that ensure the increase of quality of provision of social and information services to population in the conditions of the digital economy. A logical continuation of the studies within the selected topic is methodological provision of determining the level of interaction processes of the system of quality management during provision of social and information services to population from the position of assessment of efficiency of the key processes of the system of quality management. In the conditions of development of the digital economy it is necessary to improve the normative and legal basis in the sphere of standardization and quality management [1, 2], which allows ensuring effectiveness of information interaction and increasing the level of coordination of territorial authorities and entrepreneurial structures in the market of social services.

References

1. Pakulin SL, Feklistova IS, Pakulina AA, Pakulina HS (2018) Resource support of sustainable socio-economic regional development. In: Proceedings of the theoretical achievements for practice: XXVIII international scientific conference, USA, Morrisville, Sept 22, 2018. Lulu Press, Morrisville, pp 29–35
2. Pakulin SL, Feklistova IS, Pakulina HS (2018) Information and communication technologies as the most important sources of innovation. In: Proceedings of VII international scientific conference “Innovations in modern world”, June, 1, 2018. SI Universum, London, pp 46–54
3. Tsypkin Y, Feklistova I (2019) Assessing the efficiency of management and land use in the agrarian sector of municipalities. In: IOP conference series: earth and environmental science, vol 274 012089. IOP Publishing. https://iopscience.iop.org/article/10.1088/1755-1315/274/1/012089/pdf. Accessed 31 Aug 2019
4. Ievsieiev AS, Pakulina AA (2018) Innovatsiina ta kreatyvna ekonomika yak umova modernizatsii natsionalnoho hospodarstva Ukrainy [Innovative and creative economy as a condition of modernization of the national economy of Ukraine]. Ekonomika ta suspilstvo: Elektronne naukove fakhove vydannia Mukachivskoho derzhavnoho universytetu [Economics and society: electronic scientific professional publication of Mukachevo state University], no 16, pp 192–200
5. Pakulin SL, Pakulina AA (2016) Upravlinnia stalym rozvytkom suchasnoho pidpryiemstva [Management of sustainable development of a modern enterprise]. Traektoriâ Nauki [Path Sci] (Electronic Journal) 3(8). http://pathofscience.org/index.php/ps/article/view/50. Accessed 31 August 2019
6. Pakulin SL, Feklistova IS, Pakulina AA, Pakulina HS (2019) Strengthening public-private partnership as a factor of sustainable regional development. In: Proceedings of the Universum View 12: international scientific-practical conference, Ukraine, Lviv, January 5, 2019. LLC “Nylan-LTD”, Vinnitsa, pp 33–37

7. Pakulin SL, Feklistova IS, Pakulina AA, Pakulina HS (2019) Sustainable development of the economic base of cities in modern conditions. In: Proceedings of the Scientific achievements of Europe: IX International scientific conference
8. Semyhina T, Koryshova H, Ivanova O (2013) Public awareness on social and rehabilitation services. UNDP Project on Support to the Social Sector Reform in Ukraine, K.I.C, Kyiv
9. Tsypkin YA, Pakulin SL, Feklistova IS (2018) The implementation of the strategy of sustainable development of the social sphere. In: Proceedings of the Innovations in modern world: VII international scientific conference, United Kingdom, London, June 1, 2018. SI Universum, London, pp 18–24
10. Feklistova IS, Pakulin SL, Pakulina AA, Pakulina HS (2019) Improvement of public-private partnership as a factor of sustainable development of territories. In: Proceedings of the Universum View 12: international scientific-practical conference (Ukraine, Lviv, January 5, 2019). LLC "Nylan-LTD", Vinnitsa, pp 115–125

Chapter 27
Place and Role of a Human in the System of Robotized Circular Reproduction in the Regional Economy

Tatyana S. Kolmykova

Abstract The circular economy is a modern concept that satisfies the conditions of development of digitization and formation of "Industry 4.0". High level of interest in the study of these problems is due to environmental degradation, as well as to the abandonment of the so-called "linear" economics, based on the industrial growth in case of unrestricted use of natural resources. Active implementation of the linear model caused major ecological problems, which include exhaustion of natural resources, emission of pollutants, greenhouse gases, the increase in volumes of rubbish and waste, as well as their poor management. As a result, the need to transform the economy in the aspect of improvement of its stability, achievement of climatic goals and conservation of global natural resources is beyond doubt. One of the top-priority tasks is to create fundamentally new workplaces and development of human capital as a key competitive advantage of countries experiencing such economic transformations. Digitization of the economy and technologies of the Fourth Industrial Revolution can open up new opportunities for the implementation of circular principles of production. The Fourth Industrial Revolution is accompanied by the blurring of distinctions between physical, digital and biological technologies. The term Industry 4.0 that was first put forward at the Hanover Fair in 2011, characterizes the process of fundamental transformation of global value-creation chains. The ultimate goal of this process is the flexible interaction of virtual and physical production systems, ensuring complete adaptation of products and the formation of new operating models. Achievement of interaction should be based on automated digital production, which is controlled in real time by intelligent systems that are in continuous contact with the external environment, going way beyond a particular company with the aim of combining of things and services in a global industrial network.

Keywords Innovations · Circular economy · Regional economy · Reproduction

JEL Code O300

T. S. Kolmykova (✉)
Southwest State University, Kursk, Russia
e-mail: t_kolmykova@mail.ru

T. Kolmykova and E. V. Kharchenko (eds.), *Digital Future Economic Growth, Social Adaptation, and Technological Perspectives*, Lecture Notes in Networks and Systems 111, https://doi.org/10.1007/978-3-030-39797-5_27

27.1 Conceptual Framework of Robotized Circular Reproduction

Each region must ensure that the emerging needs of the population are satisfied by operating with its inherent socioeconomic potential, intraregional and interregional connections. In order to do this, it is necessary to organize the circulation of financial, labor and tangible/material resources forming the regional reproduction process. At the same time, some resources are allocated for the creation of conditions for reproduction, while other resources are allocated for ensuring the functioning of the reproduction process. In general terms, the process of regional reproduction is a result of the combination of territorial and sectoral differentiation of labor. We concur with the authors, who believe that it can be considered as a set of continuously repeating successive phases over time: production, distribution, exchange and consumption (production and personal) [1].

The continuity of the reproduction process is ensured by a number of factors and is directly dependent on rational use of natural resources the main properties of which are limitedness and non-regenerability. In addition, each specific region is unique, which is due to different natural climatic conditions. For regions with deficient or absent stocks of natural resources, the possibility of economic development through the application of alternative production principles opens up. In the limelight of improvement of stability of the economic growth, these circumstances charge local authorities with fundamentally new tasks, associated with the rational use of natural resources and expansion of the possibilities for their reuse.

Supporters of the concept of the circular economy claim that it its closed and restorative in nature [2]. The circular reproduction process is based on:

- enhanced control over the stocks of natural resources, including renewable resources, with a view to conserving the natural capital of territories;
- optimization of consumption processes in the aspect of manufacture and sales of products which conforms to the high level of reuse;
- establishing, preventing, or minimizing negative effects of production with a view to improving its economic and ecological efficiency [3, 4].

Circular reproduction is based on closed supply chains that maximize the added value during the full life cycle of a product and dynamically restore its value over a relatively long time period. The ideal format for such a chain is the observance of the zero waste principle and coverage of the maximum available share of social reproduction sectors. At the same time, it should be emphasized that recycling and reuse of production waste is not a single task of implementation of principles of the circular economy. It also focuses on the technological, organizational, environmental and social innovations applied throughout the entire cycle of creation of value, including both the sustainable design of a product and the prevention of waste generation.

Additional opportunities for the introduction of the circular economy are provided by a new round of industrial development. The distribution of the Internet of Things contributing to the implementation of innovations of a circular nature holds a special place in this process. According to Rose and other authors, the Internet of Things is characterized by the enhancement of networking capabilities and computational capability of facilities, devices, sensors and other items that are usually not considered to be computers [5]. As a result, there is a decrease in the cost of sensing technologies and the spread of networks to which each component of production will eventually be connected. The information obtained will allow us to register the place of origin of the product, the method of its production and the amount of resources and energy spent. It is this information that forms the basis of the implementation of principles of the circular economy into the reproduction process. With its help, enterprises, regions, countries and the entire world community will be able to more efficiently create, restore and relocate the resources used.

Another aspect of Industry 4.0 is the development of robotization of production processes. Implementation of robots, according to analysts, allows replacing a human in the course of heavy works and being in extreme conditions hazardous to life and health, as well as significantly increase labor productivity, improve the quality of products and reduce their cost [6]. The increase in performance requires the implementation of principles of circular production. Robotized production chains will make it possible to achieve the stated goals as smoothly as possible. In the course of the Fourth Industrial Revolution, in addition to automatic and adaptive industrial robots, third-generation robots with artificial intelligence are being introduced. It is they that are fully integrated in the global Internet of Things, in a way that facilitates strengthening of control over production processes in terms of minimization of consumption of virgin raw materials and the increase in volumes of processed resources. This approach will significantly reduce generation of waste that is ultimately sent for waste burial, reducing the area of landfill sites and unauthorized landfills at that. Ideally, the utilization of robots with artificial intelligence and enhancement of digitization of reproduction processes will result in the closure of the production cycle and zero-waste production.

27.2 Foreign Experience of the Implementation of Principles of Robotic Circular Reproduction

The correlation and the structure of the advantages and possible risks of the introduction of the circular economy are different for each specific territory, which gives rise to the need for differentiation of the means of its implementation. In addition, specific features of socioeconomic priorities and environmental culture of society has significant impact on these processes.

Of course, countries with a strong economy, advanced technology and a corporate culture are characterized by greater potential for the introduction of the circular economy. However, it is here that there is an increased risk of the nominal use of circular models with a view to gaining additional access to markets and guarantees of preservation of positions held, while developing countries with lower quality of life are capable of the maximally reasonable application of principles of circular reproduction. For example, the presence of circular models in production processes of developing countries is for the most part associated with the segregation and reuse of waste, which allows obtaining the so called "growing points" of opportunities for sustainable development of the territory. According to researchers, this contributes to the promotion of various forms of innovation by the business community, government, and other stakeholders [7].

Please note that even in the integration associations of countries, in the European Union in particular, each country is characterized by its own specific features of the implementation of principles of robotic circular reproduction. For example, in Germany with its powerful economy, the basis of circular reproduction is formed through material flows and the availability of materials, in the Netherlands—through innovations in materials and business models. It is natural that influential countries of the EU, such as the United Kingdom, Germany and France, occupy leading positions in terms of assessment of the development of principles of robotic circular reproduction with higher volume of investment, patents and workplaces in circular economy sectors. It is Germany that ranks first in the circular economy rating 2018 in terms of the number of patents associated with the circular economy, significantly getting ahead of France in this indicator (1260 patents against 542). Besides, Germany and the United Kingdom are leaders in terms of "circular" investment, in many ways superior to other EU members [8].

The highly efficient application of principles of circular reproduction is characteristic of the countries of Asia region. Thus, in Japan, a law was passed to promote the efficient use of resources as early as in 2000. At present, Japanese recycling figures are extraordinary: the country recycles 98% of its metals. In accordance with disposal legislation, the majority of electrical and electronic products are recycled; the share of recovered materials is about 90%. At the same time, many of these materials are returned to the production of the same type of product, which exactly conforms to principles of the circular economy.

In China, the circular economy began to develop within the program of industrial ecology, which deals with matters of how wastes of one company can become resources for another company. As noted by Benton and Hazel [9], the country has formed the legislative framework of the circular economy at present, the concepts of sustainable design and extended producer responsibility are actively developed, which is indicative of significant progress in this area.

27.3 The Urgent Character of Implementation of Robotic Circular Reproduction in Russian Regions

The urgent character of transition of Russian regions to the circular economy model is stipulated by environmental challenges facing the country [10]. One of prerequisites for the implementation of principles of robotic circular reproduction is high amount of waste and low level of their disposal and neutralization (Table 27.1).

A negative trend of an increase in volumes of waste generated during production and consumption can be observed. At the same time, the slowdown of the rate of growth of this indicator could be observed until 2015. However, this indicator has been steadily increasing since 2016, amounting to 7266.1 million tons in 2018, which is 16.81% more than in the previous period. For the period from 2011 till 2018, the share of disposed and neutralized production and consumption waste has increased from 46.26 to 52.55%. This is largely due to additional efforts by the state to improve the waste management system, within the framework of which the list of possible forms of recycling was expanded and the standards for disposal were established in the production of individual commodities.

Today, Russian regions face certain barriers preventing the transition to robotic circular reproduction, which are associated with the complicated innovative development of the country, economic and sociocultural barriers. Experts [11] note that they include, in particular:

Table 27.1 Generation, disposal and neutralization of production and consumption waste in the Russian Federation, 2011–2018

Indicator	2011	2012	2013	2014	2015	2016	2017	2018
Generation of production and consumption waste, million tons	4303.3	5007.9	5152.8	5168.3	5060.2	5441.3	6220.6	7266.1
as a percentage of the prior period	115.22	116.37	102.89	100.30	97.91	107.53	114.32	116.81
Disposal and neutralization of production and consumption waste, million tons	1990.7	2348.1	2043.6	2357.2	2685.1	3243.7	3264.6	3818.4
as a percentage of the prior period	114.53	117.95	87.03	115.35	113.91	120.80	100.64	116.96
The share of disposed and neutralized production and consumption waste, %	46.26	46.89	39.66	45.61	53.06	59.61	52.48	52.55

- high material and power intensity, as well as low resource efficiency of the vast majority of production works;
- the domination of the materials sector in the structure of the Russian economy, intensified through active state support;
- the emergence of difficulties in financing of modernization of the economy and attraction of additional investment;
- non-acceptance of the prospect of cooperation for the long-run period by the business community, its unwillingness to work in "business clusters".
- low level of awareness and understanding of the consequences of poor ecology.

In order to overcome these barriers, the national authorities of the country conduct an active modernization policy of the national economy, which is aimed at diversification of exports, implementation of innovations and the stable economic growth. Development and acceptance for fulfilment of the "Strategy for science and technology development of the Russian Federation until 2035", "The basics of state policy in environmental development of the Russian Federation until 2030", as well as the "Environmental Security Strategy in the Russian Federation until 2025" are of particular significance for the development of principles of circular reproduction.

27.4 Transformation of the Role and the Place of a Human Under Conditions of Robotized Circular Reproduction

The symbiosis of the digital transformation of the economy and the implementation of principles of robotic circular reproduction will allow creating new opportunities for entrepreneurship, improvement of quality of social services; improve the system of education and professional competencies. At the same time, digitization and robotization of production works leads to job cuts, wage stagnation, and increased inequality in remuneration of labor. According to the forecast analysis, more than five million people will lose their jobs due to new production technologies and global robotization. The most significant staff reductions will affect office and administrative employees, as well as the production personnel of individual industries. At the same time, it is expected that the specified losses will be compensated due to the two-million growth of the number of engineering, financial and computer specialties. In addition, the implementation of principles of circular reproduction will provide new additional workplaces in the area of recycling and product life extension.

It should be emphasized that a reduction in the share of human labor against the background of implementation of global robotization will be maximally manifested in developing countries that experience shortage of groundbreaking technologies, offset by affordable man power. Eventually, many large enterprises will be returned to the United States and the EU, and developing countries will lose an important industrial resource.

Changes in the quantity and the structure of jobs are not the only consequences of the digital transformation of the economy and the implementation of principles of robotic circular reproduction that are significant for a human. The development of technologies will provoke a gap between income from labor and capital, which will lead to an increase in inequality among workers. The owners of intellectual and physical capital (developers, shareholders, investors) will get the maximum benefit from the upcoming changes. The employees with a low level of education and skills will become non-demanded over time in connection with the substitution of such labor with robotic resources.

In this context, it is obvious that governmental authorities already need to take actions to restructure the education system and train specialists, modernize the infrastructure, and create new workplaces. Changes in terms of development and implementation of a new progressive tax legislation are also needed in the regulatory field.

The implementation of the concept of robotized circular reproduction makes its adjustments in the processes of formation and development of human potential. The [12] declares that in the immediate future new technologies will make it possible to overcome the current biological limitations:

- devices on the human body that are part of the Internet of Things will transmit context-sensitive information that relates directly to our emotions, increasing their controllability;
- exoskeletons and prostheses connected to the brain will increase the physical strength and endurance of a human, return mobility to the elderly and people with disabilities;
- sensors and computers built into contact lenses and permanent implants will allow a human to hear whispers behind the walls, give a natural night vision and allow immersing in virtual and augmented realities;
- nontropic drugs will expand the cognitive abilities of a human and transform work and education.

Robotization which penetrates increasingly deeper into human life will soon expand its application horizons and will be integrated into new industries and activities. As a result, the risks of active implementation of robots are quite obvious. They are associated with potential unemployment, growing inequality and the exclusion of certain categories of citizens from the economy. Another negative consequence is the increased polarization of population of regions by such parameters as education, competency, age and sex. The elderly people of working age are not so responsive to environmental changes and are less prone to the development of new technologies. As for the polarization of society by the parameter of gender, this trend can be explained by the excess of the share of men over the share of women, employed in potentially less robotized production sectors. Differences in the degree of adaptation to the robotic circular reproduction of regions will contribute to the formation of zones of social tension, increased interregional migration flows, as well as decreased stability of the budget system of certain territories.

27.5 Conclusions and Recommendations

Transition to robotic circular reproduction in regional economic systems will make it possible to improve stability of their development, enhance environmental protection and minimize emissions of harmful substances. In order to improve efficiency of the implementation of principles of robotic circular reproduction in the Russian regions, additional measures are needed in terms of improvement of environmental legislation; state support for companies that implement principles of the circular economy; provision of incentives to scientific research activity in the circular economy; internal cooperation and integration of Russian companies into global networks of ecologically responsible business; public awareness of the health and environmental hazards posed by the current production concept.

That said, the high spatial differentiation of territories by the rate of adaptation and implementation of new technologies may result in the formation of so-called "old-industry" regions, having many acute social and economic problems, including ever-increasing unemployment. In order to overcome such situations, there is a need for a detailed analysis and operational assessment of interregional distinctions. Regional and federal authorities should implement the policy of "intelligent" specialty, develop measures aimed at reducing vulnerability of certain territories. In regions with high potential for the implementation of systems of robotized circular reproduction and digitization, an active stimulative innovation policy is required, that would be focused on formation and development of the innovation infrastructure, support for scientific and educational activity, and technology entrepreneurship. The attraction of investment in high tech and knowledge-intensive types of activities with mass implementation of systems for retraining, development of robots and software, as well as circular principles of closed-loop and non-waste production will contribute to the increased sustainable development of territories with strong concentration of manufacturers.

References

1. Lomovtseva OA (2010) Features of the regional reproduction process. In: Lomovtseva OA, Sopina NA (eds) Nauchnye Vedomosti. Series History. Politology. Economics. Computer Sciences, No 13 (84), pp 5–14
2. Pakhomova NV (2017) Transition to the circular economy and closed supply chains as a factor of sustainable development. In: Pakhomova NV, Rikhter KK, Vetrova MA (eds) Bulletin of the St. Petersburg State University. Economics 33(2):244–268
3. Ellen MacArthur Foundation (2015) Circularity indicators: an approach to measuring circularity (methodology). https://www.ellenmacarthurfoundation.org/programmes/insight/circularity-indicators
4. Foundation: Cowes, UK, 2015. https://www.ellenmacarthurfoundation.org/assets/downloads/publications/
5. Rose C, Eldridge S, Chapin L (2015) The internet of things: a brief overview. Questions and problems of using the Internet on a more global scale. https://www.internetsociety.org/wp-content/uploads/2015/10/report-InternetOfThings-20151221-ru.pdf

6. Myl'nik VV (2014) Robotization of industrial production based on artificial intelligence. In: Myl'nik VV, Myl'nik AV (eds). Organizator Proizvodstva 3:5–11
7. Chatham House (2017) A wider circle? The circular economy in developing countries. Briefing. The Royal Institute of International Affairs—Energy, Environment and Resources Department, December 2017
8. Shershunovich E (2018) Towards the circular economy: assessing the progress of Belarus in the system of countries. In: Shershunovich E, Tochitskaya I (eds) EC. BEROC Green Economy Policy Paper Series. PP GE no. 3
9. Benton D, Hazel J (2015) The circular economy in Japan. March 2015. The Institution of Environment Sciences. https://www.the-ies.org/analysis/circular-economy-japan
10. Kolmykova TS, Ostimuk OV, Obukhova AS (2017) Role of institutional investors in the resource provision of the region's economic. Growth J Adv Res Law Econ VIII, 8(30):2439–2448
11. Gerasimenko D (2017) The circular economy in Russia in the context of sustainable development goals of the UN and the year of ecology. In: Gerasimenko D, Nikolaeva I (eds). International Centre for Trade and Sustainable Development
12. Program for the development of the digital economy of the Russian Federation until 2035. Access mode: http://www.consultant.ru

Chapter 28
Sociodynamic Trends in Public Relations as Historical Inevitability

Tatyana A. Zhuravleva, Irina V. Skobliakova, Lina L. Garina, Irina O. Trubina and Innara R. Lyapina

Abstract The scientific article is based on modern historical features of social development trends. The authors adhere to the principles of dialectics and materialism. The scientific article identifies the priorities and bases of property relations. The purpose of the scientific article is the formation of sociodynamic trends in social relations as a historical inevitability. To achieve the goal it is necessary: to consider the theoretical aspects of social development trends, to single out the economic component of the sociodynamic cycle of social relations, to form the staged development of the sociodynamic cycle. The research methods are the theoretical analysis method, the method of historical stageization, the cyclic grouping method. The basis of the study is the theoretical elaboration of the issue of the sociodynamic cycle of social relations.

Keywords Sociodynamic trends · Sociodynamic cycles · Dialectics · Social reproduction · Structure of social needs · Large conjuncture cycles

JEL Codes B15 · B24 · P4

T. A. Zhuravleva (✉) · I. V. Skobliakova · L. L. Garina · I. O. Trubina · I. R. Lyapina
Orel State University, Orel, Russia
e-mail: orel-osu@mail.ru

I. V. Skobliakova
e-mail: my-orel-57@mail.ru

L. L. Garina
e-mail: 1278orel@mail.ru

I. O. Trubina
e-mail: super-ya-57@mail.ru

I. R. Lyapina
e-mail: innara_lapina@mail.ru

T. Kolmykova and E. V. Kharchenko (eds.), *Digital Future Economic Growth, Social Adaptation, and Technological Perspectives*, Lecture Notes in Networks and Systems 111, https://doi.org/10.1007/978-3-030-39797-5_28

28.1 Introduction

The dynamics of the development of socio-economic systems is determined by changes in characteristics. Processes are in the nature of individual and associated entities and institutions.

A number of socioeconomic trends began to appear in public life today. Firstly, the migration scales of displacements are observed empirically. The masses of the able-bodied population move to economically stronger regions. The masses of the working-age population consolidate the qualification and industry affiliation of the territories. The masses of the able-bodied population form large urban agglomerations. Secondly, the tendencies of depopulation of human resources are observed in the country. The birth rates of the first and second children per woman continue to fall rapidly. Mortality at working age is increasing in scale. The processes are typical for urban areas. The conditions of the processes are comfort and the change of the traditional way of life transforms the birth of a child into a burden. In rural areas, identical processes are observed. These processes lead to deserted spaces. Thirdly, information technology is spreading faster. Fourthly, market relations subordinated education and health. Market relations reinforce the processes of hereditary poverty and poor health. Fifthly, information and communication technologies form consumer and utilitarian thinking. The young generation, which is actively involved in the field of trade in the qualification positions of sellers after learning at school, can't do anything. Sixthly, the role of the traditional family is declining. The trends are driven by the computer industrial revolution. Women gained self-realization opportunities to an even greater extent. Many men have replaced traditionally male sectors of labor application. The main conclusion is the reluctance to take responsibility for the family.

The list is not complete in presenting the situation. The most important trends are observed in the processes.

28.2 Methodology

Representatives of the scientific community (K. Marx, L. von Mises, V. Oyken, C. Polanyi, F. von Hayek, J. Schumpeter) studied the topic from different perspectives. The evolution of socio-economic life is most fully revealed from the perspective of a systematic approach to the objects of change. The study is part of the methodology of dialectical analysis and cyclic dynamics.

28.3 Results

Person is the owner of factors of production. Production factors are expressed in the form of productive forces. Factors are expressed through productive forces, the interaction of person and nature in the process of social reproduction. Economic relations are the interaction between economic entities. Productive forces aren't limited to those quality characteristics. Productive forces consider the combination of man, tools and objects of labor. K. Marx in his books "Economic and Philosophical Manuscripts" says: "Person is an integral part of nature, interacts with nature, has the ability to reproduce such". Person is endowed with natural forces, vitality. Person is an active natural being. In nature, there are "objects of human needs". Needs are essential items for the conduct and approval of its essential forces [1]. The productive forces of individuals include the forces of social beings. These forces interact in society. These forces become real forces only in communication and in the mutual connection of these individuals [2].

The productive forces of individuals consist of the natural productive forces of individuals and the social productive forces of real individuals. The interaction of people with nature determines the development of actual individuals themselves. This process contributes to the creation of new forces, the improvement of communication methods, the formation of new needs, a new language. The development of the forces of individuals appears as a historically determined process of improving its natural and social qualities.

The forces of activity of individuals are embodied in the products of material and spiritual activity. According to K. Marx, productive forces include those that are ready-made and created by human activity [2]. Person refers to the material conditions of life as natural conditions. This is called person processing of nature [2].

The role of public institutions is high in the historical and evolutionary processes of development. Public institutions influence the development of specific processes. The main role belongs to the development of the economic system. Processes have a decisive influence on the nature of the development and interaction of various spheres of society.

The economic system is a unity of interacting economic systems. Economic systems are formed in the process of production, distribution, exchange and consumption, wealth. In the economic system, commodity exchange relations play a decisive role for development.

Under the influence of the development of the social division of labor, the decomposition of the community and the formation of private property took place. The social division of labor is based on the relationship of domination and submission. The division of labor affects the process of establishing and developing relations between the owners of the means of production and employees. The division of labor is a source of a high degree of growth of productive forces. The division of labor influences the formation of alienation and self-reliance with the simultaneous mutual dependence of these individuals.

The dialectical laws of the development of being and production relations presuppose the separate functions of production processes. The data suggest cooperation of labor by all methods of production.

The labor cooperation is implemented in the economic system. The labor cooperation unites isolated operations and has a direct impact on the development of productive forces. The higher the degree of realization of labor productivity, the greater the extent of the economy.

The cooperation of labor also affects the processes of distribution, exchange and consumption. Processes are indirect in nature. Processes differ from their direct influence in production processes. Trends and content of economic development form a system of social relations. Tugan–Baranovsky M. I. in the study "Periodic Industrial Crises" determines the need for the distribution of social labor. The social labor is divided into proportions and correlation between the cooperation of people and the distribution of their activities. The author points to the condition for preventing economic crises [3].

In economic relations, there are collective and individual processes. Processes determines the nature of people's activities in society. For example, the nature of people's activities is the social, the political, the moral, the spiritual. Theoretical reasoning confirms the nature of the economic system. The economic system includes the social form of ownership of the means of production. The social form of property is determined on the means of production.

Historical trends and practices have confirmed the imperfect nature of capitalist economies. The economy isn't efficient with a predominant form of state ownership. The evolution of development is characterized by a successive change in individualistic and social trends. The interests of a large part of the population are important.

The collective production of the primitive communal system gradually penetrates the division of labor. In the process, the collective nature of production and appropriation changes. The division of labor enhances the productive power of society. The division of labor leads to distribution. The importance of social order. The social system is characterized by the organization of industrial relations. The system of social order includes a certain distribution system. Distribution relations correspond to historically defined forms of the production process [4].

The distribution of commodities is based on material processes. Material processes include the distribution of means of production. Material processes characterize the distribution of members of society by industry. Distribution determines the organization of production and expresses the nature of the mode of production.

The causal relationship is related to market views. In market relations, competition is determined. Demand and supply are formed on the basis of competition. The process of exchange is associated with the production process. The exchange process is secondary to the social production. Exchange without production doesn't exist.

The development of the division of labor in the process of exchange leads to a reduction in costs. Costs reduce exchange time in market research. This sequence leads to the isolation of transport, banking and insurance. Warehouses are in general use.

The division of labor leads to a division of the production process. Processes have an impact on the division of production between cities with an orientation to their industry. Exchange facilitated the concentration and centralization of capital. The capital subjugated production. The cyclical nature of production is enhanced by trading capital. The capital solves the contradictions between production and consumption. Cyclical production is observed within industries and across the national economy.

Production and consumption collide through interests. Production and consumption are revealed through supply and demand. The final stage of the process is the growth of social wealth. Important is the satisfaction of consumer demand. Overproduction of goods is the cause of crises [4].

The structure of needs changes under the influence of factors. The main factors are globalization and internationalization of production, the formation of a digital economy, mass automation, computerization and robotization of human life processes, entry into the labor market of generations X and Z. Disproportions in reproduction are the cause of modern economic cycles and crises.

Concentration processes accompany the development of social production. New technologies require scientific development. Scientific developments are being introduced into mass production. Mass production is subject to motivational reasons for the specifics of business entities.

The evolutionary dynamics of systems is revealed in the processes of inheritance and variability. In the economic system, the functions of social systems are manifested. Culture is manifested in the objective conditions of life. Culture provides the adaptation of the economic system in connection with significant changes. Intergenerational communication arises through acceptance and adaptation to change. The processes contribute to the formation of the way of thinking and worldview. Processes act on individuals in relation to the material world, social and institutional environment. The decisive role in history belongs to the activities of people under objective conditions. Theory and practice ignore the objective determinants of activity. People get happiness for 10 years. Then people experience economic and social chaos [5]. At the center of the processes of variability and heredity is man.

Trends in changes are designated as non-adaptive excess forms of system activity [6]. In critical periods of the development of systems, the degree of non-adaptive excess forms of population activity increases [7]. Activity is manifested in the form of biological disasters and social crises.

The processes of socio-economic dynamics are revealed in specific quantitative indicators of individual economic systems. Qualitative characteristics interact with other systems. Trends are formed in the process of interaction of social development.

The depravity of the existing economic system lies in the production system. Laws affect the consumption system. The production system is subject to the consumption system. The production system influences the production system. Consumption laws are aimed at changing industrial processes. The causes of crises aren't enough to look for in certain areas of the economic system. The causes of crises are caused by the totality of the phenomena of the social economy. The reason for the recurrence of crises is social in nature. Crises can stretch for a long period. Crisis can be contracted depending on economic changes, political changes and other changes [3].

The systems approach is interlinked by economic processes. Economic processes are limited in time. Qualitative and quantitative changes in economic elements can be considered separately. The economic reality is a stream of continuous and manifold qualitative and quantitative changes [8]. The development of the economic system seems to be an irreversible process of transition from one stage or stage to another.

The trends in the variability of socio-economic development are facilitated by the institutions of society. Institutions determine written and generally accepted rules of behavior in society. Institutions consolidate the processes of interaction between business entities. Informal norms and rules interact with the political system, the social system and the economic system.

Periods of upward waves of large cycles are richer in large social upheavals and upheavals in society. The downward wave is weaker [8]. Economic correctness acts as a causal dependence of phenomena [8]. Wars and social upheavals are included in the process of developing large cycles. Large cycles are the initial forces for the development of large cycles. The scientist shows the principles of consistency. The principles of consistency are related and fundamental. Social disasters don't occur suddenly and from scratch (Fig. 28.1).

Wars arise on the basis of increasing the pace and tension of economic life. The periods have different conjuncture. The economic struggle for markets and raw materials escalates annually. Social upheavals arise during the period of rapid onslaught of new economic forces [8]. Socio-economic factors lead to social catastrophes. Social disasters are the consequences of a new stage in the social division of labor.

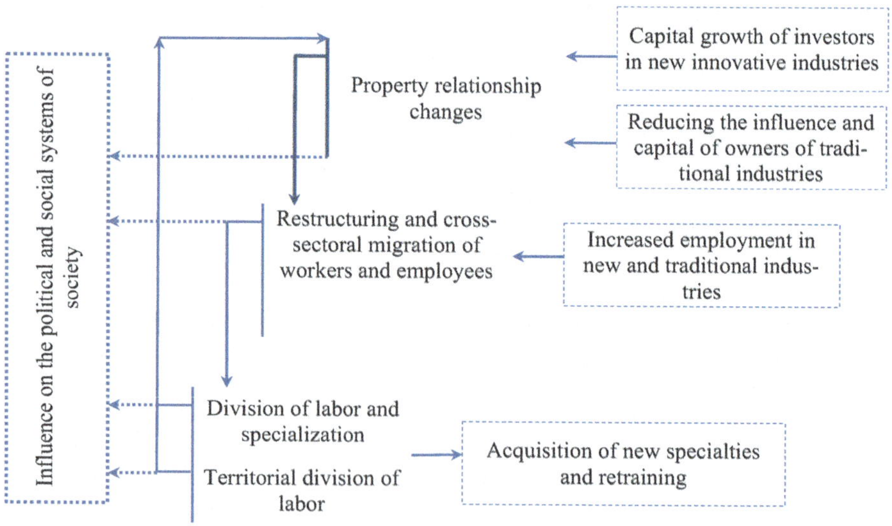

Fig. 28.1 The economic component of the socio-dynamic cycle of public relations

Table 28.1 Stages of development of the sociodynamic cycle

At the first stage	In the second stage	In the third stage
Change in mindset. The individual is related to the natural and social environment. The result of the selection process is the selection of individuals. Factors form the prevailing types of relationships and spiritual position. Changing economic institutions leads to the adaptation of habits and individual people to a changing environment	The individual contains two opposite tendencies of the evolving system: conservation and change. It reflects the evolutionary processes of developing systems. The individual makes changes. Changes are a necessary part of the overall socio-economic development	Discoveries and inventions correspond to the level of the social division of labor. The division of labor contributes to the implementation of engineering and technology in society. The division of labor becomes the result of the growth and complexity of technology. The division of labor complicates the structure of material production and the structure of society as a whole

Military conflicts on the territory of Russia may arise due to a change in world leader in the period 2024–2025. The basis of the conflict is connected with the confrontation between the USA and Russia. The global trend is pulling up under the theory of terrorist threats. Theories are based on economics and a new stage in changes in property relations. In the United States, striking workers lynched black strikebreakers a hundred years ago. This conflict was called racism. This conflict was called the struggle for survival. The labor and the capital are the socio-political public background of the conflict [9].

Changes have a significant impact on the political, social and other systems of society. The main factors are property relations and the system of production relations. In this process, restructuring and intersectoral and territorial migration of the population are advocated. Changes in the private division of labor have an impact on public institutions. Changes affect the institutional structure.

The development of the sociodynamic cycle takes place in several stages. Elements of this mechanism are closely related (Table 28.1).

28.4 Conclusions

The new situation leads to a change in the system of relations between society and individuals. Changes determine the development of the society itself. Changes require a new adaptation to changing conditions. The structure of society is changing through the scrambling organization of social development. Sociodynamic trends are preceded by social trends. Elites split and begin to share the views of most. Social disasters aren't inevitable in this situation. Conflicts are divided by the degree of tension character. For example, a war, a revolution, a putsch. Changes lead to the transformation of economic institutions. The paradigm indicates the beginning of the formation of a new cycle of the socio-economic development.

Acknowledgements This article was prepared with financial support of RFBR under the project 19-010-00124A "The theory of cyclical development of the modern economy".

References

1. Marx K, Engels F (1981) Composition, 2nd edn. Polytizdat, Moscow, pp 162–163
2. Marx K, Engels F (1966) Selected works in three volumes. Politizdat, Moscow, p 69
3. Tugan-Baranovsky MI (1923) Periodic industrial crises. History of English crises. General theory of crises. Publishing House of the Book Cooperative Partnership under the Smolgubkom of the RCP, Smolensk, p 282
4. Melnik MS (2014) The polysystemic paradigm in the theory of economic cycles: the dissertation of a doctor of economic sciences
5. Kovalzon MYa (1985) Materialistic understanding of history and activity issues. Quest Philos 3:33–36
6. Asmolov AG (1982) Basic principles of psychological analysis in activity theory. Quest Philos 2:14–27
7. Petrovsky VA (1981) Toward an understanding of personality in psychology. Psychol Issues 2:40–56
8. Kondratiev ND (2002) Big business cycles and foresight theory. Economics, Moscow, pp 24–25
9. Yamshchikova TN, Lukyanchikova TN (2017) Labor and capital: labor conflicts or social partnership. Economist 5:77–88

Chapter 29
The Role of Small and Medium Businesses in the Formation of Pleasure Economy

Tatyana A. Zabaznova, Elena V. Patsiuk, Natalya V. Shchukina, Svetlana E. Karpushova and Olga A. Surkova

Abstract *Purpose*: This paper is aimed at determining the role of small and medium businesses in the formation of pleasure economy. *Design/methodology/approach*: This paper describes the regression analysis method, which is used to determine the influence of the share of small and medium businesses on the level of happiness in the economy. In addition, the authors use the method of experiment in the form of a survey which is intended to determine the differences in the level of happiness of small and medium businesses compared to that of big business. *Findings*: Significant differences have been identified in the needs of big, small and medium business owners. When doing big business, its owners have the needs that are referred to in this paper as "lucratively inclined", because their satisfaction makes only business owners happy and in many cases causes costs for stakeholders. In contrast, small and medium business owners have fundamentally different needs referred to as "not lucratively inclined", because their satisfaction makes business owners happy and does not cause any costs for stakeholders, but rather can provide benefits for them. *Originality/value*: it has been substantiated that small and medium businesses have significant potential to help build pleasure economy, no less than that of big business. However, as exemplified by present-day Russia, an adverse market environment forces small and medium business owners to abandon the idea of doing business that brings pleasure with a view to making it break-even and maintaining their market positions. As a result, currently small and medium businesses are of little importance in building pleasure economy. A set of recommendations has been put forward that will help maintain the efficiency and competitiveness of small and medium businesses

T. A. Zabaznova (✉) · E. V. Patsiuk · N. V. Shchukina · S. E. Karpushova · O. A. Surkova
Volgograd State Technical University, Volgograd, Russia
e-mail: tazabaznova@yandex.ru

E. V. Patsiuk
e-mail: elenapatsyuk@yandex.ru

N. V. Shchukina
e-mail: schuka-amurskay@yandex.ru

S. E. Karpushova
e-mail: sfkse@yandex.ru

T. Kolmykova and E. V. Kharchenko (eds.), *Digital Future Economic Growth, Social Adaptation, and Technological Perspectives*, Lecture Notes in Networks and Systems 111, https://doi.org/10.1007/978-3-030-39797-5_29

while bringing pleasure to its owners. This will provide a significant increase in the contribution of small and medium businesses to the formation of pleasure economy.

Keywords Small and medium businesses · Big business · Entrepreneurial activity · Russia · Needs · Happiness · Pleasure economy

JEL Codes D21 · D22 · D61 · D62 · D63 · D64 · D92 · J28 · L26 · M21

29.1 Introduction

Pleasure economy is an economic and social system, the entities of which are happy (take pleasure) in meeting their needs in the process of economic activity. However, in the same manner as the needs can be different—mass or unique—the methods of becoming happy while meeting these needs are highly differentiated. Since business is one of the most common types of economic activity, it is essential in the formation of pleasure economy. The most general classification of business—by its size—implies segregation of small and medium businesses from the big business.

Although the activity of small and medium businesses has important economic significance associated with accelerated economic growth and innovative development of the economy, the possibilities of taking pleasure in doing these businesses appear to be limited. Thus, the amount of revenue (and profits, accordingly) of small and medium businesses is usually lower as compared to that of the big business; moreover, small and medium business entities are often unprofitable, especially during start-ups. Small and medium businesses almost never possess large market power—instead, in most cases, it largely cannot have a visible effect on the market.

Small and medium businesses have little to no capabilities to influence state regulation of the market. The prospects of transnationalization of business while maintaining its small or medium size are illusive even with the electronic form of its organization. In addition, small and medium businesses often do not have sufficient available resources (primarily financial) for the implementation of their own development projects (for example, innovative), which impedes the realization of its potential and the strengthening of its market positions. Small and medium businesses are characterized by an increased level of risk and, accordingly, reduced stability, leading to its more frequent crises as compared to the big business.

These aspects fundamentally differentiate small and medium businesses from the big business which may satisfy the needs of its owners, which include the needs for high incomes, for exercising their "power" over the market and lobbying their interests through communication with state regulators, as well as for expanding and developing their business. In contrast to the big business, small and medium businesses seem to be unable to bring pleasure, since the only obvious need that it is able to satisfy is the unlocking of entrepreneurial talent (self-fulfillment).

The remaining needs that are allegedly typical for most entrepreneurs, can make small and medium business owners unhappy because of dissatisfaction of these needs, as well as in case of a highly probable business crisis. For this reason, we put forward the working hypothesis that small and medium businesses play a small, or even negative (hindering) part in the formation of pleasure economy. This paper is aimed at determining the role of small and medium businesses in the formation of pleasure economy.

29.2 Materials and Method

Conceptual opportunities and practical experience of taking pleasure in doing business are presented in the writings of researchers such as Battersby [1], Belhabib et al. [2], Broome [3], Cikic et al. [4], Dirks [5], Horton and Sloan [6], Malaquias and Hwang [7], Manning and Bungay [8]. The theory and practice of doing small and medium business is reviewed in publications by scholars such as Buffa et al. [9], Caldera et al. [10], Mejri et al. [11], Müller [12], Popkova [13], Pritvorova et al. [14], Sergi and Scanlon [15].

However, the specifics of taking pleasure in doing small and medium businesses has been understudied. We use the following methods to study it within the scope of our work:

1. Regression analysis method, which is used to determine the influence of the share of small and medium businesses on the level of happiness in the economy;
2. Method of experiment in the form of a survey which is intended to determine the differences in the level of happiness of small and medium businesses compared to that of big business.

29.3 Results

We have selected the TOP 10 countries of the world with the highest share of small and medium businesses in the structure of entrepreneurship as the objects of our regression analysis (Fig. 29.1).

As can be seen from Fig. 29.1, the highest share of small and medium businesses in the structure of entrepreneurship in 2018 can be observed in the Czech Republic (98%), Italy (97%) and Spain (97%). Russia ranks fourth in terms of the number of entrepreneurs; the share of small and medium businesses in its structure is 95%. Happiness index values in selected countries in 2018 are shown in Fig. 29.2.

As can be seen from Fig. 29.2, the highest happiness index value among selected countries in 2018 can be observed in the United States (6.892 points), and the lowest happiness index value among selected countries in 2018 can be observed in Turkey (5.373 points). In Russia, happiness index value was 5.648 points. The results of regression analysis are shown in Fig. 29.3.

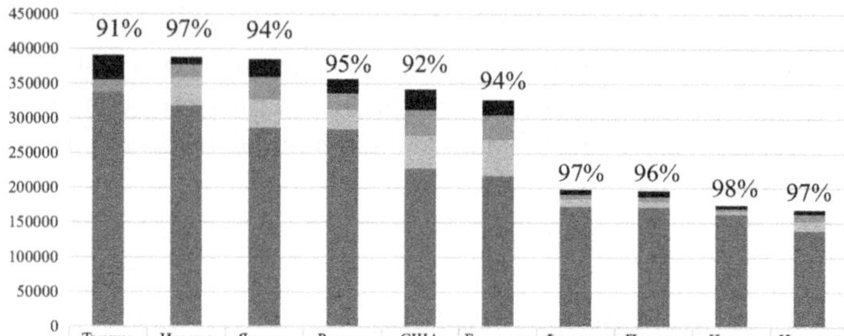

	Турция	Италия	Япония	Россия	США	Бразилия	Франция	Польша	Чехия	Испания
■ Остальной бизнес, шт.	34640	9727	24733	19604	28737	20859	5741	7977	3934	5529
■ 20-49 работников (средний бизнес), шт.	18037	19194	32915	23787	37495	36051	7128	7394	3932	10538
■ 10-19 работников (малый бизнес), шт.	0	39924	40011	28495	46635	52029	10765	7789	4483	13685
■ 1-9 работников (микро-бизнес), шт.	338347	319021	287122	284981	229045	218056	174023	172907	163076	138965

Fig. 29.1 Share of small and medium businesses in the structure of entrepreneurship in selected countries in 2018, %. *Source* Compiled and calculated by the authors based on the data of OECD [16]

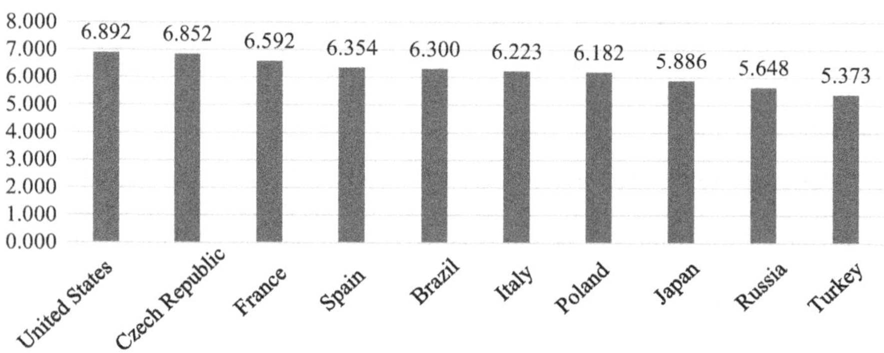

Fig. 29.2 Happiness index in selected countries in 2018, points 0–10. *Source* Compiled by the authors based on the data of Helliwell et al. [17]

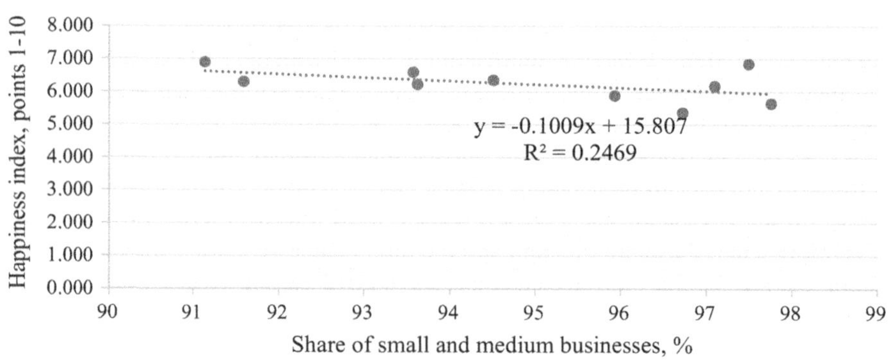

Fig. 29.3 Regression curve of dependence of happiness index value on the share of small and medium businesses in the structure of entrepreneurship. *Source* Calculated and plotted by the authors

According to the regression curve that is plotted in Fig. 29.3, the increase in the share of small and medium businesses in the structure of entrepreneurship by 1% results in reduction of happiness index value by 0.1009 points (the correlation is weak: 24.69%). As a result, currently small and medium businesses do not bring any pleasure to their owners, and their influence on the formation of pleasure economy is insignificant. In order to explain the logic of mathematical (econometric) calculations, we shall take the results of our survey that was conducted among 2500 big business owners and 2500 small and medium business owners in Russia in August 2019 (Figs. 29.4, 29.5 and 29.6).

As is shown in Fig. 29.4, big and medium business owners (7.64 points) take much more pleasure in doing business as compared to small and medium business owners (4.32 points).

As is shown in Fig. 29.5, aspects 1–6 (6.55 points on average) turned out to be the most significant for the big business: realization of profit (9.81 points), power in the

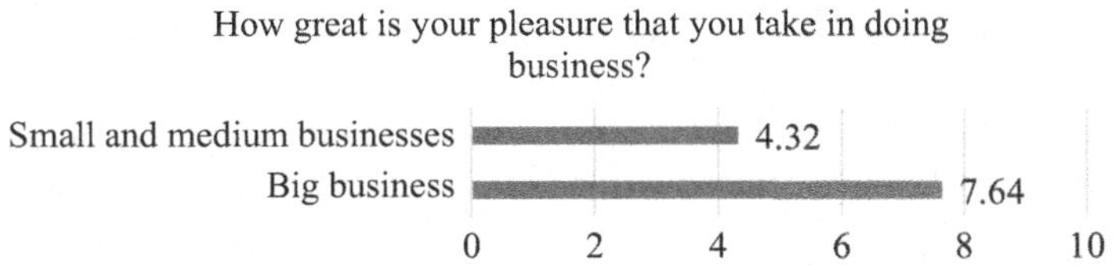

Fig. 29.4 Pleasure taken by entrepreneurs in doing business of various scale, points 1–10. *Source* Calculated and plotted by the authors

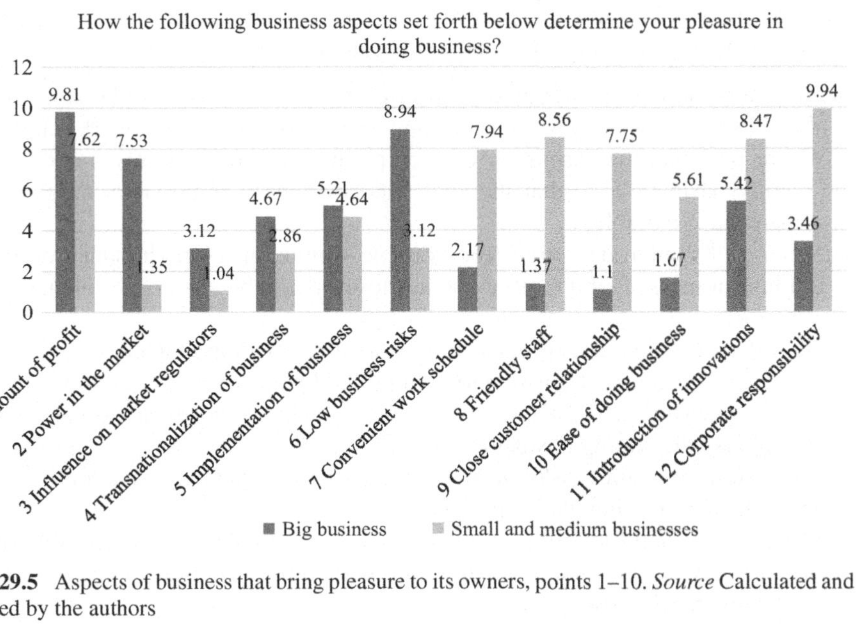

Fig. 29.5 Aspects of business that bring pleasure to its owners, points 1–10. *Source* Calculated and plotted by the authors

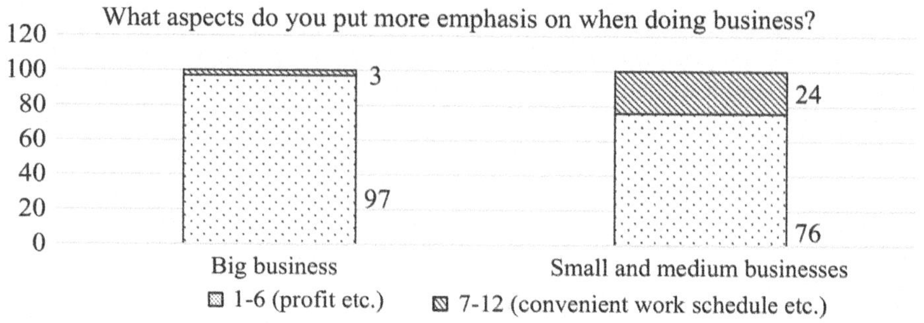

Fig. 29.6 Structure of priorities in doing big, small and medium businesses, %. *Source* Calculated and plotted by the authors

market (7.53 points), influence on market regulators (3.12 points), implementation of business development projects (4.67 points) and low business risks (8.94 points). As contrasted to it, aspects 7–12 turned out to be the most significant for small and medium businesses (8.05 points on average): convenient work schedule (7.94 points), friendly staff (8.56 points), close customer relationship (7.75 points), ease of doing business (5.61 points), introduction of innovations (8.47 points), and corporate responsibility (9.94 points).

As is shown in Fig. 29.6, aspects 7–12 (profit etc.) turned out to be top-priority both for big, small and medium businesses: 97% and 76% respectively. Hence, small and medium business owners conduct entrepreneurial activity to the prejudice of their own interest, which ensures breakeven result, but prevents them from taking pleasure in doing business. In order to solve this problem, we provide the following recommendations for small and medium business owners:

- The transition to the electronic form of doing business, if possible, since it will allow an entrepreneur to work according to the most convenient work schedule;
- Where possible, the formation of a corporate team from relatives and friends ("family business"), so that the staff was as friendly as possible, and working climate was favorable;
- The use of digital marketing (for example, communications in social networks and instant messengers) to establish as close as possible ("warm") relations with consumers;
- The fullest possible use of opportunities to simplify doing business provided by the state (for example, transition to a simplified financial reporting regime, including tax reporting and taxation practices), to avoid unnecessary organizational difficulties when doing business;
- The insurance against risks in order to maintain corporate sustainability in the implementation of innovative projects;
- The inclusion of information on the social and environmental responsibility of the enterprise in its corporate reporting and marketing communications, in order to inform all stakeholders of it.

The above recommendations are designed to ensure satisfaction of the needs of small and medium business owners along with the needs of society, thereby increasing the competitiveness of business and provide entrepreneurs with the opportunity to enjoy while realizing a profit from business.

29.4 Conclusion

The results of research have disproved this hypothesis—the role of small and medium businesses in the formation of pleasure economy is not obvious and requires in-depth study with emphasis on unique needs of equal business owners. As has been initially expected, when doing small and medium business, the opportunities for meeting the mass needs that are typical for most extrepreneurs—realization of profit, exercise of power in the market, exercise of influence on state regulators, transnationalization of business, implementation of large-scale business development projects (the need for self-expression, self-fulfillment), maintaining its stability (the need for security),—are limited as compared to the big business. These needs can be conveniently referred to as "lucratively inclined", because their satisfaction makes only business owners happy and in many cases causes costs (negative externalities) for stakeholders, for example, deterioration of environment as a result of expanded activities of large manufacturing enterprises.

At the same time, we have found that small and medium business owners have fundamentally different needs—they can be conveniently referred to as "not lucratively inclined", because their satisfaction makes business owners happy and does not cause any costs for stakeholders, but rather can provide benefits for them (positive externalities), for example, improvement of the quality of life of society through the implementation of initiatives of corporate responsibility of business. These needs include convenient work schedule, friendly staff, close customer relationship, ease of doing business (the need for convenience), introduction of innovations, and corporate responsibility (the need for self-expression, self-fulfillment).

In this regard, small and medium businesses have significant potential to help build pleasure economy, no less than that of big business. However, as exemplified by present-day Russia, an adverse market environment (aggressive competition, reliance on sanctions—penalties, withdrawal of licences—in case of state regulation) forces small and medium business owners to abandon the idea of doing business that brings pleasure with a view to making it break-even and maintaining their market positions. As a result, currently small and medium businesses are of little importance in building pleasure economy in Russia, and, probably, in many other countries of the world. With due consideration of identified specific needs of small and medium business owners, we have put forward a set of recommendations that will help maintain the efficiency and competitiveness of small and medium businesses while bringing pleasure to its owners. This will provide a significant increase in the contribution of small and medium businesses to the formation of pleasure economy.

References

1. Battersby ME (2015) Business and pleasure: attending trade shows is a valuable part of business strategy-with tax benefits. Spec Fabr Rev 2015(1):28–39
2. Belhabib D, Campredon P, Lazar N, Kane EA, Pauly D (2016) Best for pleasure, not for business: Evaluating recreational marine fisheries in West Africa using unconventional sources of data. Palgrave Commun 2(1):15–50
3. Broome B (2018) Mixing business and pleasure. Arch Rec 206(4):88–92
4. Cikic J, Jovanovic T, Nedeljkovic M (2018) Business and/or pleasure—gender (in)equalities in rural tourism in Vojvodina. J Agric Sci Technol 20(7):1341–1352
5. Dirks D (2012) Business and pleasure. Confrontation 112(1):132–145
6. Horton S, Sloan D (2016) Accessibility for business and pleasure. Interactions 23(1):80–84
7. Malaquias RF, Hwang Y (2017) Mixing business and pleasure: empirical implications for trust in mobile banking. J Electron Commer Res 18(3):212–224
8. Manning E, Bungay V (2017) Business before pleasure': the golden rule of sex work, payment schedules and gendered experiences of violence. Cult Health Sex 19(3):338–351
9. Buffa F, Franch M, Rizio D (2018) Environmental management practices for sustainable business models in small and medium sized hotel enterprises. J Clean Prod 194(1):656–664
10. Caldera HTS, Desha C, Dawes L (2018) Exploring the characteristics of sustainable business practice in small and medium-sized enterprises: experiences from the Australian manufacturing industry. J Clean Prod 177(1):338–349
11. Mejri K, MacVaugh JA, Tsagdis D (2018) Knowledge configurations of small and medium-sized knowledge-intensive firms in a developing economy: a knowledge-based view of business-to-business internationalization. Ind Mark Manag 71(1):160–170
12. Müller JM (2019) Business model innovation in small- and medium-sized enterprises: Strategies for industry 4.0 providers and users. J Manuf Technol Manag 2(1):34–46
13. Popkova EG, Morozova IA, Litvinova TN (2017) New challenges for human capital from the positions of its infrastructural role in the system of entrepreneurship. In: Human capital: perspectives, challenges and future directions. Nova Science Publishers, Inc., New York, pp 257–275
14. Pritvorova T, Tasbulatova B, Petrenko E (2018) Possibilities of blitz-psychograms as a tool for human resource management in the supporting system of hardiness of company. Entrep Sustain Issues 6(2):840–853. https://doi.org/10.9770/jesi.2018.6.2(25)
15. Sergi BS, Scanlon CC (eds) (2019) Entrepreneurship and development in the 21st century. Emerald Publishing, Bingley, UK
16. OECD (2019) Enterprises by business size [online]. https://data.oecd.org/entrepreneur/enterprises-by-business-size.htm. Accessed 7 Sept 2019
17. Helliwell JF, Layard R, Sachs JD (2019) World happiness report—2019 [online]. https://worldhappiness.report/ed/2018/. Accessed 7 Sept 2019

Chapter 30
Overview of Perspective Educational Services of the "Green" Digital Future: Online, Lifelong and Remote Learning

Oksana M. Makhalina, Victor N. Makhalin and Anatoliy B. Yaroshchuk

Abstract The article studies certain organizational issues related to implementation of the Federal project "Personnel for the digital economy". The authors formulate the goals and tasks of the project and specific features of the "green" digital economy, perform an overview of the existing directions in the educational process, and substantiate the creative directions of digitization of education. Lifelong learning, remote, and online learning are considered in detail. Recommendations for improvement of the process of personnel training for the digital economy are offered. Pursuant to the Decree of the President of the Russian Federation dated May 7, 2018, No. 204 "Regarding national goals and strategic tasks of development of the Russian Federation until 2024" [1], the Council for strategic development and national projects with the President of the Russian Federation adopted the passport of the national program "Digital economy of the Russian Federation" [2], which includes six federal projects: Normative regulation of the digital environment; Information security; Information infrastructure; Digital technologies; Personnel for the digital economy; Digital state management. The period of the program is 2018–2024 [3]. Here we dwell only on the federal project "Personnel for the digital economy". 82 organizations participated in preparation of the project—from federal ministries and state corporations to commercial structures (Rosatom, Rostelekom, Sberbank, Krok, Yandex, etc.), and educational and public organizations. The government committee on usage of information technologies for improvement of quality of life and terms of entrepreneurial activities adopted the plan of measures for "Personnel and education" of the program "Digital economy of the Russian Federation". The measures of this plan are aimed at implementing a range of the key directions of development of the system of education: update of contents, creation of necessary modern infrastructure, training of personnel for the work in the system, their advanced training, and creation

O. M. Makhalina (✉) · V. N. Makhalin · A. B. Yaroshchuk
State University of Management, Moscow, Russia
e-mail: moxanam@mail.ru

V. N. Makhalin
e-mail: mahalinviktor@mail.ru

A. B. Yaroshchuk
e-mail: yab.58@mail.ru

T. Kolmykova and E. V. Kharchenko (eds.), *Digital Future Economic Growth, Social Adaptation, and Technological Perspectives*, Lecture Notes in Networks and Systems 111, https://doi.org/10.1007/978-3-030-39797-5_30

of the most effective mechanism of the sphere management. These key directions could be specified by grouping separate measures into goals or tasks, which should have quantitative, qualitative, and time characteristics [4].

Keywords "Green" digital economy · Lifelong learning · Online learning · Digital certificates · Digital universities

30.1 Creative Directions of Digitization of Education in Russia

30.1.1 Goals and Tasks of Improvement of Education in the Digital Economy

The planned indicators in the Federal project "Personnel and education" envisage achievement of the following indicators by 2024: number of graduates of universities for specialty "IT"—120,000 per year; number of graduates of universities and establishments of secondary vocational education who have certain competencies in the sphere of IT at the medium level—800,000 per year; share of population with digital skills—40% [3].

International studies show that most countries face the lack of highly technical specialists. According to Korn Ferry, their deficit will constitute 85,2 million by 2030. Most (74%) managers of the largest organizations of the world think that technologies in the future will bring larger contribution into creation of added value of business than human factor. Highly skilled specialists will be especially valuable [5].

The number of IT specialists who graduate from Russian universities cannot fully satisfy the market's needs: the deficit is one million people. Besides, it should be taken into account that demand for them grows by 25–30% annually. How could this demand be satisfied? Which educational establishments should be created? What is to be changed in the existing system of personnel training? When solving these problems, it is necessary to use the following provisions:

- knowledge and technologies become obsolete very quickly, so it is necessary to develop new educational programs for students, which will allow for quick receipt of new knowledge and competencies;
- it is necessary to change the requirements to training and selection of lecturers and teachers for specialists, emphasizing the involvement of successful business managers in this process [6];
- when evaluating the level of specialist's qualification, it is necessary to avoid the control of the process of training and to focus only on control of the received result—i.e., to evaluate the received practical skills and knowledge;
- increase of the number of specialists on IT in Russia by 2024 by two times should take place not only by means of the youth but also by means of experienced workers of other specialties.

Besides, when solving the issues of personnel provision of the digital economy the fact that society enters the age of the "green" digital economy will be reflected in saving the expenditures of labor resources, change of the set of classic professions, and improvement of labor relations. Let us consider some of them:

- the largest personnel changes will influence the employees of traditional industrial companies, which labor will be controlled by AI;
- professions connected to unskilled and monotonous labor will disappear;
- development of the IT sphere will be accelerated. IT workers might have a monopoly in the Russian labor market;
- demand for highly-specialized specialists grows, and requirements to experience of work and the set of competencies increase;
- most of labor relations will become remote. Employers will save on offices and HR, and employees will be able to have several jobs simultaneously;
- by 2020, the number of elderly people will reach 40 million, and the "silver economy" is aimed at adapting the people of "pre-retirement" and retirement age to the modern realia.

There are a lot of various methods of obtaining knowledge. Effectiveness of these methods for development of creative skills is different. For the modern education to conform to the needs of effective creative development of companies, it is necessary to use creative education.

The notion "creative education" is widely used in the Western and Russian scientific thought and has a specific practical content. The most important feature of creative education is motivation. Human consciously tries to raise the level of his knowledge, which influences the level of his intellectual development and the possibility of occupying a competitive position in the labor market [7].

Creative education is education that is oriented at development of human's creative abilities and establishment of a setting for innovations, which includes analysis of the problems and possible variants of activities. This education motivates independent consideration of reality, self-cognition, individuality, and turning knowledge into potential of thinking and self-development [8, 9].

30.1.2 Lifelong Learning (Lifelong Learning)

Creative education belongs to special, together with lifelong, education, which is treated in different literature sources as "Lifelong learning", "Education for adults", "Lifelong vocational education", etc. The systemic factor of lifelong learning is development of human as a personality and subject of activities and communication in the course of the whole life. Lifelong learning allows human to form and satisfy his spiritual needs, develop skills with the help of different forms of education, and to conduct self-education. In the social policy of the state, lifelong learning creates favorable conditions for general professional development of human, provides a mechanism of reproduction of professional and cultural potential, and stimulates the development of public production [10].

The decisive role in provision of growth of a national economy will belong to the modern and high-quality training of specialists for satisfying the needs of science, economy, society on the whole, and separate citizens. In the conditions of quick accumulation of knowledge and update of technologies, this means that one of the most important tasks is provision of lifelong learning, which will provide each member of community with a chance of constant improvement, creative and professional development, expansion of knowledge, skills, and competencies, and information security in the course of the whole life [11]. Organization of lifelong learning could be presented in the following sequence:

- in schools—searching and supporting talented youth and stimulation of development of the quality of teaching math, IT, and technologies;
- in universities (professional education)—increase of training in IT specialties. Integration of the basic competencies of the digital economy into educational programs;
- in supplementary vocational education—personal digital certificates for development of competencies of the digital economy. Specialized programs of increase of qualification and advanced training for certain target groups;
- in mass online learning—programs of development of digital literacy for the key specialties, including for elderly people.

Let us consider two forms of lifelong creative education: supplementary vocational education and mass online learning. Organization of supplementary vocational education is based on the state system of personal digital certificates, which will allow one million people (in 2019–2024) to take training in the basic competencies of the digital economy. Certificate is financial assets provided by the state to children and adults for studying the state and directions development of digital technologies for provision of competitiveness in the labor market. This certificate's equivalent is RUB 10,000. Apart from the system of personal certificates for advanced training of specialists, at least 10 million people in 2019–2024 will take training on various programs of development of digital literacy [3]. The first 5,000 free certificates will be issued until the end of 2019. Another 300,000 certificates will be issued in 2020. Certificates will be given primarily to adults who want to acquire digital competencies. The certificates could be used for supplementary vocational education [12].

An important element of the system of lifelong learning is online learning, which allows everyone—regardless of work schedule, location, and financial opportunities—to take training. That's why the segment of online education is developing very quickly in the Russian segment of Internet, with a lot of courses for students of different ages. Depending on the level of training, skills, experience of work, and free time, potential students could select the course that is the best for them and that will allow mastering the competencies that are necessary for working in the sphere of the digital economy [10]. The program "Digital economy" sets the growth of the number of people which master at least 50% of educational programs in the form of mass open online courses. The number should be 100,000 by 2022, and 1 million by 2025 [13].

A lot of countries work on personnel training for new economy. The experts of the World Economic Forum analyzed 15 developed and developing economies and forecasted that 7.1 million jobs, of which 70% are office and administrative ones, will disappear by 2020. This is noted in the report "Russia 2025: from personnel to talents" of the Boston Consulting Group (BCG). However, new technologies not only replaced the human's work but also changed its structure and created new employment. According to the WEF, 2 million new jobs could appear in 15 studied countries by 2020 [9, 14].

30.2 Remote and Online Learning in the Age of Digitization in Russia

30.2.1 Remote Learning (e-Learning)

Remote learning (e-learning) is a decent alternative to the traditional learning. It is preferred in the corporate and state spheres, as it is the only method of quick learning with minimal expenditures. Remote learning is conducted with the help of PC's and telecommunication technologies and means. The subject of remote education is not anywhere near the pedagogue or educational resources. Let us dwell on the history of development and advantages and problems of remote education.

Usage of Internet and merging of the advantages of remote education and opportunities of telecommunications allowed for a new approach to solving the issues of studying based on a comprehensive personal experience with the help of innovative models of training and the world experience in the sphere of education. A guarantee of effectiveness of this form of learning consists in structural changes of students' personal experience. A decisive mechanism of such change is transformation of the essential landmarks of personality. The models of remote education that are based on usage of Internet technologies are offered by universities and colleges in the technical and humanitarian spheres. The main advantages of remote education are as follows:

- freedom in selecting the location, time, and speed of learning, which makes remote learning very popular among the users that cannot take intramural studies;
- lower cost of training—by 32–45% on average, according to Cedar Group;
- higher effectiveness—according to Cedar Group, learning time reduces by 35–45%, and speed of mastering of the material grows by 15–25%;
- allows increasing the quality of education by means of wide usage of international educational resources and increase of the share of independent mastering of material.
- early mastering of the skills with information and communication technologies, which allows increasing the effectiveness of using knowledge in economy.

- allows solving the task of training of a large number of an organization's employees in a minimal period of time, with optimal expenditures.
- remote and intramural learning should not be opposed to each other—these are different but mutually supplementing forms of education.

30.2.2 Online Learning

The total volume of revenues from educational resources in the world in 2018 constituted more than USD 4.5 trillion; according to the forecasts, it could reach USD 7 trillion. The share of online learning is 3%—USD 165 billion. Annual growth of revenues from remote education is 5%. Based on this indicator, it is possible to compile a forecast of increase of revenues from online education for the future five years—USD 210 billion in 2023 [15].

As for the market of online learning in the Russian Federation, it is still at the stage of its development. The volume of the educational market in 2016 constituted RUB 1.8 trillion. The online segment of RUB 20.7 billion accounted for 1.1%. According to the forecasts, it will reach 2.6% by 2021. Remote education in Russia—sphere of language training and mastering supplementary vocational skills—account for 7% of the existing programs. It is rather popular among school pupils, as it is an additional possibility of studying certain disciplines (2.7% of all programs).

Mass open online courses appeared in the sphere of remote provision in 2008, but they became really popular in 2012. Mass open online courses are a separate methodology of remote education, which includes a large complex of different activities with the following advantages: interactive character, useful connections, quick feedback and assessment, best lecturers in the world, free nature of the courses, new specialties, structured educational material, free schedule, and mixed system of online studies. This system allows obtaining higher education in any university of the world and raising qualification or taking advanced training [16].

30.3 Digital Universities and Companies in Digital Training of Personnel

30.3.1 The Concept of Digital University

Several Russian universities are to receive grants for creation and functioning of the centers for developing the models of a digital university. The first competitions for such grants will take place in the second part of the year. Financing for five centers will be provided in 2019, and financing for fifteen more centers will be provided in 2020. The grant's size—RUB 100 million [17].

The concept of the model of a digital university is the information system of managing the university, online support for the educational process, the key competencies of the digital economy, and managing the educational process based on the individual educational trajectory [17]. University of National Technological Initiative "20.35" is the first Russian university that ensures professional development of a human in the digital economy. It is oriented at training of leaders of companies and specialist that work in new global markets. The program will include such specialties as "Data-based management", "Leaders of communities", and "Technological leaders". The University of National Technological Initiative "20.35" (center of competencies for specialty "Personnel for the digital economy") emphasizes that by 2024 the elements of the models of digital university should be implemented in all Russian universities. Each student has to have access to the required educational content, effective technologies of studies, and digital support services.

The basic model of competencies for the digital economy is to be developed by the end of the year. It has to conform to the interests of business and the labor market. This new tool is to allow actualizing the state standards in the sphere of education. University of National Technological Initiative "20.35" analyzed which skills and knowledge are in demand in 50 leading companies and distinguished 211 competencies that are divided into nine blocks. The most popular block for the employers is "thinking".

30.3.2 Companies in Digital Training of Personnel

More and more universities, together with IT companies, start special programs for training of digital specialists. Let us consider these activities by the example of Microsoft. Microsoft YouthSpark's goal is to provide young people with necessary conditions for obtaining new knowledge, opportunities for employment, and development of their own projects. Microsoft Student Partners is a global community of student leaders who desire to share their technological knowledge with colleagues. Also, there is an educational initiative Microsoft Imagine Academy—free access to educational programs and certification for students and lecturers in IT, efficiency, and other specialties.

Microsoft program "Your course: IT for youth" has been active in Russia for more than six years now. Within this program, volunteers, coaches, and pedagogues in regions—on the basis of schools, libraries, and universities—conducted more than 28,000 free events, courses, and workshops aimed at stimulating young people in career guidance in the IT sphere, basic programming, increase of computer literacy, and development of digital skills. The project's events covered 5,300 schools, with more than 1,300,000 pupils aged 6–18 [18].

30.4 Conclusions

For the concept of lifelong learning to become a norm for regular change of professions, the structure of online learning has to develop and the society's attitude to learning has to change. While the first task is directly connected to development of online platforms, software, and digitization of content, the second task is connected to development of human's internal motivation for studies. The approaches to motivation in education consist not only in providing citizens with conditions for studying but also in selecting the approach for each age group. This is due to the fact that there are no parameters of selection of citizens who could take reeducation by digital certificates. That's why objective self-assessment of the level of professional qualification, digital literacy, skills with languages, and desire to receive a new job, transfer to a job in another sphere, or change the place of residence are motivation of readiness for studies.

References

1. Decree of the President of the Russian Federation dated May 7, 2018, No. 204 "Regarding national goals and strategic tasks of development of the Russian Federation until 2024". Garant RU. Information and law portal [E-source]. Access: https://www.garant.ru/products/ipo/prime/doc/71837200/. Accessed 25 March 2019
2. Program "Digital economy of the Russian Federation". Government of the Russian Federation, Decree dated July 12, 2017, No 1632-r [E-source]. Access: http://static.government.ru/media/files/9gFM4FHj4PsB79I5v7yLVuPgu4bvR7M0.pdf. Accessed 28 March 2019
3. Passport of the national program "Digital economy of the Russian Federation". Adopted by the Decree of the Presidium of the Council with the President of the Russian Federation for strategic development and national projects dated December 24, 2018 [E-source]. Access: http://government.ru/projects/selection/743/. Accessed 28 March 2019
4. Makhalina OM, Makhalin VN (2017) Organizational, economic and financial problems of the digital economy. In: Materials of the VI international scientific and practical conference "Formation and implementation of the strategy of sustainable economic development of the Russian Federation", Penza
5. Kafidulina NN (2018) Digitization as a trend: growth points for the Russian education. In: Kafidulina NN (ed) Interact Educ 2:9
6. Shpunt Y (2018) It's all about the people. In: Shpunt Y (ed). Standard 5:12
7. Zhuravlev VA (2008) Creative thinking, creative management, and innovative development of society (Parts 1 and 2). Creat Econ 4:3–8; 5:51–55
8. Korotkov EM (2006) Studying the management systems. Academic Prospect, Mir
9. Zaytseva OV (2009) Lifelong learning: the main notions and definitions. Bull TSPU 7(85)
10. Klyucharev GA (2014) Lifelong learning—stimulus for human development and factor of socio-economic inequalities. In: Klyucharev GA, Didenko AV, Latov YV, Latova NV (eds) CSFM, 433 pp
11. Makhalina OM (2018) Managing challenges and threats in the digital economy of Russia. In: Makhalin VN (ed) Management. Scientific and practical journal, No 2. State University of Management Publ., pp 57–61
12. Gusenko M (2018) The government to issue personal certificates for payment for digital education of children and adults. Russian gazette. Federal issue No 57(7520),

March 19, 2018 [E-source]. Access: https://rg.ru/2018/03/19/ekspert-predskazal-izmeneniia-na-rossijskom-rynke-truda.html. Accessed 02 April 2019
13. Pelevena N (2017) Catching emotions. Russian gazette. Special issue No 230(7396), October 11, 2017 [E-source]. Access: https://rg.ru/2017/10/11/universitet-nti-otkroetsia-7-noiabria.html. Accessed 02 April 2019
14. Yaroshchuk AB (2019) State support for formation of quality of human capital in the agrarian sector of economy in the leading countries of the world. In: Collection of the articles of the international scientific and practical conference: "Agrarian economy in the conditions of globalization and integration"
15. Veselitsky O, Khabarov D (2019) Tendencies of development of remote education in Russia and abroad. Colloquium-J 3(27). https://cyberleninka.ru/article/n/tendentsii-razvitiya-distantsionnogo-obrazovaniya-v-rossii-i-za-rubezhom. Accessed 01 July 2019
16. DIPLOM DISTANT^{RU} [E-source]. Access: https://diplom-distant.ru/2014/03/24/distantsionnoe-obrazovanie-tendentsi/. Accessed 01 July 2019
17. Noskova E. New teaching. Russian gazette. Economics. Special issue No 122(7880) [E-source]. Access: https://rg.ru/2019/06/05/vuzy-rossii-poluchat-granty-na-razrabotku-modeli-cifrovogo-universiteta.html. Accessed 30 June 2019
18. Kamneva K. Stairs into the future. Russian gazette. Economics. Special issue No 122(7880) [E-source]. Access: https://rg.ru/2019/06/05/cifrovizaciia-ekonomiki-potrebuet-kvalificirovannyh-specialistov.html. Accessed 30 June 2019

Chapter 31
Ethical and Legal Issues of Genetic Research of Indigenous Small-Numbered Peoples

Olga E. Finogentova, Olesya V. Belaya and Juliana A. Kitsai

Abstract The article raises ethical and legal issues in conducting genetic research. The article raises ethical and legal issues in conducting genetic research. If the subject of genetic research is the DNA profiles of indigenous small-numbered peoples, then the solution of these problems is of particular importance. Possibilities for editing human DNA endanger the existence of entire populations of people, preserving their identity, and ensuring continuity between generations. *Purpose*: The purpose of the study is to identify the main problems of both ethical and legal nature in conducting genetic research. The scope of the study is that during its conduct, an analysis of Russian and international regulatory acts governing the conduct of genetic research, modification of the human genome, and protection of the rights of indigenous peoples was carried out. The problem of determining the limits of intervention in the profile of human DNA during experiments, the possibility of securing the rights to the results of genetic studies and the genes themselves to specific individuals is raised. The problem of determining the legal nature of genes and the human genome as an object of civil rights is investigated. *Design/methodology/approach*: The methodological basis of the study was a combination of general scientific and private scientific methods, such as: methods of legal modeling and forecasting, analysis, structural-functional, dialectical-materialistic, comparative legal, regulatory and logical, systemic methods. *Results*: The results of the study are the identification and formulation of certain ethical and legal problems of genetic research, the inadmissibility of interference in the human genome, even the probable result of which could be the threat of losing a person's identity, transforming a person or creating a person of a "new kind". The main sociological and humanistic consequences of the lack of proper attention by the State and society to the problems of the conservation of indigenous peoples are determined. Separate proposals are formulated to identify and resolve ethical and legal problems of research and experiments on the profile of human DNA, as well as securing rights to the results of such studies.

O. E. Finogentova · O. V. Belaya (✉) · J. A. Kitsai
Immanuel Kant Baltic Federal University, Kaliningrad, Russia
e-mail: Whiteolesya@mail.ru

T. Kolmykova and E. V. Kharchenko (eds.), *Digital Future Economic Growth, Social Adaptation, and Technological Perspectives*, Lecture Notes in Networks and Systems 111, https://doi.org/10.1007/978-3-030-39797-5_31

Keywords Bioethics · Genetic information · Genome research · Indigenous small-numbered peoples · Patenting · Ethics

JEL Classification Codes K110 · K150 · K380

31.1 Introduction

More than a century after its inception, genetics gained the ability to solve one of the most important and most complex tasks, which can greatly affect the further development of human civilization. This is an international project for decoding human DNA (Human Genome Organization). Russia has its own program that solves these same problems, which is called the "Human Genome".

Scientific progress in medical science is currently inconceivable without genetic research and biotechnology. On the one hand, the achievements of modern genetics open new horizons for humanity. And, this is not only the possibility of diagnosing, preventing and treating serious hereditary diseases, but also the possibility, by editing the genome, to create a person with a predetermined set of characteristics, this is the ability to edit the human fetus.

In its turn, these opportunities cause ethical and legal problems for scientists, changing our millennia-old ideas about human uniqueness, identity, free will, and raising the problem of the sacredness of the human soul. What can fighting against bad genes ultimately lead to? What will it result? Will it lead to genetic segregation? How can the genetic portrait of humanity change? Another question is the future of human as a population. What will a human's struggle with "bad" genes lead to? Law and ethics face new challenges that need to be addressed.

31.2 Materials and Method

During the study was used a combination of general scientific and private scientific methods such as: methods of legal modeling and forecasting, analysis, structural-functional, dialectical-materialistic, systemic comparative legal, regulatory and logical methods.

31.3 Results

Academician Bayev A. A. wrote: "… The human genome is not only a fundamental scientific problem, but also a major social phenomenon, both financial and industrial. The study of the genome has reached such a state that both humanities involved in issues of philosophy, sociology, law, and religious figures, and in general the public, should finally focuses on issues of bioethics" [1].

The ideas of creating a superhuman by overcoming his given natural parameters are not new; they were expressed by Friedrich Nietzsche in the 19th century, and in the 20th century they found its continuation in supporters of the philosophy of transhumanism. At the moment, the scientific community is very vigilant about experiments in this area, since no one can accurately predict the consequences of a change in one gene and subsequent inheritance of artificially modified DNA. On November 26th, 2018, Chinese researcher He Jiankui said that he was able, using the CRISPR/Cas9 system, to edit the genomes of seven pairs of embryos during reproductive treatment, resulting in children with altered DNA with lifelong immunity to HIV. This statement caused a shock and an extremely negative reaction from the scientific community, all experiments with the human genome were temporarily banned in the PRC, and an investigation was launched into the activities of this scientist. One hundred twenty two chinese scientists wrote an open letter in which they accused the scientist of opening the Pandora's Box and urged the government to drastically limit work with human genes.

G. Onishchenko commenting on this event noted: "… our country has adopted a law on a moratorium on experiments with the human genome. The genome is 3.6 billion years old, and it really can be edited, but who gave the right to man to do this? The further science goes, the more deeply it goes into the innermost of human nature, the more acute are the questions connected with ethics, with the right to such experiments. … Presently such experiments require national and international regulation." (URL: https://radiosputnik.ria.ru/20181127/1533640732.html?in=t, [2]).

Now, one of the most important tasks of the international scientific community is to protect the gene pool of humanity from risky experiments, but at the same time not to discard (as, unfortunately, it was with the Soviet science earlier) invaluable information on the diagnosis, prevention and treatment of hereditary diseases. The opinion of Siluyanova I. V., stating that: "today the development of biomedical technologies requires the creation of an ethical and legislative concept of clinical practice and scientific research", should be supported [3].

The ethical and legal problems of genetic research are becoming especially relevant when deciding on the conservation of entire populations of people. And here the tasks are set not only in the plane of preserving the identity of certain groups of people, but also in general, preserving their life. In this regard, it seems appropriate to dwell on the analysis of individual issues related to indigenous small-numbered peoples. At first glance, it might seem that enormous attention is paid to protecting the social rights of indigenous small peoples. However, unfortunately, to date, a huge number of citizens of most regions not only do not know anything about indigenous peoples, moreover, they focus on this category of citizens only in the context of high-profile situations covered in the media. The number of indigenous small-numbered peoples is also unknown, and therefore, in 2019, a census of indigenous peoples of the North, Siberia and the Far East was launched in Russia in pilot mode. (URL: https://news.rambler.ru/other/39444791-perepis-malochislennyh-narodov-severa-sibiri-i-dalnego-vostoka-nachnetsya-v-2019-godu, [4]).

The United Nations General Assembly adopted a resolution proclaiming 2019 as the International Year of Indigenous Languages, as part of addressing world endangered languages, as well as focusing on social issues where social change is paramount Society for Indigenous Peoples (URL: http://etnic.ru/news/2019-god-obyavlen-oon-mezhdunarodnym-godom-yazykov-korennyh-narodov.html, [5]).

Current legislation requires constant updating, and therefore, there is a need for new solutions, but in the process regarding indigenous small-numbered peoples, it is sometimes extremely difficult to do due to the lack of not only relevant information that is necessary, but also information in general. In this regard, the high-profile discussion in 2019 in Russia began as part of the consideration of the draft Federal Law "On Amendments to the Federal Law" On Guarantees of the Rights of Indigenous Minorities of the Russian Federation "regarding the establishment of an accounting procedure for persons belonging to indigenous small-numbered peoples". Some conditions of this bill are justifiably called virtually unrealizable. In our opinion, the point of view of the expert of the Center for Assistance to Indigenous Peoples of the North (Moscow), a member of the International Working Group on Indigenous Issues of IWGIA—Olga Anufrievna Murashko is justified. In accordance with her point of view, certain provisions of this bill "create unforeseen legal conflicts ..., for example, in terms of specifying the condition that the family genealogy should be drawn up" with written consent and provide "documents containing information about the nationality of the applicant's relative... and also documents confirming the relationship of the applicant with the indicated person ...", that is, each member of the family genealogy needs to get to the notary public or to the archive, which makes the conditions under the bill is nearly impossible, given the remoteness and transport inaccessibility of the settlements of indigenous small-numbered peoples from the central regional archives (URL: https://arctic-consult.com/archives/5736, [6]).

Social tension is evident in other issues in this area. Often, indigenous peoples have to survive, choosing between the provision of financing or territories, which is impossible. Reality shows that for survival, indigenous peoples need to enter into the terms of commercial auctions and competitions, that is, compete with commercial companies, etc. Victory is possible in this case only, for example, if the impossible happens and all competing applications are withdrawn.

The 18th Session on Indigenous Issues was held from 22 April to 3 May 2019 in New York, at the headquarters of the United Nations. The main theme of the session was "Traditional knowledge: formation, transmission and protection" (URL: https://kmns.ru/blog/2019/04/19/cs-kmns-v-nju-jorke-sostoitsja-18-ja-sessija-postojannogo-foruma-oon-po-voprosam-korennyh-narodov, [7]). This platform could become a field to attract attention to a large number of social problems of the indigenous peoples of the Russian Federation. After all, speaking at the United Nations, indigenous peoples from around the world talk about their desire to draw attention to the problems of indigenous small-numbered peoples. It is imperative that they can bring their voices to the world community in the expectation that the opinion of the world community will affect governments and help them solve problems. But analyzing the speeches of the representatives of the Russian Federation, it is clear that the social sphere of indigenous peoples was practically not affected.

Many processes taking place today, without drawing attention to the problems of indigenous peoples, could become irreversible. Most of the social problems are not raised or are stated narrowly, but they certainly exist and become insoluble. It is extremely difficult for the young generation, who understands the magnitude of the tasks that require radical measures, to support their small groups, becoming "foreigners".

Indigenous peoples should not disappear and shrink at such scale. This main task, but without proper maintenance and development of the social sphere, is becoming practically impossible.

In the field of protecting the rights of indigenous peoples from the position of their genetic identity, attention should also be paid to the regulation of these issues not only at the national level, but also on the international level. Currently, there are already a number of fundamental international documents regulating research in the area of the human genome: Convention for the Protection of Dignity and Human Rights in Connection with the Application of Biology and Medicine ETS No. 164 (concluded in Oviedo on 04/04/1997); Universal Declaration on Bioethics and Human Rights (adopted by acclamation on 10/19/2005 at the 33rd session of the General Conference of UNESCO); The Helsinki Declaration "Ethical Principles of Medical Research with the Involvement of a Person" (adopted by the 18th General Assembly of the WWA (Helsinki, Finland, June 1964). Laws have been enacted to regulate such research at the state level. Thus, the Russian Federation was adopted at the end of 1996 Federal Law of 05.07.1996 No. 86-ФЗ "On State Regulation in the Field of Genetic Engineering"; in 1995, the French Law of August 6, 2004 No. 2004-800 "On Bioethics" was adopted; the Federal Law 07/31/2009, "On genetic research on people"; (as amended on 08/07/2013); Austria has the Law on Genetic Engineering. And, nevertheless, there remains a very large circle of unresolved issues, especially in the field of the human genome. technology research requires a whole complex of specific legal relationships that differ significantly from relationships within other biomedical technologies [8]. In particular, there are no legal norms that would regulate and determine the legal status of the embryo, there are no mechanisms for its protection in case of genetic experiments [9]. Existing international legislation only proclaims general principles and marks a further vector for the development of genomic research. Laws have been adopted to regulate such studies at the state level. So in the Russian Federation, at the end of 1996, the Federal Law of July 5, 1996 No. 86-FZ "On State Regulation in the Field of Genetic Engineering" was adopted; in 1995, the French Law of 06.08.2004 No. 2004-800 "On Bioethics" was adopted; in the Federal Republic of Germany the Law of July 31, 2009 on genetic studies in humans is in force; (as amended on 08/07/2013); In Austria, the Law on Genetic Engineering is in force. And, nevertheless, there remains the range of unresolved issues, especially in the field of the human genome. Whereas the technologies associated with his research require a set of specific legal relations, which significantly differ from the legal relations of other biomedical technologies [8]. In particular, there are no legal norms that would regulate and determine the legal status of the embryo; there are no mechanisms for its protection in case of genetic experiments [9]. Existing international legislation only proclaims general principles and provides direction for the development of genomic research.

The development of genomic research and the problems that they brought to life both in the field of social and philosophical knowledge led to the formation of a new science—bioethics, which was formed within traditional ethical limits. Its main task, according to V. R. Potter (the "godfather" of bioethics) is "the combination of a system of biological knowledge with knowledge of a system of human values" [10]. But, it must be understood that in order for bioethics could have the character of a truly practical philosophy, it must be based on the existing state of things. In bioethics, problems of a biological, social, ethical, and psychological nature are intertwined that underlie human life.

An acute problem of modern genomic research is the problem of the possible use of genetic information stored in biobanks for discriminatory purposes, which was highlighted at the twentieth session of the UNESCO International Committee on Bioethics (IBC) held in Seoul in June 2013, in which was participated more than a hundred people from 33 countries. The information stored in biobanks can be used to incite racial prejudice and discrimination of certain groups on the population. So, the sickle cell anemia gene, in particular among African Americans, can be considered a racial trait, or aggressive behavior, which according to recent studies is inherited [11], and is associated with the content of certain hormones in the blood.

Another problem is that large corporations such as Celera Genomics, which invest millions of dollars in genomic research, aim to make a profit by patenting the fragments of human DNA they have studied. So, in October 1994, the company TIGR, announced that it has a library of 35,000 DNA fragments, but this data will not be made public by the scientific community if the company does not recognize the exclusive right to them.

But how ethical is patenting genes? On the one hand, the patent institution is intended to secure the exclusive rights to the subject of patenting to its developers. This is a peculiar way of protecting their development rights, given the fact that the appearance of the final result was preceded by lengthy and costly research, testing, approbation of the results.

On the other hand, often in research developers use not only their own developments, but studies published in various publications and on different resources. In this regard, the question is the possibility of establishing an exclusive and sole author (group of individuals) for one or another development, especially if it concerns such a complex area as genomic research. For example, in the development and subsequent patenting of the BRCA1genes and BRCA2 genes, Myriad Genetics used the results of oncological studies of 2,000 patients conducted at the London Institute (URL: https://expert.ru/2013/06/18/nichi-genyi, [12]). Could Myriad Genetics do it without the results of these studies? More likely—yes, to conduct their own research, but this would significantly lead to a rise in the cost of the research process and an increase in the timing of their research.

There is a problem of a different nature. But what is a gene as an object of legal protection? Many researchers are puzzled by this question, and they have not developed a unified approach.

A. N. Levushkin believes that genes and the genome are a special object of civil rights and have a special civil law regime. At the same time, the author notes that the implementation of the civil legal regime of genes is not only legal, but also moral and ethical [13].

Mokhov A. A., Yavorsky A. N. conclude that genes by their legal nature belong to objects of intellectual property. In a more detailed analysis of which of the objects of intellectual property genes can be attributed, the authors make an ambiguous conclusion about the possibility of interpreting genes as inventions and as analogues of computer programs. In the latter case, genes are a natural program, unnamed by the legislator, with changes made by a human in cases of gene modification and without such changes [14].

Bogdanova E. E. explores the legal nature of a gene in the context of genetic information [15]. However, the Russian legislator is very inconsistent in this matter. Starting with the terminology, since the terms "gene" and "genome" are often used as synonymous concepts, and ending with that, the existing legal acts do not clearly define the legal nature of genetic information. So, in paragraph 3 of Art. 1 of Federal Law dated 03.12.2008, No. 242-ФЗ "On State Genomic Registration in the Russian Federation", genomic information is classified as personal data. On the other hand, in the norms of the Federal Law of July 27, 2006 No. 152-ФЗ "On Personal Data", the concept of genomic or genetic information is not contained, although a special type of personal data is indicated—biometric data, based on the definition of which it can be assumed that Genomic information can be attributed to biometric personal data as information about the biological characteristics of a person, allowing to establish his personality.

The question arose, however, how appropriate is it to relate information about the human genome exclusively to his personal data, because this kind of data carries information not only about this person, but also about his ancestors, but what else can be more important—about his descendants. As noted in the Decree of the Constitutional Court of the Russian Federation of June 16, 2015 No. 15-P, information on the origin of the child can be indispensable in determining biological relationships between people, identifying hereditary diseases, preventing marriages between close blood relatives, etc.

In this regard, it seems that the assignment of genomic information exclusively to the personal data of one person is incorrect. In addition, the imperfection of mechanisms for protecting genomic information makes it possible to reflect on the issue of establishing acceptable limits for using information about the human genome as evidence. Back in 2009, scientists from Israel conducted an experiment in which they showed the possibility of falsifying evidence based on human DNA data. The authors of the development were able to create samples of saliva and blood of a particular person without receiving biomaterial from him. Thus, scientists have demonstrated that knowing the DNA profile of a particular person, it is possible to artificially create a sample that is completely identical to that person's DNA without using its tissue (URL: https://www.geneticsandsociety.org/article/dna-evidence-can-be-fabricated-scientists-show, [16]).

On this subject, the words of American judges Luis Brandeis and Samuel Warren particularly meaningful: "we have come to understand the right to privacy, where the right to life is the right to be left alone" become especially meaningful [17].

31.4 Conclusions/Recommendations

Research has led to the conclusions:

(1) The problems of genetic research are not only legal, but ethical in nature. Often these two areas are so interlinked that they cannot be distinguished. In cases where there is no proper legal regulation of the issues of genetic experiments on the human genome, ethical principles that are international in nature come first. Currently relevant such science as bioethics.
(2) It is necessary to maintain a balance of interests of the parties in the behavior of genetic research: on the one hand, it is necessary to protect the gene pool of humanity from risky experiments, and on the other hand, do not create unjustified barriers to the development of science in the study of the human genome, diagnosis, and treatment of hereditary diseases.
(3) Despite the existing fundamental international documents, which only proclaim the general principles governing research in the area of the human genome, as well as the adoption of such acts at the national level in many States, a number of issues remain without proper legal regulation, for example, determining the legal status of the embryo, lack of mechanisms its protection in case of genetic experiments.
(4) An acute problem of current genomic research of an ethical and legal nature is the problem of the possible use of genetic information stored in biobanks for discriminatory purposes, to incite racial prejudice and discrimination of certain groups on the population, the potential falsification of evidence by artificially creating DNA profiles of specific individuals.
(5) There is a problem of determining the ethical feasibility of patenting genes. And here also the need to maintain a balance of interests of all process participants. On the one hand, the time, financial, labor costs of companies carrying out genetic research must be protected, and the patent institution helps to achieve these objectives. On the other hand, the granting of exclusive rights to the results of genetic researches to certain individuals and possible restrictions on their part in relation to other individuals could constitute an obstacle to the development of science, improve the results, reduce errors in the reliability of the results and more.
(6) Currently, in the scientific community there is uncertainty in determining the legal nature of the human gene and genome. Recognizing the diversity of approaches in this matter, the authors cannot agree with the opinion that genetic

information can be classified as personal data, as information characterizing only a specific person, since this kind of data carries information not only about this person, but also about his ancestors, and what else could be more important—about his descendants.

(7) In genetic research particular attention should be paid to ensuring and protecting the rights of indigenous small-numbered peoples. Risk experiments with a human DNA profile can lead to the threat of preserving entire populations of people, ensuring their identity. Indigenous peoples are currently at risk of extinction. Thus, paramount concern is required both from the side of specific states and the entire world community to ensure and protect their rights. In this case, it is necessary to take a set of measures to protect their rights in the social sphere, ensure an adequate standard of living, and take into account the characteristics of each group of people. In this area, measures are needed to systematize and modernize the legislative regulation of the social sphere related to the provision and protection of the rights of indigenous peoples; the creation of legal mechanisms to take into account the views of indigenous small-numbered peoples in decision-making, and more.

By all means, experiments with the genome of people from populations that could even potentially endanger their existence should be under a complete and uncontested ban.

References

1. Bayev AA (1995) The human genome: some ethical and legal problems of the present and future. Man 2:9
2. Onishchenko G (2018) Experiments with the human genome require international control. https://radiosputnik.ria.ru/20181127/1533640732.html?in=t. Accessed 12 Oct 2019
3. Siluyanova IV (1998) Modern medicine and Orthodoxy. Moscow Compound Publishing House of the Holy Trinity St. Sergius Lavra, Moscow
4. The census of the small peoples of the North, Siberia and the Far East will begin in 2019 (2018). https://news.rambler.ru/other/39444791-perepis-malochislennyh-narodov-severa-sibiri-i-dalnego-vostoka-nachnetsya-v-2019-godu. Accessed 12 Oct 2019
5. 2019 is declared by the UN as the International Year of Indigenous Languages (2019). http://etnic.ru/news/2019-god-obyavlen-oon-mezhdunarodnym-godom-yazykov-korennyh-narodov.html/. Accessed 15 Oct 2019
6. Murashko OA (2017) The indigenous peoples registry violates the Constitution of the Russian Federation and international law. https://arctic-consult.com/archives/5736. Accessed 12 Oct 2019
7. The 18th session of the UN Permanent Forum on Indigenous Issues will be held in New York (2019). https://kmns.ru/blog/2019/04/19/cs-kmns-v-nju-jorke-sostoitsja-18-ja-sessija-postojannogo-foruma-oon-po-voprosam-korennyh-narodov. Accessed 15 Oct 2019
8. Isaeva EL, Batasheva MA (2016) Ethical and legal aspects of intervention in the human genome. http://web.snauka.ru/issues/2016/12/75296. Accessed 15 Oct 2019
9. Ermolaeva YuN, Krasovsky VS, Sidorova ES (2016) Legal protection of the unborn life of the embryo in the light of modern reproductive technologies. Postgrad Stud 4(Part 1):25
10. Potter VR (1995) In: Reich WT (ed) Encyclopedia of bioethics. Macmillan Pub. Co., New York, p 21

11. Siluyanova IV (2002) Ethics of genetics and the “genetics” of ethics. Vestnik RSMU 4(25):71
12. Mankov S (2013) Drawn genes. https://expert.ru/2013/06/18/nichi-genyi. Accessed 15 Oct 2019
13. Levushkin AN (2019) Civil regime of genes as objects of civil rights. Lex Russ 6(151):105
14. Mokhov AA, Yavorsky AN (2018) Genes and other formations based on genes as objects of intellectual property law. Civil Law 4:28
15. Bogdanova EE (2019) Legal issues and risks of the genetic revolution: genetic information and discrimination. Lex Russ 6(151):18–29
16. Pollack A (2009) DNA evidence can be fabricated, scientists show. https://www.geneticsandsociety.org/article/dna-evidence-can-be-fabricated-scientists-show. Accessed 12 Oct 2019
17. Ajunwa I (2014) Genetic testing meets big data: tort and contract law issues. Ohio State Law J

Chapter 32
Integrating Sustainable Development Issues in the English Language Classroom and Developing Eco-Linguistic Competence of University Students

Irina N. Korenetskaya, Olga V. Lantikova and Nataliia S. Semenova

Abstract In the era of globalization, the ecological situation on our planet is undergoing catastrophic changes, which are brought about by such factors as an increase in the population's needs for services and goods, overconsumption, vigorous production activity, and irrational use of natural resources. A crucial role in education, in connection with the current environmental situation, must be assigned to environmental education, as well as to its forms and methods. The purpose of this study is to substantiate and integrate the ideas of sustainable development in a foreign language teaching. The material presents the idea of global environmental education, including the unity of goal-setting, content, process and technological components. The ambitious outcome of such a foreign language education should be an environmentally and linguistically competent person, that is developed via the course of simulation games, workshops, creative activity of designing a "smart home", a "green city", and publishing the brochures with environmentally friendly rules. This competence will allow students to respond to environmental challenges by offering solutions for the local community, as well as presenting the ideas at international events during academic mobility programs.

Keywords Competence · Environmentally friendly behavior · Foreign language teaching · Project work · Sustainable development

JEL Classification Code I 230 Higher Education; Research Institutions

32.1 Introduction

In September 2019, Russian Prime Minister Dmitry Medvedev announced the signing of a Government decree that Russia is committing to the Paris climate agreement,

I. N. Korenetskaya (✉) · O. V. Lantikova · N. S. Semenova
Pskov State University, Lenin's Sq. 2, Pskov 180000, Russia
e-mail: irinallo@mail.ru

T. Kolmykova and E. V. Kharchenko (eds.), *Digital Future Economic Growth, Social Adaptation, and Technological Perspectives*, Lecture Notes in Networks and Systems 111, https://doi.org/10.1007/978-3-030-39797-5_32

thus making Russia the 187-th state participating in this global environmental project. The topic of ecology always causes tough discussions at the international level. On September 23, 2019, during the UN environmental summit, the 16-year-old Greta Tunberg, a girl from Sweden, delivered a speech on the upcoming end of the world, caused by global environmental threats. Greta is an environmental activist who does not attend school on Fridays to fight climate warming. She did not fly to the ecological summit in New York by plane, since it pollutes the air, but sailed across the Atlantic on a sailing yacht in which the second engine was sealed for environmental reasons. Before New York, Greta spoke in the European Parliament at the World Economic Forum in Davos, and also met with B. Obama, J. Soros, the Pope, J.-K. Junker.

We should have some insight whether Greta actually improves the climate. The specialists, to put it in a nutshell, do not take the girl seriously, as well as politicians. However, Greta's performances produce a bunch of effects and cause existential fear among the younger generation. In order not to feed the flame, parents and teachers are responsible for the formation of values for the young generation and should protect the young and unstable consciousness from the ideas of global catastrophes and the end of the world, but at the same time—to explain in a comprehensive way the difficulties that exist in the field of environmental protection, and also to work out the mechanisms through which the younger generation will be able to protect the environment from negative impacts on it.

The leading principles of the modern educational paradigm are anthropology, humanization, eco-friendly education. According to the Socio-Ecological Dictionary "greening is a concept that reveals the process of penetration of the ecological approach, environmental principles into various types and spheres of human life. The essence of this phenomenon in science is to identify and examine the relationships that exist between the object being studied by a particular natural science or socio-economic science and its environment" [1]. The eco-humanitarian education paradigm includes the components such as environmental culture and environmental competence.

This article justifies the need for such a competence as environmental-linguistic, combining the foundations of environmental and linguistic education, in the development of which the foreign language discipline has great potential in the system of teaching a foreign language.

The teacher designs such an educational environment in the context of which the student masters the current global problems; develops educational modules of the discipline, which include the political, economic, social, cultural challenges of our time; provides emotional support and motivates environmentally friendly behavior, not only talking about problems, but also showing specific examples of their solutions that work effectively, in a neighboring area [2].

A foreign language teacher works not only with grammar, phonetics, lexical units of the studied language, but also develops the so-called Soft-Skills, designs teamwork, cooperation, teaches to think and feel, develops empathy, receptivity, inspiration, co-participation, introduces how inequality causes poverty, how intolerance leads to violence, how poverty negates greed, how a way of thinking and feeling can cure.

There can be no quick solution to environmental problems. Nevertheless, it is the teacher who can play the role of a decisive factor in the formation of a green model of student behavior and help them become acquainted with the goals and objectives of sustainable development, which were identified in September 2015 at the Summit on Sustainable Development, including the fight against poverty and hunger, inequality and injustice, support for the quality of education, a healthy lifestyle, the struggle for clean water, energy, decent conditions for work and economic growth, the balance between economic development and the preservation of the environment.

32.2 Materials and Method

In order to develop a nature-friendly type of thinking based on humanism and the greening of human life, as well as the development of creative thinking and social responsibility within the foreign language classroom, it is possible to integrate global issues of sustainable development into the format of an interactive series of classes using infographics, creating photo galleries, and participating in simulation games to develop an environmental bill, designing eco-houses and eco-cities, participating in a quest on the territory of the local forest with all its diverse flora and fauna [3].

The advantages of this training format are:

- formation of students' ideas about global environmental problems that can change their lives;
- formation of language skills so that students can speak (discuss and understand the opponent's point of view and express their own opinion reasonably and adequately) about environmental problems which are crucial for the entire world community;
- development of critical thinking and problem-solving skills;
- development of social skills in teamwork to solve a common problem;
- development of a positive attitude, tolerance and respect for the opinion of another person;
- development of an internal belief that it is an individual responsibility to develop concepts of environmental conservation and protection [4].

Stages of the foreign language lesson:

1. Shaping of sustainable development goals and their interpretation;
2. Ranking of goals in order of importance;
3. Development of conventions for global sustainable development goals;
4. Identifying the interconnections of the global sustainable development goals and developing an action plan;
5. Analyzing 17 global sustainable development goals (URL: https://www.un.org/sustainabledevelopment, [5]).

Here is an example of framing a lesson on working with global sustainable development goals.

The purpose of the lesson:

– to work out the global sustainable development goals and rank the discussed goals.

Description of the lesson:

Students speak about what they consider to be global problems and conclude that the global goals of sustainable development are accepted by the entire international community for the benefit of the development of all mankind and the environment by 2030. Working in pairs, students match the goal and value (Table 32.1).

After that, in mini-groups students are asked to rank existing goals by importance and provide sufficient reasoning. In pairs, students design a symbol for each of the goals presented in Table 32.1, which would in a schematic form reflect the semantic content of the concept. And further, students are working on designing the

Table 32.1 Global sustainable development goals and their interpretation

Goal 1: No poverty	To ensure that countries and organisations work together to achieve the goals
Goal 2: No hunger	To ensure that everyone has enough money to live
Goal 3: Good health	To ensure that everyone has enough food to eat
Goal 4: Decent education	To ensure that everyone has the right to medical care
Goal 5: Gender equality	To ensure that everyone has the opportunity to go to school
Goal 6: Clean water and sanitation	To ensure that girls and women have the same rights and opportunities as boys and men
Goal 7: Renewable energy	To ensure that everyone has safe water and healthy hygiene
Goal 8: Good jobs and economic growth	To ensure that everyone has access to reliable, sustainable energy
Goal 9: Innovation and infrastructure	To ensure that everyone has the opportunity to work
Goal 10: Reduced inequalities	To ensure that industry is responsible in protecting people and the environment
Goal 11: Sustainable cities and communities	To ensure that countries have laws and societies that protect everyone equally
Goal 12: Responsible consumption	To ensure that cities, towns and villages have water and electricity, and are clean and safe
Goal 13: Climate action	To ensure the reduction of waste, for example by recycling paper or glass
Goal 14: Life below water	To ensure action to slow down the impact of global warming
Goal 15: Life on land	To ensure action to protect oceans and seas
Goal 16: Peace and justice	To ensure action to protect the environment and endangered animals
Goal 17: Partnerships for the goals	To ensure the end of violence and war and give everyone access to justice

booklet "The World We Want" with illustrative stories about the problems of over-consumption, "smart shopping", the proper collection of garbage, the need to turn off the electricity and water, if not used, to use recycled packaging, and the benefits of walking and traveling to the university by a bicycle or electric board rather than by car, etc. (URL: http://www.greenliving.lovetoknow.com, [6]).

Internet information sites come to the forefront in informing a mass audience. Students are invited to analyze the principles of the BBC international channel and the priority areas for sustainable development presented on it, iconographic symbols that reflect the key ideas of sustainable development. Students answer these questions by skimming the article (URL: https://www.bbc.co.uk/responsibility/environment, [7]).

Facing the downside of the problem while reading the articles of the Air Force website (URL: https://www.bbc.com/news/science-environment-36913655, [8]), students determine the direction of environmental conservation, highlight the problem described by M. Kinver, and suggest ways to solve it. During the discussion, students compose a "fishbone" on the topic of the article, fixing the problem and its solutions.

Another important sustainable development issue is energy consumption and energy resources. To make students curious about that we can encourage them to take pictures of energy sources at home, and at the lesson they analyze whether these sources are renewable or not. The students get as well concerned about how energy is used, and whether there is over consumption [9] (Table 32.2).

Students using recyclable materials create the design of a "smart home" with energy conservation.

Working in mini-groups (with the distribution of roles for environmentalists, designers and citizens), students are invited to work on a project to design the rules of the "green city" (Table 32.3). A group of environmentalists is developing directions in which there should be some changes in the city. The design team is working on materials and resources that should be used to ensure that the city meets the goals

Table 32.2 Energy-consuming issues

Issues	What to look for?	What to do?
Lights	Are lights off when you or members of your family aren't in a room? Do you use energy-efficient light bulbs?	
Appliances	Do your appliances have Energy Star ratings? Do you switch them off/close them when you don't need them?	
Other devices	Are your computers, TVs and game consoles on stand by or off when you aren't using them? Do you remember to unplug your phone charger?	
Water	Do you leave the tap running? Do you take short showers or long baths? Do you use the washing machine more than three times a week?	
Insulation	Are there any old windows or doors in your house? Is there anything you can do to improve them?	
Heating/Cooling	Do you keep the windows/doors closed/open when the air conditioning or heating is on?	

Table 32.3 Directions of work on environmental conservation

Traffic (numbers of cars/buses)	Green areas	Buildings	Transport

of sustainable development. "Citizens" are developing rules for "environmentally friendly" behavior [10].

As a result for a series of lessons, aimed at the developing of ecological-linguistic competence, the eco-quest can be organized. Within the framework of this event, it is necessary to create conditions, encouraging students' personal conviction of the need for environmental conservation and environmental protection.

Objectives:

- increasing the level of individual responsibility when choosing his or her behavior in relation to nature and the environment;
- development of "environmentally friendly" behavior skills;
- development of problem-solving skills, including social skills in resolving issues in a team.

Stages:

1. Activation of acquired knowledge on eco-friendly behavior in the framework of the "fight" against garbage;
2. The solution of practical problems;
3. Presentation of the "happy garbage" legislation;
4. Summing up, choosing the best project.

Plan of the lesson:

Students are divided into three teams and they enter the room, in which "Garbage" begins to chase them ("Garbage" is presented by disguised students in special clothes with various garbage samples attached to them). "Garbage" complains that everyone considers it to be nasty and unnecessary, but in fact, they simply do not know how to do it or they are too lazy to use it again, and demands that it should be properly disassembled into containers. After completing the assignment of "Garbage", the teams design a portrait of the garbage, consisting of the typical contents of the garbage bin of a city dweller. Teams get points for speed, correct choices, original drawings. Students also consider the ways how the garbage can be recycled (the first part of an itinerary list is completed).

As a second task, the teams receive a card, which presents raw materials and new products from this raw material. To get the second part of the itinerary list, the teams need to estimate the amount of waste material (Table 32.4).

Keeping in mind that, for example, newspapers, handouts, cardboard boxes can be recycled into toilet paper, eco-wool, boxes, cardboard; pack—into books, notebooks,

Table 32.4 The card for calculating the amount of raw materials in the quest

Waste material	Product of recycling	Correct answer
_____ kg of newspapers	10 toilet rolls	1
_____ plastic bottles	1 fleece jacket	25
_____ aluminum cans	1 kids bike	400
_____ wine bottles	1 a package of glass wool for heat isolation of a country house	400

albums, pens; plastic into—building and decoration materials, fuel, furniture, plastic film, carpet fiber, pillow fillers and insulation for jackets; glass bottles, cans—in decor items, new containers, mosaics, concrete components; metal waste—into new metal products (cans, etc.); organic waste—in compost for fertilizing the soil, the teams compile an itinerary list for the item, selected at the beginning. The itinerary list is the path for production of a new product (created by teams) from its birth stage to eco-recycling and rearranging into a new useful product. Students need to draw and prescribe the stages, paths with possible difficulties, obstacles and ways to overcome them.

Teams present the results of their activities, the jury (students who played the role of garbage) evaluates and selects the best work.

32.3 Results

Sustainable development issues are not sufficiently addressed at the local level. Only a discussion at the international level will allow us to quickly and efficiently respond to the challenges of our time. According to the Article 12 of the UN Paris Convention, the level of international relations is defined as follows: the parties cooperate in the implementation of measures to enhance education, training, public awareness, public participation and public access to information on climate change issues, recognizing the importance of these steps (URL: http://unfccc.int/files/meetings/paris_nov_2015/application/pdf/paris_agreement_russian_.pdf, [11]).

Cooperation in any field, especially in the field of international interaction, requires foreign language communication skills, negotiation, discussion. Therefore, knowledge of a foreign language is an indispensable component of the establishment of such cooperation.

Modeling the framework of a foreign language teaching with integrations of the global goals of sustainable development provides the students with the formation of knowledge, skills that will allow them to be conscious about environmental problems and present their opinion and ways of solving problems outside the local community.

Foreign language competence together with socio-cultural and environmental one ensure not only comprehensive communication, but also adequate environmental-saving behavior. In addition, these competencies will prevent intercultural conflict and misunderstanding.

The foreign language classroom in the format of an environmental module will allow a university student to make a feasible contribution to the development of the local community and take a step towards solving the global problems of mankind, offering solutions to these problems at venues of different levels: from volunteer movement at the university to reducing harmful emissions into atmosphere by choosing a bicycle, electric board or hiking, or organizing separate collection of garbage in their area, before participating in international events. Therefore, environmental and linguistic competence is a key factor in effective international interaction and a guide for the ideas of sustainable development from one community to another.

32.4 Conclusions/Recommendations

The development of ecological and linguistic competence in a foreign language classroom should not be local and episodic in nature. It is necessary to ensure effective international cooperation, constant involvement in the studied problem, participation in modern relevant projects.

Thus, the acquired ecological and linguistic knowledge, acquired in the process of skills will not be reproductive or conditionally productive in nature, and will not be limited only by the space of the classroom, but will become a solid basis for carrying out their ideas for sustainable development at the global community level.

In Pskov region, there are currently two projects, which are actively promoted and implemented by the Pskov Administration, in particular, the Committee for the Implementation of Cross-Border Cooperation and Tourism Programs. It is the project "Multimodal cities. Transition to urban sustainable transport systems" which is implemented from October 2017 to September 2020. The project involves the development and implementation of strategic approaches and integrated solutions to ensure the effective and safe functioning of urban transport systems, including solutions in the field of development of cycling and other non-motorized types of movements. The project partners are eight European countries: Germany, Poland, Sweden, Denmark, Finland, Estonia, Latvia, Lithuania. Students' participation in the project allows them to generate and discuss ideas for a more efficient use of urban space, to offer their vision of a compact transport infrastructure, as well as the arrangement of the space of the historical center of Pskov. The holding of the 39-th International Hanseatic Days in Pskov in 2019 once again clearly demonstrated the relevance of the environmental task posed to the city and the possibility of solving it with the involvement of international experience.

The project ER 65, launched in April this year, "Improving biological diversity and increasing public awareness of environmental protection in the regional centers of Russia and Estonia," aims to increase public awareness in the field of energy-efficient practices. Until April 2022, students have the opportunity to participate in master classes and training seminars, discuss and implement their ideas on creating a more attractive and biologically diverse urban environment of Pskov and Vyru, and make their contribution to the creation of an ecologically sustainable urban area. Since one of the project activities is "the creation of methodological and informational materials on improving the quality of education in the field of environmental protection for the education of children and their families", students who study in the field of pedagogical education may be included in this active development and testing at school during the school practice period.

We believe that the student community, entering an active adult life, can benefit from such an experience of international cooperation in solving environmental problems, which once again confirms the importance of integrating sustainable development ideas into the process of teaching a foreign language and the need to form the ecological and linguistic competence of university students.

References

1. Ремизов ИН (2002) Социально-экологический словарь. Былина, Москва, с. 185
2. Hosagrahar J (2012) Culture: a driver and an enabler of sustainable development. Thematic think piece for UNESCO. Developed for the UN system task team on the post-2015 UN development agenda
3. Ochs EA (1987) Socio-cultural perspective. In: Hickmann M (ed) Social and functional approaches to language and thought. Academic Press Inc., New York, pp 305–319
4. Stevenson N (2003) cultural citizenship: cosmopolitan questions. Open University Press, Berkshire
5. Global needs for sustainable development (2019). https://www.un.org/sustainabledevelopment/. Accessed 4 Sept 2019
6. Ketcham S (2019) Developing sustainable communities. http://www.greenliving.lovetoknow.com. Accessed 5 Oct 2019
7. Hall T (2019) Environmental sustainability: our strategy. https://www.bbc.co.uk/responsibility/environment. Accessed 10 Oct 2019
8. Kinver M (2016) Civic pride can help sustain urban biodiversity. https://www.bbc.com/news/science-environment-36913655. Accessed 5 Sept 2019
9. Urry J (1995) Consuming places. Routledge, London
10. Skye J (2018) Why is sustainable development important? http://greenliving.lovetoknow.com. Accessed 20 Sept 2019
11. Paris agreement (2015). http://unfccc.int/files/meetings/paris_nov_2015/application/pdf/paris_agreement_russian_.pdf. Accessed 22 Sept 2019

Chapter 33
Human Microbiome as a Bio-indicator of the Ecological State of the Environment

Nina V. Bugero, Svetlana M. Aleksandrova and Natalya A. Ilyina

Abstract Currently, the relevant area of environmental research is the study of the species composition of biotic communities and its organization laws. To date, the research have been conducted to describe the biocenose species composition for terrestrial and aquatic ecosystems. However, with respect to symbiotic communities of micro-biocenoses of various biotopes of the human body, this issue is still relevant, which is due to the important role of the microbiome in macroorganism activity, as well as to universally observed changes in microorganisms ecology, resulting in an increase in the specific gravity of diseases caused by opportunistic microbes. The above stated was the case to study large intestine microecology state among 462 people living in different climatic and ecological regions of St. Petersburg. The data were submitted on high rates of bacterization by *Blastocystis* species of individuals living in the study region. In the topodems of the city with ecological distress (Kirovsky district), the infection rate of the population with *Blastocystis* species is higher than in areas with a relatively favourable ecological situation (Vasileostrovsky, Kurortny districts). Being in the intestine, blastocysts are involved in the formation of the microbiocenosis of this biotope. When disrupting microorganisms balance, these causative pathogens promote favourable conditions to develop pathological processes. In recent years, the protozoa etiological role in the development of intestinal dysbiosis in chronic diseases of gastroenteric tract has been made. At that, the ability to populate a particular ecological niche depends on the certain biological properties existence in bacteria. The identification of persistent properties existence in the isolated strains of blastocysts revealed a wide spread of antilysocyme activity (ALA) and the dependence of its index intensity on the source of strain isolation. A direct correlation was found between the persistent activity of blastocysts and the depth of intestinal microbiocenosis disorders. These materials make it possible to use the biological properties of microorganisms survival to study the mechanisms of microsymbiocenosis formation in the large intestine biotope and are a theoretical

N. V. Bugero (✉) · S. M. Aleksandrova · N. A. Ilyina
Pskov State University, Pskov, Russia
e-mail: bugero@mail.ru

S. M. Aleksandrova
e-mail: superkandidat@rambler.ru

T. Kolmykova and E. V. Kharchenko (eds.), *Digital Future Economic Growth, Social Adaptation, and Technological Perspectives*, Lecture Notes in Networks and Systems 111, https://doi.org/10.1007/978-3-030-39797-5_33

background to develop a method of screening diagnosis of intestinal dysbiosis based on the determination of ALA clinical isolates of blastocysts. The presence of blastocyst in human microbiome, together with a community of other microorganisms, can be considered as bioindicators of the ecological state of the environment.

Keywords *Blastocystis* species · Bio-indicator · Virulence · Microbiome · Microecological disorders of the intestine · Persistence

JEL-Code Q5

33.1 Introduction

Environmental protection is one of the most popular problems of our time. The very phenomenon of environmental pollution is not new to Russia, as can be seen from the annual State reports on the environmental situation in the Russian Federation. Scientific and technological progress and increased anthropogenic pressure on the environment inevitably exacerbate the environmental situation, which has reached the maximum peak (crisis) in the most developed countries [1].

An environmental crisis is defined as imbalance in ecological systems and in the relationship of human society with nature. It is characterized by the fact that a human, society and the state are unable to counteract the tendency of environmental deterioration [2].

The most important issue of environmental protection is the quantitative assessment of the impact of air, water and soil pollutants on the population health condition [3]. This assessment allows to obtain parameters to characterize the consequences of adverse effects on the animal and plant world, including on the population health conditions, as well as to assess the effectiveness of measures to protect the biosphere from pollution [4].

One of the fields of environmental research is the study of the species composition of biotic communities and laws of its organization [5]. To date, species composition of biotic communities for terrestrial and aquatic ecosystems is known, but the symbiotic communities of microbiocenoses of various human biotopes have not been sufficiently studied. In this regard, the issue of microbiota importance in the life of the macro-organism, the role of universally observed changes in the ecology of microorganisms, which have led to an increase in the specific weight of diseases caused by opportunistic microbes, is interesting [6].

The environmental quality largely determines the level of the population health. This refers to a wide range of environmental factors of both natural and technogenic origin and is primarily related to urbanization processes, which in turn leads to large volumes of industrial, household and other wastes entering the environment. Epidemiological studies show that even at low levels of these effects, negative changes may develop in the human body. Healthy people with tried ND TRUE adaptation mechanisms adjust to a changing environment, it is done with greater difficulty by the body if there is a particular pathology.

A "new" protozoal infection, blastocystosis, caused by parasitizing of the predominantly protozoa *Blastocystis* spp., in large intestine is now widely discussed [7]. The analysis of literature data showed high incidence of blastocysts among patients with chronic diseases of the digestive system and with the history of recurrent diarrhoeal syndrome for up to one month, as well as about the possible schizogony of *Blastocystis* spp. in case of tissue infection, which is confirmed by the detection of schizonts in cloned organs: in the intestine, in the mucosa of urogenital organs, in the lymphoid tissue of tonsils [8]. It is known that the influence of environmental factors modifies not only the macro-organism defences, but also changes the microorganism properties. However, if the effects of the external environment on the host organism are now sufficiently well studied, the dynamics of the biological properties of microorganisms under the influence of anthropogenic environmental factors remains unstudied.

In view of the above said, the study of human intestinal microbiome as a bio-indicator of the ecological state of the environment becomes relevant, which is the goal of the presented paper.

33.2 Materials and Methods of Research

Blastocysts revealing was carried out during microbiological analysis of defecations of practically healthy people. 462 people living in different anthropogenic burden areas of St. Petersburg were examined. The unequal environmental performance of the urban environment has led to the identification of three environmental zones due not only to their natural conditions, but also to the level of technogenic pollution, the main sources of which in southern topodems (Kirovskiy, Moscow districts) are metallurgical enterprises, plants of heavy mechanical engineering and reinforced concrete products, shipbuilding plant "Severnaya Verf," in western districts (Vasileostrovskiy, Primorsky district)—a plant of translucent structures, chemical industry enterprises, brewing company "Baltika," in the northern parts of the city (Kurortny district)—machine tools plant named after S. P. Voskov.

The first group was made of the 135 ($29.22 \pm 1.7\%$) examined people living in the southern areas of the city, the second group was 151 ($32.69 \pm 2.3\%$) (western district), and the third group was 176 ($38.10 \pm 2.6\%$) residents of the northern topodem of the city. The state of intestinal microbiocenosis in the subjects was assessed in accordance with generally accepted recommendations. Suresh culture medium was used to cultivate blastocysts. In order to study the anti-lysozyme activity (ALA) of microorganisms, the photometric method proposed by Bukharin et al. [9] was used.

The condition of microcenosis, quantitative and quality assessment of its composition were carried out with the use of quantitative and high-quality isolation method of types and options of microorganisms on the basis of the order of the Russian Ministry of Health of 09.06.2003 No. 231 "About the approval of the industry standard "Patient management protocol. Intestinal dysbacteriosis" (All-union standard 91500.11.0004-2003)". The classification made by Kuvayeva and Ladodo [10] was used to assess the intensity of disorders in the composition of large intestine microflora.

The data is processed using the application package for Microsoft Excel.

33.3 Results

The research done have shown that out of the 462 persons examined, blastocysts were found in 327 (70.78 ± 1.2%) residents of St. Petersburg. When studying age structure of the examined it was found that more than 50% of the infested persons were people over 50, 28%—people of 35–45 years, 15%—people of 25–35 years, 4.5% children of 7–14 years.

The study of the blastocyst attack of the citizens, depending on the place of residence, revealed its inequality in different climatic-ecological areas of the city (Fig. 33.1).

The maximum invasion figures are recorded in the ecologically disadvantaged Kirovsk district, where the indicator was (47.92 ± 0.2%) of the population. The next most intense blastocyst prevalence is Vasileostrovsky district (32.08 ± 0.4%), and the minimum values of the studied indicator were observed in Resort district (13.33 ± 0.2%).

The study of the morphological properties of the isolated blastocysts revealed two main forms—vacuolar and granular, as well as the cysts of these protozoa. The

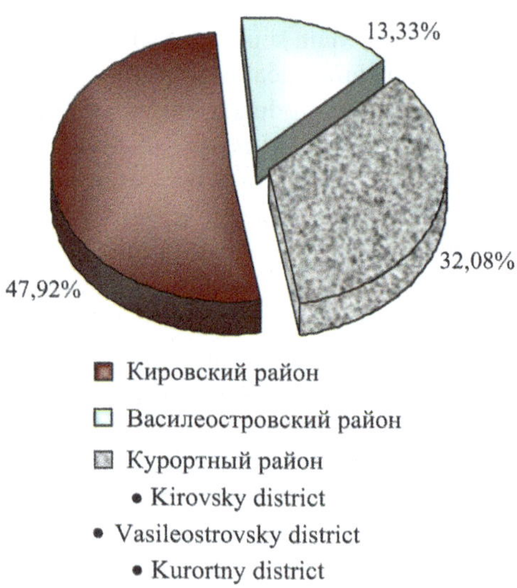

Fig. 33.1 Attack rates by the protozoa *Blastocystis* spp. of residents of St. Petersburg various districts

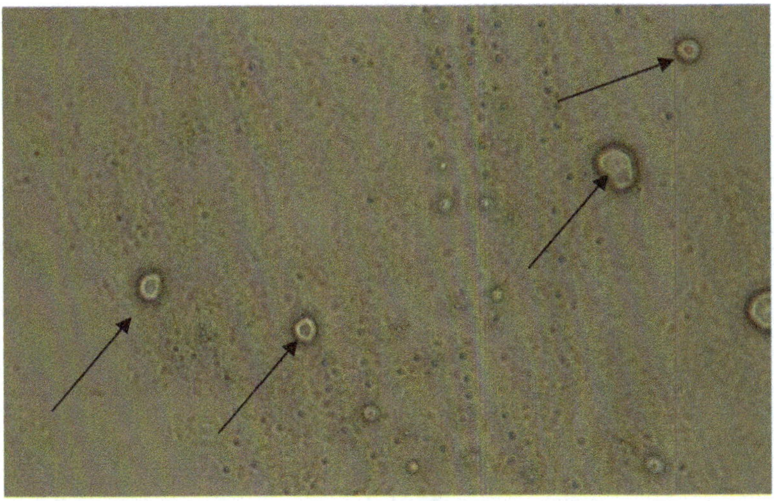

Fig. 33.2 *Blastocystis* spp. in a fecal extract sample (staining × 1000)

samples were dominated by 6–15 μm (65.0 ± 2.7%) vacuolar blastocysts. Large blastocysts (larger than 15 μm in diameter) were found in 63.4 ± 2.9% of the examined (Fig. 33.2).

Blastocyst examination with a scanning microscope revealed oval-shaped cells arranged as clusters or alone (Fig. 33.3).

When examining large intestine microecology, a change in its composition was found in the subjects of the examined groups. The dominant position in the intestinal biocenosis of individuals infested with blastocysts was occupied by anaerobic microorganisms, the qualitative and quantitative composition of which was diverse (Table 33.1).

Further interpretation of the results showed that the presence of protozoan blastocysts in the intestinal microflora in the vast majority of cases is accompanied by a violation of the intestinal microbiocenosis, which is characterized with a decrease in the representatives of the normal flora (bifidobacteria, lactobacilli, bacteroids) and an increase in opportunistic microorganisms (enterococci, klebsiella, staphylococcus).

According to the classification of intestinal dysbiosis, proposed by Kuvaeva and Ladod [10], degree I (DI) dysbacteriosis was revealed in 63 (19.26 ± 0.3%) individuals infested with blastocysts, DII—in 171 people (52.29 ± 1.6%), and degree III dysbacteriosis was diagnosed in 22.63 ± 2.3% (74) cases. Disorders in microecology of the IV degree of severity, characterized by complete imbalance in the composition of the microflora, were found in 2.75 ± 0.2% [9] of the people infested with blastocysts. Normocenosis was observed in 10 subjects (3.07 ± 0.7%).

Further on, the intestinal endoflora of individuals infested with blastocysts living in different areas of the technogenic load of the city was studied. The first group consisted of 101 people living in Kirovsky district of the city, the second—109 people

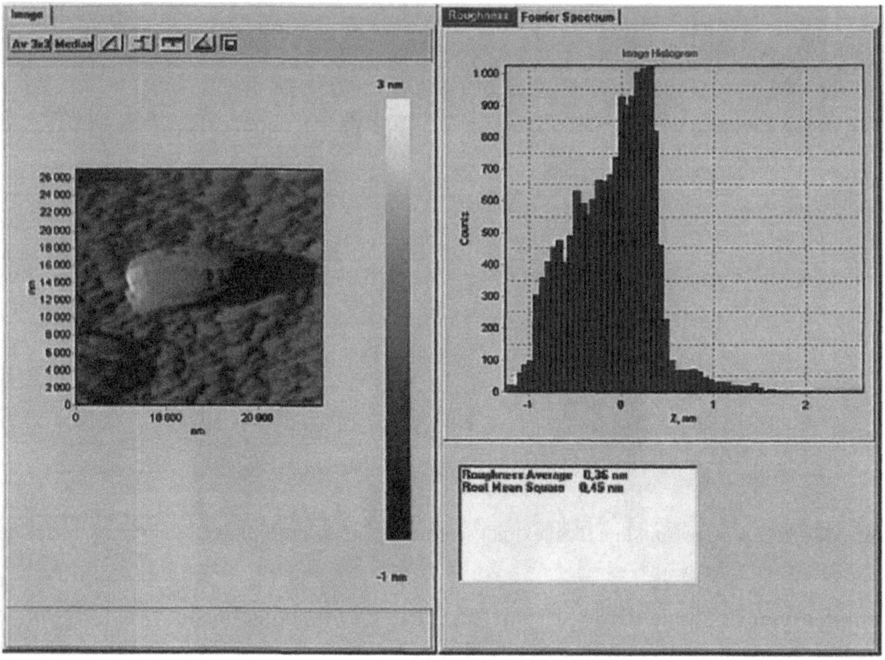

Fig. 33.3 Scanning electronic microscopy and bar graph "NANOEDUCATOR" *Blastocystis* spp. in culture

Table 33.1 Qualitative and quantitative composition of intestinal biocenosis of the examined

Group	Infested with Blastocyst			Healthy		
Microorganisms	abs.	%	CFU/g	abs.	%	CFU/g
Lactobacillus	256	78.28	7.0 ± 0.2	127	94.07	8.7 ± 0.2
Bifidobacterium	234	71.56	6.4 ± 0.1	118	87.40	8.8 ± 0.2
Escherichia	283	86.54	6.8 ± 0.4	132	97.78	7.4 ± 0.1
Bacteroides	327	100	8.5 ± 0.1	133	98.51	9.8 ± 0.3
Enterococcus	156	47.70	7.6 ± 0.2	23	17.03	4.7 ± 0.4
Proteus	78	23.85	6.0 ± 0.7	18	13.33	3.8 ± 0.2
Klebsiella	69	21.10	6.8 ± 0.2	37	27.40	4.2 ± 0.4
Enterobacter	45	13.76	3.9 ± 0.4	8	5.92	3.7 ± 0.1
Staphylococcus	87	26.60	6.2 ± 0.3	43	31.85	4.3 ± 0.2
Candida	112	34.25	6.4 ± 0.2	52	38.51	3.7 ± 0.1
Clostridium	67	20.48	6.1 ± 0.6	26	19.26	6.7 ± 0.8
Total	327			135		
R			≤0.05			≤0.05

who are residents of Vasileostrovsky district and the third group—117 examined, living in Kurortny district of St. Petersburg.

Microecological intestinal disorders analysis, taking into account the dependence on the residence of people showed that the residents of Kirovsky district of the city suffer from the deepest dysbiotic changes in the intestine, which have intestinal dysbiosis of the fourth degree, characterized by a deep imbalance of intestinal microbiocenosis with a change in the quantitative relations of the main groups of microorganisms (bifidobacteria, lactobacilli).

Dysbiosis average values are observed in individuals living in the western region (Vasileostrovsky district) of the city, low values of dysbiotic disorders were recorded in the northern (Kurortny district) topodem (Table 33.2). The most common variant of intestinal biocenosis of the examined groups was dysbiosis of the II degree, which was found in the south in 43.22 ± 2.5% (75 people), in the west—32.61 ± 1.3% (60 people) and in the north—17.12 ± 0.9% of the cases (39 examined).

To study the persistent properties aimed at the degradation of the host resistance mechanisms, we revealed the antilysocyme activity (ALA) in 327 strains of protozoa *Blastocystis* spp., isolated among people living in different areas of technogenic load in St. Petersburg.

Out of the 327 blastocyst strains studied, 178 (54.43 ± 2.7%) had ALA. The level of ALA in *Blastocystis* spp., excreted from fecal masses of the people living in different ecological areas of the city varied. To examine the persistent characteristics of blastocysts, 3 groups of protozoa were identified: the first included strains with a low ALA level of up to 2 μg/ml inclusive, the second with an average level of 3–4 μg/ml and the third with a high level of 5 μg/ml or over. The proportion of

Table 33.2 Qualitative and quantitative composition of the intestinal microflora of individuals infested with blastocysts, depending on the residence topodem

Topodem	South		West		North	
Microorganisms	%	lg M ± m	%	lg M ± m	%	lg M ± m
Lactobacillus	48.03	6.6 + 0.2	78.85	6.8 + 0.2	86.36	6.8 + 0.2
Bifidobacterium	61.18	6.7 ± 0.3	87.82	6.3 ± 0.3	88.96	6.5 ± 0.3
Escherichia	81.58	8.7 ± 0.2	92.31	8.7 ± 0.3	73.38	8.2 ± 0.2
Bacteroides	92.11	8.5 ± 0.1	94.87	8.5 ± 0.1	94.81	8.5 ± 0.1
Enterococcus	64.47	7.3 ± 0.2	46.79	6.4 ± 0.2	42.86	6.1 ± 0.2
Proteus	50.00	6.4 ± 0.7	46.15	5.2 ± 0.1	35.06	5.8 ± 0.3
Klebsiella	58.55	7.2 ± 0.2	43.59	6.3 ± 0.2	14.94	5.8 ± 0.2
Enterobacter	36.84	3.9 ± 0.4	8.33	3.3 ± 0.1	11.04	3.9 ± 0.2
Staphylococcus	73.68	6.8 ± 0.3	48.72	5.8 ± 0.1	30.52	5.3 ± 0.1
Candida	60.52	6.0 ± 0.2	34.62	5.7 ± 0.3	35.71	4.8 ± 0.2
Clostridium	22.37	6.2 ± 0.1	23.08	6.0 ± 0.1	8.44	5.5 ± 0.3
Total	101		109		117	
K			≤0.05			≤0.05

Table 33.3 Antilysocyme activity prevalence in *Blastocystis* spp.

Groups	Residence district	Total strains number	The number of strains with ALA	%
1	Kurortny district	79	53	67.09 ± 2.7
2	Vasileostrovsky district	103	41	39.81 ± 2.1
3	Kirovsky district	145	84	57.93 ± 3.6

strains with low ALA was 35.7 ± 2.5% (63 strains), with an average level of 40.3 ± 3.2% (72 strains) and 24.0 ± 1.6% (43 strains) with a high level of ALA.

Further research showed that ALA protozoan strains level varied depending on the area of the residence. In this regard, the studied blastocyst strains were divided into 3 groups. The first group included blastocysts isolated from fecal masses of the people living in the southern region of the city—145 strains, the second group (western region)—103 strains, and the third group (northern topodem of the city)—79 strains of blastocysts. The antilysocyme activity prevalence of *Blastocystis* spp. in the study groups is presented in Table 33.3.

The analysis of the research showed that the maximum number of blastocyst strains—145 (57.93 ± 3.6%)—was found in fecal masses of the residents living in the environmentally disadvantaged Kirovsky district. In the same topodem of the city, the isolated blastocyst strains had the highest antilysocyme activity. The analysis of the study of ALA indices in these groups of the subjects is presented in Table 33.4.

Table 33.4 shows that the antilysocyme activity of blastocysts isolated among residents of the northern and western regions of the city (Vasileostrovsky, Kurortny district, respectively) had low (up to 2 μg/ml) and average (3–4 μg/ml) values,

Table 33.4 Antilysocyme activity level of *Blastocystis* spp. among people, living in St. Petersburg districts with different ecology

ALA values (μg/ml)	Groups examined					
	Northern district		Western district		Southern district	
	The number of strains with ALA	%	The number of strains with ALA	%	The number of strains with ALA	%
Up to 2 mcg/ml	41	77.36 ± 3.7	25	60.98 ± 3.4	18	21.43 ± 0.7
3–4 mcg/ml	12	22.64 ± 1.7	16	39.02 ± 1.8	37	44.05 ± 3.5
5 mcg/ml and more	–	–	–	–	29	34.52 ± 2.3

whereas high rates (5 μg/ml or more) were completely absent. High ALA values (5 μg/ml or more) were observed only in protozoa found among the residents of the southern topodem of the city (Kirovsky district); its indices amounted to 34.52 ± 2.3% (29 strains).

The prevalence of ALA in *Blastocystis* spp. speaks of their high colonization potential in the large intestine. The researchers V. M. Bondarenko, V. G. Likhoded (2008) found out that many bacteria, such as *Klebsiella, Staphylococcus, Clostridia,* with persistence factors, determine the severity and course of dysbiotic intestinal disorders. In this regard, the study of the antilysocyme activity of blastocysts isolated among individuals with dysbiotic disorders of the intestinal microflora and the influence of these properties on the intestinal flora condition is of great interest.

In view of this, all the blastocyst strains under study were divided into 5 groups depending on the severity of dysbiotic disorders among the subjects.

Dysbacteriosis of the first degree (DI) was detected in 63 (19.26 ± 3.2%) individuals infested with blastocysts, DII in 171 people (52.29 ± 2.7%), and dysbacteriosis of the third degree was diagnosed in 22.63 ± 1.2% (74) of the cases. Microecology disorders of the IV degree of severity, characterized by complete imbalance in the microflora composition, were found in 2.75 ± 0.5% [9] of invasive blastocysts. The fifth group consisted of the individuals with intestinal normobiosis (Table 33.5).

The research showed that at dysbiosis of I and II degree of severity, the number of blastocyst strains with high ALA rates had low degree of incidence or were completely absent. Out of 26 strains of protozoa isolated among the subjects with I degree of dysbiosis, none of the isolates had a high level of the property under study. At II degree, the incidence degree of AL property with high values was 8.57 ± 1.2%. At the same time, with III and especially IV degrees of dysbiosis, the prevalence of strains with a high level of the property sharply increased from 97.75 ± 6.4 to 100% for the strains with ALA.

The study of the blastocyst strains persistent characteristics isolated from the fecal masses of the people who do not have obvious dysbiotic disorders showed the prevalence of protozoa strains with a set of mild signs of persistence.

The fact that ALA strains with medium and high characteristic values were found in blastocyst strains points at a significant pathogenetic role of this factor in the

Table 33.5 The prevalence of ALA protozoa *Blastocystis* spp. at intestinal dysbiosis

The degree of intestinal dysbiosis among the subjects	The number of the subjects (n)	The incidence degree of intestinal dysbiosis (%)	The proportion of strains with ALA (n = 327)	
			abs.	%
I	63	19.26 ± 3.2	26	14.60 ± 3.5
II	171	52.29 ± 2.7	66	37.07 ± 3.7
III	74	22.63 ± 1.2	71	39.88 ± 6.4
VI	9	2.75 ± 0.5	9	100
Normocenosis	10	3.07 ± 0.6	6	3.37 ± 2.3
p				$p < 0.05$

development of intestinal dysbiosis. The high level of the characteristic under study shows, apparently, the etiological significance of the blastocysts at dysbiotic conditions occurrence. Thus, an increase in the persistent characteristics of blastocysts with the rise of the seriousness of intestinal microecological disorders was identified. The revealed connection between ALA of protozoa and the seriousness of disorders in the microbiocenosis of the subjects makes it possible to consider protozoa *Blastocystis* spp. and antilysocyme activity as bioindicators of the dysbiotic process.

33.4 Conclusion

1. High infestation rates by protozoa *Blastocystis* spp. were identified among residents of various climatic and ecological districts of St. Petersburg.
2. It was revealed that in the topodems of the city with ecological distress (Kirovsky district), the population infestation with blastocysts is higher than in areas with a relatively favourable ecological situation (Vasileostrovsky, Kurortny districts).
3. It was found that blastocystic invasion is accompanied by compositional rearrangement in human intestine microbiome with varying degrees of severity of dysbiotic conditions.
4. ALA was identified in 178 ($54.43 \pm 2.7\%$) of the studied blastocyst strains, which allows them to persist for a long time in the macroorganism, thereby reduce the adaptive and reparative potencies of the host cells, and result in destructive changes in the intestinal microbiome and determine the nature of dysbiotic manifestations.
5. The microecology study of the human intestine in different technogenic load areas of St. Petersburg may be a prerequisite to use *Blastocystis* spp. together with other microbiocenosis as bio-indicators of the ecological state of the environment.

References

1. Vishnyakov Ya, Zozulya P, Zozulya A, Kiselev S (2014) Environmental protection. Academy, Moscow
2. Kukin P, Kolesnikov E, Kolesnikova T (2015) Environmental impact assessment. Security expertise. Yurait, Moscow
3. Melekhova OP, Egorova EI, Evseeva TI et al (2007) Biological environmental control: bioindication and biotesting. Academy, Moscow
4. Gelashvili DB, Dmitriev AI, Iudin DI et al (2008) Multifractal analysis of the species structure of small mammalian communities of the Nizhny Novgorod Volga Region. Ecology 6:456–461
5. Yasoveev MG, Strekh NL, Kakareka EV, Shevtsova NS (2013) Environmental monitoring and ecological expertise. INFRA-M, Moscow
6. Shenderov B (2001) Medical microbial ecology and functional nutrition. Volume III: Probiotics and functional nutrition. Grant, Moscow
7. Belova LM (1995) Fauna Blastocystis. Parasitology 29(3):208–213

8. Gritsyuk OV, Prodeus TV, Maksimova MS, Sadykova VD (2016) On the variability of isolates of Blastocystis species maintained in vitro. Med Parasitol Parasit Dis 3:12–18
9. Bukharin OV, Valyshev AV, Gilmutdinova FG (2006) Ecology of human microorganisms. Ural Branch of the Russian Academy of Sciences, Ekaterinburg
10. Kuvaeva IB, Ladodo KS (1991) Microecological and immune disorders in children. Medicine, Moscow

Chapter 34
Interdisciplinary Network Project as a Factor of the Ecologization of Youth Consciousness

Svetlana N. Vodneva, Elena A. Smirnova, Irina I. Khrapchenkova, Galina G. Maslova and Julia V. Michailova

Abstract Scientific and technological progress, globalization, information technologies advances have shown that the environmental component is an integral part of the development of human society. Modern socioeconomic and technological challenges impose special requirements on the education system and educational outcomes, which should drastically transform the worldview of a young person changing the vector of their value orientations towards ecologization of consciousness and greening of lifestyles and activities. The purpose of this publication is to focus on the opportunities and prospects of interdisciplinary young people's activities that are implemented as an international networking project on environmental issues. The methodological basis of the study are conceptual provisions of the interdisciplinary approach, ecologization of education, and the project method. The article presents the description and results of the interdisciplinary environmental network project entitled "Ecological Paths Through the Hanseatic League of the New Time" implemented in anticipation of the XXXIX International Hanseatic Days of the New Time, which took place in Pskov in June 2019. One can state that such a practice-oriented approach in teaching has resulted in changing the young people's perception of regional environmental problems. They have got steadily motivated in taking actions aimed at improving the environmental situation.

Keywords Interdisciplinarity · Project activities · Network project · Ecologization · Environmental awareness

JEL Classification Code I 200

S. N. Vodneva (✉) · E. A. Smirnova · I. I. Khrapchenkova · G. G. Maslova
Pskov State University, Pskov, Russia
e-mail: wodnewa@yandex.ru

J. V. Michailova
Secondary School No. 11, Pskov, Russia

T. Kolmykova and E. V. Kharchenko (eds.), *Digital Future Economic Growth, Social Adaptation, and Technological Perspectives*, Lecture Notes in Networks and Systems 111, https://doi.org/10.1007/978-3-030-39797-5_34

34.1 Introduction

The issues of efficient use of natural resources posing no danger to the environment are cutting-edge in the modern world nowadays. The solution to numerous environmental problems depends not only on specialists whose professional activities are directly related to the environment, but also on each individual.

For many decades the international community has been actively discussing strategies that can prevent the global destruction of the world. Special attention is paid to the integration of environmental education and aspects of international economic and sociocultural policies.

The concept of environmental protection, adopted at the UN Conference in Rio de Janeiro in 1992, and the decisions of the World Summit on Sustainable Development, which took place in Johannesburg in 2002, played an important role in establishing and developing the paradigm of environmental education.

Awareness and understanding of environmental problems and their global negative impact on society led to emerging such a phenomenon as "education for sustainable development", which implies specific methodological approaches to the goals and content of education, to the choice of means and technologies in teaching various fields of knowledge. The significance of education for sustainable development was proved at the 37th Session of the UNESCO General Conference in 2015, when the Global Program of Action on Education for Sustainable Development was adopted. The Program highlights the importance of changing the educational context so that everyone can contribute to sustainable development and have the opportunity to take actions in order to ensure preservation of natural environment and greening of economy.

Thus, the international recommendations call for balanced incorporation in the educational content of the most important components necessary for sustainable development, namely: economics, socialization, ecology. Moreover, environmental education becomes value-oriented, transdisciplinary and integrative, and aimed at developing a common system of awareness of the scientific worldview, as well as creating new ecologically-oriented values [1].

The result of such an education is a person thinking and acting in favor of sustainable development of the society.

34.2 Methodology

The special significance of environmental education in the modern world is highlighted in various legislative acts, such as Federal Laws "On Education in the Russian Federation" and "On Environmental Protection". In addition to that, conceptual documents "The Concept of General Environmental Education for Sustainable Development" and "The Basics of State Policy in the Field of Environmental Development of the Russian Federation for the Period until 2030" were also developed.

Among the main principles that contribute to sustainable development, A. P. Kushnareva singles out "variability of forms and methods, continuity, an interdisciplinary approach, taking into account regional characteristics, a practice-oriented focus, and competence in environmental decision-making" [2].

In the context of modern environmental problems, the term "ecologization" is actively used. In the scientific literature this phenomenon is interpreted as a concept that "reveals incorporating the ecological approach and environmental principles into various types and spheres of human life" [3], and "as a more versatile, more systematic than before approach to the objective world and a true awareness of the role of nature in human life" [4]. Thus, the main dominant feature of ecologization is the harmonious interaction of the triad "man – society – nature".

According to the academician of the Russian Academy of Science, Rector of the Siberian Federal University Eugene A. Vaganov the term "ecologization" refers, first of all, to the worldview and consciousness. Moreover, the researcher claims that ecologization of consciousness is a part of the general human culture. One cannot but agree to this statement, as today ecologization of consciousness of a young person is one of the global trends in modern education.

Analyzing various points of view on ecologization V. E. Chernikova comes to the conclusion that "ecologization of consciousness implies rising a completely new vision of the relationship among natural, cultural, and social aspects of human life, which requires a new worldview from a person" [5]. This idea emphasizes the special role of education in promoting ecologization of consciousness, which stems from environmental awareness based on knowledge about nature and the ability to implement this knowledge in practice. Ecologization of consciousness implies active development of critical thinking and high cognition involvement.

It is known that cognition is the leading type of human activity. The cognition of a young man is characterized by a special susceptibility, the ability to process and absorb a huge flow of information. In this period critical thinking, the desire to give one's own assessment of various phenomena and reasonably explain a particular phenomenon develop very actively. Thereby, the originality of thinking is being formed. That is why period of schooling is crucial for setting basic value orientations aimed at getting young people aware of the interrelations between the person and the world around him.

One of the planned educational outcomes indicated in the modern federal educational standards is "the formation of ecological culture, the content of which is to realize a responsible, respectful attitude to the environment, as well as to gain experience in environmentally oriented reflective assessment and practical activities in life situations" [6].

However, some researchers state that "education remains ill-equipped technologically". It is often focused on the application of methods specific to the Sciences, while the potential of socio-humanitarian knowledge, methods of collaboration, the resource of child-adult community interaction, and multicultural exchange are not applied sufficiently [7]. We share the opinion of A. I. Vasilchenko and T. B. Igonina, who claim that the prerequisite for building a common culture, including the environmental education of students, is the integration of the natural science and humanitarian knowledge synthesis in the content of education [8].

Motivation and involvement in solving environmental problems must be formed and developed throughout the entire period of life. Of particular importance is the development of environmental awareness at the stage of school education. The educational paradigm nowadays is characterized by a shift in emphasis from monodisciplinarity to interdisciplinarity. Interdisciplinarity is seen today as a new requirement and a prerequisite for the existence of the modern world, which is determined by globalization, informatization, and rethinking of existing relationships in nature. It contributes to developing a new picture of the world, which can provide responses to urgent challenges of our time. Interdisciplinarity is assessed now as a special kind of knowledge comprising more than one discipline in need to solve the central problem of research.

Interdisciplinarity as an approach most clearly fits the system of environmental education, which implies a widespread application of methods and techniques that ensure the integration of disciplinary knowledge in teaching. One of such methods promoting environmental awareness is interdisciplinary project activity. Currently, the project method is an indispensible component of the modern educational paradigm and has been actively used for a long time in theory and practice of teaching various academic subjects.

The basis of the modern domestic concept of project-based education was built up by Eugenia S. Polat, who developed the conceptual apparatus, methodology, and substantive characteristics of this approach. The researcher defines the project technology as "a way to achieve a didactic goal through the detailed development of a significant for students problem, which should end up with a practical outcome, designed in one or another form of final product" [9, 10].

The project activities can be characterized by following distinguished features: a significant task of research requiring interdisciplinarity and a need for investigations for its fulfillment; students' autonomy; practice-oriented direction of tasks; precise outline of the project content; the need for doing research and making public presentation of the result. The Russian methodology presents different project typologies. One of them is proposed by Svetlana E. Kaplina, who figures out the following types of projects: informational or orientational, practice-oriented or applied, investigating, role-playing, creative [11].

The dominant trend of educational area is the active implementation of information and communication technologies in teaching, including the process of environmental education for sustainable development. New trends change project technology and give it the impetus for advancing: project activities are more often organized in the form of network. The didactic peculiarity of the project's network model is that it creates a real (rather than virtual) activity area and it is focused not on the transfer of ready-made knowledge and thinking algorithms, but on the students' active acquisition of the activity methods and their productive self-realization [2].

We follow A. P. Kushnareva in her definition of network projects, which she determines as "the form of organizing learning in which the achievement of the goal is carried out through horizontal partnership interaction of entities supported by telecommunication means (in particular, the Internet). A network educational project is a qualitatively new information and educational technology that allows a teenager to take personal initiative and build his own track of creative development, since self-presentation on the Web is possible only by means of a creative product" [2].

Practical experience shows that at present quite a few educational institutions have been actively involved in organizing and conducting network projects specifically on environmental issues ("Cleaning Fish", "Water for Life", "Global Thinking – Hands around the Earth", "Environmental Commonwealth", " Nature Detectives", "Green School Program", etc.). However, most of them are monodisciplinary and cover the Natural Sciences (biology, geography, ecology) to be carried out.

An interdisciplinary network project offers additional opportunities for a more complete set of goals and is an effective training form for solving integrated objectives—environmental awareness and foreign language competence.

This format of project activities fosters:

- the development of communicative competence in Foreign and Russian languages;
- the capability to represent your country in your native language surroundings as well as in a foreign language milieu;
- the ability to carry out cognitive and investigating creative activity in the native and foreign languages;
- social skills to interact in a group;
- the ability to use the information and educational area for representing urgent environmental problems of the region.

Recently, school and university teachers have become increasingly aware that successive links must be built between the school and the university. Such links are provided by participating in joint projects. Project activities completely correspond to the concept of the National Project "Education", according to which by the end of 2024 twelve million children will participate in open online lessons that are implemented considering the experience of the "Projectoria" open lessons series aimed at early career guidance [12].

34.3 Results

Pskov State University has got experience in organizing and conducting joint projects with schools. Among successful projects aimed at ecologization of youth consciousness there is a joint network project organized under the auspices of the Goethe Institute: "Ecological Paths on the Territory of the Hanseatic League of the New Time." The network project was attended by teachers and students of the Pskov Secondary School No. 11, the Kaliningrad Regional Public Institution "Natural Heritage", the Gymnasium "Vector" in Zelenogradsk (the Kalinin Region), the Foreign Language School "In Down-Town" in Tallinn (Estonia). The project was implemented with the participation of the organization "Auslandgesellschaft" of the city of Lübeck (Germany). The number of participants surpassed 80 people from different countries and regions of Russia. The results of the study were presented at student scientific conferences held by Pskov State University.

The general objective of the project is to describe the unique nature reserves and landscapes of the Pskov Land, to state facts about amazing natural sites of the region, and to provide information on the possibilities of recreation and exploration of these territories in the anticipation of the events of the Hansa of New Time, which were held in the Pskov Region in June, 2019.

It's worth noting that the Hanseatic League of the New Age has become a union that can unite people who are creative, not indifferent to history, to the nature of their native land, a union that opens the borders for exploring traditions, language and culture of neighboring countries. This union is able to find new issues for mutual understanding, for implementing new projects, formulating new tasks and ways to fulfill them.

The strategic objective of the project was to bring together people from different countries, different ages, and speaking different languages. And the task of uniting representatives of different cultures and social groups was to be solved by young people living on these territories, loving their lands, exploring and preserving them for future generations. As the organizers of the project note, it is crucial to see the surrounding beauties with your own eyes, to find and feel what unites people, to learn to respect otherness and be proud of your own identity. The main thing is to protect, value the world and nature, and understand that this world can be preserved only by collaborative efforts.

One of the practical objectives of the project was a comprising of "ecological itineraries (eco-itineraries)". The eco-itinerary is a track that passes through various natural objects, such as landmarks, specific ecological systems and habitats, objects of environmental, cultural, and other values. Besides, this track must be specially equipped to obtain information about these objects.

Implementing the project showed that the creation of eco-itineraries contributes to the solution of environmental, educational, and socioeconomic problems. It can foster the development of the region due to a number of reasons. For example, the project proved that appeared alternative types of recreation focused on new values—savoring nature, acquiring new environmental knowledge, getting aware of the landscape, historical, and cultural features of the region. Participation in the project oriented to ecologization of youth consciousness manifested that working out and arranging ecological itineraries is one of the forms of developing the ecological worldview.

Visiting eco-itineraries helps, on the one hand, unwind and recharge and, on the other hand, increase public awareness of valuable nature areas, as well as attract people's attention to the problems of preserving natural heritage of their native land, protecting biodiversity, and rational nature management.

While working on the project, the participants discovered that in case the ecological destinations are located in specially protected natural areas, the major stream of tourists is to be directed to specially equipped routes. This measure will result in taking off pressure away from natural complexes. Moreover, new eco-itineraries can increase the attractiveness of this terrain as a place of tourist destination. The project participants proved that the arrangement of eco-itineraries involves the local population in the tourism sector and environmental education, providing an incentive to preserve and restore natural and historical places. It was also confirmed that the appropriate arrangement of ecological itineraries can expand the inclusive environment for physically-challenged people and engage them in ecological activities.

The implemented project falls into the type of interdisciplinary network research. The main objectives of this joint project are the following:

- widening general background of students;
- generating sustainable interest to the surrounding world, to the national and local cultures, and to the history and traditions of neighboring states;
- extending knowledge in various study fields:
  - Geography: extending knowledge about one's own country, acquaintance with the geography and culture of neighboring countries (in this case Russia, Estonia, Poland, Latvia);
  - Ecology: developing the ecological worldview, deepening knowledge of the environment and its conservation;
  - History: getting aware of the history of the Hanseatic League, active involvement into "Creating a New History of the Hanseatic Cities";
  - Foreign languages (Russian, German): developing foreign language communication skills, in-depth study of the relevant to the project terminology;
  - Intercultural Communication: building up and strengthening friendships through partner cooperation and enhancing motivation to effective interaction and communication;
  - Ecotourism: active involvement of schoolchildren from different countries in creating new attractive tourist itineraries in their homeland in order to popularize the eco-facilities of their own country, as well as obtaining reliable first-hand information about ecotourism in participating countries.

One of the practical project outcomes was working out a navigation calendar "On the Ecological Routes of the Hanseatic League of the New Time" in the languages of the Hanseatic states: Russian and German. The navigation calendar contains the description of 12 nature destinations that corresponds to the number of months in a year. The descriptions were provided by the project teams: four descriptions were submitted by Russia, two descriptions—by the Baltic States and two—by Poland.

The network project was a carefully planned students' activity which continued throughout the school year. The students were supplied with a package of training materials and instructions. The project work was supervised by teachers and project designers. Conducting the joint research and exchanging materials the project teams from different countries communicated via the Internet in a foreign language. The results of the project were highly evaluated as it became one of the prize-winners at the Russian National Contest "School for Ecology: Thinking, Investigating, Acting" and at the International Contest "Green Globe".

The project preparation, organization and complementation showed that while planning a project one should provide students with tasks and assignments that presuppose versatile approaches to their fulfillment. The offered tasks should be oriented to investigating a specific issue or a social phenomenon, an event, a problem from different aspects, which will enable students to produce a new knowledge in a form of any creative product.

Reflecting on the project work the participants noted that they were exposed to the opportunity to work in teams, autonomously substantiate and make decisions, take responsibility and initiative, manage themselves, and try on different social and public roles.

34.4 Conclusions/Recommendations

The resource potential of interdisciplinary network projects is very wide. This type of research is very popular today at all levels of education, since it provides unlimited opportunities for solving many problems. Such projects can be interpreted as a technology applied in the framework of Content and Language Integrated Learning (CLIL). The implemented project vividly demonstrates the four components CLIL as an approach where students learn a subject and a second language at the same time includes: content, communication, cognition, culture. The project participants studied a socially important *content*, namely environmental problems and nature preservation; they had to formulate their ideas, conclusions, and new information using a foreign language as a tool in *communication*; the general objective of the project is to develop ecologization of consciousness of young people, which is impossible without developing critical thinking skills and *cognition* involvement; and, finally, discussing issues and presenting the project results students were exposed to alternative perspectives and shared understandings, which deepened their awareness of otherness and self as representatives of different *cultures*.

Educational priorities are changing towards students' autonomy and self-study, integration, and interdisciplinarity. There is a distinct tendency to prepare young people for thinking globally and acting locally. A young person with an ecologically oriented consciousness will be able to harmonize the relationship between man and nature.

References

1. Stepanov SA (2014) Pedagogicheskoye obespecheniye ekologicheskogo razvitiya strany [Pedagogical support of the country's ecological development]. http://www.priroda.ru/reviews/detail.php?ID=10917
2. Kushnareva AP (2006) Setevyye ekologicheskiye proyekty kak sredstvo razvitiya tvorcheskoy aktivnosti podrostkov [Network environmental projects as a means of developing creativity of adolescents], Ulan-Ude, 189 pp

3. Dzyatkovskaya EN, Uvarova EG (2013) Ekologizatsiya predmetnogo soderzhaniya kak porozhdeniye yego novykh znacheniy i smyslov [Ecologization of content as producing new meanings and symbols]. Vestnik Moskovskoy gosudarstvennoy akademii delovogo administrirovaniya. Seriya: Filosofskiye, sotsial'nyye i yestestvennyye nauki [Bull Mosc State Acad Bus Administration Ser: Philos Soc Nat Sci] 2–3(21–23). ecoedu-cis.iseu.bsu.by › wp-content › uploads › 2017/10
4. Reymers NF (1990) Prirodopol'zovaniye: Slovar' – spravochnik [Nature-use: dictionary]. Mysl, Moskva, 637 pp
5. Chernikova VE (2017) Fenomen ekologizatsii cheloveka v kontekste sovremennykh ekologicheskikh vyzovov. Kontekst i refleksiya: filosofiya o mire i cheloveke [The phenomenon of human ecologization in the context of modern ecological challenges. Context Reflect: Philos World Man] 6(3A):161–169. https://elibrary.ru/item.asp?id=30612060
6. Kontseptsiya federal'nykh gosudarstvennykh obrazovatel'nykh standartov obshchego obrazovaniya/ A.M. Kondakov [i dr.]; pod red. A.M. Kondakova, A.A. Kuznetsova. [Kondakov AM et al. The concept of federal state educational standards of general education; under the editorship of A.M. Kondakova, A.A. Kuznetsova]. Prosveshcheniye [Education]. – 2008. — 39 pp
7. Andryukhina LM, Fadeyeva NYu, De Negri Zherar (2017) Razvitiye ekologicheskoy kul'tury studentov v protsesse mezhkul'turnoy kommunikatsii na inostrannom yazyke [Development of the environmental culture of students in the process of intercultural communication in a foreign language]. Obrazovaniye i nauka [Educ Sci] 19(8):47–74. https://doi.org/10.17853/1994-5639-2017-8-47-74
8. Vasilchenko AI, Igonina TV (2014) Realizatsiya kontseptsii obshchego ekologicheskogo obrazovaniya v interesakh ustoychivogo razvitiya v usloviyakh vvedeniya FGOS [Implementation of the concept of general environmental education for sustainable development in the context of the introduction of FSES]. Vestnik Kemerovskogo gosudarstvennogo universiteta [Bull Kemerovo State Univ] 4-2(60):83–85. https://elibrary.ru/item.asp?id=22777584
9. Polat ES. Metod proyektov. Laboratoriya distantsionnogo obucheniya. [Project method. Distance learning laboratory]. http://schools.keldysh.ru/labmro/lib/polat2.htm
10. Polat ES (2007) Sovremennyye pedagogicheskiye i informatsionnyye tekhnologii v sisteme obrazovaniya: ucheb.posobiye dlya studentov vyssh. ucheb. zavedeniy [Modern pedagogical and information technologies in the education system: textbook for students of higher. Learning institutions] Izdatel'skiy tsentr "Akademiya" [Publishing Center "Academy"]
11. Kaplina SE (2012) Metod mezhdistsiplinarnogo ekologicheskogo proyektirovaniya [The method of interdisciplinary environmental design]. Uspekhi sovremennogo yestestvoznaniya [Adv Mod Nat Sci] 11-2:58–60. http://natural-sciences.ru/ru/article/view?id=31168
12. Natsional'nyy proyekt «Obrazovaniye» (2019) [National project "Education"]. https://edu.gov.ru/

Chapter 35
The Element of Fire in Dialectal Linguistic Worldview

Natalia V. Bolshakova and Olga Kh. Gladkova

Abstract The article is devoted to the theme of fire and the related theme of conflagration in traditional folk culture and language. The relevance of the topic is due to the central position of the concept of "fire" in the linguistic worldview. The introduction of new regional material into the study allowed us to deepen our understanding of the ambivalent nature of the fire phenomenon. In the work on exhaustive empirical material extracted from regional sources, the conceptualization of conflagration, embodied in the Pskov toponymicon, is analyzed. Over 40 toponyms have been identified on this material, marking more than 70 objects on the territory of the Pskov region. The quantitative composition of toponyms motivated by the image of the fire indicates the large-scale nature of the elements of fire as a destructive force. The linguistic worldview, including the concepts of fire and conflagration, cannot be fully revealed without taking into account the mythological representations that exist in the minds of holders of traditional culture. The article reveals how the interpenetration of rational and mythological characterizes human consciousness and behavior in extreme conditions. The study is conducted using an interdisciplinary approach that integrates linguistic, linguoculturological, ethnolinguistic research methods, which allows us to present a holistic worldview in the unity and interconnection of its categories.

Keywords Linguocultural concept of fire · Conflagrations · Toponymy · Linguistic worldview

JEL Code A 130 Relation of Economics to Social Values

> I'm not on my own, but the strongest of all;
> they care for me and are scared of me (fire).
> V. I. Dahl

N. V. Bolshakova (✉) · O. Kh. Gladkova
Pskov State University, Pskov, Russia
e-mail: bolshakova55@yandex.ru

O. Kh. Gladkova
e-mail: olga.lexter@gmail.com

T. Kolmykova and E. V. Kharchenko (eds.), *Digital Future Economic Growth, Social Adaptation, and Technological Perspectives*, Lecture Notes in Networks and Systems 111, https://doi.org/10.1007/978-3-030-39797-5_35

35.1 Introduction

Fire as a universal reality and its symbolic embodiment occupies one of the central places in physical and spiritual world of a man. Being among the primary elements of the universe, along with water, air, earth, fire has a significant potential for linguistic representation. A number of studies, including monographs are devoted to the lexico-semantic "portrait" of fire on Russian material [20, 21]. "The nationwide and dialectic lexical units of the Russian language that "nominate the burning process, its stages and components" are considered. Secondly, the derivatives of the roots with the initial meaning of "fire", "be on fire", "burn", "smoulder", which are at the first stage of derivation" are analyzed [21, p. 4]. A study of the linguistic representation of the image of fire in a comparative way (in Russian and Turkish languages) reveals universal and nationally determined components of the linguocultural continuum at the level of vocabulary and phraseology [13]. In the folk epos on the material of the Karachay-Balkarian language, it was revealed that "fire, as one of the elements of the universe, is sacred, and at the same time, all-devouring and giving birth" [2, p. 76].

As the above mentioned studies and our material show, the relevance of addressing to the topic of fire in its linguistic aspect is determined not only by the central position of this phenomenon, but also by the complexity and multidimensional nature of its structure of linguistic worldview with the key concept "fire". Each aspect of its consideration, especially with the involvement of new language material enables to deepen the understanding of this natural and cultural phenomenon.

The aim of the article is to disclose the properties of fire as an element, reflected in traditional folk culture and language. The introduction of new data—the materials of Pskov folk dialects and folklore—allows us to discover regional features of this phenomenon in the context of other sources. The specificity of the work is one of the perspectives, namely, the manifestation of the highest degree of fire intensity, reflected in fires as a social disaster.

35.2 Methodology

The methodological base of the study is formed by the theoretical postulates that have been developed in linguistics and related disciplines about the connection between language and culture, about the systematic organization of language units, and about anthropocentrism of the language. The work is based on the principal points of cognitive linguistics, according to which a person perceives the world around him and conceptualizes reality using linguistic means (in this case, lexical ones), as a result of which a linguistic worldview of society is formed (in our work we are talking about bearers of traditional Russian culture in its regional variant).

The following methods are used in the work: system-structural—in establishing paradigmatic relations between lexemes, component analysis—in identifying the semantic structure of a word, lexicographic method—in analyzing vocabulary definitions in various explanatory dictionaries (normative, dialect) and identifying elements of contextual semantics.

An introduced interdisciplinary approach integrates linguistic, linguoculturological, ethnolinguistic research methods. It allows us to present a holistic worldview in the unity and interconnection of its categories.

35.3 Perception of the Dual Nature of Fire in Traditional Culture and Folk Speech

"Fire" as a phenomenon of the anthropocentric paradigm is an ambivalent category: "on the one hand, it is a formidable element that brings death and destruction; on the other, the element of light and heat, the source of life" [6, p. 513]. Folk wisdom embodied the binary character of the fire symbolism in a saying-riddle, used as an epigraph to this work: "I'm not on my own, but the strongest of all; they care for me and are scared of me (fire)" [11, p. 644]. Indeed, the source of the fire can be a natural phenomenon (lightning, self-ignition of something dry from heat) or a person's activity—conscious (making or lighting fire), careless or criminal. The perception of fire—as a blessing for a person or as a disaster—is determined by these reasons.

Let us turn to the very lexical meaning of the word *fire*. The anthropocentric approach to the semantics of a word reveals the unequal relevance of the semantic features of its lexical meaning. The meaning traditionally formulated in the definition dictionaries includes the generic seme "gases" with two most important for perception semes of heat and light: "incandescent gases". In addition to this "bundle" of semes (the first archiseme), the dictionary definition indicates the effect of "burning". However, if in some dictionaries the seme "burning" is presented as a visualization of this action ("around a burning object"—[16,1982 II, p. 587]), then, for example, in the "Pskov Regional Dictionary with Historical Data" the seme "burning" is transmitted as a causation, i.e. the conditioning of fire by the action of combustion is emphasized: "fire"—'incandescent gases generated by combustion' [14, 2011 22, p. 524]. This distinctive detail of semantics seems to be significant. In our opinion, the second version of the definition (in PRD) is more relevant, because focuses on the burning subject itself, which, from the anthropocentric point of view, is perceived as the second dominant (second archiseme) in the structure of lexical meaning. In the Dahl's Dictionary, where, as far as is known, specific techniques for the semantization of the word are used, "fire" is immediately interpreted as "a burning phenomenon", only then signs of heat and light are indicated, while the physical nature of the fire ("hot gas") is far from a leading role [11, p. 644]. Indeed, the ordinary consciousness perceives fire tactilely—as heat and visually—as light, while the object of burning acts as an indicator of the duality of perception of fire: the very nature of the object and its loss or death determine a sign of harmfulness of the fire.

At the same time, as follows from a comparative analysis of dictionary meanings, the marked duality of the perception of fire in the definition zone was not reflected.

In accordance with the specifics of the linguistic dictionary, in most such cases, pragmatic information can follow from the speech uses of the word, i.e. from a definitively formed illustrative area of a dictionary entry. Indeed, in the "Pskov Regional Dictionary", quotes first appear reflecting the use of fire in the oven for heat and cooking, and then a quote is given about the fire: "I na damá agón' pirikínulsya" (The fire has spread to the houses) [14, 2011 22, p. 524]. Thus, the analysis of contextual semantics in the dialect dictionary reveals the ambivalent image of fire existing in the human mind in the linguistic worldview.

It is important to emphasize that dictionary interpretations, in addition to the definitive-descriptive part, contain a reference to the synonym "flame". Meanwhile, "fire" and "flame" are not identical synonyms in language, which is also revealed not at the level of archisemes of lexical meaning, but in the pragmatic part. So, in the "Pskov Regional Dictionary" of seven illustrations, six describe exactly the situation of fire, for example: "Ya sizhý, na pryalke rabótayu, vdruk i svéta net. Ya [okno] otkrýla, plámya. "Kol', shkóla garít"; "Yeynyy-t" dom fso f plámi" (I sit on a spinning wheel and work. Suddenly there is no light. I opened the window and saw the flame. The school was on fire. Her house is all on fire.) [14, 2016 26, p. 215]. Thus, "flame" as 'fire rising above a burning object' (ibid.) dominates in the dialectic discourse, objectifying mainly the highest manifestation of the power of fire—conflagration.

This article does not pose the task of linguistic analysis of the entire semantic and semantic derivational paradigm of the word "fire" in folk speech, including its Pskov regional part. This task was already realized on the Russian language materials [20–22]. It is important for us to identify how the destructive power of fire is reflected in popular perception.

35.4 Conflagration as the Highest Manifestation of the Power of Fire

Conflagration in its harmful force and the damage caused, is an unconditional natural disaster. "Conflagration"—'a strong flame, enveloping and destroying everything that can burn, creating danger to people's lives, as well as burning itself, the destruction of smth. by fire' [16, 1984 III, p. 234]. Sources designed to give an idea of the traditional worldvew of the Slavs describe the cultural concept of "conflagration" as one of the natural phenomena and the elements, defining it as "a natural disaster, a manifestation of a dangerous, hostile power of fire" [5, p. 109]; as "a natural disaster sent to people as God's punishment for their sins" [18, p. 373].

35.5 The Ideogram "Conflagration" in a Toponymic Portrait of the Area

"The toponym reflects knowledge of a fragment of geographical space. This knowledge is objectified in the internal form of the name ..." [7, p. 26]. The author recalls the regularity of the appearance of names, according to which "when nominating the objects of the surrounding reality, a person marks with a name not all observed (or imaginary) phenomena, but only those that for one reason or another have become priority for it" (ibid., pp. 32–33).

The regional toponymic material studied by us has shown that the element of conflagration is one of the unconditional semantic dominants in the Pskov toponymy. Our study also includes a comparative analysis of toponymic dictionaries and directories of various territories, which is not introduced in this publication due to the overloaded list of sources. If not to take into consideration the linguistic issues (types of word formation, system of formants), then the overall picture of the prevalence of "burnt" settlements as a whole is similar to the data of the regional sources [1, 3, 14, 1967 1, pp. 27–41].

So, according to the data of only the Novosokolnichesky district of the modern Pskov region, 8 topographic objects are registered, the names of which reflect the semantics of burning: villages Zazhogino, Zhogovo, Pozhary; urochishche Zagar'ye, urochishche Gorelaya Gora, urochishche Zharki; urochishche Pal'tso; lake Pozharnoye (named after the village that existed here). Toponymy data indicate the extent of natural disasters in the specified area due to **conflagrations**. The author of the dictionary, referring to historical sources, notes that significant territories were covered by severe fires in the region: the villages of Zazhogino, Molodi, Pozhary "form a strip in the territory <...> of a region extending from west to east for more than 10 km. Most likely, this is a trace of an ancient ecological disaster, a grandiose forest fire that swept over a vast territory and left a mark even in toponymy" [3].

Indirect data can presumably be attributed to the above list and the toponym "Pechische" from this online dictionary, as the corresponding appellative in dialectic speech matters (among others) 'abandoned manor' [14, 2016 26, pp. 114, 117].

Series of given topographic objects is connected with the name of urochishche. There is no unity in the understanding of this term by different specialists, but in the common Russian language "urochishche" means a site that differs from the other surrounding area (forest in the field; swamp, meadow in the forest, etc.)' [16, 1984 IV, p. 514]. The dialectic speech retains this all-Russian meaning as obsolete: "In the old days massive forest area was called urochishche". It is important to emphasize that the people's memory also illustrates the fire as cause of urochishche: "burnt urochishche"—'burnt forest' (examples from the card-catalogue PRD).

The toponymic landscape of the Pskov region as a whole, without significant differences in the regions within the territory, is characterized by a small group of placenames motivated by the image of burning and fire. For completeness, we present the entire list of such names, mainly, villages (no indication of the district in the Pskov region is given; the number of fixations is indicated in brackets next to the toponym if there are more than one unit): Apal'ovo, Garino, Garitsy, Gorelaya Budka, Gorelaya Roshcha, Gorelik, Goreliki, Gorelitsy, Gorelovo, Goren'ye, Gor'yanki, Dymokor'ye, Zhar, Zharki (5), Zharkovo, Zharok, Zhary, Zhegalovo, Zhegi, Zheglitsy, Zheglovo, Zhegovo (2), Zhegulikha, Zhogovo, Zhogori, Zhuguri, Zagar'ye (2), Negorelitsy, Ognyannikovo, Palitsy (3), Palichino, Plam'ya, Pogorelka (13), Pogorelovo (4), Pogoreloye (2), Pogorel'tsevo (2), Pogorelovo 1-ye, Podgar'ye, Pozhar, Pozhary (3), Pozherevitsy, Smagino [1].

So according to this source, in the Pskov region there are only 69 objects, the names of which contain 42 names with the root semantics. We have ventured to suggest that these impressive numbers are inconclusive. On the one hand, of course, in later registries a smaller number of names was recorded. It is connected with the disappearance of the deserted villages. But on the other hand, (upon closer examination), the composition of toponyms can be increased taking into account the dialect semantics of the root in the name. So the toponym "Pyshkovo" (Zhogovsky), an obviously verbal formation, can be presumably added to this list. In Pskov dialects the verbs "pykhat'", "pykhnut'" (puff) in their semantics are associated with intense burning: "Zatopíla péchku, a sópat ["sopot"—'fire, flame'.—auth.] tak y pýsha" (I've lit the stove, the fire is puffing); "Skóra pýkhnula i zgaréla" (Quickly puffed and burnt) (examples from the card-catalogue PRD). In the course of a continuous sampling of toponymic material, the Pskov toponym "Greshina Gora" was discovered, which is also possible to introduce into the material under the study. In Russian dialects, derivatives with the root "greh-" (sin) mean the semantics of fire: "greshina"—'fire': "In case of any greshina, drag everything to hemp"; "greshitsa"—'burn (about the fire)': "greshitsa—this is what they say about the fire" [15, 1972 7, p. 138]. Thus, both in initial semantics and in structure, the toponym "Greshina Gora" fits into the system of names under consideration; urochishche "Gorelaya Gora" [3].

There are other toponyms in the list dating back to dialectisms. So "Smagino" is semantically connected with the word "smaga"—'heat, fire, flame' [11, 1982 IV, p. 230]. The toponym "Dymokor'ye", without a doubt, is a transformation of the name "Dymokur'ye", where the phonetic substitution of the vowels "y" with "o" has also led to a change in the internal form of the word. "Dymokur'ye" is a word associated with suffixation with "dymokur"—a word in the standard dictionary marked as regional—'a fuming fire, kindled to protect against mosquitoes, midges, etc.' [16, 1981 I, p. 459].

The Pskov Folklore-Ethnographic Archive, which is quantitative and substantial, along with materials that are diverse in the genre, also contains toponymic legends that capture a popular interpretation of toponyms. It is well known that the popular interpretation of the word origin does not always correspond to historical reality. However, in the following example, with a high degree of probability it is possible to explain the name of the settlement as associated with the fire: "[Interviewer: Why is your village called Pozherevitsy?] It was a long time ago. I wasn't born yet. My dad told this. There was a village. It was named Pozherevitsy. <...> The village burnt. Here one house was on fire (dad told us), one house on the edge was on fire—everyone ran to the Pozherevitsy village, that's all. From there a jackdaw came flying and sat on this side. The whole village burnt down. And the village was called Pozherevitsy. So they was Pozherevitsy. It was a big fire. <…>" [19, p. 89]. In the modern Russian language (both in literary and in dialect), an adjective which could serve as a motivator for the toponym "Pozherevitsy" was lost, while historical material indicates the presence of such an adjective in the past: zheraviy"—'hot, red-hot' in short form "Coal zherav" (red-hot) [17, 1978 5, p. 76]. The word "zherav"—'burning, red-hot' also marks the etymological dictionary [12, 1986 II, p. 47]. It is significant that the story not only interprets the origin of the toponym, but also reflects the ancient mythological idea of where the bird will fly in the fire, the flame will spread there after it: "From there a jackdaw came flying and sat on this side. The whole village vymakhalo [i.e. burnt out.—auth.]".

The last thing to dwell on in the general review of the Pskov toponymy, united by the ideogram "fire", is the systematization of the roots that are the carriers of this meaning. The semantics of burning and the highest degree of manifestation of its intensity are transmitted in toponyms by the following roots: "гар-/гор-/жар-", "жег-/жог-", "плам-", "смаг-". In the Pskov toponymy, there was only one derivative with the root of the word "fire"—"Ognyannikovo", apparently because this was prevented by the duality of fire semantics and pragmatics noted above both as a phenomenon and as a cultural concept, while in semantics of conflagration an extremely negative side is reflected. Particularly noteworthy are toponymic formations with the root "пал-": Apalevo, Palitsy (3), Palichino. There is a specificity in the semantics conveyed by words with the root "пал": unlike formations with other roots, they convey controlled burning, «пал-»:—'burning of forest areas to prepare them for arable land under slash-and-burn agriculture' [16, 1984 III, p. 12].

For linguistic analysis of toponyms, the study of the morphemic structure and derivational models of onyms is of its value, however, this study, is focused on the root of the word as the main part of the word.

35.6 Results

The detailed study revealed a special role of dialect material in describing the meaningful and especially symbolic content of the concept of fire. An analysis of the informativity of the dictionary entry in the "Pskov Regional Dictionary of Historical Data" from the definition to the illustration has shown how lexicographic technology reveals the ambivalent image of fire in the dialect linguistic worldview.

The very subject of the conference determined the appeal to the natural and social phenomenon of fire as the highest degree of power of manifestation of fire and its impact on nature and man.

A striking consequence of strong forest fires is a change in the landscape and the emergence of so-called urochishche. The "Goreloye urochishche", often having the appearance of a forested elevation (cf. the Gorelaya Gora tract), serves other purposes—it is used by locals as a marker of boundaries between the territories.

The conceptualization of fire in toponymicon is especially vividly embodied: fires remain in the memory of local residents for a long time, which is reflected in toponymy. In the course of work, by continuous sampling from regional sources, it was possible to identify a significant range of toponyms, which indicate a fire by their internal form. More than 40 names are marked on this basis for more than 70 objects, in the main, villages of the Pskov region. The most frequent is the toponym "Pogorelka" (recorded 13 times in 7 districts), followed by the degree of frequency: "Zharki" (5 fixations), "Pogorelovo" (4), "Pozhary" (3), "Zagar'ye", "Pogoreloye", "Pogorel'tsevo" (2 times).

Placenames with a fire are secondary, appearing to replace some previous names that have shades of tragic natural disaster. So it is fair to ask about the reasons for the old name of the village being erased from memory.

We suppose that in folk tradition, the renaming of a burned-down village is apotropic in nature—it is the result of a ban typologically similar to a magical ban on building housing "na pechishche"—that is on the site of a burnt house. The choice of the place of new construction in traditional culture is associated with a number of ritual actions and beliefs. So, "a dwelling cannot be built <…> on the site of a house burned by the "pyarunami"<…>" [4, p. 35]. However, apparently, other causes of the fire, not only from lightning, but also due to arson, form a strong tradition to choose a place for new housing away from a burned-out house in order to avoid another **conflagration**. So, in the collection of proverbs, sayings and riddles of V. I. Dahl, in the section "Sense – Foolishness," we read: "Shrewd peasant set a hut on pechishche" [10, p. 450]. Foolishness, shortsightedness of a peasant who does not comply with generally accepted norms is subjected to condemnation and mockery. Returning to the secondary names of the burnt settlements, we consider the same imperative to abandon the old name, as well as the former place of residence after the fire. Both phenomena are a single taboo system that has a protective character.

35.7 Conclusion

The linguistic worldview as a system of representations and values determines the attitude to reality and determines the actions of a person—the bearer of traditional culture. Considering the two phenomena—fire and conflagration—in interconnection, we were convinced that the rational and mythological in the consciousness and behavior of a man are in close interpenetration. The study reveals the extent of fires in the Pskov region, which is reflected in the system of toponyms. Historical evidence also points to this (see cited work: [3]). The speed of fire spread, the lack of sufficient fire-extinguishing means in the past—all these objective factors determined the extent of fires. At the same time, in accordance with the mythological ideas of the peasants, the "fight against fire" contradicted the very essence of understanding fire as a punishment sent to a person for sins, which implied that extinguishing a fire is a sin; according to a legend, it is also impossible to extinguish a fire with water, because from this it will spread even more, it can be stewed with milk, kvass, throwing Easter eggs, a willow into the fire, enclosing an illuminated place with an icon [5, pp. 110–111, 18].

Thus, the scale of fires in the past was not only indicated by objective reasons, but also to the peculiarities of the mythological consciousness of the traditional culture bearers.

In the future, the further involvement of dialectal speech material in the study of the concept of fire seems to be significant. Experts have noticed that when the fires are conditionally divided into natural, social, and man-made, the classification "excludes such an important kind of fires as military fires, which make up a significant percentage of human and material losses" [9, p. 97]. The Pskov archive contains a lot of materials currently published [8] that reveal the theme of war and the occupation of Pskov territory, where "fire" as a socio-historical concept is one of the dominant ones.

The studies of the binary opposition "fire" and "water" on Pskov material, the conceptualization of religious and mythological ideas about fire in the folk language and folklore are also significant. All this will make it possible reveal more deeply the picture of the world of a "man of tradition" by means of linguistic representation.

References

1. Administrativno Territorial'noye deleniye Pskovskoy oblasti [Administrative territorial division of the Pskov region] (1988) Reference book: vol 2, 2nd ed. Lenizdat, Leningrad
2. Ahmatova MA (2016) "Yazykovyye reprezentanty kontsepta "ogon'" v tekstakh karachayevo-balkarskogo nartskogo eposa" [Linguistic representatives of the concept of "fire" in the texts of the Karachay-Balkarian Nart epic literature], Filologicheskiye nauki. Voprosy teorii i praktik, Gamota, Tambov, No 1(55), vol 2, pp 76–79. www.gramota.net/materials/2/2016/1-2/. Accessed 01 Oct 2019
3. Alekseev U (2011) Slovar' toponimov Novosokol'nicheskogo rayona Pskovskoy oblasti» [Dictionary of toponyms of the Novosokol'niki region]. https://www.endic.ru/toponyms_pskov/. Accessed 29 Aug 2019
4. Bayburin AK (1983) Zhilishche v predstavleniyakh vostochnykh slavyan [Housing in the representations of Eastern Slavs]. Nauka, Moscow
5. Belova OV (2004) "Pozhar", v slovare Slavyanskiye drevnosti ["Conflagration" in dictionary Slavic encientry]: ethnolinguistic dictionary in 5 vol, vol 4. Mezhdunarodnyye otnosheniya, Moscow, pp 109–112
6. Belova OV, Uzeneva ES (2004) "Ogon", v slovare Slavyanskiye drevnosti ["Fire" in dictionary Slavic encientry] vol 3. Mezhdunarodnyye otnosheniya, Moscow, pp 513–519
7. Berezovich EL (1998) Toponimiya Russkogo Severa [Toponymy of the Russian North]: ethnolinguistic studies. Uralskiy universitet, Ekaterinburg
8. Bolshakova NV (ed) (2016) Velikaya Otechestvennaya voyna v zerkale narodnoy rechi i fol'klora [The Great Patriotic War in the mirror of folk speech and folklore]. LOGOS, Pskov
9. Chernih AA (2010) Pervyye pozhary i ikh proiskhozhdeniye [First fires and their origin]. Izvestiya Irkutskogo gosudarstvennogo universiteta 1(4):97–105. http://isu.ru/izvestia; https://cyberleninka.ru/article/n/pervye-pozhary-i-ih-proishozhdenie
10. Dahl VI (1957) Poslovitsy russkogo naroda [Proverbs of the Russian people], Khudozhestvennaya literature, Moscow (Dahl's Proverbs)
11. Dahl VI (1978–1982) Tolkovyy slovar' zhivogo velikorusskogo yazyka [Definition dictionary of the Russian language]. Russkiy yazyk, Moscow (Dahl's Dictionary)
12. Fasmer M (1986–1987) Etimologicheskiy slovar' russkogo yazyka [Etymological dictionary of the Russian language], Progress, Moscow

13. Khayrullina PH, Turkhan H (2015) Reprezentatsiya obrazov ognya v yazykovoy kartine mira: sopostavitel'nyy aspect [The representation of fire in linguistic worldview]. Sovremennyye problemy nauki i obrazovaniya, No 1-1. http://science-education.ru/ru/article/view?id=19483. Accessed 07 Oct 2019
14. Pskovskiy oblastnoy slovar' s istoricheskimi dannymi [Pskov regional dictionary of historical data] (1967–2017), Leningrad/St. Petersburg (PRD)
15. Slovar' russkikh narodnykh govorov [Dictionary of Russian folk dialects] (1965–2016), Nauka, Moscow; Leningrad/St. Petersburg (DRFD)
16. Slovar' russkogo yazyka [Russian language dictionary] (1980–1984) Russkiy yazyk, Moscow (RLD)
17. Slovar' russkogo yazyka XI–XVII c. [Dictionary of the Russian language of the XI–XVII centuries] (1975–2015), Nauka – Azbukovnik, Moscow (DRL XI–XVII)
18. Toporkov AL (2002) "Pozhar", v slovare Slavyanskaya mifologiya ["Fire" in dictionary Slavic miphology]. Encyclopedia, Mezhdunarodnyye otnosheniya, Moscow, p 373
19. Urchuk LA (2014) Predaniya Pskovskoy oblasti. Pskovskiy gosudarstvenniy universitet, Pskov
20. Verhoturova KS (2007) "Molniya" v zerkale dialektnoy nominatsi [Lightning in the mirror of dialect nomination]. Onomastika i dialektnaya leksika, Ekaterinburg, pp 87–92
21. Verhoturova KS (2008) Ogon' v zerkale russkogo yazyka [Fire in the mirror of the Russian language]: Abstract of PhD thesis. Ekaterinburg
22. Verhoturova KS (2008) Ogon' i voda: struktura semanticheskikh poley (na materiale leksiki ural'skikh govorov) [Fire and water: structure of semantic fields (based on vocabulary from the Urals dialects] Izvestiya Ural'skogo universiteta 59:137–153

Chapter 36
Establishing an Effective System for the Introduction of Vending Technologies in the Consumer Market and in University Catering

Marina Belyaeva, Aleksandr Gazhur, Vitaliy Perov, Samuel Ali Malazi and Olga Lukhovskaya

Abstract The issue of healthy nutrition and its provision is the most urgent in science and practice both of Russia and foreign countries. Especially it concerns universities, including the objectives of ensuring quality-price ratio. The article presents the following scientific findings obtained by the authors, namely classification international approaches to university catering management; the analysis of vending application in the university's catering system both at the international and domestic levels, monitoring of opinion on catering in the consumer market and university catering through vending machines; development of set lunches and breakfasts. Within practical recommendations, we have proposed to adjust the university catering system and have justified the introduction of a new vending machine for the provision of set meals that are patented by the author's staff of the department of the restaurant business of Plekhanov Russian University of Economics.

Keywords Assessment of university catering · Vending machines · Healthy nutrition · Diets · Set meal · Sales automation · Vending business · Consumer demand · Market

M. Belyaeva (✉) · A. Gazhur · V. Perov · S. A. Malazi · O. Lukhovskaya
Plekhanov Russian University Economics, Moscow, Russia
e-mail: Belyaevamar@mail.ru

A. Gazhur
e-mail: gajour@mail.ru

V. Perov
e-mail: perov-vitaliy@bk.ru

S. A. Malazi
e-mail: samemalazi@gmail.com

O. Lukhovskaya
e-mail: ollga37@yandex.ru

T. Kolmykova and E. V. Kharchenko (eds.), *Digital Future Economic Growth, Social Adaptation, and Technological Perspectives*, Lecture Notes in Networks and Systems 111, https://doi.org/10.1007/978-3-030-39797-5_36

36.1 Introduction

In Russia, there is an acute problem of healthy nutrition in educational institutions. So, the student union is involved in developing standards of social nutrition (range of dishes, their composition, energy value).

Student's nutrition is mostly non-scheduled, "fast", 1–2 times a day. The majority do not visit the canteen. Fast food containing a large amount of fats, table salt, modified components, flavors, colorants, etc. is growing popular among students. Besides, American scientists have stated the principles: Fast food is not good. Good food isn't fast. Unhealthy nutrition leads to obesity, diabetes, cardiovascular disease, cerebral vascular accident. These diseases can be prevented if you keep a healthy lifestyle and nutrition, that's why good nutrition is an essential part of student's healthy lifestyle [1].

Today, the diseases are more often diagnosed among young people, respectively, it should be noted the risks of the gastrointestinal, oncological and other diseases [2]. They should be prevented with the help of specially-designated foods that contain physiologically active ingredients, namely dietary fiber, peptides, amino acids, vitamins, and microelements. Regular and balanced nutrition arranged following various parameters, including the development of a more informative menu, diet computer, and Web applications can fix the negative consequences to a large extent [3].

Vending machines play a significant role in ensuring effective nutrition in universities. These studies have shown the unchallenged role of vending technologies in the optimization of the university catering system.

36.2 Methodology

The research methodology of catering forms at educational institutions is based on the diets developed for students (set breakfasts and lunches) and the automation of their sales [4, 5], as well as the design of vending machines for the provision of set meals and various foods. The development of the vending business relies on the use of retail automata for sales of a wide range of foods of the FMCG (Fast Moving Consumer Goods) group [6].

It is methodologically expedient to classify students as a special segment of the catering system at educational institutions. Student years are saturated, diverse, but overstraining the nervous system. The load, especially during sessions, is growing significantly and reaching 15–16 h a day. Long-term sleep deprivation, intense brain work, distortion of the daily regime and nutrition can lead to neuropsychic breakdowns. An integrated approach to development of an effective system for introduction of vending technologies on the consumer market [5] and in university catering involves an online technology technique for monitoring of consumer demand; methodology for calculation of dietary structure calories; methods and recommendations for food sales through vending machines.

36.3 Findings

1. Assessing the performance of the catering system in Plekhanov Russian University of Economics.

We have studied the needs of students of Plekhanov Russian University Economics through online technology Google Forms (questionnaires).

Plekhanov Russian University of Economics is one of the leading universities in the country. Over 70 thousand students are studying at all universities, and over 20 thousand students—at the principal university in Moscow. The number of faculty staff of the principal university is over 1800 tutors. Catering system of Plekhanov Russian University of Economics includes several canteens, cafes, buffets, a restaurant and various vending machines located in the educational blocks of the university. For assessment of the performance of the catering system, we polled students under a sample number of approximately 170 people. Segmentation of the sample showed that 84% of the respondents are women. Age groups: 23%—up to 18 years old, 4%—18 years old, 71%—19–25 years old, 0.6%—26–30 years old. The social status of 63% of respondents was average, 26%—above average, about 7%—high, 3.7%—below average. The results showed that 43% of respondents use catering services within Plekhanov Russian University of Economic, 21% of respondents visit food facilities several times a month, and 29% of students—several times a week, 7% of respondents do not buy food at the university, but use the services of other enterprises, or bring own food.

The analysis showed that the reasons could be the inconvenient location of catering facilities; high prices; a poor range of foods; low-quality consumer properties of products; limited time and breaks between classes. Making the question "What else catering facility would you like to see at the university?", we get the following answers: machines with frozen yogurts; healthy food cafe; dessert cafe; mini pizzeria. Surveying the target audience of mainly 18–30 years old in the Plekhanov Russian University of Economics we find out that 43% of respondents assess foods by quality and range, by location of food court (42%); some make an assessment regardless of the location the food court at the University, and other answers (about 1.0%) (Fig. 36.1).

Most of the respondents (65%) were satisfied with the quality of foods sold at food courts of Plekhanov Russian University of Economic.

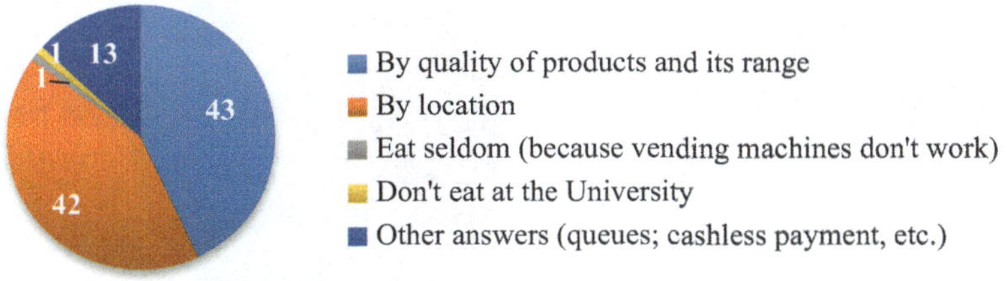

Fig. 36.1 Distribution of answers to the question "By what principle do you choose the food court at Plekhanov Russian University of Economics?"

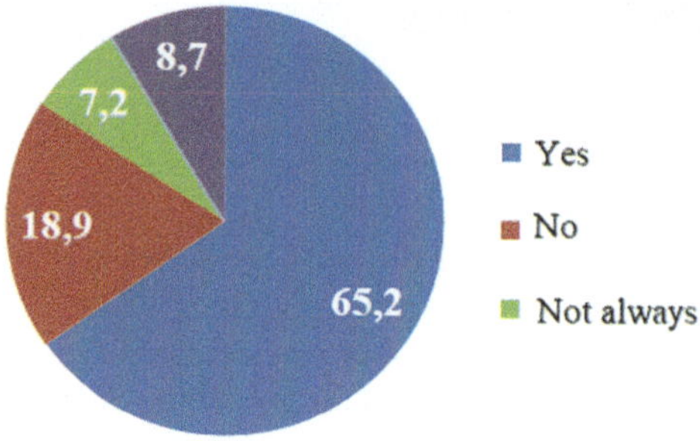

Fig. 36.2 Distribution of answers to the question "Are you satisfied with the quality of foods sold at food courts of Plekhanov Russian University of Economics?"

In 1.2% of cases, foods did not always meet the needs of the audience. Approximately 0.6% of respondents gave a fifty-fifty answer. There were other answers, but their shares less than 0.6% (Fig. 36.2).

The range of foods met respondents' expectations: 62.2% of the positive answers from the total number, 32.3% replied negatively and noted that high-calorie snacks and some similar foods should not be included in the menu (Fig. 36.3). As for the most commonly used set of products, respondents prefer hot dishes, salads, and drinks. About 18% of consumers choose soups, 24%—pastries, 0.6% would like sandwiches, rice snacks, sushi, and 17% would prefer the confectionery products. Most of the respondents prefer individually choose dishes, some would choose set lunches/breakfasts and drinks. There were opinions on the free choice of dishes, as well as the choice on the buffet type, set lunches and breakfasts preferably with national dishes of international cuisines, for example, soups, hot (second course), confectionery, sour-milk drinks.

One of the important factors is the social one, i.e. the price of the set meal. As it appeared, the preferred range of prices is 160 rub. (nearly 41% of answers), 200 rubles (23%), 220 rubles (7%), etc. The average price is 183 rub. (Fig. 36.4). There were opinions that complexes containing soups, hot and cold dishes, snacks and salads, drinks, and desserts in various combinations are required. No more than 1% of respondents remarked the attractiveness of dishes, such as pizza, pasta, sushi, non-spice dietary foods (Fig. 36.5).

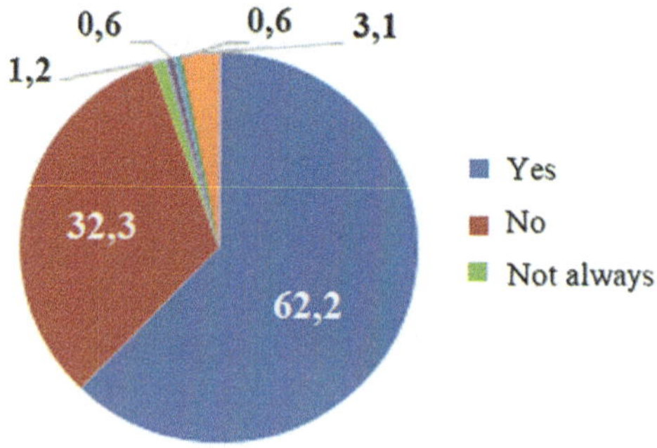

Fig. 36.3 Distribution of answers to the question "Are you satisfied with the range of foods sold at food courts of Plekhanov Russian University of Economics?"

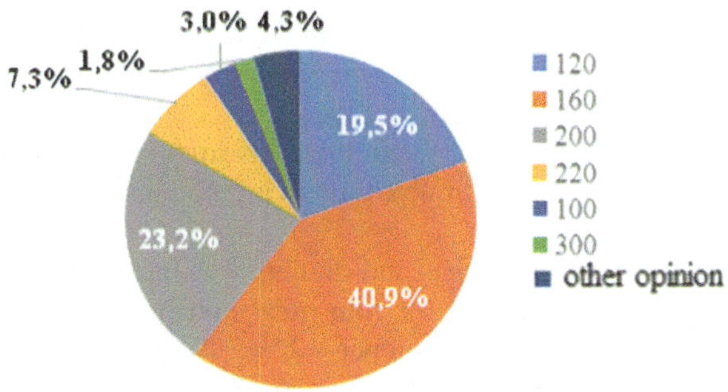

Fig. 36.4 Distribution of answers to the question "Preferred price for a set dinner"

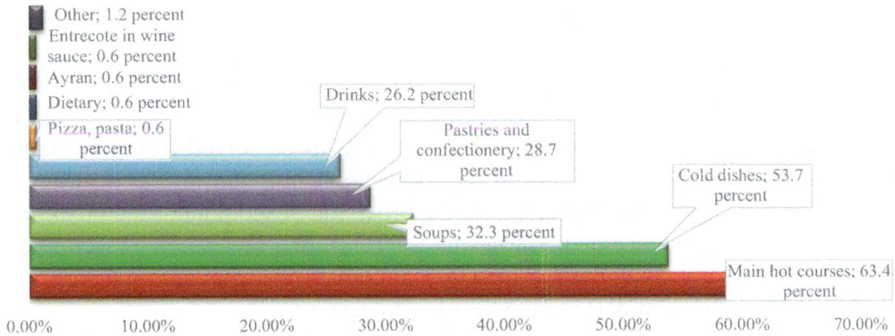

Fig. 36.5 Distribution of answers to the question "What traditional dishes and foods of national cuisines would you like to be offered?"

Okroshka, gazpacho; Wellington meat, American pie; shepherd's pie; Korean carrot and asparagus salad; Rahat lokum, churchkhela; hot mulled wine, kvass, fermented apple juice were mentioned in some questionnaires as preferred dishes of national cuisines. The results showed a certain demand for Japanese cuisine: onigiri, rolls, sushi. Also, there is a demand for the accompanying goods, such as wet wipes, chewing gum, mint-flavored toothpicks, etc.

Analysis of data on catering system automation made it possible to establish that 30.5% of respondents did not use vending machines for various reasons, approximately 15% utilized them seldom (several times a week) or rarely (several times a month—about 42%), 8%—often (1 time per day), the share of other answers was insignificant (0.6%). Making a question "What kind of vending machines would you like to see in Plekhanov Russian University of Economic?", we get that 77% of students answered "vending machines with pizza" and set meals.

2. Development of set dinners and breakfasts.

Based on questionnaires' results, we developed set dinners lunches and breakfasts that include national cuisine dishes in accordance with standards for nutritional qualities, biological value and student's diet calories, which are intended to be sold in plastic trays through a developed vending machine with a built-in microwave oven using IR, microwave and combined heating methods.

Plekhanov Russian University of Economics has studied the issues of developing product quality, the most suitable variable ratio of ingredients of animal and vegetable origin [7, 8].

The development of set breakfasts and dinners was based on the distribution of calories of the daily diet (percent) for four-meal nutrition: breakfast—25%, dinner—35%, afternoon snack—15%, supper—25%. According to five-meal nutrition: breakfast—20%, lunch—10–15%, dinner—30–35%, afternoon snack—10–15%, supper—20%. The calorie content of the daily diet of girls 18–40-year old women is 2400 kcal. When developing a set dinner under three-meal nutrition, the calories were distributed as follows: breakfast—25% (600 kcal); dinner—45% (1080 kcal); supper—35% (840 kcal).

3. Development of a vending machine project adapted for the provision of set dinners and breakfasts for students.

The issue of providing healthy nutrition is offered to be solved through vending machines that serve the consumer at any convenient time during the learning. A project of a vending machine adapted for the provision of set dinners and breakfasts for students has been developed. The design solutions for a new vending machine are based on the principle of combined heat flow patterns from different heat sources. A combined oven has been designed to heat set dinners.

The designed vending machine patented under No. 2017137620 РФ as of November 27, 2017, by Belyaeva et al. [9, 10].

36.4 Discussion

The Russian history of the introduction of vending starts in 1898 with chocolate vending machines in St. Petersburg [11]. In the 50s vending machines were installed for sales of sandwiches, notebooks, pencils, matches, cigarettes, bread, sunflower-seed oil, milk, kerosene, coffee, tea, beer and, of course, all-known soda water. The demand for such services is constantly growing.

Some monographs [7, 8, 12] are devoted to the nutritional qualities and biological value of foodstuffs by the example of semi-finished meat products, the development of diet plans for students (mainly set breakfasts and dinners). M. A. Belyaeva in his articles (2018, 2019) submits the research findings of the main properties of model minced meat systems with high nutritional and biological values, additives of animal and vegetable origin, dairy products, melted butter and their inclusion into the diet provided through vending machines [13, 14].

Russian vending includes sixty thousand machines with 4 bln. rub. of annual turnover. Payback is 2 years on average with initial investments from 100 thousand to 1 million rub. An assessment of the potential consumer market shows a ten-fold boost over 5 years; it is based on the fact that the share of vending in Russia now is about 1% of retail trade, and in the European Union—5% on average. Over 60% of the consumer market in Russia belongs to coffee machines, another 25%—snack machines, the rest of the market are machines for the sale of drinks, juices, soda, hot food, hygiene items, newspapers, lenses, etc. [15].

In the leading countries of the world, from 500 (Russia) to 230 (Japan) students fall per 10 thousand inhabitants, which is a tremendous figure (up to 5% of the population) [16–18]. Students need healthy and inexpensive nutrition, and various food courts at universities provide such an opportunity. Catering is designed to satisfy sound physiological needs under GOST 31985-2013 "Catering services. Terms and definitions" (November 03, 2017). Recently, the issue of utilization of vending machines in universities for the arrangement of healthy nutrition, which is cost-effective, has been resolved [19].

In Italy, providing the choice of foods using vending machines, including the importance of available healthy food, as well as consumer awareness is an urgent task. Consumers must be provided with information in aright way (special label), which reduces nutritional risks, in particular, obesity factors [20].

In Finland, the impact of gender and age of consumers, type of food on the structure of foodstuffs for young consumers have been studied [21]. The findings revealed that if the product is not attractive in itself, then its origin (local) can increase the perceived quality.

A comparative analysis of the catering services in famous universities of the EU and the USA showed that Humboldt-Universität zu Berlin (HU Berlin) has introduced a mensa system, which is an electronic media covering various canteens and cafeterias. The student issues a card and gets a discount on the university catering services. The site contains information on prices, composition and availability of goods at the university's food outlets, and feedback section where you can write a comment on the food quality.

Taking into account students' opinions, you can adjust the menu; vegetarian dishes are also included to satisfy nutritional needs to the fullest extent. The card allows the student to eat food in different universities. The system is working throughout Berlin, which is convenient and cost-effective. At the university, drinks and foods are also sold in vending machines.

In the EU Business School, cafes and cafeterias are prevailing. There is a discount card, but only for the students of this university. The university mainly provides healthy food to maintain a healthy lifestyle. Students can get full information on the menu on the site and also give feedback.

University of Calabria (Universita della Calabria) uses about 6–8 vending machines. Learning is going from 9:00 to 18:00 and vending machines, snack machines, coffee machines are working during this period. The latter is popular among employees and students due to the high culture of consuming this beverage in Italy [22].

The catering system at the University of Applied Sciences Würzburg-Schweinfurt (FHWS) is represented by two vending machines placed by a cascade principle, which ensures centralized sales of a wide range of products [23].

The campus of FLS St Peter's University (New York) is located in Jersey City, which is a part of the New York Metropolitan Area. For students living in the dormitory, the nutrition is provided in the dormitory's canteen under a full board system on weekdays and a half board on weekends. Drinks during dinner and supper are not generally included in the cost.

The canteen at the Aristotle University of Thessaloniki is only for students. And groups of people who learn only the language are not students, therefore they eat during the first month of study at their own expense; then they are allowed to visit the canteen. The canteen nutrition is three-meal a day from Monday to Sunday: breakfast (8:00–8:30), dinner (12:30–16:00, supper (18:30–21:00). The daily breakfast menu is coffee (or milk, tea, juice); bread (or toasts) with butter and marmalade; corn flakes (or cupcake, boiled egg, omelet, cheese). The dinner and supper menu is made up for one week with three options of the meal for dinner and two options of the meal for supper every day. Dishes are different for three weeks and the nutrition is balanced.

Nutrition at universities across the world is very diverse: the traditional types are canteens, cafeterias, the board system in dormitories, vending machines, discounts in one or another form [17, 18]. Analysis of demand for the product range and the quality of catering in leading universities of Moscow was carried out by the questionnaire method. The respondents were students and university staff of Moscow State University named after M. V. Lomonosov; The First Moscow State Medical University named after I. M. Sechenov; Russian State Humanitarian University; Financial University under the Government of the Russian Federation; Research University of Higher School of Economics; Russian State University of Oil and Gas named after I. M. Gubkin; Plekhanov Russian University of Economics.

All these universities are in the top 20 universities of Moscow or the Russian Federation according to various ratings. The sampling limitation for 7 universities was recognized by the authors of the article quite representative for this study.[1] Taking into consideration the review of the data to carry out fundamental research and development, the authors conducted research aimed at surveying public opinion on the quality and safety of foodstuffs and culinary, their range presented in the catering system of the leading Russian Economic University and various international universities. Marketing research was carried out using the questionnaire method.

36.5 Substantiation

Research findings showed that the catering system at Russian and international universities id distinguished by the extremely differentiated range of foodstuffs. Bearing in view a huge number of students in the population structure of developed countries, as well as stressful working conditions, including sessions and final certification, the introduction of a healthy nutrition is an extremely urgent issue that is being discussed today at the level of the Russian Student Union developing a standard of social dinner for students.

---

[1] https://raexpert.ru/rankings/vuz/vuz_2017/ и др.

Most universities use technology close to fast food that does not satisfy students in terms of environmental friendliness, quality, and range [24] and the problem is solved with help of vending technologies. The advantage of the vending machines is the ease of installation and operation, possible location in the zones of students' maximum concentration, the effective use of short periods between classes, as well as sales of set breakfasts and dinners. The introduction of a new vending machine for the sale of set dinners in the Plekhanov Russian University of Economics will solve the problem of providing students of evening-time training programs with food. Vending technologies create the most suitable nutritional conditions at any time and a developed vending machine with a built-in microwave oven allows heating culinary food.

References

1. Horacek TM, Erdman MB, Byrd-Bredbenner C et al (2013) Assessment of the dining environment on and near the campus of fifteen post-secondary institutions. Public Health Nutr 16(7):1186–1196
2. Cohen JE, Wakefield CE, Cohn RJ (2016) Nutritional interventions for survivors of childhood cancer. Cochrane Database Syst Rev 8(22)
3. Kim E, Tang LR, Meusel C, Gupta M (2018) Optimization of menu-labeling formats to drive healthy dining: an eye-tracking study. Int J Hosp Manag 70:37–48
4. Lukhovskaya OK, Kochetkova TS, Gur'eva OYu, Fomina NV (2019) Diagnostic assessment and strategic solutions of the development of the regional commercial network. Bull Altai Acad Econ Law 7-1:74–84
5. Lukhovskaya OK, Gur'eva OYu, Perov VI, Malova IV, Kochetkova TS (2019) Conceptual approaches to determining, diagnostics, and forecasting the region's consumer market; Specifics of decision making in modern business systems: regularities and tendencies. First edition 2019 Copyright © 2019 Emerald Publishing Limited Howard House, Wagon Lane, Bingley BD16 1WA, UK
6. Vending Business in Russia: unknown yesterday, challenging today, promising tomorrow (November 03, 2017). http://moneymakerfactory.ru
7. Belyaeva MA (2015) System analysis of technologies and business processes in meat production. Publishing House of Russian Economic University named after G.V. Plekhanov
8. Belyaeva MA (2016) Multi-criteria optimization of heat treatment of semi-finished meat products using modern electrophysical heating methods. Russians
9. Belyaeva MA et al (2018) A vending machine with a built-in microwave oven for the provision of set-meal for students. Patent for invention No. 2674708, Russia, application No 2017137620 as of December 12, 2018
10. Belyaeva MA (2018) Dietary structure with melted butter for students. Food Industry, No 7, pp 63–67
11. Nazarov I (2017) Vending: a history of emergence and development (November 03, 2017). https://www.shopolog.ru/metodichka
12. Belyaeva MA (2017) Optimization of the nutritional and biological value of semi-finished meat products during the heat treatment to provide the population with high-quality food. Russians
13. Al-Nabusli AA (2017) Modeling the combined effect of NaCl and pH against Cronobacter spp. using response surface methodology. In: Al-Nabusli AA, Osaili TM, Mahmoud KZ et al (eds). J Food Saf 37
14. Jiang S (2017) Assessment of self-produced PCR methods for the detection of toxoplasma gondii DNA in meat. In: Jiang S, He Y, Zhang Y et al (eds). J Food Saf 37(3)

15. Vending Machines of the USSR (November 03, 2017)
16. The main trends of education's development in the modern world (November 03, 2017) (UNESCO). https://www.unipage.net/en/student_statistics
17. Gazhur AA, Gridina AA, Mishina NA, Pleva DAQ (2017) Pizza vending machines as an effective means of ensuring good nutrition for students. Used in the university food analysis of the structure and problems of the food enterprises. In: "Innovations: prospects, challenges, results". Proceedings of the 5th international research and practice conference at Plekhanov Russian University of Economics. West-Ost-Publishing House, Berlin, 485 pp, pp 398–409. ISBN 978-386297-017-9 (Germany)
18. Gazhur AA, Kireeva EA, Zobneva EV (2017) Analysis of the structure and problems of the food enterprises at the university. In: "Innovations: prospects, challenges, results". Proceedings of the 5th international research and practice conference at Plekhanov Russian University of Economics. West-Ost-Publishing House, Berlin, 485 pp, pp 151–167. ISBN 978-386297-017-9 (Germany)
19. Viana J, Leonard SA, Kitay B, Ansel D et al (2018) Healthier vending machines in a university setting: effective and financially sustainable. Appetite 121:263–267. https://doi.org/10.1016/j.appet.2017.11.0.094
20. Rosi A, Zerbini C, Pellegrini N, Scazzina F et al (2017) How to improve food choices through vending machines: the importance of healthy food availability and consumers' awareness. Food Qual Prefer 62:262–269. https://doi.org/10.1016/j.foodqual.2017.05.008
21. Kumpulainen T, Vanio A, Sandell M, Hopia A (2018) The effect of gender, age and product type on the origin induced food product experience among young consumers in Finland. Appetite 123:101–107. https://doi.org/10.1016/j.appet.2017.12.01.011
22. Official Website of the University of Calabria. Universita della Calabria (October 10, 2017). https://www.unical.it/portale. Accessed 11 Oct 2017
23. Official Website of the University of Applied Sciences Würzburg-Schwanfurt (FHWS). www.international.fhws.de/en/. Accessed 21 Oct 2016
24. Ananiev MA, Lukhovskaya OK, Vasilchuk ES (2017) Regional food supply management system of the consumer market in the modern russian economy. Springer International Publishing AG, Cham, Switzerland, pp 97–104

Chapter 37
The Price of Pleasure: The Challenges and Threats of the Digital Economy

Evgeniya K. Karpunina, Sergey N. Yashin, Alexey S. Molchan, Ekaterina V. Lisova and Elena A. Derkacheva

Abstract The article attempts to study the concept "pleasure" from the standpoint of different scientific schools and through the prism of the Utility Theory. The factors determining the possibility of obtaining pleasure by a person are analyzed. It is determined that the effects of the digital economy (the attention of the target audience rather than the product or service itself, becomes a scarce resource; network effect, consisting in an increase in the number of connections; the change the relationship between different types of transaction costs) will transform the classical model of the process of choice and consumption of goods and increase the importance of "pleasure". Services and information technologies covering the process of human labor, the sphere of education and leisure, daily life and communication, as well as improve the process of choice and consumption of goods and create conditions for maximizing pleasure and reducing discomfort. It is proved that despite the obvious advantages of modern digital technologies and the expansion of the range of opportunities for pleasure, there are information, economic and psychological threats and dangers.

E. K. Karpunina (✉)
Tambov State University Named After G.R. Derzhavin, Tambov, Russia
e-mail: egenkak@mail.ru

S. N. Yashin
National Research Nizhny Novgorod State University, N. I. Lobachevsky, Nizhny Novgorod, Russia
e-mail: jashinsn@yandex.ru

A. S. Molchan · E. A. Derkacheva
Kuban State Technological University, Krasnodar, Russia
e-mail: molchan.alexey@gmail.com

E. A. Derkacheva
e-mail: derkacheva_elena@bk.ru

E. V. Lisova
Financial University Under the Government of the Russian Federation Branch in Lipetsk, Lipetsk, Russia
e-mail: lisovaekaterina78@gmail.com

T. Kolmykova and E. V. Kharchenko (eds.), *Digital Future Economic Growth, Social Adaptation, and Technological Perspectives*, Lecture Notes in Networks and Systems 111, https://doi.org/10.1007/978-3-030-39797-5_37

Keywords Pleasure · Digital economy · Utility theory · Behavioral economics · Economic security · Threats · Consumption

JEL Classification Codes O3 · P46

37.1 Introduction

The scale of digital changes in the economic system is extraordinary in the entire observed history of humanity, it is comparable, perhaps, only with the first industrial revolution, the emergence of mass production and the mass use of new energy sources. These changes create enormous opportunities that can dramatically change the way of life, the centers of value creation, the efficiency of individual sectors of the economy, the line of global leaders with new ideas of management processes.

Traditionally, the key goal of all economic activities is meeting people's needs through the maximization of utility for each individual within the possession of a certain stock of goods.

However, behind the economic side of the process of meeting needs lies the desire to achieve pleasure, which is more difficult to understand and evaluate. It is the desire to achieve pleasure in the digital economy that creates the motivation for manufacturers to improve, develop new productive technologies and create "smart" services.

Digital technologies are modifying the usual mechanisms of choice and consumption, creating new enhanced opportunities to meeting needs and accelerating these processes.

However, in the race for pleasure, a person should not forget about the dangers and threats of modern technologies that can make the "price of pleasure" unreasonably high.

37.2 Methodology

Let's define, first of all, the terminology of the study.

The etymology of the concept "pleasure" allows equating it with the term "satisfaction" and revealing a positive emotion, usually arising from pleasant sensations, experiences or as a result of meeting some needs.

Depending on the set of needs of each individual and their changes over time, will change and the range of sources of pleasure.

According to Pareto, the criterion of optimality is not the General maximization of utility, but its maximization for each individual within the possession of a certain initial stock of goods [6]. Methodologically, consumer behavior itself proceeds from the theory of marginal utility. A consumer buys goods and services to satisfy their needs, that is, they wants to gain a certain utility from them. Utility, therefore, is the pleasure or satisfaction that consumers get from the goods and services they purchase.

Bentham and other early advocates of utilitarianism, considered pleasure and the exclusion of suffering as the meaning of human activity and they connected pleasure and utility. Nevertheless, advocates of this trend point out that for different people, pleasure and utility are comparable and additive, that is, can be summed up in a kind of "common happiness of all" [5].

Besides the maximum utility of consumption, a person can enjoy work, quality environment, relationships with others, status in society, housing conditions, public order and a sense of security, and so on. In each of these aspects a person will feel more or less satisfied [24].

Applied research into patterns of consumer behavior proves that it is more correct to consider the concept of "pleasure" from the standpoint of behavioral Economics, at the intersection of both spiritual and material. In other words, the change in physical quantities in the outside world affects consumer perception [2, 16, 31].

Supporters of behavioral Economics argue that there are measurable physical quantities (the power of light, the amount of money, and so on) that are subjectively perceived by man (brightness, value), and along with them there are psychophysical laws that connect subjective sensations in the observer's brain with objective values of quantities in the material world [8]. Therefore, an integral part of a person's pleasure will be his psychophysical state, that is, sensations controlled by a certain part of the brain and responsible for creating a positive emotional background [20].

Freud Z. described the principle of pleasure as the desire of the psyche to reduce tension to a minimum level [11].

In his study, Kahneman D. showed that emotional well-being increases with income only up to a certain threshold ($75,000 per year), and identified a group of factors that affect human satisfaction, the possibility of obtaining pleasure and achieving a state of happiness [15]. So, one of the factors that gives joy and pleasure, he called "quality of work", which is understood not so much the level of income, which is important, but the sense of satisfaction that receives from work. At the same time Kahneman D. notes that boring repetitive tasks can bring little joy, while self-employment or work on creative skilled jobs, provide an opportunity for higher self-esteem and getting more pleasure from work [15]. For example, the majority of Americans selected daily in the Gallup-Healthways Well-Being Index survey feel happier and receive more pay on weekends, to an extent that depends on the social context at work and outside. Even the "weekend effect" disappears for those who work in a high-trust workplace, who consider their boss more of a partner than a boss, and maintain their social life on weekdays [13].

Danish scientists proved that "the desire to obtain new information is so powerfully fueled by the internal mechanisms of the brain, producing dopamine, that the pleasure of them outweighs any other pleasure" [14]. Therefore, the process of learning triggers a chemical reaction in people who express the highest need to search for sensational actions [9].

Similar in content and nature of the impact on the human body are social interactions, as a result of which there is not only the collection of new and useful information about the world, but also getting emotional pleasure from pleasant communication.

Kahneman D. points to another factor that determines the possibility of obtaining human pleasure-leisure. The words of another author Scitovsky T. reinforce the arguments Kahneman: "If a higher income is obtained at the expense of free time, it may negatively affect the level of happiness" [27].

Scitovsky T. gives a larger value than the quantity of leisure and its quality, seeing efficiently organised leisure activities, a source of great opportunities for pleasure. Scitovsky criticizes modern economic values, and explains that people should spend money and time on things and activities that don't adapt (beautiful scenery, meeting friends), and who have the ability "to fascinate and to provide some degree of satisfaction and pleasure." The acquisition of things, and the waste of time in those goods to which one adapts, entails in time the extinction of pleasure [27].

The well-being of family members is an important factor in obtaining satisfaction and pleasure for a person. Parents can get real pleasure when they invest to the education of their own children and they get the results in the form of their high-paying and prestigious work, their success. It is incommensurable with income growth. Good housing, low pollution, a quality natural environment, confidence, self-esteem, religious beliefs can also be a way to enjoy and increase one's satisfaction with one's own life.

In addition, according to M. Friedman, "people should be able to do what they like, what they enjoy, and give up what they do not like". In other words, a person has the "freedom to choose" [12]. This is the idea that if the person maintain and develop the possibility of realizing freedom, he will become more satisfied.

This once again confirms the fact that the concept "pleasure" is multifaceted and manifests kinesthetic (bodily), aesthetic, intellectual and emotional nature.

Measuring pleasure and life satisfaction are challenging because they have a normative subjective nature. The researchers argue that it is possible to make the economy more relevant to real life, if it consider the deeper factors that affect the quality of life.

For example, Bruno S. Frey identifies three main areas that allow for significant changes in the economy [10].

Thus, if we introduce a measure of subjective life satisfaction or pleasure into the field of Economics, we can empirically analyze the deviation from rationality due to the incorrect prediction of the utility of future consumption. It is also interesting to study the overestimation of satisfaction from higher income in the future and underestimation of utility from non-essential aspects of life (friendship and social relations). The study of economic entities with indicators of life satisfaction indicates the value of goods and services for consumers and social relations. Indeed, people want to benefit from income. They hope to participate in public relations, to self-define, to use their own competence to get more pleasure. Research on pleasure and happiness shows that raising incomes is not an effective way to increase utility: much of the increase in the utility of higher income dissipates quickly and encourages people to pursue even higher income.

37.3 Results

Will the price of pleasure change in the rapidly developing digital economy? The answer lies in the very nature of the digital economy and the changes it brings [7, 17, 29, 32].

The fundamental assumption that there is a qualitative shift in the way companies compete, management methods and business organization helps to decipher the logic of digital development. On the one hand, this shift has already known elements (mass production, mass marketing, mass organizations), on the other hand, it has new and unexplored components-flexible production, niche marketing, network organizations [35]. These changes allow companies to put into practice advanced technologies for collecting and processing information, stimulate them to find new ways to study consumer preferences, to form perfect mechanisms for managing consumer behavior on the basis of pleasure research.

We will show on the example of the utility theory, what changes determine the process of obtaining pleasure in the digital economy [33].

According to the theory, the amount of pleasure that a person receives in the process of consuming a good, that is, its usefulness depends on the urgency and intensity of the need that the good satisfies. The intensity of the need also affects the degree of pleasure received by a person from the consumption of goods. Need can take the form of vague desire, conscious longing, agonizing anxiety, overwhelming passion. The intensity of the need depends on its importance to the person, the place of the need in their overall structure, as well as the length of the waiting period. The pleasure of satisfying a need will be greater the stronger the need is expressed and the greater the utility of the good (Fig. 37.1) (created by Rodachev and Gagarin [26]).

Figure 37.1 shows a classical model of choice and consumption of goods: a consumer's choice is based on the knowledge of the utility; the evaluation of utility is

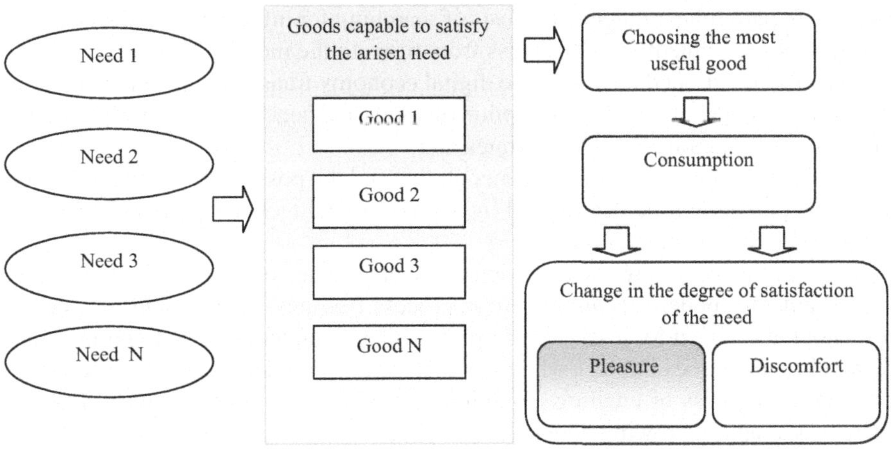

Fig. 37.1 Pleasure and its place in the process of choice and consumption of goods

subjective and individual; the dependence of the degree of satisfaction on the amount of consumed goods can have differ; the utility of the goods can be long-term; a person may receive discomfort in the process of consumption [26].

In the digital economy, companies at all levels of the corporate landscape, in every industry and field, are trying to adapt to unfamiliar circumstances and new conditions of competition.

The problem is how to trade digital products and services at near-optimal prices and how to maintain the attention of the target audience [4].

It is the attention of the target audience, rather than the product or service itself, that becomes a scarce resource in the digital economy. This follows from the idempotence of digital products, that is, the properties of an object or operation when applied again to an object give the same result as the first [23]. It is noteworthy that this scarce resource belongs to the consumer of digital goods and services, not their manufacturer [19].

The digital economy is also characterized by a network effect, which consists not only and not so much in gaining benefits from the collection of personal information about customers, but in increasing the number of connections. If there is one participant in the network, the effect is zero, with the growth of the network, this effect and, therefore, the utility of the network itself grows as the square of the number of participants [18].

The other side of the issue of network effects in the digital economy is the volume and value of content circulating on the web. After all, communication is interesting not by itself, but thanks to the content that can be transmitted or received through it. This adds another dimension to the digital economy, as much depends on the speed and content of traffic, as well as on the ability of the economic agent to benefit from it all [19].

Another important discovery in the study of consumer and producer behavior in the digital economy is the changing relationship between different types of transaction costs under the influence of digital technologies, which leads to changes in business models. In particular, reducing the costs of searching for information and concluding contracts leads to the flow of business from firms to the media [32].

The aforementioned effects of the digital economy transform the classical model of the process of choice and consumption of goods and hence increase the importance of "pleasure". Figure 37.2 demonstrates that.

Indeed, the architecture of consumer choice and the possibility of obtaining pleasure by a person will be determined by a set of digital technologies available to all participants of this interaction.

In the digital economy, the approach to the emergence of the need itself is changing. Thanks to modern technology, this process becomes manageable. If a person does not have a need for a service or product, then companies begin to purposefully form this need. For example, a while ago there was no need to use the Internet. However, in the process of creating conditions for connection and use of the Internet, there was a rapidly growing demand for connection, that is, there was a formation of the need for this service. Another example of the formation of consumer needs can be the experience of Apple. The consumer did not feel the need for Apple products

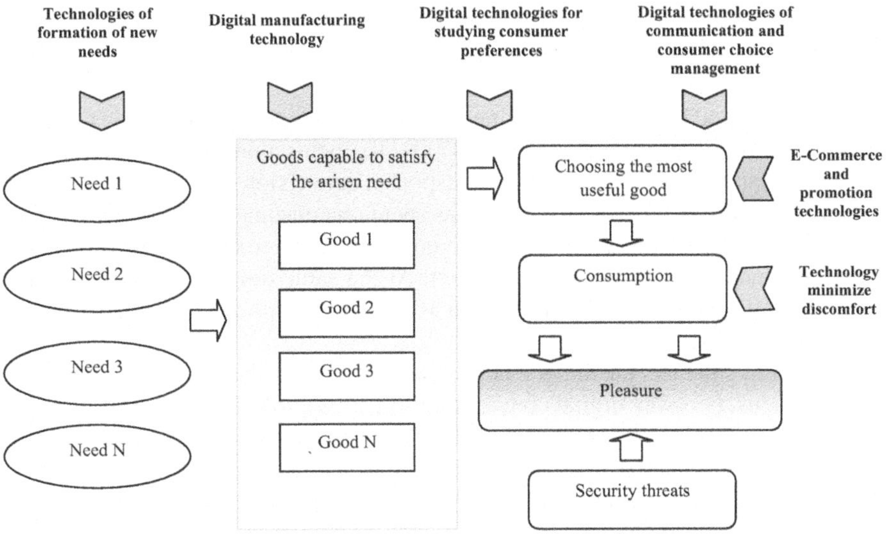

Fig. 37.2 Pleasure and its place in the process of choice and consumption of goods in the digital economy (created by authors)

until the release of its first IPhone. He did not suspect the possibility of its occurrence. However, an effective strategy to promote the brand and win the loyalty of Apple's customers has created a feeling among consumers that they are buying a premium quality and status product. This product brings joy to the consumer and aesthetic pleasure from the purchase for a long time. During this time, a large group of consumers has formed who are loyal only to Apple products and services [21].

The formation of consumer preferences through the creation of new experiences and management of memories, as well as the organization of production of goods and services that can meet emerging needs, become even more manageable in the context of digitalization. This is possible due to the active involvement of consumers in the process of creating and promoting goods and services through digital technologies. The participation and involvement of consumers through active two-way interaction with the manufacturer and purposefully structured communication process are essential from the point of view of value creation.

According to James Brian Quinn, "enhanced intelligence and its chief intermediary, service technology, are reshaping not only service industries but also manufacturing…, the General patterns of a country's economic growth, national and regional employment patterns, and position …. in world politics and international competition" [25].

Communication becomes a channel for studying the vital needs and preferences of consumers, their attitude to certain products and services of the company, as well as a way of delivering pleasure to their consumer. For example, companies like Boeing, Xerox, Apple and Motorola involve customers in their new product development. Their "smart" products, such as self-diagnostic elevators, computers, copiers, use the information processing power of the microchip to detect the probability of failures before they occur, thereby expanding the company's reach across the sales boundary into product service.

Companies find themselves in an increasingly furious race to get people to choose the most useful good among their "brands" [3]. To do this, they build communication based on personalization that increases response, conversion and revenue [22].

Statista estimates that $1.78 trillion was spent on consumer goods alone in 2018. The audience of e-commerce has grown significantly: more than 2.8 billion people worldwide make purchases on the Internet. At the same time, the average revenue per user (ARPU) increased by more than 10% compared to 2017 and amounted to 634 us dollars [28].

Companies can meet customer's new expectations, give them pleasure, "by creating continuous, two-way digital conversations that resemble customer conversations with a salesperson or service partner, conversations that treat them as individuals rather than a set of users". According to Accenture research, 75% of consumers believe they will find value in creating and managing a "needs profile", making interactions more personal and improving their brand experience.

While in the industrial economy companies had to divide consumers into "active" and "inactive", and inactive customers simply fell out of their target audience, in the digital environment, companies use the art of communication to re-activate inactive customers. To do this, use all sorts of digital services (VKontakte, Facebook, Instagram and other),[1] telephone and mobile communication facilities, Internet channels, CRM systems, mobile applications, readiness tracking systems and much more. Crowdfunding platforms and platforms for finding financing allow consumers to feel involved in the creation of the product, to influence the production and to be the first to get access to the novelty already at the stage of the primary concept. Uber and Airbnb were the most famous in this regard [1]. The growing number of connected devices and the digitalization of business processes are generating a stream of data that shapes consumer behavior and trends. Services such as Uber have emerged thanks to easy access to geolocation and maps and have greatly simplified the relationship between drivers and passengers.

37.4 Conclusions/Recommendations

The described services and information technologies cover the processes of human labor activity, the spheres of education and leisure, daily life and communication, as well as improve the process of choice and consumption of goods and create conditions for maximizing pleasure and reducing discomfort while consumption.

[1]At the beginning of 2019, the number of users of social networks worldwide reached almost 3.5 billion (according to popular social platforms in more than 230 countries and territories). In the last 12 months alone, 288 million new social media accounts have been registered, bringing the global social media penetration rate to 45% [34].

However, despite the obvious advantages of modern digital technologies and expanding the range of opportunities for pleasure, there are informational, economic and psychological threats and dangers.

This rapid development of digital technologies that can quickly satisfy human needs, leads to an acceleration of the process of emergence of new needs, and, consequently, to the reduction in the time of enjoyment. In such a scenario, the syndrome of "frustration" may soon develop, as the benefits will cease to fully fulfill their purpose. This is certainly a psychological threat, however, it can cause real economic damage to all subjects of economic relations and disrupt the balance of supply and demand.

Another real threat to people in their aspirations to enjoy consumption through digital platforms and messengers is the leakage of information about personal data [17]. Consumers who are holders of Bank cards are often the target of cyber fraud, which brings them direct financial losses in the form of loss of personal funds. Another serious menace to the economic security of the individual is cybercrime with the use of virtual wallets, virtual exchangers, the sale of business packages, twin sites with fraudulent requisites, SMS fraud, attacks on systems massively used in everyday life (Internet of things). Digital technologies are opening the door to totalitarian surveillance that limits or undermines people's legitimate privacy rights, raising issues of data abuse and the political abuse of surveillance. In addition, a specific security threat is the emergence of Internet addiction due to the virtualization of "live" communication, which is associated with "electronic network non-freedom" and the emergence of unproductive time, reduced productivity of human labor, loss of expertise.

All these threats can cause direct financial damage to the consumer and deprive him of the possibility of obtaining pleasure in the process of consuming goods and services. Therefore, the urgent task of the state is to create a strict legislative regulation of operations in the information space and clearly defined in the legal acts of the evidence base of offenses. The process of legal regulation of the Internet at the international level began in 1990, when UNCITRAL developed and adopted a number of fundamental documents: the model law on electronic Commerce (1996); the model law on electronic signatures (2001); the Legal guide on electronic transfer of funds (1987), Recommendations on the legal value of computer records (1985). The PLO Convention on the use of electronic communications in international contracts (2005) established the legal status of electronic signature and the possibility of concluding commercial transactions through the exchange of electronic messages. The Council of Europe also adopted: the Convention on the protection of individuals with regard to the automatic processing of personal data (1981); the Convention on crime in the field of computer information (2001); the Declaration on freedom of communication on the Internet and others. Legal regulation of the Internet within the European Union contributed to the adoption of legal acts such as Directive of the European Parliament on the protection of consumers in respect of distance contracts (1997); Directive of the European Parliament concerning the processing of personal data and the protection of privacy in the telecommunications sector (1997); Directive of the European Parliament on the legal aspects of information society services

(2000), European Parliament Resolution on the safe use of the Internet and new online technologies (2004) and others [30]. It is necessary to adapt these documents to the specifics of each country and develop rules and mechanisms for the functioning of the digital economy in order to ensure the financial and psychological well-being of consumers.

References

1. Abzhanbekov A (2015) Rukovodstvo po vyzhivaniyu v cifrovoj srede. https://rb.ru/opinion/v-cifrovoj-srede/. Accessed 10 Aug 2019
2. Arieli D (2010) Predskazuemaya irracional'nost': skrytye sily, opredelyayushchie nashi resheniya. Mann, Ivanov i Ferber, Moskva
3. Armano D (2007) It's the conversation economy, stupid. https://www.bloomberg.com/news/articles/2007-04-09/its-the-conversation-economy-stupidbusinessweek-business-news-stock-market-and-financial-advice. Accessed 17 May 2019
4. Ba S, Stallaert J, Whinston A (2001) Optimal investment in knowledge within a firm using a market mechanism. Manag Sci 47(9):1203–1219
5. Bentham J (1843) The works of Jeremy Bentham, published under the superintendence of his executor, John Bowring, William Tait, Edinburgh
6. Blaug M (1994) Ekonomicheskaya teoriya blagosostoyaniya Pareto. Delo, Moskva
7. Brynjolfsson E, Kahin B (eds) (2000) Understanding the digital economy. The MIT Press, Cambridge
8. Drogobyckij IN (2018) Povedencheskaya ekonomika: sushchnost' i etapy stanovleniya. Strategicheskie resheniya i risk menedzhment 1(106):26–32
9. Farde L, Gustavsson J, Jönsson E (1997) D2 dopamine receptors and personality traits. Nature 385:590
10. Frej B (2008) Schast'e: Revolyuciya v ekonomike. MIT Press, Cambridge
11. Frejd Z (1925) Po tu storonu principa udovol'stviya. Sovremennye problemy, Moskva
12. Fridman M, Fridman R (2007) Svoboda vybirat': nasha poziciya. Novoe izdatel'stvo, Moskva
13. Gallup Inc., Healthways Inc. (2014) State of global well-being. Results of the Gallup–Healthways Global well-being index
14. Gjedde A, Kumakura Y, Cumming P, Linnet J, Møller A (2010) Inverted-U-shaped correlation between dopamine receptor availability in striatum and sensation seeking. https://www.pnas.org/content/early/2010/02/03/0912319107?sid=52fbdbc8-9fa9-41c4-bc68-e97aee14d16a. Accessed 12 June 2019
15. Kaneman D (2006) Karty ogranichennoj racional'nosti: psihologiya dlya povedencheskoj ekonomiki, working paper. Fakul'tet psihologii Prinstonskogo universiteta, N'yu-Dzhersi, SSHA
16. Kaneman D (2016) Dumaj medlenno…reshaj bystro. ATS, Moskva
17. Karpunina E, Gorbunova O, Moiseev S, Cheremisina T (2019) Resistance is not hopeless…regarding the policy of countering information threats of economic security. In: Vision 2020: education excellence and management of innovations through sustainable economic competitive advantage. Proceedings of the 33nd International Business Information Management Association (IBIMA) conference in Granada, Spain, 2019, pp 2679–2687
18. Kozyrev AN (2011) Modelirovanie NTP, uporyadochennost' i cifrovaya ekonomika. Ekonomika i matematicheskie metody 47(4):131–142
19. Kozyrev A (2018) Matematicheskie modeli i prognozirovanie v cifrovoj ekonomike. https://medium.com/cemi-ras. Accessed 12 Sept 2019
20. Kuzina S (2007) Uchenye nashli u cheloveka centr udovol'stviya. https://www.tambov.kp.ru/daily/23971.3/73478/. Accessed 19 Aug 2019

21. Lajfhaker (2013) 3 sostavlyayushchie marketingovoj strategii Apple, kotorye sdelali ee uspeshnoj. https://lifehacker.ru/3-sostavlyayushhie-marketingovoj-strategii-apple-kotorye-sdelali-ee-uspeshnoj/. Accessed 10 Aug 2019
22. Okoli A (2018) Why brands must start listening in the conversation economy. https://dma.org.uk/article/why-brands-must-start-listening-in-the-conversation-economy. Accessed 3 Sept 2019
23. Peirce B (1870) Linear associative algebra, working paper. The National Academy of Sciences, Washington City
24. Pigu A (1985) Ekonomicheskaya teoriya blagosostoyaniya. Progress, Moskva
25. Quinn J (1992) Intelligent enterprise: a knowledge and service based paradigm for industry. Simon and Schuster, New York
26. Rogachev A, Gagarin A (2012) Kriticheskij analiz polozhenij teorii poleznosti. Izvestiya Nizhnevolzhskogo agrouniversitetskogo kompleksa 1(25):1–6
27. Scitovsky T (1992) The joyless economy: the psychology of human satisfaction. Oxford University Press, Oxford
28. Statista (2019) Digital market outlook. https://www.statista.com/outlook/digital-markets. Accessed 2 Sept 2019
29. Strelkova IA (2018) Cifrovaya ekonomika: novye vozmozhnosti i ugrozy dlya razvitiya mirovogo hozyajstva. Ekonomika. Nalogi. Pravo 2:226–232
30. Studme (2017) Международно-правовое регулирование сети Интернет. https://studme.org/91046/pravo/mezhdunarodno-pravovoe_regulirovanie_seti_internet. Accessed 23 Sept 2018
31. Tajler R, Sanstejk K (2018) Arhitektura vybora: kak uluchshit' nashi resheniya o zdorov'e, blagosostoyanii i schast'e. Mann, Ivanov i Ferber, Moskva
32. Tapscott D (2014) The digital economy anniversary edition: rethinking promise and peril in the age of networked intelligence. McGraw-Hill, New York
33. Viner D (1993) Koncepciya poleznosti v teorii cennosti i ee kritiki", v knige. In: Gal'perin V (ed) Vekhi ekonomicheskoj mysli. Tom 1, Ekonomicheskaya shkola, Sankt-Peterburg, S. 78–116
34. We Are Social, Hootsuite (2019) The global digital report 2019. https://wearesocial.com/global-digital-report-2019. Accessed 29 Aug 2019
35. Webber A (1993) What's so new about the new economy? https://hbr.org/1993/01/whats-so-new-about-the-new-economy. Accessed 17 Aug 2019

Chapter 38
Interregional Differentiation of Russian Regions by Levels of Educational Potential and Government Spending on the Education System

M. V. Moroshkina and O. V. Potasheva

Abstract Regional development of the Russian regions is characterized by a high level of heterogeneity, which is observed in almost all economic and social indicators. The level of development of human capital is an important factor in the social development of the territory. The main component of this capital is the level of education of the individual. The purpose of the study is to assess the inter-level differentiation of Russian regions by the level of the educational potential of the population and assess the level of expenses of the region on the education system. The information base of the study is presented by Rosstat data. Based on them, an array of data from the Russian regions was formed for economic and mathematical analysis, this allowed us to determine the importance of regional heterogeneity and to study its dynamics in the period 2005–2015. Studies have shown that the difference in regions occurs on the back of the reduction in the proportion of young people aged 15–24 years with the level of secondary education. We carried out a study of the structure of the regional budget expenditures on education. The heterogeneity is connected with the features of the regional budget and social policy. The results of the study can be used in the formation of strategic documents for the development of regions and the country.

Keywords Interregional differentiation · Regional economy · Education expenditures · Educational potential

JEL-Code R1 · R51

M. V. Moroshkina (✉) · O. V. Potasheva
Karelian Research Centre of the RAS, Institute of Economics, Petrozavodsk, Russia
e-mail: maribel74@mail.ru

O. V. Potasheva
e-mail: lelyapotasheva@yandex.ru

T. Kolmykova and E. V. Kharchenko (eds.), *Digital Future Economic Growth, Social Adaptation, and Technological Perspectives*, Lecture Notes in Networks and Systems 111, https://doi.org/10.1007/978-3-030-39797-5_38

38.1 Introduction

Russian regions have a high level of heterogeneity. This is noted by researchers according to the results of the study and analysis of various social and economic indicators. The uneven regional development is a consequence of the uneven distribution of resources, which ensure stable economic growth. Most researchers determine the resource dependence of regional development on the manufacturing sector, the resource component and the presence of a developed financial sector. The quality of conditions and the multi-level education system are of no small importance in the socio-economic development of the region. They form the labor power for the economic growth of the territory. Formation of labor power from the position of investment financial return on human capital is a complex and multidimensional process, which is provided by various mechanisms of government. The regional education system is one of the most important components of this process, since the educational potential of the territory provides the productive economy sector with specialists who allow pursuing a policy of active growth.

A distinctive feature of the educational services sector today is the need for financial investments with a previously understood time lag of return of 10–15 years. Moreover, an increasing amount of financial investments is required to ensure the conditions for the organization of educational activity in educational organizations. Evaluation of the financial investments of the state in the formation of educational potential at the regional level of management will allow us to analyze the possibility of increasing the competitiveness of the regions through the influence of the development of the sphere of educational services and the level of education achieved by the population. The research subject is the Russian regions, and the subject of the research is regional differentiation in terms of the level of educational potential and government spending on financial support for the development of educational services.

We will consider the differences in financial investments in the education system, which will allow us to study the factors of heterogeneity in the formation of educational potential by regions based on the analysis of financial investments, and this, in turn, will determine the reason for the high position in the ranking of some regions and low of others.

The results of this study can be used to conduct a comprehensive assessment of regional development and to write various program and strategic documents at the federal and regional levels.

38.2 Literature Review

The heterogeneous development of the Russian regions is the subject of research by many scientists. However, the directions by which the level of interregional differentiation of regions is analyzed do not always coincide. Most economists are interested in studying the heterogeneity of regions by economic indicators, such as GRP, factory output, investment volume, etc. [1]. The tools they proposed for the study of interregional differentiation also vary widely in methods and approaches. Some researchers consider the process of regional differentiation as a result of determining differentiation coefficients for various economic and social indicators [2]. Others use more complex mathematical approaches, determining the level of regression dependence or convergent processes [3]. Studies on regional differentiation are conducted for different economic entities. The definition of convergence processes for national economies of the world is isolated in comparisons of economic development [4]. No less relevant is the issue of regional inequality within one state.

Russian regions have a specificity of unevenness, determined by a number of economic and social indicators [5]. The history of economic issues study of regional unevenness is quite diverse: starting with the work of A. Lesh in 1959 [6] on the distribution of productive forces, who, within the framework of a theoretical approach, considered the process of uneven distribution of production resources that affect the increase in spatial heterogeneity. There are modern studies of agglomeration distribution processes, for example, Krugman [7] in the framework of a theoretical study formed the concept of agglomeration and directions of influence on spatial heterogeneity.

The question of the social differentiation of regions in their works is considered by many domestic and foreign scientists [8–10]. Most often, the results of measuring differentiation are associated with regional differences in the demographic situation, which reflect the processes of successive changes in fertility and mortality, an aging population, and migration labor flaw from peripheral regions to economically active centers. However, the results of these studies do not fully reveal regional differences in the age structure of the population from the standpoint of preservation and accumulation of human potential, in particular, in terms of education, and professional competence of the able-bodied population.

It is possible to assess the differentiation of Russian regions from the perspective of two factors of influence—demographic and educational. This is possible using data from the All-Russian Population Census, the methodology of which represents a regional section of the data on the age structure of the population by the level of education achieved. This set of indicators fairly objectively shows the number of certain groups of peers, that is, it allows you to compare and identify regions with different educational potential levels of the population, including youth—the main resource for the development of the regional economy [11].

A study of the structure of the expenditure side of the consolidated budget of the constituent entities of the Russian Federation allows us to understand the scale and causes of regional differences in financing social expenditures of the state, including the education system, because the share of federal and regional budget expenditures on education has occupied 11% in the last decade [12].

38.3 Methodology

The heterogeneity of the Russian regions is considered and proved using a wide range of tools and methods. Most researchers use the methodological base to assess the level of interregional differentiation within the framework of classical models. An assessment of the unevenness of regional development can be considered as a result of the analysis of a number of economic and social indicators.

In modern conditions of economic systems functioning, great importance is given to the development of human potential and assessment of its quality. One of the directions for assessing human potential is the level of education of the population, which can be determined using different indicators and their components.

In the framework of this study, an analysis of the heterogeneity of regional development of the Russian regions will be considered from the perspective of the formation of the educational potential of young generations of the population.

To measure the educational potential of the young generation in the region, Rosstat data on the achieved level of formal education on the generational-age structure of the population aged 15–45 are used. The choice of the boundaries of the age criterion is justified by the age classification adopted by the World Health Organization. The methodology for calculating the educational potential of the young generation is based on the data of the All-Russian population census in 2002 and 2010, according to statistical information in the context of the constituent entities of the Russian Federation on the achieved formal level of education of citizens of a certain age structure registered in the region. In order to simplify the processing of statistical data, we have determined that the aggregate of peers in a period of 10 years is one generation. Thus, we were able to distinguish three young generations: 15–24 years old, 25–34 years old and 35–44 years old.

The educational potential of young generations in the regions of Russia is assessed according to two main criteria: (1) by the number of young people per 10,000 people (2) by the rate of growth of the number of generations for the period between censuses. As a result, a typology of regions from 4 groups was developed by the rating method according to the level of educational potential of young generations.

Russian regions are not equally provided with educational resources, as a result, the quality of educational services is significantly different from region to region. In the process of assessing the development of the sphere of educational services, the availability of funds allocated for the development of the regional education system is of great importance.

The statistical data availability on the expenditures of the consolidated budgets of the constituent entities of the Russian Federation for social and cultural events, and in particular for the education system, allows us to analyze regional differences in the amount of financial expenses. Rosstat provides an information database for the period 2005–2015, which served as the basis for the study.

In the study, Russian regions were analyzed by the consolidated budget expenses of the subject for education as part of the costs of social and cultural events. The data obtained were compared, and the results were ranked with the definition of a group of leading regions, the share of educational expenses in which is of the highest importance and lagging behind.

38.4 Results

Based on the methodology of ranking the regions according to the main criteria (coefficient of the number of representative groups, growth coefficient), an information-analytical database has been formed. It is necessary to assess regional differences according to the results of censuses of Russia in 2002 and 2010 according to the level of accumulation of educational potential by young generations in the regions of the country. As part of the consideration of the impact of government spending on education, it is interesting to estimate the number of young generations in the region with the achieved level of secondary education due to the fact that most of the regional budget expenditures (from 12 to 16%) are expenditures on supporting the activities of the level of secondary education.

During the period between the censuses in 2002 and 2010, against the backdrop of a demographic wave of decline in the total population, a tendency towards a significant reduction in the share of young people with only secondary general education was manifested. So, the regions with the smallest number of people who received only school education in 2010 were: the Republic of Mordovia, the Republic of Komi, the Chuvash Republic, the Kirov and Nizhny Novgorod Regions. St. Petersburg stands out among them in the group; it is obvious that the presence of a large number of organizations of secondary and higher professional education influences the increase in the level of education of the population.

A group was recorded from regions where the proportion of people with secondary general education is very high and increases by 15–20% compared to 2002. These regions included national republics (Chechen, Altai, Dagestan, Mari El, Tuva, Adygea, Bashkortostan), the regions of the NWFD (Leningrad, Pskov, Kaliningrad regions), Novosibirsk region, Stavropol Territory.

This part of the analysis showed that the compiled ranking of the regions makes it possible to determine those regions in which the absence of a developed multi-level education system is a circumstance in which young people study in vocational education organizations in other regions of the country. A specialist rarely returns, having received an education, as he seeks to get a prestigious job and improve his living conditions. On the other hand, using this rating, it is possible to determine the regions where the situation on the regional labor market has developed so that low-skilled workers are most in demand and the return on investment in raising their educational level is very low.

At the next stage of the study, based on the statistical information base, an array of panel data was compiled, which includes the expenses of the consolidated budgets of the constituent entities of the Russian Federation in 2005 and in the period 2010–2015. The statistical information represents the expenditure part of the budget-forming document on expenditures on the social sphere, including education.

In the calculations of the study, relative indicators were used to assess the percentage of education costs of total spending on social and cultural events in the regions of Russia.

Based on the formula (38.1), the percentage of education expenses was calculated.

$$\mathrm{Ratio}_{\mathrm{ed}} = (\text{expenses} * 100)/\text{cultural events} \quad (38.1)$$

The calculations made it possible to identify the regions that have the largest share of educational expenses. The assessment in difference was analyzed from the position of identifying the leading and lagging regions (see Table 38.1).

It is worth noting that the results in the table are rather unstable and there are practically no regions with a stable leading or stable lagging position. This fact allows us to say that throughout the study period the financial situation of the region changed significantly, the provision of social services was carried out on a residual basis with the involvement of federal budget subsidies. Conversely, economically active, wealthy regions increased the share of education spending, thereby ensuring the development of the educational services sector.

Table 38.1 The ratio of spending on education among socio-cultural events

Year	Leaders	Lagging behind
2005	Republic of Sakha (Yakutia), Chukotka Autonomous Okrug, Altai Republic, Kamchatka Region, Republic of Tuva	Republic of Ingushetia, Republic of Mordovia, Chechen Republic, Omsk Region
2010	Nenets Autonomous Okrug, Chukotka Autonomous Okrug, Republic of Tuva, Republic of Dagestan, Republic of Sakha (Yakutia)	Republic of Mordovia, Chechen Republic, Omsk Region, Samara Region
2011	Nenets Autonomous District, Republic of Dagestan, Republic of Tuva, Republic of Tuva, Republic of Altai, Republic of Tuva, Chukotka Autonomous Region	Republic of Mordovia, Penza Region, Chechen Republic, Omsk Region, Astrakhan Region
2012	Republic of Dagestan, Republic of Altai, Republic of Tuva, Republic of Sakha (Yakutia), Chita Region	Republic of Mordovia, Pskov Region, Karachay-Cherkess Republic, Penza Region
2013	Republic of Dagestan, Republic of Altai, Republic of Sakha (Yakutia), Udmurt Republic, Chita Region	Jewish Autonomous Region, Amur Region, Pskov Region, Penza Region
2014	Republic of Sakha (Yakutia), Chita Region, Republic of Tatarstan, Udmurt Republic, Belgorod Region	Jewish Autonomous Region, Republic of Ingushetia, Republic of Mordovia, Amur Region
2015	Republic of Sakha (Yakutia), Udmurt Republic, Chukotka Autonomous Okrug, Belgorod Oblast, Yaroslavl Oblast	Republic of Ingushetia, Pskov Region, Samara Region, Altai Territory

Source Rosstat data, authors' calculations

Another direction of calculations is to determine the ratio of the share of educational expenses in the total amount of expenditure obligations of the subject. An analysis of this ratio was carried out on the basis of statistical data from Rosstat using formula (38.2). It is worth noting that the group of leaders in its composition is represented by the mining regions—the Republic of Yakutia, the Nenets Autonomous Okrug, etc. Along with economically developed regions, the group of leaders in terms of the ratio of spending on education among social and cultural events also includes regions that have low indicators of economic growth—the Republic of Tyva, the Republic of Dagestan.

$$\mathrm{Ratio}_{\mathrm{ed}} = (\mathrm{expenses} * 100)/\mathrm{expenses}_{\mathrm{gen}} \tag{38.2}$$

The study and the calculation of coefficient (38.2) made it possible to single out a group of regions that have a high share of educational expenses (Table 38.2).

According to the results of calculations, groups of regions that have a high share of educational expenses and a very low one were determined. As a result of the study,

Table 38.2 Education expenses ratio

Year	Leaders	Lagging behind
2005	Aginsky Buryat Autonomous Okrug, Komi-Permyak Autonomous Okrug, Ust-Ordynsky Buryat Autonomous Okrug, Koryak Autonomous Okrug, Jewish Autonomous Oblast	Moscow Region, Khanty-Mansiysk Autonomous Okrug—Yugra, Republic of Tatarstan, Tyumen Region
2010	Nenets Autonomous Okrug, Republic of Kalmykia, Jewish Autonomous Oblast, Altai Republic, Chukotka Autonomous Okrug	Moscow Region, Republic of Tatarstan, Krasnodar Territory, Krasnoyarsk Territory
2011	Nenets Autonomous Okrug, Jewish Autonomous Oblast, Republic of Kalmykia, Republic of Altai, Republic of Tyva	Moscow Region, Krasnodar Territory, Republic of Tatarstan, Krasnoyarsk Territory
2012	Jewish Autonomous Region, Republic of Kalmykia, Republic of Altai, Nenets Autonomous District, Republic of Adygea	Moscow Region, Krasnodar Territory, Republic of Tatarstan, Khanty-Mansi Autonomous Okrug—Ugra
2013	Republic of Kalmykia, Altai Republic, Nenets Autonomous Okrug, Republic of Adygea, Jewish Autonomous Oblast	Moscow Region, Krasnodar Territory, Khanty-Mansi Autonomous Area—Ugra, Sverdlovsk Region
2014	Republic of Kalmykia, Republic of Adygea, Jewish Autonomous Oblast, Nenets Autonomous Okrug, Karachay-Cherkess Republic	Moscow Region, Krasnodar Territory, Khanty-Mansi Autonomous Area—Ugra, Sverdlovsk Region
2015	Republic of Kalmykia, Jewish Autonomous Oblast, Republic of Altai, Nenets Autonomous Okrug, Sevastopol	Moscow Region, Krasnodar Territory, Khanty-Mansiysk Autonomous Okrug—Ugra, Sakhalin Oblast

Source Authors' calculations

the instability of spending on education was revealed, which leads to instability in the composition of groups of regional leaders and outsiders. The study made it possible to identify the leading territories that have a high share of educational expenses, for example, the Altai Republic and the Republic of Kalmykia, and the lagging regions—Moscow Region, Krasnodar Territory. according to the size of the share, depending on the demographic characteristics and the characteristics of the regional budget policy.

38.5 Conclusion

In the period between censuses in Russia, it was noted that there was a significant reduction in the proportion of young people aged 15–24 years with the level of secondary education reached. Moreover, in a number of country regions, the proportion of young people aged 25–34 years who have only completed school education is increasing.

A study of the structure of education spending showed that the regions that were in the lagging group are resource-rich and most of their spending is directed towards the development of the industrial sector.

The heterogeneity is associated with the features of the regional budget policy. The higher the population of the region, the greater the share of expenses and the higher the financial responsibility of regional authorities for education development and the formation of human capital in the region.

This work was carried out as part of the RFBR scientific project No. 19-010-00549 “Inter-regional differentiation of Russian regions” and the state assignment of the IE KSC RAS (No. AAAA-A19-119010990087-1).

References

1. Kolomak EA (2013) Uneven spatial development in Russia: explanations for a new economic geography. Voprosy ekonomiki 2:132–150 (in Russian)
2. Granberg AG, Suslov VI, Suspicin SA (2007) Multi-regional systems: economic and mathematical research. Novosibirsk, Sib. Nauk. ed, 371 pp (in Russian)
3. Barro PJ, Sala-J-Martin X (1995) Economic growth, 200 pp
4. Latham W (1977) Needless complexity in the identification of industrial complexes: a reply. J Reg Sci 17:459–461
5. Pinkovetska YuS (2013) Assessment of the convergence of performance indicators of small enterprises between regions of the Russian Federation. Bull Pac State Econ Univ 2(66):34–51 (in Russian)
6. Bream A (1959) Geographical location of the economy. Foreign Literature Publishing House, p 455 (in Russian)
7. Krugman P (2005) Space: the last frontier. Spat Econ 3:121–126

8. Karachurina LB (2006) Interregional migration and socio-economic differentiation of the space of modern Russia Problems of forecasting, 3. https://cyberleninka.ru/article/n/mezhregionalnaya-migratsiya-i-sotsialno-ekonomicheskaya-differentsiatsiya-prostranstva-sovremennoy-rossii
9. Lyytikäinen L, Kemppainen T (2016) Regional inequalities in self-rated health in Russia: what is the role of social and economic capital? Soc Sci Med 161:92–99
10. Zubarevich NV, Safronov SG (2013) Inequality of socio-economic development of regions and cities of Russia in the 2000s: growth or decline? Soc Sci Mod 6:15–26 (In Russian)
11. Potasheva OV, Moroshkina MV (2018) Statistical assessment of regional differences of the constituent entities of the Russian Federation in terms of educational potential of young generations of the population. Stat Econ 15(2):38–48
12. Zubarevich NV, Gorina EA (2015) Social expenses in Russia: federal and regional budgets, Moscow

Chapter 39
Human Language in the Digital Age

I. Borisova

Abstract The aim of the research is to investigate how the advent of digital technology influences the social sphere of the society, its language in particular. One of the tendencies in current communication is its active shift to the virtual reality. The generation of "digital natives" creates their own language to serve the Internet needs. It bring into existence the phenomenon of "digital creolization" which is specifically characterized by using different techniques of occasional word-formation. Applying the methods of accumulation, comparison, description and statistical analysis of the valid examples, the author finds out the most widely spread word-forming techniques, namely: patterning, holophrase, creation, and interlacing. The phenomenon under study could have both negative and positive consequences: on the one hand, it makes the literary norm of the language more unstable and eventually can result in misunderstanding between some layers of the society and, in its extreme form, in the loss of some social experience gained; on the other hand, this phenomenon is guided by the traditional word-forming techniques, which testifies to the fact that the vital conservatism of the language mechanisms will prevent it from being distorted to a fatal extent. Conducting further detailed research in this field will enable scientists to generate some practical recommendations useful for those who feel responsible for building the dependable "linguistic bridge" for smooth transfer of the experience gained to future generations.

Keywords New communicational paradigm · Written conversational speech · Word-formation · Patterning · Holophrase · Creation · Interlacing

JEL-Code Y80

I. Borisova (✉)
Financial University Under the Government of the Russian Federation, Moscow City University, Moscow, Russia
e-mail: ivb66@yandex.ru

T. Kolmykova and E. V. Kharchenko (eds.), *Digital Future Economic Growth, Social Adaptation, and Technological Perspectives*, Lecture Notes in Networks and Systems 111, https://doi.org/10.1007/978-3-030-39797-5_39

39.1 Introduction

To understand how Russia might take full advantage of its pivotal position within the world economy it is necessary to analyze thoroughly all spheres of its current development. No exception should be made for the Russian language—a powerful resource for increasing the country's social and cultural potential. The national cultural identity realized through and cemented with its language, as a primary means of communication, has always been the basis for the country's prosperity and the platform for creating new productive ideas on the international arena.

Considering the problems of any particular language traditionally starts with the review of some theses typical of human language in general. Human language is a bridge between generations which provides vitality and sustainability for the existing society. Digitalization permeates vast areas of human life. It bears both opportunities and risks for human coexistence, the consequences of which cannot adequately be assessed in many areas today. Yet one thing is certain: digitalization is accompanied by profound processes of change. "Recent developments in information and communication technology are sometimes compared to Gutenberg's invention of the printing press" [1]. The advent of digital technology symbolizes passing to a new paradigm in human communication. The language bridge does not prove to be as steady and reliable as before. The new Z-generation, the so-called "digital natives", imprudently loosen their ties with X- and Y-generations through developing their own language serving the digital reality. It is quite relevant to mention here that, according to Marc Prensky, "digital natives" are "native speakers of the digital language of computers, video games and the Internet." And those who "were not born into the digital world but have <…> become fascinated by and adopted many or most aspects of the new technology are digital immigrants" [14].

The first attempt to form a new linguistic discipline, having the language peculiarities of the digital age as its object, was made by Professor David Crystal in his monograph "Language and the Internet". He called this discipline "Internet Linguistics" arguing the idea of the Internet's being "a quintessence of the drastic processes observed in today's world" [5].

Jilliana B. Enteen goes further in calling the linguistic changes taking place in the Internet "digital creolization". She explains the introduced term as an umbrella one that "refers to digital networked communication" [8]. The depth of the changes meant can be seen through the definition of the "creolization" phenomenon found in the Cambridge dictionary: "the process of languages mixing to produce new ones". It is also remarked that the term is used especially to "refer to mixtures of local languages with European languages" [3]. So, using the term "creolization" for describing the language matter, being currently born in the virtual reality, underlines its coherence with globalization and integration processes that are cardinal for understanding the prospects of human civilization development. And it is not surprising that the Internet language is perceived as "the most advanced medium" and "associated with the most up-to-date forms of culture existence" [9].

What changes does human language undergo nowadays? How tangible is diversification between real and virtual languages? Should the current trends be treated as positive or negative? Do they pose any threat to the national identity? Could scientists display proactiveness in facing that sort of challenges? Doubtlessly, these questions are very acute for the modern scientific community. We have done our research with these and related questions in view in hope to make some mute points clear and in this way to make a difference to establishing new would be norms of our native language.

According to David Crystal, the Internet enables us to perform three main functions, namely: "to link the sites comprising the World Wide Web, to send electronic mail between private mailboxes, and to permit groups of people to engage in continuous discussion in chatrooms or by instant messaging" [6]. Tim Shortis states that there is "considerable creativity and diversity on the part of the users in the ways they deploy the vernacular resources of Txt and there is a longstanding historical basis for such practices" [15].

On the one hand, the Internet communicants' messages possess all oral speech characteristics: spontaneity, linearity, immediacy. On the other hand, it is certain to be written speech. Following Ye. Litnevskaya, we prefer operating the term "written conversational speech" which is realized as "a certain mixture of written literary and oral conversational languages" [12].

The dynamics and change rate at different language levels are not the same. The most flexible level turns out to be the lexical one, because it is exposed to numerous social changes. Therefore, the national wordage is constantly developing, which causes a lesser imperativeness and a greater openness of the lexical norm.

The object of our current research is the samples of the Russian written conversational speech represented in the Internet posts, comments in social networks, blogs and forums.

The research subject is the Internet communicants' occasional word-formation.

The novelty of the research is obvious as the process of creating new words in the World Wide Web is momentary and contextual one, and, as a result, it could be regarded as a certain "snapshot of eternity". Such snapshots, when scientifically and logically treated, could give a unique integral picture of the analyzed phenomenon and reveal its prospective potential in revising the existing norms of the Russian language. Such "mosaic" approach in studying the Internet sources seems to be valid due to the fact that the World Wide Web is "a wondrous source of information remarkable for diversity and abundance of linguistic material, its actuality, that is its continual renovation" [2].

39.2 Methodology

At the first stage of our research, we moved from the abstract to the concrete, i.e. rested upon the theoretical postulates existing in the area under study as reflected in the scientific papers of Russian and foreign investigators (see the references).

At the second stage, using the accumulated theoretical base, we applied the traditional methods relevant to collecting and verifying empirical data, namely: accumulation, comparison, description, and statistical analysis of the valid examples.

At the concluding stage, we summarized the results obtained through intent-analysis, which contributed greatly to demonstrating the Internet users' specific communicative intentions and their priorities in occasional word-formation.

The analysis was conducted on the basis of the discourse approach, which made it possible to consider both linguistic and extra linguistic factors influencing the Internet users' communicative strategies and tactics.

39.3 Results

It is known that traditional forms of communication were represented in two strict communication models, namely:

- a particular sender and a particular receiver(s)—interpersonal communication;
- a particular sender and a mass anonymous receiver—mass communication.

In the Internet, this opposition does not work. A blog commentary or a forum post is both a response to a particular statement and a text appealing to all communication participants, the latter often being much more important than the former [7].

Such deformation, alongside with the other peculiar features of the Internet communication—interactivity, anonymity, democratic mode, hypertext resources—lift all restrictions, lead to stylistic syncretism, and, as a result, to shifting the 'norm' concept. Under these circumstances, a spontaneous stream of word formation arises, bringing into existence not a usual variant (when new lexical units are built according to the rules and mechanisms fixed by the norm), but an arbitrary, i.e. occasional variant (one-time, in a definite context).

While selecting the materials for our research we depended on the signs of an occasional word, formulated by A. Lykov:

(1) belonging to speech (in our case—to a particular type of written conversational speech),
(2) creativity (non-reproducibility),
(3) word-forming derivativeness,
(4) abnormality,
(5) functional expendability,
(6) expressivity,
(7) nominative optionality,
(8) synchronic-diachronic diffusivity,
(9) individual belonging [13].

It should be noted that there exist numerous typologies of the occasional techniques and their names in word-formation. The most recent ones worthy of mentioning here are those of Zemskaja [16] and Khromova [11]. However, practically all scientists, including the above-mentioned ones, distinguish such word-forming techniques as:

- holophrase (using a word combination or a sentence as the basis for a newly-formed word),
- hendiadys (adding a rhymed word),
- blending (junction of words or their parts into the whole),
- creation (using an invented, non-semantic root),
- tmesis (inserting a new unit into the existing word),
- substitution (changing a morpheme of the original word by the other one, drawing the analogy with the structure of a concrete word),
- interlacing (creating a new word by omitting one link of the word-formation chain),
- metathesis (rearranging the word components),
- affix emancipation (using an affix as an independent lexical unit),
- pun (using paronyms),
- patterning (building a new word according to the existing model).

With this classification in view, we have analyzed 4,136 samples of the Internet written conversational speech. As a result, the following techniques have been found out (from most to least widely used):

(1) patterning (гастрабайтерша, маникюрша, бабец, худыш, химозный <соус>, фаст фуды, маргинальщина, головняк, шедеврище, запойно-быдляцкий <юмор>, тупизм, крышняк, позитивненько, примитивщина, <она> орнула, термояд, резидентишка, дибилизация, подсмысел, уговорщики, стесняшки, имбецилизм, <воспринимает> ненавистно, <тонна> косметоса, бабень, раскабанеет, красавэлло, стаканяга, молодчага, спрогрессировал, <поеду> лекционировать, симпатизанты, отъезжант, <я> вангую, замахала <своим пиаром>, <это ж надо таким> зомбаком <быть>, суррогатка, <вам там на> мейлрушечке <и бегемоты хорошо поют>, отъезжант, <что>, антислава, <он не самый> карточный <футбольный судья>, <кто у нас тут самый> МЭШобразный, <сейчас эпоха> неорабства – нераб <сам должен обеспечивать свое содержание>, <ну не могут же достойные люди так> беспределить, <мамы> аутят, <пляжи> майамщины или калифорнийщины, <бывшая> виагрянка, <они даже не в браке, а в> отношеньках, <с точки зрения> русскости, <спустит свою> жизку <в унитаз>);

(2) holophrase (всегосамадобившиеся, околоподъездная, яжемамки, полунепоймикакая <внешность>, самоудалился, быдлозал, нивжисть, шутки-самосмейки, камедиклабовское <творчество>, быдлоящик, хайпогенератор, чтотакполучилось, камедиклабовец, среднедушевое <потребление спирта>, более-менее-алкашам <на потеху>, бабдедульки, быдло-алко-голубой <огонек>, лечениетаблетками, королизолотойголос, <если б мы несли> разумноедоброевечное, <тошнит от этих> пугачевыхгалкиныхкиркоровых, <абсурдные

боевые потери –> краснаяармиявсехсильней, кустанаерожденная <бабушка>, <дочка> супер-пупер-главного-дяди-которому-нельзя-говорить-нет, женщина-на-таблеточках, НеПодругаНеймара, <а родственникам можно даже номер> ихтамнетной <могилы не сообщать>, всеминарукахнекогданосимая <заслуженная матрона>, <каждый может говорить, а не> оратьблеятьманипулировать, <министр> «никтонепросилвасрожать», <все эти> «я-же-из-добрых-побуждений» <люди>);

(3) creation (<новый бумер купит и> ушатает <его>, <женщины мужиков порой до развода> запиливают, <когда тебя> отстрапонила <твоя баба>, <он бы> балакал <только для себя>, <скоро вообще> ёкнется <умом>, <чо бы ещё> замастырить, каплоухая <крыса>, далдон, утырки, спокусил, <модная> приспособа, изврат <в голове>, <писать в стиле> мурмурмур, вякальник <свой надо закрыть>, вандефульно <сделал передачу партнеру>, <срочно приступили к> делитованию <своих изображений>, проститня, пардоньте, <поток> алармистских <новостей>);

(4) interlacing (пирсингнованные, безцеллюлитные, микояновский <майонез>, деграданский, поиздерганнее, быдловатый, недовольствоваться, антироссийскость, лекционировать, симпатизанты, уфотошопленный, сдизайнерила, <пусть они тут> поволонтерят, акционист, <проблема> профилактирования, распиаренные, <хоть> заобижайтесь, <вы че, серьезно, или так> троллируете?);

(5) blending (фитоняшки, Пугалкины, совокупляторы <совокупляться + аниматоры>, путиновыборы, диванодавы <интернетные>, Инста-мамочки, уродительский <комитет>, нефтепапа, <фрики и> звездоразы, самовостребованные <люди>);

(6) tmesis (недомужик, недомодель, <в полную боевую готовность приводится армия, которая полностью> не боеготова, недоджет, недополитика);

(7) substitution (<как чуть> перебавил, <крышу рвёт>, поклинник – <поклонник клиники (не больницы)>, Мартин-Кристо <сын Киркорова>, <пир во время> Чечни);

(8) hendiadys (бурчать-ворчать-устать, Трамп-парарам).

As we can see, the most widely used techniques turned out to be patterning, holophrase, creation, interlacing; the least widely used ones—blending, tmesis, substitution, hendiadys. The examples of metathesis, affix emancipation and pun have not been discovered.

In the contexts under study, the authors' building occasional words has different functions:

(1) to increase artistic expressiveness;
(2) to save speech efforts;
(3) to display one's creativity;

(4) to liven up the conversation;
(5) to provoke the interlocutors;
(6) to make the conversation more intimate;
(7) to demonstrate one's awareness of the subject;
(8) to protest against the fixed norms, etc.

Such amazing functional diversity gives the grounds for characterizing the Internet as "the most freedom-loving and democratic channel of human communication", the zone of "increased emotional saturation" that "stimulates a linguistic personality to realize oneself with various means", and the Internet communicant as a "playing, experimenting, creative" personality [10].

39.4 Conclusions

The issues revealed drives us to some important conclusion:

(1) The digital age brings into existence a new communicational paradigm;
(2) The most indicative communicational trends can be observed in the Internet;
(3) The Internet language is represented in a specific type of speech—"written conversational";
(4) This speech is marked by intensive process of occasional word-formation;
(5) On the one hand, this phenomenon can be considered as a negative one (it makes the literary norm of the language more unstable and eventually can result in misunderstanding between some layers of the society and, in its extreme form, in the loss of some social experience gained);
(6) On the other hand, this phenomenon is guided by the traditional word-forming techniques, which testifies to the fact that the vital conservatism of the language mechanisms will prevent it from being distorted to a fatal extent.

Linguists has always been interested in the state of the language in its synchronic and diachronic aspects. Today, we can hear the opinions that the issues concerning the language in the digital age are an example of "the overlap between popular culture and literacy", and they are "a worrying trend indeed" [4]. But we strongly believe that conducting further detailed research in this field will enable scientists to generate some practical recommendations useful for those who feel responsible for building the dependable "linguistic bridge" for smooth transfer of the experience gained to future generations.

Perspective investigations could be undertaken in the field of Internet speech transformation from the point of view of its arbitrariness or non-arbitrariness. On the one hand, the Internet-communicants' speech is obviously rich in ordinary mistakes and misprints, but, on the other hand, the intentional changes of the word graphic image are observed.

This idea could be illustrated by erratives (as cacography manifestation)—words or word collocations purposefully distorted by the language bearers, who are aware of the literary norm, for achieving some expressive goals. Writing words with capital letters, stretching words, inserting smileys and emoticons (pictograms containing printing characters and symbolizing various emotions), etc. may also be treated as intentional changes.

In syntax, special attention should be paid to the phenomenon of parceling (an emphatic construction with intentional breaking up of a coherent text into some punctuationally independent segments).

It would be interesting to analyze the selection of Internet speech samples characterized as creolized texts (consisting of both verbal and visual components—pictures, photos, audio and video fragments).

Studying these aspects would provide a solid ground for some hypotheses relating to the influence of virtual communication (linguistic, psychological, moral and aesthetic) on the real communication, would help us reveal both positive and negative sides of these communication types interference, and work out, if necessary, some relevant practical recommendations.

References

1. Ananiadou S, McNaught J, Thompson P (2011) The English language in the digital age. The Meta-Net White Paper Series, UK, 43 pp, p 4
2. Balabas NN (2015) Ispol'zovanije sovremennykh informatsionnykh tekhnologij pri izucheniji materialov autentichnoj frantsuzskoj pressy. In: Dialog yazykov, kul'tur i literatur v professional'no orientirovannom i filologicheskom aspektakh: Sbornik nauchnykh statej po materialam mezhdunarodnoj nauchnoj konferentsii, Moscow, MGPU, pp 164–168, p 165
3. Cambridge Dictionary. https://dictionary.cambridge.org/dictionary/english/creolization. Accessed 17 Nov 2018
4. Carrington V (2005) Txting: the end of civilization (again)? Camb J Educ 35(2):161–175, p 171
5. Crystal D (2006) Language and the Internet. Cambridge University Press, 257 pp, pp 20–23
6. Crystal D (2007) How language works. Avery, Penguin Group (USA) Inc., New York, 502 pp, p 153
7. Dubina LV (2013) Tendentsii razvitiya russkogo yazyka v usloviyakh Interten-kommunikatsii. In: Vestnik TGPU, Tomsk, №2(130), pp 177–181, p 178
8. Enteen JB (2010) Virtual English. Queer Internets and digital creolization. Routledge, Taylor & Francis Group, New York & London, 210 pp, p 40
9. Gusejnov GCh (2014) Russkij yazyk i novyje tekhnologii. In: Akhmatovoj MV, Belika VI (eds). Novoje literaturnoje obozrenije, Moscow, 256 pp, p 11
10. Karpova TB (2008) Okkazional'noje v seti Internet. In: Stereotipnost' i tvorchestvo v tekste, Mezhvuzovskij sbornik nauchnykh trudov. Vyp. 12. – Perm', PGNIU, pp 197–203, p 197
11. Khromova SA (2010) Individual'no-avtorskoje slovotvorchestvo v jego otnoshenii k jazykovym slovoobrazovatel'nym standartam. Dinamika semantiko-slovoobrazovatel'nykh podsistem russkogo jazyka. Saratov, Nauchnaja kniga, 327 pp
12. Litnevskaja YeI (2011) Pis'mennyje formy razgovornoj rechi (K postanovke problemy), Monografija. MAKS Press, Moscow, 304 pp
13. Lykov AG (1976) Sovremennaya russkaya leksikologiya (russkoye okkazional'noye slovo). Vysshaya shkola, Moscow, 120 pp, p 13

14. Prensky M (2001) Digital natives, digital immigrants, on the horizon, vol 9, No 5. MCB University Press, pp 1–6, pp 1–2
15. Shortis T (2016) The language of ICT—information and communication technology. Routledge, London, 128 pp, p 5
16. Zemskaja YeA (2012) Sovremennyj russkij yazyk. Slovoobrazovanije. Nauka, Moscow, 368 pp

Chapter 40
Assessment of Asynchronous Development in the Implementation of the Main Areas of Regional Policy

Yulia V. Vertakova, Maria G. Klevtsova and Yulia S. Polozhentseva

Abstract *Purpose*: The current stage of development of the socio-economic systems of the Russian regions is characterized by significant asynchrony, which is perceived in two ways, on the one hand, as an obstacle to sustainable development, and on the other, as a source of growth, due to the equalization of the level of socio-economic development and the accelerated development of lagging regions. *Design/methodology/approach*: These circumstances prove the relevance of this study aimed at developing the fundamental foundations of the formation and implementation of tools and methods to smooth out the asynchronism of regional development, which will ensure a progressively balanced socio-economic development of the Russian regions and the country as a whole. The methodological basis of the study was composed of comparative and graphoanalytic methods, methods of statistical processing of information characterizing regional economic complexes as complex socio-economic systems. *Findings*: The aim of the study is to assess the asynchronism of the development of regions on the basis of an integrated indicator in economic and social contexts and to substantiate the areas of managerial impact on various groups of regions. The article shows the relevance of developing state and regional measures to smooth out asynchrony and reduce differences in the levels of socio-economic development of regions. *Originality/value*: The study assessed the levels of socio-economic development of regions in order to identify asynchrony in their development, and also based on the application of rating technology, grouped regions to determine the main directions for reducing asynchrony. The results obtained in the study develop management theory in socio-economic systems, in particular, the level of development of regions and the degree of their asynchrony.

Y. V. Vertakova (✉) · M. G. Klevtsova · Y. S. Polozhentseva
Southwest State University, Kursk, Russia
e-mail: vertakova7@yandex.ru

M. G. Klevtsova
e-mail: klevtsovam@mail.ru

Y. S. Polozhentseva
e-mail: polojenceva84@mail.ru

T. Kolmykova and E. V. Kharchenko (eds.), *Digital Future Economic Growth, Social Adaptation, and Technological Perspectives*, Lecture Notes in Networks and Systems 111, https://doi.org/10.1007/978-3-030-39797-5_40

Keywords Regional development · Asynchrony · Socio-economic development · Regional policy

JEL Code R12 · R58 · I38

40.1 Introduction

The current Russian economic policy, focused on achieving sustainable economic growth and improving the standard and quality of life of the population, however, in recent years the problems of balanced development have become much more acute due to the consequences of the global financial crisis, the sanctions regime, the negative dynamics of world oil prices, and the aggravation of political confrontation with a number of developed countries, an unbalanced structure of the economy, low innovation and technological activity, demographic imbalance and other factors. One of the factors that negatively and comprehensively affect socio-economic development and economic growth in modern Russia is the excessive asynchrony of the economic space and the insufficient development of the tools to overcome it. Significant asynchrony in the levels of socio-economic development and living conditions of the population of the regions can be overcome by developing and implementing tools and methods for smoothing the spatial differentiation of regions aimed at ensuring effective economic regulation of regional development that can ensure close interaction of all participants in public relations on its territory.

The study of the economic space as a whole and the reasons for its asynchronous development were carried out by Fujita and Thisse [7], Fujita et al. [8]. The dependence of the level of development asynchronism on the specialization of the territory was studied by such economists as Cimoli et al. [4], Foray [6], Abdel-Rahman [1], Bodrunov et al. [3]. The relationship between structural changes and differentiation of the level of regional development was considered by Behrens and Robert-Nicoud [2], Dietrich [5]. Gonchar [10], Vertakova et al. [11] also analyzed the differentiation of territories in the Russian Federation.

State policy oriented towards an open economy does not allow for sustainable innovative development of numerous regions of the country, contributing to an increase in the quality of life of the population—reducing the actual level of unemployment to natural, raising incomes, improving education systems, healthcare, housing, etc [9]. In connection with this justifies the need to use program-targeted management of the regional development process, which is based on the development and phased implementation comprehensive target programs, the basis of which are organizational and managerial, scientific, technical and other comprehensive measures aimed at the effective implementation of the distributed functional and financial responsibilities of the program participants, meeting the deadlines for their implementation. Comprehensive targeted programs should be based on the external and internal resources of the regions, taking into account the national and demographic features of their development, aimed at realizing the formulated strategic goal of the socio-economic development of a particular region of the country. It is assumed that the sources of its financing can serve as state and municipal budgets, funds of commercial and public organizations, private deposits on a contractual basis and others.

Thus, one of the main directions of regional and state policy is to reduce the asynchrony of development and differentiation of territories, maintaining the stability and balance of the economic space of the state.

40.2 Methodology

We will form a system of indicators that includes 36 primary statistical indicators that will ensure not only all the most important components of the level of socio-economic development of the regions to assess the imbalances in the ratio of the level of socio-economic development of the regions, but also the state of the sectors of social infrastructure and industry.

According to the value of a comprehensive assessment in the form of an average rank, it is proposed to use the following classification of regions: the first group includes regions with a low level of development asynchronism; the second group includes regions with an average level of development asynchronism; the third group is regions with a high level of development asynchrony. Integral ratings of the level of asynchrony of territory development are calculated by the formula:

$$S_i = \frac{1}{n_i}\sum_{i=1}^{n}\frac{A_{ij}}{\max_{i=1}^{m} A_{ij}}, \quad (40.1)$$

where S-the level of asynchrony of the development of the territory, A_{ij} is the value of the j-ro indicator in the i-th region; m is the number of regions; n is the number of indicators; n_i—the number of indicators for a specific region.

For the convolution of primary statistical indicators to assess the level of asynchrony, the minimum (I_{min}) and maximum ratings (I_{max}). Based on this, a scale of assessments of the level of asynchrony is compiled: the values of and in the interval—$I_{min} - I$; average—$I_1 - I_2$; low—$I_2 - I_{max}$.

$$I_1 = \frac{3I_{min} + I_{max}}{4} \quad (40.2)$$

$$I_2 = \frac{I_{min} + 3I_{max}}{4} \quad (40.3)$$

Based on the obtained quantitative values of integral ratings, groups of regions were identified.

40.3 Results

The problems of development asynchronism are characteristic of large multi-regional economic systems with a high level of initial heterogeneity of the economic space. To reduce the level of asynchrony of regions, the main task of federal government bodies is to develop methods and tools to smooth the level of socio-economic development of regions.

The growth of the economic potential of the socio-economic system can be ensured by improving the organizational structures of management and increasing the stability of relations between the various elements of the economic potential of the regions. To solve a specific problem, it is necessary to create an adequate system of "flexible management", short-term and long-term planning and forecasting of scientific and technical projects, promising forms of organization of production and efficient use of economic resources, capable of timely restructuring under the influence of changes in external and internal factors, and quickly respond to market conditions. As such a "flexible system" it is advisable to use program-targeted planning and management, which involves the development and implementation of targeted integrated programs.

Most Russian regions are experiencing significant problems with the state of production and social infrastructure. On the one hand, it requires modernization, on the other hand, additional construction of new facilities is necessary taking into account the need to ensure economic growth. Countries that are actively introducing the mechanism of public-private partnership in creating industrial and social infrastructure have managed to create a legal framework, develop economic incentives and moral principles that allow the use of private capital in solving public problems. This allows these countries to channel part of the budget funds saved through the development of public-private partnerships into social problems and improving the living standards of the population.

The need to develop and successfully implement regulatory legal acts aimed at managing the differentiation of the economic space of regions in Russia is obvious to the current government. To date, in the process of developing a system of federal strategic planning documents that, to one degree or another, suggests what goals and objectives will determine in a strategic perspective the main vector of the regional policy of the Russian state.

Significant interregional imbalances and low rates of development of regional economies are becoming a real threat to the country's security. The importance and significance of this factor is confirmed by the provisions of the Economic Security Strategy of the Russian Federation, introduced on May 13, 2017 by Decree of the President of the Russian Federation No. 208 "On the Economic Security Strategy of the Russian Federation for the period until 2030".

In 2018, in his message to the Federal Assembly, the President outlined the following key tasks in the social sphere:

- achieve sustainable long-term growth in incomes and halve poverty in the next 6 years;
- implement measures to increase the demographic growth of the country;
- increase the size of the pension, perform regular indexing, in order to ultimately achieve a reduction in the gap between salaries and pensions.
- prepare a special program to support and improve the quality of life of the older generation.

As the main tool to overcome asynchrony in the development of the economic space, currently used, it is the redistribution of available resources, which is carried out either to support the "lagging" regions, equalize their position, or to stimulate and accelerate the development of those regions that are recognized "poles of growth".

The insufficient effectiveness of measures to overcome asynchrony is due to the emphasis on redistribution tools, to the detriment of the use of measures that actually stimulate the development. In the current economic model, entrepreneurship is a key factor in economic growth, which plays a decisive role in economic, in particular, spatial development. The regions of Russia are seriously distinguished from each other by their endowment with material, labor and natural resources. But this resource potential does not always correlate with the level of actual economic activity.

One of the priority measures for the development of regions is to reduce asynchrony of development, which is associated with the specialization of territories and overcoming depression. In this case, there are three main areas for the development of regional policy objectives.

(1) Support for recipient regions and reducing the level of development asynchrony;
(2) Changing the structure of regions and increasing their competitiveness;
(3) Maintaining the territorial integrity of the country and strengthening ties in creating a single economic space.

Thus, based on a generalization of the available domestic and foreign experience in implementing the regional policy of overcoming asynchrony in the development of regions, the following main stages can be distinguished:

1. stage—an integrated assessment of the level of asynchrony of the socio-economic development of the regions.
2. stage—grouping of regions depending on the achieved level of socio-economic development.
3. stage—substantiation of the directions for reducing the asynchrony of the socio-economic development of the regions, depending on the group.

At the first stage, based on private indicators characterizing the assessment of the level of development of the region (monitoring economic development, social and environmental development) and assessing the balance of its development, we calculated the integral indicator "Level of development asynchronism". Figure 40.1 shows the main private statistical indicators used in the calculation of the integral indicator.

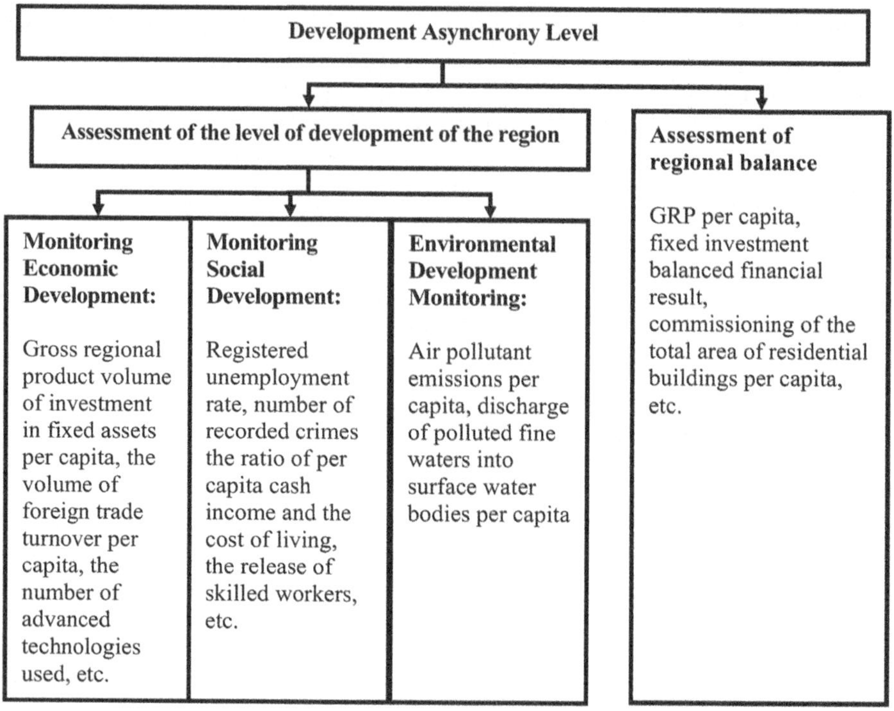

Fig. 40.1 A system of indicators characterizing the level of asynchrony of regional development

Based on the convolution of private statistical indicators (Fig. 40.1), we calculated integral quantitative ratings showing the level of asynchronism of regional development (Fig. 40.2).

At the second stage, a grouping of regions was carried out according to the level of development asynchronism. So, from the position of economic development, three subgroups are distinguished: 1. with a high level of asynchrony: Bryansk, Vladimir, Ivanovo, Kostroma, Oryol, Ryazan regions; subgroup: 2. with an average level of asynchrony: Voronezh, Kursk, Smolensk, Tambov, Tver, Tula, Yaroslavl regions; subgroup: 3. with a low level of asynchrony: Belgorod, Kaluga, Lipetsk, Moscow regions. From the point of view of social development, we distinguished the following three subgroups: 1. with a high level of asynchrony: Lipetsk region; subgroup: 2. with an average level of asynchrony: Belgorod, Kostroma, Tambov, Tula regions; subgroup: 3. with a low level of asynchrony: Belgorod, Voronezh, Ivanovo, Kaluga, Kursk, Moscow, Oryol, Ryazan, Smolensk, Tver, Yaroslavl regions.

To develop recommendations for managing structural and dynamic parameters, integrated regional ratings were calculated from the perspective of assessing the asynchrony of economic and social development, groups of regions were identified, their classification by levels of managerial impact was identified: (1) regions of a priority level of structural transformation management (recipient regions); (2) regions with

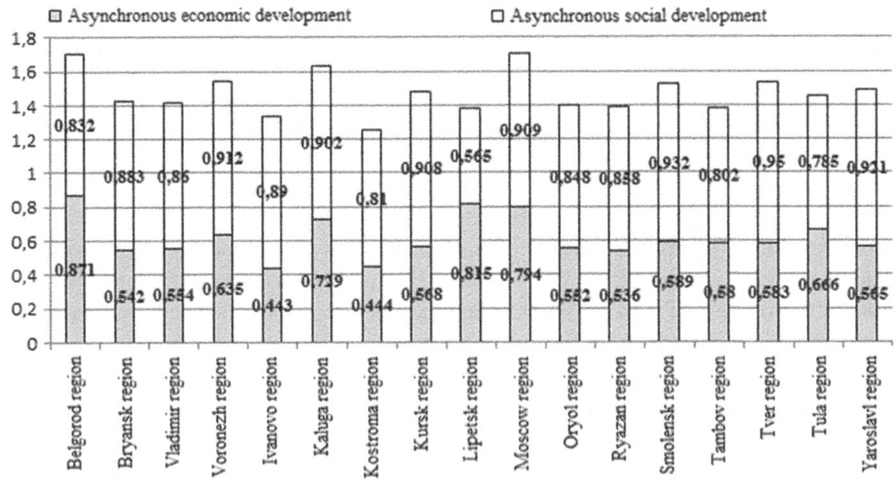

Fig. 40.2 Assessment of the level of asynchronism of regional development based on an integrated quantitative rating

stimulating management of structural transformation (evenly developing regions); (3) regions with supporting management (donor regions). The ratings of the social and economic development of the regions of the Russian Federation were also evaluated, which made it possible to identify asynchrony in development associated, in particular, with uneven structural transformations.

Based on a generalization of economic and social development, we have identified general groups of regions in terms of development asynchronism: 1. with a high level of asynchrony (recipient regions): Vladimir, Ivanovo, Kostroma, Ryazan regions; subgroup: 2. with an average level of asynchrony (evenly developing regions): Voronezh, Kursk, Lipetsk, Oryol, Smolensk, Tambov, Tver, Tula, Yaroslavl regions; subgroup: 3. with a low level of asynchrony (donor regions): Belgorod, Kaluga, Moscow region.

The third stage includes the substantiation of the directions of reducing the asynchrony of the socio-economic development of the regions, which should be combined with the main objectives of the regional development in close interconnection of economic and social development. Simplifying the situation and using a purely "economic" approach, it can be noted that the number and quality of life of the population of a particular territory have a direct impact on the level of asynchrony in the development of regions. Thus, the reduction in the asynchrony of the regional economic development of Russia is directly related to a more even population of the country and the support of the subjects of the federation with a low population density, in order to curb the outflow of the population.

The imbalance in regional investment in Russia is characterized by an increased deviation from the regions with the largest investment, there is a significant discrepancy in the degree of investment of the donor and "recipient" regions, which should be eliminated in order to unite the country's economic space.

Institutional factors of investment attractiveness of Russian regions are crucial for ensuring their economic growth and transforming them into growth poles by attracting investment. The influence of institutional factors is especially sensitive in the constantly changing external environment of any investment project, which is already reflected in the practice of evaluating them. The institutional factors of investment attractiveness of the region include: macroeconomic factors, regional development factors, regulatory factors, organizational, socio-environmental factors. The combined impact of the totality of institutional factors forms the institutional attractiveness of the region, linking together the subjects of investment activity: investors, intermediaries, recipients and authorities. It is also necessary to create organizational structures that are directly involved in investor problems. The presence of an investor in the territory, as a rule, involves the implementation of often irrevocable investments, therefore, the government should provide direct support.

The system of practical recommendations for reducing asynchrony in managing regional development, which includes recommendations for developing programmatic, strategic, regulatory and methodological documents regulating the strategic development of Russian regions taking into account the structural balance of their sectoral, territorial, reproductive and dynamic characteristics.

40.4 Conclusion

In the formation of an effectively developing economic space, the main attention should be paid to the transformation of the practice of managing regional development and its orientation towards priority areas of development in the medium and long term. Thus, the assessment of the integral quantitative ratings of the regions from the position of asynchrony of economic and social development made it possible to distinguish groups according to the level of managerial impact: with a low level of asynchrony—with a priority level of management of structural transformations; with an average level of asynchrony—with a stimulating control of structural transformations and a high level of asynchrony—with a supporting type of control.

The development of an optimal structure of the regional economy, adapted to modern conditions of reindustrialization and transformation of territorial economic systems, will allow solving a number of interrelated problems: imbalance and asynchronism of regional development, which ultimately give rise to a low level of productivity of the regional economic system and subsidization of this region; the absence of a stable economic system that does not allow the region to self-support and develop, which is reflected in a decrease in the quality of life in this territory and an increase in the poverty threshold, migration outflow of the population to more economically prosperous regions; decline in fertility in the regions and the quality of human capital.

References

1. Abdel-Rahman H (1988) Product differentiation, monopolistic competition and city size. J Urban Econ 18:69–86
2. Behrens K, Robert-Nicoud F (2014) Survival of the fittest in cities: urbanisation and inequality. Econ J 124(581):1371–1400
3. Bodrunov S, Plotnikov V, Vertakova Y (2017) Technological development as a factor of ensuring the national security. In: Vision 2020: sustainable economic development, innovation management, and global growth, vol I–IX, pp 2666–2674
4. Cimoli M, Dosi G, Stiglitz JE (2015) The rationale for industrial and innovation policy. Intereconomics 3:126–132
5. Dietrich A (2012) Does growth structural change, or is it the other way around? A dynamic panel data analysis for seven OECD countries. Empir Econ 43:915–944
6. Foray D (2015) Smart specialization: opportunities and challenges for regional innovation policy. Routledge, London
7. Fujita M, Thisse J-F (2002) Economics of agglomeration. Cities, industrial location, and regional growth. Cambridge University Press, Cambridge, MA
8. Fujita M, Krugman P, Venables AJ (1999) The spatial economy. Cities, regions and international trade. MIT Press, Cambridge, MA
9. Glaeser EL, Kallal HD, Scheinkman JA, Schleifer A (1992) Growth in cities. J Polit Econ 100:1126–1152
10. Gonchar K (2010) Urban agglomeration effects and company productivity in Russia: empirical evidence based on manufacturing industry survey. In: 14th annual conference of the international society for new institutional economics. http://extranet.sioe.org/uploads/isnie2010/gonchar.doc
11. Vertakova Y, Polozhentseva Y, Klevtsova M (2017) Transformation of structural dynamic parameters of foreign trade relations when changing geopolitical situation, 2017. In: Proceedings of the 29th international business information management association conference—education excellence and innovation management through vision 2020: from regional development sustainability to global economic growth, pp 1901–1909

Chapter 41
A Recipe for Pleasure in the Digital Economy: Managing Risk, Managing the Future

Evgeniya K. Karpunina, Ekaterina V. Kharchenko, Alexander I. Khorev, Svetlana V. Plyasova and Anna P. Anufrieva

Abstract Improving the quality of human life, the comfort of his existence and maximizing pleasure should become unconditional targets of digital development. The purpose of the research is to substantiate the fact that in the digital economy changes the content and nature of the created consumer value and its utility. On the one hand, this creates new enhanced opportunities for a person to enjoy, and, on the other hand, can significantly limit his future well-being due to system vulnerabilities and security threats. The authors revealed the features of the digital economy that determine the change in the content and nature of the created consumer value and its utility. It is proved that in the conditions of the digital economy the process of value creation shifts from a commodity-and firm-centric position to a personalized consumer experience. Models of creation of consumer value and satisfaction of requirements are analyzed. The features of their application in the digital environment are highlighted. A comparison of opportunities and threats in the system of relations "customer-manufacturer", it is proved that the maximization of pleasure and maximum satisfaction of customer is achieved by implementing a model for joint value creation and purposeful management of security risks.

E. K. Karpunina (✉)
Tambov State University named after G.R. Derzhavin, Tambov, Russia
e-mail: egenkak@mail.ru

E. V. Kharchenko
Southwest State University, Kursk, Russia
e-mail: samofalovae@mail.ru

A. I. Khorev
Voronezh State University of Engineering Technology, Voronezh, Russia
e-mail: al.khorev@gmail.com

S. V. Plyasova
Moscow University for Industry and Finance "Synergy", Moscow, Russia
e-mail: splyasova@synergy.ru

A. P. Anufrieva
Kuban State Technological University, Krasnodar, Russia
e-mail: gubskaya-a@mail.ru

T. Kolmykova and E. V. Kharchenko (eds.), *Digital Future Economic Growth, Social Adaptation, and Technological Perspectives*, Lecture Notes in Networks and Systems 111, https://doi.org/10.1007/978-3-030-39797-5_41

Keywords Pleasure · Digital economy · Risks · Economic security · Consumer behavior · Value creation

JEL Classification Codes D11 · D22 · O33

41.1 Introduction

The use of information and communication technologies and digital platforms in the modern world creates conditions for simplifying the processes of production of goods and services and their delivery to the end user. In many countries there is a steady growth of level and quality of life, enhanced comfort of existence, offers many new opportunities for the realization of human potential, meet their needs through breakthroughs in digital development.

At the same time, in real life there is a dying off of traditional economic models, the emergence of new approaches to creating consumer value, the transformation of habitual human behavior in the process of meeting their own needs, taking into account the new threats and dangers of digitalization.

New formulas and models adapted to the conditions of the new digital environment will describe the well-being of a person in the digital world, getting pleasure from the consumption of goods and services. Therefore, today, special attention should be paid to the study of changing consumer behavior and purposeful management of security risks for the benefits of digitalization in the future.

41.2 Methodology

The scientific hypothesis of the research is the assumption that in the digital economy, the content and nature of the created consumer value and its utility changes, and the institutional nature of the modern consumer is transformed. On the one hand, this creates new enhanced opportunities for a person to enjoy, and, on the other hand, can significantly limit his future well-being due to system vulnerabilities and security threats.

Economic science has many examples of how economic systems cope with new challenges and go through processes of transformation [1, 2]. The formation of the digital economy in this context is a natural stage in the development of society.

The analysis of the research results allows the authors to explain the specifics of digital transformations how:

- formation of knowledge possessed by management and staff, the main driver of progress [3];
- changing of all kinds of relations to time and space under the influence of digital communication and due to quantum leap up in volume [4];

- active development and implementation of information and communication technologies (ICT), covering a variety of software, hardware and related processes that change the "personality of the individual, society", politics and Economics are almost completely dependent on them [5];
- changing the way people interact with each other and with their natural environment [6];
- transformation of forms and methods of interrelations between economic entities in the corporate sphere [7];
- the emergence of new forms of business, the transfer of business from traditional firms to the media, the exclusion of intermediaries and direct interaction between consumer and supplier [3];
- strengthening the role of digital technologies and big data in achieving new opportunities and ways to shape the future [8].

According to the authors, the digital economy should be identified with the "creation of new value". In this paper, it is proposed to understand the digital economy as a set of all those market segments where added value is created with the help of digital (information) technologies. Automation of production processes, robotics, intensive development of human capital, the introduction of sales systems, the use of communication technologies to control consumer behavior, the development and use of advanced materials, the development of digital e-Commerce technologies, the expansion of virtual banking capital markets, the development of artificial intelligence and the Internet of things contribute to the creation of added value.

The authors explain the essence of digital transformations using the laws of dialectics (the law of unity and struggle of opposites, the law of mutual transition of quantitative and qualitative changes, the law of negation) [9].

For example, we can observe the manifestation of the law of unity and struggle of opposites at all levels of competition between companies-producers of digital technologies. Opposite tendencies easily get on in the logic of development of digital technologies (strengthening of a role and importance of the person in the digital environment entails emergence of dangers for its existence, including information character). The digital economy is based on the realization of human potential, at the same time machines and technologies displace people from production processes, etc.

The increase in the growth of computing power of the computer leads to its qualitatively new capabilities, the growth of mobile network bandwidth stimulates the change of generations of phones with qualitatively new capabilities. This illustrates the operation of the law "the transition of quantity into quality" [10]. The law "negation of negation" leads deep into the digital technologies themselves, when some materials in the process of evolution are replaced by others (the first negation), and the further course of development of digital technologies requires the return of previously used materials (negation of negation).

In this work, by applying the methods of economic analysis and synthesis, induction and deduction, the authors prove that under the influence of these changes, the institutional nature of the modern consumer and the content of the created consumer value and its utility are transformed. It develops well known theories of consumer behavior and value creation [11–14], and allows us to explain with the help of their methodology why the possibilities of pleasure (as the ultimate goal of human consumption) are changing in the new digital environment.

However, the digital economy could limit human's striving for maximization of pleasure by means of the system's vulnerability [15, 16] and the emergence of information, economic and social security threats [17].

The information base of the research is statistical and analytical materials of international databases and consulting agencies, including the European commission, World Economic Forum, Experian Information Solutions, Check Point Software Technologies, Statista, materials of scientific conferences, periodicals, scientific monographs.

41.3 Results

Statistics, materials of international reports and ratings indicate the expansion of the digital economy at the present time. We will make their systematization in the context of disclosure of the declared problems.

Firstly, the intensive development of information and communication technologies, which is manifested in the expansion of individual access to various communication channels (broadband, mobile, Internet [18]), using them in everyday life and forming ecosystems in business (ecosystem of platforms, ecosystem of innovations, ecosystem of interests, ecosystem of Commerce, ecosystem of things) [16].

According to Statista, at the beginning of 2019, 4.1 billion Internet users are registered in the world, which exceeds 53% of the World's population (this figure is 3.42 billion more than the number of Network users registered at the end of 2016). There are 333.8 million domain names registered worldwide [19]. For example, the digital economy and society index (DESI) for 2018 shows that fixed broadband is available to 98% of Europeans, and 80% of European homes are covered by high-speed broadband (at least 30 Mbit/s). 81% of Europeans go online regularly (at least once a week), which is 2 percentage points more than in 2017 [20].

Secondly, the growth in the value added of the ICT sector reflects the development of innovative digital technologies and related processes that contribute to the efficiency of the business sector on a global scale.

For the EU group alone, the value added of the ICT sector amounted to 632 billion euros in 2017, thanks to the introduction of digital technologies by European enterprises, such as the use of business software for electronic information exchange (from 26% in 2013 to 34% of enterprises in 2017), the sending of electronic invoices (from 10% in 2013 to 18% of enterprises in 2017) or the use of social networks to interact with customers and partners (from 15% in 2013 to 21% of enterprises in 2017). Finland, Estonia and Denmark are leaders in the use of digital public services among European countries [20]. At the global level, the share of value added of the

ICT sector in the GDP of European countries lags behind the main competitors - Japan, the United States, China.

Thirdly, the shift of traditional business models to media and the rapid development of e-Commerce.

According to the latest research by Digital Market Outlook company Statista, e-Commerce spending in 2018 increased by 14% compared to the same period in 2017. The company estimates that $ 1.78 trillion was spent on consumer goods alone for the whole of 2018. The audience of E-commerce has grown significantly: more than 2.8 billion people around the world make purchases on the Internet. At the same time, the average income per user (ARPU) increased by more than 10% compared to 2017 and amounted to 634 US dollars [19].

Fourthly, the priority of users' digital skills in the context of personal needs and professional business development.

Today, digital skills are crucial both for access to the labour market and for taking advantage of the ongoing digital transformation. The availability of digital skills multiplies the cost of human capital as a factor of production. The proportion of people with basic digital skills ranges from 29% in Bulgaria and Romania (despite notable progress in both these countries in 2017) to 85% in Luxembourg and 79% in the Netherlands. Between 2007 and 2017, the number of ICT professionals employed in the EU increased by 36%, compared to a 3.2% increase in total employment. Almost 83% of all ICT professionals employed in the EU in 2017 were men and about 62% had a higher education. The largest share of ICT professionals in total employment was recorded in Finland (6.8%), Sweden (6.6%) and Estonia (5.6%); the lowest in Greece (1.6%), Romania (2.1%) and Portugal (2.2%) [21].

These trends and the impressive scale of development of the digital economy draw research interest to the question of creating "consumer value": does it remain unchanged with the advent of digital technologies or is it characterized by any dynamics? Does the use of digital technologies increase customer satisfaction? What are his chances of getting an increase in pleasure from consumption?

From the standpoint of economic science, the concept of "consumer value" of goods, as a rule, is identified with the concepts of "value" and "utility" [22].

Thus, from the point of view of the labor theory of value, labor costs for the production of goods formed the basis of consumer value [23]. The views of the theory of marginal utility [24, 25] suggest that the value of goods is determined by the subjective (consumer) assessment of their usefulness and rarity. In accordance with the law of diminishing utility, marginal utility determines the value of a commodity and its market price. A. Marshall theoretically proved that the value of goods is equally determined by the utility and cost of production [23].

Shet J., Newman B. and Gross B. point to the relationship of consumer behavior with the value orientations of the individual. This allows us to identify five types of values that determine consumer choice, the ability to meet the needs of the individual and get them pleasure [26]: functional value (perceived utility of a good having functional or physical properties); social value (perceived utility of a good due to its Association with a social group); emotional value (perceived utility of a good due to its ability to excite feelings or affective reactions); epistemic value (perceived utility of a good due to its ability to excite curiosity, create novelty and/or satisfy the desire for knowledge); conditional value (the perceived utility of a good due to the specific situation in which the consumer acts).

Smith J. and Colgate M. place special emphasis on the possibility of obtaining satisfaction and pleasure in the process of consumer choice and consumption itself. They distinguish the following types of consumer value: functional value (the degree to which the product has the desired characteristics, useful or performs a certain function); hedonic value (the degree to which the product creates appropriate impressions, causes certain feelings and emotions in the consumer); symbolic value (the degree of psychological connection of the consumer with the product); cost and costs (maximizing benefits to the consumer by minimizing costs incurred in the process of buying or using a product) [27].

Thus, the consumer value of a product will be higher if it has the desired characteristics, carries utility for the consumer, creates appropriate impressions, causes certain feelings and emotions, establishes a psychological connection between the consumer and the product, and maximizes the benefit for the consumer through minimizing costs.

In the digital environment, there are more opportunities to ensure the growth of consumer value of goods.

Both the company and the consumer can now create value together [12–14].

This is more effective from the standpoint of the concept of relationship marketing by achieving the following advantages: the creation of a virtually new good for the individual consumer and the distribution of benefits from this activity occurs between all participants in the interaction; individuals can act not only as consumers, but also in determining the desired good (the good is created together with buyers, not for them); to satisfy the consumer's desire, the company forms and coordinates all business processes; is a continuous collaboration of the consumer and the producer in real time; there are prospects of establishing mutually beneficial long-term relationships through cooperation of the company with regular customers; strengthening relationships within the company, with contractors and end users [28].

Thus, in the digital economy, value and the process of creating it are shifting from a commodity- and firm-centric position to a personalized consumer experience. In such a space, informed, active, community-based (usually on the Internet) consumers with greater market power are increasingly involved in creating consumer value together with the company [12]. This model of consumer value creation is based on the concept of service-dominant logic proposed by Vargo and Lasch [13].

In accordance with this concept, there has been a transformation of the nature of marketing from commodity-centric (focusing on the production of tangible goods), to service-centric, where the focus is on intangible assets, exchange processes, as well as the whole range of relationships between producer and consumer.

In the implementation of the model of joint value creation, the consumer value of the product is manifested when it is used. The consumer is always a co-Creator of value. The company does not create value, but a commodity as a carrier of potential value, which is then extracted from it by the consumer. The provision of services is always focused on the end user and takes place with his direct participation. Value is always unique to a particular consumer and is determined by the consumer experience and the situation in which the process of consumption is carried out [22].

We also agree with E. Gumesson's statement that in a digital environment, "buyers do not purchase goods or services, they acquire offers as a set of services that create value" [29].

Thus, the use of information technology to study consumer preferences, establish feedback between the manufacturer and the consumer, which allow you to instantly track changes in the tastes and desires of the consumer and adjust the characteristics of the product to changing requirements is an example of joint value creation. For example, Sony is in dialogue with consumers who helped develop the PlayStation. Intel, Microsoft, Nokia are attracting consumers to the development of new network support devices and software for networks and cell phones. Consumers contribute to the debate both from the technical side and in terms of their expectations and views on value.

Thus, in our opinion, the joint creation of the consumer value of the goods leads to a multiple increase in the probability of satisfying consumer needs and getting pleasure from consumption (as a positive emotion from pleasant sensations as a result of satisfying any needs).

Moreover, in the conditions of digitalization, the use of the value creation model directly by the consumer (when the role of the company is reduced only to the resource provision of this process and the mobilization of the consumer to create value for himself) becomes quite real [30, 31]. That is, the value is created not only by the company, as it is established in the literature on management [30], and the actions taken by the consumer affect the value created, for example, the value of the goods.

Following the meaning of the concept of value in use, it can be argued that value creation is the creation by the consumer of value in use during use, where value is socially constructed through experience [32–35]. Consumers' individual and social experiences with resources, processes (and/or their outcomes) become the core of value creation. Value in use is not only accumulated from past and current experiences, but can also be represented in future experiences [36, 37].

For example, a person cannot derive value from using a car without access to gasoline and without a combination of these resources, value in use cannot arise from driving a car. Thus, through interaction with consumers, the company participates in the value creation process and becomes its co-Creator. At the same time, consumers purchase goods and services, they no longer pursue the goal of subsequent interaction with the company to create value [38].

The consumer may also purchase goods and services as ready value created by the company ("value in exchange", that is, goods or services created by the company and purchased by the consumer) [31]. In addition, value in exchange does not necessarily exist at one given point in time, but instead can exist as multiple singular entities. For example, a family vacation may involve multiple choices (purchases) at different points during the vacation, where potential value exists in each purchase situation, as multiple, separate objects embedded in the products or services in question.

Thus, the classical approach to the creation of a "value chain" (according to the concept of Porter M., each firm carries out certain activities aimed at the development, production, marketing, delivery and support of their goods), which involves the creation of consumer value of goods/services by the company for the consumer without his participation [39, 40] in the digital economy is gradually.

However, in the digital environment may be a dualism: on the background of a significant expansion of capacity to meet the needs of the person and the fun can start the process of reducing the time period of satisfaction and quality due to the acceleration of the processes of emergence themselves needs (availability of information, speed of response and organizational flexibility producers) and a strong range of instruments of promotion of the product to the final consumer.

In addition, new models of value creation, developing in a certain scenario, may encounter risks. Risk is the reverse side of normal development and only by referring to the reverse side of the normal form can we recognize it as a form.

Let us analyze the correlation of opportunities and risks in the process of creating consumer value and meeting the needs that arise in the digital economy (Table 41.1).

Thus, the classical model of creating a "value chain" in the conditions of digitalization provides a minimum number of opportunities for the consumer, maximizing his satisfaction and pleasure, against the background of unconditional advantages for the manufacturer. When creating value together, conditions are created to maximize utility and pleasure for the consumer, and the nature of risks and threats is, first of all, information-qualification nature, since the dependence of the results of value creation on information and communication systems and digital competencies of employees of companies is strong. In terms of applying the model of value creation by the consumer the burden of responsibility is shifted on the shoulders of the user, thereby endangering the actual existence of the producers, and the nature of threats to the consumer is directly connected with its own capacity, psychological stability, and security systems and understand the reality and virtuality of events.

41.4 Conclusions/Recommendations

As a result of the study, the following conclusions were obtained.

Firstly, the features of the digital economy that determine the change in the content and nature of the created consumer value and its utility are disclosed.

Secondly, the models of creating consumer value and meeting needs are analyzed, and the features of their application in the digital environment are highlighted.

Thirdly, on the basis of comparison of opportunities and threats for the producer and the consumer it is proved that the maximization of pleasure and maximum satisfaction of customer is achieved by implementing a model for joint value creation. The empowerment to meet human needs in the digital environment is possible due to the implementation of the model of value creation by the consumer, however, due to the vulnerabilities and threats to security, its use may limit future human well-being.

Table 41.1 Opportunities and risks in the process of creating consumer value and meeting the needs of the digital economy (created by the authors)

Model of value creation	Opportunities	Risks and consequences
The classical model of creating a "value chain"	• Maximization of the profit of the manufacturer • Customer satisfaction at the basic level • Achieving balance in the national economy	• Low level of correlation between producer and consumer, leading to dissatisfaction with consumer queries [22] • Lack of feedback mechanisms between the consumer and the manufacturer, leading to the risk of loss of income of the company • Security risks of management system (low professional level of managers; violations of labor discipline; outflow of qualified personnel; low level of accounting and analytical work and lack of control system) [41] • Financial risks (choice of unreliable partners and investors; unavailability of credit resources; low liquidity, solvency, financial stability; limited sales market and low level of marketing) • Information technology risks (failure of computer equipment; lack of information technology in accounting and analytical work; low technical and technological level of production, obsolescence of fixed assets) • Organizational and psychological risks (low level of corporate culture development, high conflictogenicity in the company, abuse of official powers, significant differentiation in staff remuneration; low innovative activity) • Legal and corruption risks • Personnel risks associated with low professionalism of specialists • Growing socio-economic instability at the macro level

(continued)

Table 41.1 (continued)

Model of value creation	Opportunities	Risks and consequences
Collaborative value creation model	• Distribution of responsibility and benefits between producer and consumer • Full satisfaction of consumer needs • Maximization of pleasure • Development of mutually beneficial long-term relationships • Enhanced coordination of business processes • Strengthening the internal integrity of the manufacturer • The formation of a personalized consumer experience • Continuous monitoring by the manufacturer of changes in consumer preferences • Meeting consumer expectations	• Unmet need of the consumer • Loss of consumer potential • Non-compliance with the quality requirements of the produced product or service • The decline in the quality of life of consumers • Failure to obtain satisfaction from the consumption of dependence of the result on information and communication channels • Risk of "electronic disintegration" • High risk of information leakage • The risk of cyber-espionage and Internet fraud • Low level of information literacy of specialists and consumers, leading to limitations of the value creation process • Reputational risk • Consumer Internet addiction • Violations of the trade secret regime • Failure of information systems of the organization of work of the manufacturer • Risk of loss of income of the manufacturer • Decrease in efficiency, risk of loss of competitive positions of the manufacturer • Bankruptcy of the manufacturer • Imbalance of supply and demand at the macro level

(continued)

Table 41.1 (continued)

Model of value creation	Opportunities	Risks and consequences
Model of value creation by the consumer	• 100% focus on consumer individual and social experience • Freedom of choice of the consumer for interaction with the producer • Shifting responsibility for value creation to the consumer • Reducing the area of responsibility of the manufacturer • Independent choice of the way of value creation by the consumer • Quick access to consumer experience at a fair price • Maximizing the utility of created goods services • Reduce transaction costs • Maximizing utility from idempotence of digital products	• The loss of consumer potential • Speeding up the process of new needs and reducing the time period of pleasure • Substitution of real pleasure with virtual one • The risk of unemployment for staff of companies as a result of declining consumer demand • Risk of loss of income of producers • Loss of the adjusted physical interaction of producers and consumers, prevalence of virtual interrelations • Psychological problems of consumers due to lack of communication • Information threats of information leakage, failures of digital communication systems and the development of forms of information fraud

References

1. Ivleva G (2003) Transformation of the economic system: review of concepts and contours of the general theory, No 10, pp 18–24
2. Klaus V (1995) Economic theory and reality of transformation processes. Probl Theory Pract Manag 6:65–69
3. Tapscott D (1995) The digital economy: promise and peril in the age of networked intelligence. McGraw-Hill, New York
4. Negroponte N (1995) Being digital. Knopf, New York
5. Schmidt E, Cohen J (2013) The new digital age: reshaping the future of people, nations, and business. Alfred A. Knopf, New York
6. Boyd D, Crawford K (2012) Critical questions for big data: provocations for a cultural, technological, and scholarly phenomenon. Inf Commun Soc 15(5):662–679
7. Zuboff S (2015) Big other: surveillance capitalism and the prospects of an information civilization. Inf Technol 30(1):75–89
8. Shah D, Cappella J, Neuman W (2015) Big data, digital media, and computational social science possibilities and perils. Am Polit Sci Rev 659(1):6–13
9. Hegel G (1970) The science of logic. In 3 volumes. Mysl', Moscow
10. Prokhorov A, Konik L (2019) Digital transformation. Analysis, trends, world experience. Al'yansprint, Moscow
11. Wiener J (1993) The concept of utility in value theory and its critics. In: Galperin V (ed) Theory of consumer behavior and demand. New Economic School, SPb, pp 78–116
12. Prahalad K, Ramaswamy V (2010) The future of competition. Creating unique value together with consumers. Publishing House "Williams", Moscow
13. Vargo S, Lash R (2006) Development of a new dominant logic of marketing. Russ J Manag 4(2):73–106
14. Payne A, Storbacka K, Frow P (2008) Managing the co-creation of value. J Acad Mark Sci 36:83–96
15. Helbing D (2012) Agent-based modeling. In: Helbing D (ed) Social self-organization. Understanding complex systems. Springer, Berlin
16. Sologubova G (2017) The phenomena of the digital economy. https://www.econ.msu.ru/sys/raw.php?o=44108&p=attachmenthttps://www.econ.msu.ru/sys/raw.php?o=44108&p=attachment. Accessed 15 Aug 2019
17. Karpunina E, Gorbunova O, Moiseev S, Cheremisina T (2019) Resistance is not hopeless…regarding the policy of countering information threats of economic security. In: Vision 2020: education excellence and management of innovations through sustainable economic competitive advantage. Proceedings of the 33nd International Business Information Management Association (IBIMA) conference in Granada, Spain, 2019, pp 2679–2687
18. ICT Data, Statistics Division Telecommunication Development, Bureau International Telecommunication and Unional (2019) The ICT development index 2019. methodology, indicators and definitions. https://www.itu.int/en/ITU/Statistics/Documents/statistics/ITU_ICT%20Development%20Index.pdf. Accessed 13 Sept 2019
19. Statista (2019) Digital market outlook. https://www.statista.com/outlook/digital-markets. Accessed 2 Sept 2019
20. European Commission (2018) Digital economy and society index for 2018. https://ec.europa.eu/digital-single-market/en/news/digital-economy-and-society-index-2018-report. Accessed 23 Oct 2019
21. ILO (2019) Global wage report 2018/19: what lies behind gender pay gaps. https://www.ilo.org/global/publications/books/WCMS_650553/lang–en/index.htm. Accessed 15 Sept 2019
22. Neganov S (2018) Processes of value creation in modern marketing theory. J Econ Theory 15(1):14–22
23. Romanov A, Nikolaeva I, Klochkov V (2015) Economic theory. Orion, Moscow
24. Boehm-Bawerk O (2009) Selected works on value, interest and capital. Eksmo, Moscow

25. Schumpeter J (1982) Theory of economic development: the study of profit, capital, credit, interest and the cycle of conjuncture. Williams House, Moscow
26. Sheth J, Newman B, Gross B (2011) Consumption values and market choices: theory and applications. South Western Publishing Company, Cincinati
27. Smith J, Colgate M (2007) Customer value creation: a practical framework. J Mark Theory Pract 1:7–23
28. Gordon J (2014) Partnership marketing. Peter, SPb
29. Gummesson E, Lusch RF, Vargo S (2010) Transitioning from service management to service-dominant logic: observations and recommendations. Int J Qual Serv Sci 2(1):8–22
30. Norman R, Ramirez R (1993) From value chain to value constellation: designing interactive strategy. Harvard Business Review, July–August, pp 65–77
31. Grönroos C, Voima P (2013) Critical service logic: making sense of value creation and co-creation. J Acad Mark Sci 41:133–150
32. Berger P, Luckmann T (1966) The social construction of reality: a treatise in the sociology of knowledge. Doubleday, New York
33. Edvardsson B, Tronvoll B, Gruber T (2011) Expanding understanding of service exchange and value co-creation: a social construction approach. J Acad Mark Sci 39(2):327–339
34. Ramaswamy V (2011) It's about human experiences…and beyond, to co-creation. Ind Mark Manag 40(2):195–196
35. Thompson C, Locander W, Pollio H (1989) Putting consumer experience back into consumer research: the philosophy and method of existential-phenomenology. J Consum Res 16(2):133–147
36. Helkkula A, Kelleher C, Pihlström M (2012) Characterizing value as an experience: implications for service researchers and managers. J Serv Res 15(1):59–75
37. Voima P, Heinonen K, Strandvik T (2010) Exploring customer value formation—a customer dominant logic perspective. Working paper, No 552, Publications of Hanken School of Economics, Helsinki, Finland
38. Grönroos C (2008) Services logic revisited: who creates value? And who co-creates? Eur Bus Rev 4:298–314
39. Porter M (2010) Competition. Publishing House "Williams", Moscow
40. Kotler F (2012) Markmeting management. Peter, SPb
41. Karpunina E, Konovalova M, Shurchkova J, Isaeva E, Abalakin A (2020) Economic security of businesses as the determinant of digital transformation strategy. In: Popkova E, Sergi B (eds) Digital economy: complexity and variety vs. rationality. Springer, Cham, pp 251–261

Chapter 42
Russia and International Humanitarian Law: History, Modernity and Prospects

Nikolay Yu. Nikolaev, Dmitry V. Garbuzov, Sergei P. Ramazanov and Daria S. Yurchenko

Abstract *Purpose*: The aim of the work is to identify Russian participation in the process of formation of international humanitarian law. *Design/methodology/approach*: While writing this article, we used the methodological principles of historicism, holism, value relationships and the historical-genetic, retrospective and historical-comparative methods based on them. *Findings*: Four stages have been identified in assessing the role and place of the Russian Empire/USSR/Russian Federation in the development of international humanitarian law and the implementation of its principles. They are contingent on a change in the international situation and on the weakening or increase in the geopolitical power of the Russian (Soviet Union; USSR) state. The first stage coincided with the beginning of the international codification of the norms of customary law of war from the middle of the 19th century and lasted until the Brest Peace of 1918. At this stage, Russian politicians and jurists played a leading role in the development of new rules for warfare. The second period is characterized by insignificant, but gradually increasing participation of the Soviet state in the formation of humanitarian law. After 1945, the third stage begins, during which the Soviet Union took an active part in the development and signing of a number of humanitarian conventions. The beginning of the fourth stage is connected with the collapse of the Soviet Union and the formation of the Russian Federation. *Originality/value*: There are four stages in Russia's participation in the development of international humanitarian law. The

N. Yu. Nikolaev (✉)
Volzhsky Polytechnical Institute (Branch), Volgograd State Technical University, Volgograd, Russia
e-mail: nikcam@mail.ru

D. V. Garbuzov · S. P. Ramazanov
Volgograd State University, Volzhsky Branch, Volzhskiy, Russia
e-mail: buendia@bk.ru

S. P. Ramazanov
e-mail: sp-ram@yandex.ru

D. S. Yurchenko
Mass-Media Institute, Russian State University for the Humanities, Moscow, Russia
e-mail: dashenka-yurchenko@mail.ru

T. Kolmykova and E. V. Kharchenko (eds.), *Digital Future Economic Growth, Social Adaptation, and Technological Perspectives*, Lecture Notes in Networks and Systems 111, https://doi.org/10.1007/978-3-030-39797-5_42

analysis of globalization is used as an important key to decipher the prospects of the events taking place.

Keywords Russia · War · Peace · Codification · Humanitarian law · International relations · Globalization · Geopolitics

JEL Code F51 · F53 · K33

42.1 Introduction

The origins of modern international humanitarian law are associated with the general evolution of international relations and the development of peacekeeping ideas in modern times [6]. Since the second half of the nineteenth century, the arms race has been increasing everywhere in the world. The bloody wars, colonial rivalry, and the struggle for economic and political influence forced not only great powers, but also small states to arm themselves. At the same time, a Western movement began to expand in Western Europe and North America, promoting the ideas of a peaceful resolution of military conflicts and general disarmament. In addition, the most important task was the struggle for the humanization of the war. Proposals for the codification of humanitarian law were seriously supported in Russian society as well, and above all among legal scholars [10]. By the middle of the XIX century in many countries, there were already rules governing the behavior of soldiers on the battlefield and in conquered territory. For example, in the Russian army some provisions of the "law of war" were enshrined in the Military Statute of Peter I (1716). However, by the middle of the XIX century, these rules of warfare were frankly outdated, requiring a radical review. The absence of clear rules of warfare led to unjustified cruelty, the use of cruel means and methods of warfare, the suffering of civilians, and the destruction of historical and cultural monuments.

42.2 Materials and Method

The article is based on the principle of historicism, which allows us to trace the development of humanitarian law in the specific historical conditions of world processes. The principle of holism (integrity) allows us to identify the stages of Russia's participation in the development of international law, taking into account its strength and influence in the world. The axiological principle allows us to highlight the real basis of the declared legal norms by various countries. Historical-genetic and retrospective methods make it possible to implement the principle of historicism in the study, in other words, consider the codification of humanitarian law and Russian participation in it in the process of formation and development. In addition, this group of methods helps to periodize the investigated problem. Using the historical-comparative method, the authors were able to identify general patterns, as well as differences in the stages of Russian participation in the formation of international humanitarian law.

42.3 Results

The Russian Empire participated in the codification of international law. Its representatives actively worked at the Geneva Conference of 1863, the decisions of which guaranteed the neutrality of medical personnel, as well as the protection of the wounded and sick during military actions. In May 1867, this international act was ratified. At the same time, Alexander II approved the charter of the Society for the Care of Wounded and Sick Warriors, which was renamed into the Russian Red Cross in 1879 [1]. The holding in Brussels of the conference in 1874 on the development of a code of laws and customs of war was an important event in the history of the humanization of hostilities. The idea of holding such a meeting was put forward by F. F. Martens [11]. The main objective of the Brussels Conference was to regulate the rights of warring parties in relation to each other and private individuals. A prominent Russian diplomat, Baron Alexander Jomini, was elected chairman of the conference. However, the draft convention of the laws of the land war proposed by Martens was not finally signed, largely due to the resistance of a number of small states and Great Britain [7].

An important event in the development of international humanitarian law was the holding of the Hague peace conference in may 1899. The initiative to convene it again came from the Russian government. The Russian government was concerned about lagging behind other great powers in the arms race. The Russian delegation was headed by Ambassador to the UK E. E. Staal, who was elected Chairman of the peace forum. Among the Russian representatives, F. F. Martens stood out. He took an active part in the development of the conference program. Public attention was focused on the discussion of the codification of international humanitarian law. The solution of these problems was engaged in the second Commission, the Chairman of which was F. F. Martens [8]. The second Commission considered the rights and duties of the occupation army. To resolve the contradictions, Martens made a number of fundamental additions to the text and proposed the adoption of the amended text in the form of a preamble to the Convention. Thus a very important provision, known in the history of law as "The Martens Clause", appeared in the final document. According to this rule, in cases not provided for by the adopted regulations, "the population and the belligerents remain under the protection and action of the principles of international law". However, the first "peace Parliament" failed to reverse the trend towards increasing international tension. The Spanish-American war that began shortly before the conference and the conflict between the British and the Boers that followed immediately after the end of the "first Parliament of the world" demonstrated numerous examples of violations of the agreements reached in 1899.

At the beginning of the 20th century, the Russian Empire continued to be one of the most prominent participants in the process of developing new rules for warfare. This was confirmed with the idea of convening a second peace conference. In September 1905, the Russian Foreign Ministry sent a circular to foreign governments with the idea of holding a new "peace parliament". Abroad, this proposal was met with approval and rated as timely. The first authorized representative of the Russian delegation, Ambassador of Russia to Paris A. I. Nelidov was elected as the chairman of the conference, and one of the commissions (on the law of the sea) was headed by F. F. Martens. The outcome of the four-month conference sessions was the adoption of thirteen conventions (on the peaceful resolution of international clashes, on the opening of hostilities, etc.).

Despite significant progress in the codification of humanitarian law, World War I fully demonstrated the reluctance of all parties to adhere to established rules. The work of the international tribunal created by the decision of the Versailles Treaty (1919) turned out to be very short and had no legal consequences [2]. Russian (or rather, already Soviet Union) participation in the post-war settlement and the establishment of new "laws of war" came down to the role of a passive observer. Meanwhile, the German side was particularly cruel to Russian prisoners of war. On the one hand, this was due to massive anti-Russian propaganda on the eve and during the war. On the other hand, this was explained by the unwillingness of the Russian authorities to make life easier for captured compatriots by concluding additional agreements with the opposing side. While, for example, England and France concluded about ten mutually binding treaties with Germany on prisoners of war (on the exchange of sick, wounded, etc.) [9].

Throughout the 1920s, the young Soviet state practically did not participate in the codification of humanitarian law. The functions of the initiators of this process were transferred to other states—the USA and the victorious countries from among the Western European states. The humanitarian initiatives put forward by Soviet diplomats in the early 1920s did not receive support from other countries. One of the reasons for the forced international legal passivity of the Soviet Union (at least in comparison with the pre-revolutionary period) was its foreign policy isolation. In addition, the Soviet leadership was generally critical of "bourgeois" pacifism. As Moscow's international authority strengthened, the USSR's participation in anti-war events became more and more significant. Since the early 1930s Soviet representatives participated in international conferences on disarmament [3]. The USSR ratified a number of conventions of a humanitarian nature, in particular "The Protocol for the Prohibition of the Use in War of Asphyxiating, Poisonous or other Gases, and of Bacteriological Methods of Warfare" (1928). However, the Soviet Union refused to participate in the 1929 Geneva Conference, as a result of which two conventions were adopted—"Convention for the Amelioration of the Condition of the Wounded and Sick in Armies in the Field" [14]. This circumstance allowed the opponents of the USSR in World War II to formally not follow the Soviet obligations of prisoner of war obligations imposed by the terms of the Geneva Convention [4].

After the end of World War II, the USSR revised its attitude to the development of international law, including in relation to prisoners of war [12]. With the participation of Soviet diplomacy, the Convention on the Protection of Cultural Property in the Event of Armed Conflict (1954), The Convention on the Prohibition of the Development, Production and Stockpiling of Bacteriological (Biological) and Toxin Weapons and on their Destruction (1972), The full title is Convention on Prohibitions or Restrictions on the Use of Certain Conventional Weapons Which May Be Deemed to Be Excessively Injurious or to Have Indiscriminate Effects (1980), etc. [5].

The collapse of the Soviet Union led to a decrease in the geopolitical influence of its successor, the Russian Federation. At the same time, Moscow selectively approaches a number of humanitarian conventions concluded at the turn of the XX–XXI centuries. Our country holds a special position on such a question as participation in the International Criminal Court. Together with other leading world powers (China, USA, India, etc.), Russia considers such international jurisdiction a limitation of state sovereignty. Therefore, Moscow decided to cease to be a party to the Rome Statute (1998), which established the International Criminal Court [13]. Now in the development of humanitarian law, there are two mutually exclusive trends. On the one hand, at the beginning of the 21st century, humanization embraced a significant range of socio-political life, protecting the rights of various categories of civilians and military personnel. On the other hand, we can observe equally obvious failures in the practical implementation of the principles of humanitarian law. The turn of the XX–XXI centuries is marked by massive examples of violation by all parties of the rights of both civilians and prisoners of war. As a result, today, compliance (not declarative, but real) even with the basic principles of humanitarian law remains a pipe dream.

42.4 Conclusion

The study allows us to distinguish 4 stages in the history of Russian participation in the formation of international humanitarian law. *The first stage* (mid-19th century—1917) is characterized by significant international legal activity of the Russian Empire, which is one of the founders of the development of new rules of warfare. *The second stage* occurred in the 1920s and mid-1940s. It is characterized by the passivity of Soviet Russia/USSR in the humanization of military conflicts, which was associated with both ideological reasons and objective difficulties in the formation of the Soviet state. *The third stage* falls on the period from the mid-1940s to the early 1990s and is determined by the results of World War II, from which the Soviet Union emerged as a great victorious power. With the collapse of the USSR, *the fourth stage* began, the most important sign of which was a marked decrease in the "international weight" of its successor, the Russian Federation. As part of the intensified geopolitical rivalry in the 2010s and a noticeable cooling in relations between Russia and the West, there has been a tendency to eliminate Russian participation in global institutions responsible for international security and arms control. Under these conditions, it is critically important to objectively evaluate and predict both the path of development of humanitarian law and Russian participation in it.

An important key to deciphering the prospects of current events is the analysis of the globalization process. The real prerequisites for the realization of globalism in the field of economy, culture, politics, ideology were formed only in the era of European modernity after the end of the industrial revolution and at the same time, the ideas of humanitarian law became popular. Moreover, while in practice European countries have long been engaged in the economic, political and territorial division of the world, then, in contrast to the bourgeois world, the communist movement appears precisely the globalist idea associated with the prospect of universal unification through the world socialist revolution. Therefore, it can be argued that at the first stage the idea of globalism was realized to a greater extent in the Soviet Union within the framework of the priority of proletarian internationalism. And only after the Second World War did the idea of Western, primarily globalism, gain strength. We can say that most of the second half of the 20th century was marked by the opposition of these two globalisms-communist (Bolshevik) and American. Communist-international and American models of globalism diverged in the declaration of their ideas on the basis for achieving human happiness. On the one hand, the basis was justice in the organization of production and distribution processes, and on the other hand, respect for democracy and human rights. However, in the methodology for the implementation of their ideas, both versions of globalism were very close. Both variants of globalism sought to spread their influence to other countries in the economic, ideological and political spheres. It is important to note that, having lost their ideological foundation as a result of the collapse of the Soviet system, the proponents of globalism in Russia have maintained a world view close to American globalism. However, the ongoing understanding of the crisis of American values, the discovery of their fake nature contributes to the formation of the idea of a multipolar world.

It is also necessary to take into account that the second model of American (Western) globalism is unfolding now. This happens due to the fact that the existing first model of globalization, based on the principles of an open free market, has lost its constructive meaning for the West, which has lost priority and delegated the initiative to the new Asian giants (China, India). In fact, they seized the initiative from the West and became the main driver of global economic growth. Having changed a lot in themselves, but at the same time keeping the traditions and features, they appeared more adapted to the global liberal market, than all other competitors, including the West. Thus, the impulse of economic market globalization, launched by the West, led to an unexpected result-instead of the process of Westernization, the world has significantly shifted and drifts towards the East.

All this makes it possible to predict (with a high degree of probability) significant changes in the globalization process. In particular, it is obvious that the existing model, based on the principles of an open free market, has lost its priority for the West. Accordingly, either a fundamental rejection of the globalization development model is possible, which does not seem realistic, since this is an objective process, or the formation of an alternative qualification model based not on the free market and competition, but on certain restrictive principles, primarily of a cultural and legal nature. The practice of "sanctions", which has been established and widely used in recent years, clearly shows that the formation of this model is already in the active phase. The philosophy of this model is quite obvious—the right to economic income, growth and prosperity is not possessed by any international player, thanks to the mere fact of its existence, but only one who meets certain criteria of the cultural and legal order. However, one must understand that the West was late with the formal legalization of this model; it missed the time when this could be done using existing legal mechanisms for making international decisions. Therefore, today the West is forced to unilaterally "push through" the recognition of this model, relying on its economic, information and power capabilities and resources. But since the time of total superiority of the West, not only in the purely power but also in the economic sphere, has already passed, an attempt to change models non-stop generates social conflict and the "war of civilizations".

Accordingly, it can be argued that the world has entered a new conflict cycle. The global West, which has lost its economic initiative and priority, but has not abandoned its ambitions, is forced right now, at this time, while it still holds primacy in international relations and institutions, to begin the revolutionary program of the new Art Nouveau. Similarly to the results of the first European Art Nouveau, it must implement such a cultural program and a type of society that will return the West an unconditional priority and at the same time drop all other societies into a new archaic. To achieve these goals, Russia and the Middle East were chosen as the object of the primary "sanctioned" and direct pressure from the West, which fully corresponds to the rational logic. If the attack is successful, then the West will receive the main sources of natural resources and eliminate the main "power" competitor (Russia). After that, it will be possible to begin to convince China to accept the new cultural and legal principles of economic globalization. It will be very difficult for China to fully comply with these principles, for example, in terms of environmental standards, and the West will regain unconditional global leadership. It can be assumed that it is in the mainstream of these processes that the formation of a new system of international humanitarian law will take place.

References

1. Abashidze AH, Bisultanov AK, Syunyaeva MD (2014) Emblema Krasnogo Kresta i Krasnogo Polumesyaca cherez prizmu otechestvennoj istorii. Sci J VolSU Seria 5 Law 3(24):134
2. Borbet'ko PV, Pushkin VI, Suharev AA (2005) Popytka provedeniya mezhdunarodnogo voennogo tribunala nad voennymi prestupnikami posle okonchaniya pervoj mirovoj vojny. Веснік Віцебскага дзяржаўнага ўніверсітэта 2(36):33–38
3. Ioffe AE (1968) Vneshnyaya politika Sovetskogo Soyuza (1928–1932). Nauka, pp 38–40, 203–204, 212–222
4. Ivanovich KB (1991) Sushchestvovavshie do sih por pravila... otmenyayutsya. Voyenno-istorichesky zhurnal 11:41
5. Lenshin SI (2009) Pravovoj rezhim vooruzhyonnyh konfliktov i mezhdunarodnoe gumanitarnoe pravo

6. Malakhov IS (2012) International human law: definition and content (politological aspect). Vestnik Kemerovskogo gosudarstvennogo universiteta 3(51):163
7. Martens FF (1879) Vostochnaya vojna i Bryussel'skaya konferenciya. 1874–1878 gg. SPb.: Tip. M-va putej soobshcheniya (A. Benke), pp 89, 125–129, 132–135, 137–139
8. Martens FF (1900) Gaagskaya konferenciya mira. Kul'turno-istoricheskij ocherk. Vestnik Evropy 3:19
9. Nagornaya OC (2010) «Drugoj voennyj opyt»: rossijskie voennoplennye Pervoj mirovoj vojny v Germanii (1914–1922). Novyj hronograf, pp 32–34
10. Nikolaev NYU (2013) Problema kodifikacii mezhdunarodnogo gumanitarnogo prava v trudah L. A. Kamarovskogo. Istoriya gosudarstva i prava 17:35–38
11. Pustogarov VV (1999) Martens, F.F.—yurist, diplomat, publicist. Voprosy istorii 8:142
12. Sidorov SG (2014) Vliyanie mirovyh vojn na razvitie mezhdunarodnogo prava o voennoplennyh. Vestnik Sci J VolSU. Seria. 4: Istoriya. Regionovedenie. Mezhdunarodnye otnosheniya 5(29):136
13. Skuratova OYU (2016) Rossiya i Rimskij statut mezhdunarodnogo ugolovnogo suda. Moskovskij zhurnal mezhdunarodnogo prava 4(104):125–137
14. Stratievskij D (2014) Sovetskie voennoplennye Vtoroj mirovoj vojny i gumanitarnoe pravo. Mogla li Moskva spasti svoih grazhdan? Zhurnal rossijskih i vostochnoevropejskih istoricheskih issledovanij 1(15):82–83

Chapter 43
Development Trends of the Marriage Contract Institution in the Digital Age

Inna Trifonova

Abstract *Purpose*: The purpose of the article is to identify the main development goals of the marriage contract institution in the era of digital technologies and determine their impact on the stability of the economy and civil relations. *Design/methodology/approach*: the authors use specific scientific legal methods, such as dogmatic, legal-technical, law interpretation, state-legal modeling, and comparative-law. *Findings*: the authors prove that (1) The digitalization of the economy has impact on the regulation of marriage and family relations; (2) A marriage contract is contradictory because, on the one hand, it cements the family ties, but on the other hand, it can destabilize civil relations; (3) Technologically conditioned civil-law regulation of a marriage contract. *Uniqueness/value*: the authors conclude that the current family law does not explicitly prohibit the use of electronic means for the conclusion of family law contracts, therefore marriage contract can be concluded in electronic form. Besides, this form will expedite and simplify the conclusion procedure for (future) spouses, especially when they far from each other. However, modern family law regulation does not allow concluding such a contract using digital means. If any economic or technological innovations can potentially strengthen the rights and freedoms in the family, as well as to reinforce their protection, then such remedies, in particular, opportunity to conclude a marriage contract in electronic form, should be enshrined in law.

Keywords Digital economy · Marriage contract · Family law · Electronic contracting

JEL Classification Codes J 12 · K12

I. Trifonova (✉)
State University of Kalmykiya, Elista, Russia
e-mail: trifon-inna@yandex.ru

T. Kolmykova and E. V. Kharchenko (eds.), *Digital Future Economic Growth, Social Adaptation, and Technological Perspectives*, Lecture Notes in Networks and Systems 111, https://doi.org/10.1007/978-3-030-39797-5_43

43.1 Introduction

The modern world is featured by the rapid development of all areas of human activity, digital technologies in particular. Modern digital technologies form a new technological environment with an acting social phenomenon such as law [6]. Moreover, digital technologies start dictating its terms which legal institutions, including civil-law ones, should be adapted to [3]. As some authors rightly point out: "Nowadays it's insufficient to state merely the impact of the technological factor on public relations. Digital technologies transform well-established state and public institutions" [9]. Surely, the introduction of all kinds of technological innovations into civil relations brings various benefits like saving time and resources, transferring typical duties from a man to machine and similar. However, the legal issues of introducing such innovations are currently poorly studied, which often causes various losses for parties to civil relations. These losses can be ones from commercial espionage or hacker attacks on the company, as well as flawed mechanisms to effect transactions using electronic communication means.

Digitalization of both the global and the Russian economy has a huge impact on all aspects of the life of a modern man, including private legal relations such as family ones.

Family relations are one of the most complex subject-matters of legal regulation for the legislator. On the one hand society, the state and civil commerce are interested in the well-being and stability of the family institution, but on the other hand, the law-maker explicitly governs only the property relations of family members and leave non-property relations to be governed by spouses themselves in accordance with their ethic, moral and (or) religious attitudes. "Property relations between family members can greatly affect both present and future of the family in particular and on the world market in general" [7].

This point is pretty reasonable, as far as Art. 13 of the Constitution of the Russian Federation prohibits any state ideology and Art. 28 recognizes freedom of conscience and religion. In this connection, any attempts to settle non-property relations of the spouses by the state will be illegal and will contradict the requirements of the Constitution of the Russian Federation.

Contracts have been the regulator of property relations in the family since ancient times. A marriage contract in most of the modern civilized countries is a common instrument for the regulation of almost all areas of family life, both concerning the property and private non-property legal relations. The lawyers suppose that "agreeing on the partition of property that had been before marriage is the most painless way to resolve possible property disputes" [2]. However, using such an effective tool in Russia is extremely low, which causes concern not only from the standpoint of balancing the marriage legal relations but also the stability of civil relations in general.

A marriage contract can be concluded by several categories of citizens such as spouses, and persons who want to contract marriage (legally capable persons who have reached marital age or minors who have got a consent).

However, pursuant to Par. 1, Art. 42 of the FC of the Russian Federation [1], only spouses can enter into a marriage contract, and citizens intending to contract marriage are not mentioned in this rule. This is due to the fact that the marriage contract becomes valid only from the date of its state registration.

43.2 Materials and Method

The research is based on specific scientific legal methods like dogmatic, legal-technical, law interpretation, state-legal modeling, and comparative-law, which make it possible to design scientifically based structure of the legal development of digital technologies when governing the institution of a marriage contract in the context of international integration.

43.3 Findings

1. *Economic digitalization influences the regulation of marriage and family relations*

The contracting parties to the marriage contract (both spouses in a registered marriage and persons entering into marriage have the right to conclusion) should be taken into account. But the second category is not stipulated in the law [8].

Thus, the marriage contract has both general and particular features. The general ones also include contracting parties. Firstly, these are persons entering into marriage. Secondly, these are married persons (spouses). Here we can agree with the opinion of some authors who consider a similar marriage contract as a suspensive condition deal [1]. Thirdly, these are former spouses bound by a marriage contract. Fourth, this is a foreign element.

The foreign element is deemed as the situation when spouses with different citizenship concluding a marriage contract under Par. 2, Art. 161 of the FC RF may choose the law of a particular country to determine their property rights and obligations. Russian law is applicable in other situations.

It should be borne in mind that if a contradiction arises between the foreign law that has been chosen and Russian law, its application will be limited in accordance pursuant to Art. 167 of FC RF.

2. *A marriage contract is contradictory because, on the one hand, it cements the family ties, but on the other hand, it can undermine civil relations*

Non-property relations and parenting cannot be the subject-matter of a marriage contract under Russian law, and therefore the marriage contract concluded under foreign law that allows such content won't be applicable in Russia since the typical feature of a marriage contract in Russian law against foreign one is sole regulation of property relations.

Spouses can change the legal regime of property both during the marriage and after its dissolution. Besides, most authors tend to consider it some kind of civil-law transaction with specific contracting parties, scope and subject-matter.

Certainly, one of the functions of a marriage contract is to avoid divorces related to property disputes, since the initial distribution of rights and obligations has a positive effect on the development of family relations. On the other hand, the essential function of a marriage contract, especially between former spouses, is to avoid property disputes after the dissolution of the marriage.

However, modern legal science often does not pay attention to the fact that a marriage contract influences not only legal relations between spouses, former spouses and persons intending to contract marriage, but also explicitly affects civil relations in general.

3. *Technologically-conditioned civil-law regulation of the marriage contract*

This is also connected to that the marriage contract can be concluded both with respect to the property of spouses at the time of contract conclusion and to property that will be purchased jointly and separately in the future.

Moreover, spouses determine in a marriage contract the scope of rights and obligations on all issues, in particular, income sharing; shared property maintenance expenses; division and utilization of property; and other property issues not specified in the legislative rule under Par. 1, Art. 42 of FC RF.

However, we see here a legal conflict that undermines civil relations in Russia. For example, civil law is applicable to family relations under Art. 3 FC RF unless it contradicts the essence of family relations. Nonetheless, Art. 40, 41, 42 SK RF does not mention anything about the termination of marriage, i.e. force of the marriage contract upon the death of a spouse, or the recognition of spouse as dead. As L. M. Pchelintseva fairly pointed out, property securing issues are not stipulated [5].

The above legal conflict from the perspective of stable civil relations requires law enforcement to independently seek a legal solution in every single case of death of one spouse who has entered into a marriage contract and, respectively, this practice is extremely heterogeneous and controversial. Today we can definitely highlight the most frequently used approaches, however, the lack of a single guideline leads to uncertainty of parties to civil relations in their counterparties.

For example, both the terms of getting a bank loan and the interest rate depends on the conclusion of a marriage contract, as far as bank has additional risks to deem the loan agreement null and void and check obligor's solvency due to the fact that disclosing the contents of the marriage contract by obligor is not compulsory.

In this regard, other factors undermine civil relations, namely, unresolved issues on the validity of a marriage contract after the spouse's death. This is especially acute when the parties to the marriage contract stipulated the elements that govern hereditary legal relations since the operating civil law contains a universal prohibition against changing the legal inheritance procedure except by drawing up a will or signing a hereditary contract. In this event, the surviving spouse may lose something owed to him/her if the marriage contract would unconditionally remain in full force and effect in all events, even upon the death of one spouse and contain the elements governing inheritance relations. In our opinion, this situation affects not only the spouses as parties to the civil relations but also their counterparties, i.e. on civil relations in general.

Probably, this factor, among other things, also does not contribute to the diffusion of the marriage contract in Russian society as a powerful tool for the regulation of family relations.

Termination of property contracts is an urgent problem not only for law enforcement practice but also for the development of legal science in general.

Some authors reckon that marriage articles between spouses are bilateral deals, and therefore are governed by civil law to a greater extent. And the grounds for amendment and termination of such contracts are identical with the procedure for disputing other deals and executed by filing statements of a claim by spouses or other interested parties and trying the case in court.

In that way, adherents of civil law application refer to Art. 4 of FC RF, which stipulates that family relations that are not regulated by family law fall under civil law if they do not contradict the essence of family relations.

We should remark that the family law establishes only two types of contracts providing for the regulation of property relations, namely marriage contract under Art. 43, 44 of the FC RF, and the alimony payment agreement introduced by Art. 101, 102 of FC RF.

Therewith, when resolving so intimate issues the law-maker invite the parties to a contract due to the economic, cultural and other differences of Russian society to independently agree on the terms of property division, alteration of the scope of property rights and obligations, as well as the procedure for amendment, termination and recognition of these contracts null and void.

Indeed, a marriage contract in itself is a powerful tool that allows both spouses avoiding disputes regarding the division of joint property, and relieving family conflicts regarding the utilization and disposal of this property, i.e. cements the family ties.

As mentioned above, the marriage contract can also destabilize civil relations, because it can be intentionally or negligently used to invalidate deals, moreover, ones made by in person by a spouse or on his behalf as well as ones made by this spouse on behalf of his legal entity.

Let's provide an example of this situation. The couple entered into a marriage contract according hereto 50% of the immovable property of a legal entity owned by the spouse is subject to transfer to another spouse upon dissolution of the marriage. The first spouse must ensure this right by getting the consent of another spouse to effect transactions or take actions that entail or could entail fall of the real value of such property at the date of the marriage contract conclusion but at the same time he/she is free to dispose of this property in any other way.

It is clear that such a condition protects the rights and legitimate interests of the second spouse, but on the other hand, it can cause serious damage to counterparties of this legal entity:

- under applicable law, the first spouse has the right not to inform the buyer of such real estate about the availability of such a condition in the marriage contract and the fact of its conclusion, even if this may entail adverse consequences for him/her;
- when the second spouse finds out that the first has made a transaction on the sale of his share in the immovable property of a legal entity, he/she will be entitled to file a lawsuit in the court to declare the purchase/sale transaction null and void on the grounds provided for in Part 2, Art. 166 of the Civil Code of the Russian Federation. Besides, there are very high chances that the court will allow the claim;
- if such a transaction is deemed null and void, the buyer will surely have the right to file an independent claim for the return of unreasonable gains; however, it will take much effort and time to get the money back. If one or both spouses act intentionally or in collusion, the buyer may not return them back at all (for example, if the legal entity is subsequently declared bankrupt).

Keeping in mind the current civil law regulation, the buyer in the above example cannot be deliberately protected by any remedies and remedial measures, since by virtue of Art. 35 of FC RF, getting the consent of the second spouse for a transaction to be made is required only if it concerns disposal of the common property of the spouses but not of the property of a legal entity owned by one of them. If it was the sale of stocks or share in OOO of the legal entity itself, then such consent would be required; however, consent is not required for the disposition of the property of the legal entity.

In our opinion, the above-stated problem can be solved using modern digital technologies [4]. This way, under Part 4, Art. 339.1 of the Civil Code of the Russian Federation, which entered into force on July 1, 2014, pledge of movable assets shall be recorded in a specially designated register on movables pledge and kept under the legislation of the Russian Federation on the notarial system. To simplify, the fact of the transfer of movable property as a pledge is subject to notarial registration in a specially designated register. The access is open around the clock to any person, i.e. when purchasing a movable thing into the property, the buyer has an objective opportunity to verify the transfer of this thing as a pledge, thereby protecting himself from a possible seizure of this thing as established by Art. 348 of the Civil Code of the Russian Federation.

In our opinion, a special register of marriage contracts, which would contain some information important for civil relations and would be accessible around the clock, can be similarly established.

We see the following information is to be made publicly available:

- surname, given name, patronymic, date of birth and taxpayer identification number, or information on another identity document of each of (future) spouses;
- date and place of the marriage contract conclusion;
- information on the notary who certified the marriage contract;

- information on whether the terms of the marriage contract can affect the legal relations between one of the spouses and third parties, while the content of the marriage contract is non-disclosed.

We reckon that the introduction of such a rule will make it possible to secure not only professional parties to civil relations but also to remove the destabilizing properties of the marriage contract institution.

As mentioned earlier, the subject matter of the marriage contract may also include a foreign element. In this situation, we are interested not in the law of the country that will be applicable to the marriage contract concluded, but the procedure for the conclusion.

This issue is especially topical both in view of the whirlwind "blurring" of state borders that allow citizens of different countries to enter into marriage, and against the backdrop of the rapid development of modern digital technologies that mitigate the consequences of separation for spouses living in different countries and give an opportunity of free online communication.

Keeping in mind that digital technologies for contract conclusion have been intensively used in the modern world, including Russia, we are interested in the possible application of this practice to the conclusion of marriage contracts.

43.4 Conclusion

1. The current family law does not explicitly prohibit the use of electronic means for the conclusion of family law contracts. However, it is particularly difficult that such a contract shall be certified by a notary public. We sum up that marriage contract can be executed in electronic form and it will expedite and simplify the procedure for (future) spouses, especially when they far from each other.
2. Regarding the possible conclusion of a marriage contract between spouses living in different countries and having different citizenships (for example, the husband is in France and the wife is in Russia), we believe that despite the sufficient development of digital technologies, the state of law, including foreign one, don't makes it possible to conclude a marriage contract in electronic form.
3. Pursuant to the current Russian legislation, a marriage contract should be certified by a notary public therefore its conclusion depends on the declaration of good will by each spouse in the presence of a notary public or an equal-status person. In view of that, even paper documents certified by a notary public of another country are not always recognized and executed in modern law enforcement practice, certification of a marriage contract by notaries of different states seems ambiguous. Nevertheless, we hope that the situation will change soon and spouses-citizens of different countries will be able to conclude a marriage contract in electronic form.
4. Under Part 4, Art. 339.1 of the Civil Code of the Russian Federation, which entered into force on July 1, 2014, pledge of movable assets shall be recorded in

a specially designated register on movables pledge and kept under the legislation of the Russian Federation on the notarial system. To simplify, the fact of the transfer of movable property as a pledge is subject to notarial registration in a specially designated register. The access is open around the clock to any person, i.e. when purchasing a movable thing into the property, the buyer has an objective opportunity to verify the transfer of this thing as a pledge, thereby protecting himself from a possible seizure of this thing as established by Art. 348 of the Civil Code of the Russian Federation. In our opinion, a special register of marriage contracts, which would contain some information important for civil relations and would be accessible around the clock, can be similarly established.

References

1. Akimov IYu (2016) Marriage contract: problems of execution. Notarius 8:26–29
2. Albikov IR (2017) The role and significance of the marriage contract in the human rights activities of the notary public. Notarius 5:43
3. Carchia AA (2019) Digital transformation law-enforcement monitoring. 1:26
4. Inshakova AO (2018) The response of law to the challenges of high technologies in innovative Russia. Legal Concept 17(4):6–15. https://doi.org/10.15688/lc.jvolsu.2018.4.1
5. Pchelintseva LM (2007) Family law of Russia: study guide, 5th ed, revised, p 113
6. Popkova EG, Inshakova AO (2019) Preface. Ubiquitous computing and the internet of things: prerequisites for the development of ICT. In: Popkova E (ed) Studies in computational intelligence, vol 826. Springer Science + Business Media, Cham, pp 5–8
7. Ralko VV (2017) Notarization of family agreements by the example of a marriage contract. Sud'ya 6:38–42
8. Stepkin SP (2018) Legal nature of the marriage contract. Fam Hous Law 2:25
9. Talapina EV (2018) Law and digitalization: new challenges and prospects. J Russ Law 2:6

Chapter 44
New Industrial Policy in a Digital Economy

Maria Kuzmina, Oksana Avdeyuk, Irina Tarasova, Svetlana Shevchenko and Tatyana Tarasova

Abstract The article analyzes and gives a general description of the law "On industrial policy in the Russian Federation". The transformation of industrial policy demands management decisions in the field of consolidating the relationship between the state and business in the process of implementation of industrial policy from public authorities. In this regard, the state should define its principled position on the problem of the distribution of powers and tasks. The scientific paper presents the results of studies reflecting the issues of the development of industrial policy in a digital economy. It reflected the theoretical provisions of industrial policy, development of directions of industrial policy in a digital economy.

Keywords Industrial policy · Digital economy · Industrial potential · Production

44.1 Introduction

Regional industrial policy takes on enormous importance in economic development of the region.

The urgent character of the study of various aspects of regional industrial policy is determined by: (1) high significance of industrial complex in regional economy;

M. Kuzmina (✉) · O. Avdeyuk · I. Tarasova · S. Shevchenko · T. Tarasova
Volgograd State Technical University, Volgograd, Russia
e-mail: mi_kuzmina@mail.ru

O. Avdeyuk
e-mail: oxal2@mail.ru

I. Tarasova
e-mail: tarasova.irina.aleks@gmail.com

S. Shevchenko
e-mail: svetashev@mail.ru

T. Tarasova
e-mail: tarasovata@gmail.com

T. Kolmykova and E. V. Kharchenko (eds.), *Digital Future Economic Growth, Social Adaptation, and Technological Perspectives*, Lecture Notes in Networks and Systems 111, https://doi.org/10.1007/978-3-030-39797-5_44

(2) the need for efficient use of the production potential of the region for the solution of social, environmental and other problems of the region.

Today, Russian economy get to a new level of business relationship, which, on the one hand, is characterized by the introduction of active integration processes into global economic system; on the other hand, Russian industry and economy are currently in severe crisis: economic sanctions, economic growth disruption, outdated manufacturing equipment. Consequently, Russia needs deliberate state industrial policy. This is a rather challenging task, since the implementation of any small-scale project in industry immediately faces the absence of the most essential components—investments, labor forces, infrastructure, etc.

44.2 The Essence and Directions of Regional Industrial Policy

Much attention is paid to issues of regional industrial policy at the domestic level, which is manifested in various regulatory legal acts.

Thus, for example, Federal Law No. 488-FZ of 31.12.2014 "On industrial policy in the Russian Federation" presents the basic definitions which reveal the essence of industrial policy: industrial policy, branch of industry; economic entities carrying out industrial activity; industrial production; industrial products; industrial infrastructure; industrial activity support infrastructure; measures aimed at promoting activities in the field of industry, etc.

In accordance with Federal Law No. 488-FZ of 31.12.2014 "On industrial policy in the Russian Federation", industrial policy is a comprehensive set of legal, economic, organizational and other measures aimed at developing the industrial potential of the Russian Federation, ensuring the manufacture of competitive industrial products.

Industrial production (industry) is a definition of a set of types of economic activity which include extraction of minerals, manufacturing activity, electric power, gas and steam supply, air-conditioning, water supply, water disposal, waste management companies, as well as pollution elimination companies based on the All-Russian Classifier of Types of Economic Activity.

Economic entities carrying out industrial activity include legal entities and individual entrepreneurs engaged in the field of industry in the Russian Federation, on the continental shelf of the Russian Federation, in the exclusive economic zone of the Russian Federation.

The goal of industrial policy in the Russian Federation is to develop high-tech competitive industry.

Directions in creating a competitive industry from the perspective of the state are as follows:

- creation and development of modern industrial infrastructure (a set of items of immovable property, transportation and communal infrastructure facilities that are required for carrying out industrial activities);

- integration of science, education and industry;
- mastering the production of innovative industrial products (goods that were produced as a result of industrial activity);
- efficient use of physical, financial, labor and natural resources;
- an increase in labor productivity;
- introduction of import-substituting, resource-saving and environmentally friendly technologies;
- upgrades of fixed production assets;
- exporting products.

At the same time, the state lays the following tasks upon itself and regions:

- to create competitive conditions for carrying out industrial activity;
- to implement measures aimed at promoting industrial activities of economic entities (activities of legal, economic and organizational nature that are aimed at achieving objectives of industrial policy and are exercised by public authorities, local self-governing authorities (profit and non-profit organizations), included into industrial activity support infrastructure):

(1) to integrate results of intellectual activity and master the manufacture of innovative industrial products;
(2) to make efficient use of resources, ensuring an increase in labor productivity, introduction of import-substituting, resource-saving and environmentally friendly technologies;
(3) to support economic entities by way of technical retooling and upgrades of fixed production assets;
(4) to support economic entities which export industrial products with high ratio of added value;

We agree with the viewpoint of Tatarkin, A. I. and Romanova, O. A., according to whom the essence of regional industrial policy is determined by the following provisions: due consideration of priorities of federal industrial policy; creation of a competitive industry in the region; clear vision of the result of implementation of regional industrial policy; promotion of increase in industrial production; efficient use of scientific and technological and innovation potential of the region; high degree of influence on results of economic development of regions [1].

Therefore, regional industrial policy must be carried out with due consideration of priorities of state industrial policy, i.e. to have the same objective and directions of its achievement (Fig. 44.1).

Competitiveness of industrial sector (a set of economic entities carrying out industrial activity within the framework of one or several classification groupings of one or several types of economic activity in accordance with the All-Russian Classifier of Types of Economic Activity) shall be understood to mean:

- the opportunity for the creation and sales of its effective, innovative and progressive goods and services in the domestic and foreign market for extended lengths of time [2].

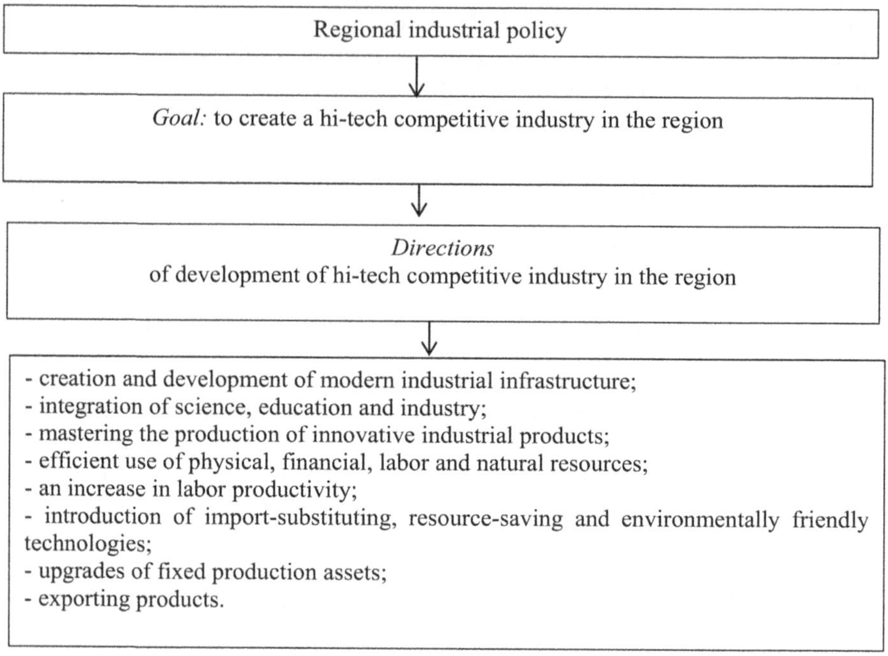

Fig. 44.1 Directions in creating a competitive industry of the region with account of national priorities (compiled by the author)

– the ability to create an increasing volume of added value (the volume and dynamics of added value per employed person) under the following conditions: improving the efficiency of production factors; ensuring the investment attractiveness of the business; developing new markets and ensuring the growth of return on business ratio [3].

Akulich, O. V. believes that the basis for the improvement of competitiveness of industrial sector consists in competitive advantages of the industrial sector [4].

Fatkhutdinov, R. A. distinguishes between external and internal competitive advantages of the industrial sector at the national level.

The author refers the following competitive advantages to external competitive advantages of the industrial sector: active state support of small and medium business; high-quality legal regulation of the economic performance; high scientific level of economic management; adequate state support of science and innovation activity; high quality management information support; low interest rates; availability of affordable and cheap resources; high-quality system of training and retraining of managerial human resources; high level of competition in all areas of activity and others [5].

The author refers the following competitive advantages to internal competitive advantages of the industrial sector: high demand for a particular product in the industry; optimal level of concentration, specialization and cooperation; high percentage of competitive staff; high-quality informational and regulatory and procedural management framework; competitive suppliers; performance of works to optimize resource efficiency; competitiveness of managers; exclusiveness of goods; high efficiency of the organization; high share of high-tech goods exports; high percentage of competitive companies and goods; certification of products and systems and others.

Abovementioned external competitive advantages of the industrial sector can only be formed through the use of efficient regional industrial policy.

Davydova, N. S. and Valova, E. V. in their research clarify the main priority tasks of regional industrial policy:

- improving competitiveness of products based on the development of new types of products, the use of new materials and technologies;
- raising the technical level of industry through the modernization of production links and selected technologies;
- formation of a competitive environment, creation of equal and predictable conditions for economic activity for enterprises in the region;
- fueling economic growth of enterprises participating in the Import Substitution Program;
- creation of loan and leasing institutions aimed at export promotion;
- competitive nature of loan institutions to ensure transparency of lending procedures;
- strengthening of technical control and introduction of modern quality systems and production certification systems;
- development of information technology [6].

According to Risin, S. I., industrial policy also implements social goals: the use of the human factor involved in industrial production; creation of favorable conditions for the increment and effective implementation of human capital [7].

Vetrova, E. N., Rokhchin, V. E. and Lapochkina, L. V. interpret regional industrial policy as a set of intentions and actions of economic entities implementing state interests relating to industrial development of regions, as well as economic interests of regions as such, municipal management structures, and large-scale manufacturing enterprises in the field of prospective development of industrial production in the region [8].

However, there are certain restrictions in the formation of regional industrial policy, which, according to Titov, K., consist in the following:

- economic growth rate of the region;
- size of the regional budget;
- characteristics of the internal competitive environment;
- own investment policy as a source of formation;
- environmental situation in the region;
- availability and condition of raw materials potential [9].

Therefore, the essence of regional industrial policy is determined by the following provisions in the aggregate:

- regional industrial policy is aimed at creating a competitive industry in the region;
- regional industrial policy includes a set of intentions and actions of economic entities of regional industrial policy;
- regional industrial policy is carried out by industrial enterprises which implement economic interests of the region with regard to industrial development of regions;
- objects of regional industrial policy is the production potential of industrial enterprises in the region; production of competitive products by manufacturing enterprises in the region; industrial development in the region (emergence of new industrial enterprises);
- directions of implementation of regional industrial policy objectives: (1) improving competitiveness of products by mastering the manufacture of innovative industrial products; strengthening of technical control and introduction of modern quality systems and production certification systems; an increase in labor productivity; integration of science, education and industry; (2) formation and use of production potential of the region through: raising the technical level of industry through the modernization of production links and selected technologies; improving the effectiveness of the use of production resource of the enterprise (innovative capacity of the region as components of industrial potential); introduction of import-substituting, resource-saving and environmentally friendly technologies; (3) industrial development of the region by means of: formation of a competitive environment, creation of equal and predictable conditions for economic activity for enterprises in the region; fueling economic growth of enterprises participating in the Import Substitution Program and export business; creation of loan and leasing institutions aimed at the promotion of industrial enterprises; creation and development of modern industrial infrastructure; ensuring the investment attractiveness of the business; developing new markets and ensuring the growth of return on business ratio.
- due consideration of several factors in the development of regional industrial policy: economic growth rate of the region; size of the regional budget; characteristics of the internal competitive environment; own investment policy of the region; environmental situation in the region; availability and condition of raw materials potential in the region.

When defining the concept of industrial policy, we stick to the following definitions.

Competitiveness of products of an industrial enterprises is a set of product properties (consumer and value) and the system for its support in the implementation and intended use, which, meeting the requirements in terms of compliance of products with real and potential requests of certain consumers from time to time, ensures its consumer appeal as compared to similar products in a particular segment of consumer's market and determines the possibility of advantageous sale of products in the competitive market along with or instead of other competing similar products [10].

Production potential of the enterprise is a level of the actual state of manufacturing resources of an enterprise in relation to the state which determines production of competitive products in sufficient quantity with minimum costs in a particular market environment [11].

Economic growth of industrial enterprises is the consistent pattern of the functioning of any industrial enterprise with a view to meeting increasing demand, which, provided that the external factors acting on it are unchanged, is manifested in: expanding operations and increasing output; improving economic indicators (revenue, profit, profitability and others). The economic growth of enterprises is ensured by means of: the growth of production factors (fixed capital, working capital, number of employees); growth in the efficiency of production factors; implementation of innovations (product-based, technological, organizational etc.) [12].

In our opinion, regional industrial policy is a comprehensive set of legal, economic, organizational and other measures aimed at developing a competitive industry in the region on the basis of: (1) improving competitiveness of products (including innovative product development) of industrial enterprises in the region; (2) development and efficient use of the production potential of industrial enterprises of the region; (3) economic growth of industrial enterprises in the region.

44.3 Orientation of Development of Industrial Policy in Russia

The main trends of development of Russian economy are associated with the concept of digital economy.

In federal documents, digital economy is defined as follows:

- a system of economic and political, social and cultural relations based on the use of digital information and communication technologies [13];
- business activity in which the key factor of production is digital data, processing large data volumes and the use of results of analysis of which, compared to conventional business patterns, allow significantly boosting the efficiency of various types of production, technologies, equipment, storage, sale, delivery of goods and services [14];
- a set of public relations emerging in the use of electronic technologies, electronic infrastructure and services, technologies for analyzing big data and forecasting in order to optimize the production, distribution, exchange, consumption and increase in the level of social and economic development of states [15];
- a new paradigm of accelerated economic development [16].

The literature on economic topics includes a lot of studies. Their authors include Aletdinova, A. S., Babkin, A. V., Burkal'tseva, D. D., Kosten', D. G., Vorobyov, Y. N., Yevtianov, D. V., Tiranov, M. V., Kovrizhnykh, A. A., Kniagina, V. N. et al.

Following the data of research results, it has been found that digital economy, unlike real economy, has the following distinctive features:

- virtual nature, digital economy exists only in the virtual world, the Internet is its main channel and space for data circulation;
- dependence on computer equipment which serves as the basis for the activity of economic entities, transition of economic entities to operation in digital format allows them to create an effective model of activity—platform—by means of network effects;
- high speed, high pace of development which should be provided by the necessary development of digital, additive, and advanced technologies;
- strategic thinking at all levels of management;
- penetration and constant updating of innovations in all areas of the digital space;
- the key factor of production is digital data; digital form of data allows significantly boosting the efficiency of various types of production, technologies, equipment, storage, sale, delivery of goods and services;
- availability of digital electronic infrastructure;
- the main tool in the creation of added value is data and software products which allow managing all technical processes;
- transition from the management of tangible assets and resources to the management of intellectual activity and innovations;
- availability of digital entrepreneurial competencies—the ability of IT-based entrepreneurs to turn ideas into actions through innovations.

Thus, we can conclude that there are cardinal differences between the digital and the real economy.

Various federal documents that have been published over the last years allow stating that industrial development priorities will be associated with digital economy in the future. The following documents can be distinguished among them:

- Decree of the President of the Russian Federation of 01.12.2016 No. 642 "On the Strategy for the scientific and technological development of the Russian Federation";
- Order of the Ministry for Economic Development of the Russian Federation of March 23, 2017 No. 132 "On approval of Methodological recommendations for developing and adjusting the strategy for social and economic development of a constituent entity of the Russian Federation and an action plan for its implementation";
- Decree of the President of the Russian Federation of May 9, 2017 No. 203 "On the Strategy for the Information Society Development in the Russian Federation for 2017–2030";
- Digital Economy Development Program in the Russian Federation until 2035, approved by the Decree of the Government of the Russian Federation of July 28, 2017;
- Decree of the President of the Russian Federation of May 7, 2018 No. 204 "On the national goals and strategic objectives of the development of the Russian Federation until 2024".

It is obvious that changes are required in industrial policy of state and region.

We shall consider the content of the above documents with a view to determining industrial policy priorities and instruments in a digital economy.

The Decree of the President of the Russian Federation of 01.12.2016 No. 642 "On the Strategy for the scientific and technological development of the Russian Federation" identifies two priorities of industrial development:

- scientific direction—creation of an efficient research and development management system, ensuring high performance and being in demand in social and economic sphere of research and development; creating conditions for research and development that would comply with modern approaches to the management of scientific, scientific and technological, and innovation activities, as well as best Russian and international practices;
- technology direction—ensuring technological renewal of industry on the basis of transition to advanced digital, intelligent production technologies, robotic systems, new materials and methods of design, creation of systems for the processing of large data volumes, machine learning and Artificial Intelligence.

We will present instruments of industrial policy by the following directions:

(1) human capital (to create opportunities for development of intellectual potential of the country):

   - targeted support of young scientists and professionals in the field of scientific, scientific and technological, and innovation activities;
   - the use of reputational mechanisms for the recognition of the scientific skill and achievements of researchers, strengthening the credibility of scientists in society;
   - creating a competitive environment open to attract world-class scientists and young talented researchers with high-level scientific findings to work in Russia;

(2) infrastructure and environment (for the creation of conditions for research and development that would comply with modern approaches to the management of scientific, scientific and technological, and innovation activities, as well as best Russian practices):

   - supporting the establishment and development of unique mega-science research facilities (international research complexes where unique, large-scale experiments are carried out, which allow achieving high-quality breakthroughs in fundamental studies), large research infrastructures in the Russian Federation;
   - ensuring access of research groups to national and international information resources;
   - simplification of procedures for the procurement of materials and samples for research and development;

- promotion of certain territories (regions) with a high concentration of research, development, innovation infrastructure, production and their connection with other constituent entities of the Russian Federation in terms of the transfer of technologies, products and services;

(3) interaction and cooperation (for the formation of an effective system of communications in the field of science, technology and innovation, increasing susceptibility of economy and society to innovations, development of knowledge-intensive business):

- system support of interaction of major companies and public authorities of the Russian Federation with small and medium-sized innovation, scientific and educational establishments, as well as their involvement in the technological update of sectors of economy and creation of new markets;
- creation of a system of state support of national companies ensuring their technological breakthrough and taking steady state position in new, emerging markets, including within the scope of the National Technological Initiative;
- implementation of the public communications policy designed to develop technological culture, innovative susceptibility of the population, and to promote significant results in science, technology and innovation, achievements of outstanding scientists, engineers, entrepreneurs, and their role in the social and economic development of the country;

(4) management and investment (to form an effective modern management system in the field of science, technology and innovation, ensuring improvement of investment attractiveness of the field of research and development):

- expanding the access of private companies to the participation in promising, commercially attractive scientific and research and technology projects owned by the federal government and creating flexible mechanisms for adapting to changing competitive business environment at all stages of the implementation of these projects;
- creation of significant tax incentives in the field of scientific, research and technology, and innovation activities;
- venture capital financing for the creation and (or) modernization of production works based on the use of Russian technologies, as well as arrangement of conditions for economic entities, credit and financial institutions and individuals that would enable their investment in research and development;

(5) cooperation and integration (for the international scientific and technological cooperation and international integration in the field of research and technology by means of mutually beneficial international interaction):

- definition of goals and the form interaction with foreign countries depending on the level of their technological development and innovative capacity;
- implementation of coordinated support measures that ensured the entrance of Russian scientific organizations, educational establishments, and production companies to global knowledge and technology markets, as well as the

proactive participation of Russia in the development of technological standards and academic formats that contribute to the enhancement of its role in the formation of new markets.

In our opinion, this document discloses the following directions of industrial policy: informational, intellectual, scientific, technological, and investment.

The Order of the Ministry for Economic Development of the Russian Federation of March 23, 2017 No. 132 "On approval of Methodological recommendations for developing and adjusting the strategy for social and economic development of a constituent entity of the Russian Federation and an action plan for its implementation" contains recommendations for the elaboration of industrial development strategies in the regions of Russia in a digital economy.

The document focuses on the following directions:

- innovative and technological development of the constituent entity of the Russian Federation;
- development of science and innovation;
- development of human capital;
- development of innovative territorial clusters;
- development of interregional and external economic relations of the constituent entity of the Russian Federation.

Table 44.1 presents some activity assessment indicators by directions.

The Decree of the President of the Russian Federation of May 9, 2017 No. 203 approved the document "The Strategy for the Information Society Development in the Russian Federation for 2017–2030" (hereinafter referred to as the Strategy).

The strategy defines the guiding principles of the Russian Federation in the field of information and communication technologies.

In particular, it is pointed out that the emergence of digital economy requires development of Russian information and communications technologies.

The focus is on the following priorities:

- human capacity development;
- emergence of digital economy: formation of the information space taking into account the needs of citizens and society in obtaining high-quality and reliable information; development of information and communication infrastructure of the Russian Federation; creation and application of Russian information and communication technologies, ensuring their competitiveness at the international level; formation of a new technological basis for development of economy and social sphere.

The Strategy points out that the state should ensure the strengthening of sectors of Russian economy in which business development with the use of information and communication technologies will provide competitive advantages to Russian organizations, ensure efficiency of production and an increase in labor productivity; to provide support for the entrance of Russian organizations into foreign markets of goods and services.

Table 44.1 Indicators of assessment of results of regional development by directions

Direction	Assessment indicators
Innovative and technological development	Gross regional product per year, preceding the previous one, based on current prices, million rubles
	Gross regional product volume per head of population
	Volume of investments in fixed assets, million rubles
	Ratio between the volume of investments in fixed assets and gross regional product, %
	Index of labor productivity, %
	Coefficient of renewal of fixed assets, %
	The number of high-performance jobs
	The ratio between the number of high-performance jobs and the average annual number of employed population
Development of science and innovation	Domestic research and development costs, billion rubles
	The growth in the number of utility patent applications, %
	Share of products of hi-tech and knowledge-intensive industries in gross regional product, %
	Percentage of organizations that implement technological innovations, in the total number of examined organizations, %
	Innovation activity of organizations
	Volume of exports of high-tech and medium-tech products, million USD
Development of human capital	Average annual number of population employed in the economy (thousand people)
	Real accrued wages of employees of organizations (dynamics, average monthly value)
Development of interregional and external economic relations of the constituent entity of the Russian Federation	Foreign trade turnover
	Exports, imports
	Growth in consolidated revenues from sales of high-tech industries intended for exports, %
	Imports (exports) of goods between constituent entities of the Russian Federation

It also lists new technologies that can be industrially used in a digital economy: Artificial Intelligence; cloud and fog computing; Internet of Things and Industrial Internet; robotics, biotechnologies and others.

The Strategy served as prerequisite for the elaboration of the Digital Economy Development Program in the Russian Federation until 2035, approved by the Decree of the Government of the Russian Federation of July 28, 2017 (hereinafter referred to as the Program).

The Program states the need for a digital transformation of economy based on the introduction of digital technologies into all areas of activity. At the same time, there is a change in the economic structure, a change in traditional markets, social relations, and public administration.

The structure of economy is changing due to the formation of more efficient economic processes provided with digital infrastructures.

The Program identifies the following industrial policy priorities:

- formation of a breakthrough structure of economic digital assets (resources) meeting economic priorities of digital economy;
- formation of approaches to the management of production branches that take into account achievement of digital economy;
- development of the digital infrastructure of industrial enterprises.

New concepts that are characteristic of digital economy appear in the formulation of priorities:

- digital asset is a specific form of property and resources, including intellectual property, investments in which increase capitalization of physical assets and ensure the growth of a flow of money; a collection of information in digital form (a set of digital products) about a real-world or virtual object, process, economic entity, or individual, which is of value and can be used for deriving added value; a set of digital products and infrastructures, the process of the use and modification of which leads to the formation of added value and new value, including value expressed in monetary form;
- digital platform is a model of business activity of parties concerned for the functioning in common communication and transaction environment in the digital market, participants of which benefit from interaction with each other;
- digital infrastructure is a series of integrated technologies and digital products based on them, providing computing, telecommunications and networking capacities and working on the digital basis, as well as a comprehensive set of infrastructures ensuring the realization of processes on the basis of digital technologies.

The instruments of industrial policy are as follows:

- improvement of federal legislation with a view to: (1) eliminating regulatory barriers that impede the use of digital technologies in the production sector, in the field of technical regulation and metrology; (2) identifying a legal mechanism to protect intellectual property rights and results of intellectual activity; (3) adapting the

technical regulation and standardization system to rapid changes and stimulating the development of digital technologies;
- the system of support and incentive measures which provide incentives for economic entities to introduce innovations and conduct research in the field of digital technologies;
- providing preferential financing terms for business, assisting in the testing and implementation of innovative solutions, grant-making for talented entrepreneurs and scientists;
- public and private partnership as a means of promoting the susceptibility of research and innovation activities to the changing nature of innovation; developing new markets and creating value by means of cooperation and coproduction;
- promotion and simplification of creation and use of digital platforms in a business environment, including the sector of small and medium-sized enterprises;
- provision of information to the public about the multiple benefits which are provided by platforms to the business on a global market scale;
- promotion of the smart specialization strategy within the framework of the innovative development of regions, involving the most efficient use of their features and the development of competitive advantages, and combining industrial, innovative and educational policies aimed at selecting the limited number of priority investment areas within the region, focusing on its strengths and comparative advantages.

The following long-term comprehensive program called National Technological Initiative (NTI) has been established within the scope of public and private partnership. Its tasks are presented below:

- support of the network (platform) architecture of new markets;
- cultivation of Russian companies that lay claim to gaining significant positions in global markets;
- providing a higher performance level in new sectors and markets;
- support of projects of end-to-end technological directions, which will have crucial impact on future markets (big data, Artificial Intelligence, quantum technologies, new and portable energy sources, new production technologies, components of robotics, wireless communications technologies, etc.).

The Decree of the President of the Russian Federation of May 7, 2018 No. 204 "On the national goals and strategic objectives of the development of the Russian Federation until 2024" identifies the following priority tasks in the industrial sphere:

- providing the industrial sector with highly qualified personnel;
- introduction of new modern technologies in industrial sector;
- export-orientedness of industrial enterprises;
- an increase in labor productivity and modernization of fixed assets.

Instruments of industrial policy:

- tax benefits to stimulate the introduction of advanced management, organizational and technology solutions to increase labor productivity and modernize fixed assets;

- formation and implementation of the system of methodological and organizational support for an increase in labor productivity in a business environment;
- reduction of regulatory and administrative restrictions hampering an increase in labor productivity;
- formation of a training system, aimed at training the essentials of an increase in labor productivity, including through the use of digital technologies and platform technology solutions.
- creation of advanced infrastructure for research and development, innovation activities, including the establishment and development of a network of the mega-science research facilities (international research complexes where unique, large-scale experiments are carried out, which allow achieving high-quality breakthroughs in fundamental studies);
- creation of a digital platform designed to promote the production and marketing activities of small and medium-sized enterprises, including individual entrepreneurs
- simplification of tax returns for entrepreneurs using cash register equipment to improve the business environment;
- promotion of competitiveness of Russian goods (works, services) in order to ensure their presence in foreign markets;
- creation of a unified system of export promotion institutions, providing for the modernization of trade missions of the Russian Federation in foreign countries;
- modernization of the system for the promotion of small and medium-sized export enterprises, including individual entrepreneurs.

Federal documents discussed above afford ground for the following conclusions:

- objects of new industrial policy: science, technologies, industrial infrastructure, human potential, interregional and international relations in the field of industry;
- the priorities of new industrial policy are influenced by new trends in digital economy (technological development trends);
- the task of state authorities (regional authorities) is to ensure the formation of competitive advantages of industrial sector in a digital economy;
- priority directions of industrial policy include scientific direction (research and development management system); technology direction (industrial renovation based on the transition to advanced digital production technologies); intellectual direction (human development); infrastructure direction (development of the digital infrastructure of industrial enterprises as a comprehensive set of technologies and digital products based on them, providing computing, telecommunications and networking capacities);
- the main instruments of new industrial policy in a digital economy: targeted support of scientific activities; public communications policy designed to develop technological culture, innovative susceptibility of the population, and to promote significant results in science, technology and innovation; tax incentives in the field of scientific, scientific and technological, and innovation activities; venture capital financing for the creation or modernization of production works; improvement of federal legislation; preferential financing terms; public and private partnership as

a means of promoting the susceptibility of research and innovation activities to the changing nature of innovation etc. (presented in Table 44.2).

The issue of new industrial policy has been discussed in the economic literature—in particular, in papers by Babkin, A. V., Burkal'tseva, D. D., Kosten', D. G., Vorobyov, Y. N, Manturov, D. V., Trofimov, O. V., Saakian, A. G. and many others.

Manturov, D. V. distinguishes the following directions in new industrial policy [17]:

- technology direction—platform management technology development—distributed digital systems which ensure the implementation of all operations that are necessary for managing the production activity on the basis of uniform standards.
- investment direction—formation of support tools for the production design, assessment of the needs of industries in obtaining public funding, identification and elimination of nonfinancial obstacles to investment, formation of a package of solutions that ensure the development of a relevant field, as well as those related with industrial production of service processes;
- infrastructure direction—solving problems of coordinating the provisions of documents of strategic industrial development, Energy Strategy, Transport Strategy of the Russian Federation, as well as development of regional strategic planning for industrial development (improvement of quality, harmonization and promotion of regional programs and strategies of industrial development);
- foreign trade direction—development of international relations and support of industrial exports; promotion of national products, use of comparative advantages of domestic production for the production upon the request of foreign partners.

In our opinion, directions of industrial policy lack such directions as scientific and intelligent.

44.4 Components of Development of Industrial Policy in a Digital Economy

Following the analysis of academic economical literature, we believe that the main directions of development of industrial policy in a digital economy are as follows:

- manufacture of producer's goods within the framework of import substitution with a view to maintaining a certain level of key competences in terms of competitiveness by world leaders in respective markets;
- improvement of the system of research organizations with a view to developing own technologies for the achievement of specified results for all imports-dependent positions;
- improvement of a training system for research organizations;
- the use of mechanisms to minimize the risks of deterioration of macroeconomic situation in the country: carrying out stabilization measures, including propping-up of demand for manufacturing products;

Table 44.2 New industrial policy in a digital economy: directions, tasks and implementation tools

Directions of new industrial policy	Tasks	Instruments of new industrial policy
Information direction	Creating an information space for industrial development in a digital economy	– Provision of information to the public about the multiple benefits of digital economy for industrial enterprises; – Creation of a special legal regime for organizations that have a certain level of computerization of processes and are willing to open them to the state; – Promoting significant results in science, technology and innovation/promote significant results in science, technology and innovation, achievements of outstanding scientists, engineers, entrepreneurs, and their role in the social and economic development of the country; – Creation of a flexible regulatory framework for the implementation of digital technologies in all spheres of life; amendments to legislative documents and the willingness of the state to participate in electronic interaction; creation of uniform technological standards (clouds, big data, biometrics, spatial data etc.); regulation of information security issues
Technology direction	Development of platform technologies necessary for managing the production activity	– Tax benefits to stimulate the introduction of advanced management, organizational and technology solutions to increase labor productivity and modernize fixed assets; – Formation and implementation of the system of methodological and organizational support for an increase in labor productivity in a business environment with a view to (1) eliminating regulatory barriers that impede the use of digital technologies in the production sector, in the field of technical regulation and metrology; (2) adapting the technical regulation and standardization system to rapid changes and stimulating the development of digital technologies; – Creation of a digital platform designed to promote the production and marketing activities of small and medium-sized enterprises, including individual entrepreneurs; – Simplification of tax returns for entrepreneurs using cash register equipment to improve the business environment; – Strengthening of technical control and introduction of modern quality systems and production certification systems

(continued)

Table 44.2 (continued)

Directions of new industrial policy	Tasks	Instruments of new industrial policy
Investment direction	Increasing the investment activity of industrial enterprises	– Amendments to legislative documents with a view to ensure the inflow of investments for development of digital technologies; – Venture capital financing for the creation and (or) modernization of production works based on the use of Russian technologies; – Financing the creation of digital platforms (out of budgetary and non-budgetary sources); – Financial support for developers of various digital platforms; – Administration of preferential loans to economic entities for investment in research and development
Infrastructure direction	Formation and development of digital infrastructure as a comprehensive set of technologies providing computing, telecommunications and networking capacities for managing production processes	– Financial and organizational support for the establishment and development of mega-science research facilities (international research complexes where unique, large-scale experiments are carried out, which allow achieving high-quality breakthroughs in fundamental studies), large research infrastructures in the Russian Federation; – Ensuring access of research groups to national and international information resources; – Simplification of procedures for the procurement of materials and samples for research and development; – Promotion of certain territories (regions) with a high concentration of research, development, innovation infrastructure, production and their connection with other constituent entities of the Russian Federation in terms of the transfer of technologies, products and services; – Creation of infrastructure, including secure communication lines and data centers; – Regulation of activities of institutions of digital economy: formation of favorable conditions for the activities of collection, transfer, storage, processing of data, and access to data generated in connection with the use of information technologies by individuals and legal entities

(continued)

Table 44.2 (continued)

Directions of new industrial policy	Tasks	Instruments of new industrial policy
Interregional and foreign economic direction	Development of interregional and international relations through mutually beneficial cooperation	– Amendments to legislative documents with account of new models of interaction of economic entities; definition of goals and the form interaction with foreign countries depending on the level of their technological development and innovative capacity; – Creation of a unified system of export promotion institutions, providing for the modernization of trade missions of the Russian Federation in foreign countries; – Modernization of the system for the promotion of small and medium-sized export enterprises, including individual entrepreneurs; – Creation of loan and leasing institutions aimed at export promotion; – Competitive nature of loan institutions to ensure transparency of lending procedures; – Promotion of national products, use of comparative advantages of domestic production for the production upon the request of foreign partners; – Implementation of coordinated support measures that ensured the entrance of Russian scientific organizations, educational establishments, and production companies to global knowledge and technology markets, as well as the proactive participation of Russia in the development of technological standards and academic formats that contribute to the enhancement of its role in the formation of new markets

(continued)

Table 44.2 (continued)

Directions of new industrial policy	Tasks	Instruments of new industrial policy
Innovative (scientific) direction	Increasing the investment activity of industrial enterprises	– Establishment of legal framework relating to the use of mechanisms of special legal regimes for the promotion of innovations and types of activity in the field of digital economy; – Provision of consulting services to organizations which help companies to implement innovations and do business with the use of digital technologies; – Establishment of subdivisions and venture capital funds for the operation with startups (carriers of breakthrough innovations out of small business), which are partners of large enterprises; – State support for companies that are carriers of development and competencies in the field of digital technologies that have an end-to-end intersectoral effect (processing and analysis of large data volumes, Artificial Intelligence and neurotechnologies, virtual and augmented reality technologies); – Establishment of innovative research and technology centers for the creation of innovations at the premises of existing higher education institutions; – Assistance in testing and implementation of innovative solutions; – Development of a network of the mega-science research facilities – Promotion of the smart specialization strategy within the framework of the innovative development of regions, involving the most efficient use of their features and the development of competitive advantages, and combining industrial, innovative and educational policies aimed at selecting priority investment areas within the region, focusing on its strengths and comparative advantages; – Providing preferential financing terms for business, assisting in the testing and implementation of innovative solutions, grant-making for talented entrepreneurs and scientists; – Public and private partnership as a means of promoting the susceptibility of research and innovation activities to the changing nature of innovation; developing new markets and creating value by means of cooperation and coproduction

(continued)

Table 44.2 (continued)

Directions of new industrial policy	Tasks	Instruments of new industrial policy
Intellectual direction	Development of intellectual potential	– Improvement of federal legislation with a view to identifying a legal mechanism to protect intellectual property rights and results of intellectual activity; – Creating a competitive environment open to attract world-class scientists and young talented researchers with high-level scientific findings to work in Russia; – Targeted support of young scientists and professionals in the field of scientific, scientific and technological, and innovation activities; – Formation of a system for training specialists for work in a digital economy: (1) professionals in description, modeling and optimization of business processes and requirements analysis; (2) professionals in data analysis and machine learning methods able to apply existing methods to solve specific business problems in various industries; (3) professionals in the creation and maintenance of the basic IT infrastructure used in digital economy; – The use of reputational mechanisms for the recognition of the scientific skill and achievements of researchers, strengthening the credibility of scientists in society

- supporting investment activity of industrial enterprises, ensuring their financial stability, availability of working capital and funds for investment in development.

Babkin, A. V. believes that priorities of new industrial policy in regions should be as follows:

- improvement of investment attractiveness of regions;
- an increase in investment activity of regions;
- development of production and social infrastructure;
- minimization of regional imbalances in the field of social and economic development of regions;
- strengthening of interregional relations;
- rational use of human potential [18].

Lenchuk, E. B. believes that intellectual direction of industrial policy is an important aspect of new industrial policy. According to the author, the primary function of this direction consists in the intelligent support of modernization policy by means of development of new research areas and competitive technologies, construction of manufacturing enterprises and creation of new production works [19].

Based on the content of federal documents and taking into consideration the division of industrial policy into directions, we have concluded that **industrial policy in a digital economy has its own components**. In our opinion, a number of directions of new industrial policy can be distinguished.

Each direction has its own tasks and instruments (Table 44.2).

Therefore, the following conclusions following the research results.

New industrial policy is a comprehensive set of legal, economic, organizational and other measures aimed at developing the industrial potential of the Russian Federation and ensuring the manufacture of competitive industrial products in a digital economy.

44.5 Conclusions

The main directions of new industrial policy of the Russian Federation in a digital economy are as follows:

- information direction (creating an information space for industrial development in a digital economy);
- technology direction (development of platform technologies necessary for managing the production activity);
- investment direction (an increase in investment activity of industrial enterprises);
- infrastructure direction (formation and development of digital infrastructure as a comprehensive set of technologies providing computing, telecommunications and networking capacities for managing production processes);
- interregional and foreign economic direction (development of interregional and international relations through mutually beneficial cooperation);
- innovative (scientific) direction (an increase in investment activity of industrial enterprises).

References

1. Tatarkin AI (2014) Industrial policy: genesis, regional specific features and regulatory support. In: Tatarkin AI, Romanova OA (eds) Ekonomika Regiona, No 2, pp 15–19
2. Krivorot'ko IA (2015) Competitiveness of the machine-building complex in the Russian Federation. Sovremennye Nauchye Issledovaniya i Innovatsii, No 10 [Electronic resource]. http://web.snauka.ru/issues/2015/10/58402
3. Sal'nikov VA (2006) Competitiveness of branches of Russian industry: present state and prospects. In: Sal'nikov VA, Galimov DI (eds) Problemy Prognozirovaniya, 1/29, p 55
4. Akulich OV (2011) Methodologic argumentation of the definition of competitiveness of the industrial sector. News-Bulletin of the Irkutsk State Academy of Economics, No 3
5. Fatkhutdinov RA (2005) Strategic competitiveness: a textbook. In: Fatkhutdinov RA (ed) Ekonomika, 504 pp
6. Davydova NS. Industrial policy at the regional level: objectives, tasks and directions for the development. In: Davydova NS, Valova EV (eds). Access mode: http://elibrary.ru/defaultx.asp
7. Risin SI (2009) Regional industrial policy: essence, target function, principles. Bulletin of the Tambov State University. Series human sciences, No 6(74), pp 88–90
8. Vetrova EN (2015) Development of regional industrial policy objectives and their harmonization with economic development objectives: methodological aspects. In: Vetrova EN, Rokhchin VE, Lapochkina LV (eds) Academic periodical of the National Research University of Information Technologies, Mechanics and Optics. Series "Economics and eco-management", No 2, pp 145
9. Titov K (2012) Regional industrial policy: objectives, tasks, and priorities. In: Titov K (ed) Problemy Teorii i Praktiki Upravleniya, No 5, pp 5–12
10. Subbotin MN (2009) Managing competitiveness of products of industrial enterprises: synopsis of thesis for a degree of Candidate of economic sciences. Izhevsk
11. Maksimtsov MM (2013) A modern approach to the concept of production potential of the enterprise. In: Maksimtsov MM, Malyshev NI, Komarov MA (eds) Transportnoe Delo Rossii, No 3, pp 87–90
12. Kashirin SV (2015) Economic growth of an industrial enterprise. UES: electronic academic periodical, No 6(78)
13. Presidential address to the Federal Assembly of the Russian Federation of 01.03.2018. Access mode: http://www.consultant.ru/document/cons_doc_LAW_291976/
14. Decree of the President of the Russian Federation of May 9, 2017 No. 203 "On the Strategy for the Information Society Development in the Russian Federation for 2017–2030". Access mode: http://www.garant.ru/products/ipo/prime/doc/71570570/#ixzz5aChCN19D
15. Decree of the Government of the Russian Federation of 28.07.2017 No. 1632-R. On approval of the program "Digital Economy of The Russian Federation". Access mode: http://www.consultant.ru/document/cons_doc_LAW_221756/
16. Prospects of receiving digital dividends in the Eurasian economic Union. World Bank Report. Access mode: http://d-russia.ru/wp-content/uploads/2016/10/Navas-SabaterJ._WB-Digital
17. Manturov DV (2018) On future industrial policy in Russia until 2018–2030. Bulletin of the Moscow State University of Foreign Affairs, No 4, pp 6–17
18. Babkin AV (2017) The emergence of digital economy in Russia: essence, specific features, technical standardization, development problems. In: Babkin AV, Burkal'tseva DD, Kosten' DG, Vorobyov YN (eds) Scientific and technical reports of Peter the Great St. Petersburg Polytechnic University. Economics, No 3, pp 9–25
19. New industrial policy of Russia in the context of ensuring technological self-sufficiency. Editor-in-chief Lenchuk, E.B. Aleteiya, St. Petersburg, 336 pp

Chapter 45
Development of Physical Education and Sports in Terms of Improving the Quality of Life of the Population in the Regions of the Russian Federation

Natalya S. Beskorovaynaya

Abstract High spatial extension and nonuniformity of economic development of constituent entities of the Russian Federation reproduce their significant differentiation by the quality of life of the population. The improvement of methodological aspects of assessment of the quality of life is a prerequisite for the formation of well-grounded state regional policy. The authors of the research have substantiated the use of an axiological approach to the analysis of the quality of life, which involves the generation of a system of value orientations of society (terminal, instrumental, political) and assessment of the degree of their achievement by relevant indicators. Judging from problem areas and the presence of statistical data, there is a well-reasoned selection of indicators of the proportion of the working-age population working out and playing sports on a regular basis, accessibility of higher education and income level, as the indicators of the quality of life in the regions. Instrumental values characterize efforts of public and municipal authorities aimed at equalizing the living conditions in constituent entities of the Russian Federation, and their efficiency—indicators of political values (absolute poverty rate). The calculation data made it possible to draw a conclusion about a large differentiation of regions of Russia by the quality of life of the population under averaged values of indicators characterizing state regulation, and significant distinctions in the number of poor people.

Keywords Quality of life · Regional economics · Development of physical education and sports · Differentiation of regions

45.1 Introduction

In contemporary society, one of the most significant value orientations is identified as the quality of human life, which is associated with health and development of creative abilities of a personality, his/her vigorous activity.

N. S. Beskorovaynaya (✉)
North Caucasus Federal University, Stavropol, Russia
e-mail: agb20047@rambler.ru

T. Kolmykova and E. V. Kharchenko (eds.), *Digital Future Economic Growth, Social Adaptation, and Technological Perspectives*, Lecture Notes in Networks and Systems 111, https://doi.org/10.1007/978-3-030-39797-5_45

Quality-of-life studies are distinguished by their multidimensionality, cross-disciplinary nature, availability of a large number of theoretical and methodological approaches to determination, assessment, and formation of mechanisms for the resolution of the problem.

The most important contemporary problems of development of Russia include a low standard of living, social inequality in many regions. Therefore, the formation of mechanisms which allow improving the quality of life in backward constituent entities of the Russian Federation becomes the most crucial task of state policy.

45.2 Literature Review

The quality of life is usually associated with certain properties of human life, as well as the degree of fulfilment of human expectations. The best-known patterns of the quality of life are used to assess, in addition to material well-being, opportunities for self-development, environmental safety, as well as health, physical and mental condition, spiritual sphere [1–4 etc.].

Researchers usually focus their attention in their studies on the most topical aspects of the quality of life, as well as on the overall integrated assessment when conducting country or regional comparisons. Attention is given to changes in social indicators, while in regional comparisons attention is given to the degree of exceedance of threshold, corresponding to the state of material, social, and spiritual poverty. A similar approach is implemented in the methods of Zubarevich [5], which provides for the calculation of the "crisis" quality of life index according to the main criteria (income level, social inequality, employment status and public health). At the next stage, indicators characterizing education, provision of basic services, and environmental safety are analyzed. In addition, Mstislavskiy, P.S. suggests that indicators of the quality of life should be assessed versus the standard needs of society [1].

In the academic literature, the concept of "binary" quality of life has become a frequent practice, in accordance with which both the degree of personal satisfaction with life, personal well-being (a subjective approach), and the parameters of the physical and social environment (objective approach) are studied. Moreover, the results of surveys and expert findings [6–8] are used for first-line estimates.

In our opinion, the use of subjective estimates in the studies of regional differentiation of the quality of life is currently unreasonable, since the results will most likely correlate to statistical figures etc. of public authorities. This is due to the dissemination of information on a lifestyle of the well-off, as well as people living in other, more developed countries and regions, in the media and on the Internet. As a result, subjective conclusions of the population about the sufficiency of own income or satisfaction with lifestyle will be more realistic than they used to be two decades ago, and come closer to objective estimates. In addition, the value of a subjective approach in quality-of-life studies reduces because the degree of satisfaction with life slightly changes with increasing welfare, partly due to a proportional increase in needs [9].

A significant large part of studies in the field of the quality of life in the regions of Russia is focused on methodological aspects. This being said, usually indicators applicable for cross-country comparisons are used to measure the quality of life in the regions. For example, the calculation of Human Development Index (HDI) is used to compare data characterizing the life span, educational level and income level [10].

For Russian regions, the levels of education and life span do not provide a complete idea of the peculiarities of living of people in a particular territory, their health and lifestyle. This also holds true for such indicators as life span and the number of people with higher education, which may be of poor quality. The demand for such education is explained both by low material security of students and the fact that a diploma of higher education does not guarantee the increase in incomes—due to the inability for graduates to get a job in their specialty, as well as relatively low wages in key branches of economy.

Taking into account the fact that the level of differentiation of regions by Human Development Index is much lower than the perceived one (0.915–0.821 in less developed regions [11]), researchers usually expand the range of assessed criteria. Thus, the calculations of the rating agency "RIA Rating" are based on indicators that are distributed by 11 groups: income level, labor market, housing conditions, security, demography, ecology, health and education, social infrastructure, level of economic development, small business development, development of the territory, and development of the transport infrastructure (a total of 65 indicators). The spread in values of the quality of life index is larger than HDI value: summary rating points of the three most developed regions (Moscow, Saint Petersburg and the Moscow Region) were 72.5–77.4, and summary rating points of the least developed regions were 25–34 points [12].

However, in our opinion, the use of a large quantity of interrelated indicators also disfigures (smooths) the overall results of calculations. In addition, when selecting statistical indicators under consideration, fundamental values (health, for example) are substituted or complemented with instrumental values (level of medical care), which also results in the formalization and smoothing of results by regions.

We believe that in the system of indicators of public health special attention should be paid to physical and mental condition of a personality, lifestyle, which depend on whether a person wants to and is able to work out and play sports. The proportion of the population working out and playing sports on a regular basis characterizes not the provision of population with infrastructure facilities (leisure facilities), but the state of health, mind-set, degree of physical and mental perfection.

45.3 The Role of Physical Education and Sports in the Improvement of the Quality of Life of the Population

The significance of state support of physical education and sports increases in the context of such current negative processes in society as increased morbidity rate, short life span, increased intensity and duration of working time. These problems cannot be solved by focusing only on the priorities of the population policy, improvement of the public health system, the regulatory framework in the field of labor relations, and social protection of the population. A complex mechanism is also required, which would create the necessary conditions for working out and playing sports on a regular basis and encourage the population to do so.

Working-out is an important element of human lifestyle, which includes such components as the balance of working and leisure time, eating patterns, family relationship, living and housing conditions, rejection of bad habits, destructive behavior. Physical activity helps reduce morbidity, including cardiovascular and oncological diseases. A research [13] showed that the fundamental reasons of the loss of health of the population are behavioral factors, the most influential of which are malnutrition, alcohol and drug use, smoking, as well as low physical activity [13]. Therefore, attention should be paid to such a means of improving public health as promotion of a healthy lifestyle.

Working out and playing sports on a regular basis contribute to promotion of health, increase in labor productivity, extension of employability.

The concept of physical education is distinguished from the concept of sports in research studies and regulatory documents; according to functional features, mass sports is distinguished from top-class sports. In the context of preservation of health of people, development of mass sport is of particular significance. In accordance with Russian law, mass sports are focused on physical education and personal development through organized or independent activities, as well as participating in physical training activities and mass sporting events. Moreover, physical education consists in character education, the development of the physical capacities of a person, with a view to shaping a healthy lifestyle, social adaptation, physical rehabilitation. The main principles of the legislation on physical education and sports include enforcement of the right of everyone to free access to physical education and sports, state assistance in the development of all sports and their components. The composition entities of physical education and sports includes, apart from physical organizations and other entities, citizens who are working out.

45.4 Factors that Influence Large-Scale Involvement in Working Out and Playing Sports

Short life span, high morbidity rates in Russia are associated not only with the imperfect system of health care and provision of drugs, but are primarily associated with the non-observance of guidelines for a healthy lifestyle of the population, especially due to an unhealthy diet and a sedentary lifestyle.

Low percentage of people working out and playing sports on a regular basis (28.2% of the total number of the population employed in economy in 2018 [14]), is due to both peculiarities of the Russian mind-set and the circumstances in the life of people.

All factors that determine a lifestyle of a human and his/her desire to work out on a regular basis, can be conveniently divided into objective and personal. Objective factors include:

- personal income level;
- availability of sufficient amount of leisure time;
- conditions and nature of employment;
- adequacy and accessibility of sports and recreation infrastructure (from the perspective of spatial location and price);
- promotion of a healthy lifestyle, outreach and awareness-raising work;
- training and education system;
- traditions, customs, social stress in society.

Personal factors should include:

- value orientations;
- culture and education.

Among all factors, much attention is usually given to accessibility of sports infrastructure facilities, as well as the sufficiency of personal income in order to be able to use the paid services of fitness and sports organizations. Low material security, high unemployment encourages people to work overtime to the detriment of their leisure time, to agree to work in harmful and heavy conditions, to monotonous work, and, in such a way, choose not to play sports. Meanwhile, it has been proven that workers engaged in rough labor, particularly need physical activity at leisure [15–17].

Traditions, customs, and immediate surroundings also influence the way of how a person manages his/her leisure time, as well as the way he/she chooses sport. Human behavior is also affected by social stress, which is associated with social inequality, moral and psychological stress during crises and economic reforms.

Value orientations of a personality, distinguished in Russian society by their expectance of social assistance from the state, a passive living position, lack of culture, are the reason for the immature comprehension of the necessities of life of a person, inability to manage their leisure time and available income. As a result, upon availability of chronic diseases, people prefer to receive treatment rather than exercise.

Monitoring that was carried out by the Ministry of Sport of the Russian Federation in 2015, showed that 32.2% of the total number of respondents above 18 are working out and playing sports more than 1 h a week, and only 9% of them are working out and playing sports more than 3–4 h a week [18]. The analysis revealed the influence of the following factors on the degree of physical activity of the population:

- gender (women are working out and playing sports less frequently compared to men);
- age of respondents—as the age of respondents increases, the duration of sports activities and the number of people working out and playing sports decreases;
- availability of higher education (people with higher education are working out and playing sports more frequently);
- professional composition: managers pay more attention to their physical training than workers (45% and 29%, respectively);
- quality of life: the higher it is, the more willing is a person to work out and play sports;
- social status of a person: people who affiliate themselves with the lower social strata are less interested in physical activities;
- size of the settlement, which, in our opinion, is due to the provision of the population with sports facilities, as well as the level of income of the population;
- internal and external motivation to work out and play sports.

The results of the survey showed that the main factors of motivation to work out and play sports are as follows:

- restoration or preservation of health (43% of respondents);
- getting pleasure from physical activity (22%);
- the opportunity to become more attractive (22%);
- stress release, improvement of emotional state (14%);
- the increase in the level of physical fitness, the development of certain of locomotive properties (12%).

The authors of Guidelines for the use of various forms of motivation to keep a healthy lifestyle point out that most respondents do not feel the need for health care activities, but tend to get pleasure from physical activity, to become good-looking, to improve their psychoemotional state [18]. It should be noted however that playing sports on a regular basis has determinative impact on health and the quality of life, while physical activity has not only health-improving, but also preventive value, improving the psychoemotional and physical condition of a person.

In order to increase the large-scale involvement in working out and playing sports, it is necessary to develop internal motivation, certain value orientations of a personality aimed at promotion of education, culture, improvement of social status and material well-being, preservation of health and active ageing. An important direction of state influence is also the creation of necessary infrastructure, ensuring its accessibility for all social strata, as well as the regulation of the system of economic and social relations with a view to maintaining the adequate income level, leisure time for self-improvement. Maintaining a balance between working life and leisure life in accordance with the concept of decent work, which has the following characteristics, is of particular significance: adequate income, availability of leisure time, respect for human rights, which gives them the right to vote and promotes social integration [19]. According to F. Kirkman, the amount of leisure time after a working day is a factor affecting the quality of work life [20].

45.5 Development of Mass Sports in Russia

Development of a healthy lifestyle of the population is one of priorities of the modern state policy. Physical and mental perfection of a person is evidenced by an increase in the number of people working out and playing sports on a regular basis. In the Russian Federation, the proportion of the population working out and playing sports on a regular basis increased from 36.8% in 2017 to 39.8% in 2018 [14]. At the same time, a similar indicator that is measured by the number of the employed population is much lower—as little as 28.2%, which is indicative of the lack of motivation to working-out in most employed people, and in some cases—the lack of necessary conditions for this (Table 45.1).

The positive dynamics of the popularity of mass sports in Russia should be pointed out. Thus, in 2018 as compared to 2013, the number of people playing mass sports which do not require heavy spending from the population increased to the greatest extent: fitness aerobics—by 47.7%, track and field—by 35.7%, swimming—by 32.3%, hockey, football, volleyball—by 36.7, 27.0%, 23.7%, respectively.

In the Federal Project "Sports is the Norm of Life" provides for bringing the value of the indicator of provision of the population with sports facilities up to 60%; currently, its value is as low as 54.2%, which is indicative of insufficient accessibility of sports facilities for the population.

In the Russian Federation, the top-class sports are financed mainly from the federal budget, and the development of physical education and mass sports is financed

Table 45.1 Main indicators of development of physical education and sports in the Russian Federation, % [14]

Indicators	2017	2018
The proportion of the population working out and playing sports on a regular basis in the total population aged 3–79	36.8	39.8
The level of provision of the population with sports facilities judging from the simultaneous visitor capacity of sports facilities	50	54.2
The proportion of people working out and playing sports in the total employed population	25.5	28.2
The proportion of people with special needs and people with disabilities working out and playing sports on a regular basis, in the total number of the population of a particular category	16	17.2

mainly from regional budgets. As a result, industry-specific indicators differ widely depending on the region.

In 2018, out of budgetary funds allocated for financing of expenditures connected with major repairs of sports facilities, regional and local budgets account for 83.0%, while investments for their restoration and construction account for 86.2%. This being said, the bulk of budget expenditures is intended for remuneration of labor (43.8%). The ratio of money out of non-budgetary sources in the total amount of funding of physical education and sports is 16.2% [21].

Region-wise, the indicators of development of physical education and sports are significantly differentiated due to distinctions in income levels and fiscal capacity of the population. Thus, in 2017, the level of provision of the population with sports facilities, judging from the simultaneous visitor capacity of sports facilities in the Belgorod region, was 60.2%, while in Saint Petersburg it was 23.2%. The proportion of middle-aged citizens working out and playing sports on a regular basis (women: 30–54 y.o.; men: 30–59 y.o.), in the total number of the population of this category varied from 2.9% in the Republic of Ingushetia to 40.0% in Krasnodar Krai, which is determined both by the presence of conditions for playing sports and by internal motivation of the population.

45.6 Assessment of Differentiation of the Quality of Life by the Regions of the Russian Federation

The regional development of physical education and sports can be assessed by the percentage of people who are working out on a regular basis. As has been noted above, the value of this indicator is a result of the impact of many factors, including creation of conditions for playing sports, as well as the general standard of culture, economic and social status of people, the presence of internal and external motivation. Physical activity, in turn, characterizes the level of health of the population in the regions to a greater extent than life span, since a healthy lifestyle helps preserving working capacity, reduce morbidity and mortality from cardiovascular and other diseases. This allows using the indicator of the number of people working out and playing sports as part of indicators characterizing the quality of life of the population.

The international research of public health for the calculation of the quality of life index uses the estimates of life span and mortality, or the information about socially significant diseases. However, these indicators are not characteristic of physical and mental condition of people, their employability, ability of senior citizens to lead an active lifestyle.

It is recommended to select the indicators for the integral assessment of the quality of life taking into account basic requirements, focusing on problem areas in the life of people, the ability to use statistical indicators recorded at regular intervals, an adequate characteristic of the state of the studied life area. In addition, the criteria should be selected based on the characteristics of the target of research [5, 7, 22], in this case—regions of the Russian Federation.

From all accounts, the main problems in the life of the population of most Russian regions are low income level, high indicators of morbidity and mortality. In addition, the features of spatial national development consist in considerate differentiation of territories by personal income, state of health, opportunities for self-development, and raising the educational level of a person.

As has been noted above, the use of a large quantity of indicators for the analysis of the quality of life in the regions smooths the results. This is also due to the mixture of indicators characterizing terminal and instrumental, as well as political values. The first values represent the goals of existence of a person (society), while instrumental values represent the methods for achieving goals, which are preferable in any circumstances.

The indicators of significant terminal values in the context of territorial communities can be certain indicators characterizing the condition of problem areas in the development of the population: health, education, income.

Value-conscious and social health pattern is treated as "a state of complete physical, mental and social well-being, and not just the absence of diseases and physical handicaps" [23, 24]. In this regard, the rates of mortality, rates of morbidity, and life span do not completely reflect the person's ability to lead a full active life. We believe that the indicator of the proportion of the population working out and playing sports on a regular basis, which constitutes a behavioral component of a value orientation is more suitable for characterizing people's health. It appears that the study of the number of working-age people working out and playing sports on a regular basis is more interesting, since the curriculum for schoolchildren and students in our country provides for compulsory physical education classes in sufficient volumes.

Social development of population in Russia is limited by such factors as inaccessibility of quality education and low remuneration of labor [25]. In our opinion, the indicator of the number of university students per 10,000 people would be the best thing to characterize the possibility to receive higher education, since due to the difference in nominal income, opportunities for living and training decline in the universities of other regions, Moscow and Saint Petersburg. Thus, the monthly average nominal accrued wages of employees in Moscow in 2017 measured to 73.8 thousand rubles, in Saint Petersburg—53.7 thousand rubles, and in constituent entities of the North Caucasian Federal District—24.4 thousand rubles, i.e. lower by 3.0 and 2.2 times respectively, whereas the minimum subsistence income in cities of federal significance is 17.6 thousand rubles and 11.9 thousand rubles respectively [26].

Another terminal value is due to material well-being of people. It has been established that one of the most pressing regional problems in Russia consists in differentiation of constituent entities of the Russian Federation by the income level. Thus, nominal personal income per capita in Moscow is 4.1 times higher than that in the Republic of Ingushetia (62.5 thousand rubles per month as compared to 15.1 thousand rubles), but their ratio amounts to 2.5 even with account of the minimum subsistence level.

In order to compare the quality of life in the regions of the Russian Federation, we have determined the Human Development Index based on three statistical indicators:

- the proportion of the working-age population (women—30–54 y.o., men—30–59 y.o.) working out and playing sports on a regular basis;
- the number of students who study under Bachelor Degree Programs, Specialist's Degree Programs, and Master's Degree Programs per 10,000 people;
- per capita money income placed against the minimum subsistence income.

The proportion of the population working out and playing sports on a regular basis is determined on the basis of selective statistical monitoring which is carried out on an annual basis using the oral enquiry method. This indicator takes into account citizens, covered by various forms of health and fitness activities: organized, independent, paid, educational, physical education on the shop floor. This being said, working out and playing sports at least 3 times a week or for 3 h a week in total are considered to be regular.

The level of achievement of instrumental goals can be represented by a set of indicators characterizing the efficiency of the use of various methods of regulation of federal and regional authorities, aimed at improving the quality of life of the population (medical service, quality of education, social support of disadvantaged groups). In this research, indicators characterizing the living conditions have been selected for the calculation of the integral index of the implementation of instrumental values:

- the level of provision of the population with sports facilities depending on the unified visitor capacity (UVC), %;
- the number of doctors of all specialties per 10,000 people;
- regional consolidated budget expenditures for social and cultural measures per capita adjusted for the minimum subsistence level, thousand rubles.

Provision of the population with sports facilities, as well as accessibility of these sports facilities, availability of competent coaches are the necessary conditions for the development of mass sports in regions, especially for working-age citizens. Out of dozens of indicators of development of the public health system, we selected the indicator which is the most important for remote areas—the number of experienced doctors of all specialties. And, finally, efforts of public authorities aimed at creating decent living conditions for people characterize budget expenditures for social and cultural measures.

In terms of sustainability of the local economic and social system, the degree of implementation of the basic political value orientations—freedom and equality [27]—is of particular significance. In this regard, for the total characteristic of differentiation of regions by the quality of life, we should give consideration to figures of political target indicators, complementing the overall picture of human development—inequality of incomes and poverty.

Usually the integral estimate of the quality of life of the population is calculated with the use of the linear scaling method which is based on the detection of reference points (minimum and maximum values) and standardization of indicators using the following formula:

$$I = \frac{D - D_{\min}}{D_{\max} - D_{\min}},$$

where I is the private index; D, $D_{\min}$, $D_{\max}$ are actual, lowest and highest indicator values.

Since this research is aimed at analyzing the correlation between indicator values of the quality of life by regions, we shall take minimum values equal to zero.

Human development index was determined as a geometric average of private indices:

$$HDI = \left(I_{Health} \cdot I_{Education} \cdot I_{Income}\right)^{1/3}.$$

In order to classify HDI values into groups, cutoff points were determined based on average values of quartiles out of distributions of private indicators. The quartiles of the empirical distribution are values which divide the sample into four groups containing identical observed numbers.

The results of calculations of indices characterizing the degree of implementation of value orientations of society in the regions of the Russian Federation, are presented in Table 45.2.

Human Development Index in regions with very high level of development is 2.3 times higher than that in the lowest group; moreover, efforts of public authorities

Table 45.2 Classification of constituent entities of the Russian Federation by the quality of life of the population, 2017 (Calculated by the author in accordance with: Regions of Russia. Regions of Russia. Socio-economic indicators. 2018: P32. Statistical Compilation/Rosstat. M., 2018.—1162 p)

Groups of regions by the value of the human development index	Number of regions	Average human development index	Average index of the implementation of instrumental values	Population poverty rate, %
Very high	15	0.695	0.457	10.3
High	25	0.544	0.495	14.3
Average	27	0.445	0.447	14.9
Low	17	0.303	0.468	21.6
Ratio between of indicators of groups with very high and low level of development	–	2.29	0.98	0.48

aimed at equalizing the quality of life on average for all groups are approximately at the same level, which is indicative by values of the index of the implementation of instrumental values. The poverty rate in regions with high quality of life is 2 times lower than in the previous group.

Group 1 with very high Human Development Index traditionally includes constituent entities of the Russian Federation with developed economy, ranking first in regional ratings: Moscow and Saint Petersburg, the Republic of Tatarstan, the Voronezh Region, Krasnodar Krai etc. The lowest lines of the rating are occupied by the Republic of Ingushetia, the Jewish Autonomous Region, the Tyva Republic, the Republic of Altai, Zabaykalsky Krai etc.

Table 45.3 shows average values of indicators for the groups of regions based on which Human Development Index was determined.

The quality of life in the regions is differentiated to the maximum extent by the proportion of working-age citizens working out and playing sports on a regular basis as a result of the influence of such factors as suitable conditions for working out and playing sports, as well as value orientations of a personality. In group 1, the value of this indicator is 3.57 times higher than in the group with low level of the human development index. According to other indicators, the degree of differentiation is lower—by the number of university students per 10,000 people—by 1.76 times, by the amount of money income per capita with account of the minimum subsistence income—by 1.65 times.

As can be seen from Table 45.4, provision of the population with sports facilities in the regions with the highest level of human development is higher than in the previous group by 16.8%. In practice, distinctions can be more significant, since this indicator disregards such characteristics as proportions, technical features, and function of

Table 45.3 Indicators characterizing the quality of life of the population in the regions of the Russian Federation (2017)

Groups of regions by the value of the human development index	Number of regions	The proportion of working-age citizens working out and playing sports on a regular basis, %	The number of university students per 10,000 people	Ratio between per capita money income and the minimum subsistence income
Very high	15	29.1	345.3	3.86
High	25	20.4	298.2	3.11
Average	27	16.3	218.2	2.89
Low	17	8.2	195.8	2.34
Ratio between of indicators of groups with very high and low level of development	–	3.57	1.76	1.65

Table 45.4 Implementation of instrumental values in regions grouped by the level of human development, 2017

Groups of regions by the value of the human development index	Number of regions	The level of provision of the population with sports facilities judging from UVC, %	The number of doctors of all specialties per 10,000 people	Consolidated budget expenditures of constituent entities of the Russian Federation for social and cultural measures per capita, thousand rubles
Very high	15	0.555	48.1	40.5
High	25	0.536	47.1	44.9
Average	27	0.560	48.1	42.2
Low	17	0.475	46.6	44.7
Ratio between of indicators of groups with very high and low level of development		1.168	1.034	0.906

sports facilities. The relative number of doctors of all specialties is also lower in the previous group by 3.1%. Regional policy of equalization of living conditions was manifested in a relatively larger amount of expenses of consolidated budgets of constituent entities of the Russian Federation for social and cultural measures in the group with low level of development by 10.4%.

We assessed the implementation of political value orientations by the poverty rate (the number of the population with money income below the level of the minimum subsistence income as a percentage of the total number of the population of a constituent entity of the Russian Federation). This indicator characterizes the prevalence of absolute poverty, which is manifested in the inability to satisfy the essential needs for food, clothing and housing with the use of current money income. The availability of a slight proportion of poor people adversely affects the sustainability of the local economic and social system, the state of the economy and the behavior pattern of the population, and is a serious problem for regional authorities. The reproduction of poverty is a result of both the deprivation and immature needs, which generate the habitude of living in poverty and social degradation of a person, his/her spiritual and social impoverishment.

Population poverty rate is also an important measure of efficiency of the state regional policy, since it is the state that is responsible for the management of distributive and redistributive processes.

Social inequality by the income level (indicator of such political value as equality) is inapplicable in the studies in the field of the quality of life in the regions: inequality is more pronounced in relatively successful regions than in backward regions; all people are relatively poor in backward regions. Thus, in Moscow and Saint Petersburg R/P 10% ratio (the relation between average levels of money income is 10% of people with the highest income and 10% of people with the lowest income) in 2017 was 16.1 and 14.9 respectively, while in the Republic of Ingushetia which is at the bottom of regional ratings it was 10.5.

Regions with low level of human development are distinguished by high poverty rate—21.6%; moreover, it is even higher in certain constituent entities of the Russian Federation: in the Tyva Republic—40.5%, the Republic of Ingushetia—32.0%, the Republic of Kalmykia—27.3%, in the Jewish Autonomous Region—24.9%.

From among indicators characterizing political values, we are concerned with the indicator of population movement (net migration rate per 10,000 people) from constituent entities of the Russian Federation with low level of development to regions with better living conditions. In the Russian Federation, there is an ongoing out migration from near-border and backward regions to central and industrially developed regions and big cities, which should be regarded as a process that negatively characterizes the regional policy of the state. The highest degree of out migration can be observed from regions with special living conditions—the Jewish Autonomous Region, the Komi Republic, Chukotka Autonomous Region, the Magadan Region (97–132 people), as well as other regions: Kalmykia (97 people), the Republic of North Ossetia-Alania, the Republic of Daghestan, the Republic of Karelia, the Novgorod Region, the Oryol Region, and the Tver Region (over 30 people).

Thus, distinctions between regions in terms of the quality of life are defined not only by the income level, but also value orientations of the population, as well as efforts of public authorities aimed at creating social infrastructure.

45.7 Conclusions

Quality-of-life studies are mainly focused on methodological aspects of measurements of the quality of life. These estimates are based on the study of main problem areas (health, education, income etc.) with account of both physical indicators and the degree of personal satisfaction with life, availability of relevant statistical indicators, judging from the goal and the subject of regulation. Estimates are completed by the ranking of countries and regions according to the Human Development Index (quality of life).

In our opinion, quality-of-life studies should utilize axiological approach, which consists in the generation of a system of value orientations of society (terminal, instrumental, political) and assessment of the degree of implementation.

In the context of assessment of territorial differentiation, indicators of implementation of terminal values characterize physical and mental aspects of life of people, instrumental—efforts of public authorities aimed at equalizing the living conditions, political—the degree of stability of local economic and social systems.

The regions of the Russian Federation judging by the quality of life in them were compared by the degree of their physical and spiritual perfection (the proportion of the working-age citizens working out and playing sports on a regular basis), accessibility of higher education (the number of students per 10,000 people), level of income as compared to the minimum subsistence income. Regular sports activities constitute a component of a healthy lifestyle, characterize value orientations of a personality, focused on self-improvement, improvement of social status and material well-being, represent the possibility and the willingness to lead a full and active life.

All factors defining the lifestyle of the population in terms of large-scale involvement in working out and playing sports, can be conveniently divided into objective (material well-being, leisure time, accessibility of sports and recreation infrastructure) and personal—value orientations, culture and education.

The absolute poverty rate, which causes degradation, spiritual and social impoverishment, inactivity of this category of the population, is of particular importance both for the economic stability of regions and for the formation of value orientations.

Human Development Index in the group of regions with very high level of development is 2.3 times higher than in the group with low level of development; moreover, efforts of public authorities aimed at equalizing the living conditions slightly differ between groups in the mean. Moreover, differentiation of territories by the quality of life is mainly formed on the basis of the possibility to receive higher education and popularity of mass sports, and to a lesser degree—based on the difference in income of the population.

Regions with the highest quality of life are also leading in terms of implementation of political values: poverty rate in them is 2.2 times lower than the average poverty rate for a group with low level; the largest migration inflow of the population from other constituent entities of the Russian Federation can be observed.

References

1. Mstislavskiy PS (2002) Issues of theories and principles of life quality analysis. Uroven' Zhizni Naseleniya Regionov Rossii, No 2, pp 5–24
2. Felce D (1997) Defining and applying the concept of the quality of life. J Intellect Disabil Res 41(2):126–132
3. Schalock RL (2004) The concept of quality of life: what we know and do not know. J Intellect Disabil Res 48(3):203–216
4. Quality of life index: the economist intelligence unit's quality of life index. http://www.economist.com/media/pdf/Quality_Of_Life.pdf
5. Zubarevich NV (2007) Social development of the Russian regions: problems and tendencies of the transition period, 3rd edn. LKI, 264 pp
6. Zarakovskiy GM (2009) The quality of life of the Russian population: psychological components. Smysl Publishing House, 319 pp
7. Stiglitz J (2016) Mismeasuring our lives: why gdp doesn't add up. In: Stiglitz J, Sen AS, Fitoussi JP (eds) The report by the commission on the measurement of economic performance and social progress. Publishing House of Gaidar Institute, 216 pp

8. Cherkashin AK (2015) Dynamic and qualimetric models of assessment of the quality of life in a family (in Russian), Cherkashin AK, Leshchenko YA, Boeva AV (eds), Narodonaselenie, No 3, pp 61–76
9. Uglanova EA (2002) Economic psychology: procedural guidelines. Yaroslavl State University, Yaroslavl, 28 pp
10. Human Development Indices and Indicators. Updated Statistical Data 2018. Published for the United Nations Development Programme. Access mode: hdr.undp.org. Accessed 29 Aug 2019
11. National Human Development Report for the Russian Federation (2016) Analytical Center under the Government of the Russian Federation
12. Quality-of-life rating of regions of the Russian Federation—2018: RIA Rating
13. Molchanova EV, Burkin MM (2018) Public health in Russia and countries of Northern Europe. Narodonaselenie 21(2):84–98
14. Progress report on the implementation of the State Program of the Russian Federation "Development of Physical Education and Sports" for 2018. Access mode: www.minsport.gov.ru. Accessed 24 May 2019
15. Bahls M, Groß S, Baumeister SE, Völzke H, Gläser S (2018) Association of domain-specific physical activity and cardiorespiratory fitness with all-cause and cause-specific mortality in two population-based cohort studies. Sci Rep 8(1). ISSN 2045-2322. https://doi.org/10.1038/s41598-018-34468-7
16. Coenen P, Huysmans MA, Holtermann A, Krause N, Mechelen W (2018) Do highly physically active workers die early? A systematic review with meta-analysis of data from 193 696 participants. Br J Sport Med 52(20):1320–1326. ISSN 1473-0480 0306-3674, 1473-0480. https://doi.org/10.1136/bjsports-2017-098540
17. Gram B, Holtermann A, Søgaard K, Sjøgaard G (2012) Effect of individualized worksite exercise training on aerobic capacity and muscle strength among construction workers—a randomized controlled intervention study. Scand J Work Environ Health 38(5):467–475. ISSN 1795-990X 0355-3140, 1795-990X. https://doi.org/10.5271/sjweh.3260
18. Guidelines for the use of various forms of motivation to keep a healthy lifestyle, the use of capabilities of physical education and sports in health promotion, self-improvement, self-education and social adaptation intended for outreach activities. Moscow, 2015. Access mode: www.minsport.gov.ru. Accessed 24 May 2019
19. Decent Work Pilot Programme. http://www.ilo.org/public/english/bureau/dwpp/. Accessed 19 Apr 2015
20. Kirkman F (1981) Who cares about job design? Some reflections on its present and future. Int J Oper Prod Manag 1(2):3–13
21. Information about physical education and sports as at December 31, 2018. Form No. 1-FK. Access mode: www.minsport.gov.ru. Accessed 24 May 2019
22. Ayvazian SA (2012) The analysis of quality of human life and lifestyle. Nauka, 432 pp
23. Kal'yu PI (1988) Essential description of the concept of health and certain issues of reorganization of the healthcare system: an overview
24. Preamble of the Constitution of the World Health Organization
25. Avraamova EM (2019) What Russians have and what they need. Ekonomicheskoe Razvitie Rossii 26(8):93–95
26. Regions of Russia (2018) Socio-economic indicators. 2018: Statistical Compilation. Rosstat, 1162 pp
27. Rokeach M (1973) The nature of human values
28. Social atlas of Russian regions: independent Institute for Social Policy. Access mode: http://atlas.socpol.ru/

Chapter 46
Main Stages of Evolution of Economic Knowledge from Antiquity to Metamodern

Svetlana L. Sazanova and Galina N. Ryazanova

Abstract The article explores the main stages in the development of economic knowledge in conjunction with the development of philosophical knowledge. The author found that ancient philosophy had a significant impact on ancient economic thought. The philosophy of Christian Aristotelianism determined the attitude of the scholastics to the economic problems of the Middle Ages. The philosophy of the Renaissance and the New Age has influenced the formation and development of economic science. Classical philosophy has become the methodological basis of classical political economy. The philosophy of positivism led to the formation of neoclassical economic theory. The methodological pluralism of metamodernism has generated a variety of trends in economic science. Based on the revealed patterns, the author predicts the further development of modern economic science.

Keywords Economic thought · Classical school · Marginalism · Neoclassical · Marxism · Evolutionary economics · Institutionalism · Neoinstitutionalism · Economic methodology · Philosophy · Positivism · Pragmatism · Rationalism · Modernism · Postmodernism · Metamodernism

JEL Code B1 · B2 · B4

S. L. Sazanova (✉) · G. N. Ryazanova
State University of Management, Moscow, Russia
e-mail: sazanova@mail.ru

G. N. Ryazanova
e-mail: ryazanovagn63@gmail.com

S. L. Sazanova
Financial University Under the Government of the Russian Federation, Moscow, Russia

T. Kolmykova and E. V. Kharchenko (eds.), *Digital Future Economic Growth, Social Adaptation, and Technological Perspectives*, Lecture Notes in Networks and Systems 111, https://doi.org/10.1007/978-3-030-39797-5_46

46.1 Introduction

In modern economic science there is no comprehensive analysis of the evolution of the development of economic knowledge in conjunction with the development of philosophical knowledge; as well as the challenges posed by the philosophy of postmodern and metamodern to economic science in general and economic theory in particular. The study of the main stages of the evolution of economic knowledge in conjunction with the development of philosophical ideas will reveal the patterns and main stages of this process, which, in turn, will reveal the logic of the development of economic science in general and economic theory in particular.

The aim of the study is to identify the general logic of the development of economic knowledge in conjunction with the development of philosophical knowledge. The scientific novelty of the study is to identify patterns of evolution of economic knowledge in general economic theory in particular, in conjunction with the development of philosophical knowledge.

The scientific problem of the study is as follows: modern economic theory is in crisis, which is expressed in the absence of a methodology that is relevant to the current state of the economic system and its subsystems, which means that there are no reliable theoretical and practical research results and recommendations for improving the management of economic systems and organizations. The solution to this problem is to improve the methodology of economic theory by identifying the relationship of philosophical and economic knowledge.

46.2 Methodology

The study is based on a synthesis of philosophical and economic methodology, general and special scientific methods. The method of rational reconstruction of science has revealed the relationship of philosophical and economic ideas. Using a causal method, factors that influenced the development of economic theory are determined. The historical method made it possible to identify patterns in the development of economic knowledge in general and economic theory, in particular, in a historical context. The method of expert assessments made it possible to carry out a comparative analysis of expert points of view of domestic and foreign scientists on the problem of the relationship between the evolution of economic and philosophical knowledge. On the basis of general scientific methods (methods of comparison, typology, scientific abstraction, analysis and synthesis, and others), the challenges of metamodern are revealed and the nature of their influence on economic science in general and economic theory in particular is revealed.

46.3 Results

Modern economic theory is the methodological and theoretical basis of economic science as a whole and its important sections such as the economic theory of organizations, the theory of entrepreneurship, the theory of economic behavior and others. Since economic theory studies the relations of economic entities in the process of relations regarding the creation, distribution and use of economic goods, it is also interconnected with other human sciences, such as philosophy, psychology, and sociology. The philosophical foundations of economic theory have been the focus of attention of domestic and foreign researchers in the history and philosophy of science for many decades. O. Ananyin, G. Kleiner, V. Polterovich, B. Erznkyan, R. Nureev, Yu. Osipov, E. Popkova, O. Sukharev, V. Tambovtsev, S. Kirdina-Chandler made a huge contribution to the study of the relationship between philosophy and economics., E. Mayburd, M. Blaug, K. Popper, T. Kun, I. Lakatosh, R. Valenchik and others, which, on the one hand, contributed to the development of the methodology of economic science and its heuristic potential, and on the other, it allowed to identify the crisis state of modern economic theory.

The crisis of scientific knowledge is manifested, in our opinion, primarily in its inability to answer the challenges of our time, and therefore, to fully realize the functions of science (cognitive, practical, philosophical, etc.).

Among the challenges of the modern world that affect the state and development of economic science, we can distinguish economic, technological, worldview, geopolitical and others. The most important of them are:

- acceleration of scientific and technical progress (digitalization, nanotechnology, etc.);
- instability (lack of sustainable development) of the economic environment;
- instability of the geopolitical situation;
- the desire for control over the natural resources, human capital and space of some countries by others;
- strengthening control over human choice in all areas of activity and even thinking (due to the development of cognitive psychology, neuroeconomics, digital technologies);
- global environmental issues.

At the same time, we are witnessing a crisis of modern humanitarian knowledge, which is expressed in the conflict of value "worlds" (values of self-expression vs. traditional values); the crisis of modern economic and managerial science (lack of effective strategies for sustainable development).

A study of the evolution of philosophical and economic ideas made it possible to establish a relationship between them. In the development of economic knowledge, we can distinguish the pre-scientific period (from ancient times to the XVIII centuries) and the scientific period (from the XVIII century to the present day). In the pre-scientific period, economic knowledge developed within the framework of religious and philosophical systems. The classical ideas of Xenophon, Plato and

Aristotle influenced the economic ideas of antiquity. The methods of cognition of economic processes and phenomena were identical to the methods of philosophical cognition: contemplation, empiricism, rationalism and Aristotle's logic. The main problems of the economic thought of the ancient period were the problem of managing the household and the state, the problem of the division of labor, the problem of money and money circulation, the problem of value and exchange, the ratio of money and subsistence economy (economics and statistics). The economic ideas of antiquity and its philosophy had a huge impact on the economic thought of the Middle Ages.

The philosophy of the scholastics was based on a synthesis of Christian dogma and the philosophy of Aristotle (Christian Aristotelism of Thomas Aquinas). The scholastics developed the idea of the natural laws of Aristotle and applied it to the world of economic phenomena and processes. Later scholastics first tried to explore economic phenomena outside of a religious or philosophical context.

The philosophy of the Renaissance (XV–XVI centuries) had a great influence on the economic views of the New Age. Distinctive features of this period were: the naturalistic philosophy of Paracelsus, Telezio and other philosophers; Montaigne's skepticism; heliocentric system of the world of Copernicus; anthropocentrism; desire to apply mathematical methods in the knowledge of nature (Nikolai Kuzansky). Under the influence of philosophical ideas of the Renaissance, mercantilism was formed—the first system of economic policy. The mercantilists did not create a holistic economic theory, but their experience-based recommendations in economic policy were quite effective. However, the empiricism of mercantilists did not allow them to reveal the essence of economic phenomena and processes. For example, they explained profit as "profit from alienation", the result of an nonequivalent exchange.

The philosophy of the New Age (XVII–XVIII centuries) presented to humanity the experimental natural sciences of F. Bacon, the mechanistic picture of the world of T. Hobbes, the rationalism of R. Descartes. The predecessors of classical political economy (W. Petty, R. Cantillon, D. Hume, J. Locke, F. Quesnay) applied mathematical and natural science methods to the study of economic processes and phenomena, since they themselves were mainly philosophers. The synthesis of rationalism and empiricism, deduction and induction ultimately led to the formation of classical political economy—the first scientific economic school.

For various schools of classical philosophy of the XVIII–XIX centuries. Empiricism, rationalism, materialism, idealism, dialectics, faith in the limitless possibilities of reason, the desire for absolute truth, a teleological approach, and essentialism were characteristic. This period is characterized by a variety of not only philosophical directions, but also areas of economic science: classical political economy, German historical and economic school, Marxism. The rationalism and universalism of English philosophy has become a hallmark of English political economy. Empiricism and materialism of German philosophy became the basis of the German historical and economic school, and the combination of Hegel's dialectical logic, Feuerbach's materialism and the achievements of English political economy formed the basis of Marxism.

The development of rationalistic methods of cognition led to the emergence in the 30–70s. XIX centuries philosophy of positivism (O. Kont), according to which the source of genuine (positive) knowledge is specific sciences and their synthetic associations, and not philosophy. Positivism suggested abandoning the knowledge of the essence of phenomena and processes and focusing on their functions. The discovery of the potential for ultimate analysis in economic research became the prerequisite of the "marginalist revolution" of the 1870s. Marginalists have proven that reliable economic knowledge can only be obtained using marginal analysis and mathematical methods.

The ideas of the philosophy of positivism (1830–1940s) were realized in neoclassical economic theory: atomism (economic individualism), formalism, relativism, economic universalism, faith in the endless progress of a market economy, economic imperialism, economic liberalism. In the XIX centuries. orthodoxy and heterodoxy of economic science are formed in an explicit form. Neoclassical economic theory is becoming the mainstream, and heterodoxy is based on Marxism [9] and the German historical and economic school.

The philosophy of modernism (70s. XIX–1st half. XX centuries) was based on the idea of endless progress, which was understood as the total submission of the world to human will. She was also characterized by relativism, holism, denial of absolute truth, evolutionism; recognition of the influence of values, cultural norms, institutions on the knowing subject. In economic science at the turn of the XIX–XX centuries. institutional theory appears. The "old" institutionalism was based on the methodology of holism (economic collectivism), relativism, evolutionism, and also the recognition of the influence of socio-economic institutions, stereotypes of thought, cultural norms on the cognizing subject [7]. However, the influence of formalism in philosophy, economic science and social thought persisted, so areas such as the Russian School of Economics and Mathematics and Keynesian economic theory gained great influence. These scientific schools had significant methodological and theoretical differences, but they were united by economic collectivism and the use of mathematical methods (aggregation methods).

Postmodernism (2nd half of the 20th–beginning of the 21st centuries) was characterized by deconstruction, irony, stylization, relativism, nihilism, discreteness, locality, fragmentation, randomness, emphasis on the game principle, anti-anthropocentrism, the absence of tough oppositions and universal ideals. The orthodoxy of the economic science of this period was the Keynesian-neoclassical synthesis (the result of the neoclassical revival). The heterodoxy of this period was composed of a wide range of directions: new and neo-institutionalism, institutional-evolutionary economic theory, and behavioral economics [4]. The intellectual opposition of postmodern philosophy has become systemic economic theory.

An analysis of the relationship between the development of philosophical and economic knowledge showed that philosophical paradigms and paradigms of economic knowledge in the XVIII–early XXI centuries were largely interconnected. But can the philosophy of metamodernism give rise to a new direction in economic theory?

Metamodernism (metamodern)—from the Greek "Metaxis" (the oscillation between two opposing concepts and the simultaneity of their use). The term was proposed by Dutch culturologists T. Vermeulen and R. der Akker in "Notes on Metadomism" (2010). In 2011, L. Turner published on his website the "Metamodern Manifesto", in which fluctuations were recognized as the natural order of things, therefore, "movement should be carried out only by fluctuations between positions with diametrically opposed ideas, acting as pulsating poles of a colossal electric machine that drives the world into action" [1, 10, 11]. What should economic theory become in the 21st century in the context of its relationship with philosophy?

What picture of the world of economic phenomena and processes can modern economic theory offer at the boundary of postmodern and metamodern? Economic theory examines the behavior of people in the process of creating, distributing and using economic goods. The economic man of the modern era was rational and sought to maximize utility. The economic man of the postmodern era was limitedly rational and sought to obtain a satisfactory result from the possible ones, institutions influenced his choice. The economic man of the metamodern era is organically irrational, therefore he constantly fluctuates between extremes, using them fully or partially to satisfy his desires. The values of economic activity have long been determined by traditional values, but from the second half of the 20th century. they are increasingly influenced by the "values of self-expression", which is fully consistent with the philosophy of metamodern.

To what extent are metadomes of economic science most compatible with such areas of economic science as behavioral economics, sharing economics, the theory of business ecosystems, the theory of "nudge", system economics, and the theory of socio-economic ecosystems? Behavioral economics is based on the hypothesis of irrational behavior; it is irrational behavior that prompts a person to make a choice between alternatives. This choice is different each time. The sharing economy denies the need for ownership of goods to conduct economic activity, but recognizes the adequacy of access to the use of goods. The rejection of ownership makes the economic entity more mobile without restricting the "fluctuations" between different options for goods, markets, and spaces. Business ecosystems (digital platforms and others) create an environment for "fluctuations" of unlimited interactions at any given time [6, 8]. And the theory of "nudge" allows us not to predict the behavior of economic entities, but to control it, initiating impulses of "fluctuations". The theory of systemic economics and the theory of socio-economic ecosystems, in our opinion, are in intellectual opposition to them [2, 3, 5]. These theories do not imply hesitation; they imply orderliness, systematicity. Discussions about how economic theory will develop in the era of metamodern should undoubtedly continue.

46.4 Conclusions

A study of the development of philosophical and economic knowledge from ancient philosophy to metamodern has revealed the influence of philosophy on the evolution of economic science at each stage of its formation and development. A retrospective ontological analysis of various areas of philosophy showed that the evolution of philosophical views on the problem of the origin of the world, its universal connections, basic characteristics and laws always influenced ideas about the origin of economic phenomena, their properties and characteristics, and the relationships between them. The economists perceptions of the cognoscibility of the world of economic phenomena and processes have changed in conjunction with the development of philosophical perceptions of the cognoscibility of the world (epistemology). The values of people's economic activities have also been rethought as social values evolved. Currently, the value of self-expression has a significant impact on the economic activities of people. Thus, the study shows that between the development of economic and philosophical knowledge there is a relationship at the ontological, epistemological, axiological levels.

Acknowledgements The study was carried out with the financial support of the Russian Foundation for Basic Research in the framework of the research project No. 19-010-00905.

References

1. Kesting S (2019) Rigor in economics means pluralism, Forum Soc Econ 48(4):383–393. https://doi.org/10.1080/07360932.2018.1432496. stefan.kesting@aut.ac.nz
2. Kleiner G (2017) From the economy of individuals to systemic economy. Voprosy Ekonomiki 8:56–74
3. Kleiner GB (2011) A new theory of economic systems and its applications. Her Russ Acad Sci 81(5):516–532
4. Kleiner GB (2018a) Three questions to political economy (An attempt of system introspection). Voprosy Ekonomiki 8:118–127
5. Kleiner GB (2018b) Humanistic management, social management and system management—the way to the management of the 21st century. Rossiiskii Zhurnal Menedzhmenta 16(2):231–252
6. Pulyaeva VN, Gibadullin AA, Kharitonova EN, Kharitonova NA (2019) Problems and prospects of electric power enterprises in monotowns. In: International scientific conference on global challenges and prospects of the modern economic development (GCPMED), vol 57. Samara State Univ Econ, Samara, Russia. Dec. 06–08, 2018, pp 801–813
7. Sazanova SL (2019) Structural modeling of the institution of Russian entrepreneurship in the spirit of "old" institutionalism. In: Popkova EG (ed) The future of the global financial system: downfall or harmony, vol 57. Lecture Notes in Networks and Systems. ©Springer Nature Switzerland AG 2019, ISC 2018, LNNS 57, pp 451–460. https://doi.org/10.1007/978-3-030-00102-5_48
8. Sazanova SL, Ryazanova GN (2019) Problems and opportunities of development of the agricultural industry of Russia from the point of view of Marxist theory. Marx and modernity: a political and economic analysis of social systems management. A collective monograph. In:

Alpidovskaya ML, Popkova EG (eds) A volume in advances in research on Russian business. Information Age Publishing Inc, 651 pp, pp 599–608. ISBN: 978-1-64113-749-2 (Paperback), 978-1-64113-750-8 (Hardcover), 978-1-64113-751-5 (ebook)
9. Sazanova SL, Sharipov FF, Dyakonova MA (2019) Spatial economics, geopolitics, and Marxism. Marx and Modernity: a Political and Economic Analysis of Social Systems Management. A collective monograph. In: Alpidovskaya ML, Popkova EG (eds) A volume in advances in research on Russian business. Information Age Publishing Inc., 651 pp, pp 279–288. ISBN: 978-1-64113-749-2 (Paperback), 978-1-64113-750-8 (Hardcover), 978-1-64113-751-5 (ebook)
10. Turner L (2011) The metamodernist manifesto. http://www.metamodernism.org/
11. Venkova AV (2018) Identification politics of metamodern art. Tomsk State Univ J Cult Stud Art Hist 32:203–213

Part III
Technological Perspectives and Future Horizons of Development of the Digital Economy

Chapter 47
Digitalization of the Russian Economy (Target-Oriented Approach): First Results, Risks and Prospects

Inna V. Mitrofanova, Svetlana G. Pyankova, Inna A. Ryabova, Larisa V. Ob'edkova and Anastasia B. Shcherbina

Abstract Since 2017 in Russia the national digital strategy is being implemented, the main aims and goals are presented in the target program "Digital economy of the Russian Federation for the period 2018–2024" which is synchronized with the "Strategy of development of informational society in the Russian Federation for the period 2017–2030". The analysis of the results of the first years of the program realization showed that by the beginning of 2019 only one out six parts of the programs which deals with the staff training has been completely implemented. Within this part two projects were launched, i.e. the competency based profile of CDO (the launch of the program Chief Data Officer) and all-Russian "Lesson of Digit" (the first digital learning and teaching support kit for secondary school students). It was identified that an important result of the realization of such a part of the program as "Digital technologies" became the making of a list of potential Russian and international leading research centers in the "cross-cutting" technologies which pretend to have the government support. Also the project of the transfer of 52 federal organizations and their territorial bodies to the accounting with the use of cloud technologies has begun. A significant lagging of Russia in the development of the digital economy from world leaders (41st rank in the world rating according to the preparedness for

I. V. Mitrofanova (✉) · I. A. Ryabova
Federal Research Centre, Southern Scientific Centre of the Russian Academy of Sciences, Rostov-on-Don, Russian Federation
e-mail: mitrofanova@volsu.ru

I. A. Ryabova
e-mail: mia05011986@yandex.ru

S. G. Pyankova
Ural State Economic University, Yekaterinburg, Russian Federation
e-mail: silen_06@list.ru

L. V. Ob'edkova · A. B. Shcherbina
Volgograd State University, Volgograd, Russian Federation
e-mail: econmanag@volsu.ru

A. B. Shcherbina
e-mail: htes@volsu.ru

T. Kolmykova and E. V. Kharchenko (eds.), *Digital Future Economic Growth, Social Adaptation, and Technological Perspectives*, Lecture Notes in Networks and Systems 111, https://doi.org/10.1007/978-3-030-39797-5_47

a digital economy and 38th rank according to the importance of economic and innovative results for the use of digital technologies) is explained by the imperfectness of legislation in a digital economy under the conditions of an unfavorable environment for business and introduction of innovations, low level of use of digital technologies by business structures.

Keywords Russia · Digital economy · Target program · Platforms and technologies · Informational society · Informational infrastructure · Informational security

47.1 Introduction

Today the digitalization process assumes a global character and the transfer to the keeping and data processing in the digital format leads to an exponential growth of the world data volume. In many countries of the world the national digital strategies have been adopted and now they are being implemented and Russia is not an exception.

In July 2017 the Government of the Russian Federation approved the program "Digital economy of the Russian Federation for 2018–2024" (hereby: "Digital Economy of the RF …") which obtained the status of the national one. The program is directed on the creation of conditions for the development of knowledge society in the country, well being growth and quality of life of citizens by means of the growth of accessibility and quality of goods and services produced in the digital economy with the use of modern digital technologies; growth of awareness and digital literacy of the population (both receiving and providing digital services including the public one); improvement of the accessibility and quality of public services for citizens including the electronic ones due to the fact that they will be provided with the help of modern digital technologies and, consequently, in the digital economy; growth of quality of healthcare and education and also security including the digital one both domestically and abroad [17].

The program "Digital Economy of the RF …" is synchronized with the "Strategy of development of informational society in the Russian Federation for the period 2017–2030" [7] stems from the fact that the digital economy presents the economic activity, the key factor of production in it are the data in digital form and they contribute to the creation of the informational space taking into account the needs of public and society in the obtaining of qualitative and reliable information, development of informational infrastructure of Russia, creation and use of Russian informational and telecommunication technologies and also the creation of a new technological basis for social and economic sphere.

The program sets aims, goals, directions and terms of realization of the main measures of public policy in the creation of necessary conditions for the development of the digital economy in Russia in which the data in digital format are a key factor of production in all spheres of social and economic activity. This is a necessary condition for competitiveness growth of the country, quality of life of the population, provision of economic growth and national sovereignty.

The main aims of the program of digitalization of the Russian economy are as follows: (1) creation of the environment of digital economy in which the digital data are the main factor of production in all spheres of social and economic activity. This environment is meant to provide an efficient interaction of the government, business, scientific and educational community and citizens; (2) elimination of existing restrictions capable of preventing the appearance and development of highly technological businesses and also the removal of new barriers both in traditional and newly established industries in technology intensive markets; (3) competitiveness growth in the world market of both separate branches of national economy and its economy in the whole.

The program provides for three levels of digital economy which interact one with another and influence the life of the country, society and every citizen of the state: 1 are platforms and technologies; 2 is the environment creating the conditions suitable for the creation and development of platforms and technologies and also for an efficient interaction of economic agents and branches of the economy; 3 are the markets and industries (spheres of activity) in which the interaction of specific agents takes place and specifically of consumers and suppliers of goods, services and works. In the result of program's implementation at least 10 national company-leaders are expected to be created. They are supposed to become technology intensive enterprises dealing with the development of "cross-cutting" technologies which are mentioned above and emerging technologies and management of digital platforms working in the global market.

The program "Digital Economy of the RF…" includes the work in six directions: "Informational infrastructure", "Informational security", "Digital technologies", "Staff for digital economy", "Normative regulation", "Digital public administration" within which the specific target indices are chosen which are expected to be achieved by the year 2024. The article analyses the first results of the implementation of the program of digitalization of the economy of Russia.

47.2 Literature Review and Methodology

For the first time the term "digital economy" was used by the Canadian scientist D. Tapscott in 1995 in his book "The Digital Economy" and it meant the economy based on digital technologies [24].

The report of the Organization for Economic Cooperation and Development (OECD) uses this term not as a separate phenomenon but it is rather used for the reflection of the process of digitalization of various sectors of the world economy including the banking, retail trade, power energy supply, transport, education, publishing, mass media, healthcare and so on [15].

Theoretical aspects of digitalization as a phenomenon and a process and also the practical aspects of the introduction of digital economy in Russia and other countries were reflected in the papers of Avdeeva [1], Biychuk [3], Bondarenko [4],

Kapranova [10], Mitrofanova [12, 13], Pyankova et al. [16], Hungerland [9], Schwab and Sala-i-Martín [20, 21] and others.

The methodology of research is based on the combination of the following methods: system, historical and logical, institutional and evolutional approaches including the particular general scientific methods of subjective and objective, factor and comparative analysis.

47.3 Results

47.3.1 Digitalization of Russian Economy: First Results

In the fall of 2017 the autonomous non profit organization "Digital Economy" was established which is devoted to the formation of business orders, monitoring of the program, assessment of the efficiency of its implementation, becoming an advisory board which determine the strategic and technical problems and also proving the cooperation between business, expert communities and public authorities at the implementation of the program "Digital Economy".

In January 2019 the non profit organization "Digital Economy" published on its web site the visual navigator through the national program "Digital Economy of the Russian Federation" which presents the visualized "Scheme of movement to the digital economy" [23] in the form of a metro map.

Recently in global companies a principally new managerial position Chief Digital Officer (CDO) has been introduced. The senior officials and top managers of the leading transnational companies point out to a critically important role of such managers. In Russia the CDO position is only appearing.

CDO as and also their teams are not only just key managers who are competent in the data management but they are also people who create a new managerial culture. The culture which welcomes the grassroots initiative and there are no rigid dogmata and operating procedures, the decision making is carried out on the basis of objective data and a preferable way of problem solution is not the spending of public funds but a mutually advantageous cooperation [5].

In the frame of the national project (program) "Digital Economy of the Russian Federation" it is expected to train 30,000 people till the end of 2019 for CDO as and 270,000 people are to be trained till the end of year 2024. Today the number of managers who are capable of managing the data in the system of public administration is 1–2%. At the same time in every federal, regional or municipal executive authority at the level of vice ministers and lower, in companies, enterprises, industrial systems the professionals are expected to work who can manage the data. In order to solve the problems of the lack of the CDO personnel the number of such specialists should grow 10 times.

The Ministry of Digital Development, Communications and Mass Media of the Russian Federation, Ministry for the Development of the Russian Far East, NTI 20.35

University and Far Eastern Federal University came to the agreement to teach and form the managerial teams for the work with data (Chief Data Officer) in government authorities and also to teach the public, to create the educational environment for public officials in the competence development necessary for the work in the conditions of digital economy [22]. In July 2018 in the Far Eastern Federal University a ten day intensive educational course "Island 10–21" took place in which businessmen, public officials and innovative entrepreneurs were taught to deal with the artificial intelligence, block chain, advanced technologies during several days. On the basis of CDO program 230 people were taught among them there were vice governors, regional ministers, heads of departments for informational technologies from 38 regions of Russia. They were taught how to gather the data about a real ongoing process in how to analyze these data, how to build the managerial system on the basis of these data which will be constantly improving the life of people.

The second project of staff training is the "Lesson of Digit", which became an example of a successful cooperation of the state and business in education, was launched in December 2018 and has already encompassed millions of schoolboys and girls. This lesson is planned to be given monthly. Besides, in February 2019 a lesson under the support of Sberbank devoted to artificial intelligence took place across the country and in the schools of the Russian language abroad.

According to the program "Digital Economy of the RF …" the number of graduates of educational organizations of higher education according to training specialization which are connected with informational and telecommunication technologies is expected to reach 120,000 by year 2024. The corresponding trend has already been formed i.e. about 80,000 people are taught in the higher educational programs devoted to the training of IT specialists and among them 27,000 student are trained for IT profession against contracts under the methodological and financial support of technological companies [27].

Within the part of the program "Informational infrastructure" the first from the planned projects is only at an active stage of development. This is the concept of the construction of narrow band Wi Fi network entitled "Internet of things".

Within the direction "Laws and regulations" the program progressed the most in comparison with other parts. The reason is that the laws and regulations is in fact the basis for other directions and requires an accelerated development. The map of measures in legal regulation includes more than 50 drafts of legal and normative acts, passed laws and subordinate legislation. In every of the mentioned directions a separate group is working for which its own aims and goals are set. One of the groups is busy with the creation of legal conditions for the formation of the sphere of electronic public data turnover. The Bank of Russia assumed the duties in the organization of the work in this sphere.

The work in this group is carried out according to two laws: the law on information and archives. According to the first one the replacing scanning is being introduced. It means the opportunity of the creation of the notarized electronic copies which the same legal force the original version with the consequent destruction of the paper original of the document. The businesses initiated this process who complain about a large number of paper documents for control of public authorities. They were supported by a large number of federal executive authorities who were found in a similar situation. The question is not about the archived papers or documents with an extended storage period but about 95% of documents whose storage life is limited by 5 years. It is planned to write into law the opportunity of creation of notorized electronic copies. It is said that it will be possible to make the notarized copied for specific types of documents with the approval of the authority responsible for the documents. The algorithm of the replacing scanning will be written into subordinate laws.

In December 2018 the list of key laws was determined whose soonest approval by the State Duma (Lower Chamber of the Russian Parliament) of the Russian Federation is considered to be a priority. The most important laws according to the opinion of the business are four drafts of law which are directed to the determination of digital rights, regulation of the crowd funding sphere, introduction of notary services online, legalization of electronic deals and smart contracts [29].

The second part of the program is the development of the procedure of migration and conversion of electronic documents which is expected to be included into the "Law on Information". The third norm which is suggested for introduction into the same law it is the question of documents' and information destruction. Today the new norms of information holding are being introduced. For instance, The "Rostelecom" company is obliged to keep the biometric personal data during 50 years. The law on "Archive keeping" does not reflect the questions of document destruction. In general in the law mentioned above the norms which forbid the destruction of long term storage documents are provided.

It is planned to make corrections into the Civil Code of the Russian Federation and a number of other laws and subordinate legal acts in part of conclusion of deals and contracts. Already today the courts consider the contracts concluded via the e-mail but they ask to capture this norm in Civil Code. The normative base for smart contracts (automated self executable contracts) is being developed separately. Today the courts are actively considering as evidence various electronic documents (electronic mail, data base). It is necessary to improve the legislative regulation in order to standardize the approaches of the courts and to make it so that the court's decisions do not depend on the qualification of a judge.

In the introduction of electronic personnel documents' turnover the work is being lead in three key directions: (1) key document turnover in the organizations, the opportunity is analyzed to make a part of documents electronic especially of a short term storage period and which do not deal with pension rights of citizen; (2) electronic documentation of labour and pension rights of citizens (electronic employment record book); introduction of electronic employment record books which was preliminary planned for the end of 2019; (3) electronic control over the administration of labour legislation and labour protection with the help of labour inspection. Here it is spoken about the necessity of working with a digital format of documents. For example, it can be the signature of the worker going on shift at a dangerous enterprise. This work is carried out by the Ministry of Labour and Social Protection of the Russian Federation.

Since April 2018 the experiment in the use of electronic personnel documents' turnover is being carried out in 10 or 12 companies including the "Russian Railways". During this experiment the electronic labour contracts are being concluded, the personnel receives the acknowledgement of the salary transfer via the e-mail, the vacation periods and temporary detachments on duty are recorded in the electronic format, record of working time, issue of work permits is made electronically. The end of this experiment was expected by the end of 2018. The main question was whether this document turnover was worth transferring to electronic format, what the advantages would be and how numerous they would be.

The correction in the law on electronic signature is being prepared, a serious revision of notary legislation is planned and it is connected with the opportunity of use of electronic documents [18].

One of the directions of the program is the "Informational security" but in 2018 the situation in this part worsened in comparison with year 2017. From January to November 2018 according to the data of the Ministry of Interior of the Russian Federation, 156,307 crimes connected with the use of computer and telecommunication technologies was recorded and the growth rate of this type of crime was 90% higher that during the similar period 2017 (82,440 crimes) [30].

Within the direction "Digital Technologies" in October 2018 the working group "Creation of research competences and scientific and technological stock" at the Autonomous Non Profit Organization "Digital Economy" summarized the results of the work for the third quarter of 2018 taking into account the transformation of the program "Digital economy of the RF…" into the homonymous national project. One of the results became a preliminary, unpublished list of potential Russian and international leading research centers in "cross-cutting" technologies which expect the government support and also one of the results is also the formation of the list of legal acts and study manuals necessary for the realization of measures within the program "Digital Technologies" in 2019 and the schedule of their approval by the Government of the Russian Federation [8].

In 2018 a project was launched implying the transfer of 52 federal executive authorities and about 1500 of their regional offices and also of organizations connected with them to a centralized book keeping on the basis of use of cloud technologies. The first "pilot" executive authority became the company "Rosakkreditatsiya" (RusAccreditation or Federal Service for Accreditation) and in 2020 the transfer to a new solution of the rest is expected to happen. The service of the cloud book keeping is based on technical solutions "1C" [26]. The centralization denotes the acceptance of a transparent presentation of assets of the whole nation, business processes and common methodology as the Federal Treasury expected to supervise the project. This technology allows managing more efficiently the public finances by means of analytical instruments.

In the program "Digital economy of the RF …" the indicators for 2016 (it is the last year for the moment of program's approval) are presented. Today there is an opportunity of comparing these indicators with the data for 2017. In 2016 the share of people using the Broad Band Internet was 18.8% and by the end of 2017 this indicator reached already 21%. In 2016 71.3% citizens of Russia used the mobile access to the Internet and in 2017 this indicator increased by 79.9% [14, 16, 28].

In the program "Digital Economy of the RF…" it is mentioned that by the beginning of year 2017 the Russian market of commercial centers of keeping and data processing grew up to 14.5 billion rubles what is 11% higher than in 2016. In the program the forecasted and real indicator (14.5 billion rubles) of the volume of the market of commercial centers for data processing is mentioned. In 2017 in the Russian data centers 34.9 thousand storage points were established what is 13% more than in 2016 [2].

Till 2016 a stable growth of "cloud" technologies market was observed or about 40% per annum. In September 2018 a new report was published according to which the Russian market of "cloud" technologies including the public and private clouds in 2017 grew by 49% and made up $663,74 million [6].

In the international rating of global competitiveness World Economic Forum's Network Readiness Index, Russia had the 43rd rank in 2016 considerably lagging behind many more competitive world economies such as: Switzerland, Singapore, USA, Netherlands, Germany, Sweden, Great Britain, Japan, HongKong and Finland. A low level of innovations and undeveloped business, also insufficiently developed public and private institutes and the financial market are "bottlenecks" for the competitiveness of Russia for global digital market. In the report for 2017 Russia already has 38th rank having advanced Panama, India, Kuwait, Poland, Lithuania [20, 21].

47.4 Prospects and Risks of Digital Economy Development in Russia

The prospects of digital economy development in Russia directly depend not only on the potential of the country from the point of view of a maximal transfer of analogous data to digital ones and this concerns not only the economy but also all types of social life, ability to use skillfully and to analyze these data and on the degree of realization of this potential and also the reduction to minimum of the possible risks which already arise and will be appearing during the realization of the program "Digital Economy of the Russian Federation …".

The prospects of the digitalization of the economy can be analyzed from two points of view. Firstly, it is important how and by whom the program is realized. Secondly, it is important how actively the population of the country is involved into this process. "Cisco" in its research published the index of the preparedness of the countries and judging from the indices, the largest part of the countries have specific target values and in these directions an active work is being carried out. For example, it is the use of Internet, in particular Broad Band Internet, smart phones, cloud technologies, quality of mathematical education; exports of top technologies, success of government in the promotion of informational and communicative technologies. Every from the indicators got the grade from 0 to 25. The preparedness of Russia was assessed at a medium level with the total grade 13.13 (Fig. 47.1).

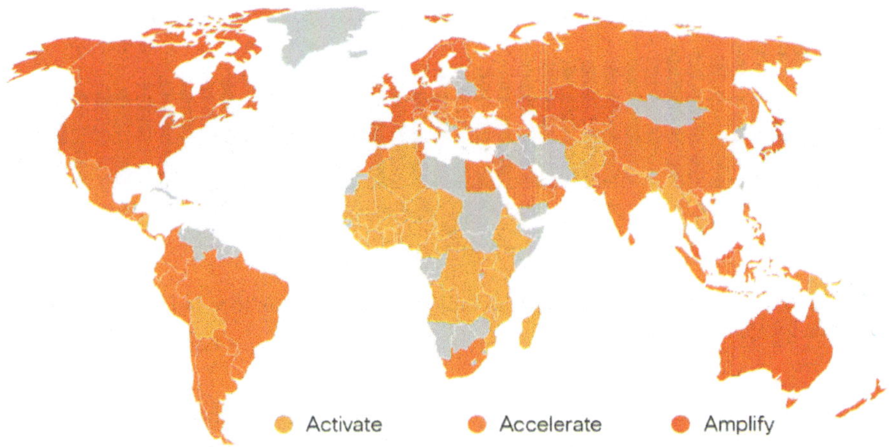

Fig. 47.1 Stages of digital preparedness [19]

The digital preparedness of the most developed the countries in the digital sphere such as for instance, Denmark, Norway was estimated in 17.27 and 17.38 correspondingly. The average index of the countries with a high level of preparedness made up 16.38; the index of the countries with a medium level is 12.49; the index of the countries with a low level of preparedness is 8.10 [19].

The determination of the group of the countries according to the level of preparedness in which Russia is found sounds as follows: a moderate stage of preparedness to digital technologies with the necessity of investments into critically important human needs, development of human capital and improvement of business environment in the region. Such countries are usually found in Asia, South America and Eastern Europe. Among the factors under analysis which are the weaknesses of modern Russia, the conditions for business and the share of youth among the population are analyzed. Thus, the share of the population of Russia aged from 0 to 19 for the beginning of year 2017 was 16.7% when in highly digitalized countries this indicator is significantly higher: in Iceland it is 26%, in Denmark it is 23% [25].

There exists a direct relationship between the share of the country's population who use the global net and the share of people of senior generation who use the Internet, the degree of involvedness into the social globalization and also with the share of citizens who shop online and at a lesser degree who access the public services via the Internet. In order to get a significant growth of population's share using the Internet in addition to the growth of the mentioned indicators it is necessary to reduce maximally the gap between the shares of Internet users among the group of people of higher education with different levels of education. The gap between the shares of users of global net according to ages should be reduced by means of increasing the number of senior age users (Fig. 47.2).

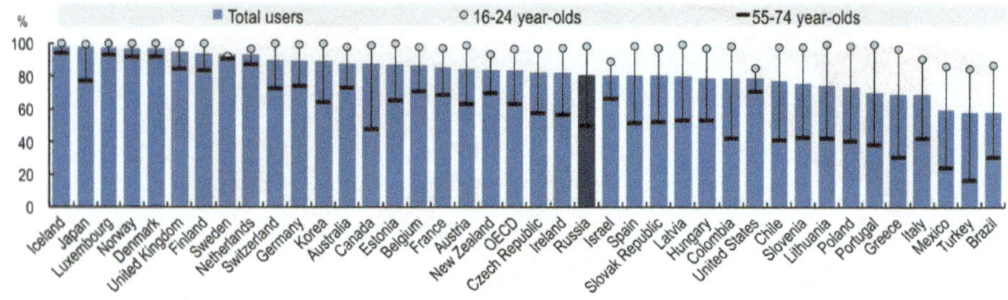

Fig. 47.2 Share of internet users in 2016 according to ages 16–24 and 55–74 [11, pp. 63–64, 15]. *The data for Russia is taken for age 55–72

According to the results of year 2016, Russia, despite a low share of elderly population (from 55 to 72) who use the Internet and also the share of the population for online shopping and online public services, in 2016 it grew up significantly in relation to the total percentage of the population using the Internet (80.8%). This fact proves that the country has serious prospects taking into account an active work in the achievement of aims and goals of the program "Digital Economic of the Russian Federation for 2018–2024".

In addition to prospects of digital economy development in Russia the risks should be taken into account. They can be connected not only with barriers which appeared during the development of the program but also with the risks which arise in the process of specific actions. Thus, for instance the program supposes the keeping of the data in the centers for data processing which should be situated in Russia and not abroad. The program planned the increase of number of such storage facilities (34.88 thousand stocks what is 13% more than in 2016) proved to be lower than expected as the time of completion of work in a number of projects in creation and development of data centers with the guaranteed level of reliability not less than 99.98% (Tier III according to the Uptime Institute classification) were postponed for the first half of year 2018. The difficulties with its timely launch are explained by a high level of projects' costs and the necessity of conservation of expensive components of engineering infrastructure which double each other in compliance with the requirements for data centers Tier III [2].

The risks are also connected with the human factor and also with the fact that the planned measure cannot be synchronized with the involved organizations. Thus, the digitalization program implies the use of the common center for data processing of the data of government authorities but the Tax Service may not agree with that.

According to the suggestion of the World Economic Forum for the assessment of countries' preparedness to digital economy the latest version of the international index of net preparedness is used (World Economic Forum's Network Readiness Index) which was presented in the report "Global Informational Technologies" for 2016 [20]. The improved index measures how well the economies of the countries use the digital technologies for the growth of their competitiveness and well being and also it assesses the factors influencing the development of digital economy.

47.5 Conclusion

Thus, by the end of 2018 the stages in one of six directions of the program "Digital economy of the Russian Federation for the period 2018–2024" were realized. In particular two projects were launched and successfully were realized which were included into the direction "Personnel in the national economy". They are the competence based CDO profile (the launch of the program Chief Data Officer) and All-Russian "Lesson of Digit" (the first digital teaching and learning kit in informatics for school students).

Within the direction "Informational infrastructure" the first from the planned projects is being developed. It is the conception in the construction of narrow band Wi-Fi nets from the "Internet of Thins". The projects should be ready by the end of December 2018.

The work in part "Legislative regulation" is connected with the "Law on Information" and "Law on Archive Keeping". In compliance with the "Law on Information" the introduction of the replacing scanning is being carried out. It is the opportunity of creation of notarized electronic copies having the same legal force as the original with a consequent destruction of the paper original document (this situation does not refer to archived documents but deals with the documents with storage term up to 5 years). The implementation of the program in part "Normative regulation" is being carried out with the lagging from the planned terms. By the end of year 2018 the corresponding amendments to these laws were not submitted into the State Duma of the Russian Federation.

The situation with the implementation of the direction "Informational security" worsened and the number of crimes committed with the use of computer and telecommunication technologies in 2018 grew.

One of the results of the implementation of the part "Digital Technologies" became a preliminary, unpublished list of potential Russian and international leading research centers in "cross-cutting" technologies which expect the government support and also one of the results is also the making of the list of legal acts and study manuals necessary for the realization of measures within the program "Digital Technologies" in 2019 and the schedule of their approval by the Government of the Russian Federation. In 2018 within the same direction a project was launched implying the transfer of 52 federal executive authorities and about 1500 of their regional offices and also of organizations connected with them to a centralized book keeping on the basis of use of cloud technologies.

Russia today has the 41st rank upon the preparedness for the digital economy with a considerable lag behind the top ten leading countries such as Singapore, Finland, Sweden, Norway, USA, Netherlands, Switzerland, Great Britain, Luxemburg, Japan. From the point of view of economic and innovative results of use of digital technologies, Russia has the 38th rank with a considerable lag from the leading countries such as Finland, Switzerland, Sweden, Israel, Singapore, Netherlands, USA. A considerable lag in the development of digital economy from the world leaders is explained by the problems in the legislation for digital economy, insufficiently favourable environment for business, development and introduction of innovations and as a result with a low level of use of digital technologies by business structures.

In addition to prospects of digital economy development which are connected with an active realization of the program "Digital Economy of the Russian Federation for 2018–2024" Russia has one the one hand a serious potential connected with the involvement of the population into the use of the results of its digitalization, but also on the other hand it has serious risks connected not only with the identified barriers which will arise during the realization of the program but also with such problems as a low share of young population of the country and insufficiently favourable business environment.

Acknowledgements The article is written within the State grant allocated for the project of the Southern Scientific Center of the Russian Academy of Sciences, project state registration number AAAA-A19-119011190184-2.

References

1. Avdeeva IL (2017) The development of the digital economy in the context of globalization: a managerial aspect. Int Sci Pract Conf World Sci 3(4 (20):57–60
2. Big data is running out of space (2018) Retrieved from https://www.rbc.ru/newspaper/2018/11/06/5bdc45019a79472ab0ecdbc2
3. Biychuk AN (2017) Digital transformation of business in the modern economy. Econ Environ 2(20):14–16
4. Bondarenko VM (2017) World outlook approach to the formation, development and implementation of the "digital economy". Mod Inf Technol IT Educ 13(1):237–251
5. CDO as a conductor of digital transformation in the regions. Economics. Data Economy Russia 2024 (2019) Retrieved from https://data-economy.ru/15102018_1
6. Cloud services (Russian market). (2018). Retrieved from http://www.tadviser.ru/index.php/Статья:Облачные_сервисы_(рынок_России)#.2A_IDC:_.D1.80.D0.BE.D1.81.D1.82_.D1.80.D1.8B.D0.BD.D0.BA.D0.B0_.D0.BD.D0.B0_49.25
7. Decree of the President of the Russian Federation dated 05.05.2017. No. 203 "On the Strategy for the Development of the Information Society of the Russian Federation for 2017–2030" (2017) Retrieved from http://pravo.gov.ru/proxy/ips/?docbody=&nd=102431687
8. "Digital technology" prepared for the start of the national project". Economics. Data Economy Russia 2024 (2018) Retrieved from https://data-economy.ru/26102018
9. Hungerland F (2015) The digital economy. Berenberg Bank und Hamburgisches WeltWirtschaftsInstitut (HWWI), Hamburg, 63 pp
10. Kapranova LD (2018) Digital economy in Russia: state and development prospects. Econ Taxes Right 2:58–69
11. Laikam KE, Abdrakhmanova GI, Gokhberg LM, Dudorova OY et al (2017) Information society in the Russian Federation: statistical digest. Research University "National University Higher School of Economics", Moscow, 328 pp
12. Mitrofanova IV, Shcherbina AB (2018) Comparative analysis of Internet use by the Russian population. Econ Yesterday, Today Tomorrow, 8(12A):158–172
13. Mitrofanova IV, Zemlyanskaya SV, Gorshkova OP, Scherbina AB (2019) Digital economy as a concept and as a phenomenon: theoretical-conceptual aspect. Econ: Yesterday, Today Tomorrow 9(5A):241–253

14. Monitoring the development of the information society in the Russian Federation (as of 10/03/2018) (2018) Retrieved from http://www.gks.ru/free_doc/new_site/business/it/monitor_rf.xls
15. OECD Digital Economy Outlook (2015). OECD Publishing, Paris, 282 pp
16. Pyankova SG, Argunova OT, Mitrofanova IA (2018) Digitization of the economy: Russian and foreign experience. Reg Economy. South Russ 3:16–25. https://doi.org/10.15688/re.volsu.2018.3.2.11
17. Program "Digital Economy of the Russian Federation for 2018–2024 (2017) Retrieved from http://static.government.ru/media/files/9gFM4FHj4PsB79I5v7yLVuPgu4bvR7M0.pdf.ю
18. Results of the first year of the implementation of the Digital Economy program (2018) Retrieved from https://12n.ru/video/6475-itogi-pervogo-goda-realizacii-programmy-cifrovaja-ekonomika.html
19. Research: Modelling an Inclusive Digital Future (2018) Retrieved from https://www.cisco.com/c/dam/assets/csr/pdf/Country-Digital-Readiness-White-Paper-US.pdf
20. Schwab K, Sala-i-Martín X (2016) The global competitiveness report 2016–2017. World Economy Forum, Geneva, p XIII
21. Schwab K, Sala-i-Martín X (2017) The global competitiveness report 2017–2018. World Economy Forum, Geneva, p IX
22. Sorokina NY, Gubarev RV (2019) Strategic prospects for the development of the coastal regions of the far east. Adv Econ Bus Manag Res 47:1118–1120
23. Scheme of movement to the digital economy. Economics. Data Economy Russia 2024 (2019) Retrieved from https://data-economy.ru/dataeconomymap
24. Tapscott D (1994) The digital economy. McGraw-Hill, 368 pp
25. Total population in Iceland from 2007 to 2017, by age group (2018) Retrieved from https://www.statista.com/statistics/594621/total-population-in-iceland-by-age/
26. The digital transformation of the state or 10 the most important events of the year according to Tadviser] (2018) Retrieved from http://www.tadviser.ru/index.php/Статья:Цифровая_трансформация_государства_–_10_главных_событий_года_по_версии_TAdviser
27. The Ministry of Science and Business has agreed to jointly train IT staff. Economics. Data Economy Russia 2024 (2019) Retrieved from https://data-economy.ru/17012019
28. The number of fixed broadband Internet subscribers per 100 population (2018) Retrieved from http://www.gks.ru/free_doc/new_site/business/it/mon-sub/1.3.8.xls
29. The Supervisory Board of ANO "Digital Economy" called the bills that require priority approval by the State Duma. Economics. Data Economy Russia 2024 (2018) Retrieved from https://data-economy.ru/04122018
30. What is the reason for a sharp increase in cybercrime (2018) Retrieved from https://news.rambler.ru/other/41475806/?utm_content=rnews&utm_medium=read_more&utm_source=copylink

Chapter 48
Study of Approaches to the Formation of the System of Business Processes at the Enterprises of Mechanical Engineering

Alexander P. Garin, Natalia S. Andryashina, Elena P. Kozlova, Svetlana N. Kuznetsova and Sergey D. Tsymbalov

Abstract In Russia, developments on the affected issues are at the stage of theoretical search and are carried out mainly in the field of organizational design, engineering, and automation of production. Until now, there is no unified theory of system organization of production through building a system of business processes. The aim of the study is to develop a system for business processes of domestic enterprises of mechanical engineering in the conditions of adaptive flexibility and reduction of product introduction time to the market. The object of research were industrial enterprises of the machine-building industry. The subject of the study is the economic relations arising in formation of the system of business processes. The novelty of the research consists in the development of theoretical provisions, methodological and practical recommendations for the formation of a system of business processes at the enterprises of mechanical engineering. The methodology consists of: development of conceptual apparatus, forecasting of development and estimation of methodology of landscape formation of business processes of the new generation. The results of the study: the author's interpretation of the category "landscape of business processes of modern industrial enterprise" is given and its components are substantiated; the algorithm of construction of the landscape of business processes; methodology of forming the landscape of business processes of the new generation has been developed. Conclusion: taking into account the developed proposals in domestic practice

A. P. Garin (✉) · N. S. Andryashina · E. P. Kozlova · S. N. Kuznetsova · S. D. Tsymbalov
Minin Nizhny Novgorod State Pedagogical University, Nizhny Novgorod, Russia
e-mail: rp_nn@mail.ru

N. S. Andryashina
e-mail: natali_andr@bk.ru

E. P. Kozlova
e-mail: elka-a89@mail.ru

S. N. Kuznetsova
e-mail: dens@52.ru

S. D. Tsymbalov
e-mail: sergey.cymbalov@mail.ru

T. Kolmykova and E. V. Kharchenko (eds.), *Digital Future Economic Growth, Social Adaptation, and Technological Perspectives*, Lecture Notes in Networks and Systems 111, https://doi.org/10.1007/978-3-030-39797-5_48

can contribute to improving the efficiency of production management in terms of the system organization of production.

Keywords Industrial development · System of business processes · Management · Mechanical engineering

JEL Code L6 · M5

48.1 Introduction

Modern machine-building production with extensive use of complex technological processes requires changing the theoretical and methodological base of production management in a system organization of production. Since the 1990s the dominant importance in scientific development of production organization problems and management of industrial enterprises is occupied by works on creation of integrated adaptive production systems and management technologies based on a complex of business processes [1].

In Russia, developments on the affected issues are at the stage of theoretical search and are carried out mainly in the field of organizational design, engineering, and automation of production. Until now, there is no unified theory of system organization of production through landscaping of business processes. The issues related to the formation of a unified concept of process management of domestic enterprises, which comprehensively covers the technology of business process design, methods, and tools for modeling the system of integrated high-tech processes. There remains a need for the scientific development of theoretical, methodological and practical approaches to the management of business processes of engineering enterprises based on the formation of an integrated landscape of business processes. The "landscape" is similar to the "system". Problems of creating a balanced complex of enterprise business processes in the production of a high-tech product have not been solved, which determined the choice of topic, the purpose of the study and its tasks.

48.2 Methodology

Theoretical and methodological basis of the research consists of concepts and hypotheses presented in scientific works of domestic and foreign authors devoted to development problems of the process management of an industrial enterprise [2–5]. The methods of systemic, structural, comparative analysis were used in the research process.

48.3 Results

Conducted in the context of the research topic, the following results were obtained.

1. The author's interpretation of the category "landscape of business processes of modern industrial enterprise" was given and its components were substantiated. Various interpretations of the "landscape of business processes" are proposed in the works of scientists related to the formation of a complex of business processes (Table 48.1).

- is a set of business processes of the top-level enterprise, as well as the interrelation between them;
- is a process map presented in the form of a SIPOC diagram (SIPOC—method of creating a simplified process model), which allows you to identify bottlenecks and visualize process, clarify its boundaries, responsibilities, improve the process and design system;

Table 48.1 Characterization of business process landscape mapping types

Process map type	The purpose of the development	Main flows	Advantages	Disadvantages
Top level process maps	Selection and structuring of all the main processes of the enterprise	Three main process streams: 1. order cycle; 2. production cycle; 3. cash flows	Standardizes the structure and main features of interaction of enterprise processes	Unable to develop in detail
SIPOC maps (process maps)	Data management	SIPOC streams that are included in one of the five components: provider; inbound threads; process; outbound threads; client	The map allows you to identify bottlenecks, visualize the process, clarify its boundaries, responsibilities, improve the process and design system	Unable to develop in detail
Value creation flow maps	Process improvement	Flows of value creation	Detailed procedures of the process in the time period are prescribed, which allows to identify time and cost savings	There is no possibility of replication

– is the basis of the enterprise management system, which shows the interaction of top-level processes in four categories: management processes, business processes, support processes (supporting processes), and measurement, analysis and improvement processes.

The essence of modern interpretations of the landscape is as follows. Landscape is a map of top-level processes that allows you to visually represent a set of selected business processes, identify and organize them, distinguish between them the relationship. The purpose of the development is data management, and in the future—the development of the system as a whole. Landscape elements are: (a) top-level business processes in four categories (management processes, business processes, support processes and processes of measurement, analysis and improvement); (b) technologies of their interaction; (c) human resource. In general, modern methodologies for enterprise landscape formation are based on the concept of PLM (Product Lifecycle Management) [6]. In accordance with this, four phases of landscape formation are distinguished:

– Inclusion of processes in the landscape—highlighting business processes in relation to other processes;
– Process definition—phase of structuring and detailing processes;
– Implementation of the process, its regulation;
– Process monitoring by setting process targets and tasks for specific processes [7].

At the same time, the modern development of high-tech production, systems of intercorporate interaction, as well as the study of the experience of successful global manufacturers such as Toyota and Volkswagen. It allowed us to make a number of changes to the map of processes, in particular—to expand the landscape of business processes of a modern industrial enterprise by integrating into its external participants:

– Suppliers of R & D—technologies;
– First-level suppliers developing and manufacturing automotive subsystems (FTS-suppliers);
– Second and subsequent tier suppliers that produce individual components [8].

In our opinion, integration of participants is possible through:

(1) Involvement of suppliers and subcontractors in a set of interrelated technological transitions within one product and assortment chain [9];
(2) Development of enterprise standards; organization of unified design and technological documentation, unified technical electronic document flow and so on;
(3) Formation of an internal supply chain with the subsequent development of an integrated supply chain of the brand within the framework of maintaining the functional product life cycle.

In the author's interpretation under the landscape of business processes of a modern industrial enterprise should be understood as an open multidimensional adaptation system, providing uniform conditions, sufficient quantity of resources [10] and balance of processes (business processes, technological processes) of all participants in the production of high-tech products. The aim of the system is to add value to the product, to form and develop adaptive production flexibility and to accelerate the process of developing new competitive products.

Algorithm of construction of the landscape of business processes of the new generation. Conducted researches allowed to reveal features of formation of business processes of large industrial enterprises within the framework of functional and process models of management, equally used on enterprises [11]. Diversity of approaches to allocation of business processes within the framework of two management models has determined the difference in algorithms of landscape of business processes formation. However, in conditions when the majority of domestic enterprises are at the stage of transition from linear functional control to process control, there is a need to develop an integrated algorithm of forming the landscape of business processes of the enterprise. The idea is to transform the format of enterprise activity by "vertical and horizontal compression" of the set of business processes, as well as design hybrid production processes. According to the proposed methodology, an algorithm of actions begins with the allocation of the "basic control", which in industrial enterprises usually acts as the business process "Production", since most industrial enterprises produce products in accordance with the technical documentation and existing business processes in the enterprise are tied to them.

The following is carried out:

1. Object modeling, involving technology analysis of consumer needs in the framework of the business process cycle. In this regard, a plan, an action program, etc. (developed the first step of the functional management model—analysis of all orders of the enterprise).
2. Complete set of "business architecture of the system"—a group of processes that are responsible for the content of production activities, implementation of its strategies, goals, plans, programs is highlighted. The aggregate of top-level processes is determined by chain: from business models to business processes to enterprise architecture.
3. "System Expansion" is a group of processes responsible for improving or developing new business processes.
4. Implementation of "incoming" production logistics. Configuration of the integrated landscape of business processes of all participants of the system is developed, and the relationship between the set of top-level processes is established.
5. Resource support. Business processes of flow control the supply and use of resources.
6. System monitoring. Effective for monitoring production processes is the DDS methodology used in a system with ARM model.
7. Change in processes.

The proposed algorithm allows to improve qualitative indicators: reduce the time of the business system reaction to environmental changes; depersonification of the management system; increase its transparency and manageability.

2. Methodology formation of business processes landscape of new generation. In general, the introduction of new technologies into production is typically carried out in three directions:

1. Business decisions on the issue of product creation in the industry;
2. Organizational and technical management systems;
3. The system of intercorporate interaction (networks of business processes).

In case of realization of the first position—technologies related to product creation in industry, and proposed methodology formation of business processes landscape of the new generation, according to our plan, should include the following steps:

Step 1: Development of the product concept—its initialization, covering the idea of creation, selection, and conceptual testing. Creation of a product is expedient through the use of systems to support the life cycle of products CALS (an industry strategy aimed at effective creation, exchange, management and use of electronic data that supports the full life cycle of the product through international standards) using PLM tools, design tools (CAD) and production tools (CAM). In particular, vendor CAD systems can be connected to customer systems, and major car manufacturers can use ticket entry systems connected to Vendor's CAM systems to solve" the problem of timeliness" ("just-in-time"). Expected result is the provision of positioning for product creation systems in the landscape of the company's business processes and establishment of their relationship with its other systems;

Step 2: Implementation of the product concept. At this stage the development, trial marketing and launch of products into mass production are carried out; the landscape of business processes of the enterprise is formed directly. It is based on a detailed study of the product by using the method "Design under the possibility of production".

The integration of internal and external business processes of participants, as intended, can be implemented using SCM (Supply Chain Management), CRM (Customer Relationship Management), VAN (information exchange system), ERP (enterprise resource planning/management). This applies not only to the totality of internal workflows of companies but also to inter-organizational systems (IOS). At the same time, the main difficulty in using the technique is the presence of gaps arising at the joints of various technological processes.

Proposed methodology for forming the landscape of new generation business processes allows us to solve a number of problems:

- Structure activities in the field of different business processes;
- Simultaneously design the product and processes, that is, complex technical systems;
- Integrate product creation systems into the landscape of business processes of the company;
- Integrate participating enterprises within the PLM concept.

48.4 Conclusion

The conducted research on formation of the landscape of business processes at the enterprise allowed to formulate the following conclusions and proposals.

1. The author's formulation of the category "landscape of business processes of a modern industrial enterprise" is given, which should be understood as an open multidimensional adaptation system that provides uniform conditions, a sufficient amount of resources and balanced processes (business processes, technological processes) of all participants in the production of a high-tech product.
2. Algorithms of identification of business processes of large industrial enterprises producing a large line of technologically complex products are developed. In functional management, the algorithm is limited to six steps, and the set of business processes is tied to technological production processes. In process management steps to highlight and develop business processes ten, and their totality is limited to either the flow of value creation or the loop of quality (product life cycle). At the stage of transition from one model to another, the algorithm is based on functional control model through selection of the "basic control" with subsequent system transformations.
3. The method formation of the landscape of business processes of modern industrial enterprise is proposed, which will allow: structure activities in the field of different business processes; simultaneously design product and processes, i.e. complex technical systems; integrate product creation systems into the landscape business processes of the company and enterprises participating in the PLM concept.

Taking into account developed proposals in real practice can contribute to improving the efficiency of production management in terms of the system organization of production.

References

1. Garina EP, Garin AP, Vinogradova SA, Batsyna YV (2019) Management of investment programs at the stage of their implementation. Financ Econ 3:464–467
2. Andriashina NS, Garin AP (2016) Evaluation of complex development of the product on the basis of modern methods of quality management (on the example of separate production). Econ Hum 4(291):74–88
3. Romanovskaya EV, Kuznetsov VP, Agafonov VP, Andrashina NS, Artemyeva MV (2018) Design and reengineering of business processes: textbook. University of Minin, Nizhny Novgorod, 200 pp
4. Parshina AA, Shpilevskaya EV, Garina EP (2018) Improving the competitiveness of domestic production through the implementation of management concepts of complex product development. Hum Socio-Econ Sci 3(100):113–118
5. Potashnik YS, Garina EP, Romanovskaya EV, Garin AP, Tsymbalov SD (2018) Determining the value of own investment capital of industrial enterprises. Adv Intell Syst Comput 622:170–178

6. Kuznetsov VP, Garina EP, Romanovskaya EV, Kuznetsova SN, Andryashina NS (2018) Organizational design and rationalization of production systems of a machine-building enterprise (by the example of the contract assembly workshop). Espacios 39(1)
7. Mizikovsky IE, Druzhilovskaya TY, Druzhilovskaya ES, Garina EP, Romanovskaya EV (2018) Accounting for costs and expenses: problems of theory and practice. Adv Intell Syst Comput 622:152–162
8. Garina EP, Garin AP, Vinogradova SA, Batsyna YV (2019) Formation of the risk management system in project practice of enterprises. Financ Econ 3:459–463
9. Kuznetsova SN, Kuznetsov VP (2016) Realization of the mechanism of support of industrial parks for small and medium-sized businesses. Electron Netw Polythematic J "Scientific Work KubSTU" 3:273–285
10. Markova SM, Narkoziev AK (2019) Methods of research of the content of professional education. Vestn Minin Univ 7(1) (26):2
11. Garina EP, Kuznetsov VP, Romanovskaya EV, Andryashina NS, Efremova AD (2018) Research and generalization of design practice of industrial product development (by the example of the domestic automotive industry). Qual—Access Success 19(S2):135–140

Chapter 49
Technique of the Information Asymmetry Assessment as One of the Important Factors Influencing the Economic Growth in the Conditions of Economy Digitalization

Natalia N. Skeeter, Natalia V. Ketko, Sergey Y. Kuznetsov, Elena M. Vitalyeva and Yuliya A. Saykina

Abstract In the modern conditions of digitalizing a large number of the economic activity types, creation of the favourable conditions for the country economic growth becomes relevant. Existent models for assessment of the economic growth include several parameters such as investment, production capacity, labor productivity and others. However, in the current situation, it seems necessary to develop an evaluation model that takes into account the qualitative factors such as the amount of information and its asymmetry (surplus, shortage, distortion, etc.), which also have a significant impact on the economic growth.

Keywords Decision support methods · Methods of the theory of fuzzy sets · Information asymmetry · The economic growth · Method for assessing the information asymmetry

JEL Code C65 · C14 · C02

N. N. Skeeter (✉) · N. V. Ketko · S. Y. Kuznetsov · E. M. Vitalyeva · Y. A. Saykina
Volgograd State Technical University, Volgograd, Russia
e-mail: ckumep@mail.ru

N. V. Ketko
e-mail: gsa-buch@list.ru

S. Y. Kuznetsov
e-mail: floyd_80@mail.ru

E. M. Vitalyeva
e-mail: lenkooooo@gmail.com

Y. A. Saykina
e-mail: yuliya1914@gmail.com

T. Kolmykova and E. V. Kharchenko (eds.), *Digital Future Economic Growth, Social Adaptation, and Technological Perspectives*, Lecture Notes in Networks and Systems 111, https://doi.org/10.1007/978-3-030-39797-5_49

49.1 Introduction

At the present time, information became a strategic resource that provides not only a competitive advantage, but also the organizations survival.

The source of any information, regardless of the information support types on which it is presented and methods of its transmission, is the person who systematizes his knowledge with digital devices into information [1].

Asymmetry of information is a specific feature in various sectors of an economy. Depending on the activity field, asymmetry of information may have certain characteristics, forms of impact and demonstrations. Detailed consideration and research of information asymmetry in the modern conditions are extremely important, as the lack of proper attention to this phenomenon can adversely affect the market participants. Asymmetry of information increases transaction costs and becomes a factor that increases the inaccuracy and uncertainty of data, which in turn prevent effective decision-making and efficient allocation of necessary resources and lead to moral risk [2]. Asymmetry in information distribution for enterprises means that a potential investor, buyer, supplier, lender, partner can't accurately determine the probability of changes trends in the financial condition of the enterprise.

There are always opportunity costs in the process of obtaining information, data is not always reliable because it is short-lived, and not every economic subject has the skills to make good usage of the necessary data. Asymmetry of information exists almost constantly, it gives the opportunity to abuse ignorance of other market participants.

The main causes of information asymmetry are the following [3–8]:

1. Complexity of access to information. There are often costs in the process of obtaining information. Therefore, the desire to obtain the necessary data should be comparable to the costs of obtaining information and the benefits of obtaining it. Additional costs if it is necessary to check false information;
2. Reliability of information. The economic environment has been changing, that's why the information, received today, may become obsolete tomorrow. Most of information cannot be collected and stored due to the lack of funds. There is a negative economic selection (preference is given to inexpensive information) [9].
3. Economic agents are unable to assess the full amount of available information, and some of the aggregate information is lost [2]. Absolutely all the market players cannot equally find, use, select, analyze all the information they encounter in the process of activity;
4. In order to assess the received information, it is necessary to have sufficient skills and knowledge, and the economic subjects are not always sufficiently qualified.

Information asymmetry is an essential part of the digital economy, but with development of the digital economy, many transactions are cheaper and faster due to the lack of intermediaries. The digital economy in Russia has been formed for a long time without much government intervention. Table 49.1 presents the prerequisites for emergence of the digital economy.

Table 49.1 Prerequisites for formation of the digital economy in Russia

Prerequisite	Substantiation
Unemployment growth	In connection with the global financial crisis, and subsequently with the systemic crisis, many enterprises left the markets or liquidated their branches in the regions, thereby left many people unemployed. Staff reductions and optimizations also became ubiquitous in the large state-owned enterprises. People refused from the previous working places and started to look for a new job, including through the Internet, using portals for job search such as hh.ru, superjob.ru etc. These portals have already formed a mobile app for added convenience in interaction between an employee and employer. Temporary and remote work can also be found through the Internet resources. Some professions do not require the mandatory presence of an employee at the place of work, only the final product is needed. With the advent of an electronic signature, mobile banking, all interactions between employee and employer in the network can be legalized. Some people with extra time free began to develop their "craft" and offer their own products and services through online shopping and social networks
Decrease of real income of people	Due to the fact that the solvency of the population has decreased and many goods have become inaccessible because of their high prices, people prefer to buy goods in online stores or on Internet platforms. It became possible to negotiate in social networks about the joint payment for the delivery of goods from stores, as well as implementation of a larger purchase in order to reduce the cost of a unit of production or obtain a greater discount. People began to consider and study new brands with the help of social media, they began to find the advantages of buying things from new manufacturers, rather than well-known brands. Moreover, even on credit, many products are no longer available

(continued)

Table 49.1 (continued)

Prerequisite	Substantiation
Increase of state commissions and fees from enterprises, increase and emergence of new excise taxes	It should be noted that this factor is negative, due to the fact that it is directly related to the growth of the shadow sector of the economy. For example, if a person wants to open a small business, he is immediately forced to think about buying a new cash register equipment, which will allow the Federal tax service to receive data on purchases in the real time, to inspect the rented building for fire safety, sanitary standards, etc. The costs incurred by the enterprise for administrative fees and deductions will increase the cost of goods and be transferred to the buyer. However, in conditions of economic instability, a small price of goods is a competitive advantage and if it is not, the company will leave the market. It will be much easier not to register a company, but to carry out its activities through the Internet. Of course—It is not suitable for every business, but for many of them this method is the beginning, before the official opening
Development and globalization of the media	Due to the modern technology, most people have the opportunity to get information from anywhere in the world, translate it in any language, distribute it and make their own value judgment
Development of the Internet and speed of information transfer	The speed and availability of information has also increased, due to the popularization of the various channels of information transmission. At the present time the skills that people have developed over the years and life experience can be transferred in a few hours by recording a video series
Social media development	The society strives for communication, exchange of knowledge and experience. In social networks there are a lot of opportunities for implementation of economic and social projects, promotion of goods and services, etc.

Source Compiled by the author

Thus, as it can be seen in the Table 49.1, not all the factors affecting development of the digital economy are positive.

49.2 Methodology

For solution of the above problem, there are special information areas of data collection, processing, storage and usage, namely Big Data, Data Science and Machine Learning. These technologies contain a set of methods designed to extract relevant and perspective information and its analysis:

1. Methods of the Data Mining class (data mining, intellectual data analysis, deep data analysis)—a set of methods for detecting previously unknown, non-trivial, practically useful knowledge in the data necessary for decision-making. Such methods, in particular, include association rule learning, classification (categorization), cluster analysis, regression analysis, detection and analysis of deviations, etc.
2. Crowdsourcing—classifying and enriching data by a wide, indeterminate range of people doing this work without entering in the labor relations.
3. Data fusion and integration—a set of techniques that integrate heterogeneous data from a variety of sources for in-depth analysis (e.g., digital signal processing, natural language processing, including tonal analysis, etc.).
4. Machine learning, including teacher-led and unsupervised learning—usage of the models based on the statistical analysis or machine learning to produce complex predictions based on the underlying models.
5. Artificial neural networks, network analysis, optimization, including genetic algorithms (genetic algorithm is heuristic search algorithms used to solve optimization problems and modeling by random selection, combination and variation of the desired parameters using mechanisms which are similar to natural selection in nature).

   - Images recognition
   - Predictive analytics
   - Simulation—a method that allows to build models that describe the processes as they would take place in reality. Simulation modeling can be considered as a kind of experimental tests
   - Spatial analysis—a class of methods that use topological, geometric, and geographic information extracted from data
   - Statistical analysis—time series analysis, A/B testing (A/B testing, split testing — marketing research method; using it, the control group of elements is compared with a set of test groups in which one or more indicators were changed, in order to find out which of the changes improve the target indicator)
   - Visualization of the analytical data—information presentation in the form of figures, diagrams, using interactive features and animations both for obtaining results and using as input data for further analysis. This is a very important stage of the big data analysis, which allows to present the most important results of the analysis in the most convenient form.

Thus, the authors propose to combine Big Data methods with the expert evaluation in their methodology (Fig. 49.1).

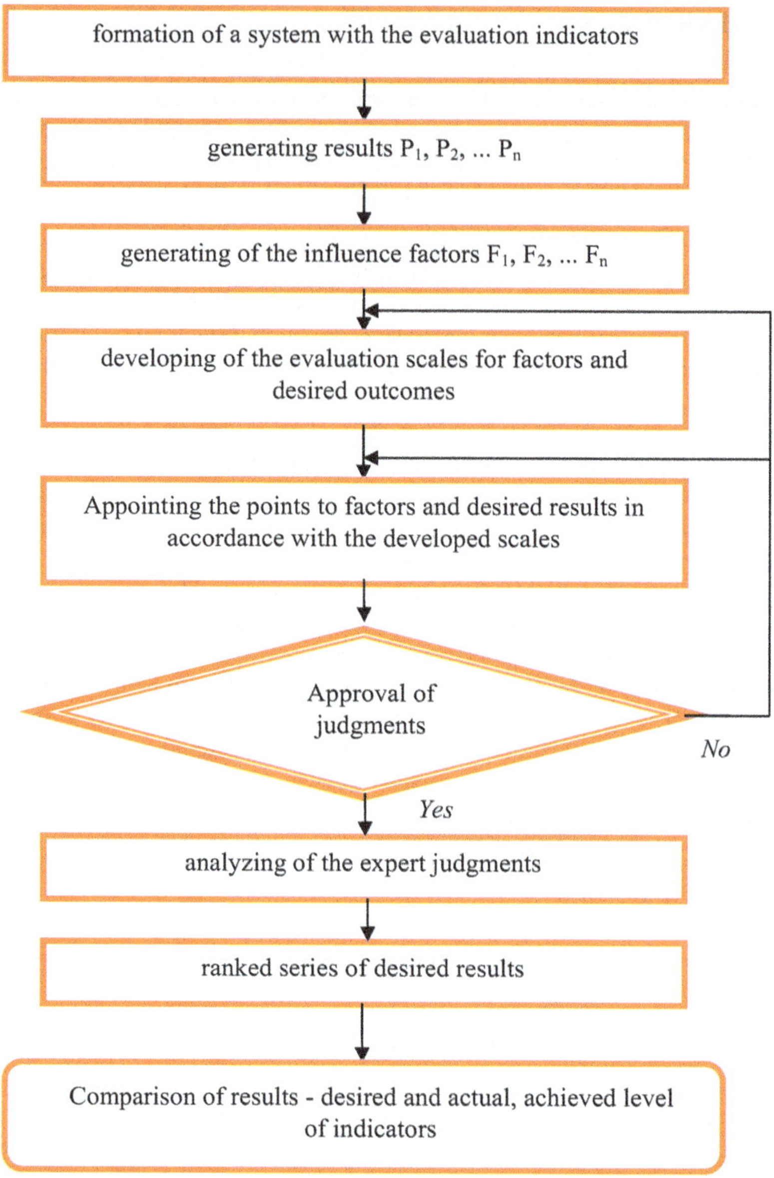

Fig. 49.1 The information asymmetry assessment algorithm. *Source* Compiled by the author

Initially, an expert commission is formed, after that the experts form the evaluation indicators characterizing the asymmetry of information:

Stage 1—formation of the evaluation indicators.
Stage 2—generation of the expected level of the indicators values.
At this stage, the experts determine the expected level that the asymmetry indicators can reach.
Stage 3—generation of the influence factors, which are necessary to assess a probability of achieving the desired level with the indicators.
Once the information about the desired outcomes of the indicators was collected, the factors, influencing the behaviour of these indicators, must be introduced for assessment to take place. Factors in the method are external and internal conditions that can enhance the positive effect of the asymmetry or, in the opposite direction, weaken or reduce it to zero.
Stage 4—formation of the basic assessment scales. Development of the point-factor scales for assessment of the results and factors.
Stage 5—appointment of a point assessment to the results and factors by the experts
Stage 6—coordination of the information about the scores by the experts. This procedure can be carried out several times as long as the experts come to a common agreement.
Stage 7—analysis of the data received from the experts. At the present time, there are many methods of the expert judgments convolution, for example, maximin convolution or minimax convolution, after which the indicator is appointed to a score, adjusted for external and internal conditions, reflecting the quantitative assessment of the desired result.

- to carry out an average assessment of each from the quality indicators:

$$m_j = \bar{m}_{ij} / \sum_{j=1,i=1}^{N,K} \bar{m}_{ij}, \tag{49.1}$$

where:

mj is the average score of the j-th qualitative indicator;

$\bar{m}_{ij}$ is the point score "weight" of the j-th qualitative indicator, set by the i-th expert;

- weighted average score:

$$m_j = \sum_{i=1}^{N} m_{ij} * a_i / \sum_{i=1}^{N} m_{ij}, \tag{49.2}$$

where:

mj—weighted average score of j-th quality indicator;

mij—the point value of the assessment j-th quality indicator by the i-th expert;
ai—the weight, reflecting the significance i-th expert.

After determining the average weighted score of the qualitative indicators, it is necessary to conduct a computer convolution of the expert information, for this purpose, determine the fuzzy ratio of dominance of one over another, which is determined by the function μD (a, b), characterizing the intensity of its dominance:

$$\mu(a, b) = M(a, b) - M(b, a), \text{ if } \quad M(a, b) \rangle M(b, a) \\ 0, \text{ otherwise,} \tag{49.3}$$

where:

μ(a, b) is a function of preferences, reflecting the intensity of the domination of a motif over the motif b;

M(a, b) is a function of agreement that motive a is preferable to employee than motive b;

M(b, a) is a function of disagreement that motive a is preferable for an employee than motive b.

Next, we calculate the intensity of dominance of each indicator by crossing all the indicators μND (a, b) for all indicators that are not dominated by any indicators.

$$1 - \max \mu_D(b, a) = 1 - \max[M(b, a) - M(a, b)] \tag{49.4}$$

where μ*(MA) is a function of the indicator belonging to the set of non—dominated indicators.

$$\mu^*(M_a^*) = \max_{a=1,\ldots,n} \mu^*(M_a) = 1 - \min\left\{\max\left[\underset{\substack{a=1,\ldots,n \\ b=1,\ldots,n \\ b\neq a}}{M}(b, a) - M(a, b)\right]\right\} \tag{49.5}$$

where:

M* a is an indicator from the set of non-dominant indicators.

Stage 8—orderliness of the results in descending order of their score.

This technique will allow more reliably and objectively assess the information asymmetry and apply the most effective Big Data technologies for the information analysis.

49.3 Results

Assessment of the information asymmetry as well as thoughtful usage of the Big Data technologies can solve the following problems:

- forecasting the market situation

- marketing and sales optimization
- product development
- management decisions
- increased productivity
- efficient logistics
- monitoring of the fixed assets.

In industrial enterprises, during the process of generating big data, the main components and parts of machines are equipped with sensors, actuators, controllers and, sometimes, inexpensive processors capable of performing fuzzy calculations. During the production process, data is continuously collected and possibly pre-processed. Analytical platforms analyze these information arrays in the real time, present the results in the most convenient form for perception and save them for future usage. Based on the analysis of the obtained data, conclusions are attained in relation to the equipment condition, its efficiency, the quality of products, the necessity for changes in technological processes, etc.

Due to the real-time information monitoring, the company's personnel can:

- reduce downtime
- improve equipment performance
- reduce equipment operating costs
- prevent accidents.

In addition, the results of the big data analysis can be used to calculate: the payback period of the equipment, the prospects of changing technological regimes, reduction or redistribution of staff—that is, to make strategic decisions on further development of an enterprise.

Despite the objective positive aspects, digitalization has also the negative consequences (Fig. 49.2).

It should also be noted that in the digital economy it will be particularly relevant to consider the information asymmetry and its impact on various economic subjects.

Changes in factors affecting business subjects lead to the following risks:

- financial;
- moral;
- production;
- loss of time, etc.

Method for assessing the information asymmetry, developed by the authors, will reduce the level of risks, caused with the information asymmetry, by means of increasing the information reliability.

The main negative effects that can be mitigated after the assessment of the information asymmetry include:

- negative economic selection;
- consumer disappointment;
- opportunistic behaviour of the market subject;
- growth of the shadow economy.

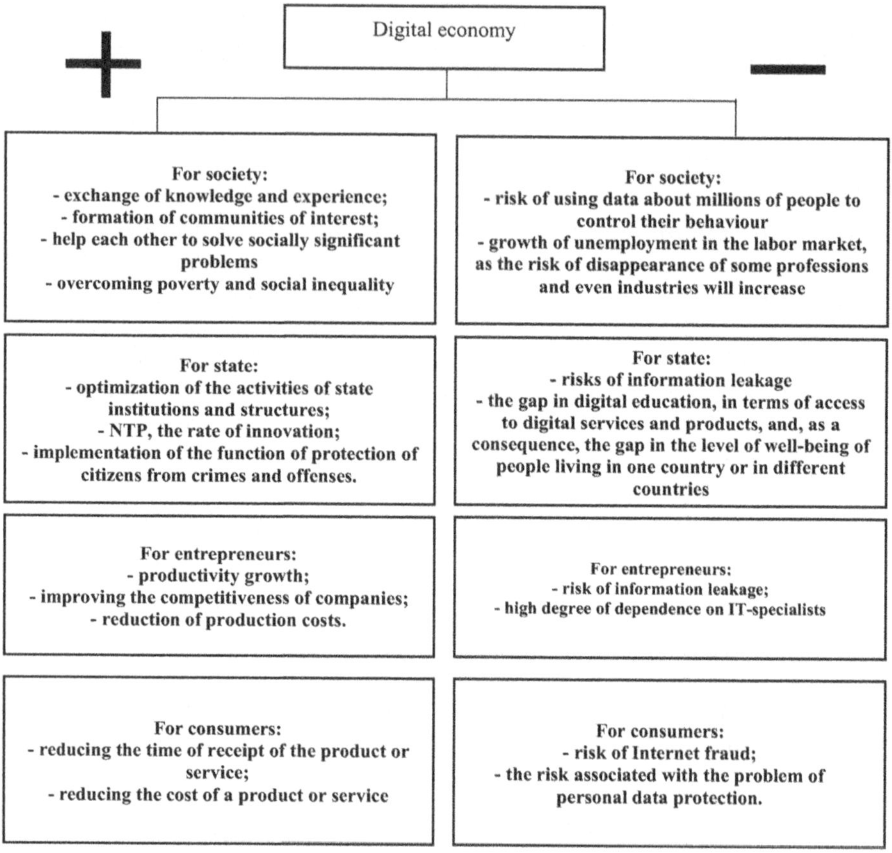

Fig. 49.2 Advantages and disadvantages of Russia's transition to the digital economy. *Source* Compiled by the author

However, it is also worth noting that the availability of more information from one of the subjects is its competitive advantage, which must be used for its own development.

49.4 Discussion

Consideration of the sources and factors affecting the economic growth of the country is a relevant theme for research, especially in conditions of economic instability and crisis phenomena. Knowledge, as a leading factor of the modern economic growth, is not only a form of existence and systematization of the results of human cognitive activity, but also the possession of proven information that allows to solve the problem [1]. In this regard, it is necessary to identify information as the most important resource of the economic growth and the structural element of each factor. Information, produced by the economic subjects acts as an additional indicator of the economic growth in development of information technologies and innovations.

The causes of information asymmetry are:

- difficulty in identifying and reaching the target audience;
- political and economic barriers;
- social and moral barriers arising from the historical facts, stable ideas about each country;
- different accessibility to information, products or services [1].

Taking into account the above reasons, it can be assumed that the impact of information asymmetry is expressed in the following—with its increase, the number of the product consumptions and services at the micro level decreases, and economic growth slows down at the macro level.

At the same time, the main signs of the slowing economic growth are:

- reduction of real investment compared to potential;
- reduction of the real trades number compared to potential ones;
- the emergence of intense socio-economic interactions;
- reduction of real aggregate demand compared to potential.

In these circumstances, it becomes important to identify possible ways to reduce the information asymmetry.

49.5 Conclusion

Information asymmetry in socially important markets, in contrast to the markets of goods and services, causes significant damage to society, therefore, requires control by a state. It is also worth noting that the society seeks to reduce the information asymmetry due to the modern technologies and the possibility of transferring information through social networks. Moreover, the form of information transfer is not only text, but also photos and videos. Through development of social media, individuals can acquire a large amount of knowledge and skills and share experiences in a short time. This trend should be maintained in organizations that operate in socially important markets.

Thus, it should be noted that the information asymmetry will be presented in any market. However, in socially important markets it should be minimized. It can be achieved through close monitoring of information by the state, non-profit organizations and people who are engaged in socially significant areas of activity (in educational, medical organizations, etc.). In the market economy, it is difficult to reduce the information asymmetry, but in transition to a digital economy, the task is simplified by the large number of channels for transmission and dissemination of information. It increases the factor of trust, and the individual task in obtaining information is reduced to verifying the identity of the person transmitting and distributing information. It should also be emphasized that the collection, analysis, processing and usage of data is a key element of the digital economy. However, Russia's position on the role and status of open data is quite contradictory: on the one hand, the benefits and potential of their using are obvious, but on the other hand, the control over data collection and storage is strengthened, and the requirements for access to the Internet are tightened. Data privacy, control over dissemination of information in the global network can, in the long term, lead to a restraint in development of the digital economy in Russia, contribute to emergence of new challenges and lead to technological lag and restraint of the socio-economic development of the country.

References

1. Popkova HG, Scientific library of dissertations and abstracts disserCat. http://www.dissercat.com/content/formirovanie-kontseptsii-novogo-kachestva-ekonomicheskogo-rosta-v-usloviyakh-globalizatsii#ixzz4p3PLuxPlwages,andwell-being
2. Akerlof GA (1970) Lemons market: quality uncertainty and market mechanism (The market for "lemons": quality uncertainty and the market mechanism). Q J Econ 84:488–500
3. Ketko NV, Akimova OE (2016) Analysing, evaluating and ranking of motivational preferences of entrepreneurs in small business development. Actual Probl Econ 1:361–369
4. Zadeh L (2002) From computing with numbers to computing with words—from manipulation of measurements to manipulation of perceptions. Int J Appl Math Comput Sci 12(3):307–324
5. Saaty TL (1977) A scaling method for priorities in hierarchical structures. J Math Psychol 15(3):234–281
6. Salo AA, Hamalainen RP (1997) On the measurement of preferences in the analytic hierarchy process. Multi-Criteria Decis Anal 6:309–319
7. Kuznetsov VN (2003) Security Eurasia-2002: encyclopedic dictionary-yearbook. The author of the idea and concept, the project manager. Book and Business, p 236
8. Sergi BS (2019) Modeling economic growth in contemporary Russia. Emerald Publishing, Bingley, UK
9. Aiello G, Giovino I, Vallone M, Catania P, Argento A (2018) A decision support system based on multisensory data fusion for sustainable greenhouse management. J Clean Prod 172:4057–4065

Chapter 50
Wide-Scale Green Digital Technologies in the Economy: Future and Nowadays

Vyacheslav Chekalin, Victor Petukhov and Denis Chepik

Abstract The paper deals with the issues of trends of the impact of green digital technologies both on the formation of economic processes and on environmental changes.

Keywords Environmental changes · Green digital technologies · Bioengineering · Green technologies · Economic structure · Fixed assets

50.1 Introduction

50.1.1 A Subsection Sample

The development of mankind is associated with environmental changes. Over the centuries, humanity has been inventing and improving technologies and techniques for survival, giving rise to various social and economic formations. Starting from the mid-XX century, current global trends are formed: increase in the size of population, life expectancy, globalization processes, development of new industries (a phenomenon that L'vov and Glaz'ev will refer to as the change of technological modes in 1985 [1]), the intensification of research and development and design and experimental activities, and an important aspect for the topic under consideration—production and consumption extensiveness.

V. Chekalin (✉) · V. Petukhov · D. Chepik
Federal Research Center of Agrarian Economy and Social Development of Rural Areas—All Russian Research Institute of Agricultural Economics, Moscow, Russian Federation
e-mail: checkalin@gmail.com

V. Petukhov
e-mail: keyneslab@gmail.com

D. Chepik
e-mail: denis_chepik@mail.ru

T. Kolmykova and E. V. Kharchenko (eds.), *Digital Future Economic Growth, Social Adaptation, and Technological Perspectives*, Lecture Notes in Networks and Systems 111, https://doi.org/10.1007/978-3-030-39797-5_50

50.2 Materials and Method

The research presents a look-back analysis of statistical data on the implementation of green digital technologies. Foreign experience of OECD and BRIC countries was analyzed. A systemic analysis of the Russian production base, basic production assets, their structure by sectors, and depreciation rate has been carried out. The economic and statistical method of research is based on the data from Rosstat and Eurasian Economic Commission.

50.3 Results

Limitedness of resources is the driver of development: a search for alternative materials, technologies, etc. is in progress. Their use changes the social lifestyle, creates new social institutions, management and regulation mechanisms. Environment (in strict accordance with Marx's ideas) influences the mind-set of people. Nowadays, in 2019, we can observe the change of technological modes. The Fourth Technological Mode, based on an internal combustion engine, electronics, plastic manufacturing, evolves into the Fifth Technological Mode based on bioengineering, information and communication technologies (ICT) and microelectronics (ME). At the same time, most of the focus is on digitization of various areas of life and their sustainability. Clean technologies are also called green.

Actually, the formation of mass green digital technologies will become the response of civilization to environmental changes associated with the Fourth Industrial Revolution.[1] How will the image of global society change? For a common person, this will be expressed in:

- increased volume of waste recycling, including separate waste collection;
- energy efficiency improvement, emergence of alternative energy sources;
- wide occurrence of digital platforms;
- involvement in digital services (public service website etc.);
- severization of requirements for narrowly-specialized education;
- expansion of telecommunication networks;
- environmental degradation;
- increase in the share of urban population;
- increased risk of information control.

The fundamental basis of the above factors is the worldwide increase in R&D costs in the structures of national balances of different countries (Table 50.1).

As can be seen from the Table, Russia is the country with the lowest R&D costs expressed in absolute terms, among selected countries and regions. The share of such costs in the GDP breakdown is shown in Table 50.2.

---

[1] It should be noted that the Fourth Industrial Revolution, also known as Industry 4.0, conforms to the concept of the Fifth Technological Mode.

Table 50.1 Gross expenditures on R&D in certain countries and regions of the world, 2002–2015, billion dollars*

Region/Country	2002	2007	2009	2011	2013	2014	2015
The entire world	790.3	1132.3	1225.5	1340.2	1477.7	1522.3	N/A
United States	277.1	359.4	373.5	383.1	396.7	405.2	420.6
European Union	206.2	251.3	262.8	278.0	282.0	N/A	N/A
Germany	56.7	69.5	73.8	81.7	83.7	86.1	87.2
United Kingdom	30.6	37.2	36.7	36.8	36.2	36.0	37.3
Japan	108.2	139.9	126.9	133.2	141.4	143.5	140.3
South Korea	22.5	38.8	44.1	55.4	64.7	69.0	69.8
Russia	15.9	22.2	24.2	23.0	23.3	24.2	24.2
China	39.2	116.0	169.4	220.6	287.5	313.2	342.5
India	12.9	31.1	36.2	42.8	N/A	N/A	42.0
Brazil	13.0	23.9	26.1	30.2	31.3	32.5	N/A

*PPP, in real terms of 2005

Sources UNESCO (2016). *Science Report 2015: Towards 2030.* Second Revised Edition. UNESCO Publishing.—P. 26; UNESCO (2010). *Science Report 2010: The Current Status of Science around the World.* Second revised edition. UNESCO Publishing.—P. 2–3; 2014–2015 гг.: UNESCO (2017). *UIS.Stat database*—http://data.uis.unesco.org/index.aspx?queryid=81

Table 50.2 The share of R&D costs in GDP in certain countries and regions of the world in 2007–2015, as a percentage of GDP

Region/Country	2007	2009	2011	2013	2014	2015
The entire world	1.57	1.65	1.65	1.65	1.69	N/A
United States	2.63	2.82	2.77	2.74	2.75	2.79
European Union	1.71	1.86	1.89	1.92	N/A	N/A
Germany	2.45	2.73	2.80	2.85	2.89	2.88
United Kingdom	1.69	1.75	1.69	1.66	1.68	1.70
Japan	3.46	3.36	3.38	3.32	3.40	3.29
South Korea	3.00	3.29	3.74	4.15	4.28	4.23
Russia	1.12	1.25	1.09	1.06	1.09	1.13
China	1.40	1.70	1.84	2.08	2.02	2.07
India	0.79	0.82	0.72	N/A	N/A	0.63
Brazil	1.11	1.15	1.20	1.15	1.17	N/A

Sources UNESCO (2016). *Science Report 2015: Towards 2030.* Second Revised Edition. UNESCO Publishing.—P. 26; UNESCO (2010). *Science Report 2010: The Current Status of Science around the World.* Second Revised Edition. UNESCO Publishing.—P. 2–3; 2014–2015 гг.: UNESCO (2017). *UIS.Stat database*—http://data.uis.unesco.org/index.aspx?queryid=81 (date of request: 10.12.2017)

As can be seen from the Table and in relative values, the share of R&D in the structure of GDP of Russia is quite small, even in comparison with the BRICS countries. This means that in the immediate future Russia will not be able to take the lead in the scientific and technological revolution. Some portion of equipment and technologies will have to be purchased abroad. Nevertheless, there is an enormous potential for the development of domestic hi tech production sector. This potential will be implemented with due consideration of expansion of mass green digital technologies. Their main fields of application will be ICT, life sciences (biomedicine, pharmaceutics etc.), logistics (transport systems, and the development of general aviation—especially relevant for Russia), material sciences (improving the quality of existing and the emergence of new ones), cosmonautics (especially in terms of development of the near-Earth space), military industry, energy.

The main task for the Russian government now is to properly modernize the existing industry and create prerequisites for working with the latest achievements of scientific-and-technological advance.

Over the past ten years, the expert community has been closely studying top-priority changes in state economic policy, occurring during modernization of the economy, noting the positive trends in the transition of the innovative economic growth. Nevertheless, the intensification of modernization processes is characterized by a rather slow response of the Russian economic system due to the diversity of structural features, which result in the slowdown of the rate of innovative development.

The time period associated with the beginning of market transformations in the country can conditionally be divided into three stages; moreover, each of these stages has its own peculiar dominant modernization components.

1991–1998—Stage 1. It had the most painful market reforms. The main feature is the institutional and structural changes in which a market environment began to take shape, and the structure of national economy was created on the basis of fundamentally new principles and grounds. In 1990, a decline in industrial production begins, then it sharply increases after the introduction of price liberalization in 1992, with its peak falling on 1994, the main reason for which is the tightening of financial policy. Due to export growth in 1995, a decline in production sector was significantly decelerated, and in some sectors, such as the chemical industry, non-ferrous and ferrous metallurgy, there is even a slight increase.

However, the decline in domestic demand has kept the trend of detrimental effect on the production level in manufacturing industries. One can say that the total amount of industrial production in the Russian Federation has decreased more than twice for the entire period from 1991 to 1998. This is indicative of a large-scale de-industrialization of the economy [2]. The default in 1998 bears evidence of inefficiency, but at the same time, there was a chance of a transition to large-scale recovery growth.

Stage 2 (1999–2007) was marked by the fact that the largest financial industrial groups were fully established in the Russian Federation. The domestic demand has increased significantly due to an increase in earnings of population, resulting in an emergence of a prerequisite to the industrial growth with the production sector, as well as wide-scale modernization in the field of fixed assets, development and implementation of new groundbreaking technologies and production works. The increasing role of the state in motivating industrial development was reflected in the formation of the institution of state-owned corporations, which were engaged with the goal of intensification of technological development and the intensification of modernization processes in the new economy.

1999–2007 became one of the most prosperous years for the industry of the Russian Federation for the entire post-war history—somewhat higher rates were recorded only in the 1950s. The remarkable thing is that the industrial production has increased almost by one third, and it has increased 1.5 times in the engineering industry. In fact, the increase occurred due to the increased efficiency of use of capital, labor and raw materials, and specifically the system of energy resources. However, the percentage of fuel and rough manufacturing complex in the structure of industry has increased, and imports dominated in the domestic market of hi-tech products and commercial equipment. This trend has been maintained to date (Table 50.3).

Please note that for the two stages under consideration, the same consistent pattern was that, in fact, the preference of economic policy was to strengthen the integration of the national economy into the world economy, which would result in an overly open economy, moreover, at a sacrifice for the long-term economic interests. According to E. B. Lenchuk, Russia did not use the opportunity to build up innovative and technological potential, thereby being "trapped in noncompetitiveness, losing to the "developing economies" in terms of cost of goods manufactured, and to the leading countries in technologic advance—in terms of its technology level" [3].

Eventually, a primitivization of industrial-engineering bonds occurred; breaking of bonds between the area of creation and implementation of innovations; "import

Table 50.3 The structure of industry of the Russian Federation, % (in actual prices)

Industries	1990	1995	2000	2005	2010	2015
Power generating industry	3.5	11.0	7.9	7.5	8.1	6.2
Extraction of fossil fuels	6.7	14.6	17.5	20.9	20.0	20.9
Extraction of minerals, except fossil fuels	0.4	0.8	0.6	2.9	2.7	2.8
Metallurgical industry	9.8	13.0	15.7	13.0	10.8	9.8
Machine-building industry	27.1	15.2	16.4	13.9	14.2	13.9
Chemical and petrochemical Industry	7.0	7.1	6.2	6.8	7.1	7.3
Forest, wood-working and paper-pulp industry	5.2	4.6	4.0	2.8	2.6	2.6
Industry of construction materials	3.3	4.3	2.4	3.3	3.0	2.6
Consumer goods industry	10.9	2.2	1.4	0.9	0.9	0.7
Food processing industry	12.1	10.6	11.1	11.5	12.0	12.4
Other industries	14.0	16.6	16.8	16.5	18.6	20.8
Total	100	100	100	100	100	100

Source Industrial production in Russia 2016: Statistical collection/Rosstat.—M., 2016.—347 p.; *Russian annual bulletin of statistics 2016*: Statistical collection/Rosstat.—M., 2016.—725 p

substitution of R&D results used by the business in case of gradual degradation of applied and fundamental science, as well as engineering activity" [4–6]. According to Bendikov, M. A. and Frolov, I. E., the vast majority of enterprises of the country "are not only capable of carrying out independent research and development works … but of obtaining their results" [7].

The fundamental difference in the third stage (from 2008 to the present day) of modernization of the national economy was the recognition of the lack of prospects of integration into the international economy, moreover, "on any terms", and the need for accelerated recovery of inner economic and technological potential as the basis of long-term development of the country. As the President of the Russian Federation Putin, V. V. noted in his program paper in January 2012, "Russia cannot afford itself having an economy that does not guarantee us stability, sovereignty, or decent well-being. We need a new economy, with competitive industry and infrastructure, with developed services sector, with efficient agriculture. An economy that would operate on a modern technological base" [8].

The creation of mass technologies, especially those that can be classified as "green", faces a number of limitations and risks in Russia. These include those that depend on the actual industrial production base and economic assets.

The dynamic patter of depreciation of basic production assets bear evidence to the fact that in general in the manufacturing sector, its main production works, as well as in several other sectors of the economy, the degree of depreciation has been consistently reducing throughout 2000s, with its minimum established by 2008 (Tables 50.4 and 50.5).

A dramatic increase in the degree of depreciation of fixed assets was caused by such processes as world economic crisis, followed by the unstable economic recovery of our country, which degenerated into prolonged stagnation. Eventually, in 2016, the total degree of depreciation reached 50%, and 54.9% in extractive industries.

The rate of deterioration of machinery and equipment is the highest. During the analyzed period, it was 50%, and in 2015 it was at its peak level of 57.6%. Moreover, in the mining and construction sectors, this indicator in 2015 exceeded 63% (64.6 and 63.2%, respectively) (Table 50.6).

The cornerstone impact on the low rates of renewal of fixed assets ensures slowdown of investment in the production sector.

The specified negative trends in the situation in the material and technical base of industry can be supplemented by estimates of underinvestment in the system infrastructure and environmental support systems. Overall, this circumstance can create strategic risks of manmade disasters for the economy and for society. Please note that in the period from 1991 till 2010, on average, for a sample of 84 developed and developing countries, the share of infrastructure costs in the global GDP was equal to 3.8% of GDP (in average annual terms) [9]. In an expert assessment of Porfir'ev, B. N., the same share of capital expenses in the Russian Federation over the same period amounted to an average of 1.2% of GDP. This situation led to domestic investment debt of about 832–900 billion dollars and, as of 2013, resulted in urgent annual investment in infrastructure of at least 6% of GDP [10].

Table 50.4 Depreciation rate of fixed assets by certain types of economic activity (at the year-end), %

Types of economic activity	2003	2005	2008	2010	2011	2012	2013	2014	2015	2016
All funds	43.7	44.1	43.6	45.7	46.3	45.9	46.3	47.9	48.8	50.2
Agriculture, hunting and forestry	46.5	42.9	32.8	33.5	34.0	35.2	36.0	37.0	38.0	38.5
Mining industry	50.9	51.7	45.6	46.8	48.4	49.6	52.3	53.0	52.8	54.9
Manufacturing sectors	46.4	44.1	41.0	42.2	42.5	43.4	43.5	44.7	45.9	47.4
Production and distribution of energy, gas and water	55.7	48.6	40.1	42.0	41.9	39.2	39.2	39.6	40.2	41.7
Construction	39.3	42.0	39.8	42.5	38.7	44.2	47.1	47.3	47.3	49.9
Wholesale and retail trade, repair	55.8	58.1	60.8	63.8	63.7	61.8	62.2	62.3	63.7	64.3
Transport and communications	23.8	25.5	31.6	35.8	37.9	39.3	40.0	42.7	44.0	44.5

**Note* Data on commercial organizations excluding small small business entities

Source Industrial production in Russia 2016: Statistical Yearbook/Rosstat.—M., 2016.—347 p.; *Rossiyskiy Statisticheskiy Sbornik 2016:* Statistical Yearbook/Rosstat.—M., 2016.—725 p

Table 50.5 Depreciation rate of fixed assets by certain manufacturing sectors (at the year-end), %

Manufacturing sectors	2003	2005	2008	2010	2011	2012	2013	2014	2015
Manufacturing sectors, in particular:	46.4	44.1	41.0	42.2	42.5	43.4	43.5	44.7	45.9
Cellulose and paper production; publishing and printing activity	40.5	41.2	39.6	42.3	41.9	44.7	43.5	49.9	50.9
Production of oil-products	N/A	36.9	N/A	41.2	41.9	41.8	40.0	44.7	47.1
Chemical production	52.9	36.7	42.9	43.8	44.4	43.6	45.3	44.6	42.2
Manufacture of rubber and plastic commodities	47.1	44.4	38.1	41.3	40.8	43.0	43.1	42.5	46.2
Manufacture of other non-metallic mineral commodities, metallurgical production and	42.9	38.5	32.9	36.2	38.4	38.2	35.8	35.2	41.0
Manufacture of finished metal products	45.2	45.0	40.2	39.9	40.9	42.1	43.7	44.1	46.6
Production of machinery and equipment	48.9	46.9	42.3	43.2	44.0	44.6	44.9	44.5	44.4
Production of electric equipment, electronic and optical equipment	50.8	46.6	45.9	47.3	47.4	47.1	46.9	45.3	42.8
Production of means of transport and equipment	55.7	51.9	51.5	49.6	49.3	48.2	47.4	48.7	47.8
Other manufacturing sectors	37.2	34.7	35.4	35.5	37.7	41.0	46.2	48.7	49.5

**Note* Data on commercial organizations excluding small small business entities

Source Industrial production in Russia 2016: Statistical Yearbook/Rosstat.—M., 2016.—347 p

Table 50.6 Deterioration rate of machinery and equipment by certain types of economic activity (at the year-end), %

Types of economic activity	2003	2005	2008	2010	2011	2012	2013	2014	2015
All funds	55.4	52.5	50.6	53.4	54.1	54.1	54.6	56.3	57.6
Agriculture, hunting and forestry	50.3	42.3	34.2	41.4	43.3	46.6	48.5	50.6	53.2
Mining industry	51.3	56.0	53.9	57.8	59.1	60.3	62.1	64.3	64.6
Manufacturing sectors	55.0	51.5	48.0	50.3	51.6	52.6	53.4	53.9	55.1
Production and distribution of energy, gas and water	н/д	н/д	н/д	43.9	42.2	42.2	43.2	43.0	44.7
Construction	49.4	50.1	49.0	52.7	53.0	53.3	58.1	59.9	63.2
Wholesale and retail trade, repair	69.9	70.8	69.3	70.8	71.9	68.3	66.6	67.9	69.1
Transport and communications	36.1	38.9	47.1	53.1	56.2	57.0	56.7	59.8	61.0

**Note* Data on commercial organizations excluding small small business entities

Source Industrial production in Russia 2016: Statistical Yearbook/Rosstat.—M., 2016.—347 p.; *Rossiyskiy Statisticheskiy Sbornik 2016:* Statistical Yearbook/Rosstat.—M., 2016–725 p

The next problem which exercises influence upon the slowdown in modernization of the Russian economy is the "discontinuity" of organizational and technological production chains and the breakage of cohesive reproduction contours of the real sector [4–6]. In the industry of the Russian Federation, zones of inefficient, superfluous production operation were formed, occupying at least 15% of total production. Technological disconjugacy in chains of finished products could be observed for a long time; what is more, the number of technologically homogeneous chains of production of finished products was small. These trends result in the reduced potential of synergetic interaction of related industries, taking away the necessary multiplier effects from the economy.

The above trends were internal factors of the process. In addition to them, there are also external factors that also have great influence on its priorities. Among these factors, the Strategy for the Scientific and Technological Development of the Russian Federation, emphasizes such factors that have an impact on great challenges for society, the economy, system of public administration, affecting significant risks and at the same time presenting new windows of opportunity and the prospects for science and technology development and modernization of the Russian economy.[2] Hidden potential for creation and implementation of wide-scale green digital technologies can be observed here. Even now, the leading countries and regions that invest heavily in R&D, have already been clearly identified. The scientific-and-technological advance will further increase man-caused impact on nature. But it is the countries that implement the concept of mass green digital technologies can reduce threats, such as morbidity and mortality, associated with environmental pollution, regional energy and food insecurity etc.[3]

The transformation of developed and dynamic economies of developing countries during the transition to a new model of industrial production (Industrial Revolution 4.0) based on technologies of the Fifth Technological Mode (ICT) and a new, Sixth Technological Mode (nanotechnologies, green technologies, additive manufacturing) creates significant strategy risks for the Russian economy due to technological inferiority.

At the same time, the anticipated transition to a new technological mode opens up the possibility of a "breakthrough" into a new technological mode and more significant positions in the future system of the global division of labor, which will ensure sustainable socioeconomic development of the country, as well as improvement of the quality of life of population [11].

In our country, there is currently no unified approach to the definition of hi-tech complex and the definition of its particular branches and producing units. The concept of green digital technologies is not yet enshrined in legislation. Therefore, the "greenness" of technologies is determined by the characteristics of the industry, special features of competitive struggle, availability, and, most importantly—accessibility

[2]Strategy for the Scientific and Technological Development of the Russian Federation. Approved by the Edict of the President of the Russian Federation No. 642 of December 1, 2016. http://sntr-rf.ru/ (Accessed: 10.10.2017).

[3]Ibid.

of breakthrough technologies (in particular, nanotechnologies). However, the complexity of statistical accounting prevents from objectively recording the dynamics and implementation of high technologies in the national economy in statistics.

50.4 Discussion

When it comes to green digital technologies, we think of a hi tech complex. Professor Folom'eva, A. N. et al. defines the high tech complex as an integral multisectoral set of scientific, knowledge-intensive, production, educational, managerial and consulting structures that provide the innovative focus of modernization, stability and safety in the world economy [12]. As a result, science and production leverage the forefront of modern research, realizing public-private partnership. When public research institutes fulfill orders from private companies, this helps to conserve and redirect resources of commercial entities to other goals, reducing costs for domestic R&D. Further development of this direction will make it possible to create and develop the intellectual property market in Russia.

However, Frolov, I. E. rightly insists on the need to distinguish features of high technology, research intensity and innovativeness of sectors of the economy and individual industries [13]. According to his classification, the hi-tech industry is characterized by the localization of results in a group of manufacturers that implement high technologies for the first time. "At the same time, a key feature of high technologies consists in the fact that in case of mass production of hi-tech products they cannot be replaced by technologies of previous generations".

As for the knowledge-intensive industries, it is assumed that the high level of R&D costs, as well as private ownership of R&D results (know-how) will be constantly maintained; therefore they are a relatively narrower concept.

50.5 Conclusion

Strategy for the modernization of the national economy in the medium term and in the long term includes two stages. This being said, the task for Stage 1 (until 2025) is based on the increased domestic demand, namely, provide the improvement of efficiency of industrial production to the level required for the export potential of non-resource industries of commercial value. The focus areas of technological modernization of the economy of the Russian Federation, with account of digital green technologies, must include:

(1) improvement of transport and logistics systems (with a focus on development of general aviation, creation of a competitive environment in the railway transportation, development of the shipping industry);

(2) development of the electrical energy complex, search for alternative energy sources, reduction of the hydrocarbon effect;
(3) low-waste and energy-efficient technologies of recycling in the extractive industries (oil and gas production, metallurgical complex and forestry) [14].

Stage 2 (after 2025) is implemented due to the increase in income from export. This being said, "the main accent should be on the transition to preventive medicine and the development of the entire complex of bio- and environmentally-oriented technologies" [15].

These strategies must be based on three priority areas [4–6]. The first area involves the *development and commercial exploitation of breakthrough technologies*, mainly the Sixth Technological Mode, providing a "level of leadership" through the achievements of domestic fundamental science based on the implementation of 10–15 major projects.[4] The second area is the renewal of basic production assets of companies participating in the fulfillment of Defense Procurement order, and their further diversification. The third area is creation of prerequisites for the modernization of *wide-scale manufacturing sectors, implementation of digital technologies that cut expenses of companies*. Moreover, the increase in the energy and resource efficiency of production should become the basis of green growth, the basis of modernization.

This circumstance is crucial for the modernization policy of the national economy, taking into account the green factor.

Acknowledgements The team of authors expresses its thanks to the administration of the Federal State Budgetary Scientific Institution Federal Research Center of Agrarian Economy and Social Development of Rural Areas—All Russian Research Institute of Agricultural Economics for help in the research in the area of digitization of green technologies. Besides, the team of authors expresses its thanks to the administration of the Institute of Economic Forecasting of the Russian Academy of Sciences for methodological support.

References

1. L'vov DS, Glaz'ev SY (1985) Theoretical and applied aspects of management of the scientific-and-technological advance. Ekonomika i Matematicheskie Metody 1:783–804
2. Bodrunov SD (2016) Things to come. A new industrial society: reloading: a monograph, 2 edn, revised and enlarged. Institute of New Industrial Development named after S. Witte, St. Petersburg, p 76
3. Lenchuk EB (2016) (Editor-in-chief) The new industrial policy of Russia in the context of technological safety. Aleteya, St. Petersburg, p 53
4. Ivanter VV (2017) Organizational and investment policy of economic growth in Russia: a monograph. Nauchny Konsultant, p 15
5. Ivanter VV (2017) Organizational and investment policy of economic growth in Russia: a monograph. Nauchny Konsultant, pp 62–63
6. Ivanter VV (2017) Organizational and investment policy of economic growth in Russia: a monograph. Nauchny Konsultant, p 168

[4]Academy Science (an interview with the President of the Russian Academy of Sciences Sergeev, A.M.) Rossiyskaya Gazeta, October 6, 2017.

7. Bendikov MA, Frolov IE (2007) The high tech industrial sector in Russia: condition, tendencies, innovative development mechanisms. Nauka, 583 pp
8. Putin VV (2012) We need a new economy. Vedomosti, January 30
9. Dobbs R, Pohl Y, Lin D-Y, Mischke J, Garemo N, Hexter J, Matzinger S, Palter R, Nanavatty R (2013) McKinsey infrastructure practice. Infrastructure productivity: how to save $1 trillion a year? McKinsey Global Institute, McKinsey Co., 100 pp
10. Ivanter VV, Ksenofontov MY (eds) (2013) Economy growth prospects of Russia: a forecast for the period until 2030: a multi-authored monograph. Ankil, pp 157–159
11. Ivanov VV, Malinetskiy GG (2017) Russia: XXI century. Breakthrough strategy, 2nd edn. Technologies, Education, Science, LENAND, 304 pp
12. Arkhangelskiy VV, Arkhangelskiy VN, Ivanov VV et al (2013) The innovative type of economic development: a textbook, Folom'eva AN (eds). Ekonomika, pp 175–178
13. Frolov IE (2007) The concept of economic and technological mechanism for accelerated development of knowledge-intensive, hi-tech sector of the economy and its theoretical foundations. Kontseptsii (1):29–57
14. Frolov IE (2011) Opportunities and challenges of modernization of Russian high tech complex. Problemy Prognozirovaniya 3:54
15. Ivanter VV, Porfir'ev BN, Shirov AA (2016) From modernization of economic policy to the qualitative economic growth. Rossiyskiy Ekonomicheskiy Zhurnal 1:15

Chapter 51
Methodological Approach to Classification of Digital Technologies According to the Criterion of Ecological Effectiveness and Sustainability

Sergey V. Shkodinsky, Inna N. Rykova and Denis Y. Taburov

Abstract *Purpose*: The purpose of this work is to substantiate the methodological approach to classification of digital technologies according to the criterion of ecological effectiveness and sustainability. *Design/methodology/approach*: The logic of the research is built on the necessity for describing the normative and legal foundations of formation of digital technologies in Russia, performing an overview of statistics in this sphere, and building a classification of digital technologies in the interests of development of "green" economy and according to the criteria of ecological effectiveness and financial sustainability and in view of the stages of products' life cycle. The research is based on the principles of objectivity and systemic character. The authors use a complex scientific approach that is based on application of the methods of systemic analysis, synthesis, and statistical grouping with detailed analysis of the initial materials of the legal, economic, and financial character. The authors use the following methods: abstract and logical, statistical grouping, and forecasting. Besides, the authors use the tools of graphic interpretation of the obtained results. *Findings*: Studying the specific features of legal regulation of the digital sphere in Russia allows concluding on the necessity for creating the systems of information security of a new generation and formation of a scientific foundation for implementation and successful functioning of digital technologies in Russia. Usage of scientifically substantiated classification of digital technologies according to the level of financial sustainability and ecological security allows determining specific measures for solving the problems of development of digital economy. Evaluation of the indicators of financial sustainability of digital technologies allows determining the problems in the segment of their application in organizations that conduct nature protection activities. *Originality/value*: The presented scientifically substantiated approach is

S. V. Shkodinsky (✉) · I. N. Rykova · D. Y. Taburov
Financial Research Institute of the Ministry of Finance of the Russian Federation, Moscow, Russia
e-mail: sh-serg@bk.ru

I. N. Rykova
e-mail: rycova@yandex.ru

D. Y. Taburov
e-mail: taburov@narod.ru

T. Kolmykova and E. V. Kharchenko (eds.), *Digital Future Economic Growth, Social Adaptation, and Technological Perspectives*, Lecture Notes in Networks and Systems 111, https://doi.org/10.1007/978-3-030-39797-5_51

very original, as it has not analogs in the scientific and special literature. The importance of the presented research consists in the possibility of further development of the studied problems and supplementing it with new approaches and criteria—based on the tendencies of development of the digital economy in the context of ecological performance and sustainability.

Keywords Digital technologies · Ecological performance · Classification · Sustainability · Effectiveness

51.1 Normative and Legal Foundations of Formation of Digital Technologies in Russia

In the modern trend of development of the Russian economy, it is expedient to distinguish the stage of digitization of technologies of production, processing of products, creation of innovations, and provision of offer of services on a new platform, which envisages comprehensive automatization and informatization of technological processes.

The task of automatization activities is one of the main directions of development of any organization. However, the issue of evaluation of effect from automatization is very important. Accessibility of information does not guarantee right decisions. For managing the infrastructure and understanding the directions of development of information technologies in an organization, it is important to determine the set of key indicators of effectiveness of the project of automatization and to use expert methods of their assessment [1].

For this, the "Strategy of development of the information society in the Russian Federation for 2017–2030" is implemented. It determines long-term policy of comprehensive transformation of the Russian economy into the digital industry.

Apart from the normative and legal aspect, a certain status of globalization of the digital economy is formed, which is peculiar for international indexing of digital technologies. Index I-DESI entered the scientific and economic turnover in 2016, when the European commission acknowledged human capital and the system of international communication as attributes of digitization of society.

At present, the digital economy becomes one of the main factors influencing the economic growth and leads to important consequences for measuring of GDP, efficiency, and well-being of households in all sectors of the economy. For successful development of the digital economy and reduction of the underrun from the leading countries, Russia has to increase personnel, intellectual, and technological advantages and form a flexible normative basis for implementing digital technologies in all spheres of life. The strategy of intensive digitization of economy and hope for its full-scale transformation, which envisages fundamental change of the state's approaches to decision making, will lead to preservation of competitiveness in the global market and achievement of positive results of socio-economic development [2].

According to the Program "The digital economy of the Russian Federation", adopted by the Decree of the Government of the Russian Federation dated July 28, 2017, No. 1632-r, the key end-to-end digital technologies include the following:

- bid data;
- neurotechnologies and AI;
- systems of distributed register;
- quantum technologies;
- new production technologies;
- industrial Internet;
- components of robototronics and sensors;
- technologies of wireless connection;
- technologies of virtual and alternate reality.

Figure 51.1 shows basic directions of development of the digital economy in Russia for the mid-term.

In the context of this research, let us note that the digital economy allows ensuring sustainability of the Earth's eco-system, transition to green economy, and restoration of destroyed areas in this sphere—e.g., by means of monitoring of green "Big Data", including online screening of the whole specter of profile indicators. The planet

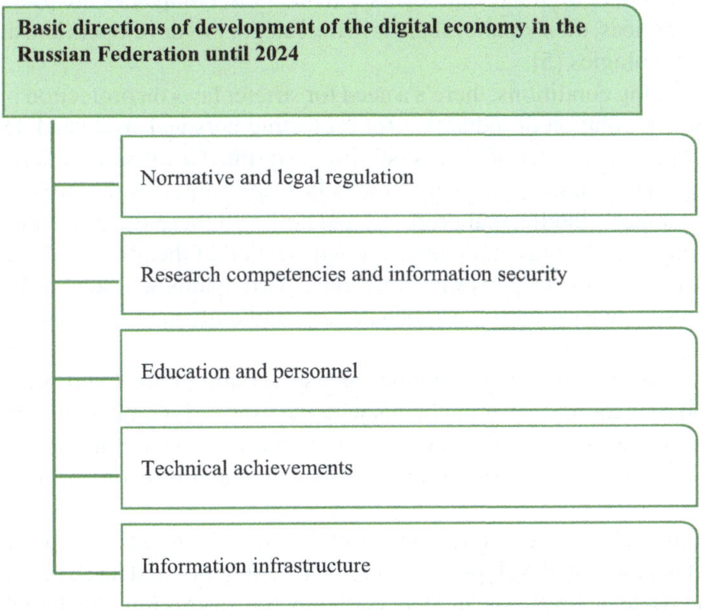

Fig. 51.1 The basic directions of development of the digital economy in the Russian Federation until 2024. *Source* Compiled by the authors based on: Decree of the Government of the Russian Federation dated July 28, 2017, No. 1632-r "Regarding the adoption of the Program 'The digital economy of the Russian Federation'". E-source. http://static.government.ru/media/files/9gFM4FHj4PsB79I5v7yLVuPgu4bvR7M0.pdf

orders determine the bio-physical borders, within which humanity could prosper. There are a lot of paths that could ensure inclusive and sustainable development in this space. These paths are challenged: different cultures that have their own needs, views, and values will look at the costs, risks, and profits differently [3].

At the same time, let us note that digital transformation of production does not cancel the necessity for improvement of the management systems. The necessity for systemic organization of joint work does not disappear. Things and robots that are connected to Internet will become a part of the interaction of humans and corporate information systems. Performers change, the algorithm of the production process improves, and the role of the system of management of industrial companies grows—as in the future competition will be between systems of management, not between goods and technologies [4].

Managing digital technologies is a complex and multi-aspect process of influence of the subject (persons who makes decisions) on the object (additive technologies, robots, the Internet of Things, etc.) for achieving the goals and mission of the socio-economic system that functions in the new digital platform "Industry 4.0" with further transformation of the object into a product or service with the set qualities, with minimization of expenditures per human/hour.

According to I. L. Tukkel, S. N. Yashin, and A. A. Ivanov, it is important that one distinguish the process of "usage" from the process of "development, dissemination, and management of digital technologies". While most companies are users of digital goods and services, their smaller share is in the category of developers and distributors of these technologies [5].

In the existing conditions, there's a need for stricter laws on protection of information and self-regulation of industry—for protecting personal space and determining the frameworks and order of access, storing, and transfer of information. Who will have the property rights, who will be legally responsible for the product containing the consumer's intellectual property, and how will these issues be solved during international deals? At last, inevitable reconfiguration of the global value chains and change of the balance of power between their participants will lead to larger problems regarding generation of cost, beneficiary, and collection of taxes and customs fees [6].

In the conditions of a large number of legal contradictions and problems with assessment of taxes and other payments with the usage of digital technologies, there is an objective necessity for the search for managerial decisions on implementing the information and technical opportunities of development of economy into the mass industry.

According to E. A. Yakovleva, mega-trends should be treated as a powerful force that predetermines the development of national economies and large global corporations at the strategic level. The modern economy transforms into the digital economy and becomes more intellectual and less material. Large scale and global character of this tendency allows assigning it to the mega-trend of the modern times with its specific problems [7].

Thus, the overview of the normative and legal acts and modern economic literature on the issues of the digital economy shows the necessity for creating the systems of information security of a new generation, increasing the qualification of personnel in the IT sphere, and creating the scientific basis for implementing and and successful functioning of digital technologies in Russia.

51.2 Statistics of Development of the Digital Economy in Russia

In the modern conditions of development of digital technologies, it is possible to observe the formation of added value chains with increased weight of the objects of intellectual property, which are treated in economy as soft assets. Distribution of gross added value between the sectors of the country's economy undergoes serious changes.

In the conditions of *digitization of measures in the ecological environment, added value is distributed not equally, and* increase of security of the ecological component of the modern business requires the achievement of the balance of interests between the robots that are applied in the eco-system and the efficiency of interaction between digital platforms and customers. That's why the managerial approach to the existing problems has to take into account the controlling and preventive measures on the optimal involvement of digital technologies into the business environment, oriented at expansion of "green" economy.

The numbers of statistics show that financing of various project measures for achievement of target landmarks of improving the ecology in Russia has a large differentiation depending on completeness and quality of the applied digital technologies (Fig. 51.2).

The numbers of budget financing show the domination of financial resources in balances of expenditures for organization of federal projects in the part of measures "Clean air", which is confirmed by a large share of investments into this system of digital transformation—RUB 500,139.8 million versus statistical minimum RUB 6,282.8 RUB, which account for ecological tourism.

Professor S. S. Nosov notes that digital technologies could eliminate turbulence, as they could stimulate quick and smart and precise strategic decisions in the conditions of the change of the external and internal economic environments.

Production of various goods becomes more robotize—traditional goods include sensors, which allow simplifying the usage of the good. One of the values in the activities of integrated structures is creation of a digital platform, which sets new professional standards, develops competition, and forms dynamic ratings of the participants of the integrated structures [8].

The change of technological paradigm and diversification of production in favor of low-cost, low-waste, and low-toxic "green" technologies took place at the end of the 20th centuries and influenced almost everyone—existing technologies and developed technologies. Thus, in developed countries, citizens and governments set the task and adopt the programs of national and international character on protection

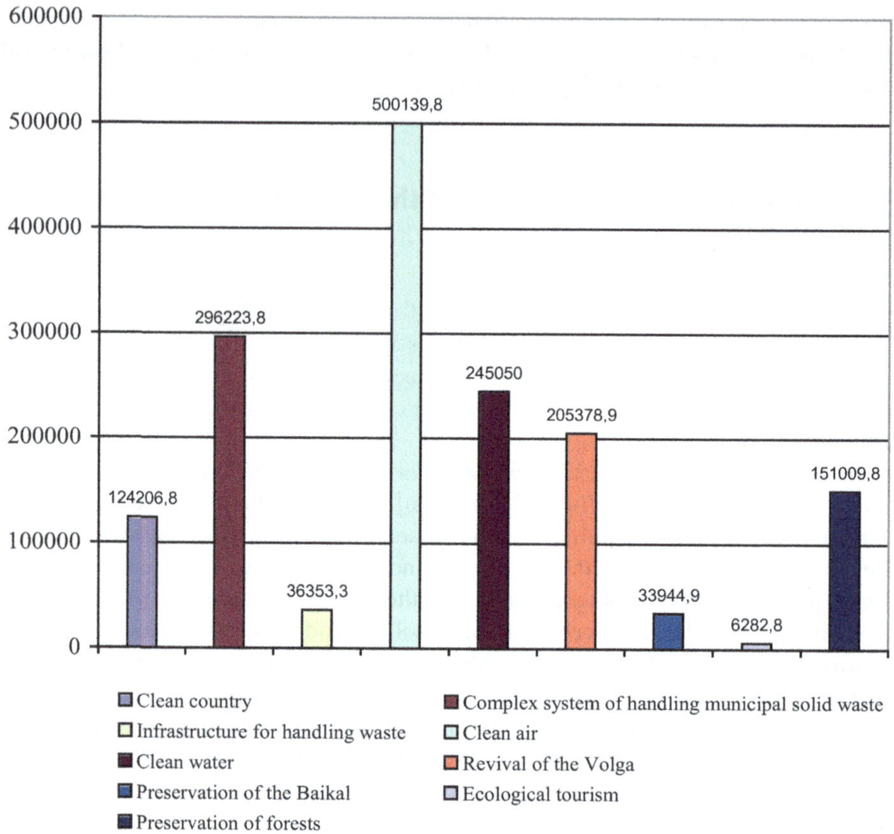

Fig. 51.2 Financing of measures for development of digital technologies for provision of ecological effectiveness of the country. *Source* Compiled by the authors for the data: Federal business network Strategy of the Russian Federation. E-source. https://strategy24.ru/rf/projects/project/view?slug=natsional-nyy-proyekt-ekologiya&category=ecology

of environment from human, having realized that it's high time to save the planet from productions that exploit and destroy nature. A response to these challenges is a new direction of science and practice—"green technologies" [9]. Analysis of digitization of economy in most countries of the world point out the technological underrun of Russia for such parameters as share of digital technologies in the volume of GDP, share of investments in the digital economy, share of innovations in "green" economy, etc. Also, there are examples of countries that are in underdeveloped sectors according to the criteria of digital technologies: Scandinavian countries, South Africa, and Egypt (Fig. 51.3).

According to Fig. 51.3, Russia is in the third group of countries of the world according to the criterion "growth rates of effectiveness of the digital economy". Together with the Russian digital economy, the group "Perspective" also includes China, India, Malaysia, Philippines, Indonesia, Chile, Mexico, and Brazil.

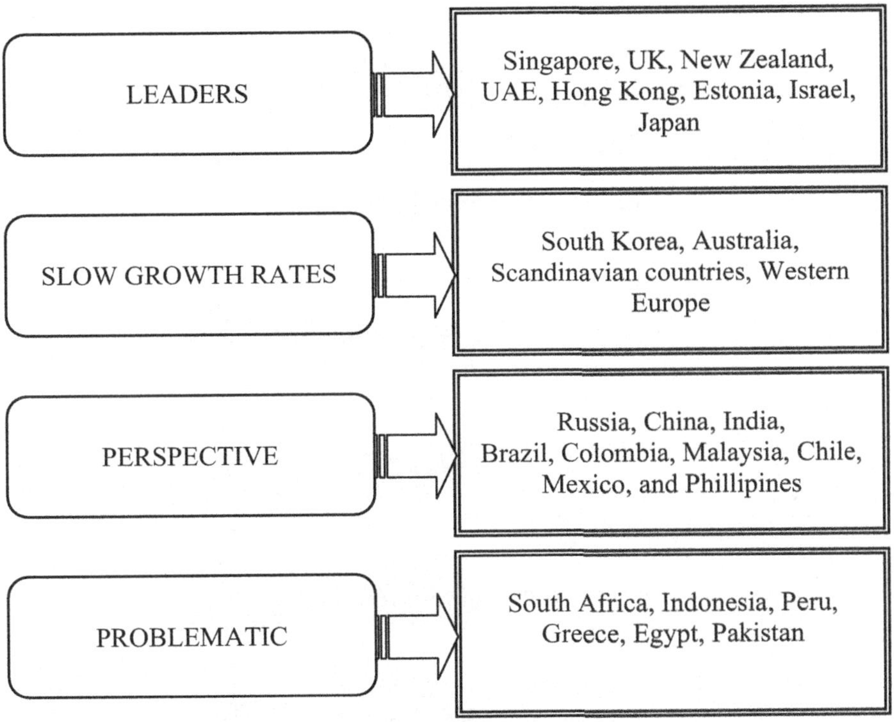

Fig. 51.3 Groups of countries according to the rates of development of the digital economy. *Source* Compiled by the authors based on: Top 10 countries with the most developed digital economy E-source. http://web-payment.ru/article/250/top-10-cifrovaya-ekonomika/ (Accessed: 30.07.2019)

Figure 51.4 shows statistics of the existing and forecast values of digital technologies, which are mastered and will be implemented in the conditions of the Russian practice.

Forecast analysis, as a type of digital technologies, has a small share in the structure of digitization of economy, as it accounts for 28%. The same value is peculiar for digital technologies "Virtual helpers".

However, in five years, the forecast analysis will achieve 39% and start competing with the following digital technologies:

- processing of queries in network language (30%);
- face recognition (31%);
- analysis of images (34%);
- machine learning (30%);
- virtual helpers (33%).

Overview of a wide circle of information materials allows supposing how the implementation of digital technologies will take place in the main spheres of the Russian industry. Technological changes allow creating and applying a range of

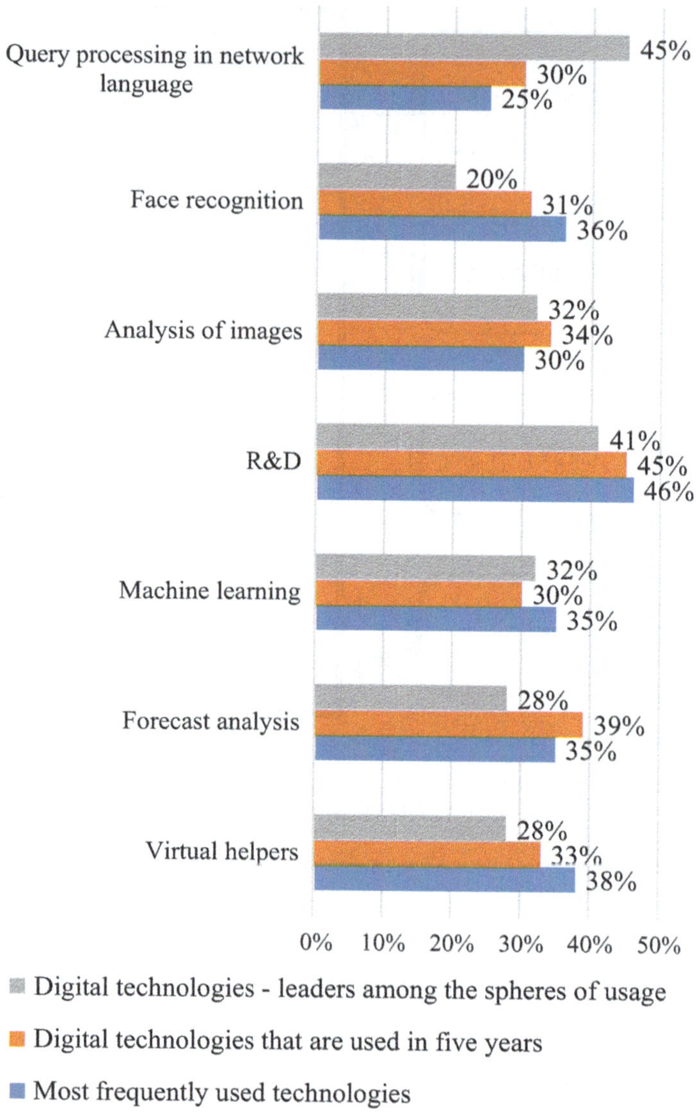

Fig. 51.4 Statistics of development of the digital economy in Russia. *Source* Compiled by the authors based on: Association of electronic communications. E-source. https://raec.ru/activity/analytics/11002/ (Accessed: 30.07.2019)

scientific and practical concepts, which allow for complex implementation of the modern digital technologies in industrial production.

For processing spheres of industry, the most perspective spheres are automatization, robotization, digitization of production (e.g., cloud technologies), application of 3D-modeling of mineral sources, systems of radar monitoring, and concepts of

Mine of the Future and Digital Mine. Processing productions, within the transition to the product platforms and the necessity to control the product's life cycle, will use analysis of Big Data, AI, and new methods of digital modeling. At the state of production, "smart" factories, additive production, precise production, and leading materials will be used. In the metallurgical sphere, a very important sphere will be production of metallurgical materials for additive production and production of materials with set qualities. As for the energy sphere, there is a tendency for disintegration of systems, transition to so called smart networks, scaling of renewable sources of energy, distributed generation, authomatization of management, technologies of energy accumulation, provision of security, and expansion of the spheres of usage of electric energy [10].

In view of the above, we suggest using the scientifically substantiated classification of digital technologies according to the level of financial sustainability and level of ecological security for specifying and solving the problems of development of the digital economy.

51.3 Scientifically Substantiated Classification of Digital Technologies for Development of the "Green" Economy

According to the national project "Ecology", there's a forecast of qualitative change of management of nature protection activities based on using the best practices in the sphere of application of digital technologies.

The period of 2019–2024 will see the implementation of a complex of measures on creation and modernization of infrastructure in the sphere of ecological security of Russia, which is regulated by the norms of 11 federal projects, including projects in the sphere of restoration of the Volga, preservation of the Baikal, and protection of unique water objects.

The budget of this national project constitutes RUB 404.1 trillion; it will be spent for the following purposes:

- provision of reduction of emissions into the atmosphere;
- creation of new nature objects, protected from various unsanctioned acts and actions;
- observation of the optimal balance of removal and reproduction of forests.

In view of the national project's tasks, large industrial centers of Russia (Fig. 51.5) will perform work on the implementation of digital technologies into the system of prevention of the negative influence on the environment.

Twelve cities in Russia acquire responsibility for involving the budget assets into the system of financing of measures for improving the favorable ecological environment in the Russian industrial areas, which have the highest level of production and are peculiar for application for complex or hazardous production and technological systems.

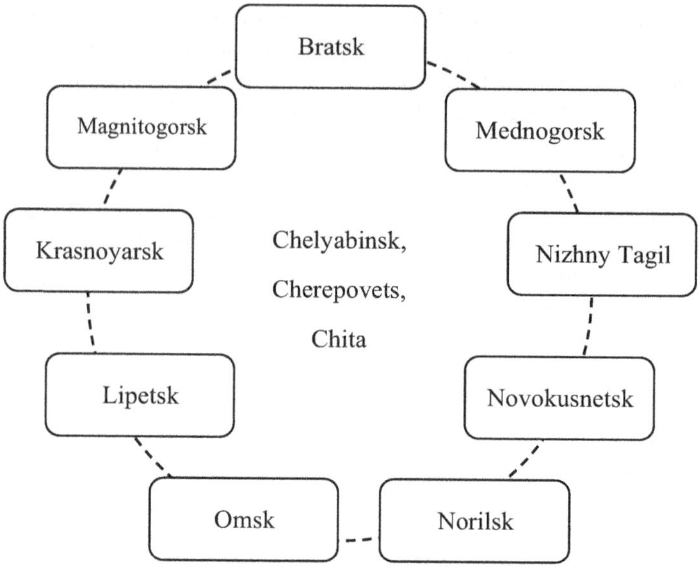

Fig. 51.5 List of industrial centers of the Russian Federation that participate in reduction of emissions. *Source* Compiled by the authors based on: Federal business network—Strategy of the Russian Federation. E-source. https://strategy24.ru/rf/projects/project/view?slug=natsional-nyy-proyekt-ekologiya&category=ecology

Synergy of economic profits from digital technologies and effects from reduction of ecological losses is achieved as a result of the following:

- reduction of irrational expenditures from usage of analog systems and devices for production of non-material products;
- minimization of negative consequences in the ecological space by means of precision and effectiveness of consumption of resources;
- optimization of spending of various types of energy during production of goods, works, and services;
- also based on growth of mobilization of digital potential for sustainability of society's ecologization.

It is necessary to pay attention to the issues of increase of quality of usage of renewable resources and implementation of the leading production systems.

Principles of classification:

- balance;
- continuity;
- multifunctionality;
- complex character of development;
- compatibility of goals and resources;
- orientation at growth of financial sustainability;

- multiple solutions for ecological effectiveness;
- eco-systemic approach;
- growth of vertical chains of added value creation;
- intellectualization;

Development of scientifically substantiated classification of digital technologies for the purpose of development of the "green" economy is one of the stages of scientific forecasting within comprehensive digitization (Fig. 51.6). In Fig. 51.6, one of the types of digital technologies according to the criterion of ecological effectiveness is the process of usage of the modern tools of precise agriculture, which is used by agricultural manufacturers. Using this type of digital technologies envisages measuring a precise volume of water resources, various types of fertilizers, and reduction of time required for operations that are performed for growing of agricultural crops. In the digital system, this process is performed based on the actions of the global positioning devices, special sensors, and photo and video cameras.

The criteria of effectiveness of digital technologies of this type are growth of efficiency and increase of crop yield in the agro-industrial complex and reduction

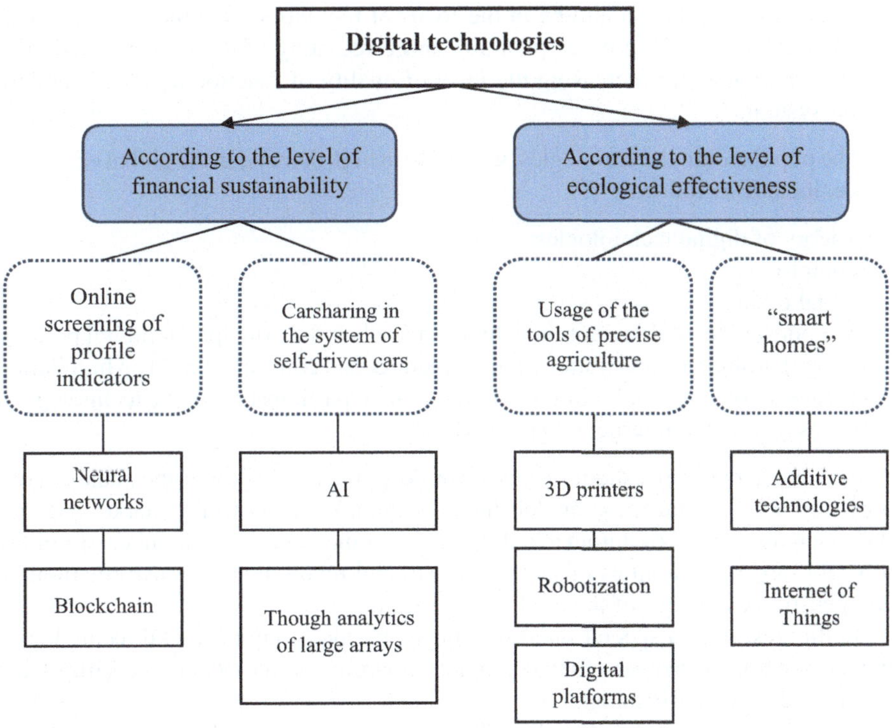

Fig. 51.6 Classification of digital technologies in the modern economic conditions. *Source* Compiled by the authors

of the volume and cost indicators of expenditures for the main types of resources: natural, material, labor, and time resources.

Orientation at growth of financial sustainability during creation of classification of digital technologies is the main principle, as in the conditions of deficit of budget resources the irrational spending of financial resources, which leads to aggravation of the financial state, and therefore, slowing down of achievement of long-term growth of financial sustainability, could be a reason of low demand for digital technologies.

Expenditures for purchase of:

- specialized equipment;
- organization of production in the conditions of additive technologies;
- 3D print;
- robototronics;
- expenditures for changing spare parts and components to expensive automatized system of information processing in case of its malfunction and investments for implementing digital technologies could exceed economic profits that are received by the economic agent from manufacturing and selling goods, works, and services with the usage of the digital platform. That's why balance of interests in the system costs-profits should be in the focus of regulation of subjects of managerial activities, which provide planning and forecasting of the implemented digital technologies for increasing the level of quality of functioning of the modern eco-system.

The popular digital technologies, which are in high demand in the global economic space, include as follows:

- synergy of digital technologies;
- robototronics;
- virtual reality;
- the Fourth industrial revolution, which is based on the principle of digital production and profitable cooperation of manufacturer and consumer based on liquidation of transaction costs and minimization of human participation in the technological processes and the Internet of Things (Fig. 51.7).

The above platforms should be used for deep transformation of the analog economy into a certain business model that could quickly react to the market changes with increased share of automatized devices and increase of the number of remote jobs, for new opportunities of market growth and revolutionary transformations in competitive advantages of the modern companies.

In this research, a special interest is posed by the group of digital technologies that ensure transformation of the ecological environment into a new specialized and specified digital eco-system.

This system is peculiar for algorithmic contacts between the intermediaries of modern business, with participation of independent (autonomous) economic agents, which function in the unified information space for achievement of the unified goals and tasks in the sphere of the "green" economy.

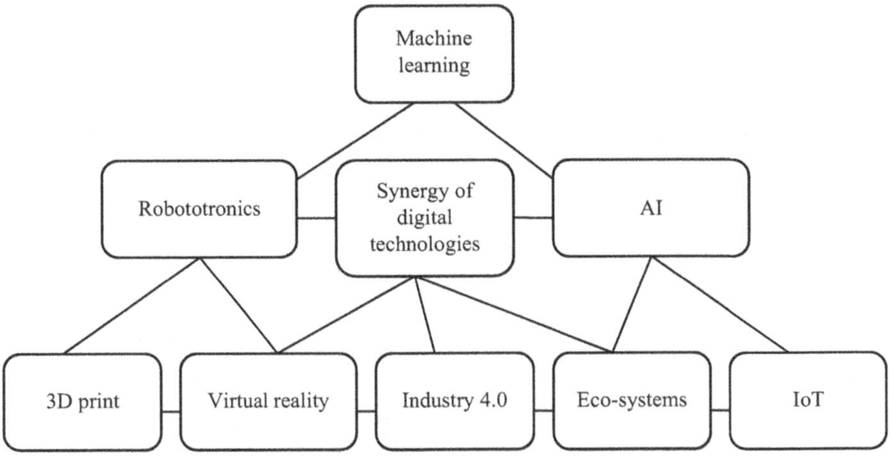

Fig. 51.7 Series of popular digital technologies in the world. *Source* developed by the authors

Within the implementation of the green economy, the top-priority investment projects, developed based on digital technologies, have to conform to the requirements of ecological effectiveness. Dynamics of estimate indicators of implementing the federal project "Digital technologies" in Russia are shown in Fig. 51.8.

We see that growth rates of expenditures for development of "end-to-end" digital technologies (300%) by 2024 will exceed growth rates of revenues from implementation of the projects (250%). This negative tendency will be visible by 2022, at the beginning of which growth of revenues will constitute 190%, and growth rate of expenditures for digital technologies will constitute 200%. This shows the adoption of ineffective decisions during forecasting of financial sustainability of digital technologies for future three years.

Table 51.1 shows dynamics of investments into fixed capital, aimed at environment protection and rational usage of nature resources.

On the whole, sustainable growth of capital investments will positively influence the application of digital technologies during processing of information on the state of lands, water resources, and air.

Increase of the number [11] of the constructions built according to the norms of the "green" economy is observed in a range of subjects of the Russian Federation, where the inflow of labor resources and direct foreign investments is formed.

The number of buildings constructed for development of the "green" economy based on certificates of quality reflects a lot of modern problems of creation of a digital business model in the sphere of development. This is characteristic of the economic agents of St. Petersburg, Leningrad Oblast, Sverdlovsk Oblast, and Samara Oblast.

The eco-system transforms from analog into digital one in the conditions of the change of the demographic situation, which predetermines the level of employment in the labor market and influences profitability of production and creation of new jobs.

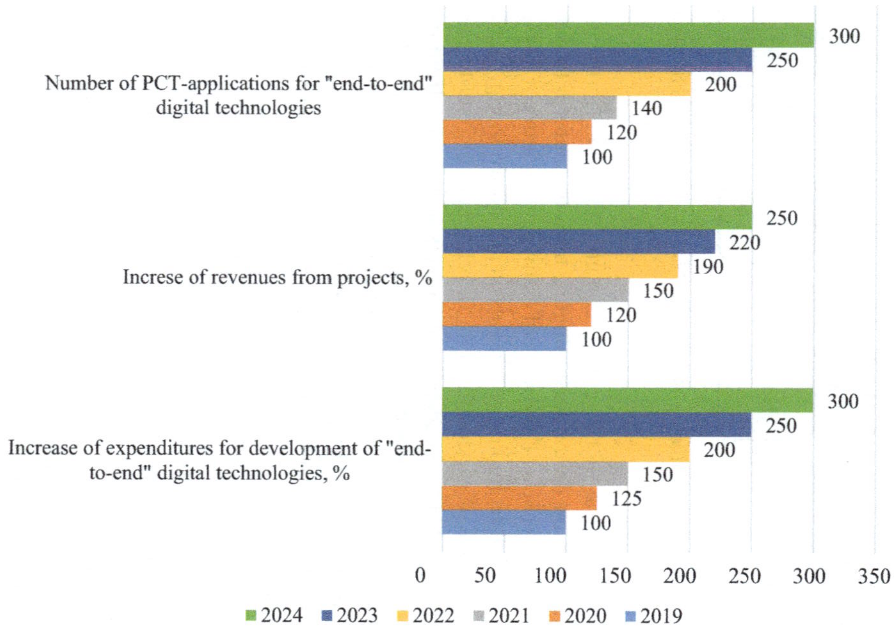

Fig. 51.8 Forecast of the federal project “Digital technologies” of Russia. *Source* compiled by the authors based on: Passport of the Federal project “Digital technologies”. E-source. https://files.data-economy.ru/Docs/Pass_Sciencetech.pdf (Accessed: 01.08.2019)

Table 51.1 Investments into fixed capital, aimed at environment protection and rational usage of nature resources in the Russian Federation in 2010–2018 (RUB million, in factual prices)

	2010	2011	2012	2013	2014	2015	2016	2017	2018
Total	89,094	95,652	116,543	123,807	158,636	151,788	139,677	154,042	157,651
Of which for protection of									
Air	26,127	27,832	34,626	41,196	55,587	40,120	40,340	60,199	65,475
Water resources	46,025	46,610	52,420	59,505	76,315	78,962	67,469	66,496	62,750
Land	9,340	13,735	19,888	13,802	14,540	15,703	12,228	10,216	10,011

Source Federal State Statistics Service. E-source. https://www.gks.ru/search?q=investments+в+основной+капитал (Accessed: 01.08.2019)

51.4 Classification of Digital Technologies According to the Criterion of Ecological Effectiveness and Financial Sustainability

For building the optimal model of classification of digital technologies, it is possible to develop a scenario of the "green" economy that takes into account the idea that 2020–2050 will see growth of investments into ecological security and sustainability of implementation of nature protection measures. Contrary to the existing scenario, development of the digital economy is modeled under the condition that the minimum level of investments is peculiar for "green" technologies.

Classification of digital technologies according to the criterion of ecological effectiveness and sustainability is presented in Table 51.2.

Based on the general principles of creation of a classification of digital technologies (automatization of decisions, possibility of network connections, application of software and hardware), let us perform analysis of their sustainability at Russian companies that have been functioning in the sphere of the "green" economy in the last three years (Table 51.3).

Assessment of the indicators of financial sustainability of digital technologies allowed determining "narrow" spots in the segment of their application in the organizations that have direct relation with operations of nature protection activities, increase of the quality of territories' ecologization, and minimization of environment pollution. Thus, in 2018, the coefficient of financial sustainability, which is calculated based on ratio of capital and balance currency, decreased by 0.1% as compared to 2016. In 2016–2018, coefficient of profitability of digital technologies (profit/assets) did not exceed the value 3.1% versus 10–15% that is sufficient for effectiveness of business development. These tendencies show the necessity for quick interference of managerial personnel with the system of planning and regulation of investments in digital technologies in view of internal potential of the analyzed companies.

Table 51.2 Classification of digital technologies according to the criterion of ecological effectiveness and financial sustainability

Type digital technologies	Criterion of classification
Internet of things	Expansion of the possibility of network connection and computational capacity of the "green economy" objects
Virtual reality	Possibility to control eco-system from a remote control center
Alternate reality	
3D modeling	Possibility to create 3D objects of forest plants. Application of a compact mathematical tool for modeling of bio-chemical processes and environment
AI	Possibility for automatization of the decision support systems in the sphere of nature use and ecology. Automatization of decisions in the sphere of options of development of the "green economy"
Software and hardware	Possibility of using the video surveillance system for the processes of nature protection activities

Source Compiled by the authors

Table 51.3 Indicators of financial sustainability of digital technologies in the sphere of the "green" economy

Indicators	Values for years			2018 to 2016, ±
	2016	2017	2018	
Financial assets/short-term obligations	2.1	2.0	2.2	+0.1
Capital/assets	3.2	3.5	4.5	+1.3
Profit/assets	2.4	2.6	3.1	+0.7
Borrowed assets/own assets	5.3	5.6	5.8	+0.5
Capital/balance currency	0.6	0.7	0.5	−0.1

Source Compiled based on the data of companies from the sphere of the "green economy"

51.5 Classification of Digital Technologies that Have a Positive Influence on the Ecological Effect in View of the Stages of Products' Life Cycle

The analysis showed that together with the problems of reduction of sustainability and profitability of investments into digital technologies into the sphere of the "green" economy there is a negative social effect, which creates problems in the Russian labor market. It is necessary to state that the value of the index of adapted net investments in % to gross national income as the indicator of effectiveness and sustainability of digital technologies is very small. As of year-end 2018, it did not exceed 9% as compared to statistical maximum of 39.2%. The analysis showed that in 2036, in case of further digital transformation, the following could be possible: increase of the share of automatization of works in human hours up to 48%. This indicator could reach 99% in 2066. These tendencies form within forecasting of the Russian policy of digitization in view of the country's participation in the global economic space [12].

Modern digital technologies, which are applied within development of the "green" economy, allow for the highest financial result in the energy sphere. Thus, in 2017, the volumes of digitization of the energy sector in the country reached the value of 1,699.9 million tons of CO_2 equivalent versus statistical minimum of 94.5 million tons of CO_2 equivalent per year.

Our analysis allowed presenting a conventional classification of digital technologies in the "green" industry, which has a positive influence on the ecological effect:

(1) technologies that ensure minimization of consumption of energy resources;
(2) technologies that stimulate reduction of CO_2 emissions, which statistics are shown in Fig. 51.10. These are tools of initial measures for digitization of the "green" economy;
(3) technologies that stimulate reduction of the volumes of pollution and waste;
(4) technologies that are aimed at nature protection measures.

Transition to ecologically sustainable economy will lead to creation of green jobs and new types of jobs, which have a very important role in ecologization of companies and the economy on the whole. Determining the number of popular green workers stimulates the specification of connection between ecological sustainability and the labor market. Combining such tools of the investment policy as market tools, rules, and state investments, subjects of the digital economy and specialized institutes could support companies and create a favorable environment that will stimulate the adoption of the practices of green jobs, investments into new green products and services, and creation of new jobs.

Jobs that stimulate the increase of quality of nature protection activities should be created in the food and processing industry, forestry and timber industry, and the fuel and energy complex.

The EU countries adopted a common program for transition to the low-carbon economy by 2050. The program contains the goals of reduction of CO_2 emissions for the sectors by 2030 and 2050 (by 40–44% and 79–82% in total, accordingly) and enumerates various measures that are necessary for achieving these and other long-term goals, including reduction of expenditures for fuel (by EUR 175–320 billion per year) and decrease of dependence on import of energy sources.

For increasing the energy effectiveness of the EU economy by 20% until 2020, it is necessary to reconsider the terms of the European Union's CO_2 Emissions Trading Scheme:

- minimizing the fixed threshold level of emissions within which the quotas are distributed;
- providing access to the market of quotas for the spheres that have not previously participated in the trade;
- Analysts emphasize large advantages of the digital economy as compared to the traditional economy;
- high demand for the leading technologies in countries and possibility to enter the world market;
- diverse and high-quality content (information, entertainment, scientific, educational, etc.), which is constantly improved depending on the users' needs;
- the quickest and most convenient access of consumers to the content;
- low expenditures—in the online system—for promotion of services and their cost, as well as higher effectiveness in the usage of resources;
- possibility to use large arrays of information, etc.

Certain contradictions between the old and new modes are inevitable, as transition from one stage to another takes place within the old model of development of the economic system, with its traditional algorithms, on the one hand, and the necessity for application of new technologies, on the other hand [13].

According to Alekseeva [14], it is expedient to conduct regular monitoring of ecological development of Russia within the global trends and to determine progress or aggravation of the situation as to certain parameters in the global ratings, depending on the conducted ecological policy.

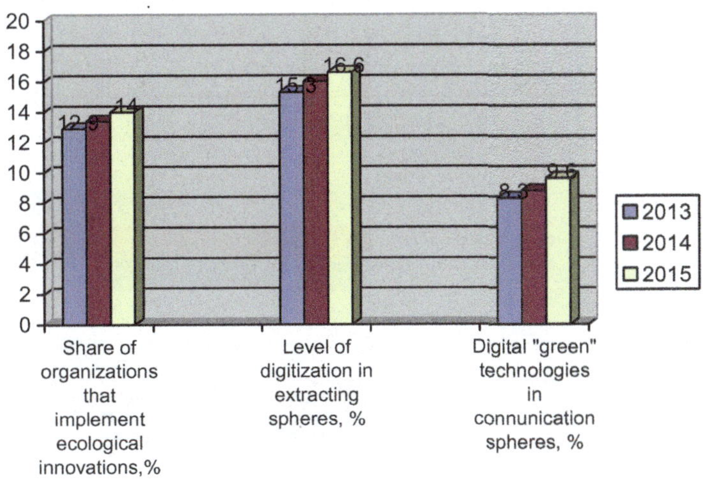

Fig. 51.9 Dynamics of digitization of production for the spheres, %. *Source* Federal State Statistics Service—"Russia in numbers"—2019. E-source. https://nangs.org/analytics/rosstat-rossiya-v-tsifrakh

Let us consider the dynamics of change of the indicators of digital transformation of the "green" economy within organization of ecological innovations. The data of statistical analysis (Fig. 51.9) show that in 2013–2015 the share of organizations that implement ecological innovations grew from 12.9 to 14%, raising the rating of investment attractiveness for Russian companies that implement digital technology into the area of the "green" economy. The sphere of communication was peculiar for increase of the share of ecological innovations from 8.3% in 2013 to 9.6% in 2015.

However, in the sphere of the information and communication services implementation of innovations is at a much lower level. Thus, in 2018 their share constituted 3.9% as compared to 15.4% in the sphere of extracting production, which achieved the statistical maximum of the analyzed indicator.

According to the performed studies, ecological innovations create demand for digital platforms in the companies that ensure everyday activities of the country's ecological system. Here it is important to ensure balance of interests between expenditures for organization of ecological business and profit from investments into digital technologies.

Thus, let us use certain statistics numbers. According to the data of Fig. 51.10, during the analyzed period with the same volume of investments—RUB 500 million—innovations into production and special ecological innovations grow every year.

Thus, 2015 showed growth of special costs by RUB 6.9 billion, or 45.7% as compared to 2013. With such dynamics, innovations in production grew by 47.3% during the same period.

Increase of the volumes of innovative services in the sphere of ecology allowed reducing the share of digitization of ecological production from 3.4 to 2.3%, which

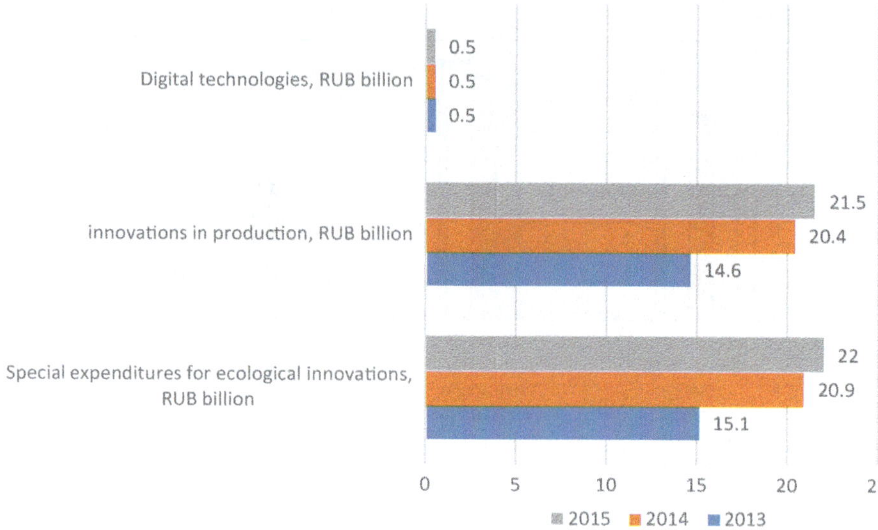

Fig. 51.10 Special and innovative costs during production of digital technologies in Russia. *Source* Federal State Statistics Service—"Russia in numbers"—2019. E-source. https://nangs.org/analytics/rosstat-rossiya-v-tsifrakh (Accessed: 01.08.2019)

negatively characterizes the economic effectiveness of the market offer of products that were manufactured with the usage of digital technologies.

Correlation analysis, which was performed by Guryeva [15], allowed determining a range of the key indicators that have maximum influence on the complex indicators of assessment of the level of development of the "green" economy, which include: gross revenues of the digital economy; index of retail turnover; birth ratio; current expenditures for environment protection; expenditures for capital repairs of fixed funds for environment protection; investments into fixed capital aimed at protection of environment; transaction costs connected to ecological innovations; volume of ecological payments; level of polluting emissions into air; share of scholars conducting R&D; share of organizations that implement technological innovations; number of the leading production technologies; investments per capita.

The largest number of private indicators belong to the ecological sphere, which proves direct interconnection between sustainability of development and ecological well-being of the territory.

Of the total number of registered organizations of smack and medium business (1,578,153), more than 29,100 legal entities are involved in the sphere of digitization of economy.

Classification of digital technologies according to the stages of product's life cycle is shown in Table 51.4.

It is necessary to pay attention to the following type of digital technologies, which is actively implemented in Russia with financial support of Finnish companies—Vocollect. It is a system of speech selection, which stimulates the growth of efficiency

Table 51.4 Classification of digital technologies according to the stages of product's life cycle

R&D	Production	Sales	Exploitation	Utilization
Robototronics	Digital source	Forecast analysis	"Smart homes"	GPS navigation
Neural networks	3D print	Internet of things	"Smart operations"	Robotization
Digital platforms	Composite materials	Blockchain	AI	Virtual reality
Deep analytics of large arrays of data	Additive technologies	Vocollect	Integrated operations	Wi-fi technologies

Source Compiled by the authors

of storage during various manipulations with goods. This type of digital technologies belongs to the stages of product's life cycle that is called sales.

This type of digital technologies of speech selection shows the possibility to reduce time that is spent for collection of the order. Data exchange between the picker and the system takes place via radio and screen of radio terminal, as well as speech commands and special set that consists of receiver, head set, and microphone.

The system tells the picker all necessary information, after which he executes all commands and reports via the microphone. The advantage of this method is using the speech interface and technologies of speech recognition which allows freeing the picker from unnecessary manipulations with the radio terminal [16].

In the course of the Fourth industrial revolution there will be created conditions for transformation of materials and resources that are used in the modern analog economy into information contours. This will allow replacing the physical weight of goods and equipment with information weight. Apart from the revolutionary changes during manufacture of products, which envisage application of additive technologies, composite materials, and neural networks, a new digital mode is formed, at which the processes of processes of production and consumption approach each other.

A perspective direction until 2024 will be online communications in the sphere of finance management, when crowdfunding, blockchain, and other modern digital platform will replace the complex tools of financing of the projects in the sphere of the "green" economy.

Apart from the above, the practice of increase of ecological effectiveness plans the development of "green" standards of communication and expansion of the specter of technologies of business analytics with application of AI.

Thus, implementation of scientifically substantiated classification of digital technologies into the practice of management of the "green" economy will allow the subjects of managerial activities to distinguish the stages of product's life cycle that will be overloaded with technological operations, which are oriented at physical weight of material, semi-finished product, and equipment, not at information weight. This will lead to excessive transaction costs and, based on the principles of classification of digital technologies—balance, compatibility of goals and resources, etc.—will allow concentrating the policy of online contacts with customers on the more rational and, thus, more profitable basis.

Rational usage of Internet resources in the system of nature protection measures and control over the state of ecology, together with long-term parameters of development of digitization of ecology in Russia, will stimulate manufacturers to select financially sustainable digital platforms from the point of view of minimization of transaction costs in "green" business. All these perspective solutions will allow for flexible management of the existing changes in the digital transformation that are connected to insufficiently regulated norms of legal responsibility for application and turnover of digital technologies, uncertainty in the sphere of import of digital platforms, which exceeds the export of the corresponding analogs by four time, and absence of guarantees and mechanisms of protection of rights of investors that are attracted for various manipulations with digital assets.

References

1. Vasilyeva EV, Deeva EA (2017) Methods of expert evaluations in applied information economy for substantiating the advantages of information systems and technologies. World New Econ 4:14–22
2. Basaev ZV (2018) Digitization of economy: Russia in the context of global transformation. World New Econ 12(4):32–38. E-source. https://doi.org/10.26794/2220-6469-2018-12-4-32-38
3. Perelet RA (2018) Ecological aspects of the digital economy. World New Econ 4:39–45
4. Konovalova GI (2019) Development of the methodology of operative management of production at an industrial company in the concept "Industry 4.0". Manag Russ Abroad 2:79–84
5. Tukkel IL, Yashin SN, Ivanov AA (2019) Digital transformation as an important part of innovative development. Innovations 3(245):45–50
6. Kondratyev V (2018) Industry 4.0 and global value chains. Issues Theory Pract Manag 6:39–48
7. Yakovleva EA (2018) Managing the intellectual resources of employees in the conditions of innovative development of the digital economy. Creat Econ 12(8):1073–1088
8. Nosova SS (2019) The digital economy in the strategy of integration of production in Russia. Econ Strat 4:18–24
9. Figovsky O, Gumarov V (2018) Green technologies. Overview of new scientific and technical developments. Sci Technol 3(336). E-source. http://www.relga.ru/Environ/WebObjects/tgu-www.woa/wa/Main?textid=5324&level1=main&level2=articles. Accessed 29 July 2019
10. Korovin G (2019) Development of the digitization processes in Russia. Ekonomist 6:38–50
11. Lyaskovskaya EA, Iminova AA (2019) Analysis of the influence of eco-economic index of the subjects of the Ural Federal District on eco-development and green construction. Bull South Ural State Univ 1:14–23
12. Krasota TG, Govako AV, Golik AS (2019) Influence of digitization of economy on the change of the structure of the labor market in the modern conditions. Financ Econ 1:487–488
13. Mezenin VG, Kudryasheva VV (2018) Digitization of economy: strategy, scales, and institutional environment. Bull. Ekaterininsky Inst. 2:19–29. E-source. https://www.niev.ru/izdatelstvo/vypusk-1-41-2018/B_1(41)_04_Мезенин_Кудряшова.pdf
14. Alekseeva NN (2019) Role of Russia in the global ecological ratings. E-source. https://ecostandardgroup.ru/about/korporativnaya-otvetstvennost/rating/Место_России_в_глобальных.Н.Н_Алексеева.pdf. Accessed 30 July 2019

15. Guryeva MA (2018) Types of regions of the Russian Federation from the position of development of the "green" economy in view of specifics of the economic space. Russ Entrep 19:2787–2810. E-source. file:///C:/Users/User/Downloads/tipologiya-regionov-rf-s-pozitsii-razvitiya-zelenoy-ekonomiki-s-uchetom-spetsifiki-ekonomicheskogo-prostranstva.pdf. Accessed 01 Aug 2019
16. Teklin BV (2019) How we implemented innovations in IT and made an invention in the sphere of personnel management. Manag Today 2(106):126–129

Chapter 52
Financial Technologies Ensuring Information Transition

Tatyana A. Yakushkin, Victoria M. Romadikova, Galina A. Gorbatkova, Valentina T. Havrel and Yuldash N. Mamengaev

Abstract *Goal*: The goal of this chapter is to substantiate the need for the transformation of the financial system for the digital transition of society. The implemented state policy of informatization of all spheres of public life dictates the need not only for resource support of informatization events, but also requires restructuring of traditional financial mechanisms of maintenance of these processes. Today's changes are impossible without a clear and efficient financing vehicle, fully adapted to the needs of digital economy and taking account of all possible complexities of remote account management. New financial instruments should be not only efficient, but also well protected from unauthorized use aimed at stealing funds from customer and state accounts. *Methodology*: in the process of the research, the principles of systematization and logical analysis of the text, the methods of graphic and tabular data analysis, horizontal and vertical analysis, the methods of generalizing and generating ideas have been used for the processing and assessment of statistical material. *Results*: In accordance with special features of a new model of ecosystem of the digital transformation of society, we need to search for new solutions both in the area of financing and in the area of finance services of occurring changes. Existing financial

T. A. Yakushkin (✉) · G. A. Gorbatkova
Bryansk Branch of the Russian Academy of National Economy and Public Administration, Bryansk, Russian Federation
e-mail: Taiayk@mail.ru

G. A. Gorbatkova
e-mail: zda101@yandex.ru

V. M. Romadikova
Kalmyk State University Named After BB Gorodovikova, Elista, Kalmykia, Russian Federation
e-mail: romadikova@mail.ru

V. T. Havrel
Bryansk Branch of REU Named After GV Plekhanov, Bryansk, Russian Federation
e-mail: khavrelvalja@gmail.com

Y. N. Mamengaev
Moscow State University Technology and Management Named After K.G. Razumovsky, Moscow, Russian Federation
e-mail: mag.mag100500@yandex.ru

T. Kolmykova and E. V. Kharchenko (eds.), *Digital Future Economic Growth, Social Adaptation, and Technological Perspectives*, Lecture Notes in Networks and Systems 111, https://doi.org/10.1007/978-3-030-39797-5_52

mechanisms do not have enough potential and capabilities for the prompt response to the needs of the financial market. All of the above gives rise to the emergence of new forms of provision of services to the customers—digital, online banks, the number and scale of which are constantly growing. *Recommendations*: The primary areas of digitalization of the financial sector that were discussed in this paper will promote accelerated transition of society to the communication platform. The development of existing and the introduction of new information tools for the provision of financial services online allows expanding capabilities of the financial sector for the provision of services for business in a digital environment. However, in order to achieve efficient use of new financial instruments, one should primarily create a new financial infrastructure that would allow keeping digital platforms operational and maintaining digital interaction.

Keywords Information society · Digital financial instruments · Budget allocations · Digital economy

JEL Codes O13 · O32 · Q13 · Q16 · P52

52.1 Introduction

The ongoing processes in the modern world demonstrate that a new approach is being formed to the consideration of the fundamentals of management in the international market, and a new redistribution of spheres of influence in the geopolitical space is taking place. In the context of universal informatization, such resources as intellectual capital, information technologies and communications, infrastructure support assume prominence. The world is moving towards a service economy, because business is increasingly providing services by means of information and communication channels of transfer of information, technologies.

The ICT market forecast defines it as growing, which is confirmed by Gartner experts. Possible hacker attacks, system hacks can not reverse the positive dynamics of market growth over the past years (Table 52.1).

The dynamic pattern of global sales of IT technologies shown in Table 52.1 proves that these instruments are in demand by all categories of users around the world. The market has been consistently growing since 2017 by 3.2–4.5% per yer. The segment of corporate software is growing most of all. According to forecasts of market analysts, by the end of 2019, the volume of sales of IT services will grow by 4.7% and amount to more than a trillion dollars, even despite the slowdown in global economic growth. In fact, society is moving towards a new round of its development; all interactions evolve into the cyberphysical space, abandoning the tangible platform and substituting it for the digital platform. But before changing to it, one should built it, ensure sustainable instruments for work and control in it, prepare other market entities to the transition to it. All of the above will require immense financial investments that are beyond the scope of the private sector; therefore, informatization becomes a strategic vector of

Table 52.1 Dynamic pattern of the global market of information technologies

Name of ICT	Sales in 2017, billion dollars	Growth, %	Sales in 2018, billion dollars	Growth, %	Sales in 2019, billion dollars	Growth, %
Data centers	181	6.4	192	6	195	1.6
Corporate software	369	10.4	405	9.9	439	8.3
Hardware	665	5.7	689	3.6	706	2.4
IT services	931	4.1	987	5.9	1034	4.7
Communications services	1392	1	1425	2.4	1442	1.2
Total	3539	3.9	3699	4.5	3816	3.2

Source Gartner: in 2019, international IT market will grow by 3.2%. Access mode: https://www.computerworld.ru/news/Gartner-v-2019-godu-mirovoy-rynok-IT-vyrastet-na-32 (Accessed: 10.10.2019)

development of the state, its formation and maintenance are laid down in the expenses of the state budget and subsidized centrally.

Financial provision of informatization of Russian society is recognized as one of strategic areas of development and is an objective of national standing. A number of regulatory documents in this area have been drawn up and approved: Information Society Development Strategy in the Russian Federation for 2017–2030; Innovation Development Strategy of the Russian Federation until 2020; The Concept of the Long-Term Socio-Economic Development of the Russian Federation until 2020; National Program "Information Society" for the period from 2011 till 2024; National Program "Digital Economy of the Russian Federation" until 2024; Information Security Doctrine of the Russian Federation and others.

According to the national program "Information society", the total amount of budget allocations for the events of interest for the period from 2011 till 2024 was 824.5 billion rubles (Fig. 52.1).

The diagram of financing of the process of formation of information society in Russia presented in Chart 52.1 shows that the volume of budgetary funds allocated for the events of interest has a positive trend, and the forecast of future development has a bottom-up linear dependency.

In fact, all government institutions today have websites through which one can receive certain services; there are governmental online service portals; the funding and financing sector provides customers with the maximum possible list of remote services; the business sector switches to cashless payments and a web ordering system; social sphere expands the online space (telemedicine, distance education system, online booking and registration systems). The population receives many operations and services remotely. In general, there is a replacement of social life with a virtual one, the maintenance of which requires new financial instruments to service these processes and operation of the entire digital economic system.

Fig. 52.1 Volume of financing according to the national program "Information society"

52.2 Materials and Method

Theoretical and applied issues of financial support for the information transition of Russian society were discussed in papers of such authors as Fedotova [1], Fedotova et al. [2], Yakunina et al. [3] who defined the primary areas of digitalization of the banking system of Russia.

The issues of legal regulation of introduction of information technologies into the financial sector are discussed in papers of the following economists: Plotnikov et al. [4], Fedotova et al. [5], Lasco et al. [6], Méndez Reátegui et al. [7], Nasir et al. [8], [9, 10], Federal Law [11].

Nevertheless, despite the abundance of publications on related topics, issues of financial support for the information transition of society are still understudied. There are many unresolved problems in the process of development of the optimization model of ecosystem of the digital transformation of the financial sector contributing to the establishment of "intelligent" and "sound" partnership relations in the processes of the use of information resources for the benefit of society. This chapter deals with these issues. The research is conducted using methods of graphical presentation of information, trend analysis, method of comparison, analogy and systematization.

52.3 Results

In recent years, the financial market has identified several key development trends within the scope of overall informatization of the social system. The main ones of them include the following: reduction of the cost of services of financial institutions, transformation of the existing business models of market participants, expansion of the range of remote financial services provided, increase in the number of operators providing payment services in addition to banks, partnership of loan institutions

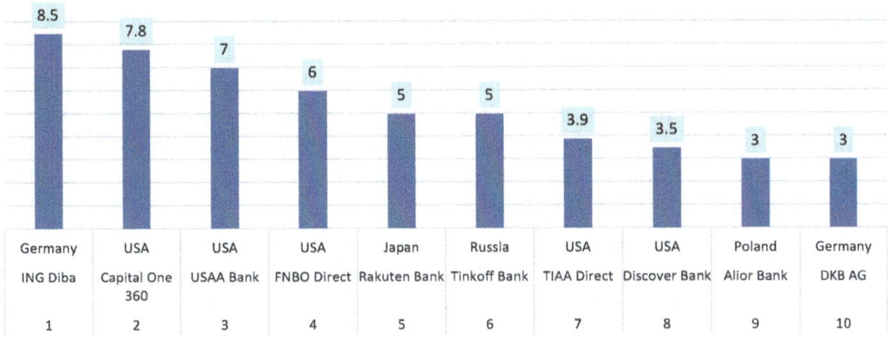

Fig. 52.2 TOP 10 of leading digital banks of the world according to the size of customer base at year-end 2018, million persons

with high-technology companies. These trends prove bear evidence of the fact that financial sector evolves and is adapted to the ongoing changes.

Modern banks all over the world are currently expanding the market of mobile and contactless payments, P2P services, digital currencies, etc. Banks are exploring the possibilities of using blockchain technologies, big data, etc. Virtual, digital banks make their appearance, handling all matters and services online (Fig. 52.2).

Digital banks are gaining more popularity and expanding their customer base. The TOP 10 banks included Russian bank "Tinkoff Bank", which has been one of the first banks to develop this service format. At its core, Tinkoff is a high-technology company with banking license, which employs up to 70% of IT professionals who create the best digital products in Russia every day. Digital banks are ahead in the financial services market due to the principles of customer-centricity, personalization of offers, and mobility. Such a bank should constantly develop according to the rate of changes in the environment; it is necessary to constantly create new financial products, be on the trend of recent changes and discoveries.

A digital bank offers most of its products and services in digitalized form with the use of digital channels. The infrastructure of such a bank is optimized for digital communications and is ready (together with a corporate culture) for a rapid change of technologies. The number of digital banks in the world is constantly growing—and the highest dynamic pattern is demonstrated by organizations that have absolutely no own offices and ATMs.

According to the data from the research, the key technologies of digital transformation in Russia were the analysis of big data (Big Data), robotization (RPA), chat bots, Optical Character Recognition (OCR), Artificial Intelligence (AI), Internet of things (IoT), Virtual and Augmented Reality (VR and AR), and blockchain. The most popular technologies in banks are AI (72%), Big Data and predictive analytics (61%), robots (56%), chat bots (56%) and optical character recognition (44%).

Modern banks are leaders in terms of the use of chat bots on an industrial scale (60% of banks use them). Financial sector was outperformed only by the telecommunications sector with its 75% figure. The third position is held by the retail sector

(50%). 45% of banks use four to 10 robots. One such robot on average replaces a minimum of 4.5 FTE (full time employees). Thus, the average market figure due to robotics is 4.5 FTE per robot. It turns out that one robot can replace 4.5 FTE; the greatest effect from the introduction of robots was noted by the heads of telecommunications companies, where one robot on average replaces more than 10 full time employees, as well as the heads of retail companies, where robotization allowed an average reduction of eight FTE per robot.

The key problem in the introduction of financial technologies into the advances sphere consists in the absence of the financial platform for remote service. In this regard, the Bank of Russia intends to take several measures in terms of strategic goals set by the state.

Thus, Russian megaregulator of financial market—the Bank of Russia defined the following long-range development goals (Fig. 52.3).

The presented development goals of the Russian financial sector call for a new digital approach to the operation of financial companies. For this reason, the existing model of management of the banking sector will be completely restructured.

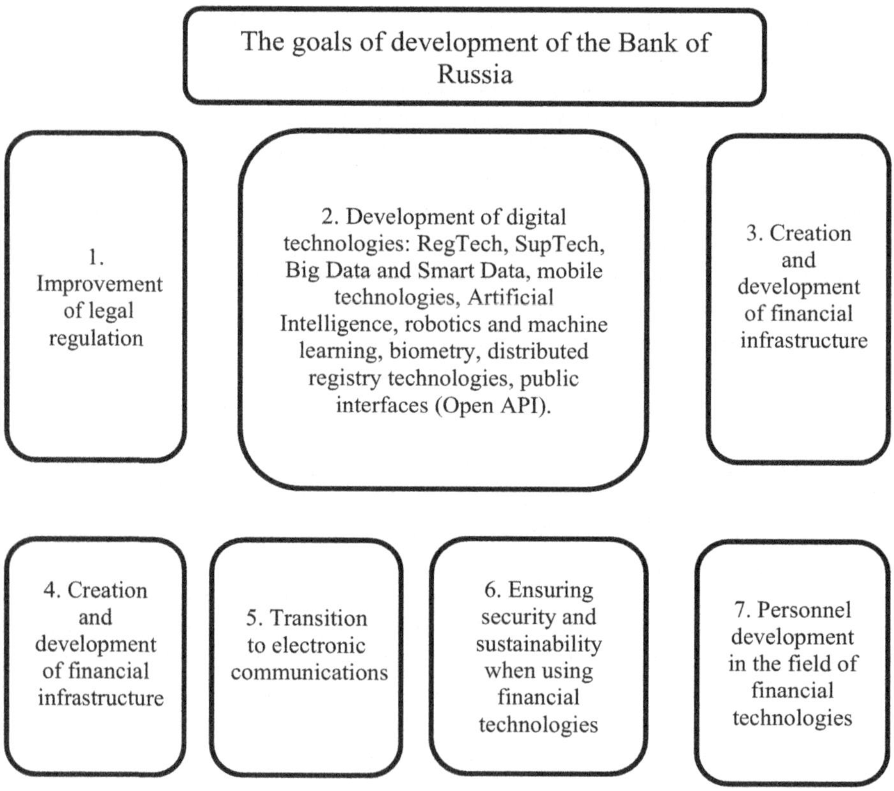

Fig. 52.3 Primary areas of the Bank of Russia in the field of financial technologies

First, legal regulation, which corresponds to the first goal of development of the financial sector. In this area, the most important tasks are as follows:

- define the legal status of the concept "digital financial technologies" which will be developed and used in the future digital financial environment, determine what means of payment will be used in this case;
- clearly indicate the basics of mining by designated institutions;
- determine the rule for attracting financial resources and cryptocurrencies.

Until these issues are resolved, there will be no clear financing vehicle in the information space.

The second area of development of the financial sector is the development and emergence of new digital technologies.

One of examples is RegTech (regulatory technology), a technology for improving the effectiveness of regulatory requirements and risk management. Usually, such technologies are used to identify customers, predict fraud by the nature of account transactions, compliance control.

Another example of promising technologies can be SupTech (supervision technologies) which are based on innovations (Big Data, machine learning, Artificial Intelligence, cloud technologies etc.). Their implementation improves regulation and supervision while simplifying administration.

In recent times, the analysts make a forecast for the prospects of application of technologies of the DARQ group, which includes distributed registry (D—DLT), Artificial Intelligence (A—Artificial intelligence), augmented reality (R—augmented reality), quantum calculations (Q—Quantum). According to 47% of respondents, Artificial Intelligence will have the greatest impact on the organizations. 19% of respondents distinguished quantum technologies as a priority, 17% of respondents emphasized the systems of distributed registry, and 15% of respondents emphasized augmented reality. This being said, 90% of managers are already experimenting with one or several DARQ technologies.

Besides, technologies for personalizing the needs and achieving a new level of digital proximity with a customer can be put to good use in the banking sector. The ability to analyze and interpret user actions, respecting their confidentiality, allows creating high-quality individual service that increases customer loyalty. 85% of bank managers believe that using digital demographics will help them better understand customer needs. Almost 30% of respondents expect an exponential increase in digital customer data that will be managed by their organization over the next 2 years.

The third area of development of the financial sector is the establishment of a new financial infrastructure which unites all participants of the digital financial environment and has additional features. Today, the existing financial infrastructure is being upgraded and supplemented with new platforms which expand its capabilities and comply with digital economy. The participants of this service and their functions are presented in Fig. 52.4. A specific feature of a new financial infrastructure consists in the presence of public interfaces (Open API), which make it possible to obtain information in systems through the use of data exchange protocols. The existence

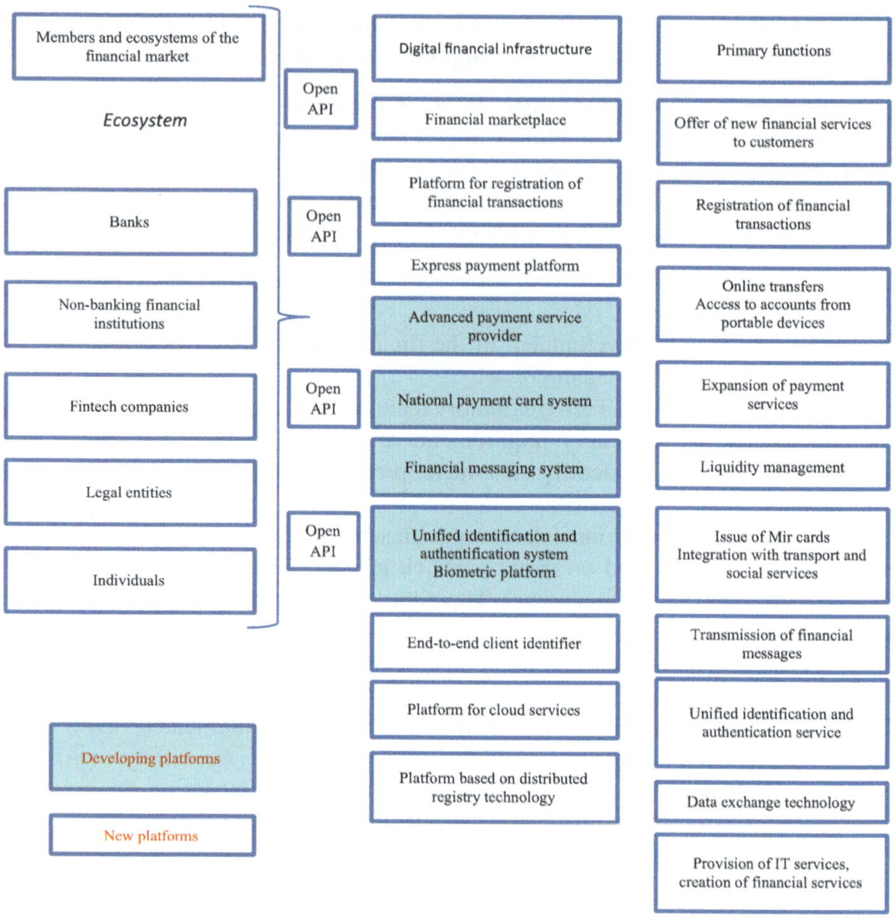

Fig. 52.4 The architecture of a new financial infrastructure

of such infrastructure will fundamentally change the financial mechanism and speed up the digitalization processes in the business environment.

The fourth area of development of the financial sector is the improvement of systems for ensuring security and sustainability when using financial technologies. Increased cyber attacks by external contractors force public authorities to carry out constant monitoring of the level of economic security of digital platforms. In order to improve technological security, the strategy for the development of the system of provision of information security (SPIS) of the Bank of Russia and information security of the banking sector and other sectors of the financial market of the Russian Federation for 2018–2022 has been developed.

All of the above activities should be carried out in strict accordance with the General strategic goal of social and economic development of the state—improvement of the quality of life of population. All transformations should be aimed at improving

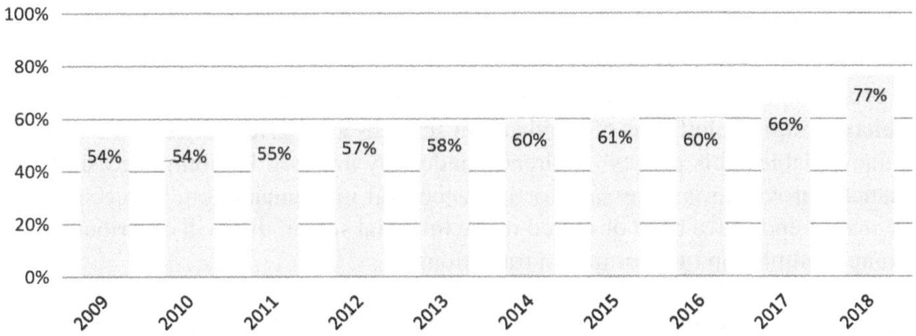

Fig. 52.5 State share in the banking sector, 2009–2018

the availability and comfort of provision of financial services to the public. The state takes on enormous importance in the digitalization of the country's economy; it is obvious that only sustainable economic development can become the basis for the development of institutions of digital economy. Against this background, there is an extremely topical issue of overcoming the negative phenomena in the financial sector, since it is the reliability of banks that is one of the most important factors determining the development of any national economic system. However, the problems of the banking sector still disturb Russian business community (Fig. 52.5).

Presented Graph 52.5 demonstrates the positive dynamic pattern of the state's share in the banking sector. This factor is an indicator of greater concentration of the policy of management of the financial sector of Russia. Granting the powers of a financial mega-regulator to the Bank of Russia narrows the boundaries of private enterprise in the financial sector. There are several reasons for that. First, the Central Bank continues to regulate the activities of commercial banks by means of rehabilitation of banking institutions, so by the early 2019 the state's share in the banking sector has already increased to 70%; moreover, according to analysts, this trend will continue for 3–4 years (the highest figure among all market sectors in Russia). Second, the aggravation of relations with the European Union and geopolitical tension stimulate the search for new national security solutions, including economic security of banking institutions. Third, the use of banking financing channels by terrorist organizations forces the government to regularly monitor the flow of financial resources to accounts with Russian banks and to demand justification for the origin of such funds. Fourth, the emerging digital platforms of online payment and remote maintenance of accounts in many cases become the targets of cyberattacks and cybercrimes, which forces us to constantly work on a system for improving not only banking, but also national security.

52.4 Conclusion

In conclusion of this research, we point out that information transition of Russian society to digital platforms of provision of services and performance of transactions is unavoidable. This process is already underway and will be completed soon, but financial provision is necessary for its successful implementation. In recent years, the main trends have been observed in the financial sector, that will contribute to the prompt completion of information transition.

The banking sector is currently lacking experts who would be the masters of new high-tech tools. The introduction of the concept "Human+" in the banks will provide every expert with the opportunity to show their worth and be successful, but this requires continuous improvement and training of employees. At the same time, it is necessary to improve the internal protection of in-house information systems from external unauthorized intrusions.

By creating ecosystems around themselves, banks must increase the stability of cyberspace in order to protect everyone. Despite the global trend towards Open Banking (banks open access to a portion of their information assets, and third parties gain access to data). In addition, the banks are not making enough efforts to ensure that their partners are worthy of the trust of end customers.

References

1. Fedotova GV (2017) Mechanisms for the improvement of information security of internet banking systems. In: Fedotova GV, Gontar' AA, Zubkova EI (eds) News bulletin of the South-Western State University. Series Economics. Sociology. Management, vol 7, no 1 (22)
2. Fedotova GV, Kovalenko OA, Malyutina TD, Glushchenko AV, Sukhinin AV (2019) Transformation of information security systems of enterprises in the context of digitization of the national economy. In: Popkova EG (eds) Ubiquitous computing and the internet of things: prerequisites for the development of ICT. Springer Nature Switzerland AG, Cham (Switzerland), pp 811–822 (Book ser. Studies in Computational Intelligence; vol. 826)
3. Yakunina IA, Sandrygailo VP, Fedotova GV (2015) An overview of economic changes in Russia during the period of sanctions. Yunost' i Znaniya – Garantiya Uspekha – 2015: collection of research papers. In: Gorokhov AA (editor-in-chief) (eds) 2nd international research-to-practice conference (October 1–2, 2015), In 2 volumes, vol 1. South-Western State University, CJSC Universitetskaya Kniga [et al.]. Kursk, pp 165–169
4. Plotnikov V, Golovko M, Fedotova G, Rukinov M (2020) Ensuring national economic security through institutional regulation of the shadow economy. In: Popkova EG, Sergi BS (eds) Digital economy: complexity and variety vs. rationality: [proceedings of the 9th national scientific and practical conference (Vladimir, Russia, April 17–18, 2019) Institute of Scientific Communications (Volgograd, Russia), Vladimir State University named after Alexander and Nikolay Stoletovs (Vladimir, Russia)]. Springer Nature Switzerland AG, Cham (Switzerland), pp 342–351. https://doi.org/10.1007/978-3-030-29586-8 (Book ser. Lecture Notes in Networks and Systems (LNNS); vol. 87)
5. Fedotova GV, Buletova NE, Ilysov RH, Chugumbaeva NN, Mandrik NV (2020) The guidelines of public regulation in terms of digitalization of the Russian economy with the industry 4.0 Tools. In: Popkova EG, Sergi BS (eds) Digital economy: complexity and variety vs.

rationality: [proceedings of the 9th national scientific and practical conference (Vladimir, Russia, April 17–18, 2019) Institute of Scientific Communications (Volgograd, Russia), Vladimir State University named after Alexander and Nikolay Stoletovs (Vladimir, Russia)]. Springer Nature Switzerland AG, Cham (Switzerland), pp 846–855. https://doi.org/10.1007/978-3-030-29586-8 (Book ser. Lecture Notes in Networks and Systems (LNNS); vol. 87)
6. Lacko R, Hurný F, Rozkošová A (2017) Conceptual framework for evaluating the efficiency of public administration. In: IDIMT 2017: digitalization in management, society and economy—25th interdisciplinary information management talks, pp 253–260
7. Méndez Reátegui R, Coca Chanalata DG, Alosilla Díaz R (2016) "The efficiency of public administration: a comparative analysis of the peruvian and ecuadorian cases from a neo institutional approach"|[La eficiencia de la administración: Un análisis comparado desde el enfoque neo institucional de los casos peruano y ecuatoriano]. Revista General de Derecho Administrativo (43):12–19
8. Nasir A, Shahzad M, Anwar S, Rashid S (2017) Digital governance: improving solid waste management through ICT reform in Punjab. In: ACM international conference proceeding series, Part F132087, 3136600
9. Edict of the President of the Russian Federation No. 203 of May 9, 2017. On the information society development strategy in the Russian Federation for 2017–2030. SPS GARANT. Access mode: https://www.garant.ru/products/ipo/prime/doc/71570570/
10. Edict of the President of the Russian Federation No. 646 of 05.12.2016. On the approval of information security doctrine of the Russian Federation. Konsultant Plius. Access mode: http://www.consultant.ru/document/cons_doc_LAW_208191/
11. Federal Law "Concerning the National Payment System" No. 161-FZ of 27.06.2011 (18.07.2017 N 176-FZ) [Electronic resource]. Access mode: http://www.consultant.ru/document/cons_doc_LAW_115625/

Chapter 53
Financial Mechanisms of Digital Transformation of Agriculture

Gilyan V. Fedotova, Yuri A. Kozenko, Tulkinbek X. Akhmedov, Tatyana E. Kozenko and Diloro N. Akabirova

Abstract The purpose of the article is to evaluate the current information changes in all spheres of the global community. Transition to the traditional spheres of the national economy to the digital platform shows the needs for a complex modernization of the sphere, the necessity for search for new approaches to production of raw materials, and growth of efficiency of agricultural companies by 2024. This goal requires the selection of the corresponding tools that conform to the modern technological requirements, which allow for correct evaluation of the effectiveness of the planned measures. For achieving the set goal, the following tasks are solved: evaluation of the dynamics of development of agriculture for the main sub-spheres of crop research and cattle breeding, analysis of the dynamics of export and import of agricultural products for the last five years, comparison of the rates of production for the main categories of raw materials, study of the departmental target program "Digital agriculture", and comparative assessment of the volumes of financing of the program's measures from various sources. The performed evaluation of the specific features of financial provision of digital transformation of the Russian agriculture is aimed at determining the tendencies of transition from the traditional approach to the technological approach in production of raw materials. The authors use the methods

G. V. Fedotova (✉)
Volga Region Research Institute of Production and Processing of Meat and Dairy Products, Volgograd State Technical University, Volgograd, Russian Federation
e-mail: g_evgeeva@mail.ru

Y. A. Kozenko
Volgograd State University, Volgograd, Russian Federation
e-mail: kozenkoja@volsu.ru

T. X. Akhmedov · D. N. Akabirova
Tashkent State Agrarian University, Tashkent, Uzbekistan
e-mail: akhmedov_tulkn@mail.ru

D. N. Akabirova
e-mail: akabirova_dilyara@mail.ru

T. E. Kozenko
Volgograd State Medical University, Volgograd, Russian Federation
e-mail: tat.kozencko@yandex.ru

T. Kolmykova and E. V. Kharchenko (eds.), *Digital Future Economic Growth, Social Adaptation, and Technological Perspectives*, Lecture Notes in Networks and Systems 111, https://doi.org/10.1007/978-3-030-39797-5_53

of normative and legal analysis and evaluation of the modern situation in the sphere of state management of digitization of the spheres of agriculture within implementation of the departmental program of digitization of agriculture. For evaluating the current situation and concluding the results of the works for the recent years, a content analysis of the results of the sphere's works is performed. Digital data are taken from open Internet sources and analyzed with the methods of financial, graphical, and trend analysis, with further formalization and systematization. In view of transition to the Fourth industrial revolution, the traditional systems of interaction and organization of the processes of production lose their importance and competitiveness. For increasing competitiveness in the Russian and global markets, companies of various spheres have to follow the modern tendencies and conduct modernization of the production basis and the system of management through implementation of digital technologies. The state is very interested in intensification of agricultural production and maximum usage of the export potential of the sphere. The program "Digital agriculture" will allow for transformation of the existing system of management and for usage of new opportunities for production of agricultural raw materials. The authors evaluate the financial provision of the implemented program of digitization of agriculture and substantiate new approaches to organization of the sphere's management.

Keywords Agriculture · Financial mechanism · Digitization · Export of agricultural products · Import of agricultural products

JEL Code G19 · O13 · O33 · P42

53.1 Introduction

State support for the spheres of agriculture and the government course at import substitution of agricultural products lead to growth of the volumes of manufactured raw materials, which are supplied to the domestic and international food markets. Russia is an agrarian country with huge agrarian territories; it can satisfy domestic demand and supply a lot of agricultural products to international food markets. The volumes of food export have been growing since 2015 (Fig. 53.1).

Evaluation of the dynamics of the Russian export (Fig. 53.1) shows sustainable growth of production of the main categories of pulses in recent five years. Crop research and cattle breeding has been raising the volumes of production. Thus, the volumes of export of meat and sub-products reached 307,000 tons—72% of growth of this category of export. For evaluating the real state of the food market, it is necessary to analyze also the dynamics of the Russian import of agricultural products. This is vividly seen in ratio of the volumes of imported and exported products of agricultural production (Fig. 53.2).

Ratio of food import/export shows simultaneous reduction of import and growth of export during the studied period. However, Russia continues importing more agricultural products, as compared to export. If the current rates of dynamics preserved,

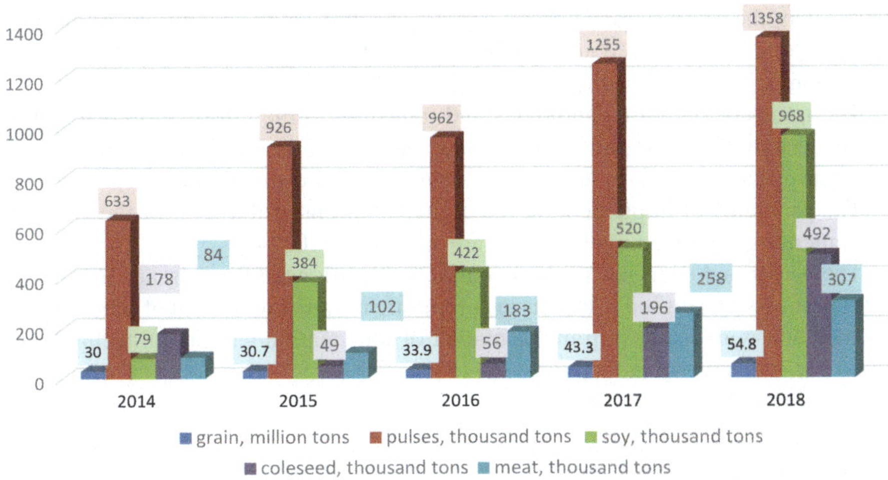

Fig. 53.1 Dynamics of export of the main agricultural products from Russia, thousand tons

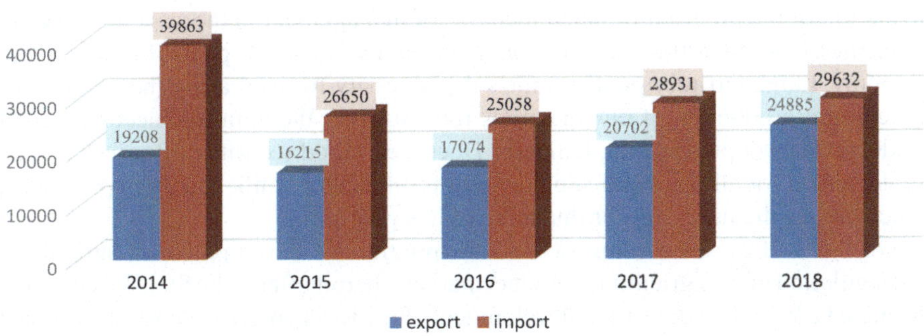

Fig. 53.2 Dynamics of import-export ratio of the main agricultural products for Russia, USD million

export will be higher than import in 2020. However, for this the state has to provide financial support for the sphere and to invest into the spheres of production of raw materials and processing of agricultural products.

53.2 Materials and Method

Experience of a lot of countries shows that digital technologies are very popular in the agricultural spheres. Food raw materials, received with application of digital solutions has the same quality as food raw materials received with the help of the traditional method—however, the economic efficiency of the former is much higher.

That's why the offered transition to digital agriculture in the conditions of growing product deficit is a justified and timely decision. This issue has been studied by a lot of economists. Let us consider some of the works.

The issues of digital technologies and transition to the information platform in the modern theories and practices of managing the national projects of development of the information and communication technologies are studied in [1–9].

The problems of implementing the digital solutions at the level of the state and implementing separate target projects are studied in [10–14].

The performed overview of the scientific publications of the Russian and foreign authors shows that digital solutions in agriculture are studied by a lot of scholars. The country that wins leadership in the market of information and digital technologies could become a world monopolist [15–18].

53.3 Results

The specific features of state support for Russian agriculture are based on implementation of the program-oriented methods of management in the system of budget financing of the main strategic directions of development of the spheres of the national economy. Large territory predetermines the necessity for increasing the volumes of agricultural production of raw materials for satisfying the domestic needs for food products. The adopted and implemented Doctrine of food security of Russia has been implemented for all main indicators, except for indicator "milk and dairy products", which causes the necessity for financing of dairy farming.

For 2019–2021—according to the State program of development of agriculture and regulation of the agricultural product markets (hereinafter—the State Program)—the main tasks and landmarks of development of the Russian agro-industrial complex are already known. The State Program has two parts: project and procedural. The main documents and directions of development of these parts are given in Fig. 53.3.

The scheme in Fig. 53.3 shows the main trends of state regulation and stimulation of development of the sphere. The documents take into account a lot of aspects of sectorial development, for strengthening of the national food security. The program envisages development of separate specialized spheres of the AIC and stimulates the investment activity in the sphere, technical and technological development of the sphere, development of the Russian export, increase of the resource basis for the light industry, and implementation of the programs of import substitution. The program shows high integration of the AIC into the Russian national economy.

As of year-end 2018, agriculture has achieved certain results (Fig. 53.4).

The graph shows that the main indicators of development of the sphere were achieved as of year-end 2018, as we have positive growth rate for the index of physical volume of investments into fixed capital—111.5%, growth rate of export—117.3%, and index of agricultural production—106.5%. Such indicators of the work of the sphere are connected to increase of the sphere of state financing and support

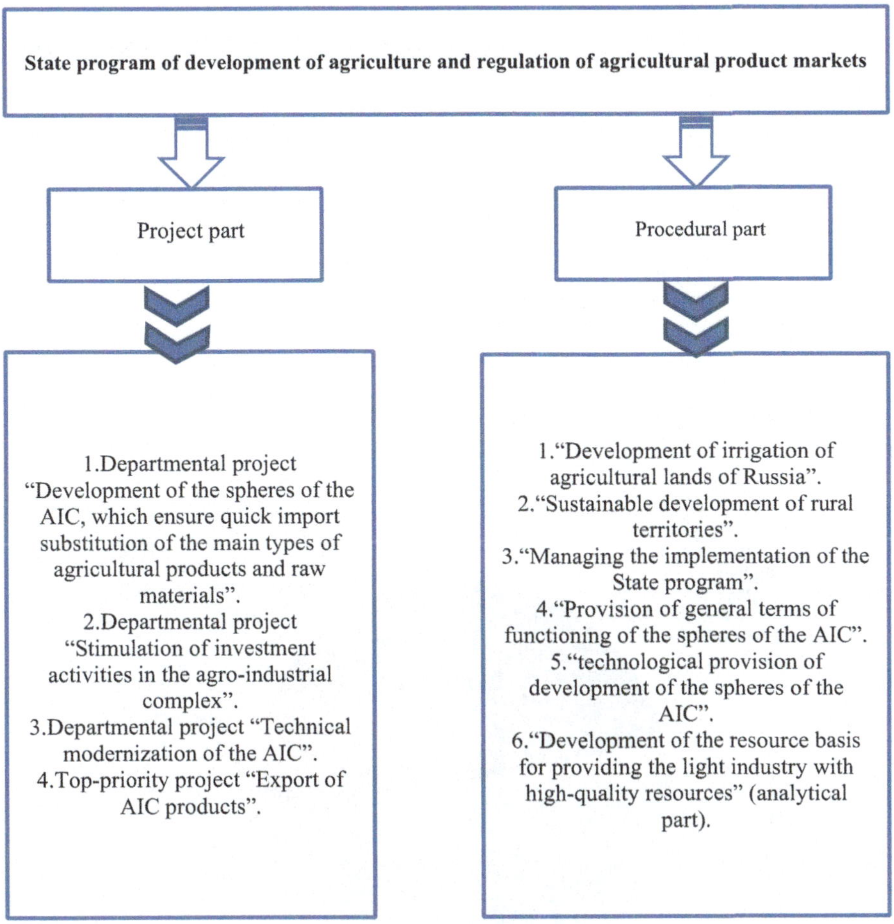

Fig. 53.3 Directions of financing of agriculture of the Russian Federation

for the sphere, which seek the goal of development of rural territories and provision of their production basis (Fig. 53.5).

Year 2018 was very good for all categories of agricultural crop research and cattle breeding. Production of vegetables was not so good, with growth of 0%, neither was production of sugar beet (−16%) and production of cereal and pulses (−13%). Other categories showed growth as compared to 2017.

State support for the sphere has led to significant results, so it is necessary to keep the volumes of financing in the future periods. Since 2019, the Government of the Russian Federation implemented certain changes: extension of the Program of development of agriculture of the Russian Federation until 2024 and correction of the program's indicators (export of AIC products should reach USD 45 billion per year and the number of the employees in the sphere should grow by 126,700). Besides,

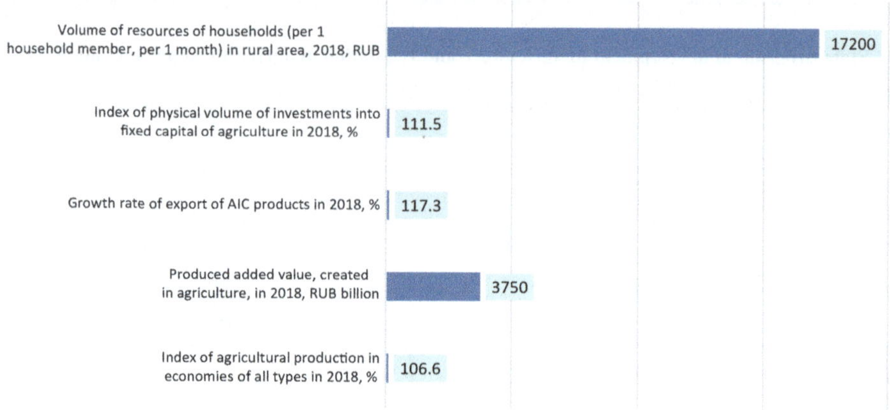

Fig. 53.4 Results of the works of the Russian agriculture in 2018

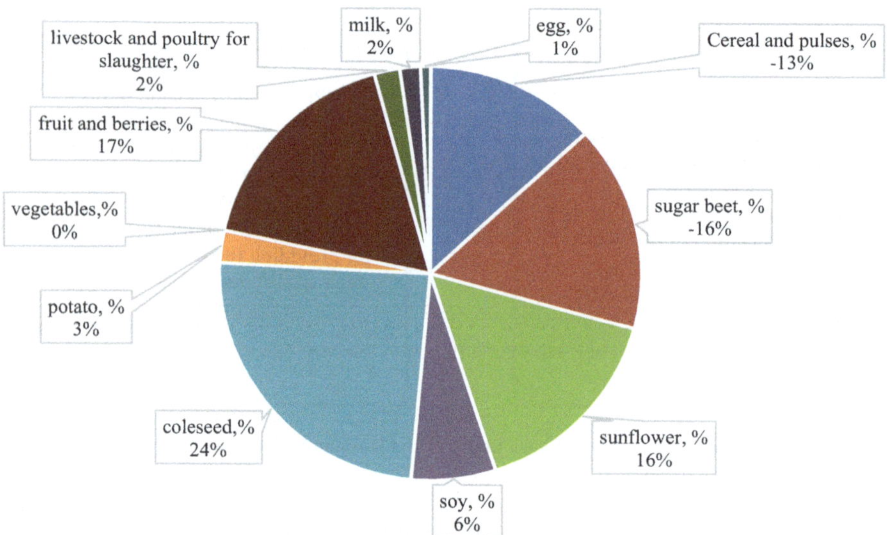

Fig. 53.5 Production of the main agricultural raw materials in 2018, as compared to 2017, %

new projects will be started: "Export of AIC products", "Creation of the system of support for farmers and agricultural cooperation", and "Digital agriculture".

Let us dwell on the project "Digital agriculture", which is to be implemented within the national policy of informatization of the Russian society. The main sense of this project is creation of the national digital platform of the AIC. The project will be implemented in three stages; let us consider each of them (Fig. 53.6).

The project is actually a plan of stage-by-stage implementation of digital technologies into the work of agricultural manufacturers of Russia. Its implementation

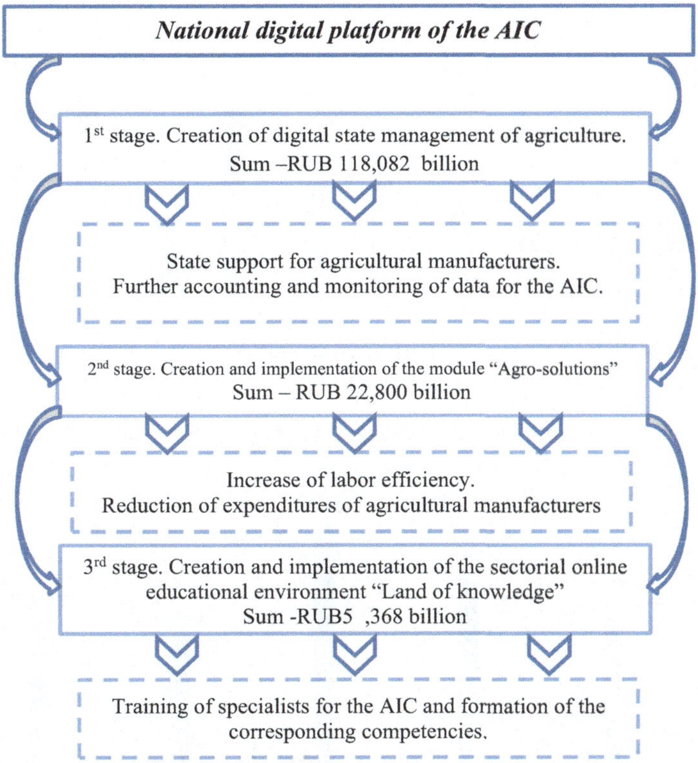

Fig. 53.6 Stages and tasks of implementing the project "Digital agriculture"

requires state support in the form of financing and legal support at all stages. Absence of experience of work in the digital environment is often a main barrier for implementing digital technologies into the practice of work. That's why the third stage—training of specialists for the spheres of the AIC—is a very important aspect of the program, which should be implemented from the start of the project.

Let us consider the volumes of financing that are necessary for successful implementation of the project (Fig. 53.7).

The sums in Fig. 53.7 show the volumes of state and non-state support for digital transformation of the Russian agriculture. The tendency and forecast of the volumes of financing are growing, which is seen in the exponential curve. At the initial state of the program the highest burden goes onto the federal budget. Then, the focuses shift, and the main financing will come from non-budget sources. Non-budget support will exceed the budget support by two times by 2024.

Let us consider the main stages of implementing the state program (Fig. 53.8).

Figure 53.8 shows the main stages of implementing the program "Digital agriculture", as well as stages of its financing. The main state support accounts for the 1st stage of the project, which will determine the work of the whole system. The volume

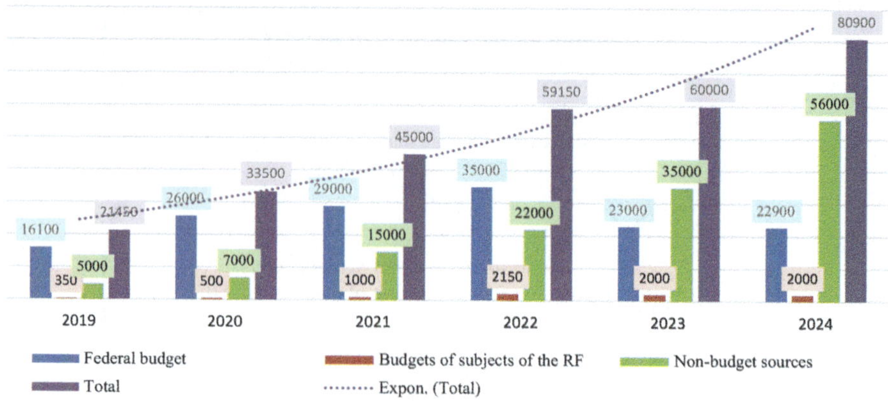

Fig. 53.7 Dynamics of the volume of financing of the project "Digital agriculture", RUB million

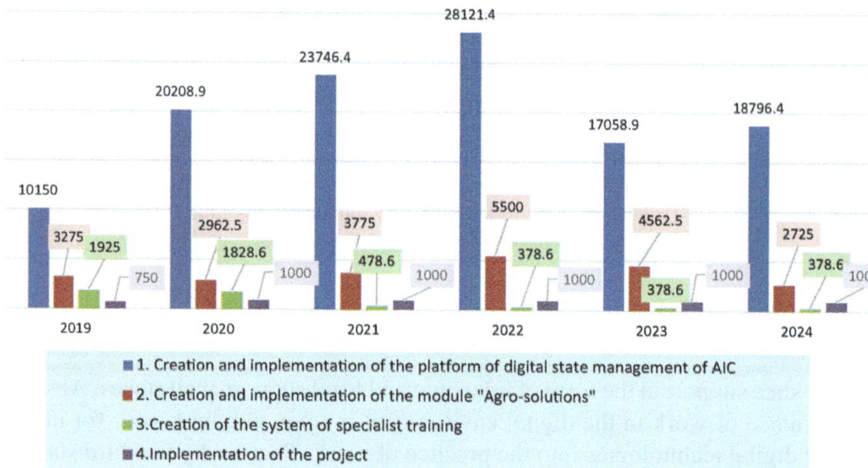

Fig. 53.8 Volumes of support for the program according to the stages of its realization

of measures of this stage is large—7 complex measures. The 2nd stage is creation of the module "Agro-solutions", which will establish interrelations and mutual dependencies between the platform participants. Financing of this stage is lower than financing of the second stage. The third stage envisages training of specialists in the sphere of IT for agriculture and creation of the training infrastructure "Land of knowledge", which will allow forming competencies with future IT agronomists and IT zootechnicians. The program envisages another direction of financing, which is connected to general management of the works on implementation of the departmental program.

Most assets for 2019–2024 will be received from the federal budget—RUB 152,000 billion; with RUB 140,000 billion from non-budget sources, and RUB 8 billion from subjects of the Russian Federation. The maximum volume of support accounts for the 1st stage of the project (RUB 118,082 billion).

Thus, according to the presented program of digital transformation of the Russian agriculture, it is planned to create and start a complex digital platform of managing all agricultural manufacturers in Russia by 2024. In case of successful implementation of the planned project, the whole system of management will be transparent, and the created interconnections within the digital platform will reduce the expenditures for search, negotiations, and evaluation of the sphere's work. The platform will allow for a clearer accounting and monitoring of agricultural resources and forecasting of the future volumes of production.

53.4 Conclusions

It should be noted that the government source at digitization of the whole economic system will allow intensifying the work of the sphere, strengthening of the national food security, ensuring a stable income from export of AIC products, and changing the opinion of Russia as a resource region. The planned measures of the program of digitization of the AIC will allow building the national digital platform and covering all agricultural manufacturers, maximizing the level of production of raw materials and increasing it by two times by 2024. The Russia's agricultural potential will be used due to active interaction in the digital environment, with reduction of expenditures of agricultural manufacturers for search for customers and sales markets. Targeted government support will stimulate the work of farmers and attract additional investments in the sphere.

Acknowledgements The reported study was funded by RFBR according to the research project No. 18-010-00103 A.

References

1. Gorlov IF, Fedotova GV, Glushchenko AV, Slozhenkina MI, Mosolova NI (2020) Digital technologies in the development of the agro-industrial complex. In: Popkova EG, Sergi BS (eds) Digital economy: complexity and variety vs. rationality: [proceedings of the 9th national scientific and practical conference (Vladimir, Russia, April 17–18, 2019)/Institute of Scientific Communications (Volgograd, Russia), Vladimir State University named after Alexander and Nikolay Stoletovs (Vladimir, Russia)]. Springer Nature Switzerland AG, Cham (Switzerland), pp 220–229. https://doi.org/10.1007/978-3-030-29586-8 (Book ser. Lecture Notes in Networks and Systems (LNNS); vol 87)
2. Fedotova GV, Gorlov IF, Glushchenko AV, Slozhenkina MI, Mosolova NI, Mosolova DA (2019) Agriculture 4.0: digital trends of development of the AIC: monograph. Volga Region Research Institute of Production and Processing of Meat and Dairy Products, Volgograd State Technical University, Plekhanov Russian University of Economics, Volgograd, 168 p

3. Glushchenko AV, Fedotova GV, Gryzunova NV, Sultanova SS, Ksenda VM (2020) Modernization of the Russian agro-industrial complex in the conditions of increase of food security. In: Popkova EG, Sergi BS (eds) Digital economy: complexity and variety vs. rationality: [proceedings of the 9th national scientific and practical conference (Vladimir, Russia, April 17–18, 2019)/Institute of Scientific Communications (Volgograd, Russia), Vladimir State University named after Alexander and Nikolay Stoletovs (Vladimir, Russia)]. Springer Nature Switzerland AG, Cham (Switzerland), pp 3–12. https://doi.org/10.1007/978-3-030-29586-8 (Book ser. Lecture Notes in Networks and Systems (LNNS); vol. 87)
4. Plotnikov V, Golovko M, Fedotova GV, Rukinov M (2020) Ensuring national economic security through institutional regulation of the shadow economy. In: Popkova EG, Sergi BS (eds) Digital economy: complexity and variety vs. rationality: [proceedings of the 9th national scientific and practical conference (Vladimir, Russia, April 17–18, 2019)/Institute of Scientific Communications (Volgograd, Russia), Vladimir State University named after Alexander and Nikolay Stoletovs (Vladimir, Russia)]. Springer Nature Switzerland AG, Cham (Switzerland), pp 342–351. https://doi.org/10.1007/978-3-030-29586-8 (Book ser. Lecture Notes in Networks and Systems (LNNS); vol. 87)
5. Shcherbina TA (2019) Digital transformation of the Russian agriculture: experience and perspectives. In: Gerasimov VI (ed) Volume: Russia: tendencies and perspectives of development. Annual edition, pp 450–453
6. Truflyak EV (2018) Results of the survey for "Digital agriculture", Truflyak EV, Kreymer AS, Kurchenko NY (eds). KubSAU, Krasnodar, 11 pp
7. Manzhosova IB (2019) Specific features of the program-oriented provision of implementing the strategy of digital modernization of agriculture. Econ, Labor Manag Agric 3(48):17–22
8. Strebkov DS, Medennikov VI, Kuznetsov IM (2019) Digital economy in agriculture. Electr Technol Electr Equip AIC 1(34):111–118
9. Kuznetsov NI, Milovanov AN, Shikhanova YA, Pototskaya LN, Monakhov SV (2019) Scientific foundations of development of the digital economy in the Russian agriculture. Bull Saratov State Socio-Econ Univ 3(77):125–129
10. Ayupov RK, Dzhumaniyazova MY (2019) Digital transformation in agriculture of Uzbekistan. Int J Humltarian Nat Sci 5–4:87–90
11. Ganieva IA (2019) Digital transformation of the Russian agriculture: consolidation of state and agri-business. Achiev Sci Technol AIC 33(4):5–7
12. Ugolnikova AM (2019) The issues of personnel provision of development of digital agriculture. Youth Sci 3:106
13. Kuznetsova GM, Safiullina AF (2019) Experience of application of information digital technologies in the modern agriculture. Mosc J Econ 9:40
14. Platonova TE (2019) The issues of development of the agro-industrial complex in view of implementing the project "Digital agriculture". Curr Probl Socio-Econ Dev Russ 1:69–73
15. Departmental project "Digital agriculture" of the Ministry of Agriculture of the Russian Federation. Access: http://mcx.ru/upload/iblock/900/900863fae06c026826a9ee43e124d058.pdf (Accessed: 02.11.2019)
16. Fedotova FV, Buletova NE, Ilysov RH, Chugumbaeva NN, Mandrik NV (2020) The guidelines of public regulation in terms of digitalization of the Russian economy with the industry 4.0 tools. In: Popkova EG, Sergi BS (eds) Digital economy: complexity and variety vs. rationality: [proceedings of the 9th National Scientific and Practical Conference (Vladimir, Russia, April 17–18, 2019)/Institute of Scientific Communications (Volgograd, Russia), Vladimir State University named after Alexander and Nikolay Stoletovs (Vladimir, Russia)]. Springer Nature Switzerland AG, Cham (Switzerland), 846–855. https://doi.org/10.1007/978-3-030-29586-8 (Book ser. Lecture Notes in Networks and Systems (LNNS); vol. 87)
17. Vectors of digital development of agriculture of the Altai Krai. Access: http://d-russia.ru/vektory-tsifrovogo-razvitiya-selskogo-hozyajstva-altajskogo-kraya.html. Accessed 01 Oct 2019
18. Klibanova YY, Kuznetsov BF (2019) Projects and developments in the sphere of digital agriculture, implemented at the faculty of energy of Irkutsk SAU. Curr Issues Agrar Sci 31:56–63

Chapter 54
Artificial Food Products: The Need for Legal Regulation

Olga V. Popova and Alla A. Serebryakova

Abstract Does the use of artificial foods affect the realization of the human right to food, a decent life? What are the risks and legal solutions to overcome and minimize the risks of artificial food use? Is there a need for legal regulation of the creation and sale of artificial food products? When answering these questions, the methods of logical-structural analysis, legal hermeneutics, comparative law, content analysis of media materials were used. First of all, it is necessary to clarify the concept of artificial food. It is important to apply the principles of prudence and precaution developed in international practice to the admission to the food market. In the Russian legislation there are the norms envisaging examination of any new products including artificial food products, however development of requirements to artificial food products is necessary. The main requirement for the use of artificial foods should be the observance of safety, quality of products, as well as the principles of equality in the right to food of every person, the maximum opportunity to choose food. Food labels must contain the composition of the product and the necessary warnings. States should strengthen the implementation of educational functions in the field of healthy and safe nutrition by developing relevant policies, provided with adequate funding.

Keywords Healthy food · Artificial meat · Artificial foods · Food safety

JEL Classification Codes K32 · K2 · K38

O. V. Popova (✉)
Department of Business Law of the Law Institute, Immanuil Kant Baltic Federal University, ul. Alexander Nevsky, 14, Kaliningrad, Russia

A. A. Serebryakova
Pskov State University, Lenin's sq. 2, Pskov 180000, Russia
e-mail: a.serebryakova@inbox.ru

O. V. Popova
ul. Alexander Nevsky, 14, Kaliningrad 236041, Russia

T. Kolmykova and E. V. Kharchenko (eds.), *Digital Future Economic Growth, Social Adaptation, and Technological Perspectives*, Lecture Notes in Networks and Systems 111, https://doi.org/10.1007/978-3-030-39797-5_54

54.1 Introduction

Modern humanity is facing a serious problem. On the one hand, the world population is growing, and the resources required for agricultural production decreases: precious fertile land built up with houses and factories; water resources are polluted and depleted; climate change; extreme weather events more frequent; reduced biodiversity. V. I. Vernadsky expressed the idea of autotrophy of mankind, which consists in the fact that humanity will be able to meet its needs regardless of the biosphere by creating artificial ecosystems, receiving artificial food and energy [1].

The concept of Vernadsky is a counterweight to the theory of Malthus [2], and the so-called "Malthus trap", into which humanity falls, increasing in number exponentially in conditions of limited resources, such a contradiction entails poverty, hunger and social upheaval.

Today, the world produces more than enough food to feed everyone, yet nearly 821 million people are chronically malnourished, and one in three people on the planet suffers from malnutrition. To meet the needs of a growing population, agricultural production should grow by about 50% by 2050 [16].

The real situation seems to be even worse. According to experts, by the end of the century the biomass of marine animals and phytoplankton will decrease, on average by 4.8% at low greenhouse gas emissions and by 17% at high, and taking into account the commercial catch of large marine animals, biomass can decrease by 48–92% even at low emissions [3], which means that the lack of food for humanity will increase catastrophically.

How to get out of the Malthus trap? Whether humanity is autotrophic? Will people be able to provide themselves with food? Currently, science is actively looking for answers to these questions from different positions: breeding, biotechnology, chemistry, demography.

And one of the areas of search in the field of food security is the search for artificial foods.

54.2 Materials and Method

The article uses the methods of logical-structural analysis, legal hermeneutics, comparative law, content analysis of media materials and scientific articles in the search for solutions to food security problems.

54.3 Results

54.3.1 *Artificial Food Products*

First of all, the concept of artificial foods is not legally defined anywhere, and researchers and the public understand by these completely different foods. For example, food products created on 3D printers. Printing of food products is carried out by mixing powdered sublimated ingredients with water, a special needle moistens the powder. This paste forms various dishes from steak to cake [17].

Such technologies allow you to create the desired taste of the product, regardless of the main ingredients. If now inks for digital food sublimate from plants and farm animals grown in the traditional way, then, apparently, the time is not far when such inks will be made on the basis of chemical compounds (proteins and carbohydrates).

Confectioners come up with recipes that enrich printed food by adding a mixture of nutrient concentrate from agar-agar, seed or mushroom spores that sprout right inside the dough. After a few days, this product can be eaten with a natural filling. These products are also created in the laboratory, and are not certified for general use [18].

Another option artificial food this the creation from a small parts of muscle tissue animal slice of meat greater volume of. Part of the muscle tissue is placed in a nutrient gel-like environment. This culture medium consists of amino acids, vitamins, salts, glucose, cells continue to divide in it, and a piece of meat increases in size.

Until recently, the cost of a kilogram of such artificially created meat was up to 300 thousand dollars, now about 90 dollars. Russian scientists have created artificial veal, which can appear on the Russian market by 2023 and will cost comparable to natural beef. It remains only a question of determining the legal regime of this product [15].

Artificial meat was seen by experts as an innovation to address food safety issues related to meat consumption, such as the growing demand for food, especially protein products. In the coming decades, artificial meat will be able to become the main alternative to traditional meat, with the potential to accurately reproduce its sensory and rheological properties [4].

Can a printed product made of natural ingredients or a gel-grown piece of veal that was originally a live calf be classified as an artificial product? Or is it necessary to refer to artificial food products only food products created from chemical ingredients?

Should genetically modified crop and livestock products be classified as artificial products [5].

In our opinion, whether ultra-processed foods or protein foods for sports nutrition should be classified as artificial foods.

The consumption of highly processed foods has traditionally been regarded as low quality food, and experts suggest educational campaigns aimed at replacing ultra-processed foods with minimally processed and unprocessed foods [6].

But the problem is that traditionally grown foods, ultra-processed, or even highly processed contain a lot of artificial ingredients [7]. Such artificial ingredients-preservatives, providing increased storage, dyes, necessary to attract buyers, can negatively affect human health, causing diseases of the digestive tract, and obesity, which in turn is associated with diseases such as diabetes or cardiovascular diseases. Therefore the quality and safety of such food products should be subject to increased control by the state in order to ensure the health of the population.

But will artificial foods be in purchasing demand? One of the articles analyzes the possibility of selling food and drinks blue. Such color is estimated by consumers as artificial for food, however, blue food and drinks are becoming more common in the world [8]. Although blue natural products exist in nature, such as blueberries, often cooks emphasize a natural source for the dye for their goods. Artificial foods may be fashionable, but in general, people intuitively regard natural foods as more useful.

Organic food, environmentally friendly food is regarded by consumers as the most useful food, organic farming is supported, carried out without the use of chemical fertilizers and pesticides [9], however, it is necessary to understand that organic foods may be unsafe, contain harmful inclusions (for example, Giardia in vegetables or brucellosis pathogens in milk) and the quality of organic products must also be ensured proper control.

Thus, it is necessary to conclude that the approach to the definition of artificial products can be ambiguous. It can be products of the chemical industry made of inorganic raw materials; biotechnological industry, and other products obtained as a result of the use of scientific and technological progress, which could not exist before these achievements became available to man.

In particular, we can talk about food products obtained as follows: artificially printed food products using 3D technologies; artificially grown food products; products using genetically engineered organisms; food products containing a large number of artificial chemical additives; food products with a significant degree of processing.

All these types of products must undergo the necessary certification procedures.

The package of artificial food products must contain information about the main consumer properties, including their composition. In particular, paragraph 2 of article 10 of the Law of the Russian Federation on consumer protection should be supplemented by the requirement to specify the percentage of chemical food additives used in the manufacture of food products, as well as the method of creating a food product [10].

54.4 Environmental and Food Safety and Artificial Food Products

Food safety now depends entirely on the suitability of agricultural land. First of all, it is important to preserve fertile land for crop production, but also for animal husbandry, the quantity and quality of agricultural land is also important, since animal feed is also crop production. Thus, one can agree with scientists who attribute the suitability of land for agricultural production to the key tools for sustainable development. At the same time, there are studies that Express reasonable concern about the preservation of land suitability in conditions of climate change [11].

The production of artificial food can reduce production costs; the involvement of agricultural land required for agricultural production; the cost of transportation of products; reducing disposal costs unrealized overdue of agricultural products, as well as the fact of reducing the negative environmental impact of unclaimed agricultural products by reducing emissions of methane and carbon dioxide.

As pay attention to the manufacturers of 3D food printers digital-food will save resources on the ingredients themselves and the cost of delivery. Sublimated products in cartridges are stored much longer than fresh ones. It is argued that the production of artificial meat instead of traditional reduces the growth of gas emissions, reduces the cost of water and electricity [4].

Of course, products grown without the use of aggressive agricultural techniques (without pesticides and chemical fertilizers), have better taste and are usually more useful, but to ensure food safety quantitatively can not. People are forced to look for ways to increase land productivity by fertilizing and controlling harmful insects. And artificial food is a way out of the food shortage situation, a way out of the Malthus trap. Humanity manifests its autotrophic quality, becoming increasingly independent of the biosphere.

However, we believe that food security lies not only in the necessary quantity, but also in the proper quality of products [12]. The quality of products should ensure the inadmissibility of harm to health when consumed, as well as a sufficient range of products for different taste, national, regional, religious, etc. preferences.

In no case should it be allowed that access to natural products was only for certain groups of people, while others could not have the right to choose such products as a result of legislative or price restrictions. That is why support for traditional forms of management, support for the peasantry and farming should be provided.

It is also important to pay attention to the fact that despite the obvious advantages of the production of artificial products for the environment, the environmental risk may be an increase in the consumption of plastic for packaging. After all, artificial foods are long lasting products, they are created, including in order to increase the shelf life of food. And therefore, the amount of packaging material will increase.

The European Union is making great efforts to prevent pollution of the environment with waste containing macro-and micro-particles of plastic [13], however, it is possible to assume with a high degree of confidence that the number of disposable packaging will increase, despite all the efforts of the new European legislation.

Thus, assessing all the advantages of the scientific and technological revolution associated with ensuring the food security of mankind through the development of new food products, the production of artificial food products can harm the environment.

However, artificial products are likely to ensure the survival of mankind in the future. But there are other issues that need to be taken into account, which can be described as human rights to a decent life, not just survival. This includes the right to live in the existing biosphere: green forests, clean air and water, the preservation of all species of existing animals. It is impossible not to pay attention to ethical and legal aspects: all categories of the population should have access to high-quality, diverse food.

54.5 Conclusions/Recommendations

We come to the conclusion that artificial foods—a heterogeneous list of products, which is replenished as a result of the use of scientific and technological progress. In particular, currently can be attributed to those and artificial products created from biological mass.

The current Russian legislation on consumer protection and on the quality and safety of food products generally regulates the creation and circulation of artificial food products.

Thus, article 16 of the Federal law on quality and safety of food products provides that when developing new food products and new technological processes for their manufacture, it is necessary to justify the requirements for the quality and safety of such food products. New food products are allowed to be manufactured after state registration [14].

In addition, food products of a new type of each name is subject to assessment (confirmation) of compliance in accordance with the technical regulations.

Food products of a new type include products not previously used by humans for food, namely: with a new or intentionally altered primary molecular structure; consisting of or isolated from microorganisms, microscopic fungi and algae, plants, isolated from animals, obtained from GMOs or using them, nanomaterials and nanotechnology products (TR CU 021/2011. Technical regulations of the Customs Union. On food safety, 2011).

In addition, food products of a new type of each name is subject to assessment (confirmation) of compliance.

It is supposed to create a unified register of food products of a new type (articles 21–29 TR CU 021/2011).

Artificial food should be classified as a new type of food product.

However, the definition of a new type of food product should contain an open list of options for the emergence of new products, in order to avoid the transfer of any new food products into circulation without proper state expertise.

Thus, a certain basis for justifying the legal regime of artificial food products and its inclusion in circulation in Russia exists, however, it is necessary to clarify both the legislation and technical regulations relating to specific requirements for quality and safety indicators of artificial food products, shelf life, packaging, labeling, information about them, as well as their production, including the conditions of manufacture and turnover, production control programs for their quality and safety, test methods and methods of disposal.

References

1. Vernadsky VI (1993) Russian cosmism: an anthology of philosophical thought. Pedagogy-Press, Moscow, pp 288–303
2. Malthus T (1895) The experience of the population law. Translation
3. Lotze HK et al. (2019) Global ensemble projections reveal trophic amplification of ocean biomass declines with climate change. PNAS 116(26):12907–12912. doi:https://doi.org/10.1073/pnas.1900194116. https://www.pnas.org/content/116/26/12907#sec-1
4. Alfieri F (2018) New food: artificial meat. Encycl Food Secur Sustain Dev 1:280–284
5. Anisimov A, Inzhieva B, Popova O, Ryzhenkov A (2019) Legal regulation of genetically modified organisms (GMOS) in agriculture: the view from Russia. Quality 20(170):123–129
6. Aguirre A, Borneo M, El Hori S, Borneo R (2019) Exploring the understanding of the term "ultra-processed foods" by young consumers. Food Res Int 115:535–540. https://doi.org/10.1016/j.foodres.2018.09.059
7. Da Louzada MLC et al (2017) Validating the usage of household food acquisition surveys to assess the consumption of ultra-processed foods: evidence from Brazil. Food Policy 72(C):112–120. https://doi.org/10.1016/j.foodpol.2017.08.017
8. Spence C (2018) What is so unappealing about blue food and drink? Int J Gastronomy Food Sci 14:1–8. https://doi.org/10.1016/j.ijgfs
9. Anisimov A, Popova O (2017) Legal problems in developing organic (enviromentally friendly) markets. Environ Policy Law 47(3–4):139–147. https://doi.org/10.3233/EPL-170031
10. Law of Russian Federation No 2300-1 of 07 Feb 1992 (2019) On protection of consumer rights. Collection of legislation of the Russian Federation, 15 Jan 1996, No 3, p 140
11. Akpoti K, Kabo-bah AT, Zwart SJ (2019) Agricultural land suitability analysis: state-of-the-art and outlooks for integration of climate change analysis. Agric Syst 173:172–208. https://doi.org/10.1016/j.agsy.2019.02.013
12. Malkhasyan AV (2014) Agrarian and legal problems of food security of the Russian Federation: abstract. Dis…. cand of law, Moscow
13. Dubovik OL (2019) New legislation of the European Union on the prevention of environmental pollution by waste containing macro-and microparticles of plastic. Int Law Int Organ 2:16–27. https://nbpublish.com/library_read_article.php?id=30107. https://doi.org/10.7256/2454-0633.2019.2.30107
14. Federal law from 02 Jan 2000 No. 29-FZ (2018) About quality and safety of food products. Collection of legislation of the Russian Federation, 10 Jan 2000, No 2, p 150
15. Online edition of RIA Novosti, available at: https://ria.ru/20190923/1558998554.html?utm_source=yxnews&utm_medium=desktop&utm_referrer=%3A%2F%2Fyandex.ru%2Fnews. Accessed 20 Oct 2019
16. Official site Food and agriculture organization OON "Global hunger continues to rise", new UN report. Available at: URL: http://www.fao.org/news/story/en/item/1152031/icode/. Accessed 20 Oct 2019
17. Official site publishing "Open system". Available at: https://www.osp.ru/os/2013/08/13037863/. Accessed 20 Oct 2019
18. Site Chloé Rutzerveld. Available at: https://www.chloerutzerveld.com/#/edible-growth-2014/. Accessed 20 Oct 2019

Chapter 55
Legal Regulation of Charitable Activities Carried Out with the Use of Digital Technologies

Roman A. Romashov and Elena A. Bryleva

Abstract *Purpose*: The purpose of the publication is to study the legal nature of charity—one of the trends of our time. Almost any state is connected with it in one way or another. At the present stage of development, we are witnessing a rapidly growing role of charity, which attracts more and more new participants and becomes more and more important. At the same time, the charity sector is changing in the digital economy, and with the use of new technological solutions this sector will develop much more efficiently, an example of such an organization in Russia is the "Rybakov Fund". *Design/methodology/approach*: The research methodology is based on the application of such research methods as the method of analysis, synthesis, deduction, induction and analogy, among the private scientific—formal-legal method, respectively, historical, logical, dialectical techniques and methods of scientific knowledge. *Findings*: Results: the article reveals the problems of charity development, in particular the lack of digitalization in this area, as in the UK. Also reflected is the requirement of time and civil society, as a social orientation of business on the one hand, and on the other the desire of legal entities actively engaged in charity to have preferences from the state: both in the tax sphere and in the sphere of moral encouragement: public recognition of merits; establishment of awards by both state and public structures. *Originality/value*: Conclusion: according to the authors of the publication, the solution to these problems is seen in the introduction of amendments to article 381 of the Tax code (part two) of 05.08.2000 No. 117–FZ (ed. of 01.05.2019) p. 29 in respect of organizations engaged in charity, as well as the adoption of special regulatory documents governing the promotion of charity.

R. A. Romashov (✉)
Petrozavodsk State University, Petrozavodsk, Russia
e-mail: romashov_tgp@mail.ru

E. A. Bryleva
Perm Institute of the Federal Service for the Execution of Sentences of Russia, Perm Branch Office of Russian, Presidential Academy of National Economy and Public Administration, Perm, Russia
e-mail: elenbrylev@yandex.ru

T. Kolmykova and E. V. Kharchenko (eds.), *Digital Future Economic Growth, Social Adaptation, and Technological Perspectives*, Lecture Notes in Networks and Systems 111, https://doi.org/10.1007/978-3-030-39797-5_55

Keywords Charity · Digital technologies · Civil society · Social orientation · Charitable foundations · Charitable organizations · Internet space · Communication · Non-profit organizations · National security · Software

JEL Code K 19

55.1 Introduction

With the use of digital technologies, everyday human life, industrial relations, the structure of the economy and education are changing, as well as new requirements for communications, computing power, information systems and services. Moreover, such previously understood as "traditional" spheres of activity undergo significant changes and reach a completely different qualitative level [1]. Digital transformation is inevitably exposed to both educational and social spheres, as providers of new solutions and solutions already working for the digital economy [2–10].

In 2018, it is estimated that more than $ 500 billion was spent on charity worldwide, with this amount increasing annually. On the one hand, it is an indicator that the charity sector is becoming more significant, attracting more new participants and social investments. At the same time, it makes sense to think about improving the effectiveness of activities in the field of charity, how to expand the experience built exclusively on people. Without technological solutions, without digital tools, many of the tasks facing the charitable community today will not be solved.

55.2 Materials and Method

The research methodology is based on the application of such research methods as the method of analysis, synthesis, deduction, induction and analogy, among the private scientific—formal-legal method, respectively, historical, logical, dialectical techniques and methods of scientific knowledge.

55.3 Results

Charity as a two-component "good" and "goodness" has always been inherent in the Russian people, especially in the days of the Empire was common among Russian merchants. Thus the Imperial surname about what in the Soviet period of time it wasn't accepted to speak, was an example, and carried out considerable donations from own means.

Modern Russian society is actively continuing the trend of charity. According to a survey conducted by the Public chamber of the Russian Federation in 2018, 53.7% of respondents occasionally give alms, transfer money via SMS, give things or products to the needy, 24.2% occasionally donate to the construction of a temple, an animal shelter, the restoration of an art monument or a valuable natural object, 16% of respondents are not engaged in charity. On a regular basis, 32% of respondents are engaged in various charitable activities: 13.8% regularly donate to a known charitable organization (Fund), transfer funds to the account of an orphanage, a nursing home, etc.; 9.6% regularly give alms/visit and provide voluntary gratuitous assistance to those in need; 4.7% regularly make donations to projects to raise funds for the construction, repair or restoration of cultural facilities; 3.9% regularly allocate to charity a certain percentage of their income, the analysis of donations shows that the most popular are assistance to patients with serious diseases (53.1%), orphans (43.7%), disabled (35.6%).

Among the trends the external environment of the charity (the donor behaviour of the population and business) significantly decrease the willingness to donate money: declining share of new regular payments on a subscription or auto payments and decreasing in overall personal budget to charity. The professional community of the charity sector today prefers to move to project activities and gradually abandons targeted assistance, is ready to raise important social problems and offer solutions to them to the state, is ready to unite to solve acute problems.

For example, in Russia the Board game "Imaginarium Dobro" created by children from orphanages—the joint project of Public Chamber of the Russian Federation and Cosmodrome Games enjoys success. 8 charitable foundations took part in the creation of the game: "Vera", "LIVE", "Sheredar", "Creation", "ship", "Nastenka", "Ronald McDonald's House" and "Rusfond". Every 100 rubles from the sale of this game go to charity funds, supervising homes for children left without parental care. All large funds today employ professional fundraisers (fundraising), attracting material or non-material resources, such as human, information, financial, legal, marketing, etc. Methods of attracting resources in fundraising: request (personal and personal); mass actions to raise funds, using agitation; grant competitions; mutually beneficial partnership; subsidies; provision of paid services; membership fees of organizations; collection of private donations; lending. Volunteering is actively developing. For example, Perm is hosting a large volunteer week for the third year in a row, when assistance is provided to cultural heritage sites, nature, children and adults in need, and each time the number of participants is growing, which indicates the growth of civil activity of the Russian population.

Another trend of charitable organizations is the use of crowdfunding (crow funding) (collective investment in a charity project). Some non-profit organizations were able to achieve self-sufficiency and became social entrepreneurs. For example, with the support of the center for innovation of the social sphere, the project "Altai string bag", which employs completely blind people, received permission to open a store at the airport of Barnaul and thus was able to pay for the work of their wards.

The new generation, generation Z, no longer triples the so-called "traditional charity." According to Cone Communications CSR Study 2017 is the year of corporate social responsibility (CSR) revision. At the same time, 63% of Americans hope that business will become a leader in promoting social and environmental change in the absence of government regulation; 78% want companies to deal with important issues of social justice; 87% will buy a product because the company protects the problem they care about, and 76% will refuse to buy the company's goods or services after learning that it supports a problem that contradicts their beliefs.

Nielsen analytical center for 2017 revealed an interesting trend in the efficiency of the company and its social orientation: 85% of global consumers and 78% of Russian pay attention to the care of the company about the environment; 80% of world and 61% of Russian prefer the products of the company, which works on social projects. Accordingly, to be benefactors is not only honorable, but also profitable. For example, Mursal Hedayat (Mursal Hedayat), included in the list of social entrepreneurs Forbes 30 under 30, in the UK founded the project Chatterbox—a platform for finding work only migrants.

One of such socially oriented on the one hand, and on the other hand "advanced" organization is "Rybakov-Fund". They find solutions to global and local social problems in a way unusual for traditional non-profit organizations, creating online platforms, algorithms and neural networks using the latest technologies. PhilTech is an inevitable trend. The Fund "arithmetic of good" has created its educational platform for orphans, operating in 31 regions (103 orphanages), on the basis of which teaches orphans. Indicator as a result, 42 people entered higher education institutions in 2015–2017, and 266 children entered secondary vocational educational institutions. This IT tool made it possible to simplify the work of employees and increase the efficiency of programs. All over the world there are already examples of successful projects that can be attributed to PhilTech. These are mobile applications, web platforms, digital services and other scalable IT solutions.

At the same time, in Russia, examples of the use of technologies are not as significant as in the UK [11], although the support of the Russian state in this direction is present. So, in 2017, through the presidential grants Fund, Ruslan Shekurov, the author of the project to create a blood donor community DonorSearch, received 3 million rubles to finalize his online platform. As noted by Alena Svetushkova, Vice President of Rybakov Fund, this is primarily due to "the lack of digital skills in management and understanding of how the digital economy affects the organizational and business models of NGOs, on the efficiency of work." (PhilTech v dejstvii).

And it affects very much: according to the same Lloyd, non-profit organizations that implement the "figure" in their activities, ten times lower costs, collect twice as much money, and spend half as much time on traditional operations. "But in addition to management, basic digital skills should penetrate into other areas of work: financial transactions, communication with audiences, creating new products and services." (PhilTech v dejstvii)

One of the modern trends is the desire of business structures to become socially responsible on the one hand, and on the other to have certain preferences, at the same time.

55.4 Conclusion

Summing up the research, we will draw the following conclusions: Unfortunately, unlike the Western legislator, Russian legislation does not fully stimulate the practice of charity. For example, in a number of Western countries, non-profit organizations engaged in charitable activities are exempt from income and value-added taxes, sometimes from corporate property taxes. As for private benefactors, they are provided with tax deductions. For example, completely charitable contributions are excluded from the tax base in Austria, Greece, Norway, a certain limit for them is set in Belgium, Canada, France and other countries [Sokol M. P.].

The obstacles to the establishment of a new charitable system in Russia can be overcome only through equal and mutually respectful cooperation between state bodies and public structures.

We also consider it possible to take into account the Russian historical traditions and foreign experience of charity in terms of stimulation by both moral and material forms of encouragement, in particular, the measures of moral encouragement include: public recognition of merits; the establishment of awards by both state and public structures; the adoption of special normative documents regulating incentives for charity and others.

To the material incentives: priority for legal entities active in charity in the awarding of public contracts and to amend article 381 of the Tax code (part two) from 05.08.2000 № 117–FZ (as amended on 01.05.2019) p. 29 in the case of organizations involved in charity work.

References

1. Inshakova EI, Ryzhenkov AY, Inshakova AO (2019) Neo-industrialization of the Russian economy: technological and digital development. In: Popkova EG (ed) Ubiquitous computing and the internet of things: prerequisites for the development of ICT, Studies in computational intelligence, vol 826. Springer Science+Business Media, Cham, pp 239–250
2. Kupriyanovskij VP et al (2017) Skills in the digital economy and challenges of the education system. Int J Open Inf Technol 5(1):19–25
3. Cone Communications research (2019) 2017 Cone Gen Z CSR study: how to speak Z. Electronic resource. http://www.conecomm.com/research-blog/2017-csr-study
4. Nielsen research Agency (2019). Electronic resource https://www.nielsen.com/ru/ru/about-us/
5. Social startup Chatterbox prepares language tutors from migrants (2019). Electronic resource https://te-st.ru/2017/03/15/chatterbox-of-migrants/
6. PhilTech in action: how technology helps charities and non-profit organizations (2019). Electronic resource https://rb.ru/longread/philtech/
7. Digital transformation of charity (2019). Electronic resource https://regnum.ru/news/society/2488449.html
8. Report on the state of civil society in the Russian Federation for 2018. https://report2018.oprf.ru/ru/
9. Sokolov IA, Kupriyanovskij VP, Namiot DE, Drozhzhinov VI, Bykov AY, Sinyagov SA, Karasev OI, Dobrynin AP (2017) State, innovation, science and talent in the measurement of the digital economy (the case of the UK). Int J Open Inf Technol 5(6):33–46

10. Sokol MP (2019) Income taxation of individuals. http://wwwHm.businesspress.EN/newspaper/article_mId_36_aId_293316.html
11. Tech Nation (2017) At the forefront of global digital innovation. Tech City UK

Chapter 56
Digital Agriculture: Opportunities for the Development of Small Agribusinesses and the Respective Problems

Nataliya Ivanova, Maxim Ovchinnikov, Maria Lata and Ivan Korabelnikov

Abstract Creation of the digital agribusiness industry ecosystem presupposes the study of innovative capacity and the full use of the modern digital platforms capabilities in farming and animal husbandry, as well as the training of the highly skilled professionals, ready to work in the conditions of the high-tech industry formation. Considering that an abnormally high proportion of the family-operated farms and small farms (99% total) and their prevalence in the structure of agricultural products is a peculiarity of Russian agribusiness industry, it is the main reason of underperformance and low salary combined with high production outlays in the agribusiness industry. The studies carried out by the authors indicate that singular attempts at implementation of digital tools in agriculture made by Russian farmers have not made a major change; the main reason is system planning and an institutional support of digitalization is required. In this regard, a hypothesis of a much-needed development of technical and economic criteria system is made. The criteria are *personnel, technical, technological, info-analytical and financial*. The fulfilment of these will allow small farms to successfully cross over to producing agricultural products using digital technologies within the time limits set by policy measures. With that in mind the authors conducted the research on personnel and innovative potential of the development of small farms on the example of the Volgograd region, identified the positive aspects, risks and opportunities for the implementation of digital processes and technologies in agricultural production.

N. Ivanova (✉) · M. Ovchinnikov · M. Lata · I. Korabelnikov
Volgograd State Agrarian University, Volgograd, Russia
e-mail: inv.74@mail.ru

M. Ovchinnikov
e-mail: oma_volgau@mail.ru

M. Lata
e-mail: mariya-lata@yandex.ru

I. Korabelnikov
e-mail: korablick.ru@mail.ru

T. Kolmykova and E. V. Kharchenko (eds.), *Digital Future Economic Growth, Social Adaptation, and Technological Perspectives*, Lecture Notes in Networks and Systems 111, https://doi.org/10.1007/978-3-030-39797-5_56

Keywords Small farms · Regional economy · Agricultural economy · Digital economy · Experimental digital farms

JEL classification *D1* · *R1* · *Q0*

56.1 Introduction

According to the National Program "Digital Economy of the Russian Federation" and the Departmental Project "Digital Agriculture" the top priorities of the Russian agribusiness industry development are the transition to "intelligent" production and the scaling of Russian holistic digital agricultural solutions, suggesting a real strategic breakthrough that can significantly increase management efficiency at the macro- and microscale.

Agriculture is an important part of the agro-industrial complex and the basis of the Russia's national food security. Agriculture has vast opportunities for modernisation, development of export potential, the implementation of new innovative solutions and management decisions.

In the process of managing their agribusiness farmers make about 40 decisions about digitalisation in a very limited time [1]. Nevertheless the current status of Russian agribusiness's digitalization raises concerns since it is characterized by an insufficient level of scientific and practical knowledge, the absence of logistics systems and agricultural market forecasting, which leads to high production costs [2]. Only a small number of agricultural goods producers have the financial capabilities to modernize machinery, use IT platforms and equipment [3].

According to the Federal State Statistic Service, the contribution of "agriculture, hunting and forestry" to GDP in 2015 amounted to 3.67 trillion rubles, 2016—3.87 trillion rubles and 2017—3.69 trillion rubles. In 2016, 119.1 million tons and in 2017 a record number of 135.4 million tons of grain was collected in the Russian Federation. According to the Federal Customs Service, in 2018 Russia was placed first in wheat exports—36.4 million tons (16). At the same time the economy of the Russian Federation lags severely behind in a number of indicators comparing to the leading economies of the USA and Germany.

In accordance with the May decrees of the Russian Federation president, the goals of the digital transformation of agriculture is to create end-to-end chains from production to consumption with deep integration into related sectors of the digital economy. The Ministry of Agriculture of Russia estimated that Russian agribusiness industry has the potential to increase agriculture production effectiveness through digital processes implementation [4].

The development and formation of a basic set of processes and methodology of digital agriculture will provide the opportunity for agricultural products production in an end-to-end digital environment "from field to market"; launch of a platform for producers that will help the formation of dynamic seasonal KPIs for crop management—"Effective hectare", "Smart greenhouse", and for animal husbandry—"Effective head", "Smart farm", creation of innovative farms as platforms for technology development and training; implementation of the "Internet of things" platforms.

56.2 Materials and Method

The authors conducted the research of agriculture digitalization using official data from the Federal State Statistic Service, Federal Customs Service, the Ministry of Agriculture of Russia and the United States. The methodological basis of the study include the National Program "Digital Economy of the Russian Federation", the Departmental Project "Digital Agriculture", the works of Russian and foreign scientists, which allowed a comprehensive and focused study of the problems and prospects of industrial digitalization.

During the research a system approach was used; as well as structural, functional and logical analysis; a method of analyzing trends, as well as a synthesis of Russian and foreign theory and practice in the application of digital technologies in the agricultural sector.

56.3 Results

Small agribusinesses continue to stay an essential part of Russian agriculture. In the structure of agricultural production in the Volgograd region small farms traditionally have strong position comparing with other regions of Russia [5]. At the same time, they fulfil a dual function: on the one hand, these farms produce up to 60% of gross agricultural output, and on the other hand small farms act as a kind of social stabilizer, ensuring the stability of rural areas, preventing the outflow of youth from villages and contributing to the solution of many other socio-economic problems. Small agribusinesses in their production and economic capabilities are certainly inferior to large agribusinesses, but at the same time small agribusinesses greatly exceed large ones in flexibility and sometimes even in production efficiency.

According to the Volgograd Statistic Service, 4800 family-operated farms, 240600 households, 124 agricultural consumer cooperatives were recorded in Volgograd region. Among small agribusinesses, "microbusinesses" prevail. Their share among small agribusinesses amounted to 60% [6].

An analysis of production and economic activity of small farms was carried out in order to determine the functional possibilities of small farms' growth in Volgograd region in the face of digital economy. The results showed a decrease in stock of tractors by 5.2%, a decrease in stock of grain combines by 2%, as well as in plows and seeders by 2.5 and 0.4% respectively in 2018, compared to 2014. In addition to this 80% of agricultural machinery is worn out, represented by obsolete samples that fail to meet the requirements of modern technologies [7].

In order to determine the innovative potential of small agribusinesses in the form of family-operated farms and their technical and economic parameters, the authors used a proprietary methodology to calculate innovative potential aimed at revealing agribusiness' innovative capabilities through performance indicators evaluation [6]:

- educational level and qualification of workers;
- use of mechanization and automation means;
- use of innovative technologies;
- use of information and digital technologies by small agribusinesses;
- the level of digitalization of small agribusinesses;
- the level of marketability of agricultural production;
- production profitability.

The information source is the survey results of a number of family-operated farm owners and small agribusinesses heads made by a scientific team from faculty of Economics of the Volgograd State Agrarian University in 2018.

The survey involved 150 owners of family-operated farms in the context of five natural areas of the region that determine the specialization of agricultural production. The first is the Steppe zone with chernozem soil, characterized by intensive grain farming and strong sunflower production. The second is the Dry-steppe zone with dark chestnut soil and has a grain growing and animal husbandry specialization. The third one is the Volga-Akhtuba floodplain, represented by the Volgograd suburban area; its specialization is vegetable and milk production together with strong poultry industry. The fourth is dry steppe zone with chestnut soil and strong grain farming and animal husbandry industry. The fifth is semi-desert zone with light chestnut soils, specialized in grain farming, growing of mustard and gourds as well as in sheep breeding [8].

Education questionnaire showed that 44% of respondents have a degree, 32% finished school and 11% have basic vocational education. The analysis showed the reasons for the preference for the chernozem zone of the Volgograd region when the innovations are being implemented.

In the process of developing a survey for owners of family-operated farms indicators of farm computer availability were identified as one of the basic indicators of farmers' digital technologies use, placed in an independent section, which includes a number of questions on the implementation of IT achievements in the small farms practices.

Questionnaire data processing showed that in farms' practices mainly Internet payment technologies and network resources for searching and acquiring equipment, seeds, young animals, etc. is used [9]. The implementation of digital technologies in the production of agricultural products itself is indeed taking place, albeit not at a fast enough pace. In particular, only 7% of respondents reported the use of specialized software products, such as "Crop History", "Meteomonitor-AGRO", "GEO-plan", and "Farm Works".

A somewhat different picture is observed in terms of the practical use of accounting and statistical reporting software by farmers, as well as processes control software. More than a quarter of the surveyed farmers admitted the use of automated control systems and accounting developed by 1C and its franchisee partners. Although in most cases we are talking about using 1C: Bookkeeping, specialized solutions built on the basis of 1C: Enterprise 8, such as, for example, "Agribusiness Management", "Satellite Monitoring", etc., are gradually becoming part of the farming.

Survey results analysis indicates that most farmers do not need such studies. According to the received data only about 10% of the family-operated owners admitted their use of such information and consulting systems.

Such indicators as the level of income and expenses, revenue, profit, profitability, etc. estimated the production, business and financial parameters of the family-operated farms activity of the Volgograd Region. Income, expenses, revenue, profitability of production, profitability of income, and some other production and financial parameters of the surveyed farms are sorted out by zones and presented in Table 56.1.

The results obtained during departmental statistical observation of family-operated farms owners survey showed that the most economically feasible and promising from the point of view of regional agribusiness digitalization are farms in the first and second zones, which have all the necessary resources for innovative development.

In studies on scientifically based zoning of Volgograd region farms was established an integrated parameter called "innovation potential index"—IPI for short, which summarized 5 main indicators: (1) qualification and educational level of family-operated farm personnel; (2) modernization level of agricultural machinery; (3) level of innovation implementation; (4) the level of computerization and IT-based management; (5) the level of profitability and marketability of production [10].

A system of weighting factors was developed in order to assess influence degree of the indicators above. The calculation of the total IPI averaged over the number of farms showed that the index reaches its maximum value (9.65) in the first zone. Based on the analysis of the natural and economic zoning of the Volgograd region, the authors concluded that the steppe zone with chernozem soil is the most suitable location for the implementation of the experimental basic family-operated farm project [8].

Dry steppe zones with chestnut and dark chestnut soil have a clear animal husbandry specialization, however, the deployment of an innovative animal husbandry facility with organic production using the yard housing method will be, in the authors' opinion, less effective than in the steppe zone with chernozem soil. Family-operated farms of the Volga-Akhtuba floodplain, engaged in vegetable growing, and a semi-desert zone with light chestnut soil specialized in poultry, sheep and goat farming also have a low index of innovative potential.

Table 56.1 Economic results of family-operated farms business activity as per one farm for 2018

Parameters	Steppe zone with chernozem soil	Dry-steppe zone with dark chestnut soil	Volga-Akhtuba floodplain	Dry steppe zone with chestnut soil	Semi-desert zone with light chestnut soils
Farm incomes received from own business activity, thousand rubles	11166	13760,2	8848	7590,22	4678,35
Revenue from the sale of agricultural products, thousand rubles	10585,3	12404,8	7673,38	6683,6	3582,15
Percentage of revenue from agricultural products in total revenue from sales of goods, products, works, services, %	94,80	90,15	86,72	88,06	76,57
Farm expenses received from own business activity, thousand rubles	8258,18	10850,4	7441,66	5965,75	4110,54
Net cost of agricultural products, thousand rubles	4731,31	5838,04	4158,03	3076,43	2368,7
The proportion of agricultural products net cost in the total share of expenses, %	57,29	53,80	55,87	51,57	57,63

(continued)

Table 56.1 (continued)

Parameters	Steppe zone with chernozem soil	Dry-steppe zone with dark chestnut soil	Volga-Akhtuba floodplain	Dry steppe zone with chestnut soil	Semi-desert zone with light chestnut soils
Tax payment amount, thousand rubles	455,692	453,478	314	319,579	111,2
Profit, thousand rubles	2452,09	2456,32	1091,99	1304,89	456,609
Profit margin, %	23,16	19,80	14,23	19,52	12,75
Production profitability, %	51,83	42,07	26,26	42,42	19,28

Data source Official website of the Federal State Statistics Service of the Russian Federation (http://www.gks.ru/)

56.4 Conclusion

The results obtained during the study allowed:

- *to systematize* the technical and economic parameters of the new generation of farms based on intelligent digital technologies, including *staffing* (training of qualified personnel capable of introducing and using digital technologies in the production and management process; *technical and technological support* (the implementation of digital technologies in agricultural production and management using software and hardware means that deliver on achieving great results on the basis of updating the reserves to increase labor productivity and reduce costs); *information and analytical support* (justification of effective tools for information and analytical support of innovative activities of basic farms); *financial support* (proposals development for state support of family-operated farms that implement digital technologies and are the starting farms for university students internships);
- *to establish* comparable efficiency parameters for small farms and agribusinesses, determined by the trend of enlargement of small agribusiness entities and a change in the commodity structure of their production, from semi-subsistence production forms to an increase in the share of individual businessmen and family-operated farms with a high level of marketability;
- *to study* the limitations and risks of digital solutions for small agribusinesses in the agro-industrial complex of the Volgograd region, taking into account the possibilities of overcoming them, which should be considered as the potential for innovative development of small farms, and their realization should be ensured

by the implementation of digital platforms such as "Smart Farm", "Smart Field", "Smart herd", "Smart greenhouse", "Smart processing", "Smart warehouse" etc.;
- *to propose* a system of state support measures by concluding smart-contracts (package solutions for easy-term loans, insurances, subsidies and grants) and integrating regional information resources that provide quick access for an unlimited authorized number of users (digital transaction support, digital warehouse receipts, goods receiving advices for processing, etc.) to increase the efficiency of interaction between participants and state with the transition to digital format;

The implementation of the proposed recommendations for the digital economy development for the research competencies and technical groundwork formation for the development of the industry will allow small agribusinesses in the Volgograd region to successfully switch to digital technologies for agricultural production within the time limits set by policy measures.

References

1. Ognivtsev SB (2019) Tsifrovizatsiya Ekonomiki i ekonomika tsifrovizatsii APK/ Mezhdunarodnyy sel'skokhozyaystvennyy zhurnal, no 2(368), 77–80
2. Ivanov VV, Ovchinnikov AS, Kochetkova OV (2019) Conceptual basis of digital transformation of the agro-industrial complex of the Volgograd region. In: Proceedings of the Lower Volga Agro University Complex 2(54), pp 18–25
3. Shchutskaya AV, Afanaseva EP, Kapustina, LV (2019) Digital farming development in Russia: regional aspect. https://doi.org/10.1007/978-3-030-11367-4_26, Retrieved from www.scopus.com
4. Karabekova AB, Osmonalieva DA, Sadyraliev ZH, Urmatayim AK (2019) The main directions of digital modernization of the agro-industrial complex of a modern region (Book Chapter). Stud Comput Intell 826, 949–955
5. Ovchinnikov AS, Balashova NN, Ivanova NV (2014) Strategy of complex development of agrarian territories and effective functioning of Volgograd oblast's agro-industrial complex in the conditions of the WTO. Econ Agric Process Co 1, 16–20
6. Popova LV, Litvinova T, Bakitko E, Loda E, Suleimanova L, Chirkina M (2015) Perspectives of the Growth of Economic Security by Clustering of Small Innovational Enterprises. Eur Res Stud J. XVIII, Special Issue, pp 163–172
7. Korabelnikov IS, Balashova NN, Korabelnikov IS, Ishkin DA (2017) Economic evaluation of the effectiveness of applying the technologies of agricultural production: a regional aspect. Bull Nizhnevolszsky Agric Univ Complex: Sci High Vocat Educ 4(48), 272–280
8. Popova LV, Lata MS (2014) Assessment of the modern state of the subjects of small economic forms in the agrarian sector of Volgograd oblast. Econ Entrep 5(46–1), 593–598
9. Kamble SS, Gunasekaran A, Parekh H, Joshi S (2019) Modeling the internet of things adoption barriers in food retail supply chains. J Retail Consum Serv 48:154–168
10. Popova LV, Lata MS, Mitrofanova IA (2016) Innovative potential of small economic forms in the agro-industrial complex. Econ: Yesterday, Today, Tomorrow 9, 353–364
11. Choudhary S, Thomas I, Bahrami M, Sumioka M (2020) Accelerating the digital transformation of business and society through composite business ecosystems. In: advances in intelligent systems and computing, vol 926. Matsue; Japan, pp 419–430
12. Kozenko Z, Kozenko Y, Kozenko K, Ivanova N, Shmyreva I (2017) Marketing instrumentarium of provision of food security

Chapter 57
Neural Networks as a Forecasting Tool in the Context of the Russian Financial Market Digitalization

Valery Aleshin, Oleg Sviridov, Inna Nekrasova and Dmitry Shevchenko

Abstract Traditional models and methods of analyzing financial markets do not always produce good results due to the inconsistency of the instruments (tools) used with the existing economic reality. Hence, the need arises to develop and use analysis methods corresponding to complex nonlinear dynamic processes occurring in the financial market. Neural network is a typical instrument in this sphere. The most important feature of this network is the parallel processing of information by all its units, the ability to learn and generalize accumulated knowledge. A large number of connections between neurons can significantly accelerate the processing of information. In addition, the network becomes resilient to error. Trained on a limited set of data, the network is able to collect the information and work well with new data. The purpose of this study is to develop an analytical system for forecasting the prices of financial assets using artificial neural networks technology. Object of testing the analytical system are stock quotes. However, the analytical system proposed by the authors consists of several neural networks, each of which makes a forecast of prices of financial assets. The system includes recurrent (feedback) neural networks with sigmoidal activation formula. This allows networks to "remember" the sequence of reactions to the same stimulus. Learning process of neural networks is carried out using the error back propagation algorithm. The study allowed to draw a conclusion about the possibility of using neural network modeling for algorithmic trading in the stock market.

Keywords Neural networks · Forecasting · Time series · Synaptic connections · Stock index

---

V. Aleshin (✉) · O. Sviridov · I. Nekrasova · D. Shevchenko
Southern Federal University, Rostov-on-Don, Russia
e-mail: vaaleshin@sfedu.ru

O. Sviridov
e-mail: oysviridov@sfedu.ru

I. Nekrasova
e-mail: inna-nekrasova@mail.ru

D. Shevchenko
e-mail: dashevchenko@sfedu.ru

T. Kolmykova and E. V. Kharchenko (eds.), *Digital Future Economic Growth, Social Adaptation, and Technological Perspectives*, Lecture Notes in Networks and Systems 111, https://doi.org/10.1007/978-3-030-39797-5_57

JEL Code C45 · E47 · G17

57.1 Introduction

Currently, neurocomputers are actively used by many financial and non-financial companies in countries with developed economies and financial markets. For example, many large American companies such as LBS Capital Management Inc. buy small neural packages and neurocomputers (up to $ 50,000) and significantly improve their trading performance on the US indices such as S&P and Nasdaq.

The relevance of using neural network modeling methods in modern financial markets for forecasting the prices of financial assets are justified by the following assumptions the authors made in the book "Handbook of Fractal Approaches for Modeling Financial Assets and Predicting Crises":

- the theoretical basis for explaining the functioning of financial markets is Fractal Market Hypothesis (FMH);
- price movement of financial assets and macroeconomic indicators of financial markets is not random (the main reason is the unequal access to publicly available information by investors);
- the all-time series of prices of financial assets are statistically self-sufficient, thus, they have the property of fractality;
- the main method for studying fractal time series is the R/S analysis, or the method of normalized amplitude;
- all prices of financial assets have the property of inertia [10].

Neural networks can compress data, highlighting the most common relationships between their parts, and express them more compactly (in a smaller dimension). The original image of the data can also be restored due to the associative memory of the neural network-if the data was damaged or noisy.

Neural networks can predict by summarizing and isolating dependencies between incoming and outgoing data. A trained network can, like any technical indicator, predict the future values of a sequence based on historical data.

Unlike classical indicators, the neural network independently adjusts the principles of data evaluation, their dependencies on each other and correct them based on successes and errors in trading. However, it is necessary to take into account the large time costs to train the network and provide the necessary infrastructure to respond to incoming data in a timely manner.

57.2 Materials

The history of neural networks dates back to the 1940s. First, the linear mathematical model of a nerve cell 'neuron' appeared in 1943 in the works of W. McCulloch and W. Pitts. They offered their own, distinctive from the future von Neumann computer architecture [7].

Artificial neural network is built on the principle of biological neural networks, that is, it copies the organization of nerve cells of a living organism and consists of artificial neurons. It depicts a mathematical model embodied in the hardware and software for complex logical calculations. Figure 57.1 below shows a model of the simplest artificial neural network [2].

In 1949, Hebb D. O. proposed a method for connecting and interacting neurons in the human brain [3]. He suggested that neural network learning depends largely on the changes in the strength of synaptic connections. According to his concept, the neural network as a result of self-learning performs the task without interference from the experimenter.

Further, in the late 1950s, Rosenblatt F. proposed a model of the perceptron and an algorithm for its training [11].

In 1969, Minsky M. and Papert S. discovered the main computational problems arising from the functioning of neural networks [8].

The problems were that single-layer neural networks could not perform "modulo-2 addition", and computers did not have sufficient computing power to efficiently process the large amount of computation that is crucial for the normal operation of neural networks.

In the 60s, American scientists noted that the multilayer perceptron (multilayer feedback neural networks) is able to carry out any mapping of input vectors at the weekend. However, during this period there was no effective algorithm for training such networks. A model of a multilayer artificial neural network is shown in Fig. 57.2 [2].

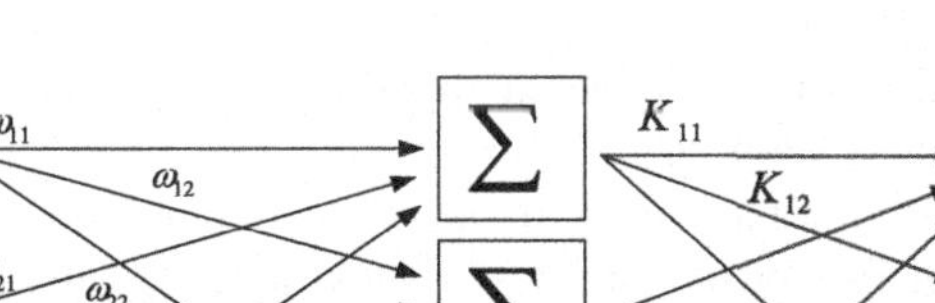

Fig. 57.1 Model of the simplest artificial neural network

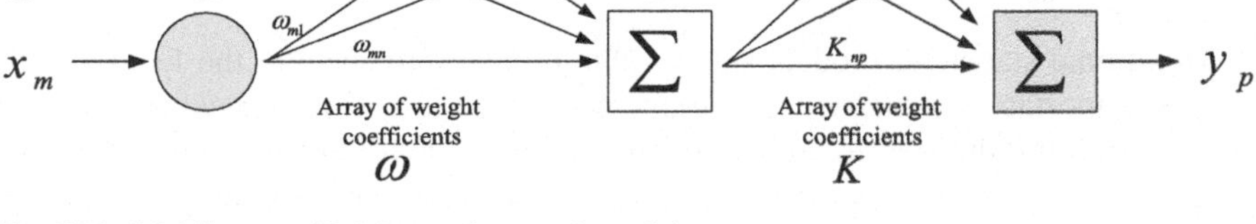

Fig. 57.2 Multilayer artificial neural network model

Since the early 1980s, after the publication of studies by T. Kohonen, Hopfield J., Hinton G. E. and other scientists, neural networks finally gained recognition and widely spread in practice.

Hopfield J. in 1982 analyzed the stability of neural networks with feedback and proposed to use them as associative memory [4]. In the same time period, Kohonen T. developed self-organizing neural networks [6].

After the publication of several works of Hinton G. in 2006, a new stage in the development of multilayer neural networks began, which continues to date [5]. Specifically, there were deep neural networks, which are as the result of the development of multilayer perceptron and integrating various paradigms of neural networks training. Thanks to the multi-layer architecture, they allow processing and analyzing a large amount of data, as well as modeling cognitive processes in various areas of human activity.

Currently, the following main trends in the development of neural networks can be distinguished.

Growth of accuracy and quality of the performed functions. Digitalization plays significant role not only in developed markets, but in emerging markets too [12]. Neural networks are predominating in many areas of the economy and has displaced the entire classical algorithm of work. That is, there are neural networks that significantly reduce the error in the calculations.

The architectural complexity of neural networks is growing, which is expressed primarily in the growth of layers. There are networks with a depth of 300 layers. Nowadays, most high-tech companies use deep neural networks to design various intelligent systems. Deep neural networks are now considered a revolutionary breakthrough in the field of artificial intelligence.

Another trend in the development of neural networks is the growth of data volumes and computing power.

57.3 Methods

The use of traditional methods for forecasting the prices of financial assets in a digital economy is not effective enough.

The limitations of standard methods are their dependence on the initial conditions and the lack of flexibility. Strict statistical assumptions about the properties of financial time series, which are set in models, limit the possibilities of methods of mathematical statistics, theory of pattern recognition, and theory of random processes.

Traditional statistical models are not able to take into account the fact that, the relative importance of individual parameters of financial time series and the factors determining them in today's dynamic financial environment dramatically and unpredictably changes over time. Furthermore, traditional approaches of forecasting have limited information and are designed to display qualitative factors or patterns in quantitative terms [9].

Due to the above reasons, the traditional methods of forecasting financial time series are replaced by new, more effective methods based on artificial intelligence. One of these methods is an artificial neural network.

For quality purposes, neural networks in the financial market use two types of data in analysis: for training and testing.

A significant advantage of neural networks is the fact that they learn on an ongoing basis due to new data and existing forecasts. Neural networks can combine technical and fundamental data, which allows them to be optimally applied.

Thus, neural networks have certain skills that allow to identify unaccounted "patterns" in the market and apply them in the preparation of forecast data, achieving the most accurate result. The disadvantage of neural networks is the use of intelligence without taking into account the possibility of human emotions influencing prices. At the same time, high volatility of quotations of certain financial assets may occur on the market due to a sudden emotional factor.

57.4 Results

In the framework of this study, we implemented a system for forecasting the rates of financial assets using the Dow Jones index as an example. It is based on 4 Elman J recurrent neural networks. Recurrent neural networks are among the most complex. Unlike direct distribution networks, they have feedbacks. Feedbacks allow for adaptive memorization of past temporal events. Recurrent networks are the most flexible tool for building nonlinear models.

The artificial Elman J. neural network is based on a multilayer perceptron. Moreover, in the Elman J. network, feedback is realized through the supply to the input layer of both the initial data and the outputs of the neurons of the hidden layer [1].

The initial data of our model are quotes (prices) of the Dow Jones index, taken from the broker terminal (company Finam). The primary time series of quotes was processed by finding the returns of the index. This is very important, because if prices, i.e. raw data, are used for forecasting with the help of neural networks, then there will be problems with network training and the accuracy of forecasts will suffer significantly.

The system is presented in the form of a neural network cluster and has the following structure of input parameters (Fig. 57.3).

Input for Artificial neural network 1—input parameters received at the input of the first neural network (profitability of the opening prices of the Dow Jones index for each business day from 05/01/2014 to 06/28/2019);

Input for Artificial neural network 2—input parameters received at the input of the second neural network (profitability of the average prices of the Dow Jones index for each business day from 05/01/2014 to 06/28/2019);

Input for Artificial neural network 3—input parameters received at the input of the third neural network (profitability of closing prices of the Dow Jones index for each business day from May 1, 2014 to June 28, 2019).

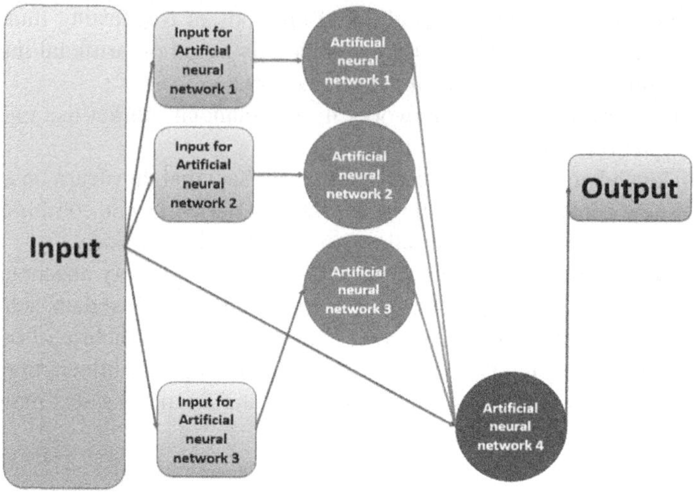

Fig. 57.3 The scheme for constructing a neural network system in the Matlab working environment. *Source* Compiled by the authors

These parameters are inputs to the 4 Elman recurrent neural networks. The fourth network (Artificial neural network 4) is a neural network that summarizes the results of previous networks and common input parameters (Output is the result of a forecast, expressed in the rise or fall of a security).

The same architecture and parameters are created for all 4 networks. Elman neural network has 42 inputs, because the input is fed by 6 converted indicators for the past 7 days. The network has 1 output, the network has 3 hidden layers, each of which uses a sigmoidal activation function, the training method is the method of error back propagation.

The network ceases its training either after 10,000 training epoch (cycles) or if the error size is zero. The execution function is set to the average absolute error. Table 57.1 below represents the results of training neural networks in the interval 01.07.19.–08.15.19.

The results of the study show that, the model based on four Elman recurrent neural networks allows a fairly accurate prediction of the future behavior of the prices of financial assets. After training on the interval from 05/01/2014 to 06/28/2019, the model showed good results from 01.07 to 08.15.2019, with a forecast accuracy of 0.8182.

Figure 57.4 shows a graph of the projected and actual values of the prices of the Dow Jones index for the interval 01.07.19–08.15.19.

Furthermore, we carried out a comparative analysis of the effectiveness of neural network methods for forecasting the prices of financial assets and traditional statistical methods: the method of exponential smoothing, the Holt model, the Holt-Winters model, ARIMA. Table 57.2 below presents the results of comparing the error of the models in predicting the prices of the Dow Jones index.

Table 57.1 Results of the trained Elman neural network on the interval 01.07.19–15.08.19

OutputTestNet4	TestDataTargetForNet4	OutputTestNet4v2	True/False
−0,0015322117	0	0	1
0,0049939115	0	1	0
−0,0005968201	0	0	1
−0,0019172296	0	0	1
−0,0003739777	0	0	1
0,0046027778	0	0	1
0,0028924760	0	1	0
0,0062378832	1	1	1
0,0011595809	0	1	0
0,0011358729	0	0	1
−0,0011109340	0	0	1
−0,0016359528	0	1	0
0,0022492815	0	0	1
−0,0021883348	0	1	0
0,0038508592	0	1	0
−0,0014752597	0	0	1
−0,1043510022	0	0	1
−0,0008888067	0	0	1
0,0019235833	0	0	1
−0,0011562541	0	0	1
−0,0005504596	0	0	1
−0,0016234676	0	0	1
−0,0132866503	0	0	1
−0,0094072053	0	0	1
−0,0021217338	1	1	1
−0,0003476335	0	0	1
0,0098496758	1	1	1
0,0005324942	0	0	1
−0,0057889545	0	0	1
0,0070703040	1	1	1
−0,0106946277	0	0	1
−0,0076843730	0	0	1
0,0083751902	1	1	1
Number of correct answers			27
Share of correct answers			0.8182

Source Compiled by the authors based on the Matlab computer program

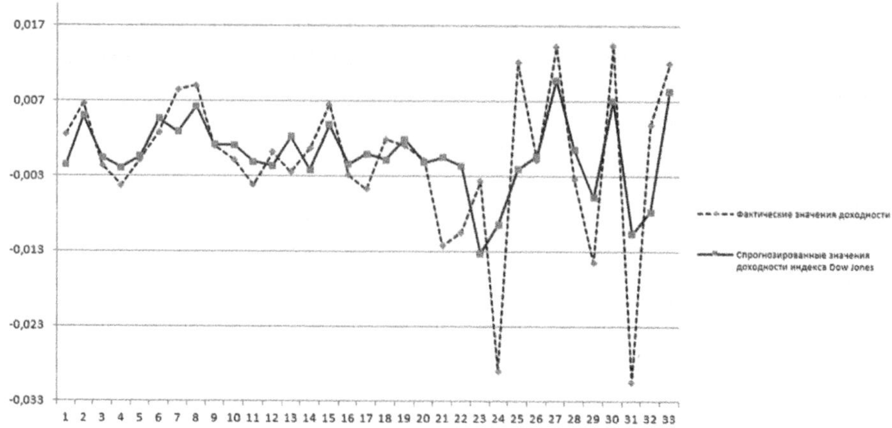

Fig. 57.4 Graph of forecast and actual values of the Dow Jones index prices in the interval 07.01.2019 – 08.15.2019. *Source* Compiled by the authors

Table 57.2 Results of comparing forecasting efficiency using traditional statistical models and neural networks

Model	Efficiency (forecast accuracy, %)
Exponential smoothing ETS	68.2
Holt model	81.78
Holt-Winters model	75
ARIMA	87
Neural network	81.82

Source Compiled by the authors

The results of the analysis presented in Table 57.2 depicts that neural networks are the second most accurate forecasts after ARIMA. This is because the training sample contains stress values, which distorts the values of complex models, but does not affect the simpler models. However, if we talk about the quality of the forecast of the trend direction of stock index quotes, then neural networks show the best result.

The authors of the article made a comparative analysis of the quality of forecasting quotes on the securities market using the ARIMA method and neural networks in a test sample. For this, two indicators were calculated: the deviation (error, E) of the forecast price value from its actual value and the accuracy of determining the trend of price movement (the percentage of predicted signs in the price movement Et). Both criteria showed the advantage of neural networks compared to method ARIMA.

The deviation of the predicted value from its actual value was calculated by the formula 57.1:

$$E = \frac{1}{N}\sum_{K=1}^{N}\frac{\left|Z_K - \widehat{Z_K}\right|}{Z_K} * 100\% \tag{57.1}$$

where N = 33 – dimension of the test sample;

Z_K – actual value;

$\widehat{Z_K}$ – forecast value.

The accuracy of determining the trend in price movements, by finding the percentage of predicted characters, was determined by formula 57.2.

$$E_t = \frac{1}{N}\sum_{n=1}^{N} t_n * 100\% \tag{57.2}$$

$$\text{where} = t_n = \begin{cases} 1, \ if\ (\widehat{Z}^{(n)} - Z^{(n-1)})(Z^{(n)} - Z^{(n-1)}) \geq 0, \\ 0, \ if\ (\widehat{Z}^{(n)} - Z^{(n-1)})(Z^{(n)} - Z^{(n-1)}) < 0 \end{cases}$$

The constructed neural network model showed a higher quality of the quotes forecast on the securities market compared to ARIMA: E = 1.1062%, Et = 81.82% for the neural network, E = 1.8703, Et = 73.41% for ARIMA.

This confirms the possibility of using the constructed model for making investment decisions in the stock market.

57.5 Recommendations and Conclusion

In the digital economy, neural networks are a tool that allows you to solve problems with an unknown algorithm in advance, initially having a strong degree of oscillation and therefore difficult to formalize. However, the use of this tool requires an accurate selection of explanatory factors and effective work with data preprocessing when using algorithmic trading on the stock market.

Neurotechnology has not yet found wide application in trading in the financial market. However, they have the potential to detect any repetitive patterns in the original data. Also, as data sources for analysis, you can use not only quotes, but, if necessary, any information, up to the phases of the moon and the Mayan calendar. At the same time, it is possible to combine both technical analysis and fundamental data as input variables.

An important advantage of neural networks in trading over mechanical trading systems is the ability to process large amounts of information, the ability to learn and adapt to market changes.

Neural networks with their current potential can predict markets better than traditional statistical methods of analysis, algorithmic advisors and indicators, but this potential has not yet been fully realized. The basics of neural network modeling need to be improved and refined.

Neural networks are good at classifying patterns, determining internal trend dynamics and behavioral cycles.

The possibility of using trained neural networks to predict the prices of financial assets confirms the existence of a fractal structure of the financial market, as well as the presence of such properties of financial assets as self-similarity and inertia.

References

1. Elman J (1990) Finding structure in time. Cogn Sci 14:179–211
2. Haykin S (2009) Neural networks and learning machines. Prentice Hall, New York, NY
3. Hebb DO (1949) The organization of behavior: a neuropsychological theory. Wiley, New York, NY
4. Hopfield JJ (1982) Neural networks and physical systems with emergent collective computational abilities. Proc Natl Acad Sci USA 79(8):2554–2558
5. Hinton GE, Osindero S, Teh Y (2006) A fast learning algorithm for deep belief nets. Neural Comput 18, 1527–1554
6. Kohonen T (1982) Self-organised formation of topologically correct feature maps. Biol Cybern 43:59–69
7. McCulloch W, Pitts W (1943) A Logical calculus of the ideas imminent in nervous activity. Bull Math Biophys 5:115–133
8. Minsky M, Papert S (1969) Perceptrons. MIT Press, Cambridge, MA
9. Nekrasova I (2015) Hurst indicator as a measure of fractal structure and long-term memory of financial markets. Bull Math Biophys 38(7–3):87–91
10. Nekrasova I, Karnaukhova O, Sviridov O (2018) Fractal properties of financial assets and forecasting financial crisis. In: Nekrasova I, Karnaukhova O, Christiansen B (eds) Handbook of fractal approaches for modeling financial assets and predicting crises. IGI Global, Hershey, PA, pp 23–41
11. Rosenblatt F (1958) The perceptron: a probabilistic model for information storage and organization in the brain. Psychol Rev 65(6):386–408
12. Shevchenko D, Bolotin Yu (2018) Actual Directions of modernization of the state regulatory function on the financial market. Mod Sci: Actual Probl Theory Pract, Series "Economics and Law", 9, 48–50

Chapter 58
Foreign Experience of Legal Regulation and Protection of Relations in the Sphere of Genome Editing

Alexander G. Blinov, Mikhail M. Lapunin, Ekaterina V. Evstifeeva and Olga A. Yakovleva

Abstract *Purpose:* to get acquainted with the experience of legal regulation and protection of relations in the sphere of human genome editing and to evaluate prospects of its use in the regulatory practice in Russia. *Design/methodology/approach:* the universal dialectical method of cognition of social phenomena and process, as well as a set of general (hermeneutical and axiological) and special (comparative legal, formal legal and modeling) scientific methods. *Findings:* the protection of human genome is essential in the digital era and DNA editing researches should be based not only on morality but also on legal standards. The world community has developed a number of fundamental documents governing researches of human genetic data and biological samples. Countries that are interested in developing human genome science are actively adopting regulations in this field. They try to create a good legal framework for scientists who are experimenting with genetic material. However, in practice, there are cases of violation of human rights and freedoms indicating the necessity of using the mechanism of criminal legal protection of such relations. *Originality/value:* editing a human genome is acceptable for therapeutic and other social purposes unless it leads to discrimination and ensures interests of an individual; international and foreign legal experience of human genome regulation demonstrates a rather cautious attitude of lawmakers to evaluation of such activities; at present the domestic legislator should pay attention to the regulatory method to influence its participants; a forming legal framework in the sphere of usage of results of scientific and technological progress will make it possible to identify areas that require the initiatives of criminal law.

A. G. Blinov (✉) · M. M. Lapunin · E. V. Evstifeeva
Saratov State Law Academy, Saratov, Russia
e-mail: blinovag2008@yandex.ru

M. M. Lapunin
e-mail: lapuninm@yandex.ru

E. V. Evstifeeva
e-mail: evstifeevak@yandex.ru

O. A. Yakovleva
Volgograd State University, Volgograd, Russia
e-mail: olgayakovlev@mail.ru

T. Kolmykova and E. V. Kharchenko (eds.), *Digital Future Economic Growth, Social Adaptation, and Technological Perspectives*, Lecture Notes in Networks and Systems 111, https://doi.org/10.1007/978-3-030-39797-5_58

Keywords Genetic information · Protection of genome · Discrimination · Bioidentification · DNA analysis

JEL Code K140 · K320 · K380

58.1 Introduction

The achievements in science and technology have enabled humanity to increase significantly the life expectancy and quality. People have got rid of many deadly diseases, expanded time frame for active work, learned to identify the predisposition of children to certain professions and sports and use effectively bioidentification. The traditional medicine needs genetic technologies for further progress.

Today the decoding of one's genome has become available to a wide range of people which not only opens great prospects for humanity but also creates new threats. For example, the results of DNA sequencing provide valuable information about an owner of genome and its relatives and evaluate genetic risk of serious diseases. But this technology has negative effects too.

58.2 Materials and Method

The universal dialectical method of cognition of social phenomena and process, a set of general (hermeneutical and axiological) and special (comparative legal, formal legal and modeling) scientific methods were used in this research paper. The content analysis of the legal literature on genetic regulation over the past five years (2015–2019) was conducted.

58.3 Results

In digital era the importance of information in the life of society is of great importance. This statement fully concerns genetics. As a human belief in high technologies is so strong that information about a person's genetic characteristics has a greater impact on his performance than genetic characteristics themselves [9]. Experiments conducted by scientists at Stanford University have shown that the very fact of studying risk of disease can cause physiological changes in a patient. The impact on human health of the abstract threat of disease has sometimes exceeded the effect associated with this disease. Scientists concluded that reporting the presence of "favorable" or "unfavorable" genes to a participant in an experiment has a stronger influence on physiological parameters than genes responsible for these factors. While providing simply data on genetic risk may increase risk, the clinicians and specialists on ethic should make cautious decisions about the justification for identifying genetic risk [10].

Thus, the genome information received from physicians has a great impact on the psyche of a modern man and effects of placebo and nocebo are important to take into account when applying the latest genetic technologies in the struggle against human diseases. On the other hand, a genome itself acts as an information subject to a special protection. When decoding a genome, the patient gives medical company private information not only about him but also about his relatives. Such data can be used as a means of influence. For example, Israeli biochemists' statements about the possibility of falsifying a DNA sample are extremely worrying. They found that due to the peculiarities of DNA expertise it's possible to imitate artificially genetic material by using information from digital genetic profile of a particular person. It's not necessary to have a person's biomaterial itself [5]. The DNA contained in the blood or saliva is removed by a centrifuge and then this space is filled with someone else genetic material. For this purpose it's enough to get access to previous results of genetic analysis of a particular person and to synthesize separate fragments of his DNA in the laboratory using this scheme. The standard genetic analysis procedure will point to this person. Deception can only be detected through the complex analysis of more DNA fragments which is not yet part of the standard forensic procedure. As a result, there may be situations where as a suspect of a crime will be a man which "bio sample" was spoofed by hackers and tossed into evidence.

It points out to fact that in the digital era the protection of human genome is important and the genome editing researches should have the legal regulation.

The world community has developed a number of fundamental documents regulating researches with human biological samples and genetic material. The key place is given to the Convention for the Protection of Human Rights and Dignity of the Human Being with regard to the Application of Biology and Medicine: Convention on Human Rights and Biomedicine (ETS No 164) (adopted by the Committee of Ministers of the Council of Europe on 19 November 1996), the Universal Declaration on the Human Genome and Human Rights (adopted at UNESCO's 29th General Conference on 11 November 1997), the International Declaration on Human Genetic Data (adopted at UNESCO's 32nd General Conference on 16 October 2003), United Nations Declaration on Human Cloning (Resolution 59/280, adopted by General Assembly on 03 August 2005), Additional Protocol to the Convention on Human Rights and Biomedicine, concerning Genetic Testing for Health Purposes (Strasbourg, 27/11/2008).

The Regulations adopted by UN, UNESCO, Council of Europe and other international organizations form an ideological basis of development of genomics and implementation of its results into daily medical practice. These regulations encourage national legislators to define the legal status of participants in biomedical research, to enhance usage of human gene information, to exclude possibility of discrimination against people on the genetic grounds, to protect human life, health, freedom and dignity by using biotechnology, to establish a responsibility for abuse of scientific research results.

Countries are encouraged to address ethical, legal and social issues that may arise in the human genome decoding and use of the information obtained in health care organizations. Any form of discrimination on the basis of a person's genetic data, the use of medical technology to choose a future child sex and human cloning should be legally prohibited. A number of appeals concern procedure for collection, processing, use and storage of individual genetic material. In order to ensure the security of human biobank, it's proposed to create a monitoring system with elements of governmental control and management [1].

Countries that are interested in the development of human genomics produce actively regulations in this sphere. The dispositive method prevails in the mechanism of legal regulation of genomic research and the application of its results in practice. For example, USA has Genetic Information Nondiscrimination Act (GINA Act). It deters employers and insurance companies from abusing genetic information. However, this law does not cover all areas. Thus, in 2012 a teenager Colman Chadam was forced to change school because of his genetic predisposition to cystic fibrosis. His parents put the boy's genetic test results on the school's medical records. Teachers reported this information to the parents of students with cystic fibrosis. Fearing for the health of their children they demanded that Chadam should leave the school. In 2013 Colman's parents sued the school. They accused the administration of violating the right to privacy [3]. However they were not able to use the provisions of the GINA Act because it applies only to employment and health insurance. It does not regulate relations in the education sphere.

China has adopted administrative rules setting out the procedures for use of assisted reproductive technologies, conditions for the rational use of human genetic resources and ethical standards for biomedical research involving humans, for example, Administrative Measures for Assisted Human Reproductive Technologies (MOH 2001); Administrative Measures for Human Genetic Resource Management (MOH 1998); Measures of Ethical Control of Biomedical Research involving People, adopted by the National Commission on Health and Family Planning of China; Ethical Principles and Technical Norms on Human Assisted Reproductive Technologies (MOH 2001, revised 2003). Rules of application of assisted reproductive technologies are seen in specialized acts, e.g. Ethical Principles and Technical Norms on Human Assisted Reproductive Technologies (MOH 2001, revised 2003). The norms formulated in these documents are aimed at development of human gene research, rational use of genetic material, protection of life, health and dignity of patients participating in biomedical research. They define the legal status of health care institutions that carry out experiments and implement their results in medical practice, establish rules for conducting biomedical research with the participation of patients and regulate usage of human genetic resources.

Experiments based on such regulations were received ambiguously by the scientific community [8]. However they have prompted scientists to pay more attention to ethical and legal aspects of human genome research. The violation of genetic research regulations by Chinese scientist He Jiankui led to the birth of the first genetically modified twins [6] and prompted a group of geneticists to propose a "global five-year moratorium" on genome editing in clinical research. This time will allow discussing technical, scientific, medical, social and ethical issues related to the modification of inherited DNA [7].

The Japanese government issued the regulation of using genetic modification technology for fertilized human embryos in the fundamental research: Guidance on Ethics for Research using Gene Modification Technological Information in the Human Embryos, 04.01.2019 (adopted by the Ministry of Education, Culture, Sports, Science and Technology in cooperation with the Ministry of Health, Labor and Welfare of Japan). The document regulates conditions of conducting experiments: institutions should have special equipment and qualified personnel to handle human embryo. The requirements for biological material including receipt, storage, use and disposal are listed. The research institutes are prohibited from transferring fertilized human embryos to other organizations. The organizational and financial aspects of research are also described.

The Japanese scientist Takashi Okamoto was accused of industrial espionage. The scientist was charged by American law enforcement agencies in Ohio in May 2001 with stealing genetic materials on Alzheimer's disease developed by the Learner Research Institute at the Cleveland Clinic Foundation in July 1999 while he worked there as a researcher. Despite the fact that there was no punishment for industrial espionage in Japan, U.S. Justice Ministry officials insisted on the extradition of Okamoto. U.S. Justice Ministry officials argued that under the Japanese law Takashi's actions could be qualified as theft and destruction of property [4]. The Supreme Court of Japan refused to extradite the scientist. This example shows that Japan's policy is aimed at protecting its high-tech interests.

The Russian Federation is planning to develop genetic technologies. It has adopted a number of programmes that demonstrate an interest in genetic engineering practices, e.g. Resolution of the Government of the Russian Federation No. 479 "On Approval of Federal Scientific and Technical Program of Development of Genetic Technologies for 2019–2027" ; Strategy of Development of Health Care of the Russian Federation till 2025 (approved by the Decree of the President of the Russian Federation No 254, 06.06.2019). Today there are several regulations concerning implementation of genetic engineering and use of its results, genomic registration, and introduction of a temporary ban on human cloning: Federal Law No. 86-FZ "On State Regulation in the Gene Engineering Sphere" , 05.07.1996; Federal Law No. 54-FZ "On Temporary Ban on Human Cloning" , 20.05.2002; Federal Law No. 242-FZ "On State Genomic Registration in the Russian Federation" , 03.12.2008. This legal framework in the field of biosecurity make contribution in improving human welfare and health, exchange of biological material and processing of genomic information.

Further development of human genome editing legislation should be focused on ensuring personal dignity as a set of physiological and social needs of a person in a society. Today the society expects from the State the legal decisions that relate to admissibility of programming parameters of the unborn child; the improvement of psycho-physiological characteristics of a person; the protection of interests of individuals with disabilities; the protection of personal data obtained in the process of genetic research; the ensuring the rights and freedoms of individual regardless of the characteristics of its genetic code, etc.

The detailed regulation of the practice of interference in the hereditary line will reveal relationship that can encourage individuals to abuse the results of scientific achievements. The rules and models of criminal law can prevent living organisms from threats to human security in genome researches. In particular, it's recommended to provide liability in the criminal law for the interference with human genome concerning dignity, the violation of the prohibition of human cloning, the usage of results of genome research without proper trials, the infringement on privacy of human gene data, the genetic research without person's consent, the illegal use of gene material and trafficking, the creation and use of gene doping [2, p. 254]. The object of legislation should determine the object of criminal law protection. In the modern circumstances the object of legal regulation yet has no clear outline, so it's early to insist on including the rules of this kind into the criminal law framework. For example, in the future it's possible to supply the criminal law with Article 137. Its disposition may look like this: "Illegal collection, storage, use, destruction or distribution (including transfer, sale) of biological samples in order to obtain information about human genome or results of human DNA analysis" .

58.4 Conclusion

The results of the research paper allow us to formulate conclusion:

- the human genome editing is acceptable not only for therapeutic but also for other social purposes if it ensures the interests of individual and does not arise discrimination in a society;
- the international and foreign experience of regulation of human genome editing indicates a rather cautious attitude of legislators towards the evaluation of genomic activity;
- the domestic lawmakers should focus on regulatory method of influence on the participants of biomedical research while forming the legal environment for genomic research;
- the legal framework being formed in the field of using the results of innovative science will reveal points of social unrest that require criminal legislative initiatives.

Acknowledgements This paper was financially supported by the Ministry of Education and Science of the Russian Federation on the program to improve the competitiveness of Peoples Friendship University of Russia (RUDN University) among the world leading research and education centers in 2016–2020.

References

1. Aleksandrova Yu (2018) Traveling documents. Mach Mech 9:28–33
2. Blinov AG (2019) International standards for regulating genetic research and their reflection in national legislation. Bull Saratov State Law Acad 3:247–255
3. Chadam J (2016) Appeal from a decision of the united states district court for the northern district of California, 4:13-cv-04129-CW, https://www.documentcloud.org/documents/2701506-Opening-Brief.html (accessed 12 Aug 2019)
4. Fesyun A. (2004) USA will not demand the extradition of a Japanese scientist, https://ria.ru/20040330/556975.html (accessed 12 Aug 2019)
5. Frumkin D, Wasserstrom A, Davidson A, Grafit A (2010) Authentication of forensic DNA samples, Forensic Sci Int: Genet 4(2):95–103
6. Xinhua News Agency (2019) Guangdong releases preliminary investigation result of gene-edited babies, http://www.xinhuanet.com/english/2019-01/21/c_137762633.htm (accessed 12 Aug 2019)
7. Lander E, Baylis F, Zhang F, Charpentier E, Berg P and specialists from seven countries call for an international governance framework (2019) Adopt a moratorium on heritable genome editing. Nature 567, 165–168
8. Reardon S (2015) Gene-editing summit supports some research in human embryos. Nature, 3rd Dec
9. Tiihonen J, Rautiainen M-R, Ollila HM, Repo-Tiihonen E, Virkkunen M, Palotie A, Pietiläinen O, Kristiansson K, Joukamaa M, Lauerma H, Saarela J, Tyni S, Vartiainen H, Paananen J, Goldman D, Paunio T (2015) Genetic background of extreme violent behavior. Mol Psychiatry 20:786–792
10. Turnwald BP, Parker Goyer J, Boles Danielle Z, Silder Amy, Delp Scott L, Crum Alia J (2019) Learning one's genetic risk changes physiology independent of actual genetic risk. Nat Hum Behav 3:48–56

Chapter 59
Neoindustrial Paradigm of Russian Corporations Integration into the Economy System Under the Conditions of Transition to the Sixth Technological Order

Olga Brizhak, Elena Klochko and Alexander Adamenko

Abstract *Purpose*: The authors of the article propose a neo-industrial paradigm for Russian corporations integration into the economic system in the context of the transition to the sixth technological paradigm. The process of corporate capital adaptation to the conditions of fierce spatial competition, the actions of mechanisms aimed at limiting access to financial and technological development resources is revealed, the specifics of the antagonism between the network and corporate forms of economic relations is reflected. *Desing/methodology/approach*: The authors sought to apply political-economic theory, theory of capital, institutional theory, dialectic and reproductive approaches, which they managed to a certain extent and made it possible to obtain significant scientific results. Basing on a synthesis of their predecessors' scientific results, the authors studied the combination of the transformation processes of the Russian economy, accompanied by contradictions of corporate capital polarization, undermining systems development and restraining the reaction of corporate capital to changes in the external environment. *Findings*: The article highlights nuclear and peripheral zones of technological transformations, identifies system barriers to the spread of network structures in raw materials economy, proposes a number of directions for implementing the neo-industrialization policy, ensuring the innovative development of Russian corporations. The use of the possibilities of fictitious capital reproduction, restoration of public confidence in the national financial market, protectionist policies implementation, import substitution and reindustrialization are proposed as a source of additional development resources in conditions of low

O. Brizhak (✉)
State University of Management, Moscow, Russia
e-mail: brizhak71@mail.ru

Kuban State University of Technology, Krasnodar, Russia

E. Klochko · A. Adamenko
Kuban State Agrarian University, Krasnodar, Russia
e-mail: magadan.79@mail.ru

A. Adamenko
e-mail: adam83@mail.ru

T. Kolmykova and E. V. Kharchenko (eds.), *Digital Future Economic Growth, Social Adaptation, and Technological Perspectives*, Lecture Notes in Networks and Systems 111, https://doi.org/10.1007/978-3-030-39797-5_59

investment attractiveness. *Originality/value*: The reliability of scientific provisions is ensured by the use of modern methods of scientific research, the use of a representative sample, factual material, generalization of methodological and practical achievements.

Keywords Corporate capital · Neo-industrial integration paradigm · Technological order · Network organization · Polarization · Commodity economy · Hicks cross

JEL Code A12 · O10 · O30 · P12 · M21 · B41 · O32 · P16 · L14 · D24

59.1 Introduction

Corporate capital, which dominates national economy, shares its destiny in all the processes of modern transformation. Together with national economy system, it is trying to integrate itself into the process of profound technological changes, adapt to the conditions of fierce spatial competition, "new protectionism", and the mechanisms of limiting access to financial and technological development resources. Deep technological shifts and related socio-economic transformations increase the unevenness of the evolutionary process, which leads to the formation of significant gaps between different economic systems, territories, spheres and public economy sectors. This fully corresponds to corporate capital reproduction, where the gaps between cycles, individual phases, interests of reproduction participants, and contradictions in the types of economic relations organization arise. In the modern economy, there has been a confrontation between two types of economic relations organization—corporate and network ones. The polarization of economic relations in the context of technological transformations, on the one hand, creates the conditions for the corporate and network forms synthesis; on the other hand, their containment, especially for Russia, which is strategically oriented towards extracting raw material rents. An analysis of these phenomena involves an appeal to cognitive resources of the polarization phenomenon, the beginning of the study of which was laid in the works.

59.2 Materials and Method

In their study the authors rely on the pillars of the following of the theories: post-industrial, systemic, transformational and creative economy theory, the theory of capital; corporation theory, neoindustrialization theory, resources of political-economic and institutional theory, competency-based and reproductive approaches. The authors, appealing to scientific foundations of classical economists such as [1, 2], etc., as well as the achievements of modern Russian researchers, such as [3–9] and others, formed the neo-industrial paradigm of Russian corporations' integration in the economic system in the context of deep technological changes, based on the possibility of fictitious capital reproduction, technological, organizational and institutional transformations, reindustrialization.

59.3 Results

The specifics of an economy acting in a "new economic reality" is the need to use the unique factors of production that form the foundation for the creative economy, based on the network model of economic relations organization. However, this does not mean that appropriate dominants should be placed between the corporate and network forms economic relations. These forms are usually actively intertwined, creating corporate networks, forming total power on the market [3]. Based on this characteristic of this process, the authors analyze the dynamics of technological transformations using the separation of these processes in the core and periphery countries [6]. The uneven technological development in core and peripheral countries creates polarization, which increasingly alienates them from each other [8, 9] and forms dual effects:

On the one hand, in developed countries, the concentration of network structures that characterize the development potential of the new economy creates the conditions for the efficient connection of corporate and network forms, the market of "total power of corporate networks". This effect allows to realize strategic advantages of network structures based on high-tech production, attracting highly qualified personnel, with relevant competencies that correspond to new technologies and organizational mechanisms [10].

On the other hand, in peripheral countries, the synthesis of corporate and network forms of organization is complicated by the lagging nature of technological transformations, and, consequently, by the slowdown of formation and distribution of network structures, which is fully related to the national economy, which is based on the predominance of corporations that focus on rental income from the sale of raw materials [4, 11]. Such a development scenario is focused on the raw material component of corporate capital and the formation of corresponding infrastructure, reducing network structures representativeness and weakening the interaction of corporate and network structures.

The raw material economy does not demand the results of deep technological shifts, innovative products as a fundamental component of its corporate sector. As a result, in the reproductive process of raw materials economy there emerges a gap, which is masked by the purchase of licenses and the supply of imported equipment, as the economy appears to be in the situation of increasing technological dependence, which weakens its competitive advantages, strengthens the protectionism policy, and restrains neo-industrialization processes at a qualitatively new level of technological development. In Russia the destruction of national technological Complex began in 1991–1992, while the period Of high prices on energy marked the peak of this process and consolidated Its results However, the long-term Crisis of 2008–2017 resulted in an urgent need to implement neo-industrialization [5, 12].

Summarizing the results of the analysis of subjects of national economy corporate sector [13], we conclude that the containment of formation and spread of network structures of raw material economy is defined by a degradation of the industrial roots of the corporate form of economic relations. This is one of the main results of corporate capital polarization in the peripheral zone of technological transformation [14]. Let's show some details.

First, this result is connected with the movement of an objective contradiction between two forms of economic relations—network and corporate, the resolution of which is a synthesis of opposites, which requires the following conditions:

- active interaction between network and corporate forms of economic relations;
- sufficient representation and the necessary level of network form development as a leading side of the interaction between network and corporate forms of economic relations [15].

But in the corporate sector of national economy, the interaction of network and corporate forms of economic relations is significantly weakened, and the representation of the network form is insufficient, so this contradiction is carried on not along the path of synthesis, but along the path of polar division of opposites, where the integrity of corporate capital is torn, divided into opposing parts, the gap between which deepens. The scenario of a corporate capital constrained transformation of the way it is integrated in the economic system is consistent and supported by the strategic orientation of the national economy towards the extraction and export of raw materials and energy.

While few high-tech network structures of national economy tend to cross-border interaction, seek to integrate more deeply into the core zone and derive additional benefits from it, corporate structures concentrated in the primary sector of the economy are limited to selling their products on the world market, attracting external investment, and also procurement of technology, equipment and organizational solutions, limited to integration in the peripheral zone [16].

Accordingly, the dominance of corporate structures integrated into the peripheral zone of the global economy in the domestic economy triggers the following processes:

- marginalization of corporations developing new technologies and manufacturing equipment;
- consolidation of an unpromising quasi-development model "sale energy and raw materials—purchase technology, equipment, and attract foreign investment" in the economic system [7].

Picture 59.1 shows the structure of participation of 600 of Russia's largest corporations in the GDP formation.

The data shown in Picture 1 indicate a low share of technology corporations in GDP, which corresponds to data on a decline in the share of high-value-added products in GDP, since the attempt to partially substitute imports turned into an accelerated growth in industries based on fairly primitive technologies and the use

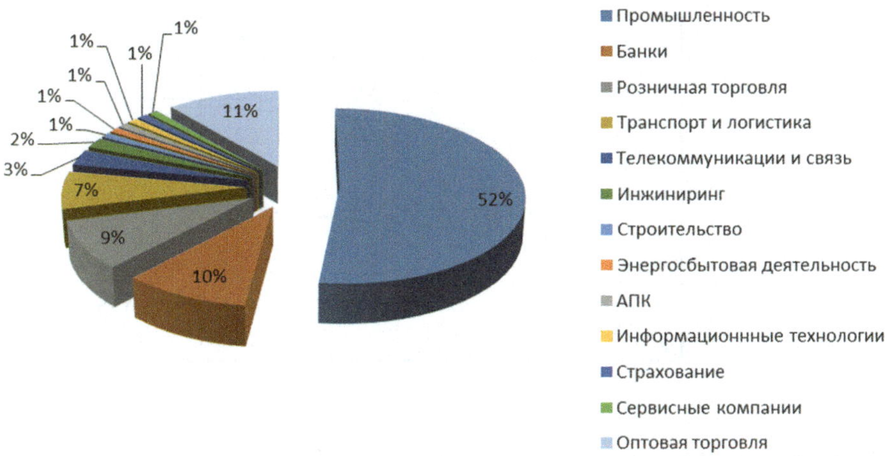

Picture 59.1 Revenue share of 600 largest corporations to GDP for 2017 in Russia by industry, % [17]

of manual labor (agriculture, recreation and tourism, the provision of traditional services, etc.) [17, 18].

Secondly, there is a widening gap in the investment attractiveness between polarized corporate capital entities. The network structures of the new economy attract additional capital, while the corporate structures of raw materials economy push it away, causing a paradoxical situation of the formation of a hidden excess capital in the national economy.

At the same time, the marked excess is performed in two ways:

- through transformation into shadow income with subsequent cross-border transition to residences with preferential tax regimes;
- through waste of capital by the owners through excessive, demonstrative consumption, which provokes a justifiable social protest amid the destruction of technological complex, loss of work and decapitalization, bleeding the economy.

In order to deepen the analysis of the consequences of the economic system technological complex destruction, we apply the heuristic potential of the method, called the "Hicks Cross" (Fig. 59.2).

In Picture 59.2, we base on the assumption of income and expenses balance under conditions of a certain stabilization of the reproduction process in the economic system, which graphically corresponds to a 45° angle of the OF line n relation to the abscissa axis.

We could also suppose that a part of the income of the participants in the economy system is provided by the consumption process, while excess incomes create savings—S. If the economy system develops, the growth in investment demand leads to a gradual decrease in the share of expenses in the total financial result of this system, which can be graphically presented by a reduction in the tilt of the BC line to the abscissa axis.

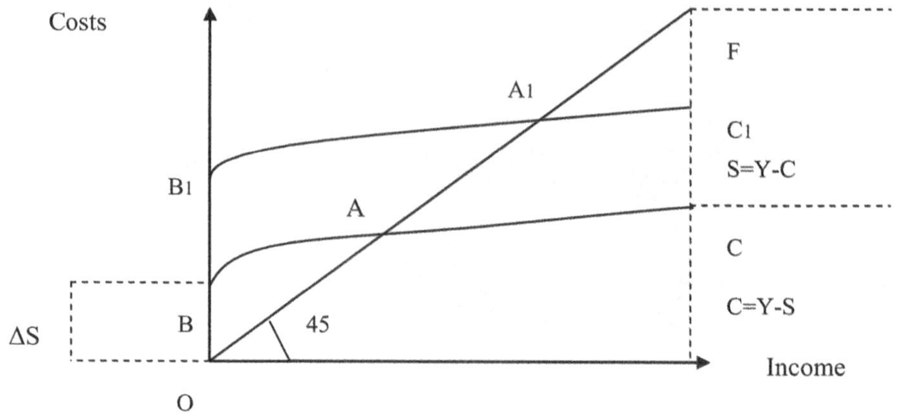

Picture 59.2 "J. Hicks cross" methodology application in the analysis of the consequences of economic system technological complex destruction [2]

Each cycle of reproduction in the economy system provides itself with funds at the expense of current income received by the participants of this system, as well as at the expense of their savings formed earlier, which is graphically represented as ΔS. Accordingly, the new cycle ensures the continuation of the saving process.

To formulate another assumption—the economy system needs an influx of development resources exceeding its ability to generate savings and transform them into investments. Then, its development requires an influx of investment resources from the external environment, which in its turn requires investment attractiveness.

But if the system of the economy has suffered a large-scale destruction of the pre-existing technological complex, then its investment attractiveness will be significantly reduced, as the infrastructure sites freed after the closure and dismantling of engineering enterprises are of interest only to real estate developers, and a technologically degraded economic system is burdened with increased risks. Moreover, such a system is forced to focus mainly on its own development resources, i.e. it is programmed for the conditions in which it appeared to be as a result of external restrictions on access to external financial and technological resources markets. It should be marked that restrictions on access to external technological markets resources for such a system is much more significant than restrictions on access to external financial markets resources due to the continuing destruction of its own technological complex.

For the analyzed system it is vital to form additional resources for the development of its reproduction, that is, graphically, the shift of the BC line to the B 1 C 1 line. If this shift happens, the system will change its equilibrium, which graphically corresponds to the movement from point A to point A 1. There is only one way to create additional development resources in conditions of low investment attractiveness—o use the opportunities that corporate capital has—opportunities to generate fictitious capital and restoring public confidence in the national financial market, undermined in the last two decades.

59.4 Conclusion

The process of neo-industrialization, which ensures the development of the economy with new technologies and equipment, ultimately contributing to overcome corporate capital negative polarization, which was studied in this article, can be supported by the implementation of the possibilities of fictitious capital generating, which will open the way to expand savings and their productive transformation into investments. We should agree with R. Nizhegorodtsev that overcoming the built-in limitations of national economy, weighed down by the phenomenon of depression, is closely related to deep technological shifts. The economic system should be ready for their implementation, which demands reconstruction of industrial complexes on a new technological basis [7].

Investment resources obtained on the basis of fictitious capital generation and aimed at developing an economy system with a destroyed industrial technological complex are not used in the reproduction process and are transformed into additional costs, hidden beneficiaries with developed mechanisms for capturing such resources being behind them; as a result, it is not the economic system that is growing, but the outflow of capital, conspicuous consumption and inflation. Accordingly, when implementing neo-industrialization, strict social and state control over the investment process is required.

Basing on the scientific results obtained in the study on the implementation of the neo-industrial paradigm of Russian corporations integration into economic system in the context of transition to the sixth technological paradigm, we will make suggestions on correcting the neo-industrialization policy in Russia, which should be aimed at:

- participation of raw materials corporations in innovative spatial systems development;
- growing innovation spaces development;
- consolidation of corporate capital connections with spatial economy systems;
- large corporations investment in innovative projects, with the government guarantees for the return of investment resources;
- public-private partnerships development;
- new growth territories creation;
- specification of connections between large corporations and small and medium enterprises;
- creation of preferential tax and credit regimes for innovative projects investors;
- updating the state's economic policy.

In the national economy, the process of neo-industrialization is in demand, ensuring its development with new technological solutions, equipment, ultimately helping to overcome corporate capital negative polarization, which will ensure network and corporate structures efficient interaction and integration, will create conditions for the gradual transformation of rigid vertical corporate mechanisms into network mechanisms, characterized by greater adaptability.

References

1. Galbraith JK (1969) New industrial society. For Scientific Libraries, Moscow, Progress, p 480
2. Hicks DR (1988) Cost and capital, Progress, 191 p
3. Buzgalin A, Kolganov A (2017) Beyond galbraith: creatosphere, Market and corporations at the beginning of the 21st century, Galbraith: Return. In: Bodrunova, SD, M. Cultural Revolution, pp. 205–207
4. Glazyev S (2009) The global economic crisis as a process of changing technological systems. Econ Issues 3:30
5. Gubanov S (2015) Major breakthrough. Neo-industrialization of Russia and vertical integration, M., Book World, pp. 70–110
6. Inshakov O (2003) "The Core of Development" in the context of a new theory of production factors. Econ Sci Mod Russ 1, 11–25
7. Nizhegorodsev RM (2012) Economy of the Inflation Gap. Alternatives 3, 146
8. Nureev R (2014) The periphery of the world economy: problems of catch-up development. Terra Econ 1, 46
9. Ryazanov V (2016) (Not) Real capitalism. Political economy of crisis and its consequences for the world economy and Russia, Moscow, Economy, 695 p
10. Buzgalin A, Kolganov A (2018) Global capital. In: vol. 2 vol. 1, Moscow, LENAND, pp. 66–88
11. Gubanov S (2016) From the export and raw materials model to the neo-industrial economic system. Russia's Econ Revival 4(46):48–59
12. Ryazanov V (2013) Time for a New Industrialization: Russia's Prospects. Economist 3
13. Brizhak OV (2016) Contribution of national corporations to the development of the national economy: monograph. TEIS, Publishing House, Moscow, p 152
14. Brizhak OV, Buyanova ME (2018) Polarization of corporate capital in the context of profound technological transformation, Volgograd state university herald. Series 3 Economy. Ecology, vol. 20 no. 1, pp. 22–29
15. Klochko E, Brizhak O (2019) Prospects of using virtual technologies in modern corporate business systems. Adv Intell Syst Comput 726:308–319
16. Ermolenko AA (2014) Resources of large corporations in the transformation of Russia's economic space. Mod Sci Thought, 2, 118–125
17. Brizhak OV (2018) The conformation of corporate capital in the conditions of systems socio-economic transformations, Krasnodar, "KubGTU", 254 p
18. Regions of Russia. Socio-economic indicators (2017). P32 Stat. Rosstat, M., Access Mode: http://www.gks.ru/free_doc/doc_2017/region/reg-pok17.pdf (call date 22.07.2019)

Chapter 60
Foreign Experience of Legal Regulation of New Financial Technologies and the Possibility of its Implementation into Russian Legislation

Alexander V. Ostroushko, Liudmila A. Bukalerova and Rosalina V. Shagieva

Abstract *Purpose*: A global understanding has now been reached that a mechanism for the long-term sustainable and secure development of financial technologies needs to be put in place to ensure their legal regulation. In the Russian Federation, this process is in its formative stage. The object of the research was public relations in the sphere of creation and use of financial technologies. The subject of the study was the normative material of a number of technologically developed foreign countries in order to identify positive experiences that can be implemented in the domestic practice of creation legal regulation and use of financial technologies. *Design/methodology/approach*: Based on the dialectical method of cognition of the surrounding reality, the research is based on a comparative analysis of the national peculiarities of the implementation of policies to regulate the basic social relations associated with the creation and use of financial technologies. *Findings*: It is noted that realizing the need to create a mechanism of legal regulation of new financial technologies the states have not fully developed a unified policy in addressing this issue. *Originality/value*: In the existing ambiguity of the selected models of legal regulation of financial technologies among foreign countries as well as on the basis of international experience positive trends have been identified on the basis of which recommendations for the improvement of domestic institutions for the regulation of financial technologies are proposed.

Keywords Financial technologies · Legal regulation · Positive experience · Implementation · Foreign countries · Improvement · Comparative analysis

---

A. V. Ostroushko (✉)
Financial University Under the Government of the Russian Federation, Moscow, Russia
e-mail: avostroushko@fa.ru

L. A. Bukalerova
Federal Center for Educational Legislation, Moscow, Russia
e-mail: l_a_buka@mail.ru

R. V. Shagieva
The State University of Management, Moscow, Russia
e-mail: shagsas@mail.ru

T. Kolmykova and E. V. Kharchenko (eds.), *Digital Future Economic Growth, Social Adaptation, and Technological Perspectives*, Lecture Notes in Networks and Systems 111, https://doi.org/10.1007/978-3-030-39797-5_60

JEL Code K23 · K24 · K34 · O38

60.1 Introduction

As practice shows that new financial technologies are being actively introduced into the modern life of the society and it has become a reality that in many countries systems which do not have the appropriate official status and outside the legal framework are being periodically developed and operated. According to N.A. Povetkina and Y.V. Ledneva, the unregulated development of the financial sector can potentially become a serious problem for financial systems not only of individual states but also of a cross-border nature which is due to the use of the Internet in the provision of financial and financial sector services [1]. There is a situation when states are not able to stop the introduction of technologies through the introduction of total bans on financial sector and in this case their activities should be reoriented to support the introduction of advanced information technologies allowing to improve the quality of services and expand their list [2] while ensuring the security of their own financial systems.

The legal regulation of financial technologies is in its formative stage in the Russian Federation in contrast to highly technologically advanced foreign countries. The reason for this is the low activity of low-budget startups in the Russian financial sector and according to Bloomchain experts, in most cases the Russian financial sector is limited to the purchase of ready-made technologies or teams that start working on new products as part of large companies.

Independent solutions which are not affiliated with any credit institution in Russia are quite rare and it is turned out that the banks to be the largest players in the financial sector in the current conditions. Banks operate according to conservative models within the existing legal prescriptions so there was no need to create an appropriate legal mechanism for regulating financial technologies in the Russian Federation in contrast to foreign countries that are rapidly implementing financial innovations as an alternative to traditional banking solutions.

The course taken in 2017 on the digital transformation of the economy had a major impact on the current situation which brought new innovative financial solutions to the market and required improvement of the legal framework for relations between the state, innovative technology companies, credit and other financial institutions as well as recipients of financial services in the creation and use of financial technologies. At the same time, the government's goal was to eliminate the need for untested and unsafe solutions in the market.

We believe that the state should solve the problem of stabilization and liberalization of the development and use of financial technologies at the same time. It has been suggested that the state will have to determine and regulate not only the general direction of development of financial technologies but also the way of development of the Internet and the IT industry in general since the provision of financial services via the Internet through the latest IT-technologies affects the interests of various social groups [3].

Certainly, such relations should be developed only in the legal field and the newly adopted norms should not contain unreasonable (excessive) prohibitions, should not meet the principles of reasonable sufficiency, should facilitate the promotion of innovations in the financial sector although such regulation should necessarily be aimed at reducing potential risks as consumers of financial services, banks, developers of financial technologies and investors.

Given the potentially global nature of financial technologies, the Russian regulator is required to carry out its activities in such a way that the country is integrated into the global international financial system in order to increase competition and reduce the risk of information crime.

The world community and foreign countries have already accumulated a certain both positive and negative experience of normative activity in this area which should be studied and used in the creation of the domestic legal mechanism for the regulation of financial technologies.

In addition, the majority of technologies are currently imported into our country and, accordingly, there are already ready-made solutions for their regulation in the developer countries which can be extended to the domestic legal practice.

All this makes it necessary to analyze international and foreign experience in the legal regulation of creation and use of financial instruments.

60.2 Materials and Method

The task was not to analyze the experience of legal regulation of any particular financial technology in the course of this work.

First of all, the experience of foreign countries is analyzed by comparing national peculiarities of policy implementation on regulation of the main social relations connected with creation and use of financial technologies. The following directions of state policies were of the greatest interest for the research which became the basis for the research program:

- The will of a particular state in relation to the development of financial technologies and the absence of key restrictions for the financial sector development;
- Availability of special government incentives for companies specializing in financial technologies;
- Issues of state support of the personnel training system for the IT-industry and the financial industry in the private sector;
- Security issues as well as related issues of digital identification of legal entities, electronic document flow, collection, storage and processing of data including personal data and etc.;

- Schemes developed by competent authorities that provide regulated and unregulated entities with the opportunity to test innovative products and/or services that are in any way related to the financial services sector;
- Regulation of digital financial assets;
- Regulation of competing technologies with traditional banking services (crowd funding, online lending, funds transfer and etc.).

The subject of the study is the documents of different states which are in the public domain on official resources on the Internet. The choice of the main countries, whose experience of legal regulation has been analyzed by us and the main directions are reflected in this article, is determined by the following criteria: firstly, these are the three world leaders in the field of financial sector which based on ratings of the Global Innovation Index, The Global Financial Centers Index, Doing Business (WorldBank) include the United States, China and the United Kingdom; secondly, the country of the continental legal system with the most extensive experience in the field of digital technology regulation is France; thirdly, the country that started to regulate financial technologies simultaneously with the Russian Federation and closely approached, according to the above ratings, the leaders of the rating for the development of the financial sector is the Federal Republic of Germany. In addition, the experience of a number of other countries was studied focusing only on their specifics without a detailed description.

60.3 Results

Studying the experience of legal regulation of digital transformation of economic relations in foreign countries, we see that since the use of financial technologies is a complex phenomenon and affects different aspects of information and economic activity, it is difficult to talk about the creation of specialized regulatory legal acts.

Focusing on the development of the financial sector in the USA, we can highlight the following features. Financial activity in the United States is regulated both by the federal government and individual states and the process of implementing financial products is more complicated and costly than in some other countries. Alternative financial platforms in the USA continue to attract a lot of interest and attention from federal and state authorities.

Based on a report by the U.S. Treasury Department, the Administration of the U.S. President Donald Trump has recognized [4] that many financial sector rules and policies are outdated and existing regulatory requirements for financial regulation are not appropriate for modern business models and financial products.

In response, the Federal Agency has decided to address the legislative uncertainties that prevent financial sector companies and data aggregators from entering into data exchange agreements. We believe that the availability of such agreements will make it possible to process the necessary information more safely and efficiently.

At the moment, the U.S. Senate is considering a bill on the introduction of a similar British "regulatory sandbox" for financial corporations, the purpose of which was to develop innovative technologies [5].

Also the bill HR56—Financial Technology Protection Act was supported in the U.S. Senate in March 2019. This bill provides for the assessment of new financial technologies (e.g., digital currencies) for the possibility of their use in terrorism and other illegal activities. The bill provides for the establishment of an Independent task force on financial technologies to combat terrorism and illicit financing which is to investigate the illegal use of new financial technologies for terrorist activities and on the basis of which to publish an annual report [6].

China has now become one of the largest markets for the financial sector in the world and in 2017 investment in this sector amounted to more than 8 billion [5]. Therefore, China's experience in regulating the financial sector is very interesting.

The Chinese government considers it necessary to increase the competitive advantage of the banking industry including protection from competition risks from the unorganized financial sector [7]. The concept of the theory of service innovation including new service concepts, interaction with new clients, new business partners, new income models, organizational innovations and technological innovations is considered to strengthen the competitive advantage of the banking industry in China.

Using a strategic approach, the Chinese financial regulator uses integrated mechanisms based on cyber risk exposure assessment to assess the capabilities of the financial institutions while also contributing to the institutionalization of cyber security among regulated firms with strategic control. For this, at the end of 2014, the Chinese government issued a decree requiring the Chinese financial institutions to "secure" and "finalize" 75% of the total amount of the financial sector used in Chinese financial institutions by 2019. The established requirements apply to all financial technology suppliers without exception so foreign companies have little chance of entering the Chinese market.

China was one of the first countries where cryptocurrency legislation was adopted back in 2017. China's legal system banned the use of ICO and imposed a total ban on both cryptocurrency exchanges and other forms of buying and exchanging cryptocurrencies. In April 2019, China classified mining as an "undesirable activity" which was found to be ineffective and should not be used to finance the country's resources.

It is important to study the experience of the United Kingdom in the regulation of financial technology development due to the fact that this country takes a leading position in Europe in the development of financial technologies [8]. London was given an unspoken status of the financial capital of Europe at the very beginning of the formation of the financial sector.

The UK has the UK Digital Economy Act 2017 which was adopted in addition to the UK Digital Economy Act 2010 (DEA).

Analyzing the legal acts and the system of state management of innovations in the UK, it is worth noting that the activities of all entities involved in the development of financial technologies are coordinated by a single entity which ensures the most effective implementation in the financial sphere.

In this regard, we fully support the opinion of M. Krylan that it is necessary to create a specialized state body in Russia similar to BEIS which will carry out public administration in the field of new financial technologies (Sandbox, U.K.).

In 2015 the United Kingdom became one of the first countries to develop a "regulatory sandbox" which is a set of rules concerning the financial sector [9]. As part of these rules, companies in the UK have the opportunity to ask the Financial Services Authority (FCA) to test new technologies such as blockchain in the sandbox. Each firm must report to the FCA during the testing period in the sandbox. If, after testing in the sandbox, a firm wants to run a service on a commercial basis, it can do so if it meets the regulatory conditions for this broader activity [10]. This experience can also be recognized as positive and there is a need in the Russian Federation to legislate the need to test new financial technologies before their distribution on a commercial basis.

Law № 2004-575 of 21.06.2004 is in force in France "On trust in the digital economy. It provides for an extensive list of changes to be made to a number of legal acts of the French Republic. Among the issues of improvement of legislation in the field of financial technologies one can mention rather serious regulation of many aspects of the activities of providers of technical services and the use of communications in the provision of financial services; careful study of issues of e-commerce including issues of taxation in the use of innovative technologies; a special role is given to the problems of legal security in the implementation of the financial instruments in the economy.

In early 2016 the French financial markets regulator AMF (Autorité des Marchés Financiers) established a special unit to assess the impact of innovative technologies on financial services, identify competition problems with traditional banking structures, create a new and adjust the existing regulatory framework for the financial sector. The reaction of the government to the financial innovations was positive and in this connection the regulation of banks' activity was simplified [11]. Also, government agencies created more favorable conditions for entrepreneurs and investors in the field of finance and activities to attract highly qualified foreign specialists in the field of financial technologies to work in the French Republic [12].

In Germany, the financial industry was not widely known until 2016 but in the last three years it has been developing at an exciting pace for innovation. This has been facilitated by the fact that activities have been supported by the developing national ecosystem of the financial sector which now unites:

- Innovative intermediaries: a number of major players and incubation programs that are developing innovations throughout Germany;
- Large German investors;
- A specially created regulatory environment that supports financial projects: The German Ministry of Finance has launched its own forum of the financial sector (FinCamp) as a space for mutual dialogue between different players;

- Industry cooperation: The Conservative German private banking association (Bundesverband deutscher Banken) has taken a clear position to make FinCamp a priority since 2016;
- Public investment: Germany's largest public bank (KfW) has launched its investment programmes since March 2016 to support growing German companies in the financial sector segment. It can provide risk capital of up to 10 million euros per company [13].

At the same time, the German regulator has a rather negative attitude towards cryptocurrencies as the German central bank has excluded the circulation of cryptocurrencies similar to the existing currencies.

It can be concluded that the national legislators of most countries are now trying to bring new developments in the field of finance under the traditional mechanism of legislative regulation. At the same time, there is a transition from a liberal attitude to financial technologies to toughening regulatory requirements while at the same time governments are trying to strike a balance between increasing the efficiency of the financial system associated with technological innovation and protecting the rights of financial market participants. This is confirmed by the fact that over the past two years there have been quite serious restrictions in the European Union concerning bans on anonymous transactions which actually puts many financial projects outside the law.

The whole world is involved in this process behind Europe. It is planned to ban in Japan any anonymous transactions, so the once liberal Japanese authorities have taken a course to stop supporting cryptocurrencies on the national stock exchanges.

South Korea is going to completely abandon digital currencies by 2020 and the Central Bank of Korea has established a separate institution to control the circulation of non-cash funds [14]. Similar rules are planned to be introduced in India by 2020 [15]. The Central Bank of India is the main regulator of the financial system which supervises commercial banks, urban cooperative banks, some financial institutions and non-bank credit institutions. Since 2015 it has introduced special licenses that granted the right of financial startups to conduct payment transactions with the consent of the regulatory authorities.

60.4 Conclusion

1. Creation and granting power to a special body (agency) or subdivision within the structure of the Ministry of Finance which will carry out public administration in the field of new financial technologies.
2. Legislative solution of the issue of trust in new financial technologies as well as legal fixation of the conceptual framework of new financial technologies.
3. Legislative consolidation of the need to create regulatory sandboxes and innovation centers for the introduction of new technologies and conducting experiments

with low risk in a strictly controlled environment with a strictly defined testing time.

4. Establishing minimum requirements and criteria for applicants to gain access to the "sandbox" with the mandatory indication of the directions of product use in Russia after receiving a positive conclusion on the results of testing.
5. Legal consolidation of the assessment of new financial technologies to ensure the safety of consumers of services as well as to counter terrorism and other illegal actions.
6. Assessing the impact of innovative technologies on financial services identifying competition problems with traditional banking structures and creating a new one in the adjustment of the existing regulatory framework to bring the legislation in line with technological innovations and eliminate discredit of traditional banking activities.
7. Creation of a favorable tax base for financial startups and introduction of a mandatory tax regime for the use of financial instruments.
8. Elaboration of the state policy at the legislative level aimed at regulation of crypto activities in the country and adoption of relevant norms regarding the use of digital financial assets and ICO.

References

1. Povetkina NA, Ledneva YuV (2018) Fintech and regtech: the boundaries of legal regulation, Law. J High Sch Econ 2:46–67
2. Ruchkina GF (2017) Banking: transitioning to a new implementation model, or fintech as a new reality. Bank Law 4:55–62
3. Bogdanovskaya IYu (2015) Network neutrality: political and legal aspects (Using the US experience as an example). Inf Law 4:39–44
4. Faden M (2019) Regulatory Sandboxes Provide Safe Spaces for Fintech Payment Services Innovation, https://www.americanexpress.com/us/foreign-exchange/articles/wcostandards-for-cross-border-e-commerce (data accessed: 12.09.2019)
5. Zhao Q, Tsai P-H, Wang J-L (2019) Improving financial service innovation strategies for enhancing China's banking industry competitive advantage during the fintech revolution: a hybrid MCDM model. Sustainability (Switzerland) 11(5):1419
6. Fintech in the USA (2019) https://iqdecision.com/finteh-v-ssha/ (data accessed: 12.09.2019)
7. Inshakova A, Goncharov A (2017) Securities as a subject of financial and economic interest and object of civil law relations. Pravo – Zhurnal vysshei shkoly ekonomiki 4, 97–110
8. Kirlan MG (2018) Domestic and foreign experience in the development of financial technologies in the field of banking services. Econ Entrep 6(95):841–846
9. Vular K (2019) Speech by the FCA director for strategy and competition, https://www.fca.org.uk/news/speeches/innovate-finance-global-summit (data accessed: 12.09.2019)
10. Regulatory sandbox (2019), U.K. Financial Conduct Authority https://www.fca.org.uk/publication/research/regulatory-sandbox.pdf (data accessed: 12.09.2019)
11. Financement participative crowdfunding (2019) https://www.amffrance.org/en_US/Acteurs-et-produits/Prestataires-financiers/Financement-participatifcrowdfunding/Cadre-reglementaire.html?langSwitch=true (data accessed: 12.09.2019)
12. Chishti S, Barberis J (2017) Fintech: a guide to the latest financial technologies. Alpina Publisher, Moscow

13. Drummer D (2016) Why German fintech is now trending https://www.law.ox.ac.uk/business-law-blog/blog/2016/06/why-german-fintech-nowtrending (data accessed: 12.09.2019)
14. Sorokina O (2019) Startups against the laws: how fintech fights against international law. Forbes
15. Romashchenko L (2019) India will ban cryptocurrencies. But he won't stop studying https://sciencepop.ru/indiya-zapretit-kriptovalyuty-no-izuchat-ne-perestanet (data accessed: 12.09.2019)

Chapter 61
Artificial Intelligence Internet Monitoring to Detect and Solve Crimes

Sergey V. Zuev

Abstract *Purpose*: The purpose of the chapter it to develop the so-called "Internet law enforcement" to monitor the Internet in order to detect and solve crimes. *Design/methodology/approach*: The author defines the features of the system, which could, without human intervention, on the principle of virus infection of computers and other technical devices that have access to the Internet, check all their contents and, when detecting marked data, signal to the quick response services about suspected crimes (committed, being committed or prepared), are determined. *Findings*: It is shown that the program "Internet law enforcement" should have various ways of checking sites, pages on social networks, hosts, disks, and various databases. Its work could be observed in real time, as well as the statistics on the results of inspections for further analysis and reporting could be accumulated. Monitoring of the Internet should take place continuously, with a quick response to the detection of suspicious transmitted or stored data. If necessary, the program should be able to block the transfer of data or the operation of a computer device until the circumstances are clarified and the law enforcement authorities remove the lock. *Originality/value*: The so-called "Internet law enforcement" program should not allow to read or guess the semantic load of phrases, interpret expressions or plots, but only highlight and fix the necessary communication segments. Such information will not violate the constitutional rights of citizens, and could be used to conduct further full-scale operational-search measures.

Keywords Artificial intelligence · Legel tech · Crime detection · Monitoring · Internet

JEL Code K24 · K42 · P37 · R28 · H41 · C45 · C55 · C61

S. V. Zuev (✉)
South Ural State University, Chelyabinsk, Russia
e-mail: zuevsergej@inbox.ru

T. Kolmykova and E. V. Kharchenko (eds.), *Digital Future Economic Growth, Social Adaptation, and Technological Perspectives*, Lecture Notes in Networks and Systems 111, https://doi.org/10.1007/978-3-030-39797-5_61

61.1 Introduction

The beginning of the digital information revolution, as a rule, is associated with the advent of electronic computers. Despite the fact that a person independently creates the necessary programs and defines tasks, computers autonomously work with numbers, accumulate, generate and transmit new information, including that which no person, no group of people could receive on their own. This becomes especially noticeable with the advent of artificial intelligence systems: all large classes of computer problems solve them without human intervention [14].

One of these tasks is to study the Internet, that is, monitoring. This term is used, as a rule, to refer to the functioning of a system that monitors the network to detect malfunctions, problem areas, failures and informs the administrator about this using e-mail, phone or other means of communication. All this allows to continuously monitor the status of the Internet, which, in turn, becomes a priority in the activities of law enforcement agencies.

The impact of the Internet on professional legal awareness is due to the emergence of new tasks for resolution in the framework of legal [6]. Active inclusion of digital technologies in the modern society has had a significant impact on everyday life and crimes [11]. The latter, committed using the communication capabilities of the Internet, are characterized by latency and difficulty in their detection and disclosure. Their perpetrators act ahead of time, applying new technologies and all complex schemes, combining with elements of encryption and thoroughly disguising the expression of their criminal intentions [4]. Law enforcement agencies are forced to respond to this, using, inter alia, operational-search forces, means and methods, including special operational-technical units, and computer programs being developed.

61.2 Materials and Method

Theoretical, practical and ethical-legal aspects of the use of artificial intelligence and Legel tech technology to monitor the Internet is emphasized in the works Sukhodolov [14], Powell [11], Broadhurst et al. [4], Aiello [1], Rosenbaum et al. [12], China [5], Borovskaya and Davydova [3], Gelin et al. [8], Nolfi et al. [10], Krichmar et al. [9], Silichev et al. [13].

61.3 Results

Currently, the Internet is considered by the Ministry of Internal Affairs, the FSB and other law enforcement agencies as a subject of study in order to obtain operationally relevant information for further testing of persons for involvement in crimes committed, being committed and prepared. In this case, we can talk about the so-called police presence on the Internet [1]. Currently, a kind of monitoring is carried out by individual employees of operational units when they personally familiarize themselves with sites, pages on social networks, during operational contacts and as a result of legendary communication. Online police presence is ultimately a rule of law activity [12].

At the same time, the question arises of the rules related to the restriction of the constitutional rights of citizens with police officers "immersing" in the private lives of citizens. Violation of the law during the implementation of operational search measures may lead to the recognition of the results as unacceptable for use in proving in criminal cases. This is because a law enforcement officer becomes aware of the privacy of the inspected person without his/her knowledge, and this kind of information can be disclosed to the public.

Thus, the need to ensure the rule of law and security of the individual, society, and the state is faced with the need of citizens for the safety of data relating to their private life. Given that through the Internet there is a constant, continuous exchange of data, the problem becomes global and, it would seem, not solved.

This problem can be solved using the technology of Legel tech and the capabilities of artificial intelligence. Legal Tech is a program, artificial intelligence and neurobots, which can be considered both individually and collectively. The widespread use of services, computer programs and algorithms allows lawyers and law firms to improve many procedures, save time, and therefore money.

Artificial intelligence is a system programmed to simulate human thinking processes in a computer. Artificial intelligence is also an area of computer science, the priority of which is the development of hardware and software tools that allow a non-programmer user to formulate and solve their traditionally considered intellectual tasks, communicating with computers in a limited subset of the natural language [3].

Legal tech is seen as a movement aimed at revising the existing approaches to solving problems in jurisprudence through modern technologies. The development of the Legal tech segment leads to a rethinking of the purpose, activities of lawyers, as well as products that simplify the work of specialists. At the same time, the development of Legal tech implies in the future a transition to a qualitatively new model of justice, which should become more accessible to millions of people through the use of various automation tools.

Today, Legal tech is a separate industry specializing in information technology support of legal activities using big data processing. The technology allows, within the framework of certain tasks, to exclude a lawyer and provide him with ready-made, well-developed data. Such work is performed by a machine, a robot, a computer, that is, an impersonal program. Using words, markers, accumulative samples of words and expressions, as well as images, audio and video clips, the system could monitor the Internet without the participation of a person and record and then notify the relevant law enforcement authorities about suspicious transactions, actions, operations, communications. Such guidance information could be used for further full-scale operational-search measures.

The Internet monitoring program, relatively speaking "Internet law enforcement", should be able to autonomously conduct checks in remote networks, having a web interface with the ability to fully control the system. Checks can be targeted when there is already information about suspicious sites and pages, and general. The results of the checks should be displayed on the screen with a graphic diagram, icons and sensors of the devices.

The proposed program could work by the type of virus infection of computers and other technical devices that have access to the Internet, check all their contents and, when detecting marked data, signal to the quick response services of the suspected crimes committed, being committed or prepared. It is fundamental that the program should not allow to read or guess the semantic load of phrases, interpret expressions or plots, but only highlight and fix the necessary communication segments.

The program could signal about detected suspicious fragments using sound, recording to the log, screen messages, as well as via e-mail or SMS messages. Everything should depend on the settings of the user-employee of the law enforcement agency responsible for monitoring.

This kind of program should have different ways of checking sites, pages on social networks, hosts, disks, various databases. Its work could be observed in real time, as well as accumulate statistics on the results of inspections for further analysis and reporting. Suspicious sites, Internet pages could fall into the so-called suspicion group, in which the degree of risk should be envisaged depending on the frequency and severity of the facts revealed.

Monitoring of the Internet should take place continuously, with a quick response to the detection of suspicious transmitted or stored data. If necessary, the program should be able to block the transfer of data or the operation of a computer device until the circumstances are clarified and the law enforcement authorities remove the lock.

Given the complexity and volume of the tasks performed, in this case it is justifiable to connect artificial intelligence (robot), which is able not only to replace a person, but also to perform work that is beyond his strength. Various data collection programs are already in use. Chat bots, which are considered symbols of Legal Tech, can serve as some kind of analogue.

The application purpose of chat bots is that they are involved in solving similar problems. They are considered as personal assistants. Chat bots are able to collect e-mail addresses from contact forms, act like viruses and worms, collect information and use it automatically.

Currently, the capabilities of artificial intelligence are widely used to study the market in business, medicine, in various search engines, including through speech and text recognition, data mining and filtering spam in email. In addition, applications are being developed that are able to recognize gestures, individual signs of voice, isolating voice from a noise stream, determine faces by emotions and non-verbal signals. Other applications have found use in robotic navigation, overcoming obstacles and setting objects.

The introduction of artificial intelligence is considered in areas with the design of expert systems and knowledge bases. The latter is understood as "a combination of data and inference rules that allow inference and meaningful information processing. In general, research on the problems of artificial intelligence in computer science is aimed at creating, developing and operating intelligent information systems, and the issues of training users and developers of such systems are solved by information technology specialists" [2].

The proposed program "Internet law enforcement" should automatically upload dialogs, analyzing them, compare the relevance of various slang expressions related to illegal actions, that is, independently replenish the list of words and phrases that are "irritants". To maintain its functionality, storage and transfer of spent information, a large amount of disk or cloud space will be required, at the moment, the closest analogy is NAS storage and disk arrays.

Information on the Internet can be both freely available and closed. Speaking about the technical aspect of network monitoring, it should be noted that both each Internet provider and any messenger have their own security protocols and methods of encrypting information.

Encryption also performs security functions, but differs from security systems. **Encryption** is a reversible transformation of information in order to conceal it from unauthorized persons, while keeping authorized users access to it. Confidentiality of information being transmitted becomes an attractive resource, and its reliability serves as a guarantee of the safety of the transmitted data. The encryption is based on the use of a key algorithm, which confirms the choice of a particular transformation from the set of possible options for this algorithm.

Users are authorized if they have a certain authentic key. There are two ways of encryption: symmetric—using one key, and asymmetric—using two keys. The whole complexity and, in fact, the task of encryption is how this process is implemented.

In general, encryption includes encrypting and decrypting.

Using encryption, three states of information security are provided: confidentiality (aimed at hiding information from unauthorized users during transmission or storage); integrity (pursuing the inadmissibility of making changes when transmitting or storing information); identifiability (allowing to authenticate the source of information and to exclude the ability of the sender of information to deny the fact that the data was sent by him/her).

Using, for example, WhatsApp, a person is confident in ensuring confidentiality due to double encryption (asymmetric method), that is, the process of transmitting information is protected by two keys. First, the message is encrypted with the first key, then the received encrypted text is processed with the second key. The bypass algorithm of this method is known and may well be used by appropriate programs to detect violations and identify the persons who committed them.

The symmetric encryption method, due to the appearance of end-to-end encryption, has lost its relevance and now rarely finds application, since there is a more secure method described above and, due to the development of hacking programs, raises doubts about its use. Of course, there are much more ways to secure data transfer; the above examples are presented as the most relevant.

Having touched upon the encryption process, decryption should also be considered. **Cryptanalysis** is the science of decrypting information and the decryption processes themselves. An attempt to disclose a specific cipher using cryptanalysis methods is called a cryptographic attack. A **cryptographic attack**, during which it was possible to reveal the cipher, is called a "hacking" or "tampering" of the cipher.

In order to read the encrypted information, the receiving side needs a key and a decryptor (a device that implements the decryption algorithm). The idea of encryption is that an attacker, having intercepted encrypted data and not having a key to them, can neither read nor change the transmitted information. In addition, in the modern cryptosystems, different keys can be used to encrypt and decrypt data. With the development of cryptanalysis, techniques have appeared that allow to decrypt closed text without a key. A crypto attack is used by running a specific program (for example, Cryptool and SubstitutionCipherDecryption.exe). In this case, the more words in the text, the greater the likelihood of their interpretation.

It seems that, despite various ways of protecting information and encryption, artificial intelligence should be able to overcome all possible barriers and monitor the entire Internet. Only in this case law and order will be ensured in the virtual space, and therefore in reality too. Perhaps this problem will be completely resolved only with the advent of neural computers.

61.4 Conclusion

Already today it is becoming clear that artificial intelligence should be able to independently select keys to decrypt suspicious information, or a special technique should be written by which it can independently identify the preconditions for committing crimes. We emphasize that we are talking about the complete exclusion of a person's ability to read a particular message, determine the meaning of a photo or video, as well as about the inadmissibility of infringement of confidential data of Internet users. In this case, the person (in this case, the operator) must remain responsible for all actions taken [8].

To achieve the progress and development of civilization, it is necessary to anticipate the consequences of technological progress and wisely introduce artificial intelligence [13].

The technical development of society necessitates the creation of more powerful computers capable of solving the most complex tasks, overcoming any obstacles to obtaining information in order to ensure security and protect society from criminal attacks. The choice between the availability of all data and the safety of society should be decided in favor of the latter, taking into account the confidentiality and interests of Internet users. It becomes possible using supercomputers, neural networks, the ability of artificial intelligence to self-study [7], as well as developing the legal framework for digital relations.

References

1. Aiello MF (2018) Policing through social networking: testing the linkage between digital and physical police practices. M Aiello. Police J 91(1):89–101
2. Artificial intelligence. Retrieved from https://ru.wikipedia.org/
3. Borovskaya EV (2018) Fundamentals of artificial intelligence. EV Borovskaya, NA Davydova M. Laboratory of Knowledge, p 127
4. Broadhurst R (2014) An analysis of the nature of groups engaged in cyber crime. R Broadhurst, P Grabosky, M Alazab, B Bouhours, St. Chon. Int J Cyber Criminol January–June 8(1):1–20
5. China D (2010) Internet policing hinges on transnational cybercrime (November 10). Retrieved from http://www.china.org.cn/business/2010-11/10/content_21310523.htm
6. Davydova ML (2016) Transformation of the legal system under the influence of the Internet. ML Davydova, VO Makarov. Bulletin of VolSU, Series 5. Jurisprudence 4(33):50
7. Fridman V (2016) The way to artificial intelligence. World Sci 3:26–33
8. Gelin R (2016) Le robot est-il l'avenir de l'homme? R Gelin, O Guilhem. Paris: La documentation Francaise
9. Krichmar JL (2005) Navigation and causal analysis in a brain-based device modeling cortical-hippocampal interactions. JL Krichmar, AK Seth, DA Nitz, JG Fleischer, GM Edelman. Neuroinformatics 3(3):197–221
10. Nolfi S (2000) Evolutionary robotics: the biology, intelligence, and technology of self-organizing machines. S Nolfi, D Floreano. MIT Press/Bradford Books, Cambridge, MA
11. Powell A (2018) Digital criminology: crime and justice in digital society. A Powell, G Stratton, R Cameron. Routledge, London
12. Rosenbaum DP (2011) Understanding community policing and legitimacy-seeking behavior in virtual reality: a national study of municipal police websites. DP Rosenbaum, LM Graziano, CD Stephens, AM Schuck. Police Q 14(1):25–47
13. Silichev D (2019) Artificial intelligence and the future of humanity. D Silichev, A Volobuev, E Kuzina. In: Popkov E (ed) Ubiquitous computing and the internet of things: prerequisites for the development of ICT, vol. 826. Research in Computational Intelligence. Springer, Cham, pp 699–706
14. Sukhodolov AP (2018) Artificial intelligence in combating crime, its forecasting, prevention and evolution. AP Sukhodolov, AM Bychkova. All-Russ Criminol J 1(6):753–766

Chapter 62
The Influence of Information and Computer Technologies and Transnational Media Corporations on Public Legal Awareness

Tatyana K. Krasilnikova, Gennady G. Egorov, Elena Yu. Dubovikova and Elena V. Ryabova

Abstract *Purpose*: The article considers specific communicative features of the Internet which have an impact on the development of public legal awareness, particularly, that of today's youth. *Design/methodology/approach*: The research methodology is based on the dialectic, historical legal, system-structural, and structure functional methods. *Findings*: The authors define the main threads in the use of Internet resources as effective agents for legal socialization. This approach should be primarily aimed at equipping the younger generation with skills which will enable them to use online information inculcating legal values. The paper shows how, with the rapid development of Internet communications and digital television, the public's awareness is left defenseless in the face of "alternative" communicative influence. *Originality/value*: While the state and society used to directly control the communicative impact of the country's mass media for centuries, since the beginning of the twenty-first century permissiveness and lack of control over subjects of mass communication have led to the situation when the younger generation is largely exposed to antisocial and criminal ideas and attitudes.

Keywords Legal culture · Legal awareness · Internet · Information technology · Public security · Media corporation

JEL Code K 100

T. K. Krasilnikova (✉) · G. G. Egorov · E. V. Ryabova
Volzhsky Branch of Volgograd State University, Volzhsky, Russian Federation
e-mail: kun_krasilnikova@mail.ru

G. G. Egorov
e-mail: egorov@vgi.volsu.ru

E. V. Ryabova
e-mail: ryabova_ev@list.ru

E. Yu. Dubovikova
National Research University MPEI, Volzhsky, Russian Federation
e-mail: dubovikovae@rambler.ru

T. Kolmykova and E. V. Kharchenko (eds.), *Digital Future Economic Growth, Social Adaptation, and Technological Perspectives*, Lecture Notes in Networks and Systems 111, https://doi.org/10.1007/978-3-030-39797-5_62

62.1 Introduction

The Internet is one of the most sought-after sources of information and, at the same time, it is a medium considerably changing its users' legal awareness. The culture of modern society is becoming increasingly dependent on IT systems. Digital information technologies are able to shape forms developing certain legal attitudes, which are necessary for any activities, both theoretical and practical, in the field of law [8].

The Science of Law has repeatedly raised the issue of the importance of any studies aimed at determining the impact of the Internet-based digital environment on its users' legal awareness. Particularly noteworthy among academic papers describing forms, means and consequences of IT influence on human consciousness are the ones by Lebedeva [6], Piskunova [9], Maurin [7], and Nadygina [3].

The impact of information technologies on various areas of public life is becoming more and more significant. The ever-increasing rate of implementing the latest information technologies, irrespective of the area of human activities, is creating a fundamentally new environment for developing socially relevant relationship, i.e. the electronic digital environment. For much of the world, as well as for Russia in particular, the transition to digital means of governing modern society through the information environment has become a priority. According to the RF President's official website, President Vladimir Putin has pointed out on several occasions that the implementation of the strategy for the development and subsequent introduction of new digital technologies in the State is one of the most essential directions for the Russian Federation (www.kreml in.ru).

62.2 Methodology

The authors of the paper used the dialectic, historical legal, system-structural, and structure functional research methods.

62.3 Results

The transition to the new society type, i.e. information society, is related to all prime areas of human activities, such as economics, politics, science, education, culture and law, which are becoming increasingly dependent on the applied digital systems.

The research into legal reality as a phenomenon determining legal awareness is particularly relevant, and it poses a whole range of serious theoretical challenges to the scholars.

The Internet influences the consciousness of large numbers of people, especially youth. The Internet-based interaction tools are of particular interest to teenagers. Network resources enable young people to communicate freely without adult scrutiny, as well as to participate in specialized interest groups dealing with various information.

A number of statistical studies have revealed that only 6% of the young respondents use online forums regularly. In our view, these respondents are mostly students, faculty and scholars who are motivated to study and pursue science. Some of them are primarily focused on research work. In this case, the digital environment is for them a basis which shapes their academic awareness. The data analysis leads to the conclusion that such respondents are interested in scientific activities, as evidenced by their research papers published on the Internet, as well as their participation in Internet-based conferences and action groups discussing various issues of state and public life. At the same time, for most of the young interviewees the intended purpose of visiting Internet resources is entertainment.

Admittedly, the Internet is merely a technological medium for information transfer. The impact caused by information systems and information in general on public and personal legal awareness is directly dependent on people's social environment, as well as their ethical and moral standing. For most users, the Internet gives the opportunity to obtain useful relevant information, which raises their legal awareness. At the same time, excessive fascination with the World Wide Web may become a cause of the users' moral degradation, which, in its turn, is a cause of distorted legal awareness [11].

The analysis of the Internet impact on citizens' legal awareness is primarily determined by their inherent value systems and world outlooks underlying their spirituality. The effective use of the Internet environment and available resources is preconditioned by the individual's developed values, attitudes and mindset allowing for a more objective perception of any information. The use of information systems as a positive set of impacts shaping people's legal awareness should be proportional to the increasing role of legal instruction in all areas involving human activities, such as family, school and professional training [1].

It is worth noting that since the early 1990s the Russian media market has been attracting the attention of transnational media corporations, which, in order to maximize their profits, reoriented communication flows so that the audience could get more entertainment (often to the detriment of its cultural standards) [5]. In the quest for ratings, various methods were used, including those that ran counter to the moral and ethical principles of Russian spirituality. The specificity of the Russians' spiritual culture should be also taken into account, as the majority of adult Russians were born and raised in the USSR, where such changed values and attitudes were alien to Soviet people. At present, most of the country's population watches entertainment programmes [8].

It should also be recognized that information perception through print and verbal channels involves considerably hard work of human consciousness and imagination, which means that it requires certain mental efforts. In contrast, emotionally coloured audio-visual images are perceived much more easily. They are largely based on subconscious psychical mechanisms, and to a much lesser extent on the conscious psyche mechanisms of the digital environment user. Emotional and sensual perception is easier than abstract and logical one. Thus, in the near future, human logical and intellectual abilities might dramatically decline, which will seriously impede progress in science and the arts, since it requires high intellectual efforts.

Another problem is to be noted: the current normative legal regulation of important public information relationship does not correspond to the dynamics of digital development. The legal norms that are supposed to regulate this area do not correspond to the pace of development in information resources. The latter require, above all, defining their legal status, which would allow for viewing them as constituent elements of current information legal systems.

Technological advances presented on the Internet, mobile technologies distributed through computer software, as well as reference, analytical and information databases have been adapted by media corporations to facilitate their use, primarily by representatives of youth groups [2].

The current government is consistently pursuing a policy of implementing all available forms of digital systems in the educational process. The introduction of various information systems makes it possible to improve qualitative indicators in the education system significantly. The use of digital technologies in classes has a positive effect on perception dynamics and practical acquisition of knowledge; and it helps to form lasing visual patterns. Various kinds of online and offline testing are aimed at intermediate and final control. The Internet environment makes it considerably easier for students to obtain information in the form of samples of past papers, such as reports, course and thesis papers and other student projects. At the same time, the development of Internet resources is also aimed at the reverse process, i.e. the detection of borrowed information, or plagiarism detection. The software system called *Antiplagiat* ('antiplagiarism') has been increasingly used in recent years. It searches for text segments borrowed from both publicly available Internet sources and a number of private databases, such as university digital environments, databases of doctoral and candidate theses and specific compilations of information that are sealed to the public. This software produces a detailed report with highlighted text selections which have been copied and specifying information about the percentage of borrowings and their sources—digital or academic databases.

The Internet-based remote communication systems provide boundless opportunities for close interaction among specialists of various levels and kinds.

Now let us review the role of social networks in the transnational media realm. A social network is an interactive multiuser environment organized through a particular website or a digital cluster group. Its content is generated by the network users according to their interests. Actually, it is possible to observe a distinctive social environment which is arranged via an abstract facility and which enables various groups of users with common interests to interact with one another.

Back in 1995 Randy Conrads founded Classmates.com, one of the first social networking services, if the modern interpretation of the term may be applied to it. The innovative project proved to be in demand with Internet users, and that year can be considered a starting point for any Internet-based social digital networks. Being one of the pioneer social networking services, the above network is still one of the most popular in the world; it has over 50 million users. Many similar services appeared later. Yet the wide usage of social networking services is generally understood to have started officially only in 2003–2004, when *MySpace* and *Facebook*, famous and popular international social networks, were launched. With these services, anyone could find a complete stranger on the site and see the chain of interim acquaintances connecting him- or herself with that person. Due to their growing popularity, social networks soon became a sort of Internet havens where anyone could find and use both technological facilities and social environment to create his or her virtual "self". In fact, every user gets the opportunity not only to communicate with the others, but also to create and subsequently share their creativity with a multimillion audience without having to bear any substantial economic or organizational costs. Social network founders focus primarily on meeting the users' need to express their individuality, which is characteristic of most people [10].

Another category of social networking service users is represented by informal youth groups whose activities often involve criminal intent. In modern society, such organizations, unlike subcultural groups of the 1980s and 1990s, are based on disregard for authoritarian charismatic leaders; they have neither centralized inner discipline nor rigid hierarchy. Such users of social networks are able to affect destructively other citizens' legal awareness. This is usually done through criminal, nihilistic or other antisocial comments on events or phenomena. Another source of negative effect on young people's sense of justice is real-life situations which might be specially selected to promote criminal views and to discredit existing moral standards which ensure public order.

For most children, using social networks is a special pleasure, and this activity takes up more and more of their free time from year to year. Minors use these digital systems to find friends, or to form and join subculture groups, thus seeking social value and aiming to prove their status through their confirmed affiliation with a certain social group. Admittedly, while choosing suitable friends for themselves, they interact actively, though digitally, with virtually unlimited number of people and interest groups. In the process, the digital majority tailors the new users' inner world to its liking and makes them take interest primarily in those issues that are of interest to the majority. In this situation, the young user is often unable to distinguish his or her own interest from the others.

It should be noted that social networks distort teenagers' attitudes and their world views, which may have a negative effect on shaping their moral values and ideals.

However, the main problem is that, as a rule, adolescents are unsupervised when they are using social networks. Parents are often quite happy that their child "is doing something useful" instead of hanging out somewhere. They hardly pay attention to any information their child is dealing with, including texts, audio or video files. Nor do they monitor the circle of the child's virtual acquaintances because it is invisible to the people not belonging to the digital environment. Teenagers may escape from the real world into the virtual one, completely neglecting their social duties, the need to do their schoolwork and acquire a vocation. The digital environment becomes so appealing to children that they feel an urgent need to stay online at any time and to respond to every post, like or comment, thus developing severe forms of addiction to social networks. At the same time, no due attention is paid to considerable harm caused both to the child's mental well-being (desocialization, agoraphobia) and to his or her physical health (visual impairment, faulty posture).

However, despite numerous disadvantages, social networks have a number of positive effects on their users. The benefits of social media in terms of their impact on a person are as follows:

- For one thing, such network services provide users with limitless opportunities for self-development. The users can watch any film even before its international release, download and listen to music they like, or have free access to articles and books of their interest. In fact, publicly available network resources allow sufficiently prepared users to master new areas of expertise with the help of numerous video tutorials.
- If the users possess necessary skills, social media may serve as a platform for setting up and developing their business.

Negative impacts of social networks on people may be summarized as follows:

- adverse health effects;
- risk of mental distress;
- desocialization;
- vocabulary degradation and deteriorated communication. Virtual interaction involves disregard of grammar and punctuation rules and simplifying speech forms to shortened slang expressions, which are meaningless to anybody except the members of a particular social group.

All this contributes to the imminent crisis of public legal awareness.

Initially, there was no intent on the part of educational institutions to develop the students' adaptation skills for interacting with available media. Nowadays, it is necessary to ensure the effective use of electronic information resources with educational potential and to prepare young people for life in the present-day information environment.

62.4 Conclusions/Recommendations

Throughout their lives, people gain social experience and obtain knowledge of the system of social ties and relationships emerging in modern information society. It has become apparent that the genesis of modern youth's legal awareness is taking place in the process of socialization and under the influence of the Internet, information, computer-based and telecommunication technologies.

The key areas of the State information policy in the current context should include:

- first and foremost, teaching adolescents to assess the social media information critically. Young people should be taught to perceive, understand and analyze information, as well as realize the consequences of social networking impact.

In order to do so, it is essential to develop independent critical thinking, since passive social networking undermines the users' civic activity, negatively affects their creative potential and, finally, results in their inability to make decisions without ready-made patterns to follow;
- secondly, teaching people to use modes of von-verbal communication through the use of technological tools [4];
- thirdly, introducing adapted information security courses in educational institutions;
- fourthly, legislating a set of measures which would require to introduce restrictions on the demonstration of misconduct and gaming activities in social media and to foster other alternative positive networking trends;
- fifthly, implementing the legislative recognition of social networks as a form of mass media and establishing legal liability of their representatives who are responsible for the dissemination of materials incompatible with legal norms.

References

1. Egorov GG (2017) The use of distance learning forms in the network interaction system (§ 1.2. Ch. 1). In: Guzeva MM (ed) Training in promising professions in the system of continuing education through the development of network interaction: monograph. Electron. text, graph. Dan. (1.08 Mb), 1 electron. opt. disk (CD-ROM). Volgograd Scientific Publishing House, Volgograd, pp 14–26
2. Egorov GG, Borzenko LE (2015) Scientific and practical approaches to the formation of general legal values of modern society in the educational process. In: Mironov YuI, Egorov GG (ed) Legislative and methodological support of the educational process in Russia: monograph. Volgograd Scientific Publishing House, Volgograd, pp 336–347
3. Gorylev AI, Nadygina EV (2006) The influence of information technology on the functions of legal consciousness. Bulletin of the Nizhny Novgorod University. N.I. Lobachevsky. Series "Law". N. Novgorod: Publishing House of UNN, 2006. Issue. 19. Problems of development and functioning of the legal system of society, pp 30–38
4. Ishkulova SH (2007) The use of mass communication in the socialization of adolescents. Educ Sch Child 9:30–31
5. Krasilnikova TK (2016) A corporation as a model of society's self-organization: domestic and foreign experience. Lawyer 9:39–41
6. Lebedeva NN (2004) Is legal information available on the Internet accessible to everyone? Russ Justice 1:53–54
7. Maurin VS (2004) Legal analysis of harmful information in the information society. Diss. … Ph.D. M., 190 p
8. Pevtsova EA (2004) Modern definitive approaches to legal culture and legal consciousness. J Russ Law 3:70–81
9. Piskunova SS (2004) Legal information as a means of forming the legal consciousness of citizens. Diss. … Ph.D. M., 180 p
10. Romashkina TV (2019) Influence of social networks of the Internet on value orientations [Electronic resource]. Socialization and re-socialization of adolescents and youth: an interdisciplinary aspect. URL http://pedagog.pspu.ru/conference/novoselova/digest/section-4/181-vliyanie-sotsialnykh-setei-intemet-na-tsennostnye-orientatsii. Date of access 24 Mar 2019

11. Tarakanov VV, Inshakova AO, Dolinskaya VV (2019) Information society, digital economy and law. In: Popkova EG (ed) Ubiquitous Computing and the Internet of Things: Prerequisites for the development of ICT. Studies in Computational Intelligence, vol 826. Springer Science + Business Media, Cham, pp 3–15

Chapter 63
Business Activities Using SAAS Techologies: Civil-Law and Tax Peculiarities

Marina Merezhkina, Tat'yana Tsytsylina and Stanislav Shiro

Abstract *Purpose*: The purpose of the article is to examine legal rules and regulatory contracts for the provision of "cloud" services and to reveal the civil-law framework corresponding to their legal nature. *Methodology*: The main research method is comparative-law one along with the general scientific dialectic method of cognition, structural analysis method and some specific scientific methods, such as formal-law, comparative analysis, grammatical and logical research methods. *Findings*: It was established that there is no direct legislative regulation and judicial interpretation concerning the legal nature of contracts for the provision of "cloud" services. As authors suppose, this type of contract is mixed with the elements of the license agreement and paid services contract. *Conclusion/recommendations*: Within the article, it is recommended to take into account the ambiguous interpretation of the legal nature of contracts for the provision of "cloud" services in their conclusion. It's necessary to enshrine grounds for the acquisition of the rights to the utilization of electronic computing machine programs by the user (customer), use the contract terms common to the license agreements for the copyright holder (contractor), as well as specify the software utilized to provide the "cloud" service. If the provision of a "cloud" service is accompanied by the rendering of services, we can recommend regulating relations through the conclusion of two agreements: a license agreement and paid services contract.

Keywords Service · License agreement · Software · "Cloud" technologies

JEL Classification code K24. Cyber Law

M. Merezhkina (✉)
Volgograd State University, Volgograd, Russia
e-mail: meregkina@mail.ru

T. Tsytsylina
Volzhsky Institute of Economics, Pedagogy and Law, Volgograd Region, Volzhsky, Russia
e-mail: meteorain@yandex.ru

S. Shiro
Volgograd State Socio-Pedagogical University, Volgograd, Russia
e-mail: stanislav_shiro@mail.ru

T. Kolmykova and E. V. Kharchenko (eds.), *Digital Future Economic Growth, Social Adaptation, and Technological Perspectives*, Lecture Notes in Networks and Systems 111, https://doi.org/10.1007/978-3-030-39797-5_63

63.1 Introduction

Modern business activity is impossible without the market for intelligent technologies. The widespread dissemination of intellectual property in commercial relations is associated with the necessity of solving the issues of proper legal regulation of these relations. The service SaaS (Software as a Service) has become widespread on the Internet due to its convenience. The idea of SaaS is that the supplier develops and manages the software by himself, providing the customer with access to the software via the Internet.

Such a service is very convenient because it is easy to use, it can significantly reduce costs against the usual distribution of electronic computing machine programs, more effectively protect software from unauthorized access, and it also allows selling user technical support services and render related services.

In Russian legal science, there are various legal approaches to the legal determination of contracts for the provision of "cloud" services.

A variety of approaches is related both to the features of such agreements and the goals of law-makers undertaking legal determination, since different approaches to legal determination have an impact on the tax and other public legal implications arising in connection with the contract concluded, put requirements to the contracting parties, its form, and other features [1–8]. A variety of approaches to determination exhibits the relevance of the research.

63.2 Methodology

The main research method is comparative-law one along with the general scientific dialectic method of cognition, structural analysis method and some specific scientific methods, such as formal-law, comparative analysis, grammatical and logical research methods.

63.3 Findings

Despite the standpoints of foreign law-makers, in particular, the German Supreme Court, authors think it is inexpedient to talk about the possibility of classifying contracts on the provision of "cloud" services for lease agreements in the context of Russian legal reality.

The most appropriate legal frameworks are a contract on paid services provision, as well as a license agreement. The basic civil-law principle of freedom of contract makes it possible to refer such contracts to as non-defined or draw them up as mixed ones.

Let's address the issue of legal determination of an agreement on the provision of "cloud" services as a license agreement.

The idea of the license agreement is that the holder of the exclusive right to the result of intellectual activity provides (or is obliged to provide) another party with the right to use such a result or such means to the extent stipulated by the agreement (Cl. 1, Art. 1235 of the Civil Code of the Russian Federation).

The working legislation of the Russian Federation does not provide for other methods of acquiring rights to the application of a electronic computing machine program, except for the conclusion of a license agreement.

Thus, the Civil Code of the Russian Federation regulates the utilization of software under a license agreement only in terms of granting rights to software copies. Disposition of the software for its intended purpose is not recognized "utilization" in copyright terms. Therefore, attributing a contract on the distant use of software to a license agreement is quite controversial.

According to A. Serov who believes that SaaS is precisely a service: "in the SaaS model, there is no subject-matter of the license agreement. Pursuant to Art. 1235 of the Civil Code of the Russian Federation, the subject-matter of the license agreement is an essential condition and the license agreement is deemed null and void without it" [9].

This finding is confirmed by the standpoint of state authorities stated in the letter of the Ministry of Finance of Russia No. 03-07-08/70744 as of November 30, 2016: "Pursuant to Subclause 1, Clause 1, Article 146 of the Tax Code of the Russian Federation (hereinafter referred to as the "Code") operations for the sale of goods (works, services) on the territory of the Russian Federation are recognized VAT taxable items. Thus, the services of providing access to computer software via the Internet rendered by a Russian organization are subject to VAT in accordance with the generally established procedure." Thus, the contract on the provision of "cloud" services in the opinion of the Ministry of Finance of Russia is a contract on the provision of information services or information processing services (in terms of tax legislation).

Assessing the opinion stated in the letter of the Ministry of Finance of Russia No. 03-07-08/70744 as of November 30, 2016, it is necessary to take into account the goals of the fiscal authorities.

If we address the legal nature of the agreement on the provision of "cloud" services as a contract on the provision of services, then V. Razuvaev's point of view seems justified: "any sober-minded person will rightly suggest that the access service is directly provided by the communications operator (provider), and the copyright holder himself in this situation provides the right to utilization of an electronic computing machine program through such an access. That is, access acts as a kind of vehicle delivering a passenger to the desired address" [10].

Judicial practice concerning contracts on the provision of "cloud" services define them either as contracts on the provision of services or as license agreements. For the sake of fairness, let us note that judgments were not trying to determine the legal nature of such contract, but were dealing with disputes on their terms and execution.

Summing up, let's remark that there is no direct legislative regulation and judicial interpretation on the issue under study. As authors suppose, this type of contract is mixed with the elements of the license agreement and paid services contract.

To outline the elements of a license agreement in the framework under study, one should take into account Article 1270 of the Civil Code of the Russian Federation pursuant hereto the utilization of an electronic computing machine program also includes its reproduction. Getting access to the program placed on a remotely located server, the user reproduces it on the screen of his monitor, that is, begins to use the program for an electronic computing machine.

63.4 Conclusion/Recommendations

When concluding a contract for the purchase of SaaS, you should watch out for the legal execution of this transaction. The contract should enshrine the grounds for the acquisition of the rights to the utilization of electronic computing machines programs. As for the copyright holder (contractor), it is necessary to use the terms common to license agreements, avoid the wordings typical for other contracts ("service", "contractor", "customer", etc.), and designations such as "cloud service", as well as clearly specify the software that is used in the provision of "cloud" service, if the contractor acts to release exercising the right to utilization of electronic computing machine programs and databases under a license agreement from VAT (Subcl. 26, Cl.2, Art. 149 of the Tax Code of the Russian Federation). If the provision of a "cloud" service is accompanied by the rendering of services, we can recommend regulating relations through the conclusion of two agreements: a license agreement and paid services contract.

References

1. The Civil Code of the Russian Federation (Part 4) No. 230-ФЗ as of December 18, 2006 (as amended on July 18, 2019). Rossiiskaya Gazeta. https://cis-legislation.com/document.fwx?rgn=14605. Accessed 20 July 2019
2. Tax Code of the Russian Federation (Part 2) No. 117-ФЗ as of August 05, 2000 (as amended on July 30, 2019) (Rev. and EXT., entered into force since September 1, 2019). Collected Legislation of the Russian Federation. https://cis-legislation.com/document.fwx?rgn=1717. Accessed 20 July 2019
3. Savel'ev A (2015) The legal nature of "cloud" services: freedom of contract, copyright, and high technology. Bull Civil Law 5:62–99
4. BGH, 11/15/2006—XII ZR 120/04. http://medien-internet-und-recht.de/pdf/vt_MIR_Dok._009-2006.pdf. Accessed 20 July 2019
5. The Decree of the RF Government No. 2036-P as of November 1st, 2013, On approval of strategy of developing IT area in the Russian Federation for 2014–2020 and until 2025. http://www.consultant.ru/document/cons_doc_LAW_154161/. Accessed 20 July 2019

6. The Decree of the RF Government No. 2602-P as of December 30, 2013, On approval of the action plan development of the IT-industry (amended and enlarged). https://base.garant.ru/70555876/. Accessed 20 July 2019
7. The Judgment ASGM in a case No. A40-85252/2015 as of July 24, 2015. http://sudact.ru/arbitral/doc/K5apbLfNbgd1/. Accessed 20 July 2019
8. The Judgment of the Commercial (Arbitrazh) Court of Zabaykalsky Krai of December 25, 2017, in Case No. A78-11563/2017. http://sudact.ru/arbitral/doc/o6t0tu78z6Qh/. ASGM Judgment in a case No. A40-101227/2017 as of November 24, 2017. http://sudact.ru/arbitral/doc/PnQrvLF0T9Lu/. Accessed 20 July 2019
9. Serov A (2011) SaaS: software or service? EJ-Lawyer 17:4
10. Razuvaev V (2010) Software as a Service. The Ezh-Lawyer, no 5, pp 142–144

Chapter 64
Digitalization and the Principal-Agent Problem in the Higher Education

Elena V. Mikhalkina, Olga A. Chernova and Anna V. Gozalova

Abstract The main aim of this article is to study the transformation of social and labor relations in the higher education in the process of the Russian economy digitalization, as well as to identify related opportunities to improve these relations because of solving the principal-agent problem. The methodology of the study is based on the principal-agent theory in with regard to the higher education, where the principal is the employer (the head of the University or structural department), and the agents are the teaching staff and employees of the University. The solution of the principal-agent problem is considered through the prism of the information asymmetry theory and its statements. The research identifies the main development tracks of social and labor relations in the higher education and proposes the digital economy tools, which allow obtaining positive effect in solving the problem of principal-agent. The article includes the conclusion and recommendations where the authors agree that the digitalization of the economy changes the content, nature and forms of the labor activity, increases the role of its individualization, and creates the new design of economic mechanisms in the personnel management system. The results of the study can be used to substantiate the strategic development directions of social and labor relations in the field of education as well as to assess the development effectiveness of these relations in the modern conditions of the Russian economy digitalization, and to develop appropriate regulatory impacts.

Keywords Social and labor relations · Information asymmetry · Digitalization · Education · Digital economy tools

JEL-codes J50 · D82

E. V. Mikhalkina (✉) · O. A. Chernova · A. V. Gozalova
Southern Federal University, Rostov-on-Don, Russian Federation
e-mail: mikhalkina_e@mail.ru

O. A. Chernova
e-mail: chernova.olga71@yandex.ru

A. V. Gozalova
e-mail: mkrtoumian12345@mail.ru

T. Kolmykova and E. V. Kharchenko (eds.), *Digital Future Economic Growth, Social Adaptation, and Technological Perspectives*, Lecture Notes in Networks and Systems 111, https://doi.org/10.1007/978-3-030-39797-5_64

64.1 Introduction

The development potential of a modern educational institution is largely determined with its ability to obtain positive effects from digitalization in all elements of the mechanism of the educational process, and in particular, those elements related to labor activity. As it is known, the principal—agent problem arises due to the actors' different degree of awareness at different levels of hierarchy, which leads to a limited rationality of their behavior. Digitalization helps to reduce information asymmetry, providing opportunities to solve this problem. There is a clear gap between the speed of digital technologies spread and the speed of adapting the regulation system in social-labor relations. This fact actualizes the objective to study the conditions and the effects of digitalization in solving the principal—agent problem.

64.2 Methodology and Literature Review

The principal-agent theory initially analyzed the management problems, which associated with the ability of employees to use information advantages for their own purposes, in large corporations [9, 13, 16]. Conventionally, the principal is the owner, who is most interested in the results of the company. An agent is a manager whose behavior, in contrast to the owner, is characterized with the desire to maximally benefit in achieving personal and professional interest.

Using the «agent's» direction of analysis in the education sphere became widespread in the early 90s of the XX century. It was caused with the active implementation of the institutional economic theory to study the processes of reforming higher education in Russia. Veretennikova N.N. explains that fact and notes, that the development of contract relations between society and universities has changed the principles of university functioning and led to the following problems:

- differences in the views of society (represented by the state) and universities on the higher education mission and its objectives in the economic system, the priority of particular functions in scientific and educational activities, the content of the orientation mechanisms to the consumers of educational services' needs and demands;
- information asymmetry, expressed in different levels of social awareness about scientific-educational functions that the University carries out [3].

As a result, the principal—agent problem arose in higher education. In this case, the principal means the state, which delegates the organization and implementation of the educational process to the agent who is the head of the educational institution. However we should notice that we can follow the chain «principal—agent» on other levels of the hierarchy. The number of links in this chain determine the organizational structure of educational institutions and the number of educational programs, etc. It means that we understand the principal as the employer-the head of the University or structural unit. He is a provider of educational services, imposes certain requirements on employees (teaching staff, employees of educational institutions), determining its scope, content, etc. Teaching staff and University staff are agents, ensuring the implementation of the educational process at its individual stages, that means in fact the implementation of set tasks. This interpretation of the principal-agent model will be the basis of this work.

Despite the existence of a large number of researches which study social and labor relations in the framework of the principal-agent theory, this problem is practically not considered in the field of education. We can note some application of this theory to the regulation of social and labor relations in universities in the works of E. Klejn and A. A. Galkin, studying the phenomenon of academic corruption [4, 8]. The public funding for research and educational institutions also mentions the principal-agent problem in education [3, 6]. The realization features of the effective contracts in achieving performance indicators in order to implement development programs for structural units (as departments, institutes, faculties, etc.) can also partly express this problem [7, 12, 14, 15].

Studying the digital scenario of education and its influence on the sphere of social and labor relations, researchers concentrated the problem field mainly on the possibilities to improve the quality of educational activities through the use of modern information technologies [1, 5], as well as on the study of new models for forming the system of employees' remuneration [2, 10]. Taking into account the obvious gap in the study of the possibilities of digitalization in improving the system of social and labor relations in the framework of the principal-agent model, the problems, considered in this article, has an important scientific and practical task.

In the context of this study, we will consider the principal—agent problem in the light of information asymmetry, implying the employer's (principal's) lack of awareness about the employee's (agent's) activities and his efforts to organize his activities. As a result, the principal's dilemma is between the need to control the agents' activities and the unwillingness to have the additional costs associated with this control. Moreover, the control process is associated with the need to collect and analyze a significant volume of information flows, which constantly complicate and do not always clearly characterize the employee's results. The dilemma of the agent is the need to follow the liabilities stipulated by the terms of the employment contract (effective contract), while trying to maximize the benefits that he has because of possessing information not known to the principal.

Digitalization, which contributes to the rapid dissemination of information, leads to the fact that the principal is able to increase control over the employee with use of modern information technologies. Consider how the principal (represented by the employer) can use the advantages of digitalization in improving the regulation system of social and labor relations in the projection on the sphere of higher education.

64.3 The Impact of Digitalization on Changing the Regulation System of Social and Labor Relations in the Field of Education

The field of social and labor relations directly reflects digitalization trends of the economic relations system, expressed in the large-scale use of information technology in solving the problems of educational institutions. We can present the most important transformation directions of social and labor relations in the higher education in the form of the following enlarged tracks:

- the University transition to project management through continuous analysis of databases and introduction of information systems;
- the implementation of prescriptive analytics principles and adaptive learning;
- higher education employees and students' development of digital competences.

In general, as R.P. Kolosova and T.A. Medvedeva noted, the system of social and labor relations in higher education changes its organizational code from hierarchical to network under the influence of digitalization. The development of interactive information interactions between varieties of distributed agents reflects this fact [11]. However, we should note that, since the main format control is not personal interaction, but interaction with use of information and communication technologies, the quality of filling the information space of the educational institution functioning largely determines the reliability of obtained information.

As a result, the participants of social and labor relations are in such conditions where they are to represent and protect their interests in a certain information space. This fact determines the need to develop tools and mechanisms to ensure the positive effects of digitalization in the field of social and labor relations.

64.4 Tools of Digital Economy in Solving «the Principal-Agent Problem»

The use of digital economy tools in solving the principal-agent problem leads, first of all, to the development of direct effects, which are expressed in reducing the costs of collecting and processing information related to the implementation of control functions. According to the World Bank's report on the digital transformation of the economy, the cost of a digital transaction is 20 times lower than the cost of using facsimile communication, 30 times cheaper than telephone communication and 50 times cheaper than live contact [17]. However, the digitalization in the field of social and labor relations in addition to economic relation causes a series of social effects directly related to the ability to solve the principal—agent problem. These social effects include increase of the remuneration fairness, the possibility of creating individual content and forms of work for the employee in accordance with their abilities and needs; the development of new forms of intra-corporate communication.

According to a number of researchers who study the problematic of social and labor relations development in the digital economy, the most essential tools for solving the principal—agent problem are those that allow providing a high level of coordination of employees' actions with the employer's interests and contributing to the formation of effective partnerships. In our opinion, such tools and technologies include the following.

The first is *the effective contract* which:

- clearly defines the effectiveness indicators of the employee's activity;
- forms a system of motivation to achieve certain results based on his individual needs;
- provides the correlation of incentives, rewards and sanctions with the results of the employee's activity;
- increases the openness of the organization social and labor relations.

Within the framework of an effective contract, a collaborative mechanism takes effect, resulting in a common vision of individual and collective goals and prospects.

The second is *blockchain technology*. Control of the activities of employees of educational institutions, as it is known, is associated with the need to collect and process a large number of information (realization of the academic load, the results of scientific, educational and other activities). This makes it difficult for employers to assess the effectiveness of scientific and pedagogical workers. The introduction of blockchain technology allows launching a transparent system of information exchange, forming an infrastructure of constantly updated data, simplifying the procedure for accounting the employee's achievements during his transition from one structural unit to another.

The third is *digital platform*. As the example we take «Electronic supervision», which allows considering citizens' applications, interacting with the employer's labor inspections, etc.; «Electronic Dean's office», which optimizes the system of interaction between the student, teacher and University administration. For efficient functioning of data platforms there are several requirements: building electronic offices and single digital deaneries to systematize the administrative and educational activities; the use of digital services as «SciVal» and «Dimensions» and others for planning and implementation of research activities; start the process of digitization of documents and the transition from paper to online.

The fourth are technologies of "smart education" as smart training, cloud technologies, which allow realizing the concept of forming individual educational trajectories of development based on the use of modern technologies that support and transform the process of interaction between student and teacher.

New systems for assessing the results of work, which propose monitoring the successful work of departments, students, teachers, the formation of tools to support and manage the infrastructure of the University "smart campus", etc.

However, the effectiveness of the use of these tools and technologies largely depends on the level of the educational institution employees' digital competence, since digitalization implies the need for changes at all levels of both the interaction system and the management system.

At the "teacher" level, the basis for the formation of digital competence and the new system development of social and labor relations should be the motivation, control and monitoring of the formed digital competence. It means the expansion of the use of digital technologies in the educational process for the introduction of ICT into everyday practice and the creation of "Smart education" system—smart training.

At the «Manager-administrator» level, digital competence implies the need to develop a University regulatory base governing the use of information management system of the University, as well as the integration of modern digital technologies in the educational process. This in turn leads to the emergence of new economic institutions that let the principal control directly the agents. At the «students» level, they should implement the task of forming basic digital competencies through the skills of using the digital economy services.

64.5 Conclusion

In that way the digitalization of the economy leads to a change in the content, nature and forms of labor activity, increasing the role of its individualization, the emergence of new economic mechanisms in the personnel management system. The digital economy factors and conditions, which contribute to the creation of new forms and approaches to the regulation of social and labor relations, strongly influence the educational activity. Digitalization of the economy poses serious challenges to traditional ideas about the content of the University employees' labor activity. That implies the need for new approaches and tools to manage social and labor interactions at various stages of their implementation, updating the monitoring system of efficiency and effectiveness.

The authors agree that the digitalization of the economy changes the content, nature and forms of the labor activity, increases the role of its individualization, and creates the new design of economic mechanisms in the personnel management system. The main prerequisites for solving the principal-agent problem in higher education are associated with the possibility of using modern information technologies and tools to reduce the level of information asymmetry about the content of the employees' labor activity. At the same time, we should take into account that the effectiveness of these tools and technologies depends largely on the level of the employees' digital competence. This fact can become a restricting factor to the use of these tools and technologies.

The results of the study can be used to substantiate the strategic development directions of social and labor relations in the field of education as well as to assess the development effectiveness of these relations in the modern conditions of the Russian economy digitalization, and to develop appropriate regulatory impacts.

References

1. Bogoslovskii VI, Busygina AL, Aniskin VN (2019) Conceptual foundations of higher education in the digital economy. Samara Sci Bull 8(1(26)):223–230. https://doi.org/10.24411/2309-4370-2019-11301
2. Borovskaia MA, Masych MA, Shevchenko IK (2013) Principles and applications of effective incentive contract in research and teaching staff in higher school. High Educ Russia 5:13–20
3. Veretennikova NV (2007) Contract relations in Russian higher education. Bull Tomsk State Univ 301:144–147
4. Galkin AA (2013) Unfavorable features of the institutional environment of higher education in Russia. Area Sci 5:32–36
5. Dneprovskaia NV (2018) Assessment of the readiness of the Russian higher education for the digital economy. Stat Econ 4:16–28. http://dx.doi.org/10.21686/2500-3925-2018-4-16-28
6. Eremenko GA (2008) The role of agency relations in science and innovation. Sci Innov Educ 4:315–330
7. Kaplan R, Norton D (2003) Balanced scorecard. From strategy to action. Olimp-Business, Moscow
8. Klejn E (2011) Corruption in Russian universities. Terra Economicus 9(1):60–70
9. Kozlova EV (2012) Economic mechanisms for identifying and limiting opportunistic behavior in Russian corporations. Bull Chelyabinsk State Univ 10(264):117–121
10. Kolosova RP (2014) Professional standards, certification of qualifications and effective contract in the system of social and labor relations. In: Problems and prospects of innovative development of the economy 2014 proceedings of the international scientific and practical conference in Saratov, Russian Federation, 2014, Publishing house "KUBiK", Saratov, pp 58–65
11. Kolosova RP, Medvedeva TA (2015) Social and labor relations in networked economy. Bull Moscow Univ 5:89–103 (Series 6.: Economy)
12. Kurbatova M, Levin S (2013) Effective contract in higher education of the Russian Federation: theoretical approaches and features of institutional design. J Inst Stud 5(1):55–80
13. Levanova LN (2010) Features of the "Principal-Agent" Relations or "Insider Capitalism" in the Corporate Governance of the Russian Enterprises. Izvestiya of Saratov University. New series. Series of Economics. Management. Right, vol 10, no 1, pp 52–58
14. Mikhalkina EV, Aleshin VA (2013) Effective contract as an institute of the social and labor relations in higher education. Terra Economicus 11(3):89–100
15. Mikhalkina EV, Skachkova LS (2013) New model concept of remuneration based on the introduction of key performance indicators in institutions of higher education. Motiv Remun 1:48–60
16. Persky YK, Dmitriev DV (2010) Interaction and interdetermination of competition processes and information asymmetry on the regional trade market. Reg Econ 1:176–182
17. World Bank (2018) Report on the development of the digital economy in Russia, September 2018, "Competing in the digital age: strategic challenges for the Russian Federation". World Bank, Washington, DC. License: Creative Commons Attribution CC BY 3.0 IGO [Electronic resource] available at: http://documents.worldbank.org/curated/pt/848071539115489168/pdf/Competing-in-the-Digital-Age-Policy-Implications-for-the-Russian-Federation-Russia-Digital-Economy-Report.pdf. Accessed 20 Sept 2019

Chapter 65
Tools for Assessing Effectiveness of Public Services as a Part of Digital Economy Development

Yulia S. Polozhentseva, Maria G. Klevtsova, Olga V. Sogacheva and Irina V. Androsova

Abstract Digitalization covers all areas of the society, including the process of interaction of public authorities with citizens. Therefore there is a current need of monitoring the quality and accessibility of public services, as well as a need to assess the efficiency of public services as a part of the development of the digital economy based on modern tools. The aim of the study is to synthesize the instrumentation that evaluates the effectiveness and ways to improve public services in the framework of the development of the digital economy. Methods of analysis and synthesis, comparative and graphic-analytical methods, methods of statistical processing of information have been used in the research. There has been conducted a statistical analysis of e-government and e-services in various countries, and the models of building e-government have been generalized. Moreover, the main problems of delivery of state and municipal services in electronic form have been systemized in the article. On the basis of the key indicators, which characterize the development of the Russian information society, the analysis of readiness for the formation of digital economy and delivery of public services in a digital format in all regions of the Russian Federation has been conducted. The research has shown that the delivery of high-quality public services based on digital technology contributes to the development of information society, openness from the authorities and the formation of an effective digital economy in many countries.

Keywords Public services · e-government · Digital economy · Information society · E-services

Y. S. Polozhentseva (✉) · M. G. Klevtsova · O. V. Sogacheva · I. V. Androsova
Southwest State University, Kursk, Russia
e-mail: polojenceva84@mail.ru

M. G. Klevtsova
e-mail: klevtsovam@mail.ru

O. V. Sogacheva
e-mail: sogachova@mail.ru

I. V. Androsova
e-mail: irinka-rusik@mail.ru

T. Kolmykova and E. V. Kharchenko (eds.), *Digital Future Economic Growth, Social Adaptation, and Technological Perspectives*, Lecture Notes in Networks and Systems 111, https://doi.org/10.1007/978-3-030-39797-5_65

JEL Code M15 · L86 · O43

65.1 Introduction

The potential of information and communication technologies and the extent of their use in the Russian system of government have increased substantially in recent years. Electronic services of commercial companies, such as e-commerce, online banking, distance education, and a number of other services, have changed the style of economic life; due to this fact the speed of social, economic and technological development is constantly increasing.

The relevance of this study lies in the fact that the positioning of the regions in the digital information environment uses a number of methodologies and tools that address the priority of economic and social problems. Through the focus area in the regions of the Russian Federation there has been implemented and developed a set of activities and projects for the modernization of the system of public service delivery in the information society based on digitalization of function space.

A number of scholars, such as Lupton [6], Garcia et al. [4], Antonelli [2], Grimes [5] were engaged in studying the features of the digital information space development, including public services.

Public service delivery system allows one to convert service delivery into electronic form at all levels—federal, regional and municipal. This is achieved by reaction with “The Integrated portal for state and municipal services” (IPSMS).

The task of developing the system of public (municipal) services delivery is aimed at improving the integration with the population and legal entities, increase of efficiency of activity of government bodies and local authorities, accessibility for the population and legal entities of information on regulations and the implementation of the state (municipal) services at every stage, verification of their execution.

The innovative nature of the research is to summarize tools for assessing the effectiveness of public services on the basis of examination their quality and accessibility of public services in the development of the digital economy [1].

65.2 Methodology

The research is based on scientific methods, including generalization of domestic and foreign experience in delivery of public services, which has shown that they are provided mainly through multifunctional centers of public and municipal services. Drawing on statistical methods, the authors have also considered and analyzed the data on the delivery of public services in the Russian Federation and abroad, as well as through the regional budgetary institution “Multifunctional Center for the delivery of public and municipal services.”

The methodological base of research also includes scientific methods: the cause-and-effect approach; the structural and functional tools for pattern identification; methods of monographic analysis, as well as classical methods of analysis and synthesis, including private-scientific methods of economic phenomena research.

The basis of the research is a systematic approach, which considers the process of formation and development of e-government in such countries as The United States of America, Canada, Great Britain, Sweden, Poland, Hungary, Malaysia, Russia, and South Korea. Furthermore, on the basis of multivariate statistical analysis there has been considered the dynamics of the key development indicators of e-government and e-services, the information society and the digital economy as a whole, for the display of the results the methodology of graphic interpretation of information has been used.

65.3 Results

In the world there are different approaches to evaluating the effectiveness of public services. Thus, in Europe ranking of countries according to the import and export of commercial services provided serves as the main tool for assessing the efficiency of public services [7].

Another tool to assess the effectiveness of public services in the countries of the world is the analysis of information society indicators, among which are: the coverage of computer, communication, insurance and financial services; number of secure Internet services; time of business registration, etc. [3].

In Russia, as key indicators for assessing the effectiveness of public services allocate indicators related to the volume of public services, their availability and the level of citizens' satisfaction with the quality of the services.

The estimation of accelerating digitalization of public services in the world is based on the UN e-government ranking (Fig. 65.1), which includes key indicators characterizing the development of informatization of society in the following areas: on-line services; ICT infrastructure and human capital. The highest growth of this indicator is observed in Denmark, which has moved up from 8th place to the first. In the Russian Federation increase of the level of e-government development is clearly seen; thus in 2018 it left the group of countries with a "high" level and joined the group of countries with "very high" level, taking the 32nd place.

On the contrary to the dynamic introduction of information technology in society, in the Russian Federation citizens still do not have a full opportunity to receive state and municipal services in electronic form. At best, users may expect to receive electronic information about the service [8].

In 2014 the Russian Federation lost eight positions in the ranking and was on the 35th place in 2016. The highest possible index of the level of electronic services development that a country can get in the state rating is equal to one. In 2016 the UK the figure had 0.9193; Germany—0.8315; Philippines—0.7948; Armenia—0.7683 and Russia—0.7215 (Fig. 65.2).

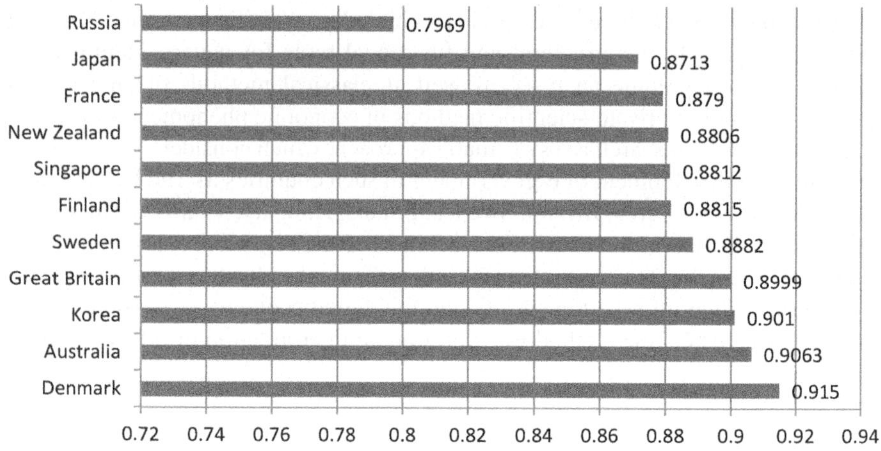

Fig. 65.1 The level of e-government development in the countries of the world according to the UN

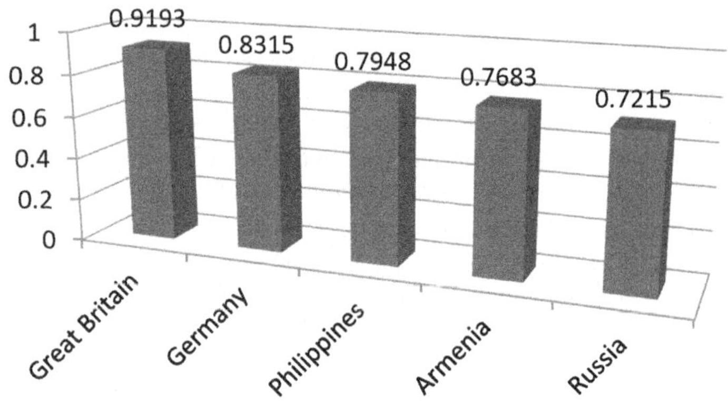

Fig. 65.2 The index of the level of electronic services development in 2016

Portal interacts in three important areas of public services delivery in electronic form [10]:

- between various levels of public authorities;
- between the government and entrepreneurs;
- between the government and citizens.

In 2015 the number of visits to this portal was 11.8 million; in 2016 –29.2 million; in 2017 –47.1 million.

Besides UN experts have grouped the countries into groups according to the models of e-government construction, which are presented in Table 65.1.

The main reason for the championship of Korea is the tangible improvement and maintenance of applications for mobile phones, which can be downloaded with a

Table 65.1 Models of e-government construction

Anglo-American	European	Asian	Russian
The United States of America, Canada, The United Kingdom of Great Britain	Continental European countries	South Korea, Singapore	Russia
This type of model is characterized by a tendency of transition from monologue of political and administrative power to communication between public employees and the public in the form of "service authority". As part of this trend, the US e-government provides services for the citizens and is accountable to them	The consumer receives e-services in real time. It means that being on the any territory of the state one can fully receive the necessary information, vote using the electronic voting system or use government services for the payment of taxes and fines.	At the core of this model is the principle of the hierarchical pyramid, typical of Asian corporate culture and specific style of public management. Therefore the delivery of public services through e-government has a multilayer system that focuses on the key problems of citizens	This model is based on the introduction of digital services and platforms, including the delivery of a variety of public and municipal services, through the implementation of easy access to various information portals and receives services in the information society

unified integrated portal of the state; it provides access to services through various channels.

South Korea aims to solve its social and public problems of low birth rates and the aging of the country, and actively respond to future social needs of citizens.

In the UK, there is e-government service called DirectGov, which provides a large number of information resources for the population. Content is generated by government departments together with the service command. Attendance of the service in a month is more than 27.5 million visits. The share of e-services in the country is 8.5% and is the leader in the use of IT-technologies.

By the level of the introduction of e-government the United Kingdom is currently on the implementation of the 4th stage. Consequently, the public can follow the work of government agencies for processing their requests, and this is the main stage of the establishment of the open, accessible and transparent government.

Among the CIS countries Kazakhstan has reached the significant achievements of e-government.

Currently in this country there are more than 2000 information, 236 interactive and transactional services (108 of them are implemented on the portal of "e-licensing" and 128—the portal of "electronic government") for citizens and businesses. On the portal of e-government in Kazakhstan (www.egov.kz) 67 facilities, 16 services, payment of 21 species of state fees, 16 types of state duties, 4 types of tax payments and payment for violation of traffic rules are represented [11].

The legislative framework, which sets requirements for the procedure of e-government (municipal) services, shows the need for a concrete definition of the term quality of the above mentioned criteria, which should cover all situations of service delivery in electronic form and direct approach to the appropriate authority, regardless of the type of service.

At this time, there have been developed and implemented such separate functional tools of e-government infrastructure as:

1. a multifunctional center of state and municipal services;
2. a unified portal of public and municipal services;
3. regional portals and registries of public services;
4. centers of public access to information about public authorities and to electronic public services;
5. call centers for citizens and enterprises.

Foreign experience (United States, Canada, United Kingdom, Sweden, etc.) shows that one of the most effective methods to ensure access for people and enterprises to the data of any public service is the portals of public services on the Internet, which are created by a country.

One way to analyze e-government services is the construction of mathematical models in the form of Petri nets [9].

Thus, the current availability of information and communication technologies is increasing in Russia; this is evidenced by a situation that in the UN e-government ranking it has risen from 59 to 27 place.

Informatization began to be perceived by regions as a way to improve the quality of public administration, means of prompt processing of requests from society and the basis for decision-making.

For a general analysis of the positioning of regions it is necessary to undertake a comprehensive assessment of the regions in all of these areas, for this purpose the authors use bubble chart, because it allows one to display three-dimensional data into two-dimensional space. The horizontal X axis displays the readiness index, the vertical Y-axis displays index of society informatization of Russian regions, calculated below in Table 65.2. The size of each bubble is proportional to the volume of goods and services in the region.

Table 65.2 The index of society informatization of Russian regions in 2018

Federal District	Society informatization index
Central	1.12
Northwestern	0.99
South	1.01
North Caucasus	1.03
Volga	1.01
Ural	1.04
Siberian	1.02
Far Eastern	1.04

It is clearly seen that the Central Federal District takes the highest place on the index of the society informatization, but on the index of readiness, it is in the middle. North-West Federal District takes the lowest place on the informatization index but on the readiness index it is slightly ahead of the Central Federal District; however it is seen from the chart that the volume of goods and services in this district is maximal. Southern Federal District holds the lowest position both on the index of readiness, and the index of the society informatization; in terms of goods and services the district occupies a middle position. North-Caucasian Federal District as well as the Southern Federal District occupies a low position on the index of the society informatization, on readiness index it occupies a middle position. Volga Federal District in terms of goods and services in this district is almost identical to the CFD, and the index of the information society is identical to the Southern Federal District, the index of readiness it is slightly ahead of the CFD. Ural Federal District in terms of goods and services in the district takes a prominent position, as well as on the indexes of readiness and society informatization. Siberian Federal District readiness index takes the highest place, and the district takes the middle position on an index of the society informatization. Far Eastern Federal District occupies a leading position in terms of goods and services, and on the indexes of readiness and society informatization it takes middle positions.

The priority of the regions is mobility, cloud computing and modernization of IT-infrastructure. Among other things, there is an increasing interest in big data, which is designed to reduce fraud and corruption.

On a 100-point scale the interval of "Digital Russia" index indicators in the first half of 2018 is in the range of 37.2–75.14 points (in 2017 this range was 26,06–70,01). This result suggests reducing the gap between leading and lagging subjects of the Russian Federation. The top ten leading regions include Moscow, Tatarstan, St. Petersburg, Khanty-Mansi Autonomous Area—Yugra, Tyumen region, Yamalo-Nenets Autonomous Area, Moscow Region, Republic of Bashkortostan, Leningrad Region, Chelyabinsk Region.

The authors have calculated the indicators of information development of the Central Black Earth economic region for the 2016–2018 in Table 65.3.

Thus, the effectiveness of the development of the digital economy of the Central Black Earth economic region for the 2016–2018 has increased, as evidenced by increased readiness to the information society index. Readiness to e-commerce index

Table 65.3 Indicators of Central Black Earth economic region assessment in a digital information space

Indicators of human factor	2016	2017	2018
Readiness for the information society index	1.96	1.99	2.18
Reflex readiness for e-commerce index	0.44	0.32	0.43
Society informatization index	2.07	2.14	2.31

Source Integrated assessment of the level of socio-economic development of regions in Russia in 2018. http://www.tchiurkounov.ru/artictes/prognos_se2018_dec2018/pril.03.xls

in 2017 fell significantly, but again in 2018 it increased to almost 2016 index, which indicates the possibility of the fail of the Central Black Earth economic region in participating in the digital economy in 2017. Society informatization index has a positive dynamics in the analyzed period, which suggests the increasing level of the information economy development in the region.

65.4 Conclusion

Development of the system of state and municipal services delivery in electronic form is complicated by the following problems:

1. Currently there is a large discrepancy between the public authorities concerning the quality and the use of information and communication technologies; the impact of the implementation of similar technologies is departmental in nature.
2. The infrastructure of open access for citizens to the websites of state institutions and other means of information and reference service are not sufficiently developed. For consumers receiving government services in a variety of situations requires personal presence in their state institutions, which leads to time-consuming and a major inconvenience.
3. The legal framework of electronic services is poorly established, as well as the standards and practices of government agencies service delivery with the support of information and communication technologies.
4. The infrastructure that provides information security of electronic forms of cooperation between public institutions, citizens and legal entities is not organized.

The difficulties that impede improvement of the effectiveness of the use of information technologies in the activities of public authorities are inter-ministerial in nature and cannot be solved at the level of individual government agencies. Implementation of program-target approach for interoperability of government systems and for the implementation of standards and technologies of their interaction is required.

Consequently, one of the top priorities, which face the Russian state authorities, is the transition to the delivery of public (municipal) services and their implementation in electronic form by the federal bodies of executive power. Another important priority is the development of infrastructure, access to state and municipal services and to information on the work of state bodies and local authorities.

In order to assess the public service, which is in electronic form, the following quality criteria are used:

- availability of data on its regulation implementation;
- availability of documents in electronic form, which are required for its acquisition;
- waiting time for a response to the application;
- convenience of the public services delivery.

Thus, the world practice shows that public services can effectively be realized abroad and are available for a large range of potential users of public services.

References

1. A comprehensive assessment of the level of socio-economic development of the regions of the Russian Federation in 2018. http://www.tchiurkounov.com/artictes/prognos_se2018_dec2018/pril.03.xls
2. Antonelli C (2003) The digital divide: understanding the economics of new information and communication technology in the global economy. Inf Econ Policy 15(2):173–199
3. Carlsson B (2004) The digital economy? what is new and what is no. Struct Change Econ Dyn 15(3):245–264
4. Garcia D, Tessone CJ, Mavrodiev P, Perony N (2014) The digital traces of bubbles: feedback cycles between socio-economic signals in the Bitcoin economy. J R Soc Interface 11(99):0623
5. Grimes S (2003) The digital economy challenge facing peripheral rural areas. Prog Hum Geogr 27(2):174–193
6. Lupton D (2013) The commodification of patient opinion: the digital patient experience economy in the age of big data. Sociol Health Illn 36(6):856–869
7. Perrons D (2004) Understanding social and spatial divisions in the new economy: new media clusters and the digital divide. Econ Geogr 80(1):45–61
8. Polyanin A, Proniayeva L, Matveyev V, Amelina A, Klevtsova M (2018) Personnel risks management in the competency approach at state service bodies 2018. In: Proceedings of the 32nd international business information management association conference, IBIMA 2018—vision 2020: sustainable economic development and application of innovation management from regional expansion to global growth, pp 2376–2391
9. Robinson C (2017) Disclosure of personal data in ecommerce: a cross-national comparison of Estonia and the United States. Telematics Inform 34(2):569–582
10. Sui DZ, Rejeski DW (2002) Environmental impacts of the emerging digital economy? the e-for-environment e-commerce. Environ Manage 29(2):155–163
11. The electronic government of the Republic of Kazakhstan. http://www.egov.kz. Date of appeal 10 Nov 2018

Chapter 66
Regional Competitiveness as a Component of the Foresight Model of Russia's Competitiveness: A Solution Within the Spatial Development Strategy

Andrey Polynev and Irina Grishina

Abstract *Purpose*: The purpose of the study is to draw competitive profiles of Russian regions to support managerial decisions to raise the competitiveness of regions. *Methodology*: The authors have suggested an approach to assessing the components of regions' competitive advantages that are important for the placement and development of various types of economic activity. The system of factors that fully displays the diversity of potential drivers of regional development and constitutes the metric footing of the competitive profile has been justified. A methodology for the quantitative assessment of factors that is based on the statistical data, including the composition and methods of representative indicators' assessment, as well as an approach to the alignment of heterogeneous regional indicators to make correct multifactor interregional comparison have been presented. *Findings*: The article submits competitive profiles developed on the back of the suggested methodology for 6 regions of Russia that differ in economic and geographical status and industry-specific pattern of the economy. Using these examples, the authors show the possibilities of revealing industry-specific priorities of their development through capitalization of regional assets such as favorable economic and geographical location and transport-logistic capacity, agro-climatic, water, forest, mineral and energy resources, cultural and historical heritage, capabilities of foreign economic relations' diversification, etc. *Value*: The methodological approach under suggestion can be used to justify industry-specific priorities of the development of Russian regions within the implementation of spatial development strategy.

Keywords Regional competitiveness · Competitive advantages · Spatial development strategy · Industry-specific priorities of regional development · A competitive profile of the region · Integrated assessment · System of indicators · Assessment methods

---

A. Polynev (✉) · I. Grishina
The Russian Foreign Trade Academy, The Russian Presidential Academy of National Economy and Public Administration, Moscow, Russian Federation
e-mail: apolynev@yandex.ru

I. Grishina
e-mail: grishinaiv@mail.ru

T. Kolmykova and E. V. Kharchenko (eds.), *Digital Future Economic Growth, Social Adaptation, and Technological Perspectives*, Lecture Notes in Networks and Systems 111, https://doi.org/10.1007/978-3-030-39797-5_66

JEL Classification Codes R12 · R58

66.1 Introduction

The competitiveness of Russia compared to other countries is objectively low today. Various comparisons show that Russia is significantly inferior to the leading countries of the European Union, USA, Japan, etc. by GDP per capita (taking into account purchasing power parity) and takes 40–60th positions, while interregional contrasts in Russia are much sharper than in the leading countries that in aggregate evidences the low efficiency of current national model for management of spatial development.

Under the adopted Strategy for Spatial Development of the Russian Federation [10] to overcome huge imbalances, the development of Russia should rely on multipolarity, i.e. growth centers of different levels (including regional and local ones) emerging through the capitalization of various territory development factors. Consequently, the emergence of competitive economies of all Russian regions is recognized as an important component of the foresight model to ensure the competitiveness of the Russian economy in general.

A large number of studies have been devoted to the issues of assessing the competitiveness of economic entities in a spatial context.

The theory of international competition by M. Porter [15] has become the most popular. M. Porter has offered to assess the competitive advantages of the economy of a particular state relying on the idea of a "national diamond" of competitiveness. The system consists of four main components, which being in a certain combination, strengthen or weaken competitive advantages of the country's economy. The studies carried out by A. Markusen [14], M. Klamut [13], et al. are also dedicated to regional competitiveness.

The main conclusions of a large-scale study performed at the request of European Commission by R. Martin et al. [12] in terms of 260 European regions have convinced that regions in the context of increasing globalization can raise their competitiveness on the back of their own identity, and regional strategies should be aimed at developing specific components of competitiveness to lay down own production base, as well as to eliminate bottlenecks.

The Competitiveness Index of the EU regions [16] is of considerable practical interest. It encompasses 263 European regions and estimates 79 indicators combined into 12 groups, including regional institutions; macroeconomic stability; infrastructure; health; basic education; higher education and lifelong learning; labor market efficiency; market capacity; technological availability of the regional economy; business development; innovations. In general, foreign research experience reflects an extensive interpretation of "regional competitiveness", which also involves basic competitive advantages traditionally attributed to the elements of the territory's socio-economic potential.

Theoretical and methodological issues of assessing regional competitiveness in Russia have been considered in different periods by R.I. Schnipper [11], Yu.K. Perskii [7], S.T. Vazhenin [2], I.P. Danilov [5], A.S. Barabanov [1], et al. In recent years, comparative assessments of the regional competitiveness have been developed by various research and consulting bodies and particular scientists: AV Group [6], the Council for the Study of Productive Forces (SOPS) [9], V.V. Pechatkin and V.A. Perfilov [8], S.A. Golovikhin and A.I. Valetov [3], et al. However, at the government level, there is still no approved methodology for the assessment of regional competitiveness that would become an information base for the justification of managerial decisions on the enhancement of the country's economic competitiveness.

In our opinion, the main objectives related to assessment of regional economy competitiveness within development and accomplishment of Strategy for Spatial Development is to reveal the total of significant factors regarded as advantages or hindrances of the economic development of each region, as well as to ensure targeted orientation of factor composition at displaying the competitive positions of the regions for localization and development of various types of economic activity within them, including specialized ones. The purpose of the study has been to identify the totality of specific significant factors of the competitive development in each region of the Russian Federation as the foundation of its future-oriented specialization, which ultimately allowed drawing competitive profiles of all Russian regions.

66.2 Materials and Methods

Shaping scientifically-based competitive profiles of Russian regions is based on estimation of their competitiveness in the global economy that is an ability of a regional economic complex to provide a dynamic and sustainable socio-economic development of the region under conservation its natural environment through mobilization of internal and external sources of economic growth, effective exercise of competitive advantages at the interregional, national and international markets for goods and services by industries of its specialization.

The competitive profile of Russian regions is a combination of significant (positive or negative) factors of their development (including competitive advantages and restrictions) that have a strong effect on the sustainability of the socio-economic development of these regions in the long term.

Each factor of the region's competitive profile can be assessed by the most representative indicators that are applied in the established practice of inter-regional comparisons. The basic methodological approach to comparative assessment of regions' competitive advantages is an adjustment (standardization) of regional figures of the corresponding statistical indicators, as well as expert scores by referring them to the average values of similar indicators across the Russian Federation. The competitiveness indicators of the regions are converted into dimensionless relative quantities, which regional values vary in relation to the average Russian unity (1.00) thereby fixing the excess of (or lag) the average Russian value in a region and displaying availability (unavailability) of favorable conditions for the development of certain types of economic activity.

Entering various factors in the competitive profile is dictated by the need to assess a wide range of significant competitive advantages of the regions from the perspective of developing different types of economic activity (Table 66.1).

For calculation of indicators, we have employed the data from Rosstat statistical reports for 2015–2017, other open data, as well as expert assessments made by authors on some indicators.

66.3 Findings

Using the proposed approaches, the authors have developed competitive profiles of all constituent entities of the Russian Federation. Figure 66.1 presents examples of competitive profiles of six regions with much different features of economic and geographical position and the industry-specific pattern of the economy, such as St. Petersburg, Krasnodar Territory, Nizhny Novgorod Region, Khanty-Mansi Autonomous Okrug, Novosibirsk Region and the Republic of Sakha (Yakutia), which have been demonstrated in the context of the corresponding average conditions across Russia.

Favorable competitive advantages of St. Petersburg are its extremely favorable geo-economic position, significant economic capacity, and a receptive consumer market, diversified economic pattern, hefty HR and tourism potential. This is the largest research and academic center and a center for the development of high technologies, a transport-logistic hub of international importance with a developed small business and ICT infrastructure. R&D, transport-logistic complex, machine-tool industry, oil refining, pharmaceutical and electrical engineering industries, a range of financial and agricultural services, and tourism should be top-priority directions which will make the greatest contribution to national competitiveness.

Along with good geo-economic location, the competitive advantages of the Krasnodar Territory are also the most favorable natural and climatic conditions in Russia, considerable recreational water and forest resources, a tourist-recreational infrastructure, relatively new tangible assets, a substantial domestic market potential, developed transport and ICT infrastructure, a high agricultural development. The development of tourism, recreation, and agro-industrial complexes; sea and pipeline transport; the production of construction materials, paper, and cardboard; engineering of specialized vehicles; paint-varnish, leather and footwear industries should be prioritized.

Competitive advantages of the Nizhny Novgorod Region include a rather favorable geo-economic location, a diversified economic pattern, a high level of innovative activity, urban area effects, relatively developed transport, and ICT infrastructure, the available water supply of the territory and relatively favorable natural and climatic conditions. The region can make its contribution through preferred development of metal-fabricating industries, the production of petroleum products, manufacturing

Table 66.1 The main components of the region's competitive profile and their indicators

Key factors of regional development	Main performances (indicators)
I. Geo-economic location and natural-resource potential	
1. Geo-economic location	available seaports (including non-frozen ones), international airports (hubs); neighboring location to countries with a receptive domestic market; proximity to international transport corridors and large regional markets for goods and services
2. Commodity and fuel-energy resources	share in the national reserves of critical commodity and fuel-energy resources
3. Water resources	the volume of water resources per capita
4. Forest resources	total timber reserves per capita
5. Agriculturally used lands	area of agriculturally used lands per capita
6. Agricultural development	agricultural products per capita
II. Comfortable living environment	
7. Climatic conditions	average annual temperature and range of average monthly temperatures; available/unavailable permanent frost; extent of favorable agricultural conditions
8. Environmental pollution	indicators of wastewater discharge and emissions of pollutants into the atmosphere (from stationary sources) per capita
9. Equipment of the housing facilities with services and utilities	the share of the total area equipped with a water supply system, sewerage system, heating, bathtubs (shower), hot water supply, gas (pipeline, liquefied) and floor electric stoves
III. Level of economic development and market capacity	
10. Gross Regional Product (GRP)	GRP per capita
11. Industrial development	volume of goods shipped (works performed, services provided) of industry per capita
12. Diversification of the economic pattern,	the share of manufacturing and the market services sector in GRP
13. Real incomes of the population	monetary incomes of the population per capita (regarding the level of purchasing power)
14. Consumer market development	retail trade turnover paid services (regarding the level of purchasing power) per capita
15. Foreign economic activity	export (to non-CIS and the CIS countries) per capita
IV. Investment and financial potential	
16. Own budget revenues	return of taxes, fees and other payments to the consolidated budget (regarding the level of purchasing power) per capita

(continued)

Table 66.1 (continued)

Key factors of regional development	Main performances (indicators)
17. Capital/labor ratio of the economy	the value of fixed assets (net of depreciation) per capita; the ratio of completely depreciated fixed assets
18. Private investment	the amount of private (off-budget) investment in fixed assets per capita
V. Infrastructure security	
19–20. Energy security of the economy and population	electric power generation per capita
	power plant capacity per capita
21. ICT infrastructure	scope of communication services provided to the population, the number of subscriber's mobile radiotelephone devices connected, the number of active subscribers of fixed broadband Internet access (per capita); the number of personal computers used for educational purposes and in educational institutions (per 1,000 students)
22. Transport infrastructure	the density of public highways and railway tracks (regarding population density under Engel's ratio)
23. Market infrastructure	the number of enterprises and organizations by type of activity "wholesale and retail trade, repair…", "hotels and restaurants", "financial activities", "real estate operations, lease and provision of services" (per 10 thousand people)
VI. Human potential	
24. The number of urban areas	the number and category of urban areas of over 5 million; 1–5 million people; 500 thousand–1 million people
25. Highly-skilled personnel	percentage of the employed population with higher and secondary vocational education
26. Human potential development	the value of the Human Development Index (HDI) and rating (position of the Russian Federation) in terms of HDI
VII. Innovative activity	
27. Patent activity	patents granted per 10,000 employed in the economy
28. Innovative goods and services	the scope of innovative goods (works, services) per 1 employed in the economy
VIII. Institutional capacity	
29. Investment attractiveness	position of the region in the National Investment Ranking
30. The development of small business	the turnover of small enterprises per capita

Source Drawn up by the authors

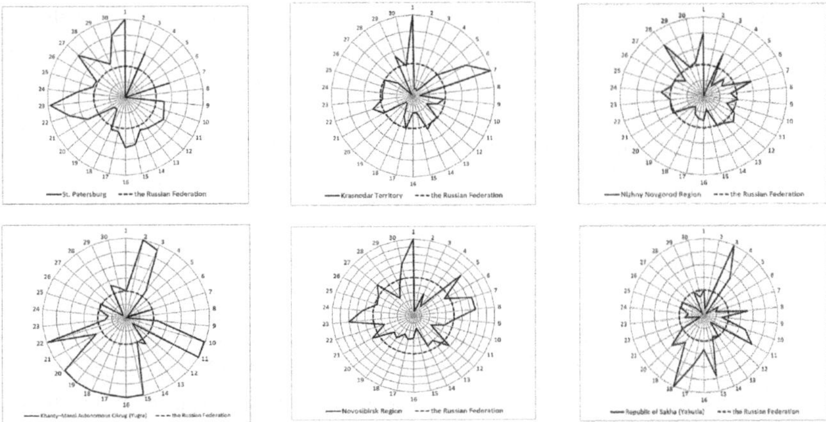

Designations: 1 – Geoeconomic location 2 – Mineral resources 3 – Water resources 4 – Timber stand 5 – Agriculturally used lands 6 – Agricultural products 7 – Natural and climatic conditions 8 – Environmental state 9 – Equipment of the housing facilities with services and utilities 10 – GRP 11 – Manufacturing 12 – Diversification of the economy 13 – Monetary income 14 – Retail trade and fee-based services 15 – Export 16 – Budget revenues 17 – Fixed assets 18 – Private investment 19 – Electric power generation 20 – Power plant capacity 21 – ICT infrastructure 22 – Transport infrastructure 23 – Market infrastructure 24 – The number of urban areas 25 – Highly-skilled personnel 26 – Human development index 27 – Patent activity 28 – Innovative goods and services 29 – National Investment Rating 30 – Small Business

Fig. 66.1 Competitive profiles of St. Petersburg, Krasnodar Territory, Nizhny Novgorod Region, Khanty–Mansi Autonomous Okrug, Novosibirsk Region, Republic of Sakha (Yakutia) (*Source* Drawn up by the authors)

of engineering constructions, development of electric power industry and transport infrastructure, R&D, and the agro-industrial complex.

The crucial competitive advantages of the Khanty-Mansi Autonomous Okrug are its richest natural-resource potential (significant reserves of oil, gas condensate, gas sulfur, substantial forest and water resources), leading positions by the main economic indicators in Russia (GRP, exports, and industrial output, private investment per capita), young labor resources, highly-skilled personnel, developed ICT infrastructure, high energy security. The development of the oil and gas complex, R&D, railway transport, woodworking, and the production of construction materials is a priority in the region.

Important competitive advantages of the Novosibirsk region are its favorable geoeconomic position on the Trans-Siberian Railway, urban area effects, high supply of water resources and agriculturally used lands, availability of multi-discipline large research and academic centers, a diversified economic pattern, developed transport and ICT infrastructure, highly developed small business, favorable environment. The main contribution of the region to national competitiveness can be reached thanks to the development of transport and financial services, metallurgical, woodworking, pulp-and-paper, pharmaceutical, agrochemical and other types of processing industries, R&D and biotechnology, tourism.

Competitive advantages in the Republic of Sakha (Yakutia) with a prominent commodity-based economy are high export and investment activity, high GRP, fiscal capacity of the region, the development of extractive industries, favorable environment, significant forest reserves, high energy and water supply of the territory, new tangible assets and labor resources, available research and academic centers. The national competitiveness of the region can be enhanced through priority-oriented development of mining production of non-ferrous metal ores, coal, diamond processing industry, energy and transport infrastructure, and R&D.

66.4 Conclusion

The system of factors under suggestion displays the system of criteria for assessing the territory's proneness to the placement of different industry-specific productions within it. Assessing the competitive advantages of each region and shaping its competitive profile is the basis for making managerial decisions on setting industry-specific priorities for the development of regional economies by representatives of federal and regional state authorities.

Setting priorities should be based, on the one hand, on the current performances of the established economic pattern of the regions, and, on the other hand, on the opportunities for targeted use of their competitive advantages, which effective exercise is associated with specific investments projects for the development of economic activity types [4]. A precise definition of breakthrough trends of industry-specific development will allow each region to make own contribution to enhance the competitiveness of Russia in the global economy.

Acknowledgements The research has been conducted in the framework of the R&D state assignment of the Russian Presidential Academy of National Economy and Public Administration. The authors express gratitude to E.V. Nesterkina for invaluable assistance in collecting data and making calculations.

List of References

1. Barabanov AS (2014) Regional competitiveness management. Institute of Socio-Economic Development of Territories of Russian Academy of Sciences. URL http://uisrussia.msu.ru/docs/nov/isedt/2014/1403589639BARABANOV.PDF. Accessed date 31 Aug 2019
2. Vazhenin ST, Zlochenko AR, Tatarkin AI (2004) Regional competitiveness situation. Reg: Econ Sociol 3:23–38
3. Golovikhin SA, Valetov AI (2015) Competitiveness of the Russian regions: rating. Basic Res 3:157–161. URL http://fundamental-research.ru/en/article/view?id=37102. Accessed date 31 Aug 2019
4. Grishina IV, Polynev AO, Shkuropat AV, Kotov AV (2018) Spatial development strategy in Russia: methodological approaches to the development of the economic block. The Russian Foreign Trade Academy, Moscow
5. Danilov IP (2007) Competitiveness of Russian regions (theoretical foundations and methodology). Kanon Plus Regional Society of Disabled People "Rehabilitation"
6. Regional Competitiveness Index AV RCI. Russia's Growth Poles. AV RCI (2016). URL https://av-group.ru/av-strategy/av-rci/. Accessed date: 20 Aug 2019

7. Perskii YuK, Kalyuzhnova N Ya, Mingaleva JA (2003) Regional competitiveness: theoretical and applied aspects. TEIS, Moscow
8. Pechatkin VV, Perfilov VA (2010) Theoretical and practical aspects of assessing the competitiveness of Russian regions. Issues Mod Econ 3:285–290
9. Polynev AO (2010) Competitive capabilities of regions: research methodology and ways to increase it. KRASAND, Moscow
10. The Strategy for Spatial Development of the Russian Federation until 2025. Decree of the government of the Russian Federation No. 207-p as of February 13, 2019 "On approval of the Strategy for the Spatial Development of the Russian Federation until 2025". URL https://government.ru/docs/35733/. Accessed date 20 May 2019
11. Schnipper RI (1991) Region: economic management methods. Novosibirsk
12. A study on the factors of regional competitiveness 2003. URL http://ec.europa.eu/regional_policy/sources/docgener/studies/pdf/3cr/competitiveness.pdf. Accessed date 20 Aug 2019
13. Klamut M (1999) Konkurencyjność regionów [Competitiveness of regions]. AE, Wroclaw
14. Markusen A (1987) Regions: economics and politics of territory. Rowman and Littlefield, Totowa, NJ
15. Porter ME (2008) On competition (updated and expanded ed.). Harvard Business School Publishing, Boston
16. The EU Regional Competitiveness Index 2016. URL http://ec.europa.eu/regional_policy/sources/docgener/work/201701_regional_competitiveness2016.pdf. Accessed date 30 Aug 2019

Chapter 67
Prospects for the Digitalization of Regional Agro-industrial Complex

Larisa Popova, Natal'ya Balashova, Vasilii Ivanov and Victor Bocharnikov

Abstract In accordance with the decrees of the President of the Russian Federation issued in May, the main goal of the agrarian policy for the medium term is to establish a highly productive export-oriented sector that would develop based on modern technologies and would be staffed with highly skilled personnel. According to the approved program measures, proactive use of digital technologies can become a huge driver for (by 3–5 times) performance of the Russian agricultural sector. To date, the share of the Russian digital economy in GDP is small. Russia ranks 15th in the world by the digitalization of the agricultural sector. Having a huge resource potential, Russia seeks to enhance its competitive positions in the agricultural market. In this regard, the authors put forward a hypothesis that the Volgograd region, being one of the leading manufacturers of agricultural products and having a huge production, resource, innovation and export potential, will be able to make a great contribution to the Russian agribusiness that will allow the Russian Federation to become a worthy competitor on the global agricultural market. To this end, the authors have identified forward-looking areas for the development of regional agribusiness in the context of digital transformation. In the course of the study, it has been found that the greatest effect from the introduction of digital tools and platforms in the agricultural sector of the region can be achieved through the formation of an industrial cluster composed of government bodies, leading enterprises, and academic institutions interested in digitalization of the industry. Valuable perspectives in this connection can be provided by the Volgograd State Agrarian University, which is capable of performing complex and knowledge-intensive development and transfer innovations at a high level, as

L. Popova (✉) · N. Balashova · V. Bocharnikov
Volgograd State Agrarian University, Volgograd, Russia
e-mail: lvpopova@bk.ru

N. Balashova
e-mail: balashova_nat@mail.ru

V. Bocharnikov
e-mail: bocharnikov_vs@mail.ru

V. Ivanov
Volgograd Region Administration, Volgograd, Russia
e-mail: mcx@volganet.ru

T. Kolmykova and E. V. Kharchenko (eds.), *Digital Future Economic Growth, Social Adaptation, and Technological Perspectives*, Lecture Notes in Networks and Systems 111, https://doi.org/10.1007/978-3-030-39797-5_67

well as training highly skilled personnel adaptive to the demands of the digital economy. Practical implications are determined by suggestions on altering the concept of regional management that assumes shifting from the process model of agribusiness to one which is an object-oriented, more stable and adaptive to transformations.

Keywords Regional economy · Agricultural economy · Digital economy · Digital tools · Management technologies in agriculture

JEL Classification Codes R11 · O13 · Q1 · Q16

67.1 Introduction

The innovative model of agricultural development in Russia is characterized by the formation of digital technological platforms, the expanded use of the knowledge-driven economy, the digital transformation of production and management systems. High-tech production is an innovative process of making competitive goods and services with a high share of added value on the target market on the back of advanced digital technologies, intellectualized instruments of labor and highly skilled labor [1].

As estimated by experts, the main constraints on the path of the digital transformation of the Russian agro-industrial complex are out-of-date technologies, unavailable specialists and teams with required competencies and skills, failure to integrate existing and new technologies, which results in dramatic transformation of the infrastructure, technology, scientific approaches, principles, methods of production arrangement [2–5].

Therefore, the digital transformation of agricultural enterprises should be based on fundamentally new high-tech approaches to production, i.e. using a multi-level matrix of target indicators and resource constraints, automation and a system of intelligent assistants to significantly reduce the time for testing and launching competitive products on the market [6].

Besides, we should note that the digital transformation of the organization of agricultural production has both pros (breakthrough digital technologies, big data analysis, unmanned vehicles, mass information services, labor robotization, etc.) and cons (dependence on borrowed technologies, the degradation of own competencies, job cuts, the elimination of particular specialties, unemployment, social tension, ambiguity in the legal field, fraud escalation, digital colonization and other negative factors arising out of the globalization of possible risks [7]).

Integrated consideration of the totality of these factors when introducing digital tools into agribusiness management will streamline state regulation of the economy by increasing its transparency and manageability, curtailing management costs and making more reasonable government decisions [8, 9]. However, according to the authors, regulatory mechanisms under consideration require independent scientific and practical conceptualization to design a digital model of regional agribusiness management.

For this purpose, by the example of the Volgograd Region the authors have explored the priorities for the establishment of a high-tech agricultural business in the region by laying conditions for higher transparency of agricultural markets, providing agricultural manufacturers with new IT business opportunities, reducing prime costs through digitization of business processes in the agro-industrial complex and transition to risk planning- and forecasting-based management that will ensure a positive economic effect and cut costs when using an integrated approach.

67.2 Materials and Method

The issues of the digital transformation of regional agriculture in this article are addressed from the perspective of examination of decrees of the President of the Russian Federation, Resolutions of the Government of the Russian Federation, information and statistical materials of federal authorities, Internet resources. The methodological background of the study has been proceedings of domestic and foreign scientists, which ensured a comprehensive and purposeful study of the problems and prospects of industry-specific digitalization.

A system approach relying on the basic methods of scientific knowledge, such as analysis and synthesis, has served a methodological background of the study. The study has been carried out with the help of a set of economic research methods, such as economic-statistical, monographic, comparative analysis, abstract-logical, computational-constructive. Depending on the problems to be solved, the authors have applied academic techniques of statistical, financial, trend data analysis, a benchmarking method, analogy, and systematization.

67.3 Findings

Nowadays, the agrarian sector of the economy in the Volgograd Region, as well as the whole Russia, is slipping into digital space and, therefore, have to form its ecosystems using digital technologies. Being one of the largest agricultural producers in the Russian Federation, the Volgograd region has a huge natural and climatic potential, which allows not only to satisfy domestic needs but also to have a rather strong effect on the formation of the food market in Russia. Volgograd region ranks 8th in the country by agricultural production output, its ratio in gross agricultural production is 2.6%. The Volgograd region has hit high and sometimes performances in particular sectors of agricultural production.

Over the past 5 years, the region has managed to dramatically transform the industry into an important sector of the region's economy. The foundation for this development was laid back in 2014 when agricultural development was defined as a priority area for the region. The Governor of the Volgograd Region A. I. Bocharov has set clear tasks on the disposal of the potential available in the agricultural sector.

Participation of science in devising the Strategy for the development of the agricultural sector of the region until 2020 played a major role in the progress. The basic provisions of the Strategy underlay the programs to be adopted and implemented [10]. The zone-based approach suggested by scientists of the agrarian university helped make government support more effective. The return on budgetary investments, in turn, allowed the region to participate in all federal programs. As a result, government support has increased.

The total amount of budget financing amounted to 22.8 billion rubles over 2014–2018. In 2014 3.5 billion rubles of government support funds were allocated to the agro-industrial complex, but in 2018 the figure was 5.9 billion rubles (1.7 times higher). The Volgograd region ranked 10th in Russia by the scope of direct government support (subsidies and grants) as of 2018 (4.2 billion rubles).

The governor's instructions on the engagement of additional funds into the region's economy through diversification of activities and growth of export supplies to foreign markets can be fulfilled through searching for new ways to raise labor productivity and reduce production costs using modern digital technologies.

Taking into account the priorities of the scientific and technological development of the agro-industrial complex of the region, the staff of the Volgograd State Agrarian University and authors commenced to develop the concept of the second stage of the Strategy for Development of the Agro-industrial Complex in the Volgograd region, the implementation of which will bring the region to a qualitatively new level thereby providing an integrated balanced solution of economic, social and environmental tasks, preserving the natural resource and historical-cultural potential of the rural settlement.

Within the national projects "International Cooperation and Export", "Digital Economy", "Small Business" and "Labor Productivity and Employment Support", scientists from the Volgograd State Agrarian University have already developed 8 project proposals for the digital transformation of the regional agro-industrial complex to be launched. In particular, the authors have taken direct participation in:

- studies on the development of digital technologies in livestock farming within the state assignment of the Ministry of Agriculture of Russia on the topic "Development of technical and economic parameters of experimental digital farms created in the basic farms of higher educational institutions";
- academic research works on the development of the Information system "Personal account of agricultural producer";
- calculation of technical and economic performances and an economic assessment of the risks of introducing digital models "Smart Field", "Smart Greenhouse" with regard to regional zoning and industry-based specialization by the example of the basic farms of the Volgograd State Agrarian University;
- joint development by the Pre-Caspian Consortium and authors of a platform for impersonal monitoring and management of transport-logistic infrastructure in the agro-industrial complex of the Volgograd region;

- joint development by the Federal Scientific Agro-Engineering Center "VIM" and authors of a distribution information and management system for command transmission to control robotic technical devices.

Within the implementation of the strategy, instructional guidelines for dispatching and aggregating data flows to create end-to-end chains from agricultural production to consumption with deep integration into related sectors of the digital economy will also be developed. Applied scientific research on the development of expert systems for control of water application rate under various irrigation methods have been carried out and digital technologies in precision farming have been planned.

It is these projects designed to scale up domestic integrated digital agro-solutions, that are interesting for investors and promising a real breakthrough in the development of the Russian agro-industrial complex since the deep integration of agribusiness with digital technologies is forward-looking from the perspective of possible competitive advantages. Thus, application of IoT technology solely as a result of the analysis of accurate and timely data [11] opens up new opportunities for enterprises for expanding the range of their services, bring them closer to their customers, sophisticating business processes [12] and differentiating their offers on the market, while ensuring a real return on investment and payback [8, 9].

The rapid introduction of digital technologies in agribusiness will require a qualitatively new level of staff competence. The staff should be able to work effectively with information flows and know modern means of establishing useful ties on the Internet, use various Internet services and tools to examine analytical reports and reviews, search for effective ways of developing new resources and markets, quickly monitor the state of the enterprise and interact online with various partners and systems for real-time control.

In framework of the Strategy for the Development of Agro-Industrial Complex of the Volgograd Region until 2030 developed today, staff of the Volgograd State Agrarian University together with the Committee on Agriculture is devising a "Comprehensive Program for the Training and Retraining of Specialists of Agricultural Enterprises" that show the demand for staff able to successfully put into practice the main development vectors. First of all, a new generation of managers of agricultural organizations who can take the level of international cooperation and export of domestic products should be trained.

In the course of the study, the authors have distributed the number of managers of agricultural organizations in the Volgograd region by the level of education and have found out that:

- about 90% of managers of large agricultural enterprises have higher education, but only 58%—agricultural (field-specific) education; 9.9%—secondary vocational education.
- 78% of managers of small agricultural enterprises have higher education (including field-specific education—57%); 17%—secondary vocational education (9.6%—field-specific education), 1.1%—basic vocational education and 3.7%—secondary education. In recent years, the category "practitioners" mentioned until 2015 has almost disappeared from statistics.

The use of digital technologies in the agro-industrial complex demands the specialists to know information systems and electronic programs that ensure the competitiveness of agricultural enterprises and their products. Currently, the regional demand for such specialists is over 80%. This especially concerns working in the State Information System for Monitoring of Agricultural Lands of the Volgograd region "AGROPORTAL" with additional modules "Economics" and "Cadastre", as well as in the State Information System of the Volgograd Region "Regional Information and Analytical System of the Volgograd Region" (RIAS) and continuously upgraded modules and versions of 1C program for accounting documentation.

This requires continuous training of industry staff with the help of an information-oriented approach, relevant progressive training methods, and modern toolkit.

Such a toolkit is engaged in training new generation specialists for digital agribusiness at Volgograd State Agrarian University. So, modern software products of over 180 domestic and foreign vendors are used in the educational process of the university.

Additional educational programs in information technology are executed at the university academies of world-renowned vendors like Microsoft, Cisco, Oracle. Given the high importance and potential of IoT technology for the agro-industrial complex, Samsung is ready to establish soon an Internet of Things Academy in the Volgograd State Agrarian University [7].

Along with HR issues of the industry development, designing a management model in the digital economy has become particularly important in the context of digitalization. Based on theoretical research findings, the authors proposed a promising algorithm for the introduction of digital tools into the practical management of regional agribusiness.

We offer the concept of objects' interaction within the target connecting network typical for the IoT technology to adapt to the management system of the regional agro-industrial complex.

The difference is that IoT exists within a single technological network and the management of the agro-industrial complex of the Volgograd region—within a single management network. Goal-setting is common both for the Internet of things, where it is impossible to establish meaningful interaction of things in multi-branch and connected Internet without a clear understanding of the strategic goal and the tactical tasks that reveal it, and for a new concept of the regional agro-industrial complex management, which should precisely define the goals of its development and functional existence.

The introduction of digital technologies in the agro-industrial complex will lead to the transformation of management concepts thanks to departure from the process business model that assumes the management of interrelated functions model, to the object one. Shifting from a functional model to a process one allows strengthening a business only in terms of business processes to achieve tactical goals, while management of business objects lets to evolve business at the level of interrelated points to execute the strategic goals of the regional agricultural sector, providing flexibility and mobility of business processes.

In this situation, enterprises of the regional agro-industrial complex will be presented as a network of interacting objects united on mutually beneficial terms into a dynamic process model under an agreed strategy to achieve a common goal. Thus, the tasks of designing and reengineering business processes are being replaced by a complicated management approach with a higher level of abstraction, which, nevertheless, is more stable and adaptive.

This fact will make it possible to develop various models of interaction between much-differing objects, for example, producers and potential customers, producers, and logistics companies, agricultural enterprises and academic institutions, competing agricultural producers. Thus, the producer can involve potential customers in designing and promoting the product; exploration of market trends and preferences by the trial-and-error method is replaced by the acquisition of big data for multi-factor intelligent analysis of market responses, and market competitors agree on issues of shared production and adoption of general industry standards.

67.4 Conclusion

The research findings allowed making a conclusion on the need for a new model to be developed to manage the regional agro-industrial complex of the Volgograd region in the context of digitalization. The success of the transition to a digital economy in the agro-industrial complex of the Volgograd Region depends on the development of its digital ecosystem based on a platform that provides the opportunity for all interested parties to work together through a single system of interconnected technologies, methods, protocols, standards, and procedures.

The industry-specific cluster established as an independent unit that unites leading enterprises and academic institutions interested in the development of the digital industry can become a cornerstone in developing an ecosystem of digital agro-industrial complex in the Volgograd region. Volgograd State Agrarian University can perform the most complex and high-tech processes of developing, operating and testing digital services, as well as procure a transfer of innovative solutions at agricultural enterprises of the Volgograd Region and digital transformation of business processes.

References

1. Amelin SV (2017) Digitalization of production as a factor of raising its efficiency. In: Proceedings of the international research and practice conference theory and practice of organizing industrial production, efficiency of organizing and managing industrial enterprises: problems and solutions. Series "theory and practice of industrial production organization (Voronezh, 14–15 Nov 2017). Voronezh State Technical University, Voronezh, pp 12–17
2. Popova LV, Lata MS (2015) Analysis of innovation risk management when implementing agricultural innovations. Scientific review. series 1: economics and law, vol 6. Economic Education, Moscow, pp 57–65

3. Kamble SS, Gunasekaran A, Parekh H, Joshi S (2019) Modeling the internet of things adoption barriers in food retail supply chains. J Retail Consum Serv 48:154–168. https://doi.org/10.1016
4. Ognivtsev SB (2019) Digitalization of the economy and economy of the agro-industrial complex digitalization. Int Agric J 2(368):77–80
5. Amelin SV, Shchetinina IV (2017) Organization of high-tech production of competitive products. FES Financ Econ Strat 12(137):25–36
6. Karabekova AB, Osmonalieva DA, Sadyraliev ZH, Urmatayim AK (2019) The main directions of digital modernization of the agro-industrial complex of a modern region. In: Studies in computational intelligence, vol 826, pp 949–955
7. Ivanov VV, Ovchinnikov AS, Kochetkova OV (2019) Conceptual basis of digital transformation of the agroindustrial complex of the Volgograd region. In: Proceedings of the Lower Volga Agrarian University Complex, vol 2, no 54, pp 18–25
8. Popova LV, Korobeinikov DA, Korobeinikova OM (2016) State regulation in the organizational-economic mechanism of agriculture. Bull Low Volga Agro-Univ Complex Sci High Educ 4(44):292–299
9. Popova LV, Lata MS, Mitrofanova IA (2016) Innovation potential of small businesses in the agro-industrial complex. Econ Yesterday, Today, Tomorrow 9:353–364
10. Ovchinnikov AS, Balashova NN, Ivanova NV (2014) Strategy for the Integrated development of rural areas and the effective performance of the agroindustrial complex of the Volgograd region in the context of the WTO. Econ Agric Proc Enterp 1:16–20
11. Choudhary S, Thomas I, Bahrami M, Sumioka M (2019) Accelerating the digital transformation of business and society through composite business ecosystems. In: Advances in intelligent systems and computing, vol 926, Matsue, Japan, pp 419–430
12. Shchutskaya AV, Afanas'eva EP, Kapustina LV (2019) Digital farming development in Russia. Regional Aspect. https://www.scopus.com

Chapter 68
Digitalization and Digital Economy: Issues of Legal Regulation and Its Prospects

Aleksei Fat'yanov, Rosalina V. Shagieva and Tat'yana Lakhtina

Abstract The relevance of issue under exploration is explained by the need for the legal regulation of the digital economy and the insufficient development of theoretical, substantive-technological and scientific-conceptual aspects of developing a base of legal concepts in terms of their application in legal arrangements. **Purpose**: The purpose of this article is to develop scientifically based and time-tested terms of digital economy in the context of a system-activity approach and their testing to establish a complex legal system that allows arranging a system of prohibitions on the processing of big data for mercenary and other sinister purposes and so on by restricting the access of various actors to these data. **Materials and Methods**: Formal-law method is a principal research method, which allows us to consider this issue as a process of deliberate and conscious acquisition of knowledge, skills, and abilities by specialists to make legal acts in the digital field only provided that the respect for rights and freedoms of man and citizen will be ensured. **Findings**: The article presents the author's concept of the digital economy, reveals legal foundations for the development of relations in a virtual environment, and offers methods of developing these relations. As it seems to the authors, it's not blockchain technology, but rather the Value Web, precisely its paradigm, which is a much more wide-scale phenomenon, will determine the development of these relations in the digital economy. Uniqueness/value: The developed theoretical principles and recommendations are aimed at adapting legislative and law-enforcement authorities to the new conditions of professional activity in the digital economy and are directed at developing a scientific and methodological framework of the legal regulation of the digital economy in legal science and practice.

A. Fat'yanov (✉)
Plekhanov Russian University of Economics, Moscow, Russia
e-mail: 561712@mail.ru

R. V. Shagieva
State University of Management, Moscow, Russia
e-mail: shagsas@mail.ru

T. Lakhtina
Moscow University of the Ministry, Moscow, Russia
e-mail: laxtin@yandex.ru

T. Kolmykova and E. V. Kharchenko (eds.), *Digital Future Economic Growth, Social Adaptation, and Technological Perspectives*, Lecture Notes in Networks and Systems 111, https://doi.org/10.1007/978-3-030-39797-5_68

Keywords Information technology · Digital economy · Digital law · Legal regulation · Protection of human rights

JEL Classification Code K 40 · M 21 · O 10

68.1 Introduction

The mankind has never been so often surprised by changes flaring as a bright star but shortly turning into commonplace for the last decades. This situation happened with the Internet, with mobile communication, digital photography and digital video, GPS and its analog GLONASS, and many others, small-scale inventions. Let us pay attention to the fact that this list encompasses only information innovations, which offer us new amenities that facilitate information exchange and acquisition of the necessary information. For example, a satellite telephone and GPS navigators allow us not to get lost in the wild places of our planet, and thereby save the lives of geologists, hunters and just tourists.

A personal computer connected to the Internet, and its simplified substitutes (tablet computers and smartphones) take a particular place among these technologies for the development of economic relations. We have been talking about the information society [1], the electronic government, electronic commerce, electronic services and even electronic money for at least two decades, but recently we have abruptly shifted to slightly different terms and replaced 'electronic' with 'digital', which indeed is more consistent with the nature of these relations. Then, some organizational and technological breakthroughs have dramatically changed a vector of development in recent times, statutorily referred to as the "digital economy".

What way can it be explained? First of all, by the development of information technologies. The most significant of them are the information and telecommunication network "Internet", its projection on a mobile communication network ("broadband Internet") and a smartphone as a mobile means to receive, transmit and process digital information.

As this combination develops, each of us has the opportunity not only to communicate from any part of the world covered by a mobile communication network but also to transmit text, graphic and video information, check e-mail, use many apps, including those important for economic relations (for instance, apps to manage your bank accounts) [2], enter into contractual relationships with carriers [3], address the public authorities and get legally relevant information, as well as public services (the State Services portal). The list of such services is dramatically expanding, and their management is getting easier [4]. All of the above is a material basis for the development of the digital economy.

68.2 Materials and Methods

Formal-law method is a principal research method, which allows us to consider this issue as a process of deliberate and conscious acquisition of knowledge, skills and abilities by specialists to make legal acts in the digital field only provided that the respect for rights and freedoms of man and citizen will be ensured [5]. This is due to the fact that there is no clear understanding of what the digital economy should be. New technological solutions, as a rule, generate a new system of terms. Some of them easily become part of the language, while the rest is used by a small number of specialists.

It is also known that legal science has a specific attitude to the concepts engaged in the legal vocabulary and statutory matter, i.e. it aims to unambiguity of definitions. Terms in the digital economy are emerging now and the purpose of this paper is to explore some new terms concerning their applicability in legal arrangements.

68.3 Findings

Let's start with the very concept of the digital economy. It is defined in the Strategy for the Development of the Information Society in the Russian Federation for 2017–2030 approved by Decree of the President of the Russian Federation No. 203 as of May 09, 2017 as follows:

> The digital economy is an economic activity with the digital data as a key factor of production, the processing of large array and using the analysis results of which can considerably increase the efficiency of various types of production, technologies, equipment, storage, sale, delivery of goods and services against traditional forms of management.

It follows from the analysis of this definition that the main method of production within the digital economy model is the evaluation of large data volumes. Such an analysis can be carried out only when the information is presented in the so-called machine-readable form, that is, in a sequence of binary numbers, that is called the digital representation of data. All modern electronic devices for storage and transmission of information ultimately present it in this sequence, regardless of the input (image, sound or text). The Strategy also defines the category "processing large data volumes", which means "combination of approaches, tools, and methods of the automated processing of structured and unstructured information coming from numerous information sources, including scattered or loosely-coupled, in volumes that cannot be processed manually during a reasonable time".

Many researchers believe that findings of the Big Data analysis open up huge opportunities for enhancement of economic relations, also by identifying and filling new niches for the production and provision of services [6]. However, such an analysis will inevitably affect most people, since their behavior, movement, and fulfillment of desires will become the object of continuous observation and system analysis, which will narrow the frames of their private life to a great extent.

It seems to us that the above definition of the "digital economy" is incomplete. At least two global-scale technologies called the Internet of Things and the Industrial Internet have the greatest impact on new economic relations. The first technology is defined in the Strategy as "the concept of a computer network connecting things (physical objects) that are equipped with built-in information technologies for interaction with each other or with the external environment without human intervention." It should be noted that something in this direction is already being implemented, for example, fully automated fire extinguishing systems, but these services will surely go further and enter every home to release us from needless troubles. And this is a huge potential for the development of the production of new things and services.

The industrial internet is defined in the Strategy as the concept of establishing information and communication infrastructures by connecting industrial devices, equipment, transducers, sensors, and industrial process control systems to the information and telecommunication network "Internet", as well as "the integration of this software-hardware with each other avoiding human intervention." Honestly, automated processes were introduced into industrial production more than a century ago. Obviously here the question is fully automated completed cycles; analysis of need, production, testing, packaging, transportation, stock-keeping, delivery to the consumer are fully automated and controlled by computer systems. A man acts as an actor who sets a common objective and as an end-user [7]. Any system with human intervention will be hardly competitive with an automated system. Except for, a man can analyze the needs more creatively.

Let us also point out that the above definition of the "digital economy" is not the only right and far from the only one. So, for example, "Introduction to the Digital Economy" [8] provides for another two definitions of this concept:

1. A digital (electronic) economy is an economy existing in a hybrid world.
2. A digital (electronic) economy is an economy a specific feature of which is the highest possible satisfaction of the needs of all its participants through the use of information, including personal one. This can be reached thanks to the development of information-communication and financial technologies, as well as available infrastructure, which together provide the opportunity for full-fledged interaction in the hybrid world of all participants of economic activity, i.e. subjects and objects of developing, distributing, exchanging and consuming goods and services.

We can give many definitions, and all of them will have something but turn out to be flawed from the standpoint of trying to cover all aspects of emerging economic relations. Several hundred categories of the economy have been defined in science over the centuries. The same situation will be with the digital economy. Nevertheless, let us try to define this concept on our own. We suppose that the ***digital economy*** *is an economic activity aimed at satisfying the needs for physical and non-physical assets through the use of mainly information and computer technologies, as well as making a profit from this activity.*

Nowadays any product can be produced in quantities that fully satisfy the needs of all who can pay it. We can state that this was an outcome of the industrial development of the economy. It seems that the next stage is the rationalization of production and consumption by more accurate analysis of the demands for certain goods and services, filling in the economic gaps with new supplies, and engaging in the economic activity the segments of the world's population that do not have real access to banking services but could become micro-investors, financing new economic initiatives (start-ups) under schemes different from rather slow and costly bank lending [9], various programs and grants [10].

Is there an alternative option to such a model for the development of economic relations? It's apparently not existing. But we can move ahead only if the rights and freedoms of a man and a citizen are fully ensured [11].

The widespread dissemination of mobile communication and broadband Internet among the population of our planet has allowed reaching a higher level of communication between people when interest groups being created on the Internet and new media celebrities appear that could not have previously become official broadcasting participants.

And these people gain high profits from posting ads in their social accounts. This phenomenon changes the familiar concept of value as mostly some kind of financial asset. The number of followers and some other factors become the value. The changing paradigm of values and non-computable capabilities of their exchange thanks to the above technologies make it possible to talk about the emergence of a new generation of the Internet as a social phenomenon, namely "Value Web". This phenomenon is also a part of the digital economy phenomenon.

Let us also point out that negative consequences inevitably arise along with the positive consequences of the qualitative development of the Web. One of them is the so-called Darknet, where people sell and buy weapons, explosives and drugs through cryptocurrency, place orders for killing people, distribute child pornography. In general, there are no limitations imposed to ensure adequate development of public relations [12]. The power of Darknet is still unclear, but it is estimated that it takes a large niche on the Internet and is not controlled by states. Moreover, this is also some kind of economy with a substantial turnover of physical and financial resources.

68.4 Conclusion

Broadening the sense of 'value' in the context of the 'Value Web' emergence led to the need to define some new concepts that attracted the attention of the authors. The first one is a 'token' translated from English as a sign, symbol, identification mark, badge [13].

A token can be called a special coin for access to the subway or a chip received in exchange for money in a casino. In non-obscene language, this concept is most often associated with ensuring information security. For example, "Information Security Concept of the Federal Tax Service", approved by order of the Federal Tax Service of

Russia No. MMB-7-4/6@ as of January 13, 2012, establishes that "encryption keys should be downloaded from the workstation of a tax authority IS user using portable data storage devices (token, smart card)." The Methodological Recommendations for Preparing and Holding the Unified State Exam at examination stations in 2017 at Educational Institutions located outside the territory of the Russian Federation" provides a mechanism for issuing cryptographic information protection means (tokens) to members of State Examination Commission.

Now let's look at the definition of this term in current and promising regulations. One of the few acts of the EAEU member states devoted to the development of the digital economy is the Decree of the President of the Republic of Belarus No. 8 as of December 21, 2017 "On the Development of the Digital Economy". In this act, the term under study is defined as "a record in the blockchain ledger or another distributed information system proving that owner of digital sign (token) has titles to civil rights' objects and (or) being a cryptocurrency."

Thus, the Belarusian law-maker interprets token as a sign acknowledging property or non-property rights or a certain number of cryptocurrency units. But, the definition of "token" should have a direct regulatory relationship with blockchain anyway. The logic of this definition suggests that it should not be used out of this relationship.

Draft law No. 419059-7 "On digital financial assets" (hereinafter referred to as "Digital Assets Draft Law") introduced to the State Duma of the Federal Assembly of the Russian Federation in 2017 has defined 'token' in a slightly different way. "It's a type of a digital asset that is issued by a legal entity or an individual entrepreneur to raise funds and is recorded in the register of digital transactions." In the opinion of this draft lawmaker, there are only two digital financial assets (cryptocurrency and token). Moreover, both have a single system feature, i.e. electronic property created using cryptographic tools.

The authors are skeptical about the fact that any property can be created in electronic form. It seems that the idea of the draft law contradicts the scientific understanding of the physical world structure.

The interpretation of token provided for in the draft law on digital assets generates a direct analogy with "share" and "bond" common to civil law. Both types of civil rights' objects in legal realities of today are mostly uncertificated securities, but actually a record in the official register about the person's title to several securities of certain face value. Replacing the official register "with a distributed ledger of digital transactions" (blockchain), we get a token. Let us recall that securities pursuant to Part 2, Art. 142 of the Civil Code of the Russian Federation are a share, a bill of exchange, a bill of sale, an investment unit of unit investment fund, a bill of lading, a bond, a check and other securities called in such a quality in the law or deemed such in the procedure established by law. All of the above indicated civil rights' objects are jointly deemed financial assets.

A comparative analysis of the Russian and Belarusian statutory approaches to 'token' shows that the Belarusian law-maker defines it verification of the owner's right to any civil rights' objects, which under the civil law include things, money and securities, other property, including property rights; works and services; undisclosed information; exclusive rights to the results of intellectual activity and means of identification by parties to civil relations, goods, works or services; intangible benefits.

As we see there a lot of objects and only some of them can be easily, figuratively speaking, "tokenized". In view of the digital economy needs, the approach of the Russian draft lawmakers seems more forward-looking to the author.

However, the draft law on digital assets considered is not the only one in the emerging legal field of the digital economy. The same year, a group of deputies of the State Duma and members of the Federation Council introduced into the Lower Chamber of the Parliament a draft law No. 424632-7 "On Amendments to Parts 1–2, 4 of the Civil Code of the Russian Federation" (hereinafter referred to as "Draft law on digital rights") to legitimize the concept "digital rights". The statutory approach to define is the following:

> In the cases provided by law, titles to objects of civil rights, except for intangible benefits, may be certified by data library (digital code or designation) existing in the information system that meets the statutory parameters of a decentralized information system. The system provides a person who has unique access to this digital code or designation with the opportunity to familiarize himself with the description of the corresponding object of civil rights anytime. The specified digital code or designation is deemed digital right.

We will try to explore this pattern of legal provisions. First of all, one should pay attention that it is purely general and subject to further specification in legislative acts for the following reasons. The first one is that the digital rights arise only in events provided for by law. The second one is that digital rights arise to any object of civil rights, except for intangible benefits. The third reason is that the key technological premise for the emergence of digital rights is their certification by data library (digital code or designation) and only in the information system that meets the statutory parameters of a decentralized information system. The fourth reason is that information technologies and technical means of a decentralized information system should provide a person with the opportunity to familiarize themselves with the description of the corresponding object of civil rights anytime. The fifth reason is that access provided to a digital code or designation must be unique. And the last reason is that digital rights become a kind of civil rights due to the amendments relating to objects of civil rights made to the Civil Code of the Russian Federation.

Let us answer the question "Are digital rights really a new and comprehensive term and can it qualitatively certify the availability and transfer of civil property rights?"

Trying to find an answer to this question, let address to the working rules of keeping the register of securities owners established by the Federal Law No. 39-ФЗ as of April 22, 1996 "On the Securities Market", Article 8 hereof contains the following regulations:

> The collection, recording, processing, storage of data constituting the register of securities owners, and the provision of information from the register of securities owners are deemed activity on keeping the above register.
>
> Only legal entities are entitled to be engaged in keeping the register of securities owners.
>
> The register of securities owners is a from-time-to-time system of records on persons who open personal accounts, securities recorded on these accounts, encumbrances on securities

> and other records under the legislation of the Russian Federation. The register holder shall carry out its activities under federal laws, regulations of the Bank of Russia, as well as the rules of register keeping to be approved by register holders. The requirements for these rules are established by the Bank of Russia.

And what then, all of the above indicated will be thwarted in the event of the emergence of "digital rights" and all these functions will be performed exclusively by the so-called "miner," i.e. a volunteer involved in ensuring the blockchain operation for some reward in cryptocurrency? In other words, proceeding from the need to execute public interest, the country will discontinue intervention in these relations. There is a big doubt because it can lead to chaos and abuses [14].

The next question is that "if the register is centralized, a citizen or legal entity has an ordinary title to securities, and if the register is decentralized, then it turns into "digital right" in a fabulous way." This is strange and inconsequent.

The definition of "digital rights" under a detailed analysis is very similar to a digital sign (token) pursuant to the Decree of the President of the Republic of Belarus No. 8 and much different from the definition of token given in the draft law on digital assets.

What will be in the event of adopting both draft laws as the federal ones? Moreover, the title to a digital financial asset will be one thing, and digital rights will be different. We have been looking after similar discrepancies between civil, tax and budget legislation for a long time. It is useless.

Unfortunately, the current edition of the draft law on digital rights has a few more flaws. Trying to find the most common definitions, its developers use categories such as "digital code or designation" and "unique access to a digital code or designation". It immediately catches an eye that digital code is extremely broad and can certify anything.

For example, a product's article and its bar code are a digital code; digital designations are most widely used in budget legislation when reporting material value movement, and so on.

Consequently, this kind of digital code should be distinguished from other categories, which has not yet been done. Meanwhile, it is clear that the question should be about cryptographic keys, which are also a sequence of characters, but not a digital code.

What is meant by 'unique access to a digital code or designation'? Most likely, it should be the possession of a private cryptographic key, since, as far as we know transactions in blockchain are confirmed by ciphers from the class of asymmetric cryptosystems having simultaneously a private (secret) key (which is always unique) and open key which is in mathematical relationship with a private one. But will all law-enforcers be able to draw such an analogy? It is unlikely because it is right but very complicated since it requires additional expertise in this area.

General recommendations:

1. Further development of relations in a virtual environment strongly needs a designation of virtual value to be elaborated, therefore the category of token apparently

can be applied in the regulatory framework. However, it should not be interpreted too broadly, let it be in the "niche" that is close to the familiar share or bond.
2. Cryptocurrencies in one or another form will inevitably be used within the digital economy, as far as this technology in the best possible way solves the problem of rapid exchange of values on the Web, including the exchange of values when introducing industrial Internet and IoT technologies. Given the above said, the regulation of these relations should be developed right now.
3. Despite that blockchain technology is promising for many relations arising in the economy and the public area (for example, in the election process and in arranging referenda, in the authorization system), but cannot be considered as something extraordinary, actually replacing the whole philosophy of digital economy. 'Value Web' seems to be a much large-scale phenomenon, and it is its paradigm that will determine the development of these relations by the authors' opinion.
4. Law as a regulatory system has been developing for a thousand years. Many legal arrangements, particularly in private law, have been transforming during this period.

Therefore, before to be including "new legal entities" into the current statutory base, they should be assessed at least from two scientific points: (a) whether they have similarities with existing legal arrangements and whether substitute them; (b) whether they conflict with the current legal rules and established doctrinal attitudes and principles. Therefore, before making them mandatory behavior rules, one should prove their apparent novelty and consistency with the operating law. Only in this event, they will be perceived as the development of the latter.

References

1. Levina EY, Masalimova AR, Kryukova NI, Grebennikov VV, Marchuk NN, Shirev DA, Renglikh KA, Shagieva RV Structure and content of e-learning information environment based on geo-information technologies. EURASIA J Math Sci Technol Educ 13(8):5020–5031
2. Sandalova V (2019) The modern state of the banking secrecy institution in the context of banking services' digitalization. In: Popkova E (ed) Ubiquitous computing and the internet of things: prerequisites for the development of ICT. Studies in computational intelligence. Springer Science + Business Media, Cham, vol 826, pp 85–93
3. Inshakova AO (2018) Law response to high tech challenges in innovative Russia. Pravovaya Paradigma 17(4):6–15
4. Inshakova E, Ryzhenkov A, Inshakova A (2019) Neo-industrialization of the Russian Economy: technological and digital development. In: Popkova E (ed) Ubiquitous computing and the internet of things: prerequisites for the development of ICT/studies in computational intelligence, vol 826. Springer Science + Business Media, Cham, pp 239–250
5. Bukalerova L, Muratkhanova M, Ostroushko A, Simonova M (2019) Protection of minors in the digital economy of the Russian Federation and the republic of Kazakhstan. Bulletin of St. Petersburg University, Pravo, no 1, PP 149–165
6. Weigend A (2018) Big data. All technology in one book. Bombora, 384 pp
7. Shitov A (2018) Will we live till the moment when money to be cancelled: what does the revolution of financial technology offer us. ARB: point of view. https://arb.ru/b2b/pointofview/dozhivem_li_my_do_otmeny_deneg_chto_nam_sulit_revolyutsiya_v_finansovykh_tekhn1-1

8. Kosheleva AV (2017) Introduction to the digital economy. VNIIGeosistem, 28 pp
9. Khasanov AK, Ostroushko AV (2019) US experience in combating insurance fraud. Leg Stud (5):29–35
10. Ng AW, Kwok BK (2017) The Emergence of fintech and cybersecurity in a global financial center: strategic approach by a regulator. J Financ Regul Complian 25(4):422–434
11. Bukalerova LA, Amandykova SK, Ostroushko AV, Avakova OV, Eliseev MA (2019) Criminal punishment of person for failure to report under russian law. Opcion 35(89):680–695
12. Bartlett D (2017) Underground internet. The dark side of the world wide web. EXMO, 352 pp
13. Goncharov AI, Goncharova MV (2019) Digital tokens in the toolkit of modern foreign trade activity of economic agents of BRICS jurisdictions. Pravovaya Paradigma 18(3):31–42
14. Ostroushko AV (2019) Adoption of foreign practice of the legal regulation of financial technologies in Russian law. Leg Stud 8

Chapter 69
Legal and Technical Forms of Developing Digital Law-Enforcement in Modern Russia: Problems and Prospects

Ksenia Shubenkova and Gennady G. Egorov

Abstract *Purpose*: The article from the perspective of an analysis of the legal and technical features of development discusses the prospects for the introduction of digital relations in civil relations. *Design/methodology/approach*: the methodological background of the study is the dialectical, system-structural, structural-functional research methods. *Findings*: the authors determine the need to distinguish between social processes and the legal forms of securing such processes, that are based on them, in the context of digitalization and attention to the peculiarities of expressing daily social relations that arose under the terms studied. Modern lawmaking and law enforcement cannot effectively fulfill its tasks without accepted and legally commented regulators aimed at determining some peculiarities common to social relations communicating with the digital environment. The state should take into account the conditions of ensuring the most favorable legal regime that develops and implements new digital technologies, as well as means for the effective use of human activity in the economic areas. The above noted creates the need for legal reasoning of the criteria of social and state demand for existing prohibitions and analysis of public law restrictions intended to reduce the risk of adverse effects both in the economic and social areas. This paper, from the perspective of analysis of international experience and development trends of this area in the country, addresses the main problems associated with the use of digital systems in civil relations. *Uniqueness/value*: the authors conclude on the importance of favorable legal regime for the emergence and development of modern technologies and their reasonable use in economic activity, the need to lift current bans, but at the same time imposition of certain public-law restrictions to prevent the adverse effects of economic digitalization and protect bona fide parties to emerging relations. The paper justifies the need to establish restrictions on the rights to cryptocurrency use, rather than bans on cryptocurrency use.

K. Shubenkova (✉) · G. G. Egorov
Volzhsky Branch of Volgograd State University, Volzhsky, Russia
e-mail: shubenkova34@yandex.ru

G. G. Egorov
e-mail: egorov@vgi.volsu.ru

T. Kolmykova and E. V. Kharchenko (eds.), *Digital Future Economic Growth, Social Adaptation, and Technological Perspectives*, Lecture Notes in Networks and Systems 111, https://doi.org/10.1007/978-3-030-39797-5_69

Keywords Digital law-enforcement · Information technology · Prospects for the development of digital relations · Issues of digitalization of civil relations · Cryptocurrency

JEL Classification Codes K 100

69.1 Introduction

Modern society is inconceivable without continuously accessible digital systems involved in the most significant areas of both law and social life. The first conscious activity of children initially faces with a variety of digital systems. The interaction of citizens in most relationships is based on network systems operating on modern computer developments.

These are the simplest devices, as well as high-level network systems, which are very close to the most familiar machine-technical indicator "artificial intelligence" by their level of development. As various authors have repeatedly noted, the development of digital technologies, as well as the widespread use of technological intermediaries (the so-called digital platforms), which allow uniting a significant number of parties making a huge number of transactions on various goods' markets through the exchange of information, suggest a fundamentally new understanding of social-economic processes [1]. In 2015, when analyzing the economic environment, D. Medvedev uses the term "new normal" to define these processes, which can be transcriptionally translated as "new normality" or "new reality" [2]. Moreover, such a reality is interpreted differently in the legal literature.

It seems reasonable that for understanding the peculiarities of the legal forms of some real social processes we should first take these real social processes as an object of study and then show the forms they can be performed.

Conditions for digitalization of real social processes is one of the possible forms of their execution. The legal form is also a condition (form) of executing real social processes, including ones that are performed in the context of digitalization of these social processes.

As for the economy, it is a complex of relevant processes (production, consumption, exchange, and distribution) that are not relating to digitalization and formalizing relations through appropriate, traditional paper forms. The legal forms of such relations, as a rule, include transactions, administrative acts, and other legal facts, legal relations in different sectors, acts of execution of such legal relations, etc.

69.2 Materials and Methods

The authors of this article have employed the dialectic, system-structural, and structural-functional research methods.

69.3 Findings

The modern digital economy often employs various electronic accounting systems united into databases, exchange systems, automated systems for the management of relevant information required to effect deals on electronic trading platforms. Social relations that have been previously made in the paper are now often digitalized. Besides, studies of the legal forms of such relations deserve particular attention. As a matter of fact, they are not much different from previous non-digital relations that are expressed in classical legal terms, which does not rule out but often mean representation of the basic peculiarities that can arise and be executed in real sectors of the modern economy [3].

Keeping in mind the aforesaid, we can state that lawyer's task is to determine the peculiarities of the legal forms of the digitally-organized society and regard its economic component, i.e. directly develop, and subsequently interpret and establish the ways to apply legal rules aimed at governing relations on digital information, and on the basis of which the legal relations under study may arise, be concluded and terminated. This means that a lawyer should solve the following problems [1]:

- determine the peculiarities of the legal status of the parties to legal relations mediating digital relations (users, operators, etc.);
- find out the peculiarities of the grounds for legal relations mediating digital relations to arise, to be altered and terminated (getting an access code to an information system, license agreements, etc.);
- establish the features of the legal regime of the parties to legal relations mediating digital ones (digital assets);
- reveal the specific aspects of the scope of legal relations mediating digital ones (rights and obligations of parties to digital relations, compliance with the terms of use of the digital system).

While identifying the ways to fix the problems highlighted above, we can state the major practical prospects. Let's address them in more detail. One should distinguish special legal status of the parties to digital relations: (a) their major status as parties to real social relations; for example, the sale/purchase relations, the legal form of which is the purchase and sale obligation (seller, buyer); (b) the peculiarities of their major status determined by a digital form of the purchase-sale relations (seller as a user of a digital information system and buyer as a user of the same digital information system). The use of a digital information system involves participation in the fulfillment of public relations of an intermediary, i.e. the operator of this system. Respectively, along with the rights and obligations of the parties that make up the scope of the major legal relations, additional rights and obligations of the parties caused by the digital form of the execution of the major legal relations arise, as well as the rights and obligations of the digital platform operator who provides services to these parties.

Specific aspects of the analysis of the grounds relating to the emergence, alteration, and termination of legal relations mediating digital ones. One should distinguish between: (a) traditional ways to express a will by person that are aimed at the emergence, alteration, and termination of real public relations, for example, purchase and sale obligations, which legal features are defined by law and an oral or written purchase/sale agreement; (b) peculiarities of will expression caused by a digital form of relations (getting an access code to an information system, transmitting information (offer, acceptance, proposal to amend or terminate an agreement in electronic-digital form), etc.).

It has been repeatedly stressed in the legal literature that the status of digitally made transactions and smart contracts should be legally defined [4], the concept of which is not reduced to the electronic contract conclusion [5]. This interpretation is controversial as far as some authors remark that in such relations there is not a new type (kind) of transaction, but just another form of its presentation (digital), i.e. we are talking about several features related to the procedure for conclusion, amendment, and termination of legal relations using digital systems (technology). It is the issues above that will be solved by the amendments made to Art. 160, 309, 432 of the Civil Code of the Russian Federation, Federal Law No. 34-ФЗ as of March 18, 2019 "Amending Parts 1–2 and Article 1124, Part 3 of the Civil Code of the Russian Federation" that are related to the peculiarities of various transaction-making ways, including various electronic and technical means (i.e. directly using modern digital technology).

At the same time, some scientists expressed the opinion that "the actual seller of goods via Internet sites is artificial intelligence effecting transactions and acting under technologically defined algorithms on behalf of the business agent" [6]. This statement is questionable both from a legal and technical point of view. It misleads one party as to another party to the transaction (is it a person, a developer, the copyright holder, or a computer complex?). We agree with the point that there is no reason to endow the machine (the object of law) with the will and the corresponding legal personality, regarding it as an agent acting on behalf of an economic agent (copyright holder of the digital complex).

As the theory of law has repeatedly highlighted, the procedure for making transactions in electronic-digital form is not a new phenomenon for the Russian economy and law. Similar transactions have been made for a long time at on-exchange tenders (stock exchanges) in electronic form by virtue of the current rules of the Federal Law No. 325-ФЗ as of November 21, 2011 "Concerning On-exchange Trading".

Regulating the legal status of objects of legal relations directly connected to the digital environment, one should distinguish [7]: (a) an object of real public relations, for example, a thing and money as objects of a sale/purchase relations which legal regime is determined by law and purchase/sale agreement (legal regime of their possession, use, disposal), (b) specific features of the legal regime of these objects caused by the digital form of the sale/purchase relations (digital asset as digital information on the object of real public relations).

Considering the status of these objects, the question on the liquid, monetary component of the relations associated with these objects has been repeatedly raised.

In fact, we are talking about digital assets. According to the Instruction of the President of the Russian Federation dated October 10, 2017, the Government of the Russian Federation and the Bank of Russia must ensure the amendments determining the status of digital technologies used in the financial sector and their concepts, including cryptocurrency, to be made into national legislation, proceeding from the fact that the ruble is the only means of payment in the Russian Federation. The Ministry of Finance of Russia has drafted the corresponding Draft Federal Law "On Digital Financial Assets" (drafted by the Ministry of Finance of the Russian Federation) (not introduced to the State Duma of the Federal Assembly of the Russian Federation, last revised on February 16, 2018), which gives the definition of cryptocurrency and the more general associated concept of digital financial asset. According to the terminology used herein, digital financial assets mean electronically represented property represented that is created with the help of available cryptographic means. Therewith, the title to such a property is confirmed by entering digital records in the register of digital transactions. Under this approach, various cryptocurrencies and tokens can be attributed to digital financial assets. From the legal point of view, such digital financial assets are not recognized as a legal payment instrument on the territory of the Russian Federation (Par. 2, Cl. 2 of the Draft Law). Along with that, cryptocurrency acts as a type of digital asset that is developed and accounted for in the distributed registry of digital transactions by the participants of this register under the rules of keeping digital transactions register (Par. 12, Cl. 2 of the Draft Law).

Exploring the approach suggested to define the legal regime of a digital asset, let's address the foreign experience of resolving these issues in national law. So, for example, the Japan Payment Services Act provides for the use of cryptocurrencies as a means of payment. Besides, Japanese legislation defines cryptocurrency as a digital abstraction with property value, which can act as a means of payment in all civil-law relations, the so-called payment for goods, rent, services, as well as the transfer of their digital analogs for virtual use. Thus, under Japanese law, cryptocurrency is understood not only as abstract digital property but also as a means of payment. There had been noted in the literature that reference to the functional importance of cryptocurrency allowed creating a favorable business environment in Japan for the development of the cryptocurrency market and engagement of investors [5]. Proceeding from the analysis of the draft law in Russia, we can state only the introduction of definitions which have not yet been elaborated at the proper theoretical level. Thus, for example, the concept of digital financial assets in the draft law has two different interpretations. On the one hand, as an electronic object-asset, and on the other hand, as a financial instrument. Therewith, reference of cryptocurrency to financial assets is a controversial issue, especially keeping in mind the ban on their use as a legal means of payment on the territory of the Russian Federation (Par. 2, Cl. 2 of the Draft Law). When determining the specific features of the scope of legal relations underlying digital ones in the theory of law, one should distinguish [8]: (a) public relations that have a tangible form of expression (the acquisition/transfer of things, money), the legal regime of which is determined by law; (b) relations exercised without tangible component, i.e. related to the acquisition and transfer of rights/obligations arising from the digital relations of rights transfer, which are complicated by the requirements to comply with the terms of digital systems use.

In the course of the discussion of the draft law "Amending Parts 1–2, 4 of the Civil Code of the Russian Federation" regarding digital rights, attention has been repeatedly drawn to the ambiguity of digital rights for a reason: whether it is an object of law, or it is an element of scope of the legal relations (legal right). It has been suggested to interpret digital law as follows: "… digital code or designation, a collection of electronic data is as a digital right belonging to a person registered in a decentralized information system. Disposing of this digital right, the holder exercises his respective civil right through information technology and technical means" [9]. The main point is to avoid confusing the object and the scope of civil law relations. It is no coincidence that in the explanatory note to the draft law digital rights are put in quotation marks and called fiction close to the essence of the security. But as for securities, there is a book-entry and uncertificated form of securities established by law and certifying obligations and other rights, which can be exercised or transferred only upon submission of such documents (book-entry securities), or the same rights set forth in security issue resolution or other act of issuing person in accordance with the requirements of the law that can be exercised and transferred in the event of compliance with the rules for regard of these rights under Art. 149 of the Civil Code of the Russian Federation (uncertificated securities) [1].

69.4 Conclusion

Summing up the research of this problem, we can agree with the statement of several authors on the need to distinguish real social processes and their legal forms, as well as these real social processes in the context of digitalization and peculiarities of legal forms of real social processes in the above conditions.

Modern lawmaking and law enforcement cannot effectively fulfill its tasks without accepted and legally commented regulators aimed at determining some peculiarities common to social relations communicating with the digital environment.

Forging conditions that ensure the most favorable legal regime for development and implementation of new digital technologies, as well as means for the effective in economic areas of human activity, require to reason social and state demand for existing prohibitions, and an analysis of public-law restrictions intended to reduce the risk of adverse effects both in the economic and social areas. The establishment of a favorable legal regime for the emergence and development of modern technologies, their reasonable use in economic activity is impossible without lifting current bans, but at the same time, imposing certain public law restrictions to prevent the adverse effects of economic digitalization and protect bona fide parties to emerging relations [10].

The statutory definition of the legal regime of cryptocurrencies based on the principles of lawful and socially acceptable use of this instrument should be commensurate with the departure from the ban on the use of cryptocurrency as a payment means.

Thus, we consider urgent to restrict the rights on the use of cryptocurrency (not the use of cryptocurrency) also as a means of payment and specificate the grounds and limits of such restrictions, which will open up opportunities for the effective use of new digital technologies for the successful development of the Russian economy in general, as mentioned in the Instruction of the President of the Russian Federation dated October 10, 2017.

References

1. Popondopulo VF (2019) Legal forms of digital relations. Yurist (Lawyer) 6:29–36
2. Medvedev DA (2015) New reality: Russia and global challenges. Econ Iss (10):8
3. Egorov GG (2018) Legal and regulatory peculiarities of implementing digital technologies in justice system of Russia. In: Mironov YI, Egorov GG (eds) Environment, economics, law: future outlook: a collection of articles from the proceedings of the research and practice conference held as part of the science week, April 2018, Publishing House of VolSU, Volgograd, 2018, pp 145–153
4. Vajpan V (2017) Foundations of the legal regulation of the digital economy. Law Econ (11):8
5. Sannikova LV, Kharitonova YuS (2018) Digital assets and technologies: some legal issues of developing a conceptual framework. Law Digit Econ 1:26–29
6. Laptev VA, Lapteva AV (2018) The methodology of developing an idea of entrepreneurship within transition to a digital economy in Russia. Bulletin of the University named after O.E. Kutafin (Moscow State Law Academy) (3):192–193
7. Tarakanov VV, Inshakova AO, Dolinskaya VV (2019) Information society, digital economy, and law. In: Popkova E (ed) Ubiquitous computing and the internet of things: prerequisites for the development of ICT/studies in computational intelligence. Springer Science + Business Media, Cham, vol 826, pp 3–15
8. Shubenkova KV (2011) On legality and legal order in a law-governed state. In: Protection of human rights by branches of Russian law: proceedings of the 4th interuniversity research and practice conference of students and young scientists, 12 Dec 2011, Volgograd, pp 90–93
9. Andreev VK (2018) On the concept of digital rights and their transferability. J Bus Corp Law 2018(2):39
10. Makarchuk NV (2018) Public law restrictions on the use of cryptocurrencies as a way to minimize the possible risks of economic digitalization. Law Digit Econ 1:21–24

Part IV
Ecological Costs of the Digital Future: Economic and Legal Foundations of Solving the Problem of Climate Change

Part IV
Ecological Costs of the Digital Future: Economic and Legal Foundations of Solving the Problem of Climate Change

Chapter 70
Reporting in the Field of Sustainable Development as a Tool of Digitalization of the Economy

Victor P. Kuznetsov, Elena P. Kozlova, Ekaterina P. Garina, Elena V. Romanovskaya and Natalia S. Andryashina

Abstract In the conditions of innovative development industrial enterprises are the central figure in the sphere of the latest developments of the intangible sphere, while their activity becomes the most significant in the process of material support of society. At the present stage of society development, one of the main aspects is the sustainable development of industry, which, in turn, will entail technical and technological progress in the economy. The purpose of this study is to analyze domestic enterprises for publication of non-financial statements, as well as to establish the reasons for which organizations prepare reports in the field of sustainable development. The methodology of this study includes empirical (observation, comparison, description), theoretical and cognitive (formalization, ascent from abstract to concrete) as well as general analysis and synthesis methods. As a result of the work, the main reasons for the publication of information were identified, as well as the factors that stimulate companies to form non-financial reporting in the field of sustainable development. A transition of most economic processes to a digital system, in our view, will expand opportunities and reduce the need to regulate the configuration of the provision of non-financial reporting entities.

V. P. Kuznetsov (✉) · E. P. Kozlova · E. P. Garina · E. V. Romanovskaya · N. S. Andryashina
Minin Nizhny Novgorod State Pedagogical University, Nizhny Novgorod, Russia
e-mail: kuzneczov-vp@mail.ru

E. P. Kozlova
e-mail: elka-a89@mail.ru

E. P. Garina
e-mail: e.p.garina@mail.ru

E. V. Romanovskaya
e-mail: alenarom@list.ru

N. S. Andryashina
e-mail: natali_andr@bk.ru

T. Kolmykova and E. V. Kharchenko (eds.), *Digital Future Economic Growth, Social Adaptation, and Technological Perspectives*, Lecture Notes in Networks and Systems 111, https://doi.org/10.1007/978-3-030-39797-5_70

Keywords Sustainable development · Reporting · Enterprise · Digitalization of economy · Industry

JEL Code L6 · M5

70.1 Introduction

Industrial production is of great importance in the basis of material security of society in conditions of innovative growth, as well as is a central figure in the field of new developments, produced by the non-material sphere. In modern conditions one of the most important aspects is the sustainable development of industry, which, in turn, will entail technical and technological progress of the economy as a whole. The choice of the path of development in the direction of sustainability has already been defined by a greater number of foreign companies. Russian organizations also consider this development as one of the priorities. The challenge of reducing negative environmental impacts and assisting in the development of local associations has been limited to a relatively small number of enterprises reporting on sustainable development. Regulation of social and environmental impact increased operational productivity, and the logical use of natural resources can focus on sustainability issues. One of the characteristic features, when companies cooperate with shareholders, investors, consumers, employees, and other stakeholders, is the provision of additional information.

Publication of non-financial statements allows to solve this problem. In Russia and around the world, quite a large number of companies make voluntary reporting. The Sustainable Development Report is a possible option for such reporting in the vast majority of countries. Information on the management of sustainable development of industry should be provided in an open format, published in such a way that it would be accessible to all objects engaged in production, be public to users both inside and outside an enterprise.

Reporting will allow you to choose directions for further actions not only for management but also for employees, partners, and investors. Detailed information and provision of it to the relevant divisions of the enterprise will allow to form certain proposals in the areas of activity in accordance with the goals of development and potential opportunities.

Researchers and practitioners, therefore, believe that one of the most relevant trends, based on the results of scientific and technological modification and determining modern economic dynamics, is the publicity of non-financial reporting as one of the tools for digitalization of the industrial economy [1, 2].

70.2 Methodology

The focus on a certain group of users should be not only on economic but also on environmental and social performance of companies. The list of persons interested in this information can be wide, so the composition of users of the reporting can be diverse and different interests and willingness to understand the received data. At the same time, there is the possibility of an appearance of parties interested in the results of the organization, an appearance of which does not depend on the desire of the company.

Therefore, non-financial reporting on sustainable development should not be provided only in the template. A balance should be maintained among all interested groups of persons while applying all possible options for providing economic, social and environmental data in the reporting process, not forgetting generally accepted principles.

Most foreign companies have already developed a strategy and methods of interaction, in this regard it is most often recommended to use already tested principles. The most popular at the international level were the principles enshrined in the following documents:

- Guidance on sustainable development developed by the Global reporting initiative (hereinafter GRI) [3];
- Account ability 1000, hereinafter AA 1000, developed by the British Institute for Social and Ethical Accountability;
- Principles of the UN Global Compact;
- IPIECA/API voluntary reporting on sustainable development in the oil and gas sector (2016);
- Basic performance indicators of the Russian Union of Industrialists and Entrepreneurs (2008).

If we talk about the Russian Federation, the standard of the Chamber of Commerce and Industry of the Russian Federation "Social reporting of enterprises and organizations registered in the Russian Federation" is most often used. It was compiled on the basis of the key principles of previously designated documents, as well as on the basis of the instructions of the state and other interested parties to business in Russia.

According to the Russian Union of Industrialists and Entrepreneurs on July 17, 2019, 176 organizations were included in the national register of non-financial reports. At the same time, there were 954 registered reports published since 2000. The reports included: environmental (EE)—83, social (SO)—327, sustainable development (ESD) reports—325, integrated reports—192, industry reports—31 (Table 70.1).

Table 70.1 Distribution of reports by industry (July 17, 2019)

Industry companies	Number companies	Number of reports				
		IO	ESD	SO	EE	Total
Oil and gas	21	6	114	9	41	166
Energy	42	96	54	45	7	191
Metallurgical and mining	20	17	46	55	3	115
Manufacture of machinery and equipment	2	12	0	0	0	12
Chemical, petrochemical, and perfumery	12	43	18	20	6	83
Woodworking, pulp and paper	5	0	2	4	18	24
Manufacture of food and other consumer goods	10	0	29	23	0	52
Telecommunications	11	7	14	22	0	40
Finance and insurance	18	7	24	68	0	97
Housing and communal services	5	2	0	14	1	17
Cement production and construction	2	0	1	0	1	2
Education and health	6	0	1	12	0	13
Transport	5	1	5	12	6	24
Retail	2	1	3	0	0	4
Other services	4	0	7	11	0	18
Non-profit organizations	7	0	7	32	0	39
Total	171	192	325	327	83	927
Industry reports	4	0	1	26	0	27
Total	175	192	327	353	83	**954**

70.3 Results

The reasons why organizations prepare sustainable development reporting are also different. It is difficult for representatives of medium and small businesses to find such information, mainly it is engaged in the largest organizations. The expectations of stakeholders regarding information openness and the desire to gain competitive advantages significantly affect their choice (Fig. 70.1).

The collection of information on the results of the company's activities, which were not previously recorded, allows for the expansion of non-financial reporting in the field of sustainable development. This information enables the company to take actions to reduce the exploitation of natural resources, increase the efficiency and modify economic activity, which allows increasing transparency, including financial and economic results. Sustainability reporting also helps the enterprise to protect

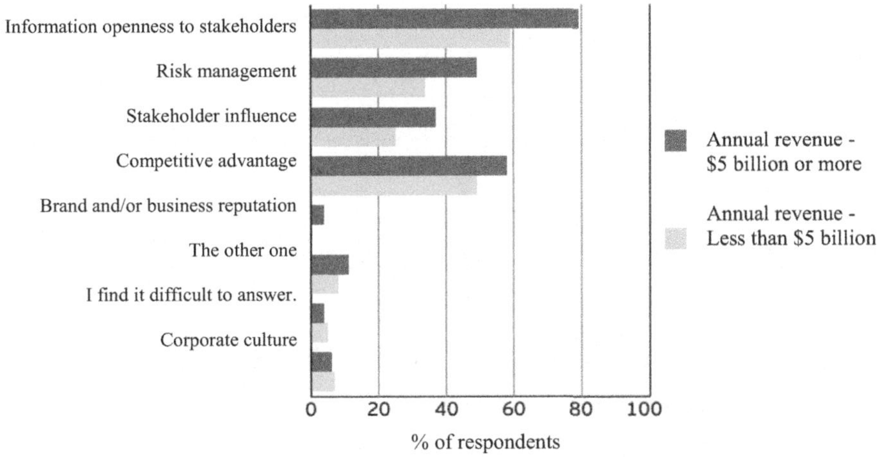

Fig. 70.1 Reasons for the preparation of statements depending on the size of the company, data for 2018 [9]

itself from economic, environmental and social risks that can significantly affect financial performance activities or shortens them. This direction of reporting will lead to an increase in the company's value, increase its social importance and financial sustainability, as well as reduce the negative impact on the environment [4, 5].

From our point of view, it is necessary to identify the factors that stimulate business to report on sustainable development. Most companies operate in the field of sustainable development but do not conduct public reports on its results. In general, reporting enterprises have a higher level of their reputation for sustainable development than non-reporting organizations (Fig. 70.2).

In the view of this study, improving reputation is an important element in the preparation of sustainable development reports and does not become the main reason for reporting.

We found that information openness to stakeholders is the main reason why companies decide to disclose information on environmental, social action and corporate governance (Fig. 70.2). The decision to prepare reports is influenced by a variety of internal and external factors:

- Agency analysis of sustainable development information [6];
- A desire of business executives, shareholders and investors to ensure that sustainable development risk management is in the right direction;
- The willingness of many communities to know how the organization manages the economic, environmental and social impacts of its activities;
- Normative and legal documents regulating the economic, environmental and social consequences of the enterprise;
- The desire for existing and potential employees to see information about the life of the organization in this field.

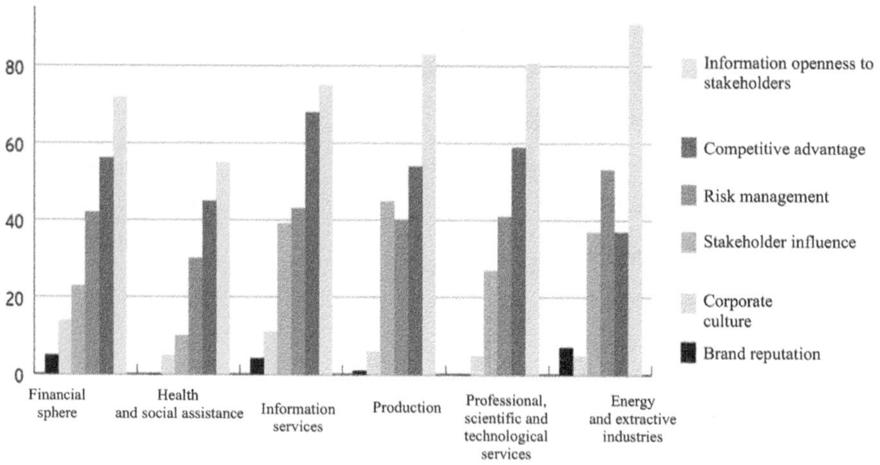

Fig. 70.2 Reasons for which companies prepare reports in the field of sustainable development, data for 2018 [9]

To generate a report, it is necessary to have a set of consecutive steps, a scheme of actions leading to the desired result. The method of formation of non-financial reporting in the field of sustainable development of the enterprise belongs to the category of management processes and is implemented in stages (Fig. 70.3).

The first stage is of a general nature, that is, it is necessary to specify the subject and object of management within the framework of the concept of sustainable enterprise development, which ensures economic and social development, not causing damage to the environment.

In the second stage, a strategy aimed at sustainable development of the organization should be developed, using the criteria and approaches set out in the international recommendations for public reporting on sustainable development.

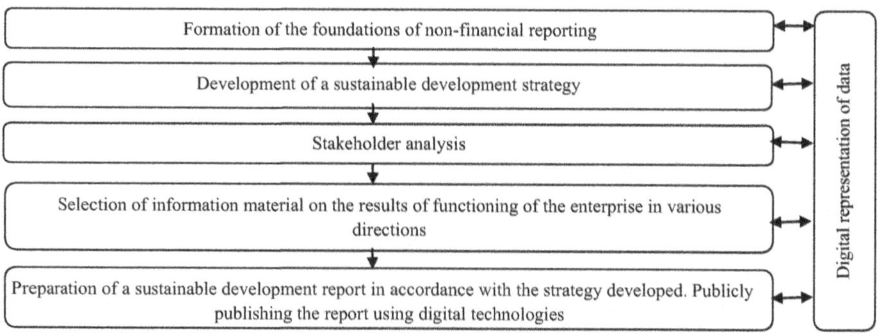

Fig. 70.3 Methodology of formation of non-financial statements

In the third stage, the range of stakeholders should be analyzed. Since non-financial statements are public, it is worth identifying the potential persons for whom the reporting will be most interesting, based on this information, should be selected the required array of data.

As part of the fourth stage, there is a selection of information material on the results of the enterprise's functioning in various directions.

The final fifth phase includes the preparation of a sustainable development report in accordance with the developed strategy. This report should be developed on the basis of Balanced Counting Chart (BSC) methodology and technology and international recommendations for open reporting on sustainable development, and disseminated among all participants in the process and be accessible to both external and internal users. Also, at this stage, it is necessary to define in what format and on which sites this report will be posted.

70.4 Conclusion

The list of types, the content of reporting at the stage of its emergence and throughout the transformation period gradually becomes an element of the events existing in society, and the forms of public provision depend on the degree of development of relations, technical and economic development of the territories and the needs of users.

For a long time, the format of reporting in writing remained the preferred format. Only with the advent and development of the informatization process, information became increasingly available in electronic form. Reliability of the information offered in it at different stages of development of society was guaranteed in different ways. The institution of an audit or state structures acted as guarantor.

A modern type of publication of non-financial statements of economic entities allows to interact with certain categories of users [7, 8]. At the same time, confirmation of its authenticity was decided through the development of an electronic signature. Today, the format of the presentation of financial statements is clearly defined. There are basic forms of such reporting, which reporting companies can modify only by detailing material information or excluding those elements for which the organization data are not available. If we talk about non-financial reporting, domestic and international standards do not regulate its format, but at the same time, the principles and rules according to which the reporting data should be are provided, define though implicit, but still format. The transition of most economic processes to a digital system, in our view, will expand opportunities and reduce the need to regulate the configuration of the provision of non-financial reporting entities.

Digitalization of the economy already at the present stage of development of society has the strongest potential, allowing, in its implementation, to achieve both companies and leading countries to strengthen their positions in basic areas of social and economic development. This transformation is based on a fundamentally new type of information and telecommunication technologies, covering and transforming all spheres of modern production and social life, although and is in the process of formation.

References

1. Garina EP, Kuznetsov VP, Romanovskaya EV, Andryashina NS, Efremova AD (2018) Research and generalization of design practice of industrial product development (by the example of the domestic automotive industry). Qual Access Success 19(S2):135–140
2. Mizikovsky IE, Druzhilovskaya TY, Druzhilovskaya ES, Garina EP, Romanovskaya EV (2018) Accounting for costs and expenses: problems of theory and practice. Adv Intell Syst Comput 622:152–162
3. Shaoul J, Stafford A, Stapleton P (2012) Accountability and corporate governance of public private partnerships. Crit Perspect Account 23(3):213–229
4. Kuznetsov VP, Romanovskaya EV, Egorova AO, Andryashina NS, Kozlova EP (2018) Approaches to developing a new product in the car building industry. Adv Intell Syst Comput 622:494–501
5. Kuznetsov VP, Garina EP, Romanovskaya EV, Kuznetsova SN, Andryashina NS (2018) Organizational design and rationalization of production systems of a machine-building enterprise (by the example of the contract assembly workshop). Espacios 39(1)
6. Campbell-Verduyn M (2015) Sustainability and the renewed semi-professionalism of leading Anglo-American financial services firms. Compet Change 19(5):355–373
7. Garina EP, Garin AP, Vinogradova SA, Batsyna YaV (2019) Management of investment programs at the stage of their implementation. Financ Econ (3):464–467
8. Garina EP, Kuznetsov VP, Garin AP, Romanovskaya EV, Andryashina NS (2019) Increasing the complexity of the product manufacture: assessment of trends and prospects of the process. Lecture notes in networks and systems, vol 57, pp 748–756
9. Chaldaeva LA, Krasikova AS, Fedchin IN (2015) Non-financial reporting in Russia: problems of formation and development. Finance and Credit. No. 13 (637)

Chapter 71
Innovative Development of the Municipal System for the Control of Municipal Solid Waste Management Processes

Yury A. Tsypkin, Roman A. Kamaev, Tatiana V. Bliznukova, Serhij L. Pakulin and Inessa S. Feklistova

Abstract The goal of the authors' research was to develop measures for the innovative development of the municipal system for the control of municipal solid waste management processes. The authors of the paper have used the systemic approach to substantiate the measures. In the course of research, the authors have identified the key features and specifics of the municipal solid waste management system. State of the art review of the theory and practice of municipal solid waste management in Russia and statistical data allowed the researchers to highlight the key features and special aspects that determine the need for the transition of municipalities from local solutions to innovative mechanisms for managing this area. Relying on the analysis of the world and domestic theory and practice, the authors of the paper have identified the areas of innovative development of mechanisms for the management of operations of municipal solid waste management companies at the municipal level. In addition, they have substantiated that further research studies on the creation and implementation of controlling centers of the holding company based on the concept of joint use centers have potential. The use of this concept will make it possible to adapt controlling methods and tools currently existing in domestic and international practice, to apply them in the field of municipal solid waste management, including the introduction of elements that take into account specific features and special aspects of the Russian market of municipal solid waste.

Y. A. Tsypkin (✉) · T. V. Bliznukova · S. L. Pakulin
Moscow State University of Land Management, Moscow, Russia
e-mail: tsypkin@valnet.ru

S. L. Pakulin
e-mail: stizapet@gmail.com

R. A. Kamaev
Moscow University of Finance and Law, Moscow, Russia
e-mail: RKamaev@fadm.gov.ru

I. S. Feklistova
The Russian Presidential Academy of National Economy and Public Administration, Oryol, Russia
e-mail: Inessa_pakulina@mail.ru

T. Kolmykova and E. V. Kharchenko (eds.), *Digital Future Economic Growth, Social Adaptation, and Technological Perspectives*, Lecture Notes in Networks and Systems 111, https://doi.org/10.1007/978-3-030-39797-5_71

Keywords Municipal solid waste · Management · Region · Efficiency · Holding

JEL Classification O18 · Q28 · Q53 · Q56 · R11

71.1 Introduction

Development of the digital economy, social and economic development of cities results in the increase of public production, more qualitative satisfaction of consumers' needs, and the increased amount of municipal waste. After the world conference on environment (Rio de Janeiro 1992), elaboration of strategic development programs started in all countries, where the most important constituent part consists in waste management problems at the national, regional and municipal levels [1]. The importance of developing theoretical and methodological principles of strategic management of complex systems stipulates the need for the development of conceptual and applied aspects of the scientific challenge of administration in the field of municipal solid waste (MSW) management [2], ensuring the transition of municipalities from conventional to modern, innovative management mechanisms that allow rebuilding the organization of management, make more efficient use of available resources, attract investment and improve the safety of residents [3]. The urgent character and complex nature of the scientific challenge under consideration have formed the basis for the selection of the topic of our research.

The goal of research was to develop measures for the innovative development of the municipal system for the control of municipal solid waste management processes. In order to achieve this goal, the following primary scientific goals were set: justify the priorities in the field of MSW management; identify the main features and specifics of the municipal system for the control of municipal solid waste management processes; identify the areas of development of mechanisms for managing operations of MSW management companies of the municipality; give scientific credence to the primary goals of authorities associated with the innovative development of the municipal system for the control of municipal solid waste management processes; identify the main principles of formation of strategic plans and programs of municipal enterprises engaged in collection, removal and disposal of municipal waste of population.

71.2 Methodology

The authors have used various methodological tools, including calculation and analytical method, economic and statistical method, and other research methods. A systemic approach was used to justify recommendations for the innovative development of the municipal system for the control of municipal solid waste management processes.

71.3 Results

The beginning of the 21st century was marked by the rampant development of industrial potential, which inevitably led to an increase in the use of raw materials and energy resources that must be transported thousands of kilometers away, to produce mass-produced goods for long and short-term use, and put waste that is generated during production and consumption, on the scrapheaps. The industrial growth results in urbanization, creates consumption cult, significantly increases the consumption of products by households and the amount of MSW [4].

Our analysis indicates the global nature of the waste management issue. Today, it can be argued that a tendency towards the development of innovative management mechanisms that provide both more efficient commercial use of waste and reduction of its adverse environmental impact, has developed in the world [5].

The complexity of the situation in the field of management of MSW generated within the territory of municipalities, requires development new financial and economic mechanisms along with improvement of management technologies and implementation of organizational arrangements which allow elaborating a methodological approach to the development of mechanisms for the innovative management of operations of MSW management companies, which is to identify and provide a scientific rationale for the development directions and the tasks of implementing these development directions. This conforms to the provisions of which are formulated in the concept of reforming of the housing and utilities sector of Russia.

In the course of research, the authors revealed key features and specifics of the MSW management system (Fig. 71.1).

Activities aimed at development of innovative municipal solid waste management mechanisms should be organized in the following areas: strengthening of control of municipal authorities over collection, removal and disposal of waste; adoption of new amendments to existing legislative acts; creation of a system for the information and methodological support of the waste management system. In order to improve the efficiency of waste management activity, funds that are received from potential investors and due to the introduction of additional charge for the uncontrolled pollution entities within the territory of municipalities, can be used. An individual source can be a disposal charge, a collection of products introduced for manufacturers and importers to promote eco-friendly recycling of packaging and out-of-use products.

In our opinion, the economic mechanism for the management of operations of municipal solid waste management companies at the municipal level, should include: the formation of a municipal division that would be responsible for the monitoring and control of the waste management system; development and application of a unified pricing policy; attracting investors for the production of recycled products.

In the context of development of the digital economy, the innovative management of operations of municipal solid waste management companies is impossible without the creation of databanks; the criteria for the segmentation of information in them may be as follows: waste generation sources; waste-tracking system; collection and

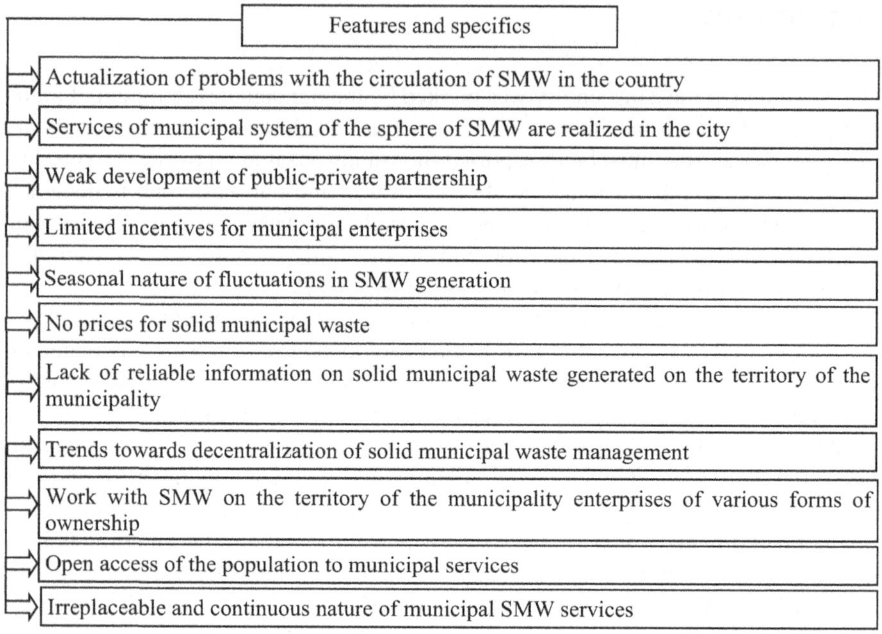

Fig. 71.1 Main features and specificity of municipal system of solid municipal waste (SMW) sphere

disposal systems; technologies and equipment for recycling; city's needs for recoverable resources and materials made of waste; ecological state of municipalities etc. Which are necessary for the development and adoption of management and business decisions, formation and development of the relevant service market.

State of the art review of the theory and practice of municipal solid waste management in Russia, on the one hand, and statistical data, on the other hand, made it possible to highlight the key features and special aspects that determine the need for the transition of municipalities from local solutions to innovative mechanisms for managing this area.

The existing environmental protection mechanisms are complex and call for various financial, scientific, technical and organizational means. The following economic tools are used in Russia:

- development of plans and allocation of financial means for environmental protection measures;
- providing a legislative framework for the limits on the use of natural resources, amount of emissions and discharges of pollutants, according to the waste dumping volume;
- setting tax and other allowances for enterprises, which implement low-waste and resource-saving technologies [2];

- introduction of insurance of cases of ecological and natural disasters, accidents and manmade disasters.

In our opinion, one of the most effective leverage was that environmental charges are levied in the form of taxes and fees associated with environmental pollution.

The following types of payments for environmental pollution are available: for air emissions, for wastewater discharge, for waste dumping.

Economic and social development of municipalities leads to an increase in public production, to better satisfaction of consumer needs and to the increased amount of municipal waste.

After the world conference on environment, most developed economies create the so-called sustainable development strategies, where the highest value is placed on the issues of nature preservation and environmental protection, and disposal of industrial, municipal solid waste and its problems are central. The following priorities have been defined almost in all of these strategies, including those that are developed at the municipal levels:

- reducing the amount of hazardous substances in generated waste;
- reducing the amount of waste;
- establishment of companies for disposal, recycling and reuse of waste.

In case of overall improvements of territories of municipalities, the following tasks should be solved:

- regular cleaning of streets, roads, and driveways,
- removal of snow in wintertime,
- removal of snow other seasonal garbage,
- elimination of unauthorized landfills,
- collection, removal and disposal of municipal waste.

The wide use of both global and domestic experience, as well as available scientific potential in the field of municipal solid waste management will provide an opportunity to improve the level and quality of the management of collection, removal and disposal of waste.

Relying on the analysis of the world and domestic theory and practice, we have identified the areas of innovative development of mechanisms for the management of operations of municipal solid waste management companies at the municipal level:

- introduction of safe and advanced resource-saving technologies into the municipal economy;
- motivation to create environmental programs and implement them at production works;
- retraining of manufacturing enterprises of the municipality that dangerously pollute the environment [6, 7];
- expansion and modernization of facilities for burial, disposal and recycling of waste;
- improvement of the waste management system.

The abovementioned areas of innovative development of mechanisms for the management of operations of companies will make it possible to improve the efficiency of the circulation process in the field of MSW management.

The consolidation of the efforts of municipal authorities, residents and business entities to develop the municipal solid waste management mechanism is an overarching challenge of our time [6, 7]. The policy of local authorities in this area is based on the use of authoritative powers of the municipality, and the right in this construct is a barrier showing what can be done and what cannot be done within the limits of the law.

Municipal policy is influenced not only by legal standards, but also such factors as: national and regional policy in this area; activity of enterprises and organizations of various forms of ownership; interests of the urban population.

Garbage accumulates and problems grow like a snowball, having a negative impact on the attitude to municipal authorities.

In order to establish the innovative MSW management system, a significant time expenditure and availability of skilled personnel is required. But the result will be significant, especially if a strategy for development of the municipality is developed [8].

The efficient solution of problems in the field of MSW management makes it possible to fulfill the primary goal of municipal politics—to improve the standards and the quality of life of population. The science-backed indicators of satisfaction of material and spiritual needs and interests of the population are the criteria of real situation of the level and the quality of life. The task of municipal authorities is to create a comfortable living environment within the territory and to manage the provision of relevant municipal services.

A complex multifactorial nature of MSW management needs a systemic approach to the resolution of arising problems, forces local self-governing authorities to use innovative management methods, which will not be able to ensure achievement of main targets and goals of development of the municipality.

Based on the analysis of domestic and foreign executive experience in this area, we have formulated the primary goals of municipal authorities in the field of municipal solid waste management:

1. Establishment of an innovative MSW management system.
2. Development of strategic and current MSW management plans and programs within the territory of the municipality.
3. Formation of the system for the monitoring of strategic and current plans and programs.
4. Improvement of efficiency of information support of the MSW management system.
5. Establishing a controlling center in the field of MSW management.

In order to create an effective system for the management of municipal solid waste management companies in the municipality, it is necessary to:

- combine strategic and current plans and programs for the development and operation of this area;

- form and coordinate goals aimed at the development of this area throughout all levels of municipality management with all stakeholders;
- develop a unified accounting and reporting planning system which allows assessing efficiency of provision of services to residents, and provide authorities and population of the municipality with planning and factual information;
- develop uniform methods and tools: determination of key indicators, development of a unified approach to the use of financial and material resources, assessment of management efficiency, assessment of municipal plans and programs etc.

The main principles of formation of municipal strategic plans and programs for collection, removal and disposal of municipal waste of population can be formulated as follows:

1. Compliance principle. It is necessary to integrate strategic plans for the social and economic development of the region and municipality [9]. The key areas of regional and municipal politics in the solution of problems of collection, removal and disposal of municipal waste must serve as a basis for the creation of plans and programs.
2. Accounting principle. Plans and programs cannot be developed without regard to special features of the existing system for the control of MSW management processes. This is not an autonomous process. Consideration must necessarily be given to plans for social and economic development of the municipality.
3. Interrelation principle. Innovative system of management of processes of collection, removal and disposal of MSW must be associated with municipal and regional plans by key indicators and terms for their achievement.
4. Interest balancing principle. The entire urban community is interested in efficient work in the field of MSW management. Efficient implementation of tasks set is impossible without active interaction with the population of the municipality, non-governmental organizations, enterprises and other entities of municipal economy and society.

Development of the strategic plan should be a special project. This project cannot be created without the necessary resources, management, definition of goals and methods for measuring results. At this stage, and essential act is the creation of universally recognized concepts of strategic planning in MSW management, such as: strategic planning system, strategic planning, participants, processes and functional strategic planning systems, strategic planning activities and documents.

Strategic plan is focused on the main areas of development of the municipality that were formed during the analysis of this area. At the same time, the points of focus are on the means of strategic programs that determine the phased plan of action aimed at implementing the primary goals and main problems associated with collection, removal and disposal of MSW [8].

The adoption of the strategy helps to create a long-term plan for the development of this area of the municipal economy, consisting of specific programs and measures implementing the areas included in the strategy.

In the context of development of the digital economy, the formation of indicators of the strategic plan requires the use of modern tools: the use of modern methods (structural and functional analysis, trend analysis, forecasting, simulation modeling, expert assessment method, optimization of economic and mathematical methods, etc.); the analysis of changes in indicators and development trends of the city; combination of different methods [10].

Strategic plans are implemented through the fulfillment of current plans and programs, ensuring the achievement of projected figures.

In the formation and implementation of plans and programs, there is a need for the feedback from population of the municipality, since the population is the main customer and consumer of MSW management services. In the context of development of the digital economy, monitoring can provide efficient feedback.

Monitoring and assessment are based on measuring performance indicators for achieving actual performance against projected performance in a certain period of time, which allows:

- amending targets and goals in case of deviation from the plan;
- assessing the efficiency and the effect of the performance of municipal programs, projects and activities;
- inform population about the progress of fulfilment of strategic and current plans and programs.

Monitoring consists of the three constituent parts: environmental monitoring; monitoring of the overall implementation of the strategic plan; monitoring of pf the implementation of current plans and programs.

The formation and implementation of current plans and programs cannot do without the informational support of the innovative management system.

The level of efficiency of the innovative management system can be assessed through residents' satisfaction indicators, and the assessment requires specific indicators.

The list of indicators is being compiled along with the formation of the municipal development strategy of the area of collection, removal and disposal of MSW. Development of strategic and tactical plans is based on the regulatory framework; therefore, the innovative management system must include indicators evaluating regulatory support of plans and programs. In the context of development of the digital economy, the system for the management of collection, removal and disposal of MSW should regulate: interaction with service users; the procedure for informing and accepting applicants; the procedure for the provision of services; interaction with other information systems; list of services; list of documents necessary and sufficient to receive a particular service; duties and fees and the procedure for their payment; appeal procedure etc.

In the context of development of the digital economy, information is an important part of the process of monitoring of the current situation and managerial decision making; therefore, creation of the information database and its constant update are extremely helpful for the chief officers of authorities.

In order to find a viable solution to the above problems that ensure the implementation of areas of development of innovative mechanisms, new forms of management arrangements in this area are required.

In order to achieve efficient implementation of directions, providing the development of innovative mechanisms for the management of operations of MSW management companies, we have identified the following primary goals:

- creation of an effective system for the management of collection, removal and disposal of municipal solid waste of population;
- development of strategic and operating plans and programs for collection, removal and disposal of municipal solid waste;
- creation of a system for monitoring of strategic and operating plans and programs;
- improvement of the effectiveness of information support of the municipal system for the management of collection, removal and disposal of municipal solid waste;
- establishing a controlling center.

The studies of organizational frameworks engaged in the management of collection, removal and disposal of municipal solid waste allowed us to find out there are no established organizational frameworks for entrepreneurial activity; therefore, additional research and development works are needed. The tendencies for the consolidation of companies providing services of municipal solid waste management significantly affect organization of management as well as competition in this services sector.

There is a strengthening of market institutions In the municipal economy, and in the context of a low level of their internationalization, competitive, hierarchical (holding) structures have competitive advantage, since organizations which form part of the holding, for the most part, cannot independently change the principles, functions and methods of functioning: identify missions, shape goals, perform functions and tasks that are generally different from the entire holding system. Under these conditions, there is a need for development of methodological approaches to interaction in the general structure of the holding and methodological provisions in the coordination of internal business structures of the holding with stakeholders. First of all, it interaction with municipal authorities is required.

In the context of the current trend towards the integration of organizations of this area, the holding must have an innovative management system, ensuring the solution of problems associated with prompt managerial decision making, cost improvement, implementation of new information technologies, monitoring of plans and programs, organizational and methodological support for management etc.

The controlling center is established as the basis for the methodological and instrumental implementation of the basic management functions, and also provides a tool for assessing the state of the holding to justify operating and strategic managerial decisions.

The need to create a Center is due to: the instability of the external and internal environment, which always requires an operational response to all kinds of changes; the need for assessment of the impact of external and internal environment through the use of monitoring; setting strategic goals and maintaining current management processes to ensure achievement of goals facing the holding.

In the context of development of the digital economy, to solve these problems, it is necessary to: develop the information database for the support of managerial decisions; provide strategic and operating planning with necessary tools; develop a system of monitoring performance indicators of the holding; develop an effective system of coordination of operations of the holding.

Controlling must provide the provision of information and satisfaction of request from the holding managers in it. The functions of information support are fulfilled in the Center and in the holding.

Controlling ensures the provision of information to the holding to measure performance. For these purposes, it is necessary to create an information system that would ensure the provision of adequate and transparent information for managerial decision making [11]. At the same time, special features of competencies of controllers are that they must have good knowledge of the state-of-the-art information technologies and their software and have the necessary knowledge and skills to build integrated information systems.

In the context of development of the digital economy, the controlling system in the innovative management of operations of municipal solid waste management companies in the municipality should be oriented towards the implementation of environmental matters, in particular:

- consideration of indicators aimed at achievement of environmental goals;
- consideration of environmental components in the development of methods for the assessment of capital investment projects;
- consolidation of economic and environmental reporting;
- use of key environmental performance indicator in the incentive system of the holding etc.

Consideration of environmental components must provide the observance of environmental standards and help the management of the holding in the achievement of strategic goals.

71.4 Conclusions/Recommendations

In the course of research, we have identified key features and specifics of the MSW management system: actualization of MSW management problems within particular territories; services in the MSW management system are provided within the territory; poor development of public and private partnership; insufficient provision of incentives to public utility companies; seasonal nature of fluctuations of MSW generation; absence of prices on MSW; lack of reliable information about MSW generated within the territory of the municipality; decentralization trends in MSW management; the companies of various types of ownership work with MSW of the municipality; open access of the population to municipal services; indispensable and uninterrupted nature of services in the field of MSW. Waste generation trends have been studied, features and specifics of the management of operations of MSW management companies have been identified and highlighted, and a methodological approach to the development of mechanisms for the innovative management of operations of MSW management companies has been put forward, which is to identify and provide a scientific rationale for the areas of development and the tasks of implementing these areas of development. This is fully consistent with the provisions of which are presented in the concept of reforming of the housing and utilities sector.

71.5 Practical Implications

In order to achieve efficient implementation of directions ensuring the innovative development of mechanisms for the municipal management of operations of MSW management companies, the following primary goals have been identified: creation of an effective system for the management of collection, removal and disposal of municipal solid waste; development of strategic and operating plans and programs for collection, removal and disposal of municipal solid waste; creation of a system for monitoring of strategic and operating plans and programs; improvement of the information support for the system for the management of collection, removal and disposal of municipal waste; establishment of a municipal controlling center. Further research studies on the creation and implementation of controlling centers of the holding based on the concept of joint use centers have potential. The use of this concept will make it possible to adapt controlling methods and tools currently existing in domestic and international practice, to apply them in the field of municipal solid waste management, including the introduction of elements that take into account specific features and special aspects of the Russian market of municipal solid waste.

References

1. Bliznukova TV, Pakulin SL, Tsypkin YA, Feklistova IS (2018) Creation of an effective system of solid domestic waste management in the territory. In: Proceedings of XXV international scientific conference "New look in world science", 1 Sept 2018. Lulu Press, Morrisville, pp 14–21
2. Pakulina AA, Pakulin SL, Pakulina HS (2018) Rational organization of financial controlling systems at the enterprise. In: Proceedings of the international scientific and practical conference "Universum View 4", 4 Nov 2018, Poltava, Ukraine. LLC "Nylan-LTD", Vinnitsa, pp 29–32
3. Feklistova IS, Pakulin SL, Pakulina AA, Pakulina HS (2019) Effective territorial strategic development in the context of changes. In: Proceedings of international scientific conference "Universum N V", San Francisco, 12 Jan 2019. Lulu Press, Morrisville, pp 24–28
4. Pakulin SL, Pakulina AA (2016) Management of sustainable development of a modern enterprise. Path Sci 3(8). http://pathofscience.org/index.php/ps/article/view/50. Accessed 31 Aug 2019

5. Pakulin SL, Tsypkin YA, Pakulina HS, Bliznukova TV (2018) The formation of the civilized enterprise in the sphere of solid domestic waste. In: Proceedings of XXV international scientific conference "New look in world science", 1 Sept 2018. Lulu Press, Morrisville, pp 45–51
6. Pakulina AA, Pakulin SL, Feklistova IS, Pakulina HS (2019) Dynamics of small business development in the municipality. In: Proceedings of I international scientific conference "Science of 2019", Berlin, Germany, 9 Feb 2019. Hamburg, tredition GmbH, pp 8–13
7. Pakulina HS, Feklistova IS, Pakulin SL, Pakulina AA (2019) Partnership between government and business as a tool for strategic management of territorial development. In: Proceedings of IV international scientific conference "Scientific achievements of the present", Berlin, Germany, 11 May 2019. tredition GmbH, Hamburg, pp 32–43
8. Tsypkin YA, Pakulina HS, Pakulin SL, Bliznukova TV (2018) Development of the optimal scheme of solid domestic waste management in the region. In: Proceedings of XXV international scientific conference "New look in world science", 1 Sept 2018. Lulu Press, Morrisville, pp 38–44
9. Tsypkin Y, Feklistova I (2019) Assessing the efficiency of management and land use in the agrarian sector of municipalities. In: IOP conference series: earth and environmental science, vol 274. IOP Publishing, p 012089. https://iopscience.iop.org/article/10.1088/1755-1315/274/1/012089/pdf. Accessed 31 Aug 2019
10. Feklistova IS, Pakulin SL, Bliznukova, TV, Tsypkin YA (2018) Optimization of the location of waste treatment facilities as a factor in the ecological and sustainable development of agglomerations. In: Proceedings of XIV International scientific conference "Science in 2018", 26 Jan 2018. Lulu Press, Morrisville, pp 29–33
11. Tsypkin YA, Feklistova IS, Pakulin SL, Bliznukova TV (2018) Ensuring sustainability of the solid domestic waste management system. In: Proceedings of XXVII international scientific conference "New step in science", 15 Sept 2018. Lulu Press, Morrisville, pp 31–38

Chapter 72
Startup Financing in the Green Economy

Anna N. Zhilkina, Marina B. Trachenko and Anastasiya V. Kozhanova

Abstract The active development of the green economy over the past decade has served as the basis for increased funding for innovative environmental startups. The paper deals with the main approaches to the definition of the green economy concept, the volume of financing has been studied, and the significance of ongoing reforms has been assessed at the global level. The authors have analyzed the market of Cleantech startups and revealed the reasons for its unstable development. The authors estimated the degree of state influence on the field of green finance, the level of interest of private investors in Cleantech startups, and developed a transitional startup financing model in the process of environmentalization of the economy. The authors have put forward possible further measures of state influence and development prospects in the transition to a new model of the international economy. Methodological basis of the research is comprised of the modern scientific methods of inquiry, including the systemic approach, logical analysis method, bibliographical analysis method, synthesis method, dialectical method, and systemic analysis method.

Keywords Startup · Green economy · Startup financing · Cleantech

72.1 Introduction

The green economy that has been growing rapidly over the last years, has dozens of interpretations. Its concept includes the ideas of many sectors of economics and philosophy, such as postmodernism, environmental economics, antiglobalism, international relations theory and many more [3].

A. N. Zhilkina (✉) · M. B. Trachenko · A. V. Kozhanova
State University of Management, Moscow, Russia
e-mail: angh@inbox.ru

M. B. Trachenko
e-mail: m.trachenko@gmail.com

A. V. Kozhanova
e-mail: a_kozhanova@bk.ru

T. Kolmykova and E. V. Kharchenko (eds.), *Digital Future Economic Growth, Social Adaptation, and Technological Perspectives*, Lecture Notes in Networks and Systems 111, https://doi.org/10.1007/978-3-030-39797-5_72

The term "green economy" was mentioned for the first time in the report called "Green Economy Plan" that was submitted to the UK Government by environmental economists in 1989 [11].

Today, there is still no single internationally agreed definition of the green economy. However, the most concise, brief and accurate is probably the definition given in the UNEP program (United Nations Environment Program): green economy is the kind of economy "which leads to improved human well-being and social equity, while significantly reducing environmental risks and ecological deficits" [17]. This definition has been cited in a number of more recent reports, including the International Labor Organization [5]. Its special feature consists in the fact that it encompasses three main components of sustainable development: ecological (environmental), social and economic. Other organizations, such as the Organization for Economic Co-operation and Development [9], use the term "green growth", which is similar to the concept of the green economy, but is slightly different from the perspective of implementation of approach. The report of OECD "On the Way Toward the Green Growth" defines "green growth" as "promotion to the economic growth and development, supported by natural assets that still provide ecological resources on which the well-being of mankind rests" [9].

It is expected that creation of the green economy without building of sustainable development in the state is ineffective and even impossible. According to the report by the UN experts "Transition to the green economy", the concept of the green economy should be considered along with a wider concept of sustainable development [16].

The report on "the required fiscal system" that was published in the end of 2016 by UNEP, points out the fundamental role of green finance in a global fiscal system. This report claims that a need grows in recent years for a sustainable fiscal system with a particular focus on green finance and due consideration of climatic risks [18]. Special emphasis in creation and development of the green economy in the document is put on state financial institutions.

Today, green economy is viewed by the global political community as an innovative process which is necessary for achieving sustainable social and economic development of the whole mankind and having significant impact on the socioeconomic and environmental challenges in the modern global economy.

72.2 Results and Discussion

Based on the foregoing, it can be concluded that the green economy is a direction of economy which provides for the increase in employment and quality of life of global population due to financial means that are intended to ensure the reduction of environmental pollution, air emissions, promoting the efficient use of energy and resources, and preventing any harm to mankind.

Business communities, responding to the challenges of governments of advanced countries, also pay significant attention to the green economy, since the basic programs for the development of green technologies provide entrepreneurs with the opportunity to earn extra profit while solving global problems. The venture capitalists, development funds, crowdfunding sites, strategic investors and business angels also participate in the formation of the green economy along with the governments of advanced countries and consumers themselves.

72.2.1 The Financing Volume of Green Technologies and Results of Cleantech Startups Market Analysis

According to the estimates of Global Environment Center [8], 400–600 billion dollars per year are needed for financing of conservation of lands, woods and waters, and more than 350 billion dollars are needed for financing of projects in the field of renewable energy resources and energy efficiency improvement. Today, less than 15% of the required capital goes to conservation, most of which comes from governmental and charitable organizations. According to the UN report for the year 2018, it will take 140–300 billion dollars for the improvement of stability of societies and economies against climatic changes and reduction of the rate of exhaustion of natural resources of the planet [15].

The brisk growth of the green bond market is demonstrative for the assessment of the interest in investment in green technologies. According to the international non-commercial organization "Climate Bonds Initiative" [1], the size of the green bond market in 2018 was 168.5 billion dollars, and in 2019 it is expected to increase to 250 billion dollars. As of today, green bonds are a widely used financing tool for all green projects which are the most common form that is used by corporate and private investors both in developed and in developing countries.

Politicians from leading countries, major corporations, strategic investors suggest the benefits of investing in technologically "green" startups. Is this part of the common strategic way of improvement of environmental situation in the world, the source of the increase in employment, a method of conservation of natural resources, using the method of increase in income or a cool trend of the technological sector of the economy, investors receive increasingly more incentives to invest in innovative companies engaged in green or clean technologies called Cleantech.

Cleantech is a general term used for the definition of technologies which optimize the use of natural resources, generate energy from renewable sources, increase efficiency and productivity, generate less waste and reduce the level of environmental pollution [13]. Cleantech helps in solving ecological problems due to the use of innovative approaches. According to the report of Cleantech Group company [2] investment in equity capital of Cleantech startups amounted to more than 55 billion dollars over the period from 2010 to 2016. According to the research center Startup Genome, the sector of Cleantech startups amounted to 2.1% of the total volume of investment in the startup economy for the period from 2015 till 2017, which approximately amounts to 48 billion dollars [13].

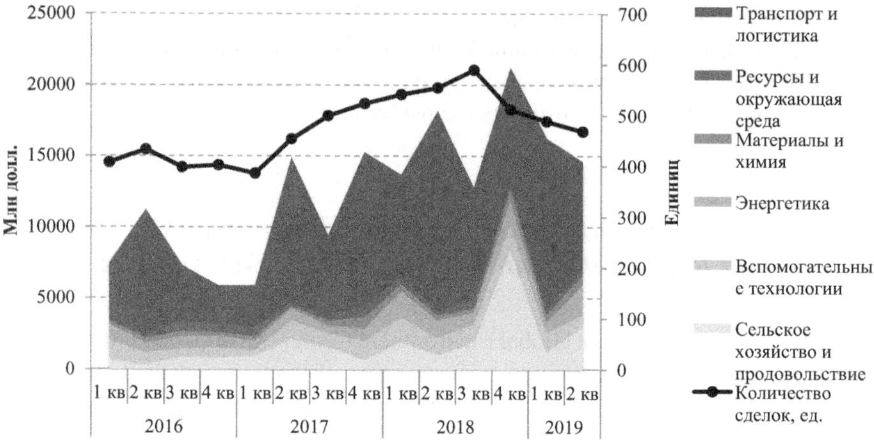

Fig. 72.1 Dynamic pattern of investment in Cleantech startups

Despite the impressive volumes, significant changes occurred over the past several years in the market of innovative Cleantech startups. Having been marked by a decline over the period of 2012–2014 (about 40% compared to 2011), the Cleantech market demonstrated that it is impossible to turn most scientific developments into commercial profit. Many investors lost their funds during this period, which reduced overall confidence in the sector.

According to English researchers of the venture market [6] in the area of Cleantech, investors are interested in the reduction of the degree of risk and the increase of the return on investment that the Cleantech market cannot offer them, since these startups require significant funding, have long development times, are uncompetitive in commodity markets, and cannot attract corporate buyers. As a result, they fail, and those that survive return capital to investors in a limited form (falling short of expectations). However, researchers point out that companies involved in Cleantech and developing software solutions were the best choice as investment facilities [6].

Therefore, the market of Cleantech startups, reoriented to software, started its systematic growth as early as in 2016 (Fig. 72.1).

Млн долл.	Million dollars
Транспорт и логистика	Transport and logistics
Ресурсы и окружающая среда	Resources and environment
Материалы и химия	Material supplies and chemistry
Энергетика	Energy production
Вспомогательные технологии	Assistive technologies
Сельское хозяйство и продовольствие	Agriculture and food supplies
Количество сделок, ед.	Number of transactions, pcs

(continued)

(continued)

Млн долл.	Million dollars
1 кв.	Q1
2 кв.	Q2
3 кв.	Q3
4 кв.	Q4
1 кв.	Q1
2 кв.	Q2
3 кв.	Q3
4 кв.	Q4
1 кв.	Q1
2 кв.	Q2
3 кв.	Q3
4 кв.	Q4
1 кв.	Q1
2 кв.	Q2

Source Compiled by the authors based on data from https://www.cleantech.com/

Today, Cleantech still remains not only an important tool for governments and companies, but also a potential growth area. In 2014, even despite the decline, World Bank estimated the general possibility of Cleantech startups at 6.4 trillion dollars in developing countries, including 1.6 trillion dollars for small and medium businesses [19].

The analysis of Cleantech startups made it possible to reveal the following features of their financing:

- dependency on state policy as compared to other process engineering sectors;
- greater volume of required initial investment in development compared with other areas;
- longtermness of return on investment;
- absent or poor financial infrastructure for development.

72.2.2 Startup Financing Features in the Green Economy

Cleantech startups are largely knowledge-intensive. This being said, the search for an effective business model, customers, as well as an effective commercialization strategy is still of great importance to them. In connection with this special feature, the state is most important in financing and promotion of startups, especially at the seed stage, providing them with subsidies and grants for R&D. In future, venture capital funds, crowdfunding sites, corporations, business angels, strategic investors, portfolio investors and private equity funds are sources of financing at later stages.

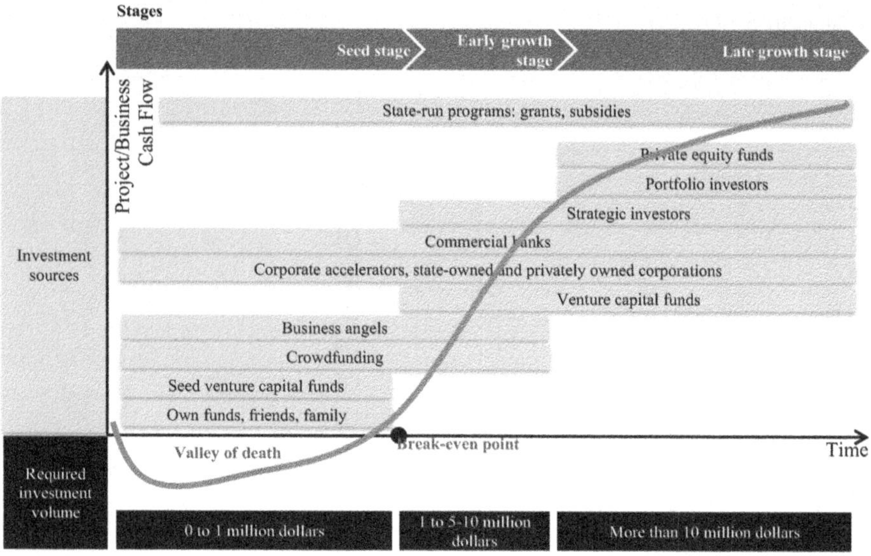

Fig. 72.2 Startup financing sources depending on the life cycle stage. *Source* Modified by the authors using [4, 7, 12]

Figure 72.2 presents an adapted model of the sources of startup financing depending on the life cycle stage, described in a certain form in papers of many authors [4, 7, 10, 12].

However, Cleantech startups do not always have access to business financing by the venture market. Most of these companies can obtain finances in grants and business incubation programs, but this funding is often limited and insufficient for development. At the same time, investors who may be interested in investing in Cleantech, focus on companies with a solid reputation. For the same reason, there are certain difficulties in obtaining bank financing. Therefore, today, many Cleantech startups are "stuck" in the valley of death (Fig. 72.2), which is almost impossible to get out of without significant financial support for such companies from the state.

"Sustainable Development Initiative of Stock Exchanges" schematically describes the structure of green finance, which is indicative of financial capacity of private investors and the major current role of public finance, in Fig. 72.3 [14].

Green financing is financing of the transition to a less polluting and less resource-intensive global economy: it is called a green transition. It is a transition from green bonds to profitability, because private finance has a number of different instruments to channel investments in environmental mitigation projects and adaptation to them [14].

There is no doubt that the growth of green finance opens up new investment opportunities for private business on the one hand and provides originators of innovative Cleantech startups with monetary means, on the other hand. Financing for

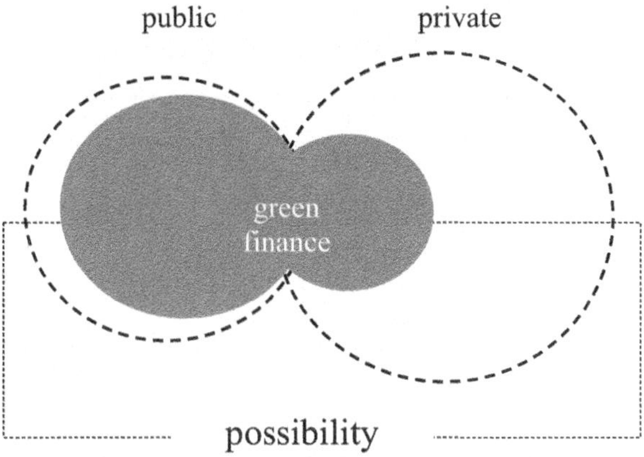

Fig. 72.3 The structure of green finance. *Source* SSE. Green finance. World investment forum 2016. www.SSEinitiative.org/green-finance

the preservation of the world ecosystem may call for a vast amount of money, and private investment may become their major source.

The main forms of financing in green finance, like in many others, are participatory and debt financing. Participatory financing is often used in the early stages of developing a project, product, or company itself, and is a form of investing in company shares in exchange for participatory interest.

Debt financing can be carried out both by means of a loan provided by the investor, and by means of the issue of bonds and bills. Unlike loans from investors, bonds are traded on the market in much larger amounts and are accessible to a wide range of investors. However, green bonds are still focused on specialized investors in a greater degree. In addition to bonds, there may also be bills, accessible to the greater number of investors, in the green finance market.

As noted above, Cleantech startups working with software, can attract investors to equity participation is much easier. Specialized venture capital funds are one of the main sources of financing in green finance. Specialized Cleantech funds can combine several projects in completely different areas, for example, water projects and projects of agriculture, but they will all be at approximately the same stage of development, at the same life cycle stage. These funds allow investors to diversify their investment portfolio, thereby reducing the risks of a complete loss of funds.

In developed countries, state institutions can fulfill the function of the insurant of investors' risks. And the latter willingly take advantage of the offered guarantees that can protect their monetary means from possible defaults.

At the domestic level, attraction of private investment in financing of Cleantech startups is a challenging task, since potential investors point out low yield and high riskiness of Cleantech. Therefore, startups at the seed stage and part of startups at the early growth stage in a greater degree depend on state financing and utilize the "subsidized" form of financing without any repayment of money (Fig. 72.4). In doing so, the state promotes development of innovative companies at the initial stage for attracting much larger volume of private investment during the following stages of startup growth.

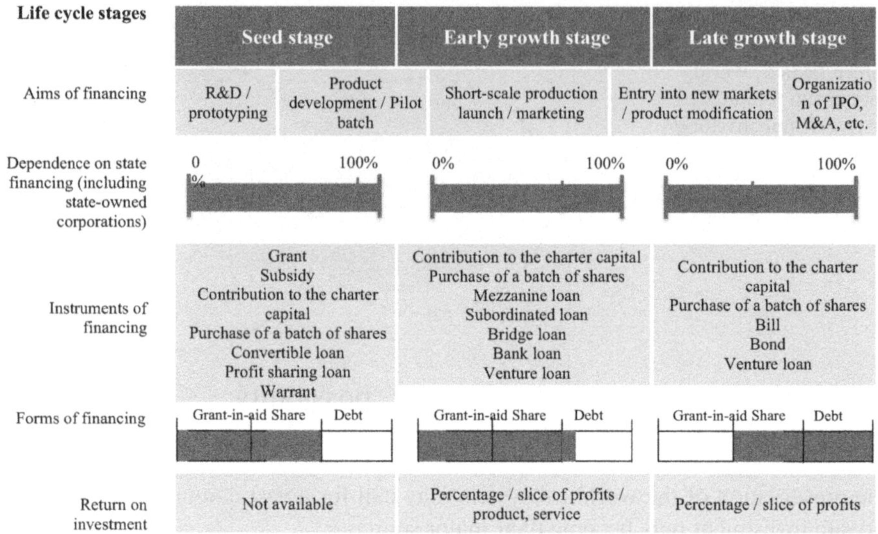

Fig. 72.4 Startup financing in the green economy at different life cycle stages. *Source* Compiled by the authors with the use of Karen and Wilson [7], Paschen [10]

Thus, the state lays the foundation for financing new projects, improving conditions for private investors. The instruments of financing include state grants, subsidies, contribution to the charter capital (of state-owned companies) and others, providing free grants and attracting private investment to financing.

According to the Global Environment Center [8] scalability and replicability are key issues in the development of the green finance market. As for now, only a few projects can be scaled beyond the threshold of 5 million dollars.

72.2.3 Green Startups Financing Model at Different Life Cycle Stages

Figure 72.5 shows an enlarged model of the current startup financing structure at different life cycle stages in the green economy as a basic model for the transitional period.

Under current conditions, the startup financing model that is shown in Fig. 72.5 assumes the presence of the state as a key participant and initiator of development of the green economy. Thus, the state has to take on the startup financing function at the seed stage almost completely, provide benefits and guarantees for private companies investing in green technologies and to rapidly develop innovative ecosystem, providing technical support to startups by creating specialized accelerators and business

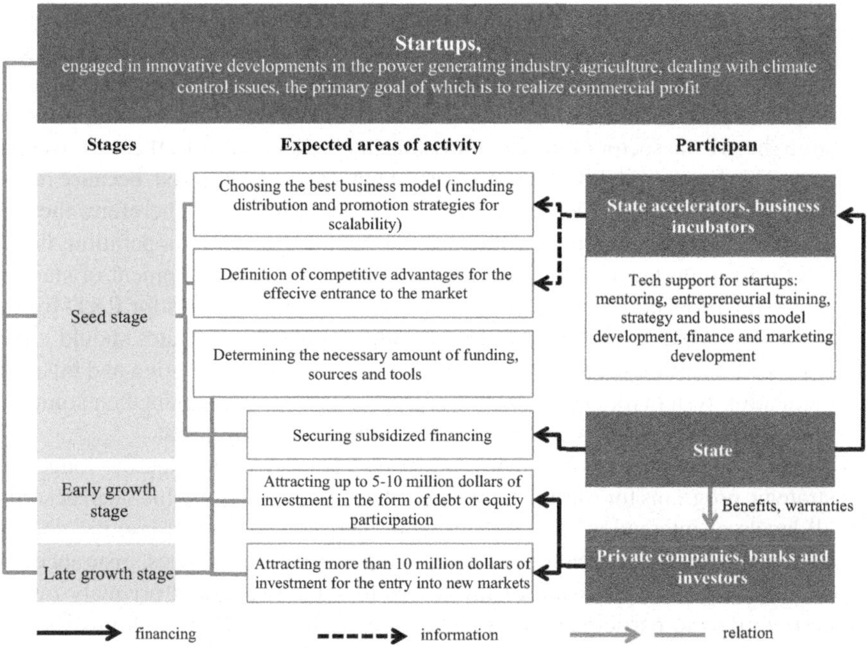

Fig. 72.5 Startup financing model in the green economy. *Source* Compiled by the authors

incubators, getting young entrepreneurs introduced with special aspects of conduct of business in the field of Cleantech technologies.

However, the proposed model should not be the only one for the support of innovative Cleantech startups. It may be that the development of financial intermediaries, such as specialized funds, will increase the scale of the launch of Cleantech startups and realization of commercial profit in the longer term. The creation of separate divisions in major corporations engaged in solution to the green economy issues will also improve the environment for the commercialization of projects. However, the role of governments in the development of these areas is key today. In order to give an impetus to these sectors to invest, the state needs to develop appropriate support programs, providing benefits, guarantees, and insurance. Active cooperation with the private sector will allow the state to do this as smoothly as possible.

72.3 Conclusion

The following conclusions can be drawn following the results of research:

1. Out of all sectors of the startup economy, Cleantech is a unique blend of politics, economics and sociology. Governments point out the need and urgency of solution of ecological problems, but there are no real tools for solving these problems

today. International reports claim that there is an urgent need, but the Cleantech sector is not yet ready for the implementation of ideas of creation of a global green future of planet. Today, Cleantech startups need significant financial resources for the implementation of their ideas; these financial resources must be attracted from the private sector of the economy than any other sectors of the startup economy. This being said, low activity of private investors is observed, because return on investment is not always expected and is too long-term. Therefore, the role of the state in creating an innovative ecosystem, a space for cooperation, is still of utmost importance in this case. The impetus for the development of startups is also the provision of more grants and subsidies to companies for R&D for the attraction of bigger private investors to further investment. States should cooperate with institutions, research laboratories, create an acceleration and business incubation system for entrepreneurs, and more actively implement their solutions in the business environment, helping to commercialize projects.

2. At the international government level, it is necessary to develop the long-term strategic programs for startup financing in the green economy, which will describe all levels of interaction with the business environment, to increase the volume of financing of green technologies, including benefits, guarantees, insurance and others, as well as provision of incentives to state-owned and privately owned corporations to participate in the implementation of support programs for new technologies.
3. For the sector of Cleantech startups, it is necessary to search for new sources of financing that would be able to withstand the long-term return on investment. They can be found among major patrons and benefactors, public and private foundations, funds, and institutions.

In general, it can be noted that global changes that need to be taken into account in the context of sustainable development, are associated with green technologies today, and the issues of development of an effective finance system for startups of this sector so far largely depend on state policies.

References

1. CBI (2019) Green bonds market 2019. https://www.climatebonds.net. Accessed: 11 Aug 2019
2. Cleantech Group (2018) The global cleantech innovation index 2017. https://wwf.fi/mediabank/9906.pdf. Accessed: 11 Aug 2019
3. Diyar S, Akparova A, Toktabayev A, Tyutunnikova M (2014) Green economy—innovation-based development of Kazakhstan. Procedia Soc Behav Sci 140:695–699
4. Dutta S (2016) Financing innovation: a complex nexus of risk & reward, MPRA paper, № 75584. https://mpra.ub.uni-muenchen.de/75584/. Accessed 11 Aug 2019
5. ILO (2018) World employment social outlook. Greening with jobs. https://www.ilo.org/weso-greening/documents/WESO_Greening_EN_web2.pdf. Accessed 7 Aug 2019
6. Gaddy B, Sivaram V, O'Sullivan F (2016) The wrong model for clean energy innovation an MIT energy initiative, working paper, MA 02139

7. Karen E, Wilson F (2013) Silva policies for seed and early stage finance, OECD science, technology and industry papers, № 9. http://doi.org/10.1787/5k3xqsf00j33-en. Accessed 08 Aug 2019
8. Krushelnytska O (2017) Introduction to green finance. World Bank, Global Environment Facility (GEF), Washington, D.C. http://documents.worldbank.org/curated/en/405891487108066678/Introduction-to-green-finance. Accessed 11 Aug 2019
9. OECD (2011) Towards green growth, p 9. https://www.oecd.org/greengrowth/48012345.pdf. Accessed 5 Aug 2019
10. Paschen J (2016) Choose wisely: crowdfunding through the stages of the startup life cycle. Bus Horiz № 60(2). http://doi.org/10.1016/j.bushor.2016.11.003. Accessed 11 Aug 2019)
11. Pearce D, Markandya A, Barbier E (1989) Blueprint for a green economy, vol 1. Earthscan
12. Salamzadeh A, Kesim HK (2015) Startup companies: life cycle and challenges. Paper at the 4th international conference on employment, education and entrepreneurship, Belgrade, EEE. http://doi.org/10.13140/RG.2.1.3624.8167. Accessed 11 Aug 2019
13. Startup Genome LLC (2018) Global startup ecosystem report 2018. https://startupgenome.com/report2017/. Accessed 11 Aug 2019
14. SSE (2016) Green finance. World investment forum 2016. www.SSEinitiative.org/green-finance. Accessed 05 Aug 2019
15. UN (2019) Environment adaptation gap report. https://www.unenvironment.org/resources/adaptation-gap-report. Accessed 11 Aug 2019
16. UN-DESA (2011) UN environment programme the transition to a green economy. https://plagiarism.repec.org/trica-papuc/trica-papuc2.pdf. Accessed 1 Aug 2019
17. UNEP (2012) What do we mean by green economy? https://wedocs.unep.org/rest/bitstreams/11027/retrieve. Accessed 11 Aug 2019
18. UNEP (2016) The financial system we need, 2nd edn. http://unepinquiry.org/wp-content/uploads/2015/11/The_Financial_System_We_Need_EN.pdf. Accessed 8 Aug 2019
19. World Bank (2014) Opportunity for 10 year timeframe starting 2014. http://www.worldbank.org/en/news/feature/2014/09/24/new-report-identifies-major-cle-an-tech-market-opportunity-for-small-businesses-in-developing-countries. Accessed 11 Aug 2019

Chapter 73
Improving the Control of Municipal Solid Waste Management Processes in the Region

Yury A. Tsypkin, Roman A. Kamaev, Tatiana V. Bliznukova, Serhij L. Pakulin and Inessa S. Feklistova

Abstract The authors of the research aimed at elaborating measures to improve the control of municipal solid waste management processes in the region. The authors of the paper have used a comprehensive approach for the substantiation of measures. The researchers substantiated fundamental principles of the development of a territorial strategy for the municipal solid waste management. The strategy development system and the process management should be based on the business integration of stakeholders and take into account the application of mechanisms of the provision of economic incentives for separate collection. The authors have proved the expediency of creation and modernization of the municipal solid waste recycling system. There is a need to form a differentiated pricing policy for the removal of refuse and to grant tax benefits for the activity of a municipal solid waste management company. The strategy must take into account the development of motivators for the application of new technologies and mechanisms at all stages of municipal solid waste management, the application of public and private partnership mechanisms, fostering environmental culture of society, raising public awareness, development of new principles of producers' liability for compliance with environmental requirements for safe disposal of products and packaging. Creation of a complex system for the collection, transportation, recycling and disposal of municipal solid waste within the territory will make it possible to: reduce environmental impact of waste, change the quality of life of population, introduce separate waste collection, increase the

Y. A. Tsypkin (✉) · T. V. Bliznukova · S. L. Pakulin
State University of Land Use Planning, Moscow, Russia
e-mail: tsypkin@valnet.ru

S. L. Pakulin
e-mail: stizapet@gmail.com

R. A. Kamaev
Moscow University of Finance and Law, Moscow, Russia
e-mail: RKamaev@fadm.gov.ru

I. S. Feklistova
Central Russian Institute of Management, Branch of RANEPA (The Russian Presidential Academy of National Economy and Public Administration), Orel, Russia
e-mail: Inessa_pakulina@mail.ru

T. Kolmykova and E. V. Kharchenko (eds.), *Digital Future Economic Growth, Social Adaptation, and Technological Perspectives*, Lecture Notes in Networks and Systems 111, https://doi.org/10.1007/978-3-030-39797-5_73

percentage of recovery of recoverable resources, set up new advanced waste recycling companies, build landfills that would meet the current technological horizon, increase the level of environmental culture of society, attract investment in the field of municipal solid waste management.

Keywords Municipal solid waste · Strategic control · Region · Control mechanism · Efficiency

JEL Classification Q28 · Q53 · Q56 · R11

73.1 Introduction

Efficient municipal solid waste (MSW) management is based on the concept of strategic control, the use of technical, economic and financial methods of long-term development, as well as psychological, social and other instruments. Waste management includes not only collection, transportation, separation, recycling, and disposal of waste, but also explanation to the public, cooperation with public and governmental organizations [1]. Many problems need to be solved in the process of control of the field of MSW management: establishment of criteria for the creation of the waste management system; how to finance changes; what instruments should be used etc. [2]. Therefore, it is necessary to intensify the development of conceptual and applied aspects of an important scientific challenge of strategic control of the field of municipal solid waste, ensuring the transition of territories of Russia from traditionally established to modern, innovative mechanisms of strategic control that would allow rebuilding the entire setup of control, using available resources in a more efficient manner, attracting investment, and raising the level of safety of people in this territory [3]. The relevance and complex nature of the phenomena considered formed the basis for the choice of the topic of our research.

The research was aimed at elaborating measures to improve the control of municipal solid waste management processes in the region. To achieve the goal, the following main scientific missions were identified: substantiate priorities in the field of MSW management; identify the key conditions in the field of MSW management; justify the primary goals on which territorial holding strategy in MSW management should be focused; substantiate the expediency of the creation of a complex system for the collection, transportation, recycling and disposal of municipal solid waste within the territory.

73.2 Methodology

The authors have used diverse methodologic tools, including calculation and analytic, economic and statistical, as well as other research methods. In order to substantiate recommendations on improvement of the control of municipal solid waste management processes in the region, a comprehensive approach was used.

73.3 Results

In the development of the system for strategic control of MSW management processes, there is a need to create a new, improved model of waste management, introduce innovative methods of management of the field of MSW disposal. Within the scope of a new strategy for the development of the field of MSW management, it is expected that regions determine approaches and alternatives to solving the MSW management problems; it depends on specifics and peculiarities of local conditions and resources. But when shaping the goals and planning the strategy, it is expedient to clearly define the priorities in the field of waste management: reducing the waste generation level; reducing the content of hazardous substance in the waste; maximum possible reuse; waste composting; waste landfilling of the remaining waste with minimum environmental damage [4].

When developing a strategy, attention should be paid to a number of critical conditions in the field of MSW management.

1. In order to achieve the maximum recovery of recoverable resources, it is necessary to introduce a mandatory system of separate (selective) MSW collection due to the following reasons: combined costs are reduced; the costs of purchasing and operating complex segregation technologies are reduced; incentives are created to reduce the amount of waste. Selective collection programs must take into account the following principle—minimization of requirements to people in terms of expended efforts results in the increase in the level of their participation [5].
2. Territorial local self-governing authorities should coordinate and oversee activities in the field of MSW management, taking into account national and regional strategies. To do this, laws must be passed that demand the following: mandatory selective waste collection; prohibition of acceptance of certain types of waste; setting waste disposal limits to streamline and control operations in the field of MSW management.
   Today, there is a topical matter of development of statutory instruments regulating the hazardous waste management regulations, and an absolute prohibition on landfilling of waste that can be reused [6].
3. Decisions made during the development of the strategy should be based on the opinion of all participants of the MSW management process, using all available local resources: human, organizational, economic and other, including:

population, public organizations, expert communities; representative and executive bodies; government supervision departments; management of MSW disposal companies (landfills, MSW recycling plants etc.); heads of the largest enterprises; mass media.
For the efficient interaction of these groups, it is necessary to create a committee of representatives (a task group), whose work should consist in an expert assessment of areas of development of the field of municipal solid waste. As global and Russian experience shows, any, even the most advanced technology is unable to solve MSW management problems without the authoritative opinion of experts, the political will of the city authorities and public support, decisions are made taking with account of local social, economic and other conditions.

4. Elaboration of principles of producers' liability for disposal of waste of their products. At the national level, in the field of MSW management, the principle of "producers' liability" should be legally enshrined, the essence of which is that the manufacturer must be jointly and severally liable for the neutralization of its products after the end of their service life. This principle can be implemented through the introduction of the disposal charge. This, in particular, will urge Russian producers to use recoverable resources in the production process as much as possible.

   - The MSW management strategy that is currently being developed, must take into account the current trends interrelated with the MSW problem we have identified based on the systemic analysis:
   - the volumes of MSW are constantly increasing both in absolute terms and in terms of each resident of Russia [7];
   - morphological composition of MSW becomes more complicated, new types of waste, containing environmentally-dangerous components, appear;
   - the existing waste collection and removal system requires improvement and modernization;
   - burying MSW in landfills and dumps fails to meet health standards and environmental safety requirements.

5. The lack of financial resources at all stages of MSW management results in decreased rate of modernization, forming the residual financing principle.
6. Service tariffs do not ensure the receipt of such an amount of funds that would guarantee the planned modernization of the waste management system, construction of landfills that meet health standards, and the rehabilitation of those operating in post-operational period. In these conditions, the pricing policy is required to ensure specified allowances, optimization of financial flows, financial independence of the holding. In order to improve the pricing policy, one should analyze the existing rates and actual expenditures on all waste management operations for compliance with their market conditions; justify the need and agree on tariff changes, with a particular focus on the fact that otherwise it would be impossible to improve the waste management system.

7. A new approach to the development and implementation of strategic plans and programs is needed. The latter should be designed for all kinds of sources of financing, in particular, take into account the possibility of entrance to the domestic and world capital markets, target federal and world funds, use international and interstate programs.
   Experience has shown that for a holding to implement a strategy in the field of MSW management, with account of its economical efficiency and competitiveness, there is a need to develop an efficient investment policy. Investment policy cannot be developed without regard to investment policy of the city [8], which includes the following areas: economical efficiency indicators, social benefit and significance, focus on the improvement of environmental situation. The implementation of capital investment project becomes easier if it was recognized in the investment market. Holding must have agency instructions from the city hall, which will allow it to act as a Party in state and interstate programs, use administrative guarantees, be a participant of official informational support etc.
8. The socially important factors in the field of MSW management must be taken into account [9]:
   - raising the level of environmental education and public awareness;
   - ensuring the housing improvement;
   - increase of income of population;
   - ensuring the sustainability of operation of city-forming enterprises.
9. The primary environmental factors in the field of MSW management should include: (1) maintenance of temporary waste storage (container yards, refuse collectors) with account of public health regulations; (2) status of equipment used for transportation and collection of waste (check and control of toxicity of exhaust gases); (3) condition of places of MSW disposal, meeting public health regulations.
10. As a rule, a short list of contractors for this type of service is due to the lack of a coordinated urban policy in the field of MSW management. City leaders should make informed decisions on involvement of private enterprises in the work in the field of MSW management, taking into account the opinion of people and public organizations. One can involve public utility companies in the region, part of the operating costs of which can be allocated from the city budget [10]. Often, there are not enough funds in the city budget, so problems can be solved in the cheapest, and not the most environmentally justified way. Private enterprises cannot be funded directly out of budget and at the same time they must comply with all environmental standards. The competition is able to improve efficiency of operation of private enterprises regardless of political changes in municipal administration. High price, certainly, is one of factors that influences the reduction of waste volume, but only in case if there is a direct relationship between the amount of fee and the amount of waste generated by people and institutions. This can be an incentive to reducing the amount of waste buried in landfills. The task of municipal authorities in the region in this

case is to provide alternatives to the landfilling of MSW and exercise control over rapid emergence of unauthorized landfills.
The presence of the legislative framework, tax benefits, raw material base and consumers contributes to the revitalization in the market of services of selective collection and recycling of waste.

11. A legal regulatory framework should be formed so that the city authorities were focused on the creation of an integral MSW management aimed at improvement and development.
Today, the authorities mostly use a formal approach aimed at monitoring from the perspective of compliance with health standards, in particular, reduction of hazardous environmental impact of waste by isolating the landfill body from groundwater. However, a much more efficient way is to control not landfills but waste delivered to landfills or waste sorting plants. In a perfect world, municipal waste of various size should be sorted and disposed of separately, using the most eco-friendly methods. In our opinion, without changing approaches to the solution of MSW management problems, a new understanding of the essence of what municipal waste is, the process of efficient strategic waste management is impossible.

The holding strategy in MSW management within the territory must be focused on the following primary goals:

- construction of new capacities in volumes sufficient for the disposal of the non-processed fraction of waste;
- implementation of technologies that make it possible to significantly reduce the generation of unseparated waste, and the maximum reuse is the top-priority goal;
- strategy reflecting the general goals of the holding should be based on a large number of additional factors, reflecting the specific nature of each municipal district of the region;
- features of the territory: area of the territory; density of population; population structure; environmental conditions etc.;
- the status of waste disposal facilities, non-compliance of the existing infrastructure with statutory requirements, the state of waste collection and transportation system, availability of waste recycling facilties, the analysis of existing MSW disposal projects;
- prospects and areas of development of the region, as well as prospects for the attraction of private investors etc.;
- new waste recycling technologies etc.

The strategic planning of the development of the regional sector of MSW management must ensure:

- relation of plans for the development of the field of MSW management to the program of social and economic development of the region, to other target programs (for example, housing and utility sector reforming programs, ecological programs, development of housing construction programs etc.);

- establishment of a hierarchy of goals and proposals for the implementation of innovative waste management methods;
- creation of the improved regulatory framework and data base;
- coordination of plans of development of the field of MSW management of the region;
- constant provision of information to the public about the development of entrepreneurial activities in the field of waste disposal.

The holding must use the following innovative instruments in the elaboration of the strategy for the development of the field of MSW management:

- preferential loans, grants and awards at the expense of budgets of constituent entities and regional authorities, environmental funds and other investment resources (for example, foreign investment);
- benefits for paying environmental fees, which can be granted as a result of construction of new, innovative MSW recycling (disposal) plants;
- tax benefits;
- prohibitions on burying MSW, that can be recycled or disposed of, in landfills;
- fee for the collection of MSW to be recycled;
- electronic waste-tracking system;
- tax on burying MSW in landfills;
- quotas for disposal or landfilling of MSW in the plants or waste landfills;
- scheme for payment or disposal of MSW on weight basis;
- provision of information services to the public, for example, at points of sale of glassware, paper, rubber, plastic bags, etc.

At present, systemic problems dominate in the economy of regions, which should primarily include the focus on the solution of short-range problems in the field of environmental safety.

The solution to long-term problems is contributed by investment elaboration of programs of development of the field of municipal solid waste management.

The implementation of programs is based on a set of organizational and methodological measures on the implementation of events that constitute a program.

Program management must be achieved by means of a technology that must include:

- organizational support: realization of procedures of planning, organization, coordination and monitoring of the entire process of implementation of programs;
- banking support: processing, granting of credit, collecting payments and settlements on credit resources, consulting. At the same time, the issues of coordinating schedules, interest rates, risks, and repayment of credits are resolved;
- technical support: rational use of funds and their payback, assessment of intermediate indicators and forecasting results of implementation of projects. Elaboration of solution for adjusting programs;
- audit support: examination of the financial condition of the program performer and the agent bank, intermediate audit transactions.

Professionals from expert, consulting and other organizations should be involved in the support of programs.

An efficient method of program management is monitoring, which provides a focus of actions on the operational determination of the difference between the actual results of implementation of the program and scheduled indicators, and the timely development of management decisions facilitating program change.

The primary goals of monitoring are as follows:

- management of established targets and standards;
- measurement of deviation of actual results from the planned indicators;
- assessment and diagnosis of deviations.

Monitoring is a coordinating system of interrelation between the creation of the database, analysis, planning and control.

The basis for the information support can be the qualitative data for each event of the program: cost, period of implementation, payoff period, economic indices, source of financing, performers.

Program and event indicators are ranked by significance level. The first level consists of projects with the most important monitored indicators. The second level consists of projects where indicators are related to the indicators of priorities of the first level. Subsequent levels are formed in a similar way. The need for ranking indicators by levels is dictated by the fact that the process of analysis and explanation of possible deviations of the actual values of indicators from the specified values in the program is facilitated.

After controlled indicators have been ranked, one should establish quantitative "standards" (indicators) which can be either absolute or floating.

Monitoring uses instruments that cover all theoretic and real-world (technical) auxiliary means used for collection, structuring, analysis and processing of information of all organizations. Theoretical means include methods, techniques, working, while real-world means include electronic data processing, computers, diverse software, and communications equipment.

Monitoring is the most important controlling mechanism, which provides a communication system that supplements but not replaces indirect contacts between various stakeholders of the project.

The monitoring of implementation of the program is aimed at provision of reliable information about results and prospects of their implementation to interested organizations implementing certain measures, for making reasoned management decisions.

Monitoring provides for the presence of the following:

- a specific meaningful program, which should be clearly structured and recorded; this is necessary to ensure control, including in the process of introduction of changes in the projects;
- a formalized reporting system that would allow demonstrating the current state of the program, comparing it with initial plans, based on common approaches and criteria;

- a system that would allow for the efficient analysis of actual indicators and trends (cost and time are the main values for analysis);
- an efficient response system that would allow for the timely correction of identified shortcomings and the elimination of negative trends within the framework of the program, up to making changes to it.

Monitoring processes ensure:

- collection of factual data and their analysis for the degree of deviation between the actual implementation and planned performance indicators;
- assessment and comparison of the achieved results with the planned results, identifying the causes and methods of influence on identified deviations;
- development of proposals that would allow carrying out activities in accordance with the adopted program, minimizing adverse consequences from deviations (identification and use of benefits from favorable deviations).

In the course of this research, we developed a method for assessment and selection of capital investment projects. This method is based on the analysis of projects on the basis of multilateral expertise, taking into account many different, contradictory project performance figures with different quantitative or qualitative characteristics. Some of these performance figures relate to the economic, environmental and social consequences of project implementation, while others demonstrate various risks occurring in the process of project implementation. The application of the developed method allows forming investment programs for the development of the field of MSW management of the region, formulating and implementing strategic development priorities in this field with account of economic, social and environmental imperatives based on the interaction of stakeholders.

Under current conditions, the vital task of creation of a complex system for the collection, transportation, recycling and disposal of waste within the territory emerges full blown.

The need for increasing the efficiency of municipal solid waste management in various countries was recognized long ago. The primary goal of creation of the municipal solid waste management system is to achieve a certain level of environmental protection indicators.

International practices bear evidence to the fact that the creation of a system should be based on the following principles: reduction of environmental impact of waste; introduction of selective collection of waste; change in the quality of life of population; the maximum possible preservation of recoverable resources for later use; construction of waste recycling plants and waste landfills based on a new technology level; increasing the level of environmental culture of society.

Implementing these principles requires a new approach to the creation of a complex MSW management system. We have identified the following blocks as the main organizational and technical components of a standard municipal solid waste management system: (1) collection of MSW; (2) delivery of MSW; (3) segregation of MSW; (4) recycling of MSW; (5) municipal solid waste landfill.

Recoverable resources, which are an essential component of waste, are almost not used in Russia, but are transported to landfills, affecting the state of the environment. The averaged morphological composition of MSW in the Oryol Region shows that segregation of recoverable resources (metal, glass, polymers, paper) will allow reducing costs for waste landfilling and increasing the resource potential of tech region. Requirements defined in the development of the MSW management system can be formulated as follows: recovering the maximum amount of recoverable resources, reducing the level of adverse effects on human health, general reduction of the amount of waste generated in the cities.

Low focus on issues of removal of refuse is the main cause of rapid emergence of unauthorized landfills. According to the current regional laws, unauthorized landfills are defined as the accumulation of industrial and municipal waste emerged as a result of their unauthorized disposal.

This phenomenon can be observed throughout Russia. It is difficult to eliminate such landfills, because they are mainly located on the plots of land, which in fact are not owned by anybody and are not controlled by any public authorities, although they are found by supervision bodies.

Due to the lack of sufficient control, the number of landfills is growing, which not only spoils the looks, but also negatively affects the environment. Even a small landfill can cause significant harm to human health, because decomposition products enter the soil, water and are released in the air in the form of volatile compounds.

This situation calls for new MSW management solutions: an efficient system for control and transportation of MSW to places of recycling and landfilling is needed.

Today, two alternatives for removal of MSW can be distinguished: direct removal and two-stage removal. The MSW is removed from container yards using special-purpose vehicles—waste collection trucks. According to the method of loading, they can be divided into front-loaded, side-loaded and rear-loaded waste collection trucks.

Selection of the type and capacity of a waste collection truck can be determined by the collection system; however, certain features should be considered:

- condition of the road bed;
- availability of sufficient space for the entrance and U-turn of vehicles;
- amount and morphological composition of generated waste.

Removal of waste using waste collection trucks (with a capacity of 12–24 cubic meters) directly to the landfill is expedient if the distance to the landfilling object does not exceed 15–30 km, otherwise transport costs will increase significantly, not to mention a decrease in turn-round of waste collection trucks.

Two-stage removal is used if the landfill is located far away from the place of collection, separation, or for additional compaction of waste. Further delivery of waste to the place of recycling or landfilling is performed with the use of waste collection trucks with removable containers 20–30 cubic meters in a compacted state.

When operating with heavy-duty waste collection trucks, one should pay attention to the following:

- overall weight of a waste collection truck (whether it does not exceed the acceptable traffic load);
- vehicle dimensions;
- compliance of emissions with environmental legislation;
- noise level;
- nuances associated with wintertime work.

Control is an essential point in the operation of the waste collection and removal system.

An integral element of efficient operation of transport at the moment is the use of a global positioning system that monitors the distance run by vehicles; fuel consumption; movement time; parking place and time; the fact of speeding; departure from the route; allows visualizing the route traveled and the event series on an electronic map. In addition to monitoring, the system helps to: optimize traffic routes; reduce the time required for waste collection and removal; promptly eliminate various emergency situations.

Experience in MSW recycling shows that there is no exclusive most efficient way or method that would meet all current norms and requirements in the field of environmental legislation, market needs, and needs of modern economy. The construction of combined waste recycling plant, capable of using waste as an energy source and at the same time produce secondary raw materials, can meet these conditions to the fullest extent (recommendations of international environmental congresses). The principle of combination of various MSW recycling methods that is used in the design and construction of recycling plants, can significantly reduce the drawbacks of each method, if it is used separately. A complex approach and the system combination of methods of segregation, heat treatment, fermentation and other processes take into account the features of morphological composition of raw materials and must ensure the minimum amount of production waste, its maximum return and environmental friendliness.

Waste with high calorific efficiency can be used as a secondary solid fuel or burnt in power plants. Advantages: reducing the amount of waste buried in landfills, reduction of expenses for waste landfilling, increased recycling rate; however, there are also disadvantages such as atmospheric emissions from combustion and increased class of residual waste.

For many years, waste storage in MSW landfills has been the only way of waste storage, but it is not environmentally and economically sound, because household waste contains many useful and hazardous substances.

Today, the technology of producing landfill gas—a gas generated as a result of biological and chemical processes of breakdown of wastes under the layer of soil—is increasingly widely used; emitted gas contains up to 60% methane and is used as local fuel. This technology is used both at existing and at closed landfills.

This gas is comparable to natural gas in terms of its properties, but its collection is a fairly profitable business.

The collection of biogas in landfills requires special engineering solutions: the presence of the insulating screen, gas collection and storage systems, and many more.

However, decomposition process lasts for many years, thus landfill is a long-term and relatively stable source of biogas. Given the proximity to industrial and urban areas, the low cost of production makes biogas a promising source of energy used for local needs. On the other hand, its use helps to ensure the purity of atmospheric air.

Landfill and waste recycling plant must be regarded as a unified system. Unprocessed waste can be disposed of in a landfill, while the generated landfill gas can be used for auxiliaries of the waste recycling plant.

73.4 Conclusions/Recommendations

Fundamental principles of the development of a strategy in the field of MSW management have been developed in the course of research. The strategy development system and the process management should be based on the business integration of stakeholders and take into account the application of mechanisms of the provision of economic incentives for separate collection; creation and modernization of the MSW recycling system; formation of a differentiated pricing policy for the removal of refuse; granting tax benefits for the activity of a MSW management company; development of motivators for the application of new technologies and mechanisms at all stages of MSW management; the application of public and private partnership mechanisms; fostering environmental culture of society; raising public awareness; development of new principles of producers' liability for compliance with environmental requirements for safe disposal of products and packaging.

73.5 Practical Implications

Creation of a complex system for the collection, transportation, recycling and disposal of MSW within the region will make ie possible to: reduce environmental impact of waste; change the quality of life of population; introduce separate waste collection; increase the percentage of recovery of recoverable resources; set up advanced waste recycling companies; build landfills that would meet the current technological horizon; increase the level of environmental culture of society; attract investment in this field.

References

1. Bliznukova TV, Pakulin SL, Tsypkin YA, Feklistova IS (2018) Creation of an effective system of solid domestic waste management in the territory. In: Proceedings of XXV international scientific conference "New look in world science", 1 Sept 2018. Lulu Press, Morrisville, pp 14–21

2. Feklistova IS, Pakulin SL, Bliznukova TV, Tsypkin YA (2018) Optimization of the location of waste treatment facilities as a factor in the ecological and sustainable development of agglomerations. In: Proceedings of XIV international scientific conference “Science in 2018”, 26 Jan 2018. Lulu Press, Morrisville, pp 29–33
3. Pakulin SL, Pakulina AA (2016) Management of sustainable development of a modern enterprise. Path of Science 3(8). https://pathofscience.org/index.php/ps/article/view/50. Accessed 31 Aug 2019
4. Pakulina HS, Feklistova IS, Pakulin SL, Pakulina AA (2019) Partnership between government and business as a tool for strategic management of territorial development. In: Proceedings of IV international scientific conference “Scientific achievements of the present”, Berlin, Germany, 11 May 2019. Hamburg, tredition GmbH, pp 32–43
5. Tsypkin YA, Feklistova IS, Pakulin SL, Bliznukova TV (2018) Ensuring sustainability of the solid domestic waste management system. In: Proceedings of XXVII international scientific conference “New step in science”, 15 Sept 2018. Lulu Press, Morrisville, pp 31–38
6. Pakulin SL, Tsypkin YA, Pakulina HS, Bliznukova TV (2018) The formation of the civilized enterprise in the sphere of solid domestic waste. In: Proceedings of XXV international scientific conference “New look in world science”, 1 Sept 2018. Lulu Press, Morrisville, pp 45–51
7. Pakulina HS, Tsypkin YA, Pakulin SL, Bliznukova TV (2018) Development of the optimal scheme of solid domestic waste management in the region. In: Proceedings of XXV international scientific conference “New look in world science”, 1 Sept 2018. Lulu Press, Morrisville, pp 38–44
8. Feklistova IS, Pakulin SL, Pakulina AA, Pakulina HS (2019) Effective territorial strategic development in the context of changes. In: Proceedings of international scientific conference “Universum N V”, San Francisco, 12 Jan 2019. Lulu Press, Morrisville, pp 24–28
9. Pakulina AA, Pakulin SL, Feklistova IS, Pakulina HS (2019) Dynamics of small business development in the municipality. In: Proceedings of I international scientific conference “Science of 2019”, Berlin, Germany, 9 Feb 2019, tredition GmbH, Hamburg, pp 8–13
10. Tsypkin Y, Feklistova I (2019) Assessing the efficiency of management and land use in the agrarian sector of municipalities. In: IOP Conference Series: Earth and Environmental Science. Sci. 274 012089. IOP Publishing. https://iopscience.iop.org/article/10.1088/1755-1315/274/1/012089/pdf. Accessed 31 Aug 2019

Chapter 74
Influence of Digitalization Processes on the Development of Agricultural Eco-Systems

Galina N. Ryazanova and Svetlana L. Sazanova

Abstract The article is devoted to problems of influence of the processes of digitalization in the agricultural sector of the Russian economy, the Methodology of research of problems of development of agricultural eco-systems based on the use of scientific quantitative and qualitative methods: the method of scientific abstraction, comparison, historical and logical methods, methods of abstract modeling. The authors concluded that the global trends of digitalization have a serious impact on the agricultural complex, in general, and on agricultural ecosystems, in particular; the key regulatory role in the formation of agricultural eco-systems belongs to the state.

Keywords Digitalization · Eco-system · Agricultural enterprises · Agricultural sector · Agriculture · Transformation of eco-systems · System approach

JEL B59 · Q38

74.1 Introduction

In the modern economic space, not only atomized economic agents of different levels of hierarchy interact, but also socio-economic eco-systems as a set of economic entities, taking into account the space-time factor and the factor of their functioning in the economic environment.

G. N. Ryazanova (✉) · S. L. Sazanova
State University of Management, Moscow, Russia
e-mail: ryazanovagn63@gmail.com

S. L. Sazanova
e-mail: sazanova@mail.ru

S. L. Sazanova
Financial University Under the Government of the Russian Federation, Moscow, Russia

T. Kolmykova and E. V. Kharchenko (eds.), *Digital Future Economic Growth, Social Adaptation, and Technological Perspectives*, Lecture Notes in Networks and Systems 111, https://doi.org/10.1007/978-3-030-39797-5_74

Currently, the environmental impact of economic entities is steadily increasing, causing both problems of adaptation of economic agents and problems of development of the socio-economic system [1]. This impact has increased with the transformation of the economic space and the advent of digital technologies. On the one hand, the interaction of agents in terms of digitalization facilitates and enhances mutual communication, on the other, presents a challenge for microeconomic agents in part the need to adapt to new conditions to retain competitive positions in the market.

The agricultural sector is one of the key sectors of the economy, most acutely experienced the negative consequences of socio-economic ecosystem of the 90-ies of the last century, in recent years shows a steady growth through the transformation of the socio-economic system of national economy at the macro and meso—level, which ensured effective interaction at the micro level as inside the agricultural sector and with related sectors of the economy.

74.2 Research Methodology

Theoretical studies in the field of socio-economic eco-systems were first proposed by J. G. Moore in "business ecosystems" [2], which was based on "biological ecosystems" by A. Tansley and gained many followers, who, based on an interdisciplinary approach cause in his writings, the penetration of the historical, political, social and other institutions in economic analysis that provides identification of the relationship of individual agents of the ecosystem, causal relationships of economic processes and phenomena and prediction of the dynamics of economic development based on the systemic analysis.

Among the followers of Moore the greatest contribution to the development of interdisciplinary research of socio-economic ecosystems to make foreign scientists: Baldwin, R. Adner, M. Jacobides, M. Castels et al. and Russian scientists: G. Kleiner, V. Dementyev, E. Ustyuzhanina et al.

The agricultural sector in Russia is one of the key in importance, since effective eco-systems of agricultural enterprises are not only important in terms of food security of the country, but also in terms of geopolitics and the preservation of state sovereignty [3]. The study of problems and transformation of eco-systems of agricultural enterprises in the era of digitalization and prospects for their development is relevant both in theoretical and practical aspects.

The research was carried out on the basis of General scientific methods, quantitative and qualitative economic research methods, historical and logical method, as well as using the method of abstract modeling.

74.3 Problems of the Post-Soviet Eco-System of the Agricultural Sector of Russia: Their Background and Consequences

The period of liberalization of the national Russian eco-system is characterized by serious consequences in all segments of the economy. The crisis has engulfed the political, social and economic spheres. The volatility of energy prices, instability of social relations, psychological tension in society have become evidence of the transformation of the current eco-system. The collapse of the Soviet Union and the attack on the "invisible hand of the market" dealt a blow to all the republics belonging to the USSR. The systemic ties that played a key role in the interaction of microeconomic agents, formed during the decades of the Soviet state's existence, were severed, meso-economic interactions were lost, and microeconomic relations were significantly reduced. This situation has affected, in particular, the horizontal and vertical links of agricultural enterprises, which led to a deep crisis in agricultural production [4].

Agricultural science in all directions ceased to meet the needs of economic practice: the scientific system of agriculture was not in demand among the new owners of soil—it was tenants, whose purpose was not to preserve and improve the natural biological eco-systems, and, therefore, the cultivation of only highly profitable row crops (corn, sunflower). These plants carried out soil fertility and, in the absence of a crop rotation system restoring the soil eco-system, led to the depletion of soil fertility of a large part of high-productivity agricultural land. Veterinary science and breeding in animal husbandry and population genetics was plunged into a deep depression. Reproductive centers were closed, breeding of new breeds of animals was not made. Research and achievements in the development of new highly productive breeds in the field of dairy farming and meat production through the use of genetics, which were in the period of the Soviet Union, in the 90s were completely lost, because agricultural farms, due to the low profitability of animal husbandry, reduced the number of all types of cattle and did not need to update the herd. During this period, sheep and poultry farming was almost completely destroyed, the number of cattle and pork production decreased at times.

The state variety-testing eco-system, which provides Soviet seed production with zoned varieties of elite, super-elite and the first reproduction, in conditions of low funding could not cope with the task of providing Russian crop production of the end of the XX century with high-quality highly productive seeds with high germination, bushiness of shoots and, as a consequence, high yields, taking into account regional soil and climatic features. Crop areas of agricultural enterprises were sown with clogged seeds with poor productivity and endurance against weather and soil conditions, which significantly reduced crop yields. More than 80% of the enterprises of agricultural complex of Russia are geographically located in the zone of risky agriculture. During the Soviet Union, land reclamation activities had a significant impact on the eco-system of agriculture. They optimized hydrological, soil and climatic deficiencies of regions, leveled soil fertility, fought water and wind erosion of fields and prevented it. In the post-Soviet period, land reclamation activities were almost not carried out at all.

Stagnation was observed in the creation, production and application of fertilizers. Industrial production of mineral fertilizers was reduced slightly, and the production of organic and organo-mineral fertilizers fell significantly.

Agricultural enterprises due to the lack of working capital, were forced to reduce the production of crop and livestock. In the agronomic process, the eco-system of crop rotations was violated, agricultural crops were cultivated on the principle of quick benefits, for example, during the acute need of sunflower oil seeds for the production of sunflower oil, the fields were sown with sunflower for two or even three years in a row. This row crop carries a significant amount of nutrients and requires mandatory rotation. However, in the conditions of aggressive agriculture crop rotations were not observed. As a result, some of the fertile land was seriously depleted, and some even went out of agricultural circulation. This happened including as a consequence of the unstructured legal field:

- the lack of clear legislative regulation of relations between landlords and tenants;
- the lack of clear land surveying of agricultural lands (profile and boundaries of fields);
- the lack of control over the environmental safety of agricultural production by the state.

The socio-economic eco-system of the meso-level in which agricultural enterprises found themselves in the late 1990s was deeply depressed. The interaction of agricultural enterprises both within the agricultural sector (horizontal and vertical links) and with enterprises of related industries were broken and unmanageable. All segments of the integration chain of the eco-system are destroyed (Fig. 74.1).

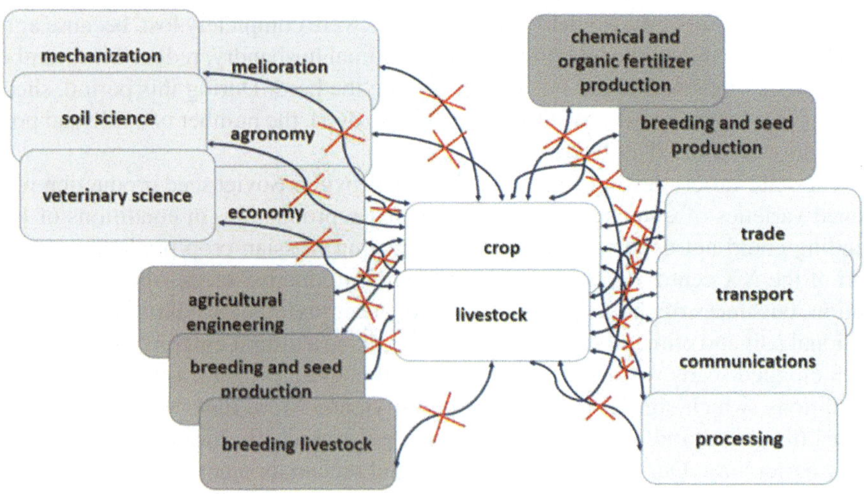

Fig. 74.1 Model of eco-system of agricultural enterprises of Russia in the 1990s

The model demonstrates the gap between the eco—system of agricultural production (crop and livestock-in the center of the model) and science (top left). Institutions of mechanization, reclamation science, agronomic research institutes, structures in charge of the economy of agriculture, could not provide adequate support to agricultural enterprises. The problems of the internal eco-system of science-weak funding, personnel problems, lack of modern equipment did not make it possible for the scientific unit to create new ideas, models of agricultural machines and trailers, new agricultural technologies.

The lack of support for agricultural machinery has led most Russian agricultural machinery production to decline. New models did not appear for a long period. Breeding works of the Soviet era in the field of seed production and animal breeding were suspended, many breeding stations and centers were closed. The vertical integration "from field to counter"—"crop production-livestock production-processing-transport logistics-sales" has collapsed.

State regulation was needed to restore the national eco-system and the internal eco-system of agricultural enterprises, to develop agricultural production, to get out of the deep crisis of the agricultural sector.

74.4 The Transformation of an Agrarian Eco-Systems and the Processes of Digitalization

Since the 2000s, the Russian Government has decided to support the agricultural sector. From this period, the stage of restoration of the eco-system within the agricultural sector and agricultural farms with a meso-level eco-system begins. The main documents aimed at the formation of structural certainty of agricultural production were in the 2000s:

- "Main directions of agri-food policy of the Government of the Russian Federation for 2001–2010" dated July 27, 2000;
- "Land code of the Russian Federation";
- Federal target program "Social development of rural areas till 2010".

It should be noted the General nature of the first normative documents and the lack of a clear target function.

The next stage is characterized by a clearer structure, the document Federal law No. 264 of 29 December 2006 "on the development of agriculture" outlines the main provisions of regulation of agricultural production, in terms of antitrust, fiscal and other aspects.

The accession of our country to the world trade organization provoked new challenges from foreign suppliers of products. What strategy of development of the country to choose in the conditions of rigid world competition? Only the introduction of innovative technologies can ensure development and sustainable economic growth [5]. One of the essential points was the need to ensure food security of the state. The targets for the development of sub-sectors of the agricultural sector are reflected in the "food security Doctrine "and in the" State program for the development of agriculture and regulation of markets for agricultural products, raw materials and food for 2013–2020". These documents indicate the indicators of support for agricultural enterprises for each target result.

Digitalization has had a major impact on the development of agricultural eco-systems, as the introduction of modern innovative technologies in the eco-systems of agricultural enterprises has made it possible to solve several management problems at the same time [6]:

- energy problems in the provision of agricultural production [7, 8];
- environmental concerns (improving soil resource productivity, weed and pest control with the least impact on the environment);
- technological problems (ensuring control over compliance with the requirements of technological operations);
- personnel problems (partial replacement of human artificial intelligence;
- problems of management of financial flows to (digitalization creates a basis for optimization of control of execution of budgets of agro-industrial enterprises) [9];
- the problem of risks of opportunistic behavior of counterparties (there are bases of unscrupulous suppliers, partner reviews) [10].

According to experts of the Russian-German dialogue, in 2018 Russia was on the 41st place in terms of readiness to digitalize business processes in the agricultural sector of the economy with a significant gap from the leading countries-leaders. Experts expect that the market of information technologies in agricultural business will grow 3 times, which will reduce the cost of production of crop products by more than 20%, which will become a driver of development of agricultural eco-systems, and a driver of further national economic growth [11].

At the same time, the development of digitalization of agricultural enterprises is already showing some results. In many regions of Russia, as a result of digital land monitoring, uncultivated lands were identified, some of which have already been put into circulation. Precision farming technologies have made it possible to ensure an optimal feeding area for each plant, limit the consumption of expensive elite seeds and increase crop yields. Artificial intelligence technologies are implemented in many livestock complexes, greenhouses and their use is growing every year.

The main problem of digitalization of agricultural eco-systems is the systematization of digital data both within the enterprise (coordination of all business processes) and integration into a single digital eco—system of the national economy, since currently the bulk of transactions (sales of agricultural products, reporting and subsidies) are carried out using digital technologies.

Digital technologies are used in the implementation of state support:

- when subsidizing investment costs for the construction of new infrastructure;
- with preferential lending at the rate of 5% from Rosselkhozbank, including small and medium-sized businesses;
- in the implementation of per-hectare payments to agricultural producers (according to the area of the agricultural economy);

- when subsidizing productive dairy farming (if the technology of milk production corresponds to the current level of development of the industry);
- when subsidizing technical means of production, if they are produced by domestic agricultural machinery;
- with the support of farms and agricultural cooperation.

Part of the subsidies is transferred using the "budget planning" system, when the agricultural enterprise downloads scans of documents and receives an electronic signature that is sufficient to sign additional agreements and financial documents. Preferential loans and equipment for leasing can also be obtained remotely by filling out a customer questionnaire on the websites of organizations and attaching the necessary documents for decision-making.

The digital transformation of institutional conditions has had a positive impact on the eco-system of agricultural enterprises, as it saves time, reduces the cost of agricultural operations, increases revenue by increasing the productivity of all segments of agricultural production. Digitalization has allowed to establish interaction of agricultural enterprises with enterprises of related industries, the model of eco-system of agricultural farms demonstrates significant positive changes-many links of the chain have been restored (Fig. 74.2).

The interaction of microeconomic agents of the agricultural sector within the meso eco-system with enterprises of related sectors, due to the regulatory function of the state, has changed significantly. Ties within the eco-system of agricultural business, in the sub-sectors of crop and livestock production have been fully restored. Many agricultural farms have expanded their own feed base to reduce costs on the main cost item, as well as reduce the risks of purchasing poor-quality feed in the field of livestock breeding for meat and dairy farming. Digitalization of business processes

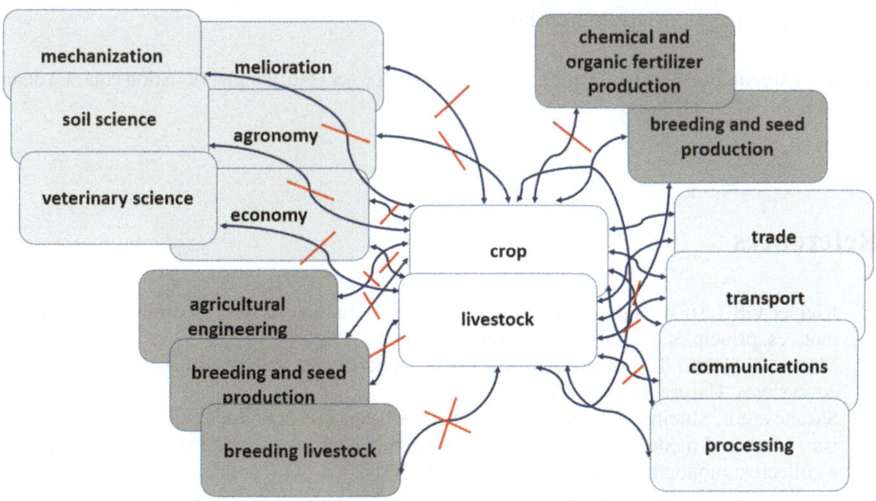

Fig. 74.2 Model of eco-system of agricultural enterprises of Russia in the 2010s

contributed to the expansion of acreage, transformation of the eco-system of crop rotations, introduction of new crops, the most nutritious feed and sederates into modern crop rotations. The latter, in turn, had a positive impact on the state of the soil and, by optimizing the eco-systems of natural biocenoses, increased soil fertility.

The production and application of chemical and organic fertilizers has expanded, the nomenclature of protection products has grown several times taking into account soil and climatic, technological conditions, vegetation conditions of different agricultural crops and other factors.

Production of agricultural self-propelled machines and adapters has stabilized [12], Russian agricultural machinery has become in demand in international markets, which indicates an increase in the level of management at production enterprises [13].

74.5 Conclusions

The theoretical analysis of the transformation of ecosystems of agricultural enterprises of Russia showed that changes in the eco-system at the macro level entail transformations at the levels of lower hierarchies.

The study of empirical material conducted in this article revealed the negative impact of imported liberal eco-systems on agricultural business. The transformation of the institutional conditions of the last decade and the digitalization of the economic space have had a positive impact on agricultural eco-systems and have a serious growth potential.

The authors of the study emphasize the need for state intervention in the formation and maintenance of effective eco-systems in the conditions of digitalization of the economy, as well as state the fact that further research prospects of agricultural eco-systems has heuristic potential.

Acknowledgements The publication was prepared on project No. 3 under donation contract dated March 1, 2019 No. 1154.

References

1. Kleiner GB (2017) System modernization of domestic enterprises: theoretical background, motives, principles. Ekonomika Regiona-Econ Reg 13(1):13–24
2. Moore JF (1996) The death of competition: leadership & strategy in the age of business ecosystems. Harper Business, New York
3. Sazanova SL, Sharipov FF, Dyakonova MA (2019) Spatial economics, geopolitics, and marxism. Marx and modernity: a political and economic analysis of social systems management. a collective monograph. In: Alpidovskaya ML, Popkova EG (eds) A volume in advances in research on russian business, vol. 651. Information Age Publishing Inc., Charlotte, pp 279–288. ISBN: 978-1-64113-749-2 (Paperback), 978-1-64113-750-8 (Hardcover), 978-1-64113-751-5 (ebook)

4. Sazanova SL, Ryazanova GN (2019) Problems and opportunities of development of the agricultural industry of russia from the point of view of marxist theory. Marx and modernity: a political and economic analysis of social systems management. A collective monograph. In: Alpidovskaya ML, Popkova EG (eds.) A volume in advances in research on russian business, vol. 651. Information Age Publishing Inc., Charlotte. pp 599–608. ISBN: 978-1-64113-749-2 (Paperback), 978-1-64113-750-8 (Hardcover), 978-1-64113-751-5 (ebook)
5. Popkova EG, Sergi BS (2018) Will industry 4.0 and other innovations impact Russia's development? In: Sergi Bruno S (ed) Exploring the future of Russia's economy and markets: towards sustainable economic development. Emerald Publishing, Bingley, pp 51–68
6. Ksenofontov AA, Ksenofontov AS, Kirpicheva MA, Trifonov PV (2019) The use of modern management technology to improve business efficiency IOP Conference Series: Materials Science and Engineering, vol. 012114. IOP Publishing Ltd, Bristol, pp 1–4
7. Kamchatova EY, Vasilieva AV, Lyasnikov NV, Dudin MN, Vysotskaya NV (2018) Energy saving management in urban economy and industry. Int J Civil Eng Technol (IJCIET) 9:1423–1429. http://www.iaeme.com/ijciet/IJCIET_Paper.asp?sno=11945
8. Arutyunov AL (2010) On the prospects of using the primary and alternative types of fuel in the agricultural industry of russia. Stud Russ Econ Dev 21(3):282–289. ISSN 1075-7007
9. Rogulenko T, Bodiaco A, Zelenov V, Ponomareva S, Mironenko V (2016) Budgeting-based organization of internal control. Int J Environ Sci Edu. 11(11):4104–4117. Scopus
10. Surkova LE (2019) Snizhenie riskov vzaimodejstviya organizacij na osnove ispol'zovaniya internet-resursov. [Reducing the risks of interaction between organizations based on the use of Internet resources] Estestvenny'e i texnicheskie nauki 2(128):193–194
11. Bogoviz AV, Sergi BS (2018) Will the circular economy be the future of Russia's growth model? In: Sergi BS (ed) Exploring the future of Russia's economy and markets: towards sustainable economic development. Emerald Publishing, Bingley, pp 125–141
12. Degtyaryova VV (2016) Problemy' i perspektivy' innovacionnogo razvitiya otrasli mashinostroeniya a period sankcij i importozameshheniya. Aktual'ny'e problemy' upravleniya–2016. [Problems and prospects of innovative development of the engineering industry during the period of sanctions and import substitution. Actual problems of management–2016. Materials of the 21st international scientific and practical conference.]. State University of Management, Moscow, pp 88–90
13. Adekola A, Sergi BS (2016) Global business management: a cross-cultural perspective. Routledge, New York

Chapter 75
On Improving the Legal Provision of Sustainable Defense-Industrial Development in the Face of Global Climate Change: Setting a Problem

Alexey V. Barkov, Yana S. Grishina, Alexander S. Kiselev and Alexey P. Sokolov

Abstract The article is dedicated to justifying the need to improve the legal provision of sustainable development of the defense industry in the face of global climate change. The aim of the study is to assess the impact of environmental ecologization on the legal provision of sustainable development of the defense industry, which can affect the competitiveness of defense enterprises. *Methodology:* The methodology of the study is based on the development of the basic idea, the essence of which is that the sustainable development of the defense-industrial complex is a modern sought-after direction of development of Russian economy, creating impact, both on market relations and the environment. The main methods used in the work are: methods of empirical cognition (comparative-legal, description, interpretation), theoretical methods of research (formal and dialectical logic, systemic method), as well as analysis, synthesis, method of induction and deduction. *Results:* The conclusion has been made that the sustainable development of the defense-industrial complex is ensured by the ecologization of legislation in this area. At the same time, the change in the course of exploitation of natural resources in industrial production, taking into account environmental factors, is ensured by environmentally responsible obligations of legal relations subjects. The possibility of realizing foreign experience in the sphere of defense-industrial complex in the reconstruction of Russian environmental-oriented model of corporate governance is justified. The aim of the new model is to achieve the competitiveness of the enterprise operating in the field of defense industry, increasing the demand for dual-use products in the regional market, due to the environmental over orientation of production and management.

A. V. Barkov (✉)
Legal Regulation of Economic Activities, Financial University of Russia, Moscow, Russia
e-mail: barkov_a_v@mail.ru

A. V. Barkov · A. S. Kiselev · A. P. Sokolov
Department of Civil Law, Russian Defense Ministry Military University, Moscow, Russia

Y. S. Grishina
Russian State University of Justice, Moscow, Russia

T. Kolmykova and E. V. Kharchenko (eds.), *Digital Future Economic Growth, Social Adaptation, and Technological Perspectives*, Lecture Notes in Networks and Systems 111, https://doi.org/10.1007/978-3-030-39797-5_75

Keywords Sustainable development · Defense-industrial complex · Ecologization of legislation

JEL Classification Codes K2

75.1 Introduction

In recent years, scientists of all countries have been hitting the nabat, warning of an inevitable environmental catastrophe, linking its approach to global climate change, caused by the rapid growth of active economic activity Humanity. The reality of the threat of warming from the greenhouse effect caused by the burning of coal, gas and oil, which leads to the imbalance of all natural systems, puts on the agenda the urgent question of what should be done to prevent climate chaos, but not to slow down scientific and technological progress? In these circumstances, the best state anti-crisis course is "sustainable development," which means harmonizing the process of economic and social change, taking into account the needs of citizens, as well as future generations, without significant damage to the biosphere.

The security of the Russian Federation, including its environmental component, is the defense-industrial complex (the defense industry), whose mission, along with national security issues, is also to ensure the quality of life Russian citizens. Consequently, the sustainable development of the defence industry, as a real sector of the economy, in the face of global climate change, largely depends on balanced state development, in which the exploitation of natural resources, direction investment, the orientation of scientific and technological progress, agreed with each other to ensure the current and subsequent increasing needs of Russians, which makes the topic of this study relevant.

75.2 Materials and Method

The methodology of the study is based on the development of the main idea of the theme of work, the essence of which is that the sustainable development of the defense-industrial complex is a modern, sought-after direction of improving the Russian economy that have a positive impact on both market relations and the environment. As a result of the methodological basis of the study, the scientific thesis is that "sustainable development of the defense-industrial complex" is a social and environmentally-oriented public-private partnership, within the framework of the which, together with the systematic extraction of profits, is being addressed by the global problems of humanity, such as the preservation of a favourable environment and wildlife.

The empirical basis of the study is the legal experience of foreign countries (Kazakhstan), which have taken a course on the sustainable development of the defense-industrial complex, where there is already a practice of legal regulation of the state defense order and formation of a defense industry based on environmental principles.

Accordingly, the activities of companies engaged in systematic cooperation on the basis of public-private partnership are also environmentally-oriented. According to the purpose of the study, empirical (comparative-legal method, description) and theoretical (formal-logical) methods of the study became the most significant.

75.3 Results

Results: The rationale for the main areas of improvement of the legal provision of sustainable development of the defense-industrial complex in Russia in the face of global climate change.

75.3.1 The Importance of the Defense-Industrial Complex for the Development of the Country's Economy

The importance of the defense industry for the development of the country's economy is great. The complex is not only a guarantor of the national security of the state, but also on the basis of its innovative developments is carried out regional economic development. The role of the defense-industrial complex cannot be overestimated, as it is an essential element in ensuring the security of the state and the quality of life of the country's population, both in peacetime and in the context of the escalating geopolitical situation. Since the mid-twentieth century, the defense-industrial complex has consistently taken a leading position in the economy.

Today, as in earlier times, the United States and its allies seek to establish global hegemony not only through economic and political influence on international processes, but primarily by rebalancing the armed forces in their favor. It is obvious that in order to maintain the functional well-being of the Russian defense industry at the necessary and sufficient level, its constant development is needed: the development and production of weapons, military equipment, scientific and technical and design and construction activities, training of qualified personnel and more.

At the same time, in the conditions of the impending environmental crisis, the development of the defense-industrial complex should be fully in line with the ideology of sustainable development, providing not only the defensive function of the state, but also the ecological. The state is obliged to set the vector of the development of the defense industry and policy, to determine the necessary level of military and technical equipment of the armed forces taking into account the geopolitical situation, environmental security, as well as socio-economic factors affecting the well-being of the country's citizens. It is clear that the sustainable development of the defense-industrial complex should be based on a well-developed and well-considered strategy that takes into account the basic ideas and essence of the concept of "sustainable development" that includes positive foreign experience in this Sphere.

However, as the scientific literature rightly notes, "at the moment there are virtually no methods, methods, criteria, measure and standard of sustainability of development in relation to the defense-industrial complex" [6]. The question arises whether the criteria for sustainable development apply to enterprises that comply with the state order for the defense industry? In order to answer this question, let us turn to the history of the problem.

75.3.2 The Historical Premise of the Adoption of the Concept of Sustainable Development at the International Level

Today is undeniable that in the twentieth century the issue of ecology became of particular importance to the entire world community. The consequences of the Second World War, the largest military-political clash in the history of mankind, the process of building industrial and scientific and technological capabilities, the arms race and the space campaigns of the superpowers made us think not only environmentalists and scientists, but also leaders of states about the negative consequences for the well-being of all. A natural event was a multilateral international dialogue, which was built during the second half of the twentieth century. The UN Environment Commission, the Rome Club, Greenpeace, the World Wildlife Fund and the International Green Cross have been operating on the international scene.

Over time, environmental organizations began to express ideas about the preservation of nature, which were transformed into a social and political agenda. Internationally, the UN Conference on the Environment (Stockholm, June 16, 1972) is adopted, guiding the development of humanity on the path of ecologization. As part of the work of the International Commission on Environment and Development in 1987, the report used the term "sustainable development" to mean development that satisfies the needs of the present, but does not jeopardize the ability of future generations to meet their own needs.

The theme of sustainable development is on a par with the security issues of heads of state. Many politicians, scientists and journalists begin to actively discuss environmental issues at the highest level. From that moment on, the term enters the scientific circulation, becoming common [1]. The logical continuation was the signing of the 1992 Declaration on Environment and Development (Rio de Janeiro, June 14, 1992). Article 1 of the Declaration states that "human care is central to sustainable development" and that the right to development must be respected in a way that adequately meets the needs of the current and future generations in the fields of development and the environment. This document is notable for the fact that it enshrines the principles and basic ideas of the concept of "sustainable development".

75.3.3 The Peculiarities of the Interpretation of the Term "Sustainable Development"

We believe that the first step is to focus on the interpretation of the term "sustainable development", this seems to be a necessary starting point for further research. It is noted that despite the wide discussion of the concept of sustainable development, there is no single approach to this concept [5]. According to L.V. Shchukina "this is due to the fact that in different languages the term finds its own meaning: in French the term means long-term development, in Italian - deserving of development, in German - long-term development, in Swedish - sustainable development, in Japanese - long-term development" [10].

A.G. Ivolga and A.A. Chaplitskaya speak about the peculiarities of the semantic meaning of translating a foreign term into Russian: "At first glance there is some inconsistency, illogicalness of the Russian version of the translation of the term "sustainable" "I'm in a state of "Sustainability" implies equilibrium, and "development" is possible only if the system is permanently out of equilibrium" [3]. Indeed, given this interpretation of the term, there is some doubt in the true understanding of the concept itself. The most accurate explanation of the meaning of the term, in our opinion, gives S.A. Suntsov, who points out the following: "The phrase "sustainable development" literally means "supportive development", with the adjective "sustainable" implies, along with sustainability, a development process that can be maintained indefinitely through the constant reproduction of its basic properties" [9]. Based on this position, it implies the possibility of reusing the resource, its "refill" in the process of industrial production of material goods without causing significant damage to the environment.

Also noteworthy is an important piece identified in A.N. Kolmakov's study: "Sustainable Development" (continuous, sustained, long-term, supported, protected development) means maintaining a balance between mutually related elements of the system: economy, social sphere and environment" [4]. We focus on this moment: resources, including natural ones, are limited, their thoughtless use in order to increase profits inevitably leads humanity to a world catastrophe. Accordingly, sustainable development involves creating parity between the needs of humanity (social sphere), commercial activities (economic), and the preservation of the world (environmental) by strictly complying with environmental principles under the law.

By systematizing information, it can be concluded that the concept of sustainable development guides the state in a way that takes into account the needs of all citizens, as well as future generations, without significant damage to the biosphere at the expense of political and legal regulation aimed at harmonizing the socio-economic and environmental spheres.

75.3.4 On the Legal Regulation of Sustainable Development in the Russian Federation

The creation of a single document (declaration), which includes the basic tenets of ecology, is primarily due to the need to change the worldview and mentality of the person, which is based on the principles of conservation of nature and animals after the ratification of the "Brazilian" Declaration, many countries have taken the path of sustainable development, and active rule-making has begun, accompanied by the adoption of the relevant regulations.

Russia was no exception: in 1996, the Russian President's decree approved the concept of the Transition of the Russian Federation to Sustainable Development, which defines the main objectives, directions and conditions of Russia's consistent transition to sustainable development, to ensure a balanced solution to socio-economic and environmental challenges and natural resource potential. Of course, the adoption of the concept does not mean immediate and rigorous implementation of the provisions. As with any other innovative solution, the implementation of sustainable development principles is difficult. O.V. Skotarenko believes that "the main problem of sustainable development is the contradiction between the thoughtless, predatory approach to meeting the needs of today, the struggle of individual groups and states for control of the world's resources - and the approach to the these needs from the point of view of social partnership, mutually beneficial for the long term interstate and interregional (both inside and supranational) relations, concern for future generations of humanity" [8]. This problem is one of the key ones, and the world powers are currently striving to solve it.

Some researchers point out that in order "to plan for the sustainable innovation development of defense-industrial complex enterprises, it is necessary to ensure the formation of a holistic management system of innovation activities of defense industry enterprises. Including: the elaboration of documents describing the main areas of technological elaboration of defense industry enterprises; the formation of management structures being responsible for the technological and innovative development of defense industry enterprises; implementation of a system of measures aimed at expanding scientific and industrial cooperation; the formation of an intellectual property management system, etc." [2]. Of course, compliance with this set of measures increases the efficiency of the defense-industrial complex. However, the proposal does not have a place to address environmental issues that are relevant to any industrial enterprise, especially in the field of defense with high production capacity. We believe that the sustainable development programs of defense-industrial enterprises should not only include the elaboration and implementation of new technologies, but also take into account the environmental aspects of the problem.

75.3.5 Ecologization of Production as a Way to Ensure the Competitiveness of Enterprises Engaged in the Field of Defense Industry

In the context of global socio-economic and environmental crises, the problem of ensuring the competitive advantage of defense industry enterprises comes first. Despite the current stabilization of the state-defense order [7], this does not mean that the enterprises of the defense-industrial complex have protected themselves from all risks. If there is a hypothetical possibility of reducing the state defense order, then inevitably there is a problem of ensuring the competitiveness of defense industry enterprises in the market. Scientists and practitioners are encouraged to introduce new technologies, promote research and development, create patent departments in enterprises and more. Nevertheless, we can state that doctrine and practice do not pay enough attention to improving the competitiveness of defense industry enterprises by greening production through innovative solutions to the problems associated with market failures. This approach can help to increase the recognition of the company's social brand in the market, eco-oriented image, demand for dual-use products from the regional consumer.

In foreign countries this practice has been found to be successful, spreading among private enterprises and entrepreneurs. We believe that for Russian enterprises, which carry out their detailed implementation of the state order to build up the defense-industrial complex, the use of environmental innovative technologies will also contribute to the use of environmental innovative technologies to make them more competitive.

75.3.6 The Problem of Legal Regulation of the Ecologization of Defense Production: The Main Areas of Improvement of Legislation Taking into Account Foreign Experience

However, the sustainable development of the defense industry is hampered by the lack of right enforcement of the defense industry. Which should be achieved by the ecologization of legislation in this area. At the same time, unlike Russia, in the allied state of Kazakhstan on March 18, 2019, the Law No. 236-V- missile defense system "On Defense Industry and State Defense Order" was adopted, providing, in particular, the processing of property, that is. Bringing unused property in a state that excludes the use of it for its intended purpose, with the extraction of raw, or any other, materials, further used in the production (manufacturing) of goods or other products, as well as changes in properties for purposes facilitate the treatment of it, reduce its volume or dangerous properties. The Legislature of the Republic of Kazakhstan has provided, one of the main areas of legal support for the development of the defense industry, environmentally friendly treatment of waste production which fully corresponds to the idea of sustainable Development. It follows that in the competition for the defense order and in its implementation, companies will be obliged to look for such technological solutions that will allow the processing of waste and residues. As a result, the effect will help to solve environmental problems, and from an economic point of view, environmental-oriented technologies will benefit the state, the competitiveness of private enterprises, and accordingly increase the quality of life for citizens. It seems that such experience can be in demand in Russia while improving the legal provision of sustainable development of the defense industry in the face of global climate change.

75.4 Conclusions

In conclusion, the uniqueness and novelty of the proposed concept of sustainable development of the defense-industrial complex on the basis of public-private partnership are that they imply a change in the course of operation natural resources in industrial production on the need to take into account environmental factors, ensuring environmentally responsible obligations of legal relations subjects. This sustainable development is achieved by the ecologization of legislation in this area.

The aim of the new corporate governance model is to achieve the competitiveness of the enterprise operating in the field of defense industry and to increase its recognition in the market by drawing public attention to the environmentally oriented nature Production. According to the new ideology, the company's mission itself is changing: in the classical model, the company focuses itself on maximizing material goods, in the modern model there is a need to be in harmony with the environment, to take into account biosphere. This changes, first of all, the paradigm of thinking. Thus, the concept of the legal provision of sustainable development of the defense-industrial complex in the face of global climate change becomes a completely new approach to addressing the pressing environmental problems that used to be aspect was not given due importance to, both in theory and in practice.

In order to ensure the competitiveness of Russian enterprises in the defense order and production of dual-use products, it is necessary, similar to the Kazakh right-wing approach, to consolidate the need at the legislative level recycling of waste production and other activities that meet modern standards and environmental safety requirements.

References

1. Danilov-Danilian VI, Piskunova NA (2015) Sustainable development: new challenges. Aspect Press, Moscow
2. Zakutnev SE, Kandybko NV (2014) Features of planning sustainable innovation development of defense-industrial complex enterprises. Natl Interes Prior Secur 27(264):29–35

3. Ivolga AG, Chaplitskaya AA (2014) Justification of the approach to the concept of sustainable development of the region's economy. Bioresour Environ Manage 1–2:151–154
4. Kolmakov AN (2013) To the issue of the ratio of sustainable development and economic growth in Russia. Social Econ Phenom Processes 7(53):72–77
5. Companiec MA (2016) The concept of sustainable development in the agricultural sector. Her Donetsk Natl Tech Univ 6(6):53–59
6. Mikhin VF, Chevgun SS (2018) Sustainable prospective development of Russia's defense and industrial complex. Econ Ind 3:235–241
7. Popov NE (2016) Sustainable development of the defense-industrial complex by improving its competitiveness. In: Management of quality and sustainable development in a changing world, Saransk, 2016, Ogarev Mordovia State University, Saransk, pp 114–119
8. Skotarenko OV (2011) The problem of sustainable development in Russia and abroad. Herald Murmansk State Tech Univ 1:34–37
9. Suntsov SA (2018) History of the concept and concept of sustainable development. Sustain Dev Sci Educ 9:15–22
10. Shchukin LV (2015) Theoretical aspects of the sustainable development of regional socio-economic systems. Pskov Reg J 21:38–50

Chapter 76
The Issue of Environmental Protection in the Regional Mass Media Space

Yulia N. Grickevich, Svetlana V. Lukyanova and Irina N. Korenetskaya

Abstract Modern economic processes require an interdisciplinary research, due to the fact that they entail crucial changes in the economic, social, political spheres, as well as in public mind, personal views and goals. The need to include the media information space analysis in multidisciplinary studies is determined by the increasing influence of the media (which have turned into the open interactive discussion platforms about all spheres of human life) on the public opinion about the economy and the environmentalism of technological progress. The authors studied the Pskov region mass media discourse regarding the situation with the "Titan-Polymer" plant, the planned construction of which caused an ambiguity in society, because of environmental issues as well. Along with political and social forces, regional mass media can play an important role in stabilization of the economic situation. The discourse analysis of mass media identifies the basic patterns for creating meanings in the information space that influence the human acceptance or rejection of new economic objects, and to determine the mechanisms of interaction between the linguistics of the text and the real situation surrounding the object. The content analysis of regional mass media publications took into account the rating scale when highlighting an informational event. The publications analysis, which resulted from the discussions of "Titan-Polymer" plant construction, identified and described the main trends in organizing the regional mass media discourse and the effect on the public opinion transformations, which can be used to stabilize the situations related to greening the technological progress.

Keywords Discourse · Economical information reason · Environmental protection · Ecological situation · Information space · Mass media · Mass media space

JEL Classification Code Q5

Y. N. Grickevich (✉) · S. V. Lukyanova · I. N. Korenetskaya
Pskov State University, Lenin's Sq. 2, Pskov 180000, Russia
e-mail: Grickevich68@yandex.ru

T. Kolmykova and E. V. Kharchenko (eds.), *Digital Future Economic Growth, Social Adaptation, and Technological Perspectives*, Lecture Notes in Networks and Systems 111, https://doi.org/10.1007/978-3-030-39797-5_76

76.1 Introduction

Linguistic research is increasingly being included in the system of interdisciplinary studies of various phenomena of reality, including the economic sphere. Among the factors that effect the development of the economy, traditionally we include the objective economic factors, the factors of the economy state management, the factors of natural wealth, investment factors, and force majeure factors. Recently, special attention has been paid to innovative factors and the human factor as well. However, the factor of the mass media effect on the economic processes is not always considered.

The emergence of a whole range of macroeconomic issues caused by the introduction of new economic objects, even high-tech ones, results into serious shift in public mind. The environmental aspect of the economy has recently come to the forefront, often turning into a discussion point that determines the future of the economic object.

The strengthening of the interaction between the informational media space and the economic and socio-political space makes it important to analyze and describe the mechanisms of shaping the public opinion about a particular segment of the economy, which significantly effects the adoption of economic and political decisions.

The authors conducted a study of the mass media discourse in the Pskov region regarding the situation with the "Titan-Polymer" plant, the planned construction of which caused public debate because of the environmental issues as well. Along with political and public forces, regional mass media played an important role in stabilizing the economic situation. After analyzing the discourse practice of regional media, the authors highlight the main trends in the organization of mass media discourse, which significantly effected the transformation of public opinion.

The complex structure of communication both in the mass media space and in the transformed mass media reality determines the relevance of including studies of coverage and interpretation of economic informational issues in mass media in a wider theoretical context. It is necessary to identify and describe the patterns and mechanisms of interaction between mass media and the state, business and society.

76.2 Materials and Method

Media linguistics as a part of linguistics that studies the functioning of language in the media in various aspects in modern linguistic science has long been recognized as a new independent branch of science. The methodological apparatus of media linguistics integrates various studies of media texts: discourse analysis, content analysis, cognitive linguistics, critical analysis, functional stylistics, linguocultural studies. As part of the study, all components and levels of media text in a combination of linguistic and extralinguistic factors are analyzed [1].

In developing the research methodology, we refered to the works of Dobrosklonskaya [1], Chicherina [2], Nazarov [3], Reznik [4], our study is based on the conclusions of Kubryakova E. S., who substantiated the cognitive-discourse approach to linguistics and linguistic objects, which involves a comprehensive interdisciplinary study, taking into account the interaction of various linguistic areas. Moreover, any study is based on considering two fundamental functions of the language—cognitive and communicative "in constant interaction and coordination with each other" [5].

76.3 Results

Economic information in the mass media, even if it is presented in the genre of a news article, not only reflects the reality, but also introduces significant additional adjustments to the mass addressee's world picture in mass media discourse. Economic reality and its image in mass media, and especially public opinion about the economic situation, do not always coincide. At the same time, the vectors of the formation of public opinion in the mass media space significantly effect the economic processes, up to the decision to the right of an economic object to exist: the broadcast mass media vision of the situation in any sector can largely determine the attitudes and behavior of people in the economy and politicians.

One of the methods of presenting information, which is clearly visible in the media, is a negative bias in covering economic issues. This can be expressed both in more frequent and expanded coverage of negative changes in the economic sphere, and in some imbalance between the positive and negative background of introducing the economic situation into the existing information. Economic events in the information flow are often arranged in such a way that even news with positive bias deliberately fits into a broader negative context. That can be perceived as a threat of forced adaptation of the new economic conditions and, consequently, transformations of traditional behavior. The mechanism of increased attention of the addressee to the problem areas of the economy, launched by the mass media, determines the replication of either the news item itself or its interpretation in a negative way, due to the fact that the demand for the content of a particular media information channel increases, which becomes economically beneficial for the media, being a segment of the economy.

Such an informational economic reason turns out to be more significant in the regional mass media space in comparison with the mass media space of a country or the world. A smaller number of reasons in the information space of the region makes each event more significant. Economically beneficial projects often encounter prejudices in society, developed by mass media, which is the starting point for the discussion in the media space, using arguments and counterarguments of both objective and subjective nature. In such a discussion, special attention should be paid to the language means by which the authors of the discourse raise or lower the status of argumentation.

Reducing the argumentation status of opponents can be achieved by concentration of units with similar semantic components, which for an unprepared mass addressee can be invisible. So, in the title, **The woman pretending to come from Omsk at a meeting in Pskov turned out to be a Pskov animator** in "Pskovskaya Pravda" on March 26, 2019, the word **to pretend** with the meaning of "to represent oneself for somebody else, to announce oneself for somebody else" [6] launches a mechanism of distrust of the previously given argumentation. During the development of the discourse, the semantic component "lie, deception, implausibility, substitution" is gradually consolidated, both due to the units in which this component is easily recognized, and due to the units where this meaning is implicitly expressed: "**The so-called** "**resident of Omsk**", who criticized the "Titan-Polymer" company on March, 23 at the meeting of the Communist Party of the Russian Federation, **turned out** to be a Pskov animator, an anonymous telegram channel "Pskovskiy Pravdorub" reports. For her services, a professional **actress** and a wedding moderator takes a reward of 10,000 rubles … ". If in an expression **the so-called**, which is used to express an ironic or contemptuous attitude towards someone or something that has some name [6], the evaluation is already contained within the meaning of the word, but in the following words the evaluation is absent: **to turn to be** as "to reveal, to appear, to come out" [6] and the **actress** as "an artist – performer in theatrical performances, in films, on television" [6], however, within the context these units reinforce the overall impression of unreliable arguments of some participants in the discussion of the economic situation. Quotation marks for the phrase "**the resident of Omsk**" contribute to the general impression of cheating. The place of lexical units with a common semantic component, relative to the compositional structure of the discourse also plays an important role in the formation mechanism of the recipient of mistrust. So, the word **myth** as "an unreliable story, fiction" [6] at the end of the text, with the background information, which is presented in the official style, highlights the necessary semantic emphasis: "The construction is still far away, but the number of **myths**, surrounding the future enterprise is growing in arithmetic progression."

The analysis of the media discourse showed that a similar method of counterargument is actively applied by various authors: "Galina Artyunina, Doctor of Medical Sciences, Professor of the Department of Fundamental Medicine and Biochemistry of Pskov State University, **debunked** "**scare stories**" about the "Titan-Polymer" plant at a meeting of the Public Chamber of the Pskov Region on May, 28": I immediately drew attention to the "**scare stories**" about the "Titan-Polymer" plant ("Pskovskaya Pravda", May 28, 2019). Introduction of the word "**a scare story**" into the context where the professionalism of the primary discourse author is underlined, and the word is used jokingly about a story, a movie that causes fear [7], and the verb "**to debunk**" in the meaning of "to deprive of respect, public recognition" [6], the use of quotation marks contribute to the realization of the same mechanism of formation the distrust of the opponents arguments and, consequently, a shift in public consciousness regarding the conflict situation.

The inclusion of even one of the lexical units that neutralize the arguments of one or another side of the discussion in the title strengthens the position, stated in the author's discourse: **to dispel** "destroy, disperse" [6]: "The head of the "Titan-Polymer" plant **dispelled** Estonians' concerns about its ecological compatibility" in the publication of the "Moskovsky Komsomolets" newspaper, dated May, 16, 2019.

The political situation, as we know, makes an impact on the economy. The representatives of different political views use the current economic event to their advantage, which is reflected in the mass media, and is interpreted in the texts and headings: "The Pskov opposition was refused a referendum on the construction of "the Titan". NICHOSI (from Russian: wow!)" in the newspaper "Moskovsky Komsomolets", dated April, 19, 2019. The lexical unit which appears in the headline is typical for the youth slang, and it removes the initial menacingly categorical impression, adding a clearly ironic subtext. Formed as a result of the spelling game of the expression "Wow!", uttered in Russian in a fluent speech, it became an expression of an extreme degree of surprise [8] with a touch of naivety, childishness, which finally got fixed in the form of an Internet meme. And hereinafter, in order to impress the addressee, to fulfill the function of influence, the transformed precedent text was used ("Pskov opposition, I brought the terrible news to your house! Call the children!"), which is a vivid way of evaluating and creating a comic effect. The above means of influencing the recipient also lead to a decrease in the communicative complexity of the tough issue.

Interpretation of economic events, on the one hand, forces the author of mass media discourse to simplify the source material, in particular by eliminating or reducing the terms used in one or another sphere of the economy, replacing objective indicators of the economy with a subjective vision of the economic situation, so that complex economic processes are comprehensive and interesting for a mass addressee. There is an inevitable and justified simplification of information in order to reduce excessive communicative complexity.

Mass media can fix in the mass addressee consciousness some necessary components of the worldview and contribute to the interests transformation (primarily economic ones) in individuals, representing them as the interests of the whole society. Due to the fact that the interests of different communities, as a rule, do not fully or partially coincide, the arguments presented by the author of the discourse as reflecting the interests of society will also differ significantly. Thus, the adjective "**anchor**" (in the paraphrase "**an anchor resident**", "**an anchor investor**") in relation to the "Titan-Polymer" is intended to emphasize the role of a new economic enterprise in the region. The adjective "**anchor**" acquires terminological economic meaning, which is already reflected in special lexicographic sources: "Anchor residents of the technopark – the branches of large companies with a long history of successful activity, some of which correspond to one large direction of science and technology development with high tech and intellectual resources application" [9].

On the other hand, in case of conflict issues, one of the ways to stabilize the information space, and therefore the economic situation, may be to return the terminological layer of linguistic resources to an economic discourse. The use of specific scientific terminology in a text determines its semantic parameters—informative, argumentative, and prognostic [10]. The desire for accuracy in the transfer of information leads to the emergence of highly specialized terms such as **ethylene**

glycol, dimethyl terephthalate, tetrahydrofuran, bio-purification, terephthalic acid, acetaldehyde, polyethylene terephthalate, etc. Herewith, the author is clearly interested in removing the misunderstanding of the contents of the publication posted on May 30, 2019, on the Pskov news site with an explanation by both direct and indirect definitions: "We remember that **ethylene glycol** is a dihydric spirit, which is classified as a hazard class III substance. We note that acetic acid, hydrogen peroxide, iodine, manganese compounds and other products that we encounter not only in production but also in everyday life also belong to this class".

Strengthening the terminological component of the text, which is economically reasoned, does not simplify the discourse and make it more comprehensible, but promotes achieving a communicative goal such as stabilization of public opinion. A large number of terms is perceived by the addressee of the discourse as an indicator of objectivity in the coverage of the event. This is especially important when the recipient of the discourse lacks personal experience and sufficient background knowledge regarding the interpreted economic processes. The mass addressee ceases to feel, or to a lesser extent, feels the effect of subjectivity in interpreting an informational case, which launches a complex mechanism of trusting the information received and including such information in the system of personal beliefs.

The degree of conflict and the length of the discussion of an economic news item depend on the fragment of the media space, where the discourse is realized. So, social networks are also actively used as an interactive discussion platform for the representatives of those branches of the economic sector, around which there are disagreements in society: "Based on citizens' appeals, the "Titan" group of companies prepared the main theses "for" and "against" the construction of the "Titan-Polymer" in "Moglino". Infographics is available in the official group of the company on social networks" (in "Pskovskaya Pravda", dated May 19, 2019).

This method of constructing a discourse, when an argument is formulated for an opponent in a form which is convenient for the author of the discourse and a counterargument is given, is known in constructing a discourse on economic issues and is quite effective in terms of influencing public consciousness. The arguments for and against are perceived by the mass addressee as an objective reflection of the situation. Voluntarily or involuntarily, such a discourse is assessed as a demonstration of strength, confidence and rightness, since problems in the field of economics are spoken openly. The chosen form of narration, a sudden shift in speech registers during the transition from the opponent's argument to their own counterargument contribute to fixing the desired image of each interested party in the recipient's mind. The use of many set expressions, typical for colloquial speech, is aimed at initially reducing the status of the arguments given by the opponent: "For "Titan" **everything is going smoothly** – they **gave a kickback** and no one will ever check anything. They are **making money hand over fist at our expense**". The Russian expression "**to go smoothly**" was originally applied to carts and generally to such tools or vehicles that were smeared before the trip, and meant, figuratively, 'will leave soon, is already set up for departure and ready to move, ready for'. We can compare it with German 'fertig', which means 'prepared', formed from the verb 'fahren', meaning 'to go' [11]. The expression "**to give a kickback**" is given in the dictionaries as barbaric colloquialism and it means 'to bribe someone by giving a bribe'. The expression '**making money hand over fist at**' is considered as slang and is used in the meaning of 'to appropriate (usually illegally, secretly) money allocated for any purpose' and is strengthened with the pretext 'at the expense of', which is used when indicating the subject, whose funds include any expenses [12].

The transition from concentrated use of stylistically lowered expressions to the units, which are characteristics for a neutral official style of presenting the information creates an impression of objectivity and reliability: "Even in Russia, banks **issue loans** only for **well-grounded business plans**. Apparently, "Titan-Polymer" has proved to them its potential profitability".

Creolisation of the text is an effective way of presenting material in modern media. In the modern media sphere, this are, first of all, the texts, in which verbal and visual components interact—photographs, drawings, infographics. The text, fixed in the format of infographics, is a universal tool for transmitting information to the media. Infographics can increase the efficiency of perceiving any data due to the logical structure and a clear picture that summarizes the facts and demonstrates the information, which is connected from part to part. Infographics can be called a special visual summary. "Infographics means reference or illustrated statistical information, presented with various visualization methods: using graphs, charts, histograms, mental maps (mindmap—memory cards), timelines (one of the basic principles of animated infographics), etc. This form is suitable to present quickly a large amount of information. It is a universal tool for popularizing conceptual information" [13]. With its help, information can be conveyed simply, brightly and beautifully. The combination of a text, image and design helps to remember terms or concepts that are usually difficult to perceive in a continuous text. Complicated specific scientific and economic information in the infographic format has a clear, capacious and beneficial presentation for the author: What is textile PET and how environmentally friendly is it? (in the Pskov news, September 6, 2019). The use of the subtitle "PET production: facts only" is subordinate to the objectives of the discourse—to present information objectively, reliably and clearly, taking into account the specifics of the transmitted knowledge.

The stabilisation of the conflict situation in the economic sphere is achieved due to the application of various techniques. First of all, this is an increase of people in the mass media discourse, whose opinion is authoritative for this case in connection with their professional or social activities: "**Chemists** have no questions about the Pskov "Titan-Polymer" plant, Larissa Nikolskaya said, an associate professor of Technosphere Security Chair, Pskov State University, on May 28, Ph.D. in Chemistry, secretary of the Pskov Public Chamber, freelance expert of the State Ecological Expertise" (see Larissa Nikolskaya, "All steps required for construction of "Titan-Polymer", Pskovskaya Pravda, May 28, 2019).

Secondly, the media broadcast a discourse, in which the polar points of view are not so clearly reflected, in the discourse there is not so much of a juxtaposition of opposing positions, but a combination of opponents' views: "13 billion of rubles of investments is an amount that the region has not seen for a long time. However, of course, there are risks. I understand well the questions to the investor. Safety is above all. But I want to emphasize: this topic is equally worrying for the administration of the Pskov region" (see Mikhail Vedernikov: "The Titan-Polymer" project can significantly affect the growth of the economy", Pskovskaya Pravda, May 30, 2019).

Thirdly, the concentrated use of terminological vocabulary even if it is unknown to an unprepared mass addressee, in order to include those stereotypes when the use of means of a terminological system is assessed as an indicator of the reliability and objectivity of the facts cited: "The expert reminded that "Titan-Polymer" plant in Pskov plans to produce **a biaxially-oriented film (BoPET)** ... Another product is **PET-granulate (polyester)**" (see Expert: I would call the fight against polymer disposable tableware—a populism, Pskovskaya Pravda, August 30, 2019).

Fourth, this is a transition from a discussion format to a format for approving a solution to a conflict situation around an economic object: **"Titan-Polymer will be opened ..."**: see Sergei Elizarov about "Titan plants": I am not a chemist or a biologist, but I listen to the professionals in "Pskovskaya Pravda", May 28, 2019.

76.4 Conclusions/Recommendations

Thus, mass media discourse is the result of complex processes when mass media in the economic space can be presented not only as one of the profitable sectors of the economy, as a link that unites all the parties that are significant for successful media activity: the state, society and advertisers, but also as a kind of regulator of economic interaction and information space. Mass media information channels have certainly become the main channels for broadcasting economic news and regulators of public opinion. Reducing the amount of negative information will reduce the degree of conflict in solving problems of an economic nature, but such a technique can only create the impression of a resolution of the conflict, especially if it concerns environmental issues that are perceived especially sharply by the population. Qualitative and sustainable changes in public relations can be formed in the mass media discourse with the help of concentration and increase in the amount of language units within a single text (terminological vocabulary, words with similar semantic components, set expressions, which are characteristics of colloquial speech). For stabilizing the situation in the economy it is necessary to shift from the format of discussion, debate to the format of assertion.

References

1. Dobrosklonskaya TG (2000) The issues of studying media texts: experience of studying the modern English media speech, Moscow
2. Chicherina NV (2007) Typology of media text as the basis of formation of media literacy. Bull A.I. Gertsen Russ State Univ 47(9):159–166
3. Nazarov MM (2000) Mass communication in the modern world: methodology of analysis and practice of research. URSS, Moscow

4. Reznik DA (2004) Regional mass media and public authorities: ways of interaction. In: Dylnov GV (ed) Saratov
5. Kubryakova ES (2001) Regarding the current tasks of word building at the brink of centuries. Materials of the international conference devoted to scientific heritage of Prof. M. D. Stepanova and its further development, Russia. Moscow State Linguistic University Publ., Moscow, pp 11–18
6. Ozhegov SI, Shvedov NY (1999) Explanatory dictionary of the Russian language, 4th edn. Azbukovnik, Moscow
7. The large explanatory dictionary of the Russian language: [BTS:A-Z] (2000). In: Kuznetsov SA (ed) St. Petersburg
8. Mselnik YA (2017) Socio-linguistic projects "Word of the year", as a source of information on neological processes in the modern Russian language. PhD thesis, Omsk
9. Explanatory dictionary "Innovative activities": terms of innovative management and adjacent spheres: (from A to Z) (2008). In: Suslov VI (ed) Siberian Scientific Publ., Novosibirsk
10. Valko NA (2006) Futurism in the newspaper text of the economic topic. PhD thesis, 10 Feb 19, Krasnodar
11. Birikh AK, Mokienko VM, Stepanova LI (1998) Dictionary of Russian phraseology. St. Petersburg
12. Efremova TF (2000) New dictionary of the Russian language. Russian language, Moscow
13. Simakova SI (2012) Infographics: visualization of digital content", Bulletin of V.N. Tatishchev Volzhsky University, No. 3. pp. 219–226

Chapter 77
Organization of Insect Pest Control in Gardens of Pskov and Its Suburbs at the Beginning of the Twentieth Century

Andrey P. Ignatiev, Oleg P. Korolkov and Svetlana F. Matsevich

Abstract *Purpose*: Solving environmental problems in agriculture is an important aspect of human activity. One of the ways to overcome these problems is chemicalization, which was used in Russia as early as in the nineteenth–early twentieth centuries. The aim of the study is to cover the history of chemical pest control in the Pskov province at the beginning of the twentieth century and to identify the problems faced by the Pskov department of the Imperial Russian Gardening Society leaders in the course of events. *Design/methodology/approach*: Using both general scientific and special historical methods, the authors present an objective picture of the agricultural pest control organized by the designated organization, its members and residents of the province. Based on materials from the local periodical press of the beginning of the twentieth century, archival sources and scientific literature, the article discusses organization of the garden crops insect pest control by the Pskov department of the Imperial Russian Gardening Society within the territory of Pskov and its suburbs in 1908–1909 on the example of apple moth ("May worm"). *Results*: The analysis of the material studied allowed us to draw conclusions about the results and practical significance of the measures taken. In general, the measures had a positive effect, since the goal—the destruction of pests—was achieved. However, it became obvious that the insecticides used in the processing of orchards are dangerous not only for apple moths, but also for the living beings around.

Keywords Apple moth · Flying squads · Gardening · Gardening societies · Insecticides · Nurseries

JEL Classification Codes N530 · Q130

A. P. Ignatiev (✉) · O. P. Korolkov
Department of Russian History, Pskov State University, Pskov, Russia
e-mail: deredevil1993@mail.ru

S. F. Matsevich
Pskov, Russia

T. Kolmykova and E. V. Kharchenko (eds.), *Digital Future Economic Growth, Social Adaptation, and Technological Perspectives*, Lecture Notes in Networks and Systems 111, https://doi.org/10.1007/978-3-030-39797-5_77

77.1 Introduction

Many environmental problems today are associated with the irrational use of natural resources and underestimation of consequences of a certain natural phenomenon, which negatively affects the development of the economy both globally and within a single country or a region. Under the conditions of current economic instability, it is important to use the achievements of civilization which would help to minimize risks and improve the situation in certain production sectors.

One of such achievements is cooperation which, in our opinion, seems to be an excellent example of the population economic life organization on the basis of principles of independence and self-government. Russian cooperation, having received an impetus for development in the post-reform period had become by 1917 one of the most advanced in the world and covered up to 50% of the population of the country.

Alongside with solving problems aimed at the increase of the population production and consumption, cooperative associations also received their first experience in solving environmental problems, related to climatic fluctuations and outbreaks of reproduction of insect pests in particular.

An example of such experience acquisition is the organization of insect pest control of Pskov department of the Imperial Russian Gardening Society (hereinafter referred to as Pskov department of IRGS) in the early twentieth century.

77.2 Materials and Method

The methodological basis of the article includes both general scientific and special historical methods. Due to the use of principles of historicism and science, an objective picture of the agricultural pest control organization by the Pskov IRGS department is presented, the conditions and factors that determine the development of measures used by the agricultural society, are considered.

The analysis and synthesis of a sufficiently large array of diverse historical sources and literature, as well as retrospective method, allowed us to present an integral picture of the problem the gardeners encountered in the period under study.

77.3 Results

The Imperial Russian Gardening Society arose in 1858 (March 25, 1858—The Highest approval of the Charter) in St. Petersburg. The main printing edition of the organization was the "Bulletin of the Imperial Russian Gardening Society", published since 1860.

Later, the Imperial Society of Gardening and Fruit Growing established its departments in 26 cities [1]. In the Pskov province, they were opened in two cities—Pskov

(1887) and Velikiye Luki (1893) [2]. In two other district cities, there operated Porkhov Society of Gardening and Fruit Growing (since 1908) and Kholm Society of Gardening (since 1909) [3]. The Pskov department was the largest of them.

The opening of the department in Pskov was facilitated by the current situation in the field of Fruit Growing in the Pskov province, where there were large landowner gardens once famous for their fruits. The idea of establishing a society to improve fruit growing in the Pskov province occurred in 1886 among the gardeners of Pskov. It was the initiative of the D.I. Ivanov—the chairman of the district council of Zemstvo.

The regulations on the Pskov department of the IRGS were approved by the Minister of State Property on the 30th of January, 1887 and it began its activities on the 10th of May, 1887 when the first meeting of its members took place [4]. The first meeting elected the Board of directors: The Chairman (President)—M.F. Essen, Vice Chairman—D.I. Ivanov, Secretary—V.S. Sedelschikov, treasurer—G.I. Sorgenfrey, and the members of the Board: V.I. Ivashevsky, N.I. Stolbushinsky and V.I. Berezsky [5, 6]. The Pskov department of the IRGS consisted of 40 members, whose contributions (three one-time payment 50 rubles each with an entrance fee of 3 rubles) amounted to 480 rubles [7].

The IRGS Board was located in the premises of the Mutual Credit Society of the Pskov County Zemstvo in the house of Gerasimov in Sergievskaya Street, and later in the Pogankin Chambers [7].

The composition of the Board was represented by people of different backgrounds, employed in various fields, including officials, retired officers, and house owners. For example, the Board included N.I. Stolbushinsky, the senior doctor of the Pskov Provincial Zemsky Hospital, M.F. Essen, the provincial representative ('glasny') from the Opochetsky district, V.I. Ivashevsky, Secretary of the County Military Service Presence office [6].

One of the prominent figures in the Pskov department of the IRGS was Dmitry Ivanovich Ivanov, who came from a family of state peasants. In 1883–1892 he headed the Pskov district Zemstvo. Dmitry Ivanovich was elected to the council for 15 years and was one of the organizers of the Pskov Society of Agriculture and its treasurer [8].

In the late nineteenth–early twentieth centuries such famous people of the Pskov region as V.V. Nazimov, K.I. Geldt, N.F. Fan der Fleet, P.D. Batov et al. were full members of Pskov department of the IRGS.

At the first meeting, the main activities of the department to improve the situation were also identified: collecting information on the situation of fruit growing and gardening in the city of Pskov and suburban settlements, purchasing a nursery for seedling growing, inviting an experienced gardener, organizing agricultural gardening exhibitions, and conducting educational work in this area.

In the same year 1887, on the initiative of Vice-President of the Pskov Department of IRGS department D.I. Ivanov, the information about the fruit growing situation in Pskov and suburban settlements was collected. According to the collected data, the orchards occupied 54 tithes and 264 square fathoms of the city land. The total number of fruit trees and bushes was calculated. There were 12,735 apple trees, 2527 pears, 5274 cherries, 5369 plums, 28,739 berry bushes in this area [7]. In addition, it turned out that 1/3 of the house owners had orchards, which occupied about 1/5 of the city's estate land.

The department leased a plot of land in the center of the city for 12 years to arrange a nursery at this place. Its total area was ¾ of tithes. The lively gardening issues discussions were often held there. By 1913 there had been nurseries in all three departments of the IRGS—Pskov, Velikoluksky and Kholmsky [9].

In 1888, a house for a gardener was built on the leased plot, greenhouse frames were acquired, fruit tree samplings and other plantings were stocked. The gardener started to work in the nursery in autumn.

At the initial stage, the level of well-groomed gardens in the Pskov province was very critically assessed. According to F. Kalaida, most of the gardens produced an unfavorable impression. The gardens of only a small number of landowners were in good condition, since their gardens were of industrial importance [7]. The Pskov department of IROS worried about slow gardening culture development in the province. One of the key reasons for this situation was almost complete absence of qualified gardeners and the low level of local population education.

Potentially, Pskov gardens could produce very good harvests. The climate and soil of the Pskov province, according to F. Kalaida, allowed cultivating many beautiful varieties of apples and pears, not inferior in quality to foreign ones for industrial purposes. At the exhibition in Pskov in 1889, "there were Aport and Titov apples more than a pound in weight, and also excellent Antonovka"—the latter being one of the most profitable apples for the industrial garden. It was noted that "the Americans admire our Russian varieties and ordered our apple and pear tree cuttings in enormous batches" [7].

In general, industrial fruit growing was a fairly profitable enterprise. The sources reported the data on quite decent earnings of the people who cared about their households. A noteworthy example is that of a priest of the village of Prudy, Father Peter Rudakov. The annual income received from his garden with an area of less than one tithe was 200–250 roubles [7]. On average, one fruit tree, whether an apple one or a pear, under the proper care, could give an income of up to 10–20 or more rubles. Therefore, the activities of the Pskov Department of IRGS were aimed at organizing measures to maintain order in the gardens, to increase fruit growing productivity, and disseminate the best gardeners' experience.

Exhibitions, courses on fruit growing and gardening, regular discussions on gardening, the work of gardening instructors, the creation of a library, experimental gardens and nurseries, and other activities were meant to play a positive role in the development of the agricultural sector of the Pskov province.

An analysis of historical sources shows that the Highest Authorities of the Russian Empire, scientists and Zemstvo leaders realized the importance of the development of Russian gardening as extremely important from an economic, social, environmental point of view, and contributed to the unification of representatives of different strata of society in the development of gardening.

The problems of gardening development and low fruit yields were associated not only with the low level of agricultural technologies at the turn of the nineteenth–twentieth centuries, but also with continuing invasions of pests and diseases.

Such a misfortune happened in 1908. By the summer of this year it had become clear that the apple trees, with a high degree of probability, would not bear fruit due to the invasion of the apple moth (codling tinea corrumpit). One of the gardeners in Pskov described the following picture: "all the apple trees were hung with May worms and entangled in a cobweb so that a healthy leaf and a germ of an apple were not visible." The Pskov department of the IRGS reported that "the gardens in the cities of Pskov province have the saddest look and stand with foliage eaten by apple moth" [10].

By virtue of the obligations undertaken, the Pskov department had to undertake several measures to combat the pests. First of all, local gardeners were introduced to the methods of apple moth destruction. Due to the efforts of the provincial government the agricultural inspector of the Department of Agriculture, a prominent specialist in applied entomology, Y.F. Schreiner, arrived on a business trip to Pskov. He was the author of numerous publications on methods of controlling pests of crops, including environmental methods using beneficial insects for the purpose of pest control [11].

Apple moth (or "May worm") is a butterfly from the family of moths, rather small and very harmful to apple trees. However, an adult insect harms the apple trees only indirectly—by laying eggs in the spring. The main damage to the orchards is caused by numerous caterpillars—the so called "May worms", which can number up to 1000 in spider webs located on tops of the trees [12].

Multiplied greatly, apple moth causes such damage to the apple trees that they do not bear fruit for two years. The problem associated with this insect encountered in Russia more than once. The dictionary of Brockhaus and Efron notes insect infestations in the Middle and Lower Volga region (in Saratov province, especially in the 60s of the nineteenth century, and also in 1894–1897) in Kursk province and in other territories [12].

Pest control could be done in several ways. The first, as Y.F. Schreiner notes, the most time-consuming way, was physical elimination of the tracks by collecting the young leaves which became brown. The second method, similar to the first in its laboriousness, was to collect cocoons by hand or "poles with a fork at the end."

The third method was the most effective and fast. Apple trees were sprayed with insecticides that were in use at the beginning of the twentieth century. Among them were Parisian greenery, Bordeaux liquid, or arsenic solution [12]. These chemicals practically guaranteed the result, allowing the gardeners to process large areas in a short time, and also not to spend as much effort as in the first two ways.

However, in 1908, the Bureau at the Pskov department of the IRGS proposed not the most effective method, but the simplest one. Some gardeners knew that Parisian greenery was effective in controlling pests, but did not use it for some reason [13]. The following measures were recommended against this "terrible fruit enemy": to collect all cobwebs from tree trunks and tree branches and damaged leaves and burn them; coat the trunks and low branches of apple trees with freshly slaked lime solution; remove all grass around the tree and dig up the ground around the tree trunk [14]. It was prescribed to carry out these measures as quickly as possible, but not later than July, 20, so that adult insects would not have time to develop. Further, in order to combat the remaining pests of fruit crops, it was planned to organize the direct destruction of apple moth caterpillars, as well as to instruct the gardeners.

For example, in November 1908, Y.F. Schreiner arrived in Pskov again and delivered three lectures on the insects found in gardens, "demonstrating the lectures with the "magic lantern" that aroused great interest among gardeners [15]. The Pskov department of IRGS itself spread information about other pests of fruit crops - the codling moth (codling fatebatur) and winter moth (hiems agrotis), etc., which also harmed orchards, but not as much as apple moth, as the year 1909 showed.

All measures taken were not entirely effective, since the problem of apple moth remained acute. It was exacerbated by difficulties of a different nature: the turnover of funds from the Pskov Department of IRGS reduction and the lack of trained gardening specialists (there was only one gardening instructor, A.P. Yakovlev, in the province).

It was obvious under such circumstances that the aforementioned organization could not cover the entire population with the planned work at gardening culture development. This is confirmed by the opinion of A.P. Yakovlev, who claimed that the population of the Pskov district is rather indifferent to gardening, and therefore often leaves their gardens without proper care: only a small part of the 700 Pskov gardens located on 70 tithes was treated with insecticides [13]. This attitude led to the excessive reproduction of the "May worm", which caused great harm to unprotected gardens [16].

The mechanical method of collecting caterpillars and apple moth eggs destruction proposed by the Pskov Department of IROS, according to A.P. Yakovlev, was effective only in small gardens and under certain conditions. He proposed that the main method of fighting "May worm" may be spraying trees with toxic substances, which, if carried out simultaneously, would be very effective both in large and small gardens.

In 1909, the potential cheapness of this operation became obvious, since the profit obtained from large gardens, could well cover the costs of processing (10–15 rubles per a tithing of a garden). At the same time, its benefit was undoubted, since many gardeners who resorted to the advice of the learned gardeners Y.F. Schreiner, P.I. Puzyrevsky and A.P. Yakovlev, were able to collect a good harvest of apples [13].

The Pskov department of IRGS was well aware of the problem and that's why coordinated actions of all the necessary specialists and sought to unite the gardeners to control pests. At the end of 1908, a special commission was elected in the person of its chairman D.I. Ivanov and four other members. Among other objectives of the department, the task of organizing a group of sprayers for the destruction of apple moth and other pests was set [15]. Despite the difficulties with funding, the idea of garden processing did not fade. The sittings of the Bureau of the Pskov Department of IRGS in autumn of 1908 and winter of 1908–1909 resulted in working out a year plan of apple moth control.

The plan was as follows. At first, it was necessary to collect information about gardeners and their gardens. Secondly, it was to address them with a proposal to make a decision about the desire or unwillingness to have their gardens processed with the assistance of the Bureau of the Pskov Department. Thirdly, it was necessary to familiarize as many gardeners as possible with information about the very pest, both through lectures and publication of brochures. Fourth, it was to provide material and technical base of spraying. And, finally, fifthly, to teach the gardeners to use machines for garden processing [17].

By the beginning of 1909, all the tasks set by the plan were completed. 700 invitations in which gardeners could declare their desire to have their gardens processed by Parisian greenery by the 1st of March, were printed and sent out. However, for some reason, only a small part of gardeners managed to meet the deadline, while the rest expressed their desire during March and April. In total, in the beginning of May 1909, 106 garden owners, owning 4739 apple trees in different parts of the city were registered. Later 20 more people who had about 1000 apple trees joined them. Thus, in total it turned out that it was necessary to process about 6000 apple trees, 5650 adults among them [17].

At the same time, in spring 1909, the Board of the Pskov department and the pest control committee formed squads for spraying the gardens. They included both the so-called agents entrusted with special devices for spraying apple trees, as well as workers who performed auxiliary functions. A total of 9 squads were formed and divided into 3 large units and 6 small ones [17]. They differed in number of people and material and technical equipment. The large squads consisted of 1 agent and 3 workers, equipped with a Pomona machine, and the small ones consisted of 1 agent and 1 worker, who had a Guld hydraulic console at their disposal [17].

Agents and workers of the "flying squads" received a small fee, which was assigned in accordance with the number of the processed trees, the number of convoys and people with them.

Several members of the company were authorized by the Bureau to oversee the work on processing the gardens. They controlled the process in different parts of the city: the center, Zapskovye, Zavelichie and in the suburban settlements. Among them there were such prominent figures as I.I. Danishevsky, D.I. Ivanov, A.E. Sholpo et al. among them [18].

The order of processing the gardens and payment were also agreed in advance. The owners were informed about the day their trees would be sprayed. The gardeners were to prepare buckets of water in advance to make the insecticide solution. The processing procedure was not free. A fee of 5 kopecks was charged for each adult tree, and 2.5 kopecks for a young one. Only the gardens of the poor were sprayed for free. It is for this reason, that the income estimate for 1909 has a deficit of 20 rubles [15].

This large-scale event was supposed to begin immediately after the flowering season, when the first caterpillars appeared. However, in 1909, the dates of processing had to be postponed due to cold weather, which slowed down all natural processes, including the appearance of May Worms. The establishment of warm weather became the beginning of the processing of gardens by Parisian greenery. The working day of each squad began at 6 a.m. and lasted until 8 p.m. with breaks for breakfast, afternoon tea and dinner. On a sunny day, processing continued until 11 a.m. and began again at 3 p.m. so that the trees would not receive a chemical burn from a combination of insecticide and high air temperature. A large unit was supposed to process large gardens with more than 30–40 trees, and a small one—from 10 to 30 [19]. At the end of the working day, the units gathered at the "station", where they kept their equipment and prepared for a new processing cycle.

Gardens processing teams carried out all the work within 10 days, with a short break on June, 7 due to rain. A total of 5144 apple trees were sprayed in 120 orchards, 4651 adult tree and 493 young ones among them [17]. The remaining trees from about 6000 fruit crops planned for spraying were partially processed by the efforts of representatives of the Pskov Agricultural Society, the farmers themselves and private individuals. Those who wanted to spray their garden on their own or repeat the procedure had the opportunity to rent a spraying device for the price of 40 kopecks per day, as well as buy the insecticide at a reduced price—35 kopecks per pound.

The results of such a large-scale event carried out in such a short period of time were recognized "quite satisfactory", and the goal of the "May worm" destruction "achieved" [17]. An obvious success was the fact that from that moment the population clearly realized usefulness of preventive measures and also became more literate in the sphere of agriculture in general.

But with all the positive aspects of this event, there were also small negative points. First, insufficient attention was paid to the countryside. Village gardens were already badly damaged by the time the work of flying squads in the city was over, and the control measures undertaken by local residents in Melekhovskaya and Logozovskaya volosts had no effect due to lasting rains.

The second negative aspect of the event was the insecticide itself. The fact is that the "Parisian greenery" is a toxic substance based on arsenic, which is not hygroscopic, that is, does not dissolve in water. Consequently, its effectiveness decreases if it rains. The other side of this is that arsenic is harmful not only to the caterpillars of the apple moth, but also to other insects. Thus, there is evidence that one of the beekeepers complained about the large number of dead bees after the pest control in the gardens. But the "Parisian greenery", apparently, also ruined at least one larger representative of the animal world. A report to the board of the Pskov IROS department pointed to the complaint of one of the gardeners whose cow got ill after eating grass in the processed garden [17]. But there are only individual cases of which we are aware.

77.4 Conclusion

In general, in 1909 the Pskov department of IROS, despite delays in choosing pest control methods and underestimating the extent of the threat to agriculture, in a short time was able to organize pest control of gardens. Thanks to the initiative professional work of the members of the Bureau of the Pskov Department of the Imperial Russian Gardening Society, prominent representatives of Russian applied entomology attraction to educational work, proper organization of garden processing and the response of Pskov population and gardeners, the undertaken measures got their due effect. At the same time, a serious environmental problem of the use of insecticides, expressed in isolated cases of death of bees and mortality of livestock was identified. It subsequently contributed to the more careful use of insecticides, and ultimately led to its replacement of "Parisian greenery" by other insecticides and fungicides and even to a complete ban on the use of "Parisian greenery" in agriculture due to its toxicity. The Pskov department of the Imperial Russian Gardening Society, like other similar public organizations acting on the basis of cooperative principles, solved not only some immediate local problems, but also worked for the future, enlightening the local population and instilling gardening culture in their minds.

References

1. Prineva LA (2005) Gardens bloomed for centuries: the history of gardening in Russia. Quarta, Moscow, p 17
2. State Archive of the Pskov Region, f. 20, c. 2953, l. 70
3. Korolkov, OP (2001) Cooperative movement in Russia and the Pskov province, Pskov, pp 238–241
4. Sapelkin VV (1910) Consumer and agricultural societies of the Russian Empire. St. Petersburg, p 444
5. Ad. The Pskov department of the Imperial Russian Gardening Society on May 10, 1887, opened its operations (1887) Bull Pskov Provincial Zemstvo 9:367
6. The memorial book of the Pskov province for 1899 (1899), Pskov, pp 24, 29
7. Kalaida F (1891) On the activities of the Pskov division of the Imperial Russian Gardening Society and on the situation of local fruit growing. Bull Pskov Provincial Zemstvo 1:37–46
8. Levin NF (1990) D. I. Ivanov—an outstanding Zemsky and public figure of the Pskov province in the Land of Pskov, ancient and modern. Pskov 1990:40–41
9. State Archive of the Pskov Region, f. 20, c. 3058, l. 111
10. To the attention of gardeners (1908) Bull Pskov Provincial Zemstvo 25:25
11. Schreiner JF (1899) Apple moth and measures to combat it. Printing House of the Publishing House, St. Petersburg, p 16
12. Encyclopedic Dictionary (1904) Apple Moth. Encyclopedic Dictionary, vol 81. Printing House of Brockhaus-Efron, St. Petersburg, pp 473–474
13. Garden owner D.I. (1909) On the May Worm (apple moth), harmful to gardening control. Bull Pskov Provincial Zemstvo 16:13–15, 17:15–16
14. Ivanov DI (1908) To the owners of gardens. Bull Pskov Provincial Zemstvo 27:8
15. Report of the Board of the Pskov Department of the Imperial Russian Gardening Society to the General Assembly of the society members in 1908 (1909) Bull Pskov Provincial Zemstvo 11:14–16
16. The needs of gardening in the Pskov district (1909) Bull Pskov Provincial Zemstvo 45:6
17. Report to the Board of the Pskov Department of the Imperial Russian Gardening Society on spraying orchards in the city of Pskov in 1909 (1909) Bull Pskov Provincial Zemstvo 47:11–13, 48:12–13
18. Ivanov DI (1909) From the Bureau of the Pskov Department of the Imperial Russian Gardening Society. Bull Pskov Provincial Zemstvo 20:10–11
19. Yakovlev AP (1909) Plan for the organization of flying squads for spraying gardens against pests. Bull Pskov Provincial Zemstvo 48:14–15

Chapter 78
The Environmental Potential of Sanitary Protection Zones and the Role of Self-Regulatory Organizations in the Construction Industry in the Projects' Development of Unified Sanitary Protection Zones

Julia G. Leskova and Olesya A. Zolotova

Abstract Based on the analysis of studies of Russian and foreign scientists-economists, legal scholars, as well as representatives of various areas of natural sciences, this article examines the legal regime of sanitary protection zones. It is shown that now the regulatory framework, which allows establishing the sanitary protection zones according to the industrial facilities classification, has been created in Russia. The requirements for limiting economic activity within the boundaries of such zones are fixed too. However, the classification of industrial enterprises, which was developed in the early 2000s, seems to be out of date and needs to be reviewed. In addition, the Russian legislation does not fix the obligation of operating organizations to green the territory of sanitary protection zones. This gap has been eliminated in some municipalities through the adoption of local self-government acts, which establish requirements for the greening extent and improvement of sanitary protection zones. At the same time, the federal legislation does not regulate the issue of developing uniform sanitary protection zones for several industrial enterprises. It is advisable to foresee the formulation of recommendations on sanitary protection zones' greening and the development of projects of unified sanitary protection zones within the framework of self-regulatory organizations (SRO) that unite entrepreneurs operating industrial facilities. It is also necessary to fix obligatory requirements in the rules and standards of SROs in the construction industry for property developers to landscape the adjacent to the construction site areas. At the same time, various compensation mechanisms for the costs of landscaping sanitary protection zones can be established in federal, regional and local legislation.

J. G. Leskova (✉) · O. A. Zolotova
Russian Presidential Academy of National Economy and Public Administration, Moscow, Russia
e-mail: yuliyleskova@yandex.ru

O. A. Zolotova
Moscow, Russia

T. Kolmykova and E. V. Kharchenko (eds.), *Digital Future Economic Growth, Social Adaptation, and Technological Perspectives*, Lecture Notes in Networks and Systems 111, https://doi.org/10.1007/978-3-030-39797-5_78

Keywords Industrial facilities · Environmental protection · Sanitary protection zones · Zones with special conditions for the territories' use · Self-regulatory organizations · State regulation

JEL Classification Codes K200 regulation and business law: general · K320 environmental, energy, health, and safety law

78.1 Introduction

The UN Climate Summit and ratification by the Russian Federation of the Paris Agreement in September of the current year again drew the attention to the problems of environmental pollution and global warming. The Paris Agreement offers a more flexible mechanism for regulating the effects on the ozone layer. It involves reducing emissions and cleaning them, technological re-equipment of operating enterprises, equipping them with more modern treatment facilities and adapting industrial production to climate change. The question arises: is it possible to achieve these goals within the framework of environmental protection mechanisms existing in Russia.

One of the ways to reduce the impact of pollution on atmospheric air is to establish sanitary protection zones (hereinafter—SPZ) around the objects, which have a harmful effect on the environment. SPZ is a protective barrier, outside of which favorable living conditions and environmental protection are ensured during normal operation of an industrial facility.

This research was carried out with the aim of studying the regulatory framework and legal regime of SPZ and identifying existing and potential opportunities for using SPZ in industrial enterprises to reduce their negative impact on the environment (hereinafter referred to as NIE).

78.2 Materials and Method

The environmental potential's study of sanitary protection zones was carried out through the works' analysis of scientists and economists (Krupina N.N., Anderberg S., Jacobsen N.B., etc.) and legal scholars (Ignatyev I.A., Luneva E.V., Nekrestyanov D., Obolenskaya M., etc.). In addition, the study was based on the works of various areas of natural sciences' representatives (Agafonova A.L., Agafonova G.V., Atkina L.I., Belkin A.S., Kipriyanova E.N., Kiselev A.V., Osipov I.V., Trutnev E.K., Scherbo A.P., Weslynne A., etc.).

The methodological basis of the study was: the dialectical method, methods of analysis and synthesis, induction and deduction, comparison and generalization. The formal legal approach revealed the significance of legal norms aimed at revealing the potential use of SPZ to achieve environmental goals.

78.3 Results

In accordance with SanPiN 2.2.1/2.1.1.1200-03 "Sanitary protection zones and sanitary classification of enterprises, structures and other objects" (14) (hereinafter—SanPiN) five hazard classes are established for existing and planned for the construction of industrial facilities and manufacturing, which are sources of harmful effects on the natural world. Apart from that, SaNPiN contains general requirements for the size of the SPZ depending on the hazard class of the enterprise (from 1000 to 50 m) and requirements for the regime of the SPZ territory. The level of exposure to adverse factors at the border and outside of the SPZ must comply with sanitary and hygienic standards. A special regime for the implementation of economic and other activities is established within the territories of the SPZ. SanPiN establishes the prohibition to place certain types of objects, such as housing construction facilities, in the sanitary protection zone, and the permition to place other types of objects.

The owner of the industrial facility specifies the size and the regime of the individual object of the SPZ in the following way:

- the hazard class of the enterprise and its approximate SPZ are determined (according to SanPiN from 1000 to 50 m),
- basing on the calculations of emissions, taking into account their dispersion, the estimated SPZ is established (the size may differ from the established SanPiN),
- the final size of the SPZ is calculated after environmental monitoring at the border of the SPZ.

After that, the project is subject to sanitary and epidemiological expertise and the decision on its approval by the Federal Service for Consumer Rights. Responsibility for the unidentified SPZ is not directly provided, but in the context of Art. 6.3 and 8.1 of the Administrative Code of the Russian Federation an enterprise may be fined for applicable sanitary rules' violation and non-compliance with environmental requirements when operating capital construction facilities.

As I.A. Ignatyeva correctly points out ([1], pp. 318–319), at present there are two procedures for determining the legal regime of SPZ—in SanPiN and in Rules No. 222 (18). This situation developed after the adoption of Federal Law dated 03.08.2018 No. 342-FL "On Amendments to the Town Planning Code of the Russian Federation and Certain Legislative Acts of the Russian Federation" (24), which established the general procedure for the existence of zones with special conditions for the territories' use. I.A. Ignatieva concludes that "SanPiN on SPZ continues to be the act, which regulates in greater detail the procedure for creating, changing SPZ, establishing its boundaries, conducting the necessary research … The Rules against this background set out more structurally the system of requirements for the SPZ project, the procedure for creating and approval of the project are also detailed" ([1], p. 327).

Russian Construction Charter already in 1857 contained a recommendation to place a "gap zone" between residential and industrial buildings. In modern legislation, the hazard class of industrial facilities and industries was first established in SanPiN in 1996, according to which the estimated dimensions of the SPZ were from 100 m to 2000 m. Then there were standards of 2000 and 2001, up to the adoption of the current SanPiNa of 2003 in subsequent editions. Despite the long regulatory period, the environmental potential of the SPZ is not fully disclosed.

First of all, the current sanitary classification of industrial enterprises positively affects the assessment of the environmental impact of these enterprises. However, it was created based on the level of development of technologies characteristic in the middle of the last century. Using more modern technologies can significantly reduce the harmful effects of the enterprise and adjust the size of the SPZ ([2], p. 51). However, even in the Rules No. 222 of 2018 the legislator did not dare to change the approach to determining the size of the SPZ, but merely specified the procedure for developing of SPZ project. In addition, when developing SPZ, it is necessary to take into account the assimilation potential of the environment [3].

Secondly, the abolition in March 2008 of the SanPiN requirements for the SPZ greening needs to be revised.

Foreign experience shows that having close dimensions to the Russian SPZ (in Europe up to 500 m, in Canada from 20 m to 1000 m, in Australia from 300 m to 2000 m), the barrier areas are subject to mandatory greening [4, 5]. According to Art. 1 of Model Environmental Code for the CIS member states (general part) (10) SPZ is a green area. In addition, the requirements for greening of SPZ are preserved in the sanitary rules for certain types of industrial facilities: railway transport (12, 13), cemeteries (15) and radiation facilities (16).

We have to admit that the issues of SPZ greening have moved from sanitary and environmental legislation to urban planning. So, according to the Federal Law of 06.10.2003 No. 131-FL "On General Principles of the Organization of Local Self-Government in the Russian Federation" (25), the organization of landscaping and greening of the territory of municipalities refers to issues of local importance. In order to implement such tasks, municipalities develop and approve the rules for improving their territories or display them in urban planning documents. At the same time, they should be guided by GOST 28329-89 "Gardening of cities. Terms and definitions" and SP 42.13330.2016. "Set of rules. Urban planning. Planning and development of urban and rural settlements. Updated edition of SNiP 2.07.01-89." According to SP 42.13330.2016, in the cities, where enterprises with SPZ of more than 1 km are located, the greening extent of the territory must be increased by at least 15%. It should be mentioned that sanitary-protective gardening also has a masking function, hiding industrial sites [6] and improving the aesthetic perception of the city.

To implement these requirements, specific municipalities (for example, Moscow (17), St. Petersburg (6), Vsevolzhsk (21), Kirovsk (19), Vologda (20), Cherpovets (22)) establish requirements for greening of SPZ as a percentage of their area and requirements for landscaping of SPZ. Thus, some municipalities have implemented the opportunity to normatively provide for landscaping of the territory of SPZ.

However, it is difficult to oblige organizations operating industrial facilities to maintain the SPZ in proper condition, since they do not have any land rights within it. On the one hand, this reduces the financial burden of operating organizations, and on the other, it deprives supervisors of the opportunity to prosecute such organizations for the failure to meet the requirements for greening and landscaping of the SPZ. It does not allow stimulating industrial facilities to reduce the size of the SPZ as well.

It should be borne in mind that the effective exploitation of perennial green spaces with their competent species selection can reach 35–40 years ([7], p. 3), which means the projected functioning of an industrial facility for such a long period and maintaining green spaces within the boundaries of the SPZ.

Nevertheless, in addition to the costs of the SPZ greening itself, the operating organization meets the costs of uprooting trees and shrubs after they have withered and cannot reflect green spaces in the SPZ as its own material asset. In such circumstances, it is necessary to normatively provide for some form of compensation of the SPZ greening costs: depreciation bonus ([7], p. 5), lowering of income tax, reduction in the amount of fees for non-profit and other environmental payments, allocation of funds to maintain green spaces in proper form.

In addition, the involvement in economic circulation can encourage the maintenance of SPZ in proper conditions. So, on the territory of the SPZ it is possible to place objects, whose operation does not contradict the SPZ regime and will allow to derive income from their use: organization and placement of paid parking lots, payment terminals, ATMs, advertising structures, dry closets, warehouses and trade pavilions and others. In this case, interest in the content in the SPZ in the form corresponding to the law also arises among public bodies that are landowners within the boundaries of such zones. As E.K. Trutnev correctly states, "the improvement of the environmental situation in the territories is, above all and among other things, carried out by the authorities, mainly local authorities, to solve the problems of planning the territories' development" [8].

Besides the regional and local legislation, it is possible to improve the SPZ regime through the rules and recommendations developed within the framework of self-regulatory organizations operating industrial facilities. They may relate to the accounting of green spaces in the territory of the SPZ, the choice of planting type, species of shrubs and trees, types of landscaping elements; development of the internal division of the SPZ into subzones for greening and construction, and operation of other facilities, minimizing operating costs and streamlining the SPZ budget. It should be especially noted that such requirements should take place in the rules and standards of self-regulatory organizations in the construction industry and relate to its members—developers of real estate.

Thirdly, in para. 3.13 of the Sanitary Regulations and Rules, it is stipulated that for groups of industrial facilities or an industrial complex, a single settlement SPZ can be established or SPZ can be created separately for each object inside of such site. N.N. Krupina positively assesses this possibility, believing that "such a partnership contributes to the integration of landscaping effects and improvement of the barrier territory" ([7], p. 9). In the EU countries, the creation of unified protective territories for industrial facilities located close to each other within the city is being implemented, while declining production costs and reducing urban pollution [4, 5].

As T.E. Bobkova [9], p. 14, correctly points out, when establishing a unified SPZ, first of all, the problem of organizational order arises: who is the initiator of the SPZ organization, how the general outline of the SPZ is developed, how is it taken into account the share of the Initial Environmental Examination (IEE) of each facility and what is the degree of material participation of each operating organization in this process. It seems that the main motivation of operating organizations to create a single SPZ may be to reduce its size and share the costs of its improvement with other enterprises. The initiator of the establishment of a single SPZ can be both local government bodies ([10], p. 121), and self-regulatory organizations created in construction industry, as well as industries, within the framework of which it is possible to develop a procedure for coordinating a single SPZ. In our opinion, when establishing minimum SPZ, it is necessary to green at least the industrial site and the territory right behind it.

It is worth agreeing with the opinion of E.K. Trutnev that "SPZ should be one of the mechanisms encouraging the owners of industrial real estate to improve technologies to reduce the SPZ" [8]. However, the proposal of D. Nekrestyanov and M. Obolenskaya [11] to protect the land rights by initiating the reduction of the SPZ in the courts seems to be controversial. We believe that the change in the SPZ size should be carried out only on the basis of the existing technological and economic capabilities of the operating organization while ensuring the achievement of environmental indicators, and not guided solely by the tasks of preserving the land rights of other right holders.

78.4 Conclusions/Recommendations

It should be concluded that protecting the population by distance is fading away: the widening gap between industrial and residential areas should be replaced by environmentally friendly production technologies. At the same time, SPZs are the effective way to ensure a favorable environment and, if properly modernized, can be applied along with the use of technological means in production. It seems to be right:

- to revise the classification of industrial facilities and general requirements for the size of their SPZ at the federal level;
- to consider at the federal level the obligation to plant greenery on the territories of SPZ and consolidate the compensation costs' mechanism for this kind of greening;
- to develop recommendations for the SPZ greening and establishing common SPZ within the framework of self-regulatory organizations.

References (Literature)

1. Ignatyev IA (2019) Use of land and land plots with electric power facilities: law and practice: study guide. Prospect, Moscow, p 368
2. Scherbo AP, Kiselev AV, Belkin AS (2010) New approaches in the examination of sanitary protection zone projects. Bulletin of the North-West State Medical University named after I.I Mechnikov 1:50–55

3. Luneva EV (2017) The legal essence of the environment' assimilation potential: statement of the problem. Lawyer 11:30–35
4. Jacobsen NB, Anderberg S (2004) Understanding the evolution of industrial symbiotic networks: the case of Kalundborg. In: Economics of industrial ecology: materials structural change, and spatial scales. MIT Press, Cambridge, pp 331–335
5. Weslynne A (2008) Understanding the organization of industrial ecosystems: a social network approach. J Ind Ecol 12(1): 34–51
6. Agafonova GV, Agafonova AL, Atkina LI, Osipov IV (2008) Sanitary protection zones of industrial facilities of Yekaterinburg and the experience of using small-leaved linden for their landscaping. Vestnik MGUL, Lesnoy vestnik 3:24–26
7. Krupina NN (2014) Actual issues of managing problematic green assets of enterprises. Financ Anal Probl Solutions 1:2–9
8. Trutnev EK (2019) Issues of establishing sanitary protection zones. Property Relat Russ Fed 6:20–36
9. Bobkova TE (2009) Establishment of sanitary protection zones for a group of industrial enterprises. ZNiSO 6:14–16
10. Vekovshinina SA, May IV, Sedusova EV, Shaidurova ES (2014) On the issue of legal regulation of the organization of sanitary protection zones of enterprises, constructions and other facilities. Vestnik KazNMU 3–1:118–122
11. Nekrestyanov D, Obolenskaya M (2015) Site protection from the sanitary zone. Corp Lawyer 6:60–64
12. Decision of the Council of Deputies of the municipal entity "City of Vsevolozhsk" of the Vsevolozhsk municipal district of the Leningrad Region dated February 20, 2018 No. 5 (as amended on May 28, 2019) "On the approval of the Rules for the improvement of the territory of the municipal formation" "City of Vsevolozhsk of the Vsevolozhsk municipal district of the Leningrad Region"//Vsevolozhsk City Life, No. 7, 02/23/2018
13. Federal Law of 06.10.2003 No. 131-ФЗ "On General Principles of the Organization of Local Self-Government in the Russian Federation"//SZ RF, 10/06/2003, No. 40, Art. 3822, Parliamentary newspaper, No. 186, 10/08/2003, WG, No. 202, 10/08/2003

Chapter 79
Sustainable Development Assessment of the NWFD Regions

Michail A. Nikolaev, Marina Yu. Makhotaeva and Denis P. Malyshev

Abstract The object of this paper is to analyse the essence of the concept of "sustainable development of a region" and develop the methodology to assess the level of sustainable development as exemplified by the NWFD regions. The approaches systematization presented in science literature show that sustainability must be considered in the unity of its economic, social and environmental aspects. This position corresponds to the traditional understanding of the regional social and economic system in the form of the triad "economy - social sphere - ecology". The paper introduces the integrated index of the economic growth environmental load. The economic component is evaluated by the GRP dynamics, the social component—by the consolidated budgets spending index of the constituent entities of the Federation for social and cultural activities, and the environmental component—by the environmental load index. As a result of the research, the NWFD regions are divided into three groups. The Komi Republic, the Arkhangelsk and Murmansk regions, and St. Petersburg have sustainable economic growth. The Vologda, Kaliningrad, Leningrad and Novgorod regions have an average level of economic growth sustainability. Sustainability underdevelopment is characteristic of the Republic of Karelia and the Pskov region. The problems of the Republic of Karelia are caused by problems in the social sphere and ecology, and the social sphere is a weak point in the Pskov region. Increase in innovative activity is considered as a direction to improve the sustainability.

Keywords Ecology · Economy · Innovation · Region · Social sphere · Sustainable development

JEL-Code R110

M. A. Nikolaev (✉) · D. P. Malyshev
Economics and Finance, Pskov State University, Pskov, Russia
e-mail: mihaelnikolaev@mail.ru

M. Yu. Makhotaeva
World Economy and International Business, Pskov State University, Pskov, Russia

T. Kolmykova and E. V. Kharchenko (eds.), *Digital Future Economic Growth, Social Adaptation, and Technological Perspectives*, Lecture Notes in Networks and Systems 111, https://doi.org/10.1007/978-3-030-39797-5_79

79.1 Introduction

The unsatisfactory dynamics of social and economic processes, growing uncertainty in the economy, as well as internal and external threats rise made it necessary to form an integrated approach to solve sustainable development problem. In addition, such factors as growing volatility of economic processes, global contradictions development, as well as natural resources shortage contribute to the actualization of sustainable development problem [4].

The main attributes of sustainable economy are: environmental sustainability, social orientation, innovation, energy efficiency, sustainable consumption and production, emphasis on knowledge, risk reduction [2]. At the same time, as mentioned in the paper [10], the prevailing in the country approach to the problem of sustainable social and ecological development is characterized by situationality, fragmentarity and low systemacity.

Thus, official documents constantly emphasize that development opportunities exhaustion within raw-materials export model is one of the main factors of unsatisfactory dynamics of the Russian economy. At that, the need for a transfer to knowledge-based innovative economy is manifested. At the same time, science financing from the Federal budget decreased from 425.3 billion roubles in 2013 to 402.7 billion roubles in 2016 and the consumer price index for the period of 2014–2016 was 132.6%, i.e. in real terms, funding for science declined by about 30%. Innovation activity indicators also have negative dynamics. Thus, the indicator "the ratio of technological innovations costs in the total volume of goods shipped, work and services performed" in industry decreased from 2.1% in 2014 to 1.8% in 2016.

When assessing the economic sustainability level, high importance of territorial factor is to be taken into account. In the paper [6], the expanded reproduction of all types of territory potential, competitive recovery of the economy and living standards of the population are highlighted as attributes of sustainable development of a region. The urgency of sustainable development research at the regional level is also determined by high differentiation level of the social and economic development of territories [5, 8].

The consequence of this is the formation of problem areas, causing low sustainability of the national economy. The issue of providing integrated and balanced development of regions is given priority in the strategic planning documents at the Federal level. So, in the National Security Strategy of the Russian Federation, one of the most important attributes of the national security is balanced, integrated and comprehensive development of the subjects of the Russian Federation. In the Strategy for Economic Security of the Russian Federation for the period till 2030, balanced spatial and regional development is pointed out as a condition for sustainable development.

79.2 Materials and Method

When conducting the research, the authors used the basic principles of classical economic theory, the theory of regional economics. The research was based on the principles of system approach to the study of the economic development of regions and the practical feasibility of the research results. To achieve the objectives of the research, such methods as structural and statistical analysis, system method, and analysis of planning documents were used in the paper. When using these methods, the system approach was observed, due to the fact that each region should be considered in the form of the triad "economy - social sphere - ecology".

For each of the subsystems, the use of relevant indicators is justified: economy—the index of physical volume of GRP for the period, the social sphere—the index of expenditures of the consolidated budgets of the constituent entities of the federation for social and cultural events, ecology—the index of environmental load. These indicators are presented in the official statistics, which assures the reliability of the research. The regions of the North-Western Federal District were considered as the study object.

79.3 Results

When studying the sustainability of territorial systems, it is necessary to proceed from the fact that a region is a complex system with specific features of the course of social and economic processes. Significant factors of sustainable development at a regional level include the existing structure of regional industrial complexes, which not only determines the type and pace of economic development, but also determines the high environmental burden and emergence of different environmental problems not only in regions, but also in the national economy [7].

In the paper [12], with the use of the Adjusted Net Savings Index (ANSI), the relationship between the structure of regional economies and the state of the environment for the Republic of Kazakhstan was investigated. Most industrial regions have average ANSI values; oil-producing regions have negative index values, and agrarian-oriented regions have ANSI above average value.

Economy specialization is the main factor of the environmental state for the Russian Federation regions. So, in the North-Western Federal District, the minimum value of the indicator "emissions of pollutants into the air emanating from stationary sources" in 2017 was observed in agro-industrial regions of the Kaliningrad and the Pskov Regions—26 and 35 thousand tons, respectively. Their lead industry is food production. At the same time, the Komi Republic that specializes in oil and gas production, had the value of 451 thousand tons, and the Vologda region, specializing in metallurgical production, had 420 thousand tons.

The paper [11] presents the integrated approach to the assessing the environmental intensity of economic development. The author identifies two stages to solve this problem. At the first stage, the economic result is evaluated with the use of such indicators as GRP, production output, etc. Next, the environmental load is calculated using industrial and municipal waste indicators, pollutant emissions into the atmosphere, dirty discharge, and the amount of consumed resources (electric or thermal energy, clean water, fuel resources, etc.). The indicator of environmental intensity of economic development is defined as the ratio of environmental and economic indicators. This methodology was used by the author to analyze the economy development of the NWFD regions in 2011–2015.

Thus, “sustainable economy” is considered by most of the authors in the unity of its economic, social and environmental aspects. This position corresponds to the traditional understanding of the regional social and economic system in the form of the triad “economy - social sphere – ecology”.

At that, the main problem is the choice of indicators characterizing each area and methods to calculate the integral indicator of the economic growth environmental load.

Within this research, we propose to evaluate the level of the sustainability of regions’ economic development on the basis of the following indicators:

- the economy—the index of physical volume of GRP for the period;
- the social sphere—the index of the consolidated budgets spending of the subjects of the Federation for social and cultural activities, including education, healthcare and social policy;
- the ecology—the environmental load index, which is determined on the basis of indicators of atmospheric pollutants emission, emanating from stationary sources and dirty discharge into surface water bodies. These indicators are presented in the official statistics and make possible to do the comprehensive and system research of both the dynamics of sustainable development and its determining factors.

The presented approach was used to analyze the level of economic growth sustainability in the regions of the North-West Federal District for the period 2011–2016 in the context of economic, social and environmental components. The economic component was estimated by the dynamics of GRP (Fig. 79.1). The maximum GRP growth index among the regions of the NWFD was in the Novgorod Region. Its economy grew by 25.1%. At that, the GRP of the Komi Republic decreased by 3.6%. Thus, the economic component of sustainable development was positive in all regions except the Komi Republic. In general, we can report a high level of the economic growth differentiation of the NWFD regions, which is considered in science literature, as well as in program documents, as a factor of low sustainability of the economy against external shocks.

The social component of the sustainable growth was estimated according to the dynamics of the consolidated budgets spending of the regions for social and cultural activities (Table 79.1). Thus, for the period under analyses, per capita expenditures on social and cultural activities in nominal terms increased in all regions. At the same time, the consumer price growth index for the period 2011–2016 amounted to 150.5%. Accounting this, in real terms, social expenditures increased only in the Leningrad region by 9.4%. At that, in the Pskov region they decreased by 30.3%.

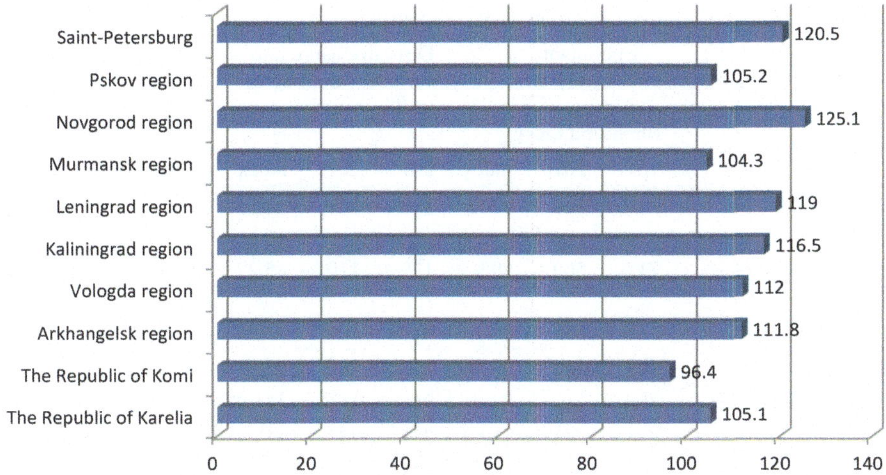

Fig. 79.1 Dynamics of gross regional product for the period 2011–2016

Table 79.1 The Russian Federation Consolidated budget spending of the subjects, per capita, rub

	2011	2016	Growth index	2016 real	Social development index
The Republic of Karelia	39779.5	42535.9	1.07	28357.3	71.3
The Republic of Komi	42787.5	61340.5	1.43	40893.7	95.6
Arkhangelsk region	36654.7	46220.2	1.26	30813.6	84.1
Vologda region	31439.2	34840.6	1.11	23227.1	73.9
Kaliningrad region	25435.1	31189.8	1.23	20793.2	81.7
Leningrad region	26550.2	43562.3	1.64	29041.5	109.4
Murmansk region	46030.1	64107.7	1.39	42738.4	92.8
Novgorod region	27100.6	33624.3	1.24	22416.2	82.7
Pskov region	27849.3	29112.6	1.05	19408.4	69.7
Saint-Petersburg	42329.9	56562.1	1.34	37708.1	89.1

Thus, the social component of the economic growth for the period 2011–2016 turned out to be negative in all regions except the Leningrad region.

Let us further consider the environmental component of the economic growth (Table 79.2). The environmental load indicator was calculated as the product of the growth indices of the two indicators presented in the table. The maximum increase in pollutant emissions into the atmosphere took place in the Novgorod region from 42 to 56 thousand tons. At the same time, the dirty discharge in the region decreased from 92 to 78 million cubic meters. The composite index that characterizes the dynamics

Table 79.2 Economic growth environmental component

	Air pollutant emissions from stationary sources (thousand tons)			Dirty discharge into surface water bodies (million cubic meters)			Environmental load dynamics
	2011	2016	index	2011	2016	index	Composite index
The Republic of Karelia	96	116	120.8	175	221	126.29	152.6
The Republic of Komi	712	569	79.9	129	110	85.27	68.1
Arkhangelsk region	373	245	65.7	375	329	87.73	57.6
Vologda region	469	441	94.0	157	147	93.63	88.0
Kaliningrad region	25	21	84.0	91	111	121.98	102.5
Leningrad region	216	243	112.5	231	272	117.75	132.5
Murmansk region	263	232	88.2	334	312	93.41	82.4
Novgorod region	42	56	133.3	92	78	84.78	113.0
Pskov region	28	33	117.9	45	37	82.22	96.9
Saint-Petersburg	69	78	113.0	1239	1093	88.22	99.7

of the environmental burden of economic growth in the region, turned out to be 113.0%.

The most adverse dynamics of the environmental load in the period under analyses was seen in the Republic of Karelia both due to the increase in pollutant emissions into the atmosphere and due to the dirty discharge. As a result, the region's composite index is 152.6%. From the point of view of the dynamics of the environmental load, the most favourable situation is in the Arkhangelsk region. It must be understood that this region, along with the Komi Republic and the Vologda region, belongs to the regions with high level of the environmental load.

The regional sustainable development index was calculated as the product of GRP indices and expenditures on social and cultural activities divided by the composite index characterizing the dynamics of the environmental load. In terms of this indicator value, regions can be divided into three groups (Fig. 79.2). The Komi Republic, the Arkhangelsk and Murmansk regions, as well as St. Petersburg have sustainable economic growth. The Vologda, Kaliningrad, Leningrad and Novgorod regions belong to the regions with average level of economic growth sustainability. Low level of sustainability is characteristic of the Republic of Karelia and the Pskov region. The problems of the Republic of Karelia are caused by problems in the social sphere and in the environment, and in the Pskov region—in the social sphere.

High performance of the Komi Republic is due to the considerable reduction in the environmental load with a slight decrease in the economic and social component. In the Arkhangelsk region, significant economic growth was accompanied by a moderate decrease in social spending and a significant reduction in the environmental load. From the point of view of the balanced economic development, it should be

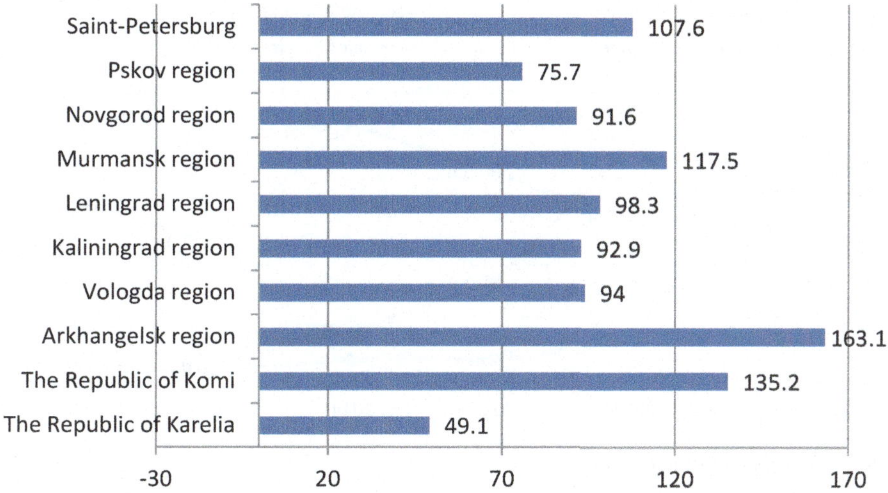

Fig. 79.2 Integrated assessment of the sustainable development

noted that no region has a positive dynamics of all three components of the sustainable development. So, the Arkhangelsk, Vologda, Murmansk and Pskov regions have positive economic and environmental components, in the Leningrad region it refers to economic and social components.

The low sustainability level in the Republic of Karelia is due to both a significant reduction in social spending and a significant increase in the environmental load. The low sustainability in the Pskov region is due to social spending by 31.3% in real terms.

The presented economic growth sustainability estimation does not take into account its innovative component, which is an important factor of the sustainable development [3]. Let us consider the dynamics of the indicator "expenditures connected with technological innovation in percentage terms of the volume of goods shipped" for the regions belonging to the group with sustainable economic growth (Fig. 79.3).

In the regions under analysis, the indicator of innovation is characterized by a high level of volatility, and in all regions, with the exception of St. Petersburg, it is at a low level. In these conditions, it is difficult for the innovative component to be included in the comprehensive assessment of the regional development sustainability. Thus, the priority direction of increasing the sustainability of the regional systems development is to improve the innovative activity level. In science literature, as well as in regional planning documents, clusters are identified as an efficient tool for sustainable development [1, 9].

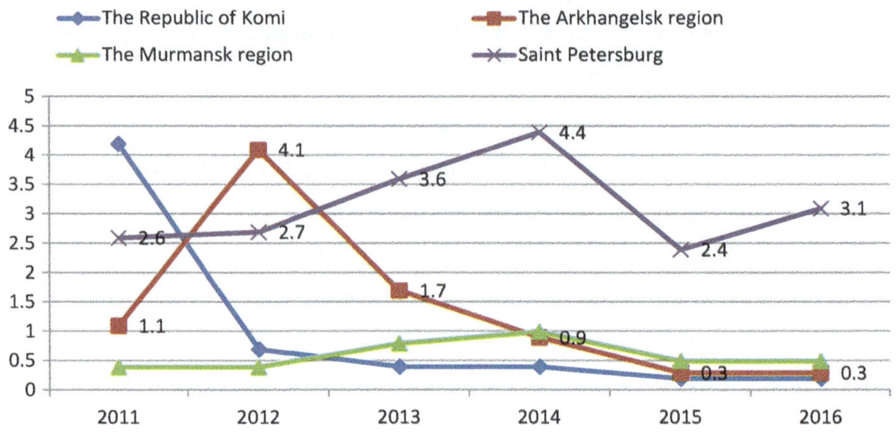

Fig. 79.3 Indicator of innovation

79.4 Conclusions

The analysis showed that the comprehensive approach to assess the sustainability of territories development prevails in science literature. The regional system is considered in the unity of its economic, social and environmental spheres. To assess the sustainability of economic growth, the integrated indicator is proposed which takes into account the development dynamics of each of these areas. The proposed methodology has been tested for the regions of the NWFD. According to the results of the research, regions are divided into groups with high, medium and low levels of economic growth sustainability.

Acknowledgements The research was carried out under the financial support of the Russian Foundation for Basic Research in the framework of a scientific project № 18-410-600004 r_a.

References

1. Akif HS, Vladyka M, Rashina E (2019) Cluster approach to sustainable development of the national economy. 37th international scientific conference on economic and social development—socio-economic problems of sustainable development, Baku, Azerbaijan, 2019, pp 14–15, 720–727
2. Bobylev SN, Kudryavtseva OV, Yakovleva YY (2015) Regional priorities of green economy. Ekonomika Regiona/Econ Reg 2:148–159
3. Bogolib T (2016) Ensuring smart, sustainable, inclusive development of the regions of the Ukraine in the conditions of structural crisis. Balt J Econ Stud 2:10–16
4. Kormanovskaya IR (2013) Ensuring sustainable development of the region: principles, concept, mechanism. Probl Mod Econ 4(48):260–266
5. Kulov AR, Gadieva AN, Fedosova EV (2019) Strategy of spatial development of the North Caucasus: theoretical and applied aspects. At conference on social and cultural transformations

in the context of modern globalism (SCTCGM), Russian Acad. Sci, Complex Res. Inst. Kh. I. Ibragimov, Groznyi, Russia, 2018, pp 971–979

6. Loktev AV, Menshchikova VI (2012) Sustainable development of the region's economy: basic conditions and support mechanism. Socio-Econ Phenom Process 10(44):115–122
7. Malykh OE, Polyanskaya IK (2018) Climatic aspects of the global economy transformation: increase of regions and municipalities' impact. At the international scientific conference on global challenges and prospects of the modern economic development (GCPMED). Samara State University of Economics, Samara, Russia, 2018, pp 1769–1776
8. Nikolaev MA, Makhotaeva MYu (2015) The role of the territorial factor in economic dynamics. Economist 3:42–49
9. Nikolaev M, Makhotaeva M, Bakumenko O, Malyshev D (2019) Tools for sustainable investment climate formation in the region. Proceedings of the international scientific and practical conference "business cooperation as resource of sustainable economic development and investment attraction" (ISPCBC 2019), 2019, pp 449–453
10. Pakhomova NV, Richter KK, Malyshkov GB (2013) Sustainable development strategy and the transition to green economy: updating priorities and mechanisms. Bull St Petersburg State Univ 4:35–54
11. Tretyakova EA (2019) Environmental intensity of economic growth in the Baltic Sea region. Balt Reg 11(1):14–28
12. Varavin EV, Kozlova MV (2018) Assessment of the development of the green economy in the region: Exemplified by the Republic of Kazakhstan. Econ Reg 14(4):1282–1297

Chapter 80
Evolution of Fire Conceptual Metaphor as Unit of Ecological Consciousness

Natalia Pitolina and Anna Kuzmichenko

Abstract The article attempts an interdisciplinary analysis of the phenomenon of fire—both as an environmental factor and as a unit of the conceptual system of man (unit of knowledge). Fire is seen as a conceptual metaphor reflecting national consciousness on the one hand and collective consciousness on the other. Presented hierarchy of conceptual metaphors of fire reflects the overall dynamics of the processes of man cognition and interaction with the world, and in the context of current global climate and economic transformations allows to recognize the conceptual metaphor of fire as the unit of ecological consciousness. The purpose of the study requires the disclosure of such concepts as environmental consciousness, environmental factor, conceptual metaphor. As an example of an ecological factor of great importance, forest fires are considered, which, depending on the dominance of the type of ecological consciousness, can act as biotic, abiotic or anthropogenic factors. The problem of the article also involves the study of fire as a mental construct (conceptual metaphor). As a result of the analysis of dictionary entries and the data of the British National Corpus, some basic properties of fire, reflecting the ambivalent properties of the element underlying the metaphorical projection, are distinguished. Conceptual metaphors of fire are subjected to categorization, as a result, basic and subordinate conceptual models are distinguished. The authors consider the significance and role of fire in the English national consciousness in terms of culture and history then taking a look at the present stage of its development. The analysis of the English-language articles from the BBC news devoted to forest fires allows us to identify a universal metaphorical model in which fire is compared to enemy. At the same time, the article presents some positive observations indicating that this model is being rethought now and new environmental conceptual metaphors come into being. It is concluded that the basic conceptual models of fire, represented by binary oppositions, are universal concepts of ordinary consciousness, but their interpretation depends on extralinguistic (including environmental) factors, which together lead to a change in the ecological paradigm of consciousness.

N. Pitolina (✉) · A. Kuzmichenko
Pskov State University, Pskov, Russia
e-mail: peanut278@yandex.ru

T. Kolmykova and E. V. Kharchenko (eds.), *Digital Future Economic Growth, Social Adaptation, and Technological Perspectives*, Lecture Notes in Networks and Systems 111, https://doi.org/10.1007/978-3-030-39797-5_80

Keywords Conceptual metaphor · Ecological conceptual metaphor · Fire · Ecological consciousness · Ecocentric ecological consciousness · Ecological factor

JEL Classification P28

80.1 Introduction

This article is written at the intersection of two sciences—linguistics and ecology, and, as a result, some observations on the evolution of the human value system have been made.

The key terms in this study are **ecological consciousness (EC), environmental factor**, and **conceptual metaphor**. Let us consider them all in more detail.

80.2 Materials and Method

Generally, **ecological consciousness** may be characterized as a form of human consciousness (both social and individual) which includes the sum-total of ideas, theories, opinions, values, motivations, intentions, competence and skills to solve economical problems without doing harm to the environment and human health. It also includes practical issues of interaction between man and environment, between society and nature [1].

It's noteworthy to say that ecological consciousness is integral to personality that can be moral, spiritual, mindful and responsible for their deeds. When shaped, ecological consciousness might serve a key to cognition and comprehension of environmental problems and necessity of their solution.

Human–Nature interaction is long-lasting, continuous, and evolving. It might depend upon epoch, religion, mentality and other factors. That is why each stage in mankind's evolution is characterized by its own type of ecological consciousness, each implying a certain set of distinctive features. Depending upon what features can dominate, there are two principal types of ecological consciousness—**anthropocentric and ecocentric**. They are opposed to each other as shown in Table 80.1.

Table 80.1 Opposition of anthropocentric ecological consciousness and ecocentric ecological consciousness

Anthropocentric EC	Ecocentric EC
• Careless attitude towards the environment	• Concern and awareness for the environment
• Opposing oneself to the world of nature	• Seeing oneself as an integral part of nature
• Viewing the environment as object of interaction	• Viewing the environment as subject of equal interaction

The next key term in the article is **environmental factor**: the element of habitat that can have a direct impact on the organism [2].

All environmental factors are divided into **biotic** (possible influence that a living organism is experiencing from the surrounding living beings), **abiotic** (possible influence that a living organism is experiencing from inanimate nature), and **anthropogenic** (possible influence that a living organism is experiencing directly from humans).

Importantly, the fore-mentioned categorization is not that strict, since one and the same possible influence that a living organism is experiencing can be related to some/or all factors.

Let's take FIRE to exemplify it.

FIRE can originally be described as an abiotic factor: fires are often started accidentally due to a combination of ideal conditions like lack of rain, hot weather, the availability of dry grass ignited by a lightning strike.

FIRE can also be a biotic environmental factor. Historically, wildfires played a positive role in the ecosystem of our planet. Being a part of "climate" in most epigeal habitat fires had been forming the history of flora since ancient times as they contributed to the change of timber species in the world's forest. Biotic biocenoses in their turn adapted to fire as well as to temperature or humidity [2].

Humans however have changed the influence of the environmental factor of fire dramatically, so that it has become rather anthropogenic. And this influence is many-faceted.

Firstly, wildfires and grassland fires would be a method of hunting and slash-and-burn agriculture. After deforestation burn areas provided bountiful harvest, but soil fertility decreased very quickly, and new forest areas had to be chopped down. The damaged areas couldn't be reconstructed again. Thus, the wide usage of slash-and-burn agriculture resulted in some obvious environmental changes including flora, fauna, soil, climate and regime of river [3].

Secondly, a large group of risks and hazards is linked with man-made disasters and military operations. The latter may also cause fires which destroy local natural ecosystems and can lead to changes in the atmosphere—saturation with greenhouse gases, soot and propellant gases that spread far beyond the region of warfare [4].

And, finally, a man's unconcerned attitude causes the degeneration of the environment (uncontrolled small fires, non-stubbed out cigarettes and matches, burning dummies, unpermitted burns and other careless actions).

To crown all, the dominance of anthropocentric EC is leading to destructive and irreversible consequences for our planet.

Having made preliminary extralinguistic observations concerning fire as an ecological factor, we will proceed to the consideration of this concept as a mental construct, which, as is obvious from the above, is an integral unit of the conceptual system of man.

The notion of a "conceptual system" is one of the basic terms in cognitive linguistics. A *conceptual system* is a mental level or mental organization where the sum-total of all the concepts given to the human mind is concentrated; their orderly union [5].

The conceptual system of man is the result of conceptualization, and one of the means of conceptualization is *metaphor*.

The process of metaphorizing a mental construct is a hypothetical field of study. At the present stage of development of linguistic science, there are several hypotheses that explain this process, namely the theory of conceptual metaphor, proposed by G. Lakoff and M. Johnson [6] and the theory of conceptual integration in the authorship of M. Turner and G. Fauconnier [7] and their followers. Let us consider in more detail the process of metaphorization in the theory of conceptual metaphor.

In the English-language scientific literature, the process of metaphorization is explained by the statement, known as "*embodied cognition thesis*".

For the first time this thesis was introduced by G. Lakoff and M. Johnson. It was formulated as follows: conceptual metaphors in general are based on the "embodied" experience of human cognition, i.e. on how the human brain and body function in interaction with the environment [6].

Conceptual metaphors represent the systemic relationship between conceptual domains: the *source-domain*, which serves as the basis for metaphorization, and the *target-domain*, which represents the object itself.

The essence of metaphorization is the transfer from the "source-domain" to the "target-domain" of those cognitive structures in terms of which the experience related to the "source-domain" was structured [6].

The cognitive theory places metaphor not into the structure of linguistic knowledge, but into the structure of a concept. At the same time, if a conceptual metaphor appears as a mental representation underlying the functioning of a language, then metaphorical language expressions become only a means of realization of conceptual metaphors.

80.3 Results

The analysis of empirical material allows us to judge the importance of fire as an environmental factor in the process of human cognitive activity and the special place of fire as a unit of knowledge in its conceptual system. Reference to the Oxford dictionary makes it possible to talk about the following, the most important components of FIRE as a mental construct:

1. flames that are **out of control and destroy buildings**, trees, etc.—*Several youths had set fire to the police car (= had made it start burning).*
2. a pile of burning fuel, such as wood or coal, used for **cooking food or heating a room**—*Sam had lit a fire to welcome us home* (Oxford Learner's Dictionary; Merriam-Webster Online Dictionary).

This semantics represents the element of FIRE as an ambivalent concept in the consciousness of man: on the one hand, it is an evil that eliminates and destroys human life and material goods necessary for man; on the other—it is something without which the physical and social development of man would be impossible. Thus, the basic binary opposition is distinguished: FIRE IS A CREATIVE POWER/FIRE IS A DESTRUCTIVE POWER.

In the language these conceptual metaphors are implemented as follows:

- FIRE IS A DESTRUCTIVE POWER:

Founded in 1637 and rebuilt after a ***disastrous fire*** *in 1824.*
The ***fire was a blazing tornado****.*

- FIRE IS A CREATIVE POWER:

Fire *too is a similar thing. It can be* ***warming …, it can be purifying, life-giving****.*
The ***fire is the sun****.*

As the analysis of the language material presented in the British National Corpus shows, another productive model is a binary opposition FIRE IS A LIVING BEING/FIRE IS A NON-LIVING BEING.

These conceptual metaphors are objectified in the language as follows:

- FIRE IS A LIVING BEING:

FIRE broke out *during a funeral at a crematorium yesterday and set the roof alight.*

Fire breaks out *again in top floor of apartment block.*

*Firemen wearing breathing apparatus were sent in****to deal with the fire*** *at Hedley-on-the-Hill, Northumberland.*

Brandy Then in June, Ivan died in a ***freak fire*** *at his seafront nursing home.*

However, we have recently had to incur extra expenditure because the road that were put in looking at one hundred were found to be too high especially for the services and place committee which throughout that it was very difficult but apparatus you know ***moving fire****.*

- FIRE IS A NON-LIVING BEING:

TERROR TARGETS: Ford and family take cover as fanatics open fire. 'It's a ***sea of fire****. Everyone has gone. Children… everyone,' she cried.*

Who's going to be the last one ***to pull*** *their hands* ***out of the fire?***

Children really are ***under fire*** *throughout the world, literally and metaphorically.*

What it is I'm ho- we're ***holding fire****.*

The analysis of stable correlations between the source-domain and the target-domain (representations of physical experience) within the framework of these basic metaphorical models allows semantic sub-categorization of the latter:

1. FIRE IS MAN: *the* ***flames crowded****;*
2. FIRE IS AN ANIMAL: *the fire was a* ***raging tiger****;*
3. FIRE IS A PART OF HUMAN'S BODY: *And something appeared to them which seemed like* ***tongues of fire****.*

The conceptual metaphor FIRE IS A NON-LIVING BEING is characterized by such connections as:

1. FIRE IS A PHYSICAL OBJECT:

POP heart-throb Jon Secada escaped death by seconds during a shoot-out in Miami. The Florida-based star, at No. 10 with Just Another Day, was out shopping when two rival street gangs ***opened fire*** *outside the store.*

Large sections of fire *under control.*

BED and Breakfast has come a long way since the days when fearsome dragon landladies ***belched fire*** *and brimstone over their terrified guests.*

2. FIRE IS LIQUID:

'It's a ***sea of fire****. Everyone has gone. Children... everyone,' she cried.*

In January he escaped ***a fire which engulfed*** *the Cheshire Home where he was staying in Greatham, Hants.*

Earlier, eight people died when ***fire rained*** *on the city's crowded streets.*

Luckily, their homes were spared as the ***fire flowed away*** *from houses.*

These conceptual metaphors seem to be universal constructs of consciousness, but can change in the process of forming a national picture of the world, as well as depend on the level of mental, spiritual, cultural development of a person and the dominant religion.

Let us consider some examples. In the Celtic mythological system, fire was seen as a powerful symbol, seed, or essence of life: while "male" deities are associated with forest fires, "female" deities are associated with the hearth. For instance, *Taranis*, the Celtic god of thunder, was associated with lightnings and fires caused by it; whereas the Irish goddess *Brigit* was considered the patroness of the home hearth [8].

In addition, fire-related rituals were extremely important to the Celts. One of the most important, *Beltane*, was held on the first of May in honor of the sun warming the earth after a long cold winter.

In Ireland fire was associated with a number of festivals. The tradition of celebrating the summer solstice is believed to date back to pre-Celtic times. Already in the days of Christianity as the dominant religion in many regions of Ireland, the eve of St. John's day, celebrated on June 23, two days after the longest day of the year at the summer solstice, was called "Bonfire Night" [8].

In Scotland, fire was considered protective and cleansing: the fire, which was carried around the house necessarily clockwise, guarded the building and its inhabitants from harm. Fire was used to protect women during childbirth, as well as young children. It was believed that fire will save the child from being kidnapped by magical creatures.

The use of fire as a means of protection and purification was observed until the mid-eighteenth century (written evidence indicates 1767). This ritual was also observed among the Germanic tribes, as noted by the outstanding collector of folklore J. Grimm [8]. It remains unclear whether it originated among the Germanic tribes and then spread among the Celts, or vice versa. This fact allows us to draw a conclusion about the dominant ecological consciousness of ancient people, who thought themselves physically and psychologically to be a part of Nature, for whom fire was a symbol of worship, a means of survival and protection.

The change of epochs, characterized by the gradual alienation of man from the natural world, the technologization and marketization of society, global climate changes on the planet, the evolution of the value system of mankind and other factors together led to a change in the ecological paradigm of consciousness, which "mirrored" the experience of human interaction with fire.

The analysis of articles from BBC News for the last two years, which are devoted to wildfires in European countries, Canada, the USA and some countries in South America, leads us to some important observations, which can be generally reduced to the following statement: "*It's a huge ecological disaster*", acting regional governor Kostas Bakoyannis told AFP.

This conclusion made is also the result of the analysis of metaphors of fire from BBC News articles (https://www.bbc.com).

The selection of stylistic contexts where the lexeme "fire" occurs allows to single out the most productive metaphorical model—FIRE IS ENEMY, which invokes a certain assumption about negative experience and complex relationship between man and fire which can be generally characterized as *hostile, conflicting, unequal, uncompromising*. The following characteristic are attributed to fire:

1. **fire can kill/destroy**: At least 80 people *have been killed in wildfires* in Greece…. *The deadly fires in Greece* are a prime example of weather being only one important factor. *More than 100 people were killed by wildfires* in Portugal only last year and *173 people died in the Black Saturday Fires* in Australia in 2009. Hundreds of people were evacuated from nearby villages as the fire broke out in the early hours of Tuesday *and ravaged the dense forest.* Watch how *flames tore through an untouched pine forest. Greece was devastated by wildfires* in 2018 that tore through pine forests surrounding Athens, *leaving 100 people dead.*
2. **fire can threaten and attack**: It tends to be those *which threaten lives, infrastructure, or natural resources* that hit the headlines. Educating people about the risks and having evacuation plans in place can also reduce *the threat to human lives and the economy.* The fires in the seaside village of Mati in July 2018 spread rapidly, *trapping people* in their homes and cars—and even some who fled to the sea.
3. **fire can damage**: Spain and Portugal, which normally have *fire damage over a greater area* than any other European countries *Greece wildfires leave blackened forests in their wake.*
4. **fire can cause fear**: These are *too big and fierce* to be stopped, irrespective of the efforts and resources thrown at them. Archive video: "*The flames were chasing us into the water*".
5. **fire can gain momentum**: This has brought unusually dry conditions which have allowed *large fires* in regions which usually see very few. Other northern European countries including Denmark, Estonia, Finland, Germany and Latvia *have seen between 20 and 200 times the normal area burned. Wildfires have blazed* across Sweden this summer, with hot weather and persistent drought cited as the main causes. Increase of "*megafires*".

Furthermore, fire is realized as a strong, mighty and unbeatable rival: *The regional governor, Costas Bakoyannis, told Skai TV the situation seemed to be improving after* ***"superhuman efforts" to control the flames*** *on Tuesday night.*

The conceptual contents of the model FIRE IS ENEMY is revealed even fuller when human actions towards fire are taken into consideration. A human has to take the challenge of the force of nature, fight with it, protect his territory and life: ***A firefighter tackles wildfires*** *in Lancashire.* ***Spanish firefighters combat a forest fire*** *in Portugal in 2017. […]* ***firefighters had "given their all" to tackle the flames****, often with little sleep and food.*

Language verbalization of the "fire" contents goes by means of wide usage of verbal and substantive lexis semantically connected with warfare (*to fight, to attack, to combat, to kill, to threaten, to die, to devastate, to evacuate, arson, aftermath, dead, rescue, victim, и т. д.*).

Interestingly, our analysis of the same articles has lead us to another conclusion which is quite *positive*, because it shows that the model FIRE IS ENEMY is being conceptually reconsidered.

Firstly, FIRE comes to be understood as an object of study. Secondly, humans are searching for ways how to safely coexist with fire and how to prevent it: *In addition to criticism of the crisis management and co-operation during the blaze, it said there were issues with* ***the preparation and prevention to avert such a catastrophe****.*

Paradoxically, being aware of the fact that man is the greatest predator and threat to the environment man has to admit at least two truths: first, the enemy, whom man is fiercely fighting with, is man-made; and second, it is man himself who can become the ultimate loser in this battle.

And here we do not mean some occasional cases of flame development caused by man's careless attitude but global and irreversible matters when forest fires are caused by climate change, which itself is the consequence of a human activity: *Natural ignition by lightning is common in some regions, such as the North America boreal forest, but in Europe* ***the vast majority of fires are human-caused****. While climate change may have made more fuel available for such fires,* ***human behaviour has also played a role. Burning question of Wildfires and Climate Change****.*

The fires triggered mass protests against the environmental policies *of President Evo Morales.*

Conclusions/recommendations. Metaphors verbalizing that new model are few. Yet they are reflecting fluent steps our consciousness is making in the direction of ecocentric ecological consciousness: ***To explain fire****, scientists and firefighters use something called the* ***"fire triangle":*** *an ignition source, plus fuel such as dry vegetation and oxygen to keep going.* ***We need to adapt and learn how to coexist with fire****.*

What is important is that there are such steps, the steps are obvious, and the steps are gaining momentum… So now we are ready to introduce novel environmental conceptual metaphors of fire.

References

1. Бегидова, С.Н., Макрушина, И.В. (2014) «Структура экологического сознания», в сборнике Вестник Адыгейского государственного университета, Серия 3: Педагогика и психология, Режим доступа: https://cyberleninka.ru/article/n/struktura-ekologicheskogo-soznaniya (дата обращения 11 октября, 2019)
2. Одум, Ю. (1986) «Экология», Мир, Москва
3. Большаков, В.Н. и др. (2000) «Экология: Учебник для вузов», Интернет инжиниринг, Москва
4. Николайкин, Н.И., Николайкина, Н. Е., Мелехова, О. П. (2003) «Экология: Учебник для вузов», Дрофа, Москва
5. Кубрякова, Е.С. и др. (1997) «Краткий словарь когнитивных терминов», Филол. ф-т МГУ им. М. В. Ломоносова, Москва
6. Lakoff G, Johnson M (1980) Metaphors we live by. The University of Chicago Press, Chicago
7. Fauconnier G, Turner M (2002) The way we think: conceptual blending and the mind's hidden complexities. Basic Books, New York
8. Monaghan P (2004) The encyclopedia of Celtic mythology and folklore. Facts on File, Inc., New York, pp. 195, 296–299, 442
9. British National Corpus, "Fire". Available at: https://www.english-corpora.org/bnc/. Accessed 14 Oct 2019
10. Burning question of wildfires and climate change (25 October 2007). Available at: https://www.bbc.co.uk/programmes/p03jrvyg. Accessed 10 Oct 2019
11. Greece wildfires leave blackened forests in their wake (14 August 2019). Available at: https://www.bbc.com/news/world-europe-49343572. Accessed 22 Sept 2019
12. Merriam-Webster Online Dictionary, "Fire". Available at: https://www.merriam-webster.com/dictionary/fire. Access 18 Sept 2019
13. Oxford Learner's Dictionary, "Fire". Available at: https://www.oxfordlernersdictionaries.com/definition/english/fire/_1?q=fire. Accessed 14 Oct 2019

Chapter 81 Military-Political History of the Pskov-Livonian Border: History of Struggle for Natural Resources in the Context of Climatic Changes of 14th–15th cc.

Elena V. Salmina, Sergey A. Salmin and Svetlana F. Matsevich

Abstract The content of the research is the study of the climatic factor influence on the intensification of the struggle for natural resources in the border of the Lord Pskov and the Livonian Confederation in the period of climatic changes in the 14th–15th centuries. The objective of the study was the identification of the fullest possible range of the causes of spontaneous cases of the military-political situation aggravation in the period under study. Russian Chronicles, Russian scribes books, Livonian Chronicles, correspondence of the Livonian officials were used, later sources were also involved. Maps of military conflicts correlated with the known periods of climatic fluctuations in the studied region are made. The most important and significant came to be struggle for lake and river fishing grounds (Piirissaar island (Zhelachka), Narva rapids, etc.), for the territory of wild bees population mass rate (Purnau) and for haymaking. To redistribute the resources of these territories in their favor, the sides used both direct seizure of the resource, and diplomatic and military pressure on the opponent. The large-scale invasions, chronologically coinciding with lean years, should be specially marked. Mapping the regular military clashes loci allowed us to determine the list of resources, the struggle for control of which took the most violent forms. It is shown that in the 14th–15th centuries the struggle for renewable resources of the border territories, largely due to the climatic factor, made the mayor content of military actions and conflicts of Pskov and the Livonian Confederation.

Keywords Military history · Renewable natural resources · Climate · Livonian Confederation · Border conflicts · Pskov · Pskov land

E. V. Salmina (✉)
Laboratory of Socio-humanitarian regionics, Pskov State University, Pskov, Russia
e-mail: muntrik102@yandex.ru

Department Archaeological Centre of Pskov Region, Pskov, Russia

S. A. Salmin
Archaeological Centre of Pskov Region, Pskov Archaeological Centre, Pskov, Russia

S. F. Matsevich
Pskov State University, Pskov, Russia

T. Kolmykova and E. V. Kharchenko (eds.), *Digital Future Economic Growth, Social Adaptation, and Technological Perspectives*, Lecture Notes in Networks and Systems 111, https://doi.org/10.1007/978-3-030-39797-5_81

JEL Codes N13 · N43 · N53

81.1 Introduction

At the beginning of the 14th century, the economies of the Baltics and the Russian North–West turned out to be in a situation of direct opposition. If during the 13th century the purpose of the expansion for both sides was to establish control over the peoples and territories of the Baltics, then after the establishment of a stable demarcation line in the 2nd half—the end of the 13th century [1] it were natural resources of border territories that became the target of the struggle.

The situation was aggravated by the fact that the time of the transition from the climatic optimum to the small ice age coincided with the formation of the Livonian economy, which put the established agricultural system of the Livonian region in a difficult position [4]. To provide the population with food resources that were minimally dependent on climatic changes became the priority.

81.2 Materials and Methodology

81.2.1 The Region of the Research

The region of the research is the basin of Pskov-Chudskoe Lake and the Velikaya and Narva Rivers, which was the zone of the Pskov State formation and functioning during the 12th–16th centuries.

Pskov-Chudskoe Lake covers the area of 3,555 km, being the fourth largest lake in Europe. Its water collection area is 47,800 km^2. It is connected to the Baltic Sea (the Gulf of Finland) by the Narva River (Narova). The lake is divided into three zones with different hydrological characteristics. The northern part of the water body is traditionally called Chudskoe Lake, the southern part—Pskov Lake, the narrow strait connecting them is called Warm Lake or Uzmen.

In the water area of Chudskoe Lake there are islands: Piirissaar (historical names—Gelachka, Porka), Ozolets, etc., mostly concentrated in the southern part of the lake. In Pskov Lake there are such large islands as Talabsky, Kamenka and Kolpina (Kolpin) [11]. In the basin of Chudskoe Lake there were located suburbs of Pskov called Kobyla and Gdov (river Gdovka, 2 km from the mouth).

Pskov-Chudskoe Lake is among the most productive freshwater bodies of Europe. Fishing in this territory has been actively practiced by the locals since the beginning of the region development to the present [6, 9, 19, 21, 22, 24, 36].

Extensive hayfields were also located in the coast areas.

The difference in geological structure of the lake shores [7, 31] made the basis for the uneven distribution of resources in the waters of the lake. The main breeding and fishing zones were the coastal territories and the islands of Pskov and Warm Lakes, and to a lesser extent the east and north-east Shores of Chudskoe Lake.

The west coast of Chudskoe Lake was mostly low-lying (territory of the sublime coast, as for example, Kallaste, occupied small area), resulting in heavy waterlogging of the area and shifting settlements inland. This in its turn prevented the full-scale exploitation of lake resources by the Livonian side [8]. The second important factor that made it difficult for the Livonians to enter the water area of Chudskoe Lake was the absence of natural bays suitable for harbors installation on its west bank.

The Velikaya River is about 430 km in length, the water collection area is about 25,000 km^2. On the main course of the Velikaya River there is the city of Pskov itself (15 km from the mouth) and also important Pskov suburbs, which arose from the 10–11th to the 15th c., such as Ostrov (58 km from the mouth) and Opochka (145 km from the mouth). On the rivers of the Velikaya River Basin there were located the suburbs of Vellie (the river Versha and the Lake Vellie), Volodimerets (the Listvenka river), Voranach (the Sorot' river), Vrev (the Vrevka river), Vybor (the Milya river), Vyshgorodok (the Ulada river), Kotelno (the Kotelinka river), Kolozhe (the Kudka river), Krasny (the Sinaya river).

The main water highway leading inland the region, the Velikaya River was also one of the Pskov-Chudskoe Basin ecosystem basis, especially for ichthyofauna migrating between the lake and the river in the periods of spawning, gaining weight and wintering.

During the period under review, the water area and the coast of the Pskov-Chudskoe Lake were regular conflict zones on both Pskov and Livonian sides. The left bank of the Velikaya River to the south of the Kudeb' river and to the Sinaya River remained one of the disputed territories, for the control over which Pskov and Livonian landsgerers fought. Prior to the 19th century, this region was a low populated and forested area, especially in the southern part, in the area of the Yavorskaya, L.V. (2014) "Meat trade or bone paving? Archaeozoological research of the Luzhsky II excavation in Pskov", in the collection "Archaeology and History of Pskov and Pskov Land. Seminar named after academician V. V. Sedov", Proceedings of the 59th session, Vol. 29. Nestor-History, Moscow-Pskov-St. Petersburg, pp. 42–55 (in Russian).

The economic importance of these territories was determined by the local biocenosis features. Due to the abundance of mixed forests and meadows presence, the Purnau area was rich in wild bees and high-grade hay, which—at low population density—allowed these resources to be used without any excessive biocenosis burden.

In general, natural conditions in the region under consideration were unfavorable for the Livonian side. The main fishing areas were located in the Pskov and North Lakes (mainly in the western part) at the eastern shore of Chudskoe Lake and the Narova River head [18, 23, 25].

All these territories were under military control of Pskov. The west shore of Chudskoe Lake, on a significant part of the length lowland and swampy, was of less interest to fishermen [5, 10]. Due to the impossibility of founding significant settlements on the western shore of Chudskoe Lake, the Livonians found themselves in a notoriously weak position.

On the contrary, having convenient operational bases at the mouth of the Velkaya River and on the Lake Islands and being able to use the mouth of the Pskova river, located within the city, as a winter base for ships, the Pskovites established almost absolute monopoly on fishery in the waters of the Pskov-Chudskoe lake water basin [12].

The similar situation was developing in the region of Purnau. With a direct access to the main water artery of the region, the Pskov suburbs could receive timely support from the metropolitan area and thus had significant benefits compared to those of Livonian competitors.

81.2.2 Chronological Scope of the Study

The period of the 14th–15th centuries was chosen for our research as the time period most fully reflecting dynamics of political and economic relations between the members of the Livonian Confederation and The Lord Pskov (The latter received significant support from the Grand Duchy of Moscow at the end of the period under review).

It is difficult to distinguish the conflicts aimed at seizing territories from those caused by the struggle for the renewable resources in the earlier period (until the 1300s), when the Pskov-Livonian border was still in the state of formation. Since the first quarter of the 14th century, the amount of information concerning the Pskov-Livonian economic relations contained in written sources has considerably increased.

81.2.3 Sources of the Study

During the research the data of the Russian chronicles (The Pskov 1st, 2nd and 3rd chronicles [15, 16], the Novgorod 4th chronicle [14], Russian scribes books (The Original scribes book of the city of Pskov and its vicinities No. 355 (1585–1587) [26]), the Livonian and Prussian chronicles [13, 33], correspondence of Livonian officials [12] were attracted. At this, reconnaissance stage of the study, we decided to limit ourselves to available published sources.

Statistical descriptions of Russian provinces in the mid-19th century, works in hydrology, political and climatic history, as well as in historical ecology were used to obtain information about natural conditions and dynamics of their changes in the region under study.

On the basis of historical source analysis, maps of military conflicts have been drawn up, and their correlation with the dates of climate fluctuations known to us in the region under study and in neighboring territories has been made. Mapping the loci of regular military clashes revealed the list of renewable resources, the struggle for the control over which, took the fiercest forms.

81.3 Results

81.3.1 Economic Side of Conflicts in Disputed Territories

The first military clash for fishing grounds took place in autumn, 1322, when the chronicles for the first time record targeted attacks on "fishermen" in the mouth of the Narva river and on the fishing settlement of Cheremest' in Berezskaya volost [15]. The attacks were probably carried out by subjects of the Danish king from northern Estonia (as it was against this territory that the Pskovites took a retaliatory campaign in winter of 1323).

Thus, the attacks on Narvian and Cheremestian fishermen may be regarded as the first attempt of the Livonians to take possession of fishing grounds ("*lovli*") in the northern and eastern part of Lake Chudskoe (Peipsi), more open to attack than the fishing grounds of Pskov and Warm Lakes.

- In 1341—the attacks on the fishing settlements in Berezskaya volost', Koushela, Remda;
- In 1348—the German attacks on the Pskov village in Prinarovye and the subsequent construction of the "town above the Norova" by the Livonians (historical Syrenets, contemporary Vasknarva) opposite the Pskov fishing settlement Skamya and subsequent destruction of this town by the Pskovites in 1349;
- In 1367—the attack of the people of the Bishop of Derpt and members of the Order on the Pskov fishermen, accompanied by the destruction of "huts and nets" [33] which led to a five year war for the island of Zhelachka;
- In 1463—attacks on fishing settlements in the area of Kobyla, Ostrovtsy, Podoleshie, Kolpin, Skamya;
- In 1480—attacks on Gdov, Ostrovtsy fishing settlements in the area of Kobyla.

These attacks were often a prelude to larger-scale military actions.

The conflict over Piirissaar—Zhelachka (1368–1480), which took a particularly violent character and turned twice into full-fledged wars (1367–1373 (?), 1461–1463) was probably triggered by changes in the hydrological regime of Lake Chudskoye [32].

The flooding of the land isthmus that had connected the former peninsula to Livonian territory turned it into an island. As the remaining Lake Islands during the period under review were controlled by Pskov, the Pskovites immediately filed their claim on Zhelachka.

The abundance of fishing yards on this island recorded in the scribes books [26] shows the high profitability of fishing on this section of Lake Chudskoe and explains why both sides showed such persistence in defending their right to own the island.

The information on the main zone of snet (*Osmerus eperlanus eperlanus m. spirinchus*) fishing at the southern and west shore of Lake Pskov [18] and the location of the island of Zhelachka on an extreme northern point of a coastal arch turned possession of the island into a guarantee of maintaining Pskov monopoly for this fish. A number of historical documents mentions that the conflict of 1367 occurred after the third case of the Pskovites' "obstruction" to Derpt and the Order fishing [29, 30, 33].

The absolute dominance of the Pskov fishermen in the basin of Lake Pskov-Chudskoe and the important role of Russian dried fish supplies for Derpt is evidenced by a number of facts.

In 1471, the Landmaster Johann Waldhaun von Gerse attempted to end the Pskov fishing near the island Piirissaar (Zhelachka) in exchange for the rejection of Livonian claims for the developed by the pskovites part of the Purnau region in the area of town Krasny [15].

The Landmaster grounded the necessity of this agreement by the need of the residence of the Livonian master of the Teutonic order which was moved to Fellin (it should be noted that the Pskov chronicle calls Fellin the capital of the Order—«*stol*» [15]) in the supply of fish. Thus, we can assume that the volume of fish production by Derpt fishermen could at best meet only their own needs.

The refusal of the Pskovites led to the overthrow of the Landmaster by the Livonian "war party" [1].

In 1480, according to the message of Derpt magistrate, the city was unable to provide "dried fish" to the troops (landsknechts) during the war with Pskov—due to the lack of supplies of this fish to Derpt from Pskov (sic!) [12].

Equally important is the message sent at the same period to Revel which informed that there were not enough ships in Derpt to transport mercenaries to Pskov and, moreover, no shipbuilders in Derpt, that made it impossible to arrange shipbuilding there.

Derpt Magistrate asks Revel about specialists in shipbuilding and driving ships and describes the need in refurbishing the city boats ("*lodigen*") [12].

The very fact that there are not enough lake vessels in itself may not directly testify to the lack of organized fishery, but without any doubt it highlights the inability of the people of Derpt to take control over any significant part of the water area of Lake Pskov-Chudskoe. The Pskovites, unlike their opponents, acted extremely decisively, and never hesitated even to use lake ships—"*nasads*" supplied with artillery for the fishermen protection [15].

The second extremely important natural resource recorded in chronicles was the high-grade "meadow" hay. The topic of cattle breeding the development in the Pskov territories during the Middle Ages has not been the subject of a special study yet, but significant amounts of osteological material originating from the Luzhsky 2–4 excavations of 2011–2013 [34, 35] and from Neznanova slide-2 2013–2014 excavation [20] allow us to claim that livestock played a much larger role in the food supply of Pskov than it was previously imagined.

The livestock farm of the city was based on cattle, besides, the increase of the number of cavalry in Pskov troops composition, traced during the 14th–15th centuries, cannot be underestimated either. Due to the natural features of the territory, the bulk of the forage came from the marshy lands giving low-grade hay. The "meadow" hay mostly came from coastal lands and meadow areas of Purnau region, which remained disputed territory [27].

Haymaking sites were largely geographically tied to fishing settlements due to general geographical timing. The haymaking on the lake shore territories has become the subject of a violent struggle since at least 1427, when it was reported that seven Estonians who had been captured "on Vybovsk" while mowing hay "on the ground of Pskov" were executed, and the harvested forage was destroyed by the Pskovites [15].

The situation was literally repeated in 1459, when the Pskov expedition led by Prince Alexander Chartorizhsky and the posadniks carried out a number of undoubtedly symbolic and even demonstrative actions on Ozolets, Zhelachka and Rozhkin. The participants of the expedition mowed grass, organized fishing "in the old fashion" built a church in the name of St. Mikhail on Zhelachka and executed several Estonians—"chyuchnov" on Rozhkin (the fishing settlement on the Livonian border). In this case, most probably these were fishermen captured in the disputed territory.

During the retaliatory march, the Germans attacked the Pskov fishermen, who were under the protection of the "*nasad*", armed with guns, in Narov, burned the church on Zhelachka and destroyed many yards in Berezskaya Volost' (east coast of Lake Chudskoe).

The acute conflict situation on the left bank of the Velikaya River (Purnau and territory of «Latygora» in Pskov Chronicles) was determined in particular by the uncertainty of the political status of these lands, which developed during the

German conquest of Latgalia [27]. The inhabitants of these territories were considered subjects of the Livonian landsgerrers, but some of them—for example, the area of Adsele were to pay tribute to Pskov, which could not but cause a conflict of interest. The conflict over Purnau resources, which began in 1427, was one of the main causes of the Russian-Livonian war of 1480–1481 [2].

Conditions for sharing use of adjacent territories were fixed by written agreements providing for both permitted fishing territories and force majeure situations of demarcation line violations [12]. Let us also emphasize the existence of individual areas officially used by both sides [17, 27].

During the Pskov-Livonian conflicts, there was revealed quite an interesting pattern—at least three times (in 1323, 1463, 1480) a full invasion and siege of Pskov or its suburbs is preceded by an unexplained attack on fishing settlements.

Thus it was the attack of the Livonians on the fishing settlements in the autumn of 1322 that provoked the arrival of Davyd Gorodensky in Pskov, the invasion of the Pskovites in northern Estonia and the two subsequent unsuccessful sieges of Pskov by the Germans (in March and May, 1323).

In 1462/1463, the main target of the Livonian invasion were fishing settlements, and the New Gorodets (Kobyla) was but only briefly blocked by the Livonian units.

When describing the events of 1480, the Pskov chronicles deliberately specify the German destruction of the "*isads*" (fishing settlements) in the area of the New Gorodets prior to the capture of Kobyla. The Russian attacks on Kastre castle (which was a very important medieval fishery centre for the lower Emajogi River and Lake Peipsi) are part of the same trend [28].

It appears to us that this sequence of events was not accidental, but was due to the following factors:

- The winters of 1322 and 1323 were extremely severe in Prussia and Livonia. In the summer of 1322, a significant number of pilgrims arrived in Prussia. The winter cold of 1323 disrupted the march of Pilgrims and Teutonic knights to Prussia, as many of the infantry men died of frost [13]. The simultaneous revitalization of the Livonian branch of the Order in May 1323 was probably due to the necessity of withdrawal of numerous allies, in need of provision, from the Livonian lands.
- The winters of 1460–1462 were extremely cold (especially the winter of 1460, when the Baltic Sea was completely frozen) [3].
- Preparation for the war of Livonia with Pskov in 1479–1480 was preceded by an extremely hot and dry summer marked by hail and thunderstorms in Western Europe [3].

Notably, in all these cases, the rout of fishing settlements can be interpreted as a reaction to the lack of provision in Livonia caused by climatic causes. The harsh winters of 1320–1322, 1460–1462, 1479, recorded for Western European and Russian territories, could have both caused the influx of pilgrims and mercenaries into Livonia and produced a severe impact on the standard of living of its population. In this situation, an attempt to seize a food resource that had not suffered from climatic disturbances seems quite reasonable.

These observations are well correlated with information on the exacerbation of the conflict situation in Lake Chudskoye region in cases of climatic crises in Western Europe.

The military clashes of 1341, 1348 and 1367 are preceded by:

- 1340—an extremely cold winter in Western Europe.
- 1347—a very cold winter, and then an extremely wet and cold summer in Denmark, which suggests a similar situation throughout the Baltic States.
- 1366—a rainy spring and a very dry summer in Western Europe [3].

The situation is emphasized by the fact that the Pskov Chronicles do not report any famine in these periods. Perhaps this is due to the development of fishing, which provided a certain level of food supply in the Pskov region, and the presence of strategic provision stock in Pskov. In any case, in 1422 and 1434, the all-Russian famine practically did not affect the Pskov land, which had large reserves in case of siege (judging by the place of concentration of products "in the crates on the Krom") [15].

81.4 Conclusion

Among the causes of military clashes in the zone of the Pskov-Livonian border are the struggle for lake and river fishing grounds (Piirissaar [Zhelachka] island, the river Narva rapids, the Eastern shore of Lake Chudskoye-Peipus), the struggle for the territory of mass wild bee population (Purnau), the struggle for haymaking, etc.

For the redistribution of resources of these territories in their favor, both sides used both direct seizure of the resource and diplomatic and military means of influencing the opponent. The situation of passive counteraction and/or joint use of land tended to aggravate in case of unfavorable climatic conjuncture.

Until the mid of 15th century, neither side made any attempt to establish permanent strongholds to protect their interests on the disputed lands, indicating a weak interest in holding the territories themselves. Both sides were interested only in the alienated resources of the micro-regions. To protect their workers, the conflicting parties used military escort.

Thus, at least once the ship supplied with artillery-"*nasad*" was used for the protection of the fishermen on the lake. In case of capture of representatives of the opposite side on the firm ground the latter were subjected to death penalty. The fishing tackles of the alien fishermen were confiscated.

Beginning from 1461, the Pskov people began to establish fortresses on the disputed lands to protect their peasant colonists, in one case at the direct request of the latter.

The Pskov fortresses which arose in the period from 1461 to 1476 were repeatedly attacked by the troops of the Livonian Confederation. In some cases the fortifications were destroyed, but no attempts to use the captured fortifications by the Livonian side were ever noted. This may be due to different approaches to the use of local bioresources.

Unlike the Pskov people who included agricultural colonization in the use of territories, the Livonians considered them exclusively as raw materials base. As a result, the purpose of military incursions into these regions by the Livonians was not so much the retention of these lands under their military control, as the expulsion of Pskov colonists and fishermen from the disputed territories.

The struggle for the land and fishing grounds was noticeably aggravated in lean years. The cases of large-scale invasions coinciding with lean years both in the territory of Russia, and in the territories of the Western Europe, Livonia and Prussia in particular, especially stand out.

References

1. Bessudnova MB (2015) Rossiya i Livoniya v konce XV veka. Istoki konflikta [Russia and Livonia in the late 15th century. Sources of the conflict]. Moscow (in Russian)
2. Bessudnova MB (2016) Specifika i dinamika razvitiya russko-livonskix protivorechij v poslednej treti XV veka [Specificity and dynamics of Russian-Livonian contradictions in

the last third of the XVth century]. Lipeczkij gosudarstvennyj pedagogicheskij universitet imeni P.P. Semenova-Tyan-Shanskogo [Lipetsk state pedagogical university named after P. P. Semenov-Tan-Shansky], Lipetsk (in Russian)

3. Borisenkov EP, Pasetsky VM (2002) Letopis' neobychajnyx yavlenij prirody za 2,5 tysyacheletiya (V v. do n.e. - XX v. n.e.) [Chronicle of extraordinary natural phenomena for 2.5 millennia (The Vth Century BC—the XX[t] century AD)]. Gidrometioizdat [Hydrometeorology Publishing House], St. Petersburg (in Russian)
4. Camenisch C, Keller KM, Salvisberg V, Amann B, Bauch M, Blumer S, Brázdil R, Brönnimann S, Büntgen U, Campbell BMS, Fernández-Donado L, Fleitmann D, Glaser R, González-Rouco F, Grosjean M, Hoffmann RC, Huhtamaa H, Joos F, Kiss A, Kotyza O, Lehner F, Luterbacher J, Maughan N, Neukom R, Novy T, Priby K, Raible CC, Riemann D, Schuh M, Slavin P, Werner JP, Wetter O (2016) The early Spörer Minimum—a period of extraordinary climate and socio-economic changes in Western and Central Europe (2016). Clim Past Discuss. https://doi.org/10.5194/cp-2016-7. 2016 manuscript under review for journal Clim. Past Published: 12 February 2016. Available at https://boris.unibe.ch/77863/1/cp-2016-7.pdf. Accessed 5 Oct 2019
5. Estlandskaya guberniya (1852) [Estland guberniya]. In: Voenno-statisticheskoe obozrenie Rossijskoj imperii [Military statistical review of the Russian Empire], vol. 7, part 3, SPb
6. Kangur A, Kangur P, Vaino V, Tambets M, Krause T, Kangur K, Pihu E (2012) Ryby i rybolovstvo [Fishes and fishing]. In: Timm T, Rauskas A, Haberman Yu, Yaani A (eds) Pskovsko-Chudskoye ozero [Pskov-Peipsi Lake]. Eesti Loodusfoto, Tartu, pp 335–354 (in Russian)
7. Kuznetsova VS (1966) Geologicheskie usloviya rajona Chudskogo ozera [Geological conditions of Lake Peipus]. In: Karaev GN (ed) Ledovoe Poboishhe 1242 g. Trudy kompleksnoj ekspedicii po utochneniyu mesta Ledovogo poboishha [The Ice Battle 1242. Proceedings of the complex expedition to clarify the place of the Ice battle]. Moscow-Leningrad, pp 122–129 (in Russian)
8. Kustin AE (1966) Nekotorye dannye o zaselenii zapadnogo poberezhya Tyoplogo Ozera [Some data on the settlement of the West coast of the Warm Lake]. In: Karaev GN (ed) Ledovoe Poboishhe 1242 g. Trudy kompleksnoj ekspedicii po utochneniyu mesta Ledovogo poboishha [The Ice Battle 1242. Proceedings of the complex expedition to clarify the place of the Ice battle]. Moscow-Leningrad, pp 98–102 (in Russian)
9. Lebedev VD (1960) Presnovodnaya chetvertichnaya ikhtiofauna Evropeiskoy chasti SSSR [The Freshwater Quaternary ichthyofauna of European part of the USSR]. Moscow (in Russian)
10. Liflyandskaya guberniya (1853) [Lifland guberniya]. In: Voenno-statisticheskoe obozrenie Rossijskoj imperii [Military statistical review of the Russian Empire], vol. 7, part 2, SPb
11. Menke U, Kangur K, Borisov V (2007) General description and topography. In: van Eerden M, Bos H, van Hulst L (eds) In the mirror of the lake. Rijkswaterstaat Centre of Water Management, Lelystad, pp 50–61
12. Pervaya Livonskaya vojna, 1480–1481 gody. Dokumenty (2019) [The First Livonian war, 1480–1481. Documents]. Saint-Petersburg
13. Peter of Dusburg (1997) Khronika zemly Prusskoj [Chronicle of the land of Prussia]. Moscow (in Russian)
14. PSRL 2000a. Novgorodskaya 4-ya letopis' [The chronicles 4 of Novgorod]. In: Nasonov AN (ed) Polnoye sobranie russkih letopisey [The full collection of Russian annals], vol 4, iss 1. Moscow (in Russian)
15. PSRL. 2000b. Pskovskie letopisi [The chronicles of Pskov]. In: Nasonov AN (ed) Polnoye sobranie russkih letopisey [The full collection of Russian annals], vol 5, iss 2. Moscow (in Russian)
16. PSRL. 2003. Pskovskie letopisi [The chronicles of Pskov]. In: Nasonov AN (ed) Polnoye sobranie russkih letopisey [The full collection of Russian annals], vol 5, iss 1. Moscow (in Russian)
17. The Pskovo-Pechersky monastery during the Livonian War (1558–1582): holdings in Estonia, ed. by Anti Selart. Tartuer historische Studien, 5. Hrsg. Dr. Olaf Mertelsmann (Hamburg: Dr. Kovač, 2016)

18. Pskovskaya guberniya (1852) [Pskov guberniya]. In: Voenno-statisticheskoe obozrenie Rossijskoj imperii [Military statistical review of the Russian Empire], vol 3, part 2. SPb
19. Ränk G (1934) Peipsi kalastusest [The Peipsi Lake fisheries]. Tartu
20. Salmin SA, Yavorskaya LV (2016) Raskop Neznanova Gorka-II v 2013/2014 gg. [Excavation of Neznanova Gorka-II in 2013/2014]. In the collection "Arheologiya i istoriya Pskova i Pskovskoj zemli. Seminar imeni akademika V.V. Sedova" [Archaeology and history of Pskov and Pskov land. Seminar named after academician V. V. Sedov]. Materials of session 61, vol 31. Moscow-Pskov-St. Petersburg, pp 43–56
21. Salmina EV (1999) Dinamika izmenenij v sostave rybnyx ulovov v srednevekovom Pskove (po rezultatam izucheniya osteologicheskix materialov) [Dynamics of changes in fish catch composition in the Medieval Pskov town (results of osteological studies)]. In: Sedov VV (ed) Drevnosti Pskova: archaeologia, istoria, architektura. Pskov, pp 178–185 (in Russian)
22. Salmina EV (2001) Etapy razvitiya organizacii rybolovnogo promysla v srednevekovom Pskove [Stages of fishing development in the medieval Pskov]. In: Tverskoy Archaeologicheskiy Sbornik [Tver archaeological collection], vol 2, iss 4. Tver, pp 234–238 (in Russian)
23. Salmina E (2015) Fishing settlements in Pskov region in the sixteenth century (according to archaeological data and written sources). Environ Archaeol: J Hum Palaeoecol. Available at http://dx.doi.org/10.1080/14614103.2015.1129716. Accessed 5 Oct 2019
24. Salmina E, Lajus Yu (2011) Fisheries in the urban environment in the North-Western part of Russia 12–20th centuries. In: Myllyntaus T, Hettula P, Manni T, Piilola J (eds) Encounters of sea and land. The 6th ESEH conference. University of Turku, Turku, Finland, 28 June–2 July 2011, pp 171–172
25. Sanktpetergurgskaya guberniya (1851) [Sanktpetergurg guberniya]. In: Voenno-statisticheskoe obozrenie Rossijskoj imperii [Military statistical review of the Russian Empire], vol 3, part 1. SPb
26. SMAMU (1913) Sbornik Moskovskogo Ministerstva Justisii. Pskov I ego prigorody [Collection of the Moscow archaive of the Ministry of Justice, vol 5. Pskov and its suburbs]. Moscow (in Russian)
27. Selart A (2015) Vladeniya Pskovo-Pecherskogo monastyrya na yuge Estonii vo vremya Livonskoj vojny [Possessions of the Pskov-Pechersky monastery in southern Estonia during the Livonian war]. In the collection Arheologiya i istoriya Pskova i Pskovskoj zemli. Seminar imeni akademika V.V. Sedova [Archaeology and history of Pskov and Pskov land. Seminar named after academician V.V. Sedov]. Materials of session 60, vol 30. Nestor-History, Moscow-Pskov-St. Petersburg, pp 132–142 (in Russian)
28. Selart A, Tvauri A, Laanelaid A (2017) Die Burg Warbeck (Kastre). Balt J Art Hist 13(Spring). Available at http://ojs.utlib.ee/index.php/bjah/article/view/BJAH.2017.13.03. Accessed 14 Oct 2019
29. von Stern C (1937) Die Pleskauer Annexion von 1476, Der einzige Grenzlandverlust unsrer mittelalterlichen Ostmark. In: Baltische Monatshefte „Ernst Plates". Riga, S 148–156
30. von Stern C (1937) Der Kleinkrieg um die Ostgrenze im 15, Jahrhundert Stern, Carl von. In: Baltische Monatshefte (1937) „Ernst Plates". Riga, S 69–79
31. Tavast E (2012) Berega Pskovsko Chudskogo ozera [Shores of Lake Peipus]. In: Timm T, Raukas A, Haberman Yu, Jaani A (eds) Pskovsko-Chudskoe ozero [Lake Peipus]. Eesti Ludusfoto, Tartu, pp 75–84
32. Tyulina TYu (1966) K voprosu o prirodny'x usloviyax v XIII v. v severnoj chasti Tyoplogo ozera [On the question of natural conditions in the XIIIth century in the Northern part of the Warm lake]. In: Karaev GN (ed) Ledovoe Poboishhe 1242 g. Trudy kompleksnoj ekspedicii po utochneniyu mesta Ledovogo poboishha [The Ice Battle 1242. Proceedings of the complex expedition to clarify the place of the Ice battle]. Moscow-Leningrad, pp 103–121 (in Russian)
33. Wartberg G (1879) Livonskaya hronika Germana Vartberga [The Livonian chronicle of Hermann Wartberg]. In: Sbornik materialov i statej po istorii Pribaltijskogo kraya [Collection of materials and articles on the history of the Baltic region]. Printing House of A. I. Lipinsky, Riga, pp 81–155

34. Yavorskaya LV (2013) «Kostnye vymostki» v drevnerusskix gorodax. «Anatomiya» odnoj arxeologicheskoj zagadki ["Bone paving" in ancient cities. "Anatomy" of one archaeological riddle]. Zool J 92(9):1179–1189 (in Russian)
35. Yavorskaya LV (2014) Torgovlya myasom ili vy'mostki iz kostej? Arxeozoologicheskie issledovaniya raskova Luzhskij II v Pskove [Meat trade or bone paving? Archaeozoological research of the Luzhsky II excavation in Pskov]. In the collection "Arxeologiya i istoriya Pskova i Pskovskoj zemli. Seminar imeni akademika V.V. Sedova" [Archaeology and history of Pskov and Pskov land. Seminar named after academician V.V. Sedov]. Proceedings of the 59th session, vol 29. Nestor-History, Moscow-Pskov-St. Petersburg, pp 42–55 (in Russian)
36. Yurtseva A, Salmina E, Galik A, Lajus D (2014) How a millennium of fishing changed fish populations: a case study of Lake Peipus and the Velikaya River (NW Russia). Aquatic Sciences. Research Across Boundaries. Available at https://link.springer.com/article/10.1007/s00027-014-0381-4. Accessed 5 Oct 2019

Chapter 82
Self-Regulation in the Field of Environmental Safety During Construction: Assessment of Environmental Sustainability and Safety of the Regions of the North–West of the Russian Federation*

Natalya A. Seraya, Darya D. Filyakina and Oksana S. Struzhinskaya

Abstract The article discusses the issues of self-regulation in the field of ecology of the regions of the North–West of the Russian Federation. The activity of self-regulatory and public organizations in the environmental sphere, both in our country and abroad has been analyzed. The analysis of the use of environmental standards and norms in the construction industry has been done. Possible recommendations have been developed on measures to adapt the economies of the regions of the North–West of Russia to large-scale climate change. The necessary conditions for the transformation of human thinking to conduct economic activity, the use of natural resources and environmental protection are determined. The current environmental situation objectively induces the creation of self-regulatory organizations in the field of ecology that will regulate environmental standards in the construction industry, and their close interaction with all economic entities. It is proposed that environmentalists

N. A. Seraya · O. S. Struzhinskaya
Civil Law Department, Kuban State Agrarian University named after I.T. Trubilin, 13, Kalinina str., Krasnodar, Russia 350044

N. A. Seraya
Civil Law Department, Lawyer, Lawyer Chamber of the Krasnodar Territory, F. Lusana St., Krasnodar, Russia 350090

D. D. Filyakina (✉)
Entrepreneurial, Labor and Corporate Law Department, Russian Presidential Academy of National Economy and Public Administration, 82, Vernadsky Avenue, Moscow, Russia 119571
e-mail: darya@filyakina.ru

Entrepreneurial, Labor and Corporate Law Department, Interdistrict Shchelkovsky Vodokanal, 1, Svirskaya St., Shchelkovo, Russia 141100

O. S. Struzhinskaya
Civil Law Department, Consultant, Pervomaisky District Court of the city of Krasnodar, 92, Kolkhoznaya St, Krasnodar, Russia 350010

T. Kolmykova and E. V. Kharchenko (eds.), *Digital Future Economic Growth, Social Adaptation, and Technological Perspectives*, Lecture Notes in Networks and Systems 111, https://doi.org/10.1007/978-3-030-39797-5_82

preparing the environmental sections for the construction business should apply the requirement of mandatory membership in an environmental SRO, which will become another important factor in determining the social component of SRO.

Keywords Business · Self-regulation · Ecology · Construction industry · Environmental self-regulation · Non-profit organizations · State environmental control

JEL Classification Codes: K290 · K320

82.1 Introduction

Adaptation to environmental instability requires an integrated approach that includes measures implemented by both federal and regional executive bodies. It is a comprehensive system of measures of an economic, legislative, political, social and regulatory nature that can contribute to the organization of a sustainable national security sector, reduce the "sensitivity" of individual economic entities to the impact of climate changes affecting the climate of the Russian Federation, the climate of neighboring states and bordering them the territory of the global oceans. Moreover, the task of the adaptation under consideration consists not only in the preventive-recovery character, but also in deriving economic benefits in connection with some favorable manifestations of climate change. Recently, researchers [8, 11] noticed an aggravation of the orientation of state regulation spheres on the environmental sustainability of the development of a positive state of the environment, as these issues are increasingly affecting the quality of life and health of citizens. The economic opportunities of individual entities, including **the regions of the North–West of Russia**, are becoming directly dependent on the environmental factor. So, in the North–West, climatic changes are expressed in the increase of average annual active temperatures and their sums, the increasing number of rainfall in the late autumn and winter periods, prolonged thaws and frosts, and the increase of natural and climatic anomalies. It is the environmental factor that serves as a criterion for the region's ability to progressive development, since, due to its indisputable features, it concerns the interests of a wide range of nature users. Thus, taking into account also the need to perform environmental functions, ecology now acts as a complex object of economic management.

The purpose of this article is an attempt to assess environmental sustainability and make recommendations on adapting the economies of the regions of the North–West of Russia to large-scale climate change, determining the conditions for the transformation of human thinking to conduct economic activities and use natural resources, first of all, the main environmental problems of such regions should be considered in our country as: Arkhangelsk, Vologda, Kaliningrad, Leningrad, Murmansk, Novgorod and Pskov regions, republic to Karelia and Komi, St. Petersburg and the Nenets Autonomous Okrug. It should be noted that enormous concentration of timber, woodworking and pulp and paper industry objects in the above regions brings enormous damage to the environment of the North–West of Russia. Negative factors such as the rapid construction of port facilities and the development of shipping (for example, in the Murmansk region), the extraction and transportation of oil (in the Arkhangelsk region), and the extraction and processing of ores and non-metallic minerals [4] also affect the ecology of these regions. Atomic energy production is extremely unfavorable for the environmental situation in the regions of the North–West of our country. The reason for the negative and unsafe environmental conditions listed is the uncontrolled and unregulated environmental activities of organizations. The global ecology damage to the region is also affected by the defense, construction, electrical engineering and mechanical engineering of the Leningrad region. The main environmental problems of the North–West, which require speedy resolution, are: constantly growing pollution of the atmosphere, water and soil; the growth of environmentally-caused morbidity, the merciless use of natural resources, inefficient management of municipal waste; insufficient attention to the development of specially protected natural areas, the preservation of urban green spaces and ensuring nuclear and radiation safety.

82.2 Materials and Method

In this work, we used both general methods of scientific research (dialectical, systemic, logical, and others) and private scientific methods—rather-legal, concrete historical, formal legal, legal modeling method, linguistic, statistical, and other research methods.

82.3 Results

According to official data, the Ministry of Health of the Russian Federation in 2019, the North–West Federal District among all regions of our country is the worst due to the high incidence rate for a number of non-communicable and oncological diseases. Specialists and scientists in the field of medicine and ecology argue that environmental factors play a fundamental role in the emergence and development of many diseases. A case in point is the quality of drinking water. In Russia, according to the low quality of drinking water in the centralized water supply systems in the regions of the North–West, the leading ones are Karelia, Komi, Arkhangelsk and Novgorod regions. Public organizations of the North–West of Russia constantly draw the attention of the authorities and the public to the facts of a direct connection between the state of public health and the ecological state of the regions.

The right of every citizen of our country to a favorable environment, reliable information about its condition and to compensation for damage caused to his health or property by environmental violations is recognized by the Constitution of the Russian

Federation in Article 42. Federal Law of January 10, 2002 No. 7-FZ "On Environmental Protection" environment" defines the legal foundations of state policy in the field of environmental protection, providing a balanced solution of socio-economic problems and maintaining a favorable environment. In accordance with paragraph 1 of Article 12 of the said Law, public associations and non-profit organizations have the right to carry out activities in the field of environmental protection, including appealing to the state authorities of the Russian Federation, state authorities of the constituent entities of the Russian Federation, local governments and judicial authorities petitions on cancellation of decisions on activities that may have a negative impact on the environment.

On November 21, 2018, the III North–West Interregional Public Environmental Conference was held on the problems of environmental safety in the regions of the North–West Federal District. Non-governmental organizations and government agencies jointly discussed urgent problems dedicated to solving numerous environmental problems in these regions, including the need for constructive interaction between government bodies, public organizations and self-regulatory organizations in the field of ecology and nature management.

Do not forget about such a global environmental problem as climate change. The current level of economic measures taken indicates the concentration of state forces predominantly on the problem of the greenhouse effect, while the problem of adaptation to climate change in the Russian Federation is beyond the framework of priority national projects and entrepreneurship development programs. Currently, only measures are taken in case of emergency situations caused by the climate change. However, the climate factor not only directly affects the economic sector and the life of the population of the Russian Federation, but also affects the system of existing international political and international economic relations. So, the Paris Agreement on the implementation of the UN Framework Convention on Climate Change (UN Framework Convention on Climate Change. Paris Agreement. Adopted by the Conference of the Parties at the twenty-first session. Paris, November 30–December 11, 2015), includes the problem of the climate change adaptation to main topics of the contractual process of international cooperation. Paris Agreement in Art. 7 establishes the requirement for participating countries to provide adaptation plans that provide for practical steps to achieve adaptation. Unfortunately, in the Russian Federation, as in many countries, such plans are not yet available. The development of such plans is envisaged, in addition to international agreements, by the Climate Doctrine of the Russian Federation (the Comprehensive Plan for the Implementation of the Climate Doctrine of the Russian Federation until 2020, approved by Decree of the Government of the Russian Federation No. 730-r of 04.25.2011). A public discussion of the draft Concept for Planning Adaptation of the Population and the Economy of the Country to Changes and Their Consequences, prepared by V. M. Kattsov and B. N. Porfiriev on behalf of the special representative of the President of the Russian Federation. This Concept distinguishes as the main directions of adaptation planning: (1) Actively informing state authorities, local self-government, entrepreneurs, investors, NGOs and the population about existing climate risks, climate impact forecasts, measures of preventive and reaction adaptation in the process of economic activity of organizations in the Russian Federation. (2) Creation of a unified information base for the purpose of monitoring indicators implemented under the Concept of measures for individual economic sectors and for the whole of the Russian Federation. (3) Creation of conditions for state control and the provision of state support in the implementation of the Concept through adopted regulatory acts, regional adaptation plans and industry standards.

Currently, **the system of environmental regulation** in Russia includes government agencies, public organizations, non-profit organizations, including a small number of self-regulatory organizations (hereinafter referred to as SRO). State environmental supervision is carried out at 2 levels: federal and regional in the form of scheduled and unscheduled inspections. Rosprirodnadzor exercises supervision at the federal level, and its territorial divisions and executive bodies of the country's subjects carry out state environmental supervision at the regional level. As for **public associations and non-profit organizations** in the field of environmental protection and ecology of the North–Western regions of our country, at present there are quite a lot of them, for example, such as the Open Coast movement, the Karelian Republican Society for the Protection of Animals and Nature Protection, Free Coast movement, Karelia's Green Association, Green Branch public organization, Northern Natural Coalition, Pechora's rescue committee, Green Wave, Friends of the Baltic coalition, Ecological Association Journalists of St. Petersburg, City Forests of Petrozavodsk, Green World. Nevertheless, the listed organizations and many others do not have the status of a self-regulatory organization in the field of ecology. The idea of creating **self-regulatory organizations** in the field of ecology is not new, it was officially discussed earlier and was proposed at the conference "On the possibility of transferring a number of functions of public administration in the field of environmental management and environmental protection to self-regulatory organizations", held by the Federation Council Committee on Natural Resources and Protection environment in 2010. Conference participants voiced that in our country there are non-profit specialized organizations in such environmental areas as: environmental insurance, environmental education, consulting, environmental audit, which can become the basis for the transition to self-regulation. Article 3 of the Federal Law dated December 1, 2007 No. 315-FZ (as amended on August 3, 2018) "On Self-Regulating Organizations" gives the concept of self-regulatory organizations, highlighting the reasons for uniting their members as uniformity in the field of their professional activity. As Denisova A. points out, the main goal that the legislator pursued when introducing the institution of self-regulation can be defined as the transition to self-organization of business in some areas of the economy. For this, the SRO is vested with public law powers to establish standards and rules for the entrepreneurial and professional activities of its members, as well as to ensure compliance with these standards and rules [3]. For certain areas of business and professional activity, the ability to provide services and perform work is realized only upon entry into the relevant organization. Thus, membership in the SRO is **mandatory**, for example, for such areas of activity as: construction, design and hydrogeological surveys, appraisal activities, financial audits and some others for which special federal laws establish **special requirements in the field of self-regulation**.

Such self-regulatory organizations have been created and currently exist, membership in which **is not mandatory** for the implementation of a certain type of professional or entrepreneurial activity, this may include such areas as: healthcare, housing and communal services, advertising, mediation, insurance and others, and in the environmental sphere at this time SRO operates waste operators. The position of Leskova Y. G. is quite fair, that non-identity of legal status of SROs based on compulsory membership and SROs based on voluntary membership is initially manifested in special functions performed by these organizations. The author emphasizes that the specifics of SROs activities based on compulsory membership determine a wider list of functions performed by them compared to the list of SRO functions based on voluntary membership. SROs with mandatory membership perform, among other things, an important function related to "admission to the profession" [6].

In the North–West region, as in all other regions of the Russian Federation, from January 1, 2019, as part of a new reform implemented by the Ministry of Natural Resources of Russia, a transition to a new system of solid municipal waste management was launched. The so-called "garbage" reform provides that each entity is obliged to develop an electronic territorial scheme for waste management, approve the regional program and standards for the accumulation of MSW and, **on a competitive basis, select a regional operator** for the management of municipal solid waste (hereinafter—MSW), who will be responsible for full cycle of MSW management. According to Rosreestr, in our Northern Capital there are registered SROs related to waste management such as SRO NP Union of Ecologists of the North–West, SRO NP SPAR. For example, on the territory of St. Petersburg, according to the results of the competitive selection process for the assignment of the status of a regional operator for MSW management and the right to conclude an agreement on the organization of activities for the management of municipal solid waste, the St. Petersburg State Unitary Enterprise "Mechanized Processing Plant" was recognized as a regional operator of household waste." Thus, each producer of municipal solid waste in St. Petersburg **is obliged to conclude** a contract for services for the management of MSW with a regional operator, and the task of a regional operator is to organize the process of handling solid municipal waste generated in the city. For non-compliance with the above algorithm of necessary actions in accordance with Article 8.2 of the Administrative Code of the Russian Federation, administrative liability is provided. According to the Federal Law of June 24, 1998 No. 89-FZ (as amended on December 25, 2018, as amended on July 19, 2019) "On Production and Consumption Wastes", it is possible to collect, transport, process, utilize, dispose and dispose of waste only with **a license**. Speaking of licensing environmental activities, it is necessary to turn to the vast experience of the construction industry, entities that are united by a self-regulation system. Using the example of self-regulation in the construction industry, it can be said with full responsibility that, **unlike licensing, the self-regulation system initially includes** the personal organizations' responsibility for the activities of their members through payments from compensation funds in cases of damage to the environment or human health. In connection with the current situation in the field of MSW, it can be concluded that licensing activities such as the collection and transportation of waste is **an "artificial administrative barrier"**. Mentioning the construction industry, where the institution of self-regulation has already been sufficiently formed and continues to develop effectively, it must be said that the construction industry is inextricably linked with the environment. Currently, the construction business is advancing at a rapid pace, including in the regions of the North–West of the country, which of course negatively affects the state of the environment. In this regard, the legislator has developed basic environmental requirements, following which it is possible to minimize the adverse environmental impact during construction activities. Environmental requirements during construction should be taken into account and strictly observed during the development of a project for the organization of construction, as well as the mandatory compliance with environmental requirements is noted in the project of work. The ecology section in the design documentation is developed at the final stages of design, when the technology, master plan, heating and ventilation, water supply, sewage and gasification are ready. Only an environmental specialist will be able to assess the environmental restrictions of the territory under consideration, sanitary and hygienic standards and notify the customer in advance of the list of environmental documentation necessary for the project and the possible need for environmental protection measures. Therefore, a member of a self-regulatory organization in the construction industry should contact an environmentalist or an appropriate organization in the field of ecology to develop the necessary environmental solutions for the project. Whether such a specialist or this organization is a member of a self-regulatory organization in the field of ecology and is not taken into account when developing and protecting sections of project documentation related to ecology. Responsibility for the quality of all project documentation, including sections of the project affecting environmental aspects, lies with the design organization—a member of the SRO and the self-regulatory organization itself. Thus, environmentalists, in fact, do not bear responsibility for the quality of their work. One wonders why this is happening. It would be fair and logically justified if the relevant requirements, for example, the requirement of mandatory membership in an environmental SRO, are applied to environmental specialists who are preparing environmental sections for the construction business. According to Article 2 of the Federal Law dated 12/01/2007 No. 315-FZ (as amended on 03/08/2018) "On Self-Regulating Organizations", self-regulation is understood as independent and initiative activity carried out by business or professional entities, the content of which is **the development and establishment of standards and rules** of the specified activity, as well as monitoring compliance with the requirements of these standards and rules. Accordingly, to develop appropriate standards and rules in the environmental sphere, to comply with them, and to bear responsibility for their activities is appropriate for environmentalists, enlisted in a specialized SRO.

According to statistics from the Ministry of Economic Development, as of January 1, 2019, the self-regulatory system of the Russian Federation includes **627 self-regulatory organizations** that combine 225,289 business and professional entities in 12 business sectors on a mandatory membership basis, and 424 self-regulatory organizations were created by market participants on conditions of voluntary membership. In the environmental sphere of the Russian Federation, there are various non-profit organizations, associations in the form of non-profit partnerships, unions and associations that, to one degree or another, **carry out certain functions (regulatory, control, etc.)** and which position themselves as self-regulatory organizations. These organizations are created in accordance with Federal Law of January 12, 1996 No. 7 "On Non-Profit Organizations." So, for example, the Union of Subsoil Users was founded in St. Petersburg, it brought together legal entities engaged in entrepreneurial activities in the field of subsoil use. We should agree with Leskova Y.G., who determines, that the range of socio-economic, political, other needs and interests that the present non-profit organizations represent is huge, and the influence of non-profit organizations on the condition of society and the state is expanding, and, most importantly, their efforts are being fruitfully deployed thanks not only to the legal ensuring human rights and freedoms, but also to the formation of entrepreneurship.

Non-profit organizations with special legal status should include self-regulatory organizations [5]. Such environmental activities are currently not regulated by law, but they are important and of great importance such as: environmental audit (checking the manufacturer's environmental documentation for compliance with the environmental legislation of the Russian Federation), environmental consulting, organization of vocational training, retraining and certification in the field of ecology and environmental management, development of normative documentation for enterprises (mainly enterprises involve outsiders to carry out such work "Professional" environmental organizations). It is definitely worth saying that, as the authors, Vafin V. V. and Zolotova O. A., correctly pointed out not all types of entrepreneurial activities associated with the use of natural resources are environmental. Entrepreneurial activities in logging or subsoil use do not aim at solving environmental and social problems, and therefore, is one of the types of activities in the field of natural resources. Moreover, the implementation of this type of activity may have a greater negative impact on the environment than a business not related to nature management. At the same time, of course, there are areas of activity that are both related to nature management and environmental protection [2].

To understand the prospect of using self-regulatory mechanisms to solve environmental problems, first of all, one needs to ask about the purpose, role and tasks of such SROs. Romanova O.A. says that in our country, there are still no prerequisites for the introduction of mandatory self-regulation of entrepreneurial and professional activities in the field of ecology and nature management, and as for voluntary self-regulation, it is now being implemented within the framework of the current legislation [7]. Is this so and why, until now, mandatory membership in environmental SROs has not been legislated? The answer to this question is primarily in the economic field, since there is such a negative aspect for small and medium enterprises upon entry into the SRO as high material costs (compensation contribution, insurance, etc.). It is also worth paying attention to the political component, namely, a very weak interaction between the SRO and state bodies. In general, self-regulation promotes confidence in the market among its participants. Moreover, this is an effective method of regulation, since SROs have deeper knowledge in their narrowly focused area, as well as are familiar with products sold by member organizations and the problems that arise. Therefore, SROs have special knowledge and ability to more effectively implement cost-effective regulatory programs. Due to the great diversity in sizes, market conditions and types of environmental problems covered by this work, it is difficult to give one recommendation for each area of self-regulation, which directly faces the development of environmental standards and rules. However, government agencies should consider self-regulation as a regulatory method while developing, reviewing, or re-evaluating management programs. The functions of SROs should be developed, maintained, and improved as much as possible, provided that appropriate accountability mechanisms are in place to ensure that the responsibilities of the regulatory body are properly implemented and that markets operate in accordance with general performance standards in the public interest. With proper application, self-regulation can lead to effective rules and regulations, their widespread compliance and adoption, timely adjustment of the rules in accordance with changing conditions, flexible and effective compliance with the rules. In addition, self-regulation can lead to significant savings of the State due to the transfer of costs from public authorities to SROs. Self-regulation is not a form of "deregulation", it is an important part of the model of effective and broad management. Therefore, government bodies and their respective working groups should consider, recognize and incorporate self-regulatory approaches into guidelines or standards developed to address specific environmental challenges. It should be agreed with Bakulina. A. A., that environmental problems can be solved not only by state regulation tools, but also within the framework of environmental initiatives of individual companies, as well as within the framework of environmental self-regulation [1]. Since the creation of SROs, many industries, where self-regulatory standards are used, have significantly improved compared to when they were under direct state control. In our opinion, it is worth highlighting three important indisputable advantages of self-regulatory organizations:

1. **Expertise**—SROs are considered experts in their fields and, therefore, know a lot about the markets in which they operate. This, in its turn, is useful for members of such organizations, as they can be invited to participate in discussions and learn more about the diverse nuances of the industry.
2. **Higher standard of behavior**—With the creation of SROs, member organizations follow a defined standard of conduct that helps promote ethical business practices. This can lead to investors becoming more confident in the market.
3. **Enhanced supervision**—SROs do not rely on citizens' taxes, but are financed by organizations under its wing. Thus, the government is able to eliminate the abovementioned need to support the state apparatus for this purpose.

The statement that, unlike many other developed countries, the system of environmental self-regulation in Russia is not developed, will be too categorical, since the function of the main regulator in the environmental sphere in many countries lies mainly with government bodies, because the issue of environmental safety is fundamental for human life. In countries such as Finland and Sweden there is a fairly high level of environmental protection and environmental safety. In many ways, it is supported by current legislation. For example, Sweden is one of the few countries in the world that has codified environmental legislation (URL: http://www.

naturvardsverket.org). It was the policy of these states that gradually, in the long run **transformed the legal awareness of the population in the environmental sphere**. The Finnish Ministry of the Environment and the Swedish Ministry of the Environment and Natural Resources occupy a fundamental place in the environmental regulation system. The issuance of environmental permits in these countries is carried out by administrative authorities.

In order to understand the main directions of activity and tasks of self-regulatory organizations in the field of ecology, let's get down to the experience of the United States and consider trade associations that were created on the initiative of business representatives. For the first time, it was in the USA, on the initiative of representatives of the chemical industry to improve and rise the level of technical safety, labor protection and the environment, in 1989, the Responsible Care Program began to operate. At present, it is actively developing, and already within the framework of this program, new directions are being created, for example, the branch Association of Chemical Industry Manufacturers (CMA) has been created. The National Association of Chemical Distributors was established in 1991, having developed a unified environmental policy, which is mandatory for implementation in all participating companies (Responsible Distribution Process), it unites companies involved in the shipment, processing and marketing of chemical products to final consumers. In addition to complying with their own environmental policies, industry participants also have the condition of interacting only with environmentally oriented suppliers. The American Forest and Paper Association was also created to improve the image of the industry in the eyes of consumers. The main document of this organization was the Sustainable Forestry Initiative (SFI). Association members, in their turn, developed a number of necessary environmental and socially responsible measures that were aimed at reforestation and resources conservation [9].

Analyzing foreign experience, obviously the fact is clear that the role of the State in the field of ecology, environmental safety, environmental protection and environmental management is fundamental, but nevertheless such important areas as environmental audit, environmental consulting, environmental impact assessment, waste collection and disposal activities, organization of vocational training, retraining and certification in the field of ecology and nature management, as well as environmental design (Waste Generation and Disposal Limits Project, draft norms, the development of waste passports, engineering and environmental studies, the program of industrial environmental monitoring, etc.) must be within the environmental self-regulation system. The introduction of a mandatory environmental audit, at least for large, city-forming enterprises, the transfer of environmental consulting and environmental regulation to a professional type of activity could streamline all activities in the environmental sphere. Since it is impossible to separate these areas into a separate regulation, it seems possible and necessary to include in the SRO standards **the compulsory nature of obtaining an environmental education**. Undoubtedly, self-regulation in the field of ecology is a fundamentally new type of regulation of professional activity, the formation of which is only at the initial stage, so we hope that due to the improvement of the quality of interaction with the relevant executive authorities and the search for the optimal balance of powers of the state and SRO, the activities of such SROs will be improved, and deficiencies will be resolved over time. It seems that already now it has become clear to state structures, the public and business that with the help of SROs in the environmental sphere it is possible to increase the responsibility of business entities for negative environmental impacts. The benefits of membership in such self-regulatory organizations are obvious, for example, if the company is a member of an SRO, then in any case it complies with environmental laws and provides its services with special attention to reducing negative environmental impacts. It can also be assumed that any company, a potential customer, is more likely to conclude a contract with an organization—a member of an SRO in the field of ecology, rather than with a similar, but not related to any partnership. Certainly, the SRO in the field of ecology will help companies—their members to improve the quality of services in environmental activities and will provide advisory services to members, as well as represent the interests of the company in resolving disputes.

It seems necessary to formulate the main social functions of self-regulatory organizations in the environmental sphere such as: improving the quality of work in the field of ecology and nature management, strengthening the responsibility of business and professional environmental participants, raising the level of property responsibility of SRO members, providing guarantees for compensation for harm caused, organizing and the exercise of a control and oversight function over the SRO members' activities.

82.4 Conclusions

Based on the stated above characteristics and facts, it seems necessary to really characterize the environmental sustainability of the regions of the North–West of Russia as weakly effective, not safe and possessing so far low potential. Nevertheless, despite the unfavorable present, there is a real opportunity to change the situation for the better with the help of the gradual development and stimulation of self-regulation processes in the field of ecology. The current environmental situation objectively necessitates the creation of self-regulatory organizations in the field of ecology and their close interaction with all subjects of various types of economic activity, including those that carry out construction activities. These organizations will perform a number of tasks, including, for example, ensuring environmental safety in the construction industry. However, in the state register of SROs, such associations are absent in this industry, and all environmental SROs are formed exclusively on a voluntary basis, which shows insufficient state and business attention to this issue. The current underestimation of the importance of the environmental factor and the social significance of natural goods leads to a misconception about the priority of resolving primarily economic, legal and social problems, and only then environmental ones.

Based on the stated above features of the adaptation process of the regional economy to climate change and environmental problems in general, we can conclude that the measures carried out at this stage are low. Formally, a policy to implement adaptation measures to climate change exists in the Russian Federation, but in reality such a policy has not been properly implemented and, in fact, the necessary comprehensive system of adaptation policy in the country remains only within the framework of concepts. We can't talk about the complete absence of this policy, since some individual measures, such as extinguishing forest fires, sanitary and epidemiological monitoring of water quality, purchasing equipment and other assistance to the agricultural sector, as well as other measures are systematically implemented, making reports in this political sector very presentable. Nevertheless, the low priority of climate policy in the Russian Federation, the disproportion of financing for the implementation of environmental policies due to the differentiation of these issues between the budgets of the Russian Federation and the regions (lack of funds in regional budgets) and inefficient administration only increase adverse climate risks.

It can be concluded that for the development of environmental self-regulation, it is necessary that the business community is motivated "from within" to collective actions to protect the environment, since self-organization, as noted above, is evolutionary in nature. Also, currently there is an acute problem of the possibility of commercializing the ecology, which is an obstacle to the development of SRO. Business needs to recognize that the negative impact on the environment today is its low economic efficiency in the future. And vice versa, investments in solving and preventing environmental problems are the basis for obtaining additional rent.

Thus, the greatest efficiency of the regulating process of the environmental sphere in the conditions of the development of the innovation-driven economy and construction is achieved through the coalition interaction of the state, business and society, which allows taking into account the interests of all economic entities.

References

Magazines

1. Baulina AA (2010) The economic attractiveness of business participation in environmental self-regulation, Research at the Faculty of Economics. Electron J 2(2):72–95
2. Denisova A (2012) The legal status of self-regulatory organizations in the Russian Federation. LRS ConsultantPlus
3. Kartamysheva ES, Ivanchenko DS (2017) The main environmental problems of the North–West of Russia. Young Sci 25:110–113
4. King A, Lenox M (2000) Industry self-regulation without sanctions the chemical industry's responsible care program. Acad Manag J 43(4):72–80
5. Leskova YG (2011) Features of the legal status of certain types of non-profit organizations (for example, self-regulatory organizations, microfinance organizations). Law Econ 1:18–23
6. Leskova YG (2011) The role and importance of the specific diversity of self-regulatory organizations in the modernization of economic and legal relations. Civ Law 2011(2):4
7. Romanova OA (2018) On the development of self-regulation in the field of environmental management as one of the factors of sustainable development of territories in the Russian federation. Lex russica 5:72–84

8. Shkiperova GT, Druzhinin, PV (2015) Assessing the Impact of Climate Changes on the Economy of Russian Regions. National Interests: Priorities and Security, no. 34, pp 43–48
9. Vafin VV, Zolotova OA (2019) Theoretical and legal aspects of the development of environmental entrepreneurship in the region. Civil Law, No. 2. LRS "Consultant Plus"

Electronic Resources

10. Paptsov AG, Shilovskaya SA, Kolesnikov AV, Amelin AV, Adadimova LYu, Surovtsev VA (2015) Adaptation of Russia's Agriculture to Global Climate Change, available at https://www.oxfam.ru/upload/iblock/f96/f9622b41f48549945438f2292f509d14.pdf. Accessed 10 Jan 2019
11. The official website of the Swedish Environmental Protection Agency, available at http://www.naturvardsverket.se/. Accessed 24 Sept 2019

Chapter 83
Richardson's Dream: Anthropological and Cultural Aspects of Global Research on Climate Change

Artem V. Verle

Abstract Global changes in the modern world are increasingly transforming social reality and human self-determination in it. In this regard, the study of the human dimension of abstract economic, political and social processes is characterized by increasing relevance as part of a comprehensive global change research (GCR). Based on the philosophical and cultural analysis of the idea of "Forecast Factory" by Lewis Fray Richardson from his work "Weather Prediction by Numerical Process" (1922), the article considers the fundamental connection of modern weather and climate sciences with the economic, social and political trends of our time. In the work, an attempt is made, firstly, to determine the basic principles of this connection in the context of the history and current state of meteorological and climatological studies and, secondly, to characterize their influence on the ideas of modern man about the world and its future. The author offers a working terminology for thematization of some features of science as an aesthetic phenomenon in the aspect of studying the designs of scientific imagination and their systemic social-anthropological effects.

Keywords Anthropology of science and technology · Modern society · Philosophy of science · Scientific imagination · History of climatology · Global change · Lewis Fray Richardson · "Forecast Factory"

JEL Classification Codes A120 relation of economics to other disciplines · A130 relation of economics to social values

83.1 Introduction

The socio-anthropological consequences of climate change include at least two aspects. Firstly, the impact of the actually changing climate on various spheres of life from everyday practices to global political and economic processes. Secondly, a change in the anthropological model and, in general, the relationship of man to the

A. V. Verle (✉)
Department of Philosophy and Theology, Pskov State University, Pskov, Russia
e-mail: artemverle@gmail.com

T. Kolmykova and E. V. Kharchenko (eds.), *Digital Future Economic Growth, Social Adaptation, and Technological Perspectives*, Lecture Notes in Networks and Systems 111, https://doi.org/10.1007/978-3-030-39797-5_83

world in connection with the transformation of the meteorological and climatological thinking and knowledge itself. In this article, we are interested in the second of these aspects. The consequence of the development of modern science, in particular meteorology and climatology, is not only a simple increase in collective knowledge, but also a systematic transformation of lifestyle, attitudes to time and space, changes in everyday practices and reactions. An analysis of this dimension of global processes is necessary as well as an examination of climate, biological or economic systems. The question of including such an analysis in the context of climatological knowledge was already posed by researchers some time ago [27], but has not yet received a systematic development.

The science of weather and climate now unites thousands of scientists representing various methods and fields of research. Recently, sociologists and researchers in various fields of humanitarian knowledge have brought new perspectives to climate research. Climate is a complex facility, the studies of which are an expressive example of relevant interdisciplinarity. The comprehension of such a phenomenon close to everyone as the weather has a history of many thousands of years. Summarizing the features that are common to all cultures, the American sociologist George Murdoch identified, in particular, such a universal feature of them as attempts to influence the weather [24]. The life of a huge atmosphere in all its vicissitudes for many centuries has been the idea of fixing mankind: from the search for a connection between weather and bodily ailments to cosmological concepts of climatic phenomena [10]. Extensive literature is devoted to the scientific research of weather and climate in the pre-instrumental and instrumental era of measurements, in which fundamental dependences are traced between the methodology, measuring equipment, relevant science and technology, on the one hand, and social ideas about the world as a whole and its meteorological conditions, on the other hand [1, 6, 7, 9, 13, 20, 25, 28, 33]. The radical transformation of meteorological thinking began a hundred years ago and entered the modern phase fifty years ago. The beginning of the process was associated with the emergence of the first models of numerical methods for analyzing the physical parameters of the atmosphere, and the continuation and rapid development with the use of electronic computers in the implementation of these methods (for example, the ECMWF and WRF models). It is indicative in the perspective of the philosophy and anthropology of science that the origins of the modern state of the global meteorological system were utopian phantasmagoria, which arose in the imagination of the English mathematician Lewis Richardson in 1911 [29]. His dream of "Forecast Factory", where tens of thousands people work on meteorological calculation, is a vivid example of scientific insight, which has absorbed both the spirit of its time, full of sci-fi projections and hopes, and the prophecy about the impersonal world of infinite automated calculations. Richardson's "Forecast Factory" has already become a subject of interest for researchers [23], but the philosophical, anthropological, and social aspects of this image and its consequences for the transformation of climatological thinking have not yet been understood in the broad perspective of the history of science, its philosophical foundations and socio-anthropological consequences. In the second half of the 20th century, planetary observation systems and computer analysis technologies for their results became completely abstract and supra-individual. Paul Edwards called this circumstance "quasi-obligatory globalism" [12].

83.2 Materials and Method

The basic theoretical and methodological principle of this work is the thematization of the humanitarian and social aspects in natural science research and applied technology. This principle has long been basic in the philosophy and history of science and technology since at least the Frankfurt School. In addition, we rely on general methodological principles developed by B. Latour, M. Callon, J. Law, A. Rip [8, 21, 30] as part of Science and Technology Studies (STS). By extending the semiotic aspects to wide non-textual reality, the interdisciplinary approach to the anthropology of science and technology is conceptually concretized. Climatological knowledge, its historical and modern practices, as well as its economic, political, and anthropological effects are considered as a system of symbolic meanings that form a kind of metatext in relation to social and natural reality as a text. The sci-fi idea of L. F. Richardson is described as a model of systemic transformations of the modern world and worldview. A comparative comprehensive analysis of the subject of research is based on the methodological principles of the collaboration of the natural, social, and humanities developed by the Paul Holm research group that proposed the RITE project (Radically Inter- and Trans-disciplinary Environments framework) [18]. This approach involves the use of the methods of the humanities associated with the tactics of understanding and imagination, not as local and situational operations, but as a basic analytical resource in the study of science and technology. The project involves a decisive reorganization of the conceptual interfaces of scientific disciplines with the goal of their actual mutual integration. Orientation to this task is dictated by a certain fluidity and preliminary nature of the used and proposed terminology, namely, such concepts as "dynamic cross-borderness", "grandeur", "uncontrollable objectivity", "theatricality", "network organon" as applied to the philosophy and anthropology of science and technology. It is also worth noting that Paul Rodaway's work on sensuous geographies [31], in which geography, corporeality, and imagination are considered in a general perspective, was important for choosing a research angle.

At the level of research tactics and conclusions are applied general scientific methods of comparison, generalization, analogy and concretization.

83.3 Results

The scientific system of meteorology in the 19th century could not be formed on the basis of only ground-based observations, since the task of a comprehensive vertical study of the atmosphere could not be solved within the limits of the available instrumental capabilities. The theoretical awareness of this circumstance happily coincided with the first practical successes in aeronautics. The romantic ideas of early ballooning found the first and immediate response precisely among meteorologists. A landmark event in this regard was the conference of the International Meteorological Organization of 1896, at which the Commission on Scientific Aeronautics (ICSA) was created. The commission, led by Hugo Hergesell, managed to establish active international collaboration of scientists in the development of systematic aerological observations [25]. From 1901 to 1912, the Commission published 22 volumes of regular observational data from 17 stations in Europe. This international meteorological web actually stopped working with the outbreak of World War I. The invention of the radiosonde in 1927 could not reanimate dying scientific research in the upper atmosphere. They will be resumed only after the Second World War and coincide with the beginning of the era of electronic computing technologies in big data processing [20]. It is significant that meteorological and climatological studies have become the forefront in the application of the capabilities of two key technological innovations of the 20th century—the development of airspace and electronic computers.

Lewis Richardson's revolutionary book "Weather Prediction by Numerical Process" [29] fell on meteorological timelessness, went unnoticed and was rediscovered in all its significance only in 1965. The book proposed, described and applied the method of integrated analysis of large meteorological data. Richardson carried out prediction calculations based on a set of parameters collected on May 20, 1910. The choice of this date was motivated by the fact that the end of the second decade of May this year was the peak of meteorological measurements due to the close passage of Halley's comet to Earth. This circumstance also has important symbolic significance, since it indicates a fundamental connection between aerophysics and astrophysics. The forecast for May 21, 1910, calculated by Richardson, was unsuccessful, although the mathematician devoted at least two years to calculating [23]. The reasons for this failure are discussed in detail in specialized literature and are associated with the peculiarities of meteorological knowledge of the time [23]. The first successful numerical weather forecast was made by American meteorologists in 1950 under the leadership of Jules Charney on the first ENIAC general purpose electronic numerical computer, designed by John von Neumann's team at the request of the U.S. Army Ballistic Research Laboratory [25]. At that time, Richardson was 68 years old, he led a solitary scientific life and was mainly engaged in the application of numerical methods in the analysis of the causes and course of military conflicts. It is worth noting that as a convinced pacifist, Richardson always feared that his research in the field of integrated meteorology and mathematical polemology could be used for non-peaceful purposes [23]. In fact, the rapid development of meteorology in the second half of the 20th century was motivated and supported precisely by military tasks related, in particular, to the provision of military aviation and missile forces. John von Neumann's position on this issue was much less cautious than Richardson's [4]. So at the junction of applied mathematics and meteorology, the problem that was one of the key for humanity at that time was actualized—this is the connection of science, in particular, nuclear physics, chemistry, biology, with new forms of war.

Let us consider, however, the image of Richardson's "Forecast Factory" as a phenomenon of sci-fi imagination, embodying those principles and intentions that determine the role of modern weather sciences in the design, fixation and maintenance of the basic principles of the modern world. We are talking about the famous Ch. 11/2 "The speed and organization of computing" in the penultimate chapter "Some remaining problems". The dream of a "Forecast Factory," with which, according to his own testimony, Richardson began his work on mathematical weather prediction, symbolically ends many years of his work [29]. Let us dwell on the analytical review of its main explicit and implicit meanings.

First, let us turn our attention to the idea, which can be previously designated as the principle of dynamic cross-borderness. Its essence can be defined as follows: overcoming borders is fundamentally more important than marking them, and marking itself is characterized by fundamental dynamism. Research in meteorology and climatology initially involves not just a supranational, but also a "supra-geographical" data collection and processing systems. For example, the circulation of air masses in the upper and middle layers of the atmosphere is indifferent to both the forms of the earth's surface and their boundaries [19]. That is, meteorology and climatology at the level of the basic methodology are essentially global projects and otherwise impossible at all. The geographical ideas of the territory, areas, natural boundaries, undeveloped and developed lands, habitats are fundamentally dynamized for meteorology and, so to speak, overcome. That is, the modern intentions of global economic and political practices were initially tested at the beginning of the instrumental era in meteorology. On the frontispiece of Richardson's book there is a map of Europe lined with black and white rectangles in the style of a chessboard, resembling the landscape of the world of the looking glass from the novel of Lewis Carroll. In a unified landscape, we are faced with a different anthropological perspective and with different trajectories of meanings [16]. The network of the geographical coordinate system in the Richardson model goes to a new level of abstraction - it turns from a linear system into a three-dimensional one.

Secondly, we single out the principle, which we arbitrarily designate as uncontrollable physical and mathematical objectivity. Modeling of states and processes in each particular case involves a complex and high-speed analysis of a huge amount of heterogeneous data, the list of which is fundamentally open, and the content is variable. In this regard, meteorology acts as a test field for the use of the most advanced supercomputers, while at the same time demonstrating their relative imperfection. The world in this context appears as a colossal system of dynamic physical parameters, the cross-correlation of which within the framework of instant conditionally static fixations is a difficult task of mathematical calculation [14]. In this regard, we are no longer talking about the methods of this calculation, but about ensuring its acceptable speed. On the tiers of the huge Richardson globe, 64,000 people work on computing—as many "computers" as Richardson calls them should be enough to predict climate conditions. As modern practice has shown, the real implementation of this task requires at least several million Richardson factories [23], which is actually being implemented on a global scale, minus the organizational theatricality, and most importantly, the human component that is present in "Forecast Factory".

The economic opportunity of Richardson's phantasmagoria looks utopian even in the perspective of a happy future, but the modern system has proved capable of its even more grandiose implementation, primarily due to the total exclusion of the human dimension from this project. Let us note, meanwhile, the exceptional instrumental nature of human participation in the Richardson model. Although he carefully surrounds his imaginary factory with "fields, houses, mountains and lakes", where computer people can "breathe freely", the content of their working time can hardly be envied.

Thirdly, it is necessary to highlight the principle that we propose to call the network organon. By this concept, we propose to understand the organization of the global spectacle and effect. The image of a modern scientific project contains the effect of the sublime, if expressed it in the language of aesthetics. That is, it is characterized by the inner significance of the phenomena, incomparable in its ideal content with the real forms of their expression. Richardson's «Forecast Factory» impresses with the grandeur of the design and the grandeur of the dream. The sublime dramaturgy of climatic processes, fixed by the colossal organism of computer calculations, betraying the prophecies about the future state of this enormous scene, is a phantasmagoric image of a scientific office-magic theater [5]. Two aspects of this principle can be distinguished here: science as a spectacle and science as a network. The idea of openwork glass-metal dome-ball, realized in the second half of the XIX—early XX centuries in the advanced achievements of engineering and construction thought—in the passages, railway stations, pavilions of the World Exhibitions, unsupported aerial structures of huge vaults, combined with the ideas of the opera house, circus, exchange buildings, reflected in Richardson's imagination. It is important to note that these designs have become symbols of a new capitalism and an emerging mass society [2]. The idea of science as a theater has its roots, at least in the 18th century, in the practice of public demonstration of scientific experiments, in the even earlier history of anatomical theaters and huge curious hollow planetarium globes, like Globe of Gottorf. The chair of the "chief conductor" of the «Forecast Factory», waving colored beams, towering in the center of the world globe, refers both to the monumental symphonic canvases by Anton Bruckner and Gustav Mahler, as well as to "Journey to the Center of the Earth" by Jules Verne and the futuristic fantasies of Camille Flammarion. It is especially necessary to emphasize, in addition, the symbolic connection of the weather theater with the exchange and the auction house. The closest constructive analogue of the imaginary Richardson model was the world's first projection planetarium, conceived by the German engineer Walther Bauersfeld in 1919 and first earned in Jena in 1923 [11]. Modern user-made visualizations of climatic processes on the Internet at the household level and scientific visualizations in the practice of national and international meteorological centers are embodiment of the entertainment that was presented in the «Forecast Factory».

The idea of science as a network is clearly expressed in Richardson's fantasy. It is worth noting that meteorology is in principle based on the effective interaction of information collection points—meteorological stations—and analytics centers for this information [17]. Moreover, the location of the stations obeys the simple principle of spatial regularity, indifferent, as already indicated, in relation to the diversity of political and geographical circumstances. In this regard, it is the forerunner of modern network technologies again.

Fourth, we single out the idea of total interconnection as a physical and informational need, integral and parametric universality. The world of climatic and meteorological processes is a world of flexible relationships, encompassing remote global determinants. Atmospheric dramas, perceived as movements of divine forces for a long history of mankind are now thought of as scientific performances in which polyparametric analytics systems replace the principle of former animistic predestination. Attempts to influence the weather are, as noted, the essence of any culture, but modern meteorological thinking is fatalistically inclined to aero- and hydrophysical necessity, which can be influenced to an extremely limited extent. Thus, modern science contains two opposing determinants: confidence in the ability to intervene in nature by re-creating and transforming its elements, and the idea of grandiose natural law-making inexorability [32]. These trends also characterize the modern economic worldview, expressed in the contradiction between the objectively determined global system and the possibilities of local and even global regulation.

Fifth, it is necessary to highlight the idea of transforming of basic anthropological relationship between man and nature. The modern computer forecasting system predicted by Richardson has transformed and continues to transform one of the basic relationships of man to the world—this is the relation of uncertainty. Another paradox is that the predictability of natural processes transforms the structure of the sensitive effects of these processes [22]. In other words, a specific bodily experience of a person surrounded by air, humidity, pressure and wind in the modern information era is in a slightly different mental situation than in pre-information eras. Mainly, the strategy of expectation is being transformed, as the "predictability time" is increasing. In general, it is noted that modern science works on a large spatial and temporal scales, giving priority to global knowledge and marginalizing local measurements and changes, that the issue of global climate change required globally averaged long-term climate information, in which spatial and temporal details of human scale play a subordinate role [14]. However, in connection with these retrospective and perspective global expectations, the local structure of specific human expectations, anxieties, and apprehensions is changing. In the Richardson system, of course, there could still be no talk of global anthropogenic climate changes and anxiety about this, but the idea of a physical and mathematical forecast of weather conditions naturally changes the world of a hypothetical subject walking in parks and fields, giving it a sensation previously unheard of confidence. In addition, a sense of individual involvement in complex global processes is becoming increasingly relevant [3]. Like one of the 64,000 "computers" at the «Forecast Factory», a modern person is individually defined in a total (if not totalitarian) system of global production of an effect that provokes, in turn, affective reactions in one's own producer [26].

83.4 Conclusions

Taming the wildness of nature through a mathematical study of its physical predictability transforms the fundamental relationship in the "man-nature" pair, where the paradigm of power as control of nature comes in place of the old European paradigm of power as a use of nature. In this sense, a hypothetical abandonment of control means a return to unpredictable savagery, but maintaining control means an apology of inexorable power. The paradoxical nature of this situation leads to paralysis of environmental projects, which on the part of global authorities are reduced to formal complication of control procedures while maintaining the fundamental principle of appropriation. A temporary balance between these trends is still possible due to the improvement of appropriating and processing technologies. Between scientific control of climate change and modern production technologies, there is a strong systemic adhesion, that is, the structural transformation of the ecology of thinking and thinking about ecology in the current paradigm seems to be a systemic regression, the real pragmatic risks of which, unfortunately, seem more real than hypothetical climatic catastrophes. In other words, modern global politics and modern global business are trapped in their own global nature, the essence of which is to expand its colossal size, but not to limit it. The request for a new climate choice, whatever its causes and circumstances, whether they are systemic hydroaerophysical or anthropogenic in nature, requires the rehabilitation of the fragile and weak, isolated and outgoing, local and situational values. The global disposition of climatological knowledge, built in the abstract paradigm of not only supra-individual, but also suprageographic mathematical-informational analysis—impersonal and global in its fundamentals, systematically interconnected with global supra-individual and suprageographic mathematical-informational processes in economics, management and politics, expresses the current situation as at least extremely controversial [15, 17, 26]. The previously understood and not only mythopoetic and metaphorical connection between meteorology and war theory should not be forgotten [4]. Richardson's "Forecast Factory", as a grandiose and global economic and management dream project, where the human function is reduced to the tasks of specific calculus within the boundaries of given parameters, looks like a prophetic culmination of a certain dystopia. Thus, Richardson turned out to be a prophet, not only in relation to mathematical weather forecasting technologies, but also in relation to that global community, where the calculation of global parameters in its abstract anonymity and machine monstrosity is capable to report about disaster, of which it is a part itself.

References

1. Behringer WA (2010) Cultural history of climate. Polity Press, Cambridge
2. Benjamin W (1999) The arcades project, trans. Howard Eiland & Kevin McLaughlin. Belknap Press, Cambridge (MA) & London

3. Bolin B (2007) History of the science and politics of climate change. Cambridge University Press, Cambridge
4. Branagan M (2013) Global warming, militarism and nonviolence: the art of active resistance. Palgrave Macmillan, London
5. Brázdil Rudolf et al (2005) Historical climatology in Europe—The state of the art. Clim Change 70:363–430
6. Brooke JL (2014) Climate change and the course of global history: a rough journey. Cambridge University Press, New York
7. Brown N (2001) History and climate change: a Eurocentric perspective. Routledge, London
8. Callon M, Law J, Rip A (1986) Mapping the dynamics of science and technology: sociology of science in the real world. Macmillan, Basingstoke
9. Campbell B (2010) Nature as historical protagonist: environment and society in pre-industrial England. The Econ Hist Rev 63:281–314
10. Carey M (2012) Climate and history: a critical review of historical climatology and climate change historiography. Wiley Interdisc Rev: Clim Change 3:233–249
11. Chartrand MR (1973) A fifty year anniversary of a two thousand year dream [The history of the Planetarium], available at. https://www.ips-planetarium.org/page/a_chartrand1973. Accessed 9 Oct 2019
12. Edwards PN (2010) A vast machine: computer models, climate data, and the politics of global warming. MIT Press, Cambridge (MA)
13. Feldman ThSh (1983) The history of meteorology, 1750–1800: a study in the quantification of experimental physics. University of California, Berkeley (CA)
14. Fleming JR (2010) Knowing global environments: new historical perspectives on the field sciences. In: Vetter Jeremy (ed) Knowing global environments: new historical perspectives on the field sciences. Rutgers University Press, Piscataway (NJ), pp 190–221
15. Guillemot H (2017) How to develop climate models? the "gamble" of improving climate model parameterizations. In: Heymann Matthias, Gramelsberger Gabriele, Mahony Martin (eds) Cultures of prediction in atmospheric and climate science: epistemic and cultural shifts in computer-based modeling and simulation. Routledge, New York (NY), pp 120–136
16. Heymann M (2010) Understanding and misunderstanding computer simulation: the case of atmospheric and climate science—an introduction. Stud Hist Philos Mod Phys 41:193–200
17. Heymann M (2010) The evolution of climate ideas and knowledge. Wiley Interdisc Rev: Clim Change 1:581–597
18. Holm P et al (2013) Collaboration between the natural, social and human sciences in global change research. Environ Sci Policy 28:25–35
19. Jasanoff Sh (2001) Image and imagination: the formation of global environmental consciousness. In: Clark A (ed) Changing the atmosphere: expert knowledge and environmental governance. MIT Press, Cambridge (MA), pp 303–337
20. Lamb HH (1995) Climate, history, and the modern world. Routledge, London
21. Latour B (1987) Science in action: how to follow scientists and engineers through society. Harvard University Press, Cambridge (MA)
22. Livingstone D (2002) Race, space and moral climatology: notes towards a genealogy. J Hist Geogr 28:159–180
23. Lynch P (2005) The emergence of numerical weather prediction. Cambridge University Press, Cambridge
24. Murdock G (1965) Culture and society. University of Pittsburgh Press, Pittsburgh (PA)
25. Nebeker F (1995) Calculating the weather: meteorology in the 20th century. Academic Press, San Diego (CA)
26. Oreskes N et al (2010) Adaptation to global warming: do climate models tell us what we need to know? Philos Sci 77:1012–1028
27. Palsson G et al (2012) Reconceptualizing the 'anthropos' in the anthropocene: integrating the social sciences and humanities in global environmental change research. Environ Sci Policy 28:3–13

28. Parker G (2013) Global crisis: war, climate change and catastrophe in the seventeenth century, Yale University Press, New Haven (CT)
29. Richardson L (1922) Weather prediction by numerical process. Cambridge University Press, Cambridge
30. Rip A (2002) Science for the 21st century. In: Tindemans P, Verrijn-Stuart A, Visser R (eds) The future of science and the humanities. Amsterdam University Press, Amsterdam, pp 99–148
31. Rodaway P (1994) Sensuous geographies: body, sense, and place. Routledge, London
32. Smith HA (2007) Disrupting the global discourse of climate change: the case of indigenous voices. In: Mary E. Pettenger (ed) The social construction of climate change: power, knowledge, norms, discourses. Ashgate Publishing Ltd., Aldershot, pp 197–216
33. Winters HA et al (2001) Battling the elements: weather and terrain in the conduct of war. Johns Hopkins University Press, Baltimore (MD)

Chapter 84
Perception of Global Warming in the Context of Studying the Psychology of Environmental Safety

Nadezhda Parfenova, Elena Mititsina and Anna Bizaeva

Abstract The article examines the psychological approach to the phenomenon of environmental safety, including the example of students' perception ($n = 557$) of a specific environmental problem—global warming. Environmental safety is understood as a human sense of its security in interaction with the natural environment, formed on the basis of the assessment of the environment as not threatening with destructive environmental impacts, as well as the search for resources to confront the emerging dangers of a natural-tech nature. Factor analysis method justified environmental safety structure presented by the following components: cognitive (representations of nature and opportunities to influence environmental problems), affective (feeling defenseless in the face of environmental threats), value-oriented (environmental values and their relationship with values in other life spheres) and behavioral (possible behavioral strategies to solve conservation problems). The study found an average level of students' concerns about global environmental issues. With the importance of environmental values, many of them have insufficient levels of cognitive and behavioral components of environmental safety. Attitudes to climate change are linked to perceptions of other global environmental problems, but are inferior to a number of them, such as waste disposal, radioactive contamination, and environment-related diseases. At the same time, students with a higher level of concern about climate warming demonstrate a stronger desire to adhere to the rules of pro-environmental behavior in their daily behavior.

Keywords Global warming · Environmental behavior · Environmental safety · Environmental values

JEL Classification Codes Z000

N. Parfenova (✉) · E. Mititsina · A. Bizaeva
Psychology and Child Development Support Chair, Pskov State University, Lenin Squire 2, Pskov 180000, Russia
e-mail: nadezhdaparfenova@yandex.ru

E. Mititsina
e-mail: pskovprof@mail.ru

T. Kolmykova and E. V. Kharchenko (eds.), *Digital Future Economic Growth, Social Adaptation, and Technological Perspectives*, Lecture Notes in Networks and Systems 111, https://doi.org/10.1007/978-3-030-39797-5_84

84.1 Introduction

According to Maslow's theory, the need for security is one of the basic needs of the individual, the dissatisfaction of which entails the frustration of the needs of a higher level—social, cognitive, creative. The conditions of instability in modern society in the recent past have created the background for the emergence of a new concept in psychological science—"safety psychology" in which the category "safety" is understood as a sense of protection of the individual from various threats, a willingness to confront dangers, the ability to prevent and resolve them.

The environmental crisis, which is one of the global problems of our time, gives rise to the special type of psychological safety—environmental, which is reflected in a person's experience of his protection in the face of impact of threatening factors in the natural environment. Its study can be promising to solve pressing issues of environmental psychology, ranging from analyzing an individual's stress reactions to extreme exposure to negative environmental factors and up to influencing the worldview and forming a nature-centred model of interaction with the environment.

Today, the term environmental safety is widely represented in the scientific literature in areas such as law, political science and engineering. In particular, the legal field focuses on environmental security as a component of national security. In the technical sciences, environmental safety studies are moving towards the creation of methods to assess negative impacts on nature in manufacturing and reduce their devastating effects. Against this background, consideration of environmental safety in the psychological context is still little developed; there are only isolated mentions about it.

Among the basic provisions of the psychology of safety the most productive for our study are the following. The idea of [1] on the need to study psychological safety in two aspects—the safety of the environment and the safety of the individual. In this understanding the psychological safety of the environment implies the presence of such characteristics, as safety protection, absence of threats and violence [1], a state of integrity manifested in the possibility of personal development, a state of stability, defined in terms of reliability, order, law [2]. These conditions of the environment cause a person to feel protected from destructive influences, lack of anxiety and fear, reduction of tension, confidence in the future [3]. Personal psychological safety is seen as a new resource of resistance to external and internal destructive influences, coping with psycho-injuring situations, behavior that promotes sustainable development, the ability to avoid and prevent threatening circumstances.

Based on the theoretical provisions outlined above, we have formulated a definition of environmental safety. This is a human sense of self-protection in interaction with the natural environment, formed on the basis of the assessment of the environment as not threatening with destructive environmental impacts, as well as the search for resources to confront the emerging nature and man-made dangers and activity to prevent and eliminate them.

The results of our own research in area of socio-psychological security in the face of the expected terrorist threat [3], confirmed a complex structure of psychological safety phenomenon consisting of cognitive, affective, value-oriented and behavioral components which in the case of environmental safety can be filled with indicators widely discussed in the field of environmental psychology.

The cognitive component is presented by such indicators as: knowledge and assessment of global environmental problems, environmental threat situations in the region, country and globe, information in the media about nature-technological threats, views about nature [4, 5]. The affective component reveals indicators such as: anxiety caused by awareness of global and regional environmental problems, feelings of insecurity in the face of man-made disasters and natural disasters, resistance to environmental threats [6]. The value-oriented component reveals indicators such as the role of the values of unpolluted environments and conservation in the structure of common values of young people, environmental attitudes and their relationship with environmental behavior [4, 7]. The behavioral component reveals indicators such as: strategies for environmental behavior, individual responsibility for the impact of their behavior on the natural environment, participation in the work of environmental organizations [8, 9].

This environmental safety structure can be used to analyze the protection of the individual at the level of perception of a particular environmental situation in which the person is immersed. Thus, in the case of climate warming, environmental safety components may include social perceptions of climate change, the impact of the media and social groups on the formation of public opinion, and the degree of concern in due to the awareness of global warming, which in turn forms certain attitudes and determines related environmental behavior [10, 11].

84.2 Materials and Method

During the empirical data stage, a questionnaire was used on students at Pskov State University (n = 557). The questionnaire included both open questions and scales tested according to psychometric reliability criteria ("Social Security," "The Role of value attitude to nature in the structure of common values of youth," "The importance of value orientations," "Attitude to Nature," "Cohesion with Nature and the Environment," "Environmental security factors," "Influence on Environmental Solutions," "Behavior strategies of environmental safety solutions".

In the empirical study, a large amount of empirical material (95 variables) was obtained using questionnaires and standardized scales for students, which was factor-analyzed by the relevant structure blocks of environmental safety. Factor analysis helped to reduce the size of the data baseline, subject to minimal loss of information. The results were a shift from a variety of initial variables to a much smaller number of generalized factors. Factor analysis has made it possible to determine structure and content characteristics of the scales. To highlight the factors, the main component method (Varimax rotation method) was used, so many factors were singled out, which own values are larger than the unit, the significance of the variable factor weights, more than 0.4, the accumulated percentages of variance by the integral index was above 50%. Subsequently, the generalized factors were analyzed by the degree of severity among students (frequency analysis). In addition, the study used methods of counting the mean, correlation analysis of Pearson.

84.3 Results

The cognitive component of youth environmental safety is based on students' ideas of the natural environment, the diversity of which can be summarized in the form of 4 factors.

1. factor, "The notion of nature as a dangerous, but also fragile, inspiring element" (19.5% variance), includes: Nature-unbridled element that can pose a danger (0.775); Nature is a fragile structure, the balance of which is easy to upset (0.753); Nature is the mastermind of art (0.523). This factor reflects a large-scale perception of nature, given its existential aesthetic value and energy potential.
2. factor, "Ideas of Nature as a Resource Source" (16.5% variance), includes: Nature—a place of rest, a source of replenishment of human forces (0.767); Nature is a source of resources for economic and industrial development, as well as personal use (0.599); Nature is a source of knowledge of the patterns of the world around us (0.475). This factor included statements reflecting the utilitarian attitude to nature, the assessment of the benefits it can give to a person, both at the individual level and at the industry level and research.
3. factor, "Ideas of Nature as a Habitat" (14.5% variance), includes: Nature—an integral human habitat (0.831); Nature is a model for the creation of new technology and technology (0.581). It can be assumed that students not only feel the inseparable connection of humanity with the natural environment, but also spread its influence on the technical sphere, which is often created by analogy with natural objects.
4. factor, "Ideas of nature as a national wealth and condition of harmonious human development" (13.7% variance), includes: Nature is a part of national wealth (0.839); Nature is a condition of harmonious human development (0.618). Based on the content of this factor, students assess nature as a prerequisite for development. At the personal level, it can help to unlock human potential and at the state level to provide a basis for economic growth and improved quality of life.

The results of the frequency analysis showed that Factor 2 "Nature as a Resource Source" is the most ambiguous for students, as there is a greater difference of opinion than other factors have. At the same time, the number of students who highly share this perception of nature is much higher than for other factors (25%), which underline the predominance of consumer attitude towards nature. Among factors 1, 3 and 4 high level of the severity displayed 17.6, 13.7 and 19.6% of respondents respectively.

The structure of the cognitive component also contains students' ideas of the potential for influencing problems conservation to ensure environmental safety.

1. factor, "Influence on decisions to protect the environment, the environment by the state, politicians and the media" (20.8% variance), includes: government policy, officials (0.840); Party and politics (0.816); sources of information (newspapers, radio, TV, Internet). Factor combines the possibilities of solving global environmental problems, relying on the resource of power, through the adoption of pro-environmental political decisions and the creation of public opinion.
2. factor, "Influence on environmental decisions environment by scientists and environmental organizations in the regions" (20.2% variance), includes: scientists, nature researchers (0.762); Volunteers (0.717); Environmental Protection (0.661); environmental organizations in the regions (0.592). The factor shows that in the minds of students a powerful force that can influence the environmental situation in the world, are communities of people who at the local level join forces to organize environmental activities.
3. factor, "Influence on environmental decisions by each person" (18.8% variance), includes: each person in his daily life (0.686); responsible business representatives (0.680); administration of industrial enterprises (0.676); strengthening the environmental component in education (0.486). The factor reflects the level of individual responsibility in decisions relating to environmental behavior in the field of daily life and management of organizations. The factor takes into account the role of environmental education in the formation of individual responsibility.

Frequency analysis data suggest a high-level prevalence in Factor 1 "The impact on environmental decisions by the state, politicians and the media" (40.7% of students). It is likely that the solution to global environmental problems they primarily associate with the implementation of environmental measures globally. At the same time, there is an underestimation of the role of people's activities at the regional and individual level. Thus, the belief that a high level of influence on the ecological system has the behavior of each individual was shared only by 17% of respondents.

The affective component of environmental safety allows assessing areas in which students experience a state of defenselessness in the face of environmental threats.

1. factor, "Defenselessness in the face of climate and man-made disasters" (22.5% variance), includes: climate change, "greenhouse effect" (0.818); overpopulation of the globe (0.715); rising sea levels (0.658); shortage of fresh water on Earth (0.617); man-made accidents and disasters (0.615); diseases caused by pollution (0.615).
2. factor, "Defenselessness in the face of natural disasters" (20.4% variance), includes: the disappearance of the ozone layer (0.764); natural disasters and cataclysms, natural hazards (0.760); irreparable extinction of species (0.685); resource deficit (non-renewable energy sources) (0.681); use of nuclear energy, radioactive contamination (0.528). The two factors described relate to the environmental threats that humanity faces at the global level or could potentially emerge in the future.
3. factor, "Defenselessness in the face of pollution and destruction of life" (16.2% variance), includes: pollution of all environments (water, air, soil) (0.822); destruction of life (deforestation, shooting of animals, etc.) (0.798); problem

of domestic and industrial waste disposal (0.672). The factor combines environmental threats that a person may face directly in their place of residence, primarily related to pollution of the natural environment.

A comparative analysis of the severity of factors indicates that students express a high degree of concern for environmental issues both locally and globally. At the same time, yet there is a tendency to underestimate environmental threats on a planetary scale (factors 1 and 2) because there are quite a large number of students (16.7 and 22.5% respectively) who do not feel any concern about this issue. Apparently, these problems are perceived by them as insensuous, time and space remote, not directly related to them. Among the environmental problems proposed to respondents for evaluation, the feeling of being defenseless due to climate change is an intermediate position in terms of severity (mean score 2.8). It is inferior to such as recycling radioactive waste, diseases related to environmental causes (mean score above 3.0), and, on the contrary, causes greater concern, than resource deficit, overpopulation (mean score is below 2.5).

Thus, students are more traumatized by problems, either already present in their real life, or can lead to irreparable consequences for the individual. In this sense, the warming of the Earth's climate, although felt directly in the form of unusual weather events, does not cause much alarm, because it does not contain a threat "here and now". These findings are supported by the distribution of answers to the direct question "How dangerous, in your opinion, is the environmental problem of climate change of the globe". 42.6% of students responded "climate change is dangerous, rather, in the long term (tens and hundreds of years), so it doesn't bother me much", 1.8% are not at all concerned about this issue. An additional correlation analysis of this indicator with variables included in other environmental safety components showed that students with a higher level of concern about climate warming perceive nature as element that can cause harm to a person ($r = 0.317$; $p < 0{,}01$) and demonstrate a more distinct intention to follow pro-environmental daily behavior ($r = 0.307$; $< 0{,}01$).

As part of the value-oriented component of environmental safety the structure of environmental values was described, which in the process of factor analysis were united into two factors. The first factor "The importance of responsibility for nature" (40.8%), includes: responsibility for nature (0.864); nature's beauty (0.862); unity of man and nature (0.706); environmental security (0.641). The second factor "The Importance of Life and the Future of the Planet" (28.5%) includes: life (0.868); future of the planet (0.764). In other words, it can be assumed that the sphere of environmental values is heterogeneous and involves two levels—the first covers a range of phenomena directly related to the human and tangible in his real life, the second—important in remote perspective and planetary scale. At the same time, the first of them is of a greater importance for young people—high indications on it are observed in 18.5% of students, and in the second—only 9.3%.

In the overall structure of values of young people, environmental values constitute a separate factor "Values of unpolluted environment and conservation for the well-being of future generations" (16.6% variance), which includes: the well-being of future generations (0.801); unpolluted environment (0.778); feeling that he created something for the protection of nature (0.663); nature's beauty (0.596); health (0.583). In addition to it the following groups of values stood out: "Values of own security, family and state" (15.8% variance), "Values of friendship and respect" (12.7% variance), "Values of freedom" (9.6% variance), "Values welfare and equality" (8.6% variance). Among these students' values, environmental values and values of material well-being and equality are represented at a higher level (20.8 and 18.9%, respectively). Security values (oneself, one's family, one's state) and social values tend to decline, possibly due to a high degree of satisfaction with the needs involved.

The behavioral component is based on youth behavior strategies for environmental safety.

1. factor, "Active propaganda and following the rules of conservation" (26.6% variance), includes: I actively advocate rules for conservation among their surroundings (family, friends, colleagues) (0.900); I personally adhere to the rules on the protection of nature (saving water, electricity, sorting garbage, etc.) (0,847); I strive to instill environmental values in my children (younger brothers). The behaviors combined in this factor relate to individual actions that take to take shape in everyday life.
2. factor, "Public activism in the environment" (23.7% variance), includes: environmental community organization (0.889); I take part in conservation as a volunteer (0.786); I think it is appropriate to participate in rallies, pickets (0.636).
   Factor involves direct participation in communities that address specific conservation challenges.
3. factor, "Social network propaganda" (20% variance), includes: I promote environmental behavior in social networks, media (0.870); I vote for those politicians or parties that pay attention to environmental issues (0.801). The behavioral strategies identified in this factor indicate a desire to participate indirectly in environmental issues by attracting the attention of large numbers of people.

Frequency analysis has shown that behavioral strategies in relation to the environment vary in the degree of severity among students. Thus, factors 2 and 3 (high levels of 25.5 and 27.8% of respondents respectively), factor 1 showed a high level in only 15.1% of students. Consequently, students prefer social behaviors in support of the natural environment over individual ones; they tend to reduce the importance of personal input to solving global environmental problems and perhaps shift responsibility to questions from themselves to society in whole.

84.4 Conclusions

Factor analysis confirmed the theoretical model of the environmental safety structure, which includes the following components: cognitive (notions of nature and opportunities to influence environmental problems for the purpose of environmental security), affective (feeling defenseless in the face of environmental threats), value-oriented (environmental values and their relationship with values in other life spheres) and behavioral (behavioral strategies for solving conservation problems).

We can talk about the manifestation of most indicators of environmental safety at different levels—personal (as individual behavior), local (at the regional level and among the nearest social environment) and global (in planetary scale). These levels reflect the extent of perception of environmental threats, the degree of representation of the problem in people's daily lives, the possibility of personal participation in solving it.

Analysis of the structural and level characteristics of environmental safety shows that students are not generally concerned about environmental problems and their level of defense against environmental threats can be defined as average. Environmental values were represented in the overall structure of values of young people on a high level, which can probably be explained by the active promotion of environmental ideas in society through the media. However, despite this, the focus on conservation at the cognitive and behavioral level in students is much weaker. In particular, they are dominated by consumer attitudes to nature as a resource and demonstrate a weak desire to follow the rules of environmental behavior in daily life. This fact needs to be taken into account in the development of environmental programs for young people.

In addition, the following discrepancy was found in the ratio of structural and level characteristics of environmental safety. The affective and value-specific components of environmental safety are more represented at the personal and local levels. Students are more concerned and vulnerable in situations they face from personal experience.

The cognitive and behavioral components are dominated by indicators that reflect the global level of perception of environmental problems and how to address them—the belief that the protection of nature should be carried out primarily at the interstate level, weak willingness to contribute personally to environmental issues.

The highlighted trends will also be true in analyzing students' attitudes towards a specific environmental problem—climate change. The indicator "Insecurity to the Earth's Climate Change" with the highest factor weight entered the factor "Defenselessness in the face of climate and man-made disasters" where it is linked to other global environmental problems—the overpopulation of the globe, rising sea levels, freshwater shortages, man-made accidents and disasters, diseases related to unfavorable environmental conditions. It raises an average level of concern among students. At the same time, students with a higher level of concern about climate warming perceive nature as an element capable of harming humans and have a more pronounced desire to adhere to environmentally protective behavior.

Relying on the empirical evidence, we see the prospects of this study in the search for internal resources of young people to overcome situations associated with the occurrence of environmental threats. We also plan to develop a psychological support program for students, including identifying problems and reorienting them in relation to the natural environment.

References

1. Bayeva IA (2002) Psychological security in education. Soyuz, St. Petersburg
2. Krasnyanskaya TM (2011) Psychology of extreme: personality, subject, security. PGLU, Piatigorsk
3. Sereda EI, Parfenova NB (2013) Psychological security of young people in the face of the expected terrorist threat. PskovSU Publishing House, Pskov
4. Yasvin VA (2000) Psychology of attitude to nature. Smysl, Moscow
5. Hong Y et al (2019) Media exposure and individuals' emergency preparedness behaviors for coping with natural and human-made disasters. J Environ Psychol 63:82–91
6. Ferrer-i-Carbonell A, Gowdy JM (2005) Environmental awareness and happiness. In: Rensselaer working papers in economics, vol 0503, pp 67–75
7. Wright M, Klyn B (1998) Environmental attitude–behavior. Correlations in 21 countries. J Empir Gen Mark Sci 3:42–60
8. Panov VI (2004) Environmental psychology: experience of building methodology. Nauka, Moscow
9. Bamberg S, Moser G (2007) Twenty years after hines, hungerford, and tomera: a new meta-analysis of psycho-social determinants of pro-environmental behavior. J Environ Psychol 27:14–25
10. Linnik TA, Nestik TA (2018) Socio psychological mechanisms for constructing social perceptions of global climate change. The ecopsychological research collection—5 on the 8th Russian conference on environmental psychology. Publishing House Ot I Do, Perm, pp 39–47
11. Hall MP, Lewis NA, Ellsworth PC (2018) Believing in climate change, but not behaving sustainably: evidence from a one-year longitudinal study. J Environ Psychol 56:55–62

Chapter 85
Climatic and Ecological Characteristics of the Region as a Resource Influencing the Motivation for Entrepreneurial Activity

Viktoriia V. Notchenko, Yuriy N. Dyatlov and Rimma V. Druzhinina

Abstract The article, in accordance with the purpose of the study, attempts to justify the possibility of considering the climatic and ecological characteristics of the region not only as conditions necessary for human activities and vital functions, but also as specific resources referred to a subgroup of free goods. The authors conducted a questionnaire survey among entrepreneurs of the Pskov region, which gives grounds to say that availability and access to resources is a motivating factor that encourages potential entrepreneurs to engage in real business activities. Access to resources also affects generating entrepreneurial ideas and, as a result, as the questionnaire showed, affects the choice of a specific type of activities. In addition, there is reason to believe that views on the climatic and ecological characteristics of the region as on specific resources will make entrepreneurs' perception of them similar to their perception of typical economic resources, that is, they will be interested not only in their availability, but also in their preservation. Considering climatic and ecological characteristics as resources will also enhance the capacity to conduct certain types of business, in particular in the field of regional environmental tourism, agriculture and forestry, which in turn will increase the motivation for business activities.

Keywords Climatic characteristics of the region · Ecological characteristics of the region · Economic resources · Motivation of entrepreneurs · Social responsibility of business

JEL Classification Codes Q290 · Q500

V. V. Notchenko (✉) · Y. N. Dyatlov · R. V. Druzhinina
Pskov State University, Pskov, Russia
e-mail: viktoria-notchenko@yandex.ru

Y. N. Dyatlov
e-mail: dyuriy@mail.ru

Y. N. Dyatlov
Academy of the FPS of Russia (Pskov branch), Pskov, Russia

T. Kolmykova and E. V. Kharchenko (eds.), *Digital Future Economic Growth, Social Adaptation, and Technological Perspectives*, Lecture Notes in Networks and Systems 111, https://doi.org/10.1007/978-3-030-39797-5_85

85.1 Introduction

In modern society, there has clearly emerged the problem of the deterioration of the environmental characteristics quality in the regions, as well as of global climatic changes, the consequences of which are generally assessed as negative for all mankind. As a result, the society entrusts the state bodies with the function to combat the negative impact of human activities on the environment, mainly in the sphere of production of goods, provision of services of an industrial nature, performance of certain works. Public organizations, whose members are the most active in the society often, pursue the cause of protecting the natural environment. And although there has been talk of social responsibility of business, including careful attitude to the human natural environment, any effective mechanisms have not been implemented yet in that area.

In this study, the authors attempt to present the ecological and climatic characteristics of the region as a resource in the preservation of which entrepreneurs (engaged in certain types of business) may be interested. In addition, the availability of these "high quality" resources may in itself be a factor that encourages entrepreneurs to become more active, or even may be a motivating factor for start-ups. These circumstances, being accepted by business representatives, can to some extent solve two problems: firstly, concerning rational and careful attitude to the natural environment and, secondly, awareness of new opportunities to use climatic and ecological characteristics of the region in business, thus providing a motivational basis for its development.

Currently, it is characteristic for the business sector in the Russian economy to be gradually occupying market "niches" and to be searching for new or underdeveloped markets. As a result, the business community in its activities faces an urgent issue of taking into account factors and new types of resources previously neglected. Although in the narrow sense economic resources are usually understood as tangible and intangible goods, and also as human abilities used for the production of goods, services, it should be noted that the benefits, in turn, can also be differentiated into economic and non-economic ones. It is customary to refer to economic benefits as goods, the number of which in open access is fewer than it is required for economic activities or vital human functions. Traditionally, free goods are those that are more widely available than it is necessary. For this reason, most scientists refer economic goods to resources—economic goods as resources represented on any market of factors of production—and this means that we can assess their supply, demand for them, can identify their equilibrium market price in a certain time period and in certain geographical boundaries.

But one must pay attention to the fact that the goods, which are, by definition, free, i.e. those for which there is no need to pay, also can be, in some sense, resources, availability and use of which can significantly affect activities of entrepreneurs, among other things increasing the motivation of entrepreneurs to engage in certain activities. Such non-economic resources include climatic and ecological characteristics of the region in which business activities are carried out.

The issue under consideration has been insufficiently investigated in the economic scientific literature, which determines the relevance of this study. In this regard, the goal of this study is to substantiate the need to investigate the climatic and ecological characteristics of the region referred to as resources, the presence of which affects the motivation of entrepreneurs. The main objectives of this study include the following ones:

- to identify the importance of the availability of resources for motivating entrepreneurs, as well as for making them aware of this importance;
- to consider climatic and ecological characteristics of the region as specific resources that influence business activities;
- to justify the need to include climatic and ecological characteristics of the region into the generalized classification of economic resources.

85.2 Methodology

The goal and objectives of this study brought about certain scientific methods. Initially, the problem statement worded as "insufficient attention from the business community to the climatic and ecological characteristics in the region referred to as resources", necessitated the deductive method application, in particular, generalizing the hypothesis that the motivation for entrepreneurial activities would grow if entrepreneurs understood the role of climatic and ecological characteristics as additional resources which are important for implementing some activities.

While studying opinions of well-known scientists, methods of analysis and synthesis, of generalizing data were applied to investigate the role of resources in human life and business activities.

To prove the importance of resources as a motivating factor, one of the empirical methods called "questioning" was applied, which was followed by methods of statistical processing of results and methods of formulating conclusions.

In general, throughout the study, the authors proceeded on the principles of the systems approach, and, in some cases, the methods of analogy and comparison were used.

85.3 Results

Let us consider the concepts of "climatic characteristics of the region" and "ecological characteristics of the region", used in the study of entrepreneurship as a type of activities.

Under the climatic characteristics of the region in this study, we will understand the long-term weather regime in a particular area, manifested in regularly changing weather patterns. The concept of "ecological characteristics" will be interpreted as a set of dynamic elements of the external natural environment of man, determining the qualitative conditions of his existence, affecting his activities and vital functions.

It should be noted that a number of research works are devoted to the study of the influence of climatic and ecological characteristics of the region on the existence and activities of man. In particular, Baryshnikov S. V., Monichev A. Y., Basurov V. A. address the problem of interdependence of economic, social, demographic and ecological characteristics of regions of Russia [1]. Kuznetsova T. I., Bychkov I. V., Batuev A. R. and others paid special attention to the ecological potential assessment [2]. Studies of interrelations between the ecological characteristics of the region and the quality of the population life are traditional [3, 4]. Meanwhile, in the scientific sources specified above there is, practically, no justification of the necessity and the possibility of considering climatic and ecological characteristics of the region as specific resources.

The climatic conditions factor, due to the lack of influence on it from economic entities, can be either made use of or not. In the latter case it is either not called for or entrepreneurs are short-sighted and are not aware of this opportunity. Speaking of the "use"-option we should note that regional, federal and municipal governments can have some impact on the ecological characteristics, while the positive impact of entrepreneurs on these characteristics is limited. In this regard, from the standpoint of a business, the ecological characteristics of the region can be considered as elements of an infrastructure, the quantity and quality of which are determined by the extent to which the authorities control the maintenance of ecological characteristics at the required level or contribute to their improvement.

Traditionally, the infrastructure is understood as a business environment formed by industries and spheres which ensure entrepreneurship functioning and development [5]. Depending on the scope of infrastructure elements distribution, the infrastructure is divided into common one (for all the business entities, e.g. roads, housing and public utility sectors), special one (i.e. depending on the type of business activities) and the infrastructure for supporting entrepreneurship [6].

A number of researchers expand this concept, distinguishing in it the institutional infrastructure of entrepreneurship, which includes, respectively, institutional actors (i.e. institutions that ensure the implementation of institutional requirements) and social institutions [7]. Examining the infrastructure of entrepreneurship, Kumpilova A. R. considers innovation infrastructure as another subsystem, to which he refers innovation and technology centers, technology incubators, technology parks, as well as information and communication systems, educational organizations, technical means for collecting, processing and transmitting information, research and production associations [8].

From the standpoint of providing conditions for fulfilling certain types of business activities ecological characteristics can be attributed to the infrastructure elements and we can identify the concept of "the environmental protection institute", which provides for regulating environmental parameters and for using ecological characteristics in certain business activities.

It follows that the traditional classification of resources can be supplemented by a group of resources called "free goods", which, in our opinion, should include the climatic and ecological characteristics of the region.

Thus, the classification of economic resources, as specified above, would include three groups: economic goods, human abilities and free goods. In the first group we should differentiate between tangible goods (natural resources, fixed assets and working capital of the enterprise) and intangible economic goods (patents, licenses, developed technologies, results of scientific, sociological, marketing and other research, energy, etc.). The group of "human abilities" should traditionally include the ability for work and the entrepreneurial ability. While into the composition of free goods as a type of economic resources, we propose to include the entrepreneurship infrastructure and free goods which we do not refer to the infrastructure.

In turn, the infrastructure of entrepreneurship can be divided into common one (for business), in which it is worthwhile to distinguish innovation, institutional and special segments (depending on the type of the entrepreneurship) and the infrastructure for supporting entrepreneurship. Based on the above conclusions, the ecological characteristics of the region can be referred to the infrastructure as a separate element.

We propose to include the climatic characteristics of the region, which have a significant impact on certain types of business activities, into the category "free goods outside the infrastructure". The same group should include information obtained from open sources, which can be applied in practical activities and can be treated as a resource that does not require any costs for its acquisition.

Certain comments are required here on including into the above classification a specific resource called "goodwill". Many authors consider goodwill as an important and specific component that determines the value of a business. At the same time, the source of goodwill is the external environment, which means that it can be purposefully formed and, in some cases, requires certain costs. Therefore, on the one hand, this type of resources can be classified as referring to "economic goods". On the other hand, for example, a positive reputation of an enterprise can be formed naturally due to their higher quality products on the market and their proper performance of contractual obligations. Therefore, in our opinion, the goodwill component can be partially attributed to the resources of the category "free goods".

In recent decades, entrepreneurship, as a special type of activities, is developing not so much quantitatively as it is doing qualitatively. The attitude of the society towards entrepreneurs representing small and medium-sized businesses is changing in a positive way. For the most part civilized methods in competition are being employed, more transparent schemes of obtaining state and municipal orders are being used. Programs and measures to improve the entrepreneurship infrastructure are being implemented. However, the role of small and medium-sized businesses in the domestic economy is not as significant as in economically and socially developed countries. In this regard, the issue of motivation for entrepreneurial activity proves to be important.

The authors of this study have conducted a survey in the Pskov region. 54 entrepreneurs were questioned and the questioning revealed that the main factors of motivation for them were as follows: "the desire to feel independent in making decisions about their activities", "the desire to manage their life" and "the desire to implement their entrepreneurial abilities", which is understandable as these motives are basic for making decisions about starting a business. 18% of the respondents indicated "availability of resources and access to resources" as a motive for entrepreneurship. At the same time, for the question "What has had a decisive influence on the choice of a particular type of business?" 28% of the respondents indicated "the availability of the sufficient number of resources or the access to resources, including personnel ones, technical ones, etc.".

As we can see, the availability of resources is an important factor determining the entrepreneurs activity. The availability of resources and the access to them, within the region specifics, as the earlier conducted survey showed, is also a factor that has a significant impact on generating entrepreneurial ideas (noted 26.7% of the respondents) [9]. However, not all the entrepreneurs are aware that the list of resources may also include "goods" because they have not been earlier included in the list as resources. In particular it concerns advantages that a region can have within the climatic and ecological aspects.

Improving ecological characteristics of the region can substantiate for additional motivation of entrepreneurs, both in terms of its increase, and in terms of forming new incentives for entrepreneurship. Therefore, the costs of preserving the natural environment of the region are not only socially significant costs, the main "impact" of which is in the increase in the quality level and in the life expectancy of the population. The cost of "ecology" can also be considered as an investment in the business sector, the return on which can result in the growth of tax revenues within the entrepreneurship revival. It can be achieved through utilizing an additional motivating factor obtained from the activities connected with the ecological characteristics of the region.

In addition, climatic and ecological characteristics turn out to be one of the factors attracting labor resources to the region. Although these characteristics are not as important as the level of average wages in the region, the cost of renting a house, the shortage of labor resources it is likely that potential migrants, other things being equal, will choose settlements located in more comfortable climatic conditions and, more significantly, with a better environment for the potential place of residence.

The climatic and ecological characteristics of the region are unlikely to affect all the activities directly. Klyuev N. N. notes that environmental conditions have affected the agricultural specialization of the regions [10]. Oborin M. S. connects the "green" tourism development with the natural resources potential, paying special attention to the environmental conditions prevailing in the region [11]. Many authors note a significant importance of climatic and ecological characteristics of the region for the activities of different sub-sectors of agriculture [12, 13].

Based on the assessment of the human influence on ecological and, in some cases, climatic characteristics, another important and obvious conclusion follows: these characteristics are subject to changes not only for objective reasons (laws of nature), but also due to human activities, including entrepreneurial activities aimed at using natural resources to produce goods that meet the needs of the society. As for the objective causes for changes the humanity can either adapt to changes in climatic and environmental conditions or resist them to the extent that the technology development and scientific knowledge allow. Changes in the state of the natural environment caused by human activities require further studies in order to find mechanisms to encourage humanity, engaged in active labor and business activities, not to damage natural resources of the planet.

Entrepreneurial activity deserves special attention as an initiative activity involving all the types of resources in a unified process of production of a social product. Thus the responsibility for decision-making for the most part lies with entrepreneurs and, to a certain extent, with state governing bodies controlling various spheres of activities. Hence, it is important to consider the attitude of the business community to ecological characteristics which should be treated as a resource.

Forming business representatives' ideas of regional climatic and ecological characteristics (as resources) can radically change their attitude to the environment. In particular, entrepreneurs operating in the sectors of eco-tourism, of construction, agriculture, of souvenirs production (reflecting the culture of the region), and planning to expand their business in the future, can actively oppose the deterioration of the environmental situation due to the activities of other business entities, as this can lead to a reduction in the tourist flow, lower housing prices and in general cause an outflow of the population from the region. Thus, concern for the conservation of environmental resources will cease to be only a function of the state and of the most active groups of the population. It should become an important task for a number of business circles.

85.4 Conclusions

1. Climatic and ecological characteristics of the region can be considered not only as factors influencing business activity, but also as specific resources, the quality of which can stimulate business activity (increasing motivation).
2. The classification of economic resources can be supplemented by the group of "free goods", which, in our opinion, should also include "business infrastructure", which, in its turn, includes, as an element, "ecological characteristics of the region" and a separate good named "climatic characteristics of the region".
3. It is obvious that the regional public authorities will not be able to exert any significant influence on climatic conditions, but ecological characteristics should be given much attention to, because they determine the most important indicators of the population's life activities, which is not only an important social priority, but also, from the economic point of view, indirectly affects the quality of labor resources in the region and in the country.
4. The cost of improving environmental characteristics of the region can be considered as investments in the development of certain types of businesses in the region as the characteristics of the business infrastructure are in this way improved, and, therefore, entrepreneurial abilities are promoted, which in turn will lead to the

growth of the gross regional product, to the increase of the tax base thus increasing budget revenues.
5. The climatic and ecological characteristics of the region are unlikely to be "used" as a resource for all the types of business activities. The activities for which the above mentioned characteristics can be considered as an important resource are as follows: regional tourism, agriculture (including crop production), fish farming, forestry, etc.
6. Climatic and ecological characteristics, along with other characteristics of the region, influence the direction and strength of migration flows. This means that it is much easier to attract labor resources to locations more "favorable" from the point of view of ecology and climate than to regions less attractive from the point of view of ecology and climate. This circumstance is also important when it is necessary to "retain" in the region highly qualified and necessary labor resources.
7. Since business activity is associated with certain risks, and factors causing them can include changes occurring in climatic conditions and, especially, in the ecological situation of the region, it is necessary to intensify the activities aimed at preserving the available resources and both public authorities and businesses should be aware of the value of these resources for the region.

Summing up, we should note that as competition on the market is intensifying and the level of entrepreneurial "literacy" is increasing there is reason to believe that the climatic and ecological characteristics of the region should be evaluated by entrepreneurs as significant for their activities, and the improvement of these characteristics under the influence of state governing bodies should be considered not only in social but also in economic aspects. The awareness of the entrepreneurs engaged in certain types of activities, of the importance of the regional climatic and ecological characteristics, as resources that influence the economic results, will lead to the formation of a new attitude towards them which should be aimed at preserving these resources not only for reasons of their civil position, but also from the perspective of their businesses. At the same time, treating the ecological and climatic characteristics of the region as resources that expand business opportunities can contribute to the growth of entrepreneurial motivation.

References

1. Baryshnikov SV, Monichev AY, Basurov VA (2008) Research of interdependence of economic, social, demographic and ecological characteristics of the regions of the Russian Federation by factor analysis methods. Bull Nizhny Novgorod Univ N I Lobachevsky 3:110–115
2. Kuznetsova TI, Bychkov IV, Batuev AR, Plyusnin VM, Ruzhnikov GM, Khmelnov AE (2011) Structural-typological characteristics and ecological potential of Baikal region geosystems. Geogr Nat Resour 4:20–29
3. Rossinskaya MV, Rossinsky NP (2012) Elements of ecological monitoring, their brief characteristics and impact on the quality of the environment and the health of the population of the region. Eng Bull Don 1(19):258–270

4. Sukhova MG (2009) Climatic and ecological characteristics of the mountain regions environment (on the example of the Russian Altai). Bull Russ Peoples Friendsh Univ Ser: Ecol Life Saf 2:116–121
5. Kosov ME (2009) Formation of small business infrastructure as a necessary condition for its development. Natl Interes Priorities Secur 5(No 4 (37)):67–71
6. Galochkina OA, Razumova IV (2012) Infrastructure of entrepreneurship: essence and types. Bull Econ Law Sociol 1:38–40
7. Yudenko MN (2009) Methodological principles of research of institutional infrastructure of entrepreneurship. Bull USTU-UPI Ser: Econ Manag 6:4–12
8. Kumpilova AR (2011) Regional management of entrepreneurship infrastructure development. Socio-Econ Phenom Process 3–4(25–26):160–163
9. Notchenko VV, Vinogradova OV (2017) Scientific and practical aspects of entrepreneurial idea generation. Bull civil Eng 3(62):314–320
10. Klyuev NN (2017) Ecological and resource "dissonance" in agriculture of Russian regions. The collection "Strategy of development of border territories: traditions and innovations". Kursk State University Press, Kursk, pp 391–398
11. Oborin MS (2018) Assessment of natural resource potential of ecological and "green" tourism of territories with medical and recreational specialization, Scientific notes of V.I. Vernadsky Crimean Federal University. Geography. Geology 4(No. 1 (70)):49–64
12. Derevenets DK (2018) Basic economic and environmental requirements for transition from zonal to adaptive landscape system of agriculture in the context of regional food security. Colloq J 9–7(20):6–10
13. Stepin AD, Ryseva TA, Kostrova GA, Utkina SV, Romanova NV (2018) Selection evaluation of hybrid populations of flax in Pskov Research Institute of Agriculture. Proc Velik Luki State Agric Acad 4:26–35

Chapter 86
Comparative Analysis of Methodological Approaches to Assessing Threats to Environmental and Economic Security

Natalia Kurepina, Tamara Tsatkhlanova, Nyuudlya Pavlova and Danara Erendzhenova

Abstract This issue is topical, since concerning present realities when economic development is characterized by an active anthropogenic impact on the environment, an assessment of threats to environmental and economic security should serve the reason for early preventive measures to be taken. The modern mechanism of environmental and economic development is quite destructive both from the standpoint of taking into account the interests of subjects of nature management and the lack of an effective system for tracking threats to environmental and economic security. This situation jeopardizes the sustainable development of human civilization, high quality of life and public health, conservation of natural systems and maintenance of proper environmental quality. The authors have examined methodological approaches to the assessment of threats to economic security and have concluded that they need to be refined. The examination allowed the authors to find out that it is necessary to revise approaches to the concept of "environmental security" as a component of economic security, and regard it as a single concept "environmental and economic security", and threats—as negative factors affecting the development of the environment, areas of social activity and production. This conclusion was based on the research of the environmental and economic security of the Republic of Kalmykia that is a complex object both in terms of natural and other types of threats and risks. Keeping in mind the necessity of assessing threats with attention to regional features of the territory, it is proposed to supplement the list of indicators and consider the criteria depending on the object's location, the level of threats and other factors.

N. Kurepina (✉) · T. Tsatkhlanova · N. Pavlova · D. Erendzhenova
Kalmyk State University Named after B.B. Gorodovikov, Elista, Russia
e-mail: kurepinanl@mail.ru

T. Tsatkhlanova
e-mail: tsatkhlanovat@mail.ru

N. Pavlova
e-mail: pavlova_nc68@mail.ru

D. Erendzhenova
e-mail: uralan-danara@yandex.ru

T. Kolmykova and E. V. Kharchenko (eds.), *Digital Future Economic Growth, Social Adaptation, and Technological Perspectives*, Lecture Notes in Networks and Systems 111, https://doi.org/10.1007/978-3-030-39797-5_86

Keywords Impact on the environment · Assessment of environmental · Economic security threats · Methodological approaches

JEL Classification Codes C100 · Q510

86.1 Introduction

The issues of conserving the natural environment and reasonable nature management in the National Security Strategy of the Russian Federation until 2020 are referred to the priorities of sustainable development and designated strategic goals of ensuring environmental safety.

Such a decision is necessitated by threats to environmental and economic security, the significant external impact of production and anthropogenic factors on the environment. Moreover, the above factors are not only; threats of a depressed natural environment that influence the quality of the population's living conditions arise. Therefore, understanding the inevitability of environmental consequences, one should make economic decisions concerning the consequences of the impact on the environment.

The dependence on the environmental factor is extremely strong in industries such as agriculture. This is associated primarily with the involvement of natural resources in the production process, and within arid climate—the land (the most vulnerable resource).

Therefore, when assessing threats to environmental and economic security, especially in regions with agricultural production as the main industry which greatly depends on natural factors, it is necessary to consider the environmental factor that in terms of an arid climate affects both the production output of agricultural enterprises and living conditions of the population.

Most of the rural territories of Kalmykia is located in the arid zone, which is characterized by high environmental stress.

Unreasonable nature management in the eastern regions of the republic, namely year-round exploitation of pastures, has led to desertification, extensive withdrawal of previously productive lands from circulation has required significant financial costs for their restoration. Land degradation has covered a significant part of the republic's territory and incrementing livestock population without account for the load on natural pastures can have disastrous consequences. Therefore, it is necessary to promptly respond to environmental and economic challenges and take preventive measures.

Given this, an early assessment of threats should become a priority, which is especially urgent for environmentally depressed regions.

Scientists have been tackling the issue of conserving the environment and limiting the anthropogenic impact for many decades. This is a very wide range of issues that are presently addressed from different points of view.

For the issues of environmental and economic security to be solved, it is necessary to assess the level of threats to specific territories depending on their types and criteria to assess and prevent them as soon as possible. The system of criteria for assessing the level of threats that is especially important for rural areas located in the severe climatic conditions of the arid region should be refined.

86.2 Research Issues

The issues of conserving the natural environment have been studied for several centuries, but at the present stage of economic development, they have become aggravated that the point is conservation of civilization. These issues have been investigated by scientists of different branches of knowledge. In our study, we have examined not only the approaches of biologists and ecologists to the issues of sustainable development of the environment but also scientists, especially economists, who consider the issue as ecological and economic safety. Nowadays, there is no single approach to the definition of environmental safety. The definition given in the Law is disputable and is interpreted by the authors from different perspectives.

Most authors associate environmental safety with global causes of the emerging threats, growth of the planet's population, pollution of the atmosphere and water resources, loss of biodiversity, and so forth. [1–3]

Recently, the issue of environmental safety has been connected to a greater extent with the conditions for ensuring the quality of living, creating a comfortable and safe environment, and preventing negative effects associated with the activities of production enterprises [4].

Undoubtedly, threats to the environmental security of the territory affect not only the population's quality of life but also the development of the region in general.

At the present stage of research dedicated to the environmental and economic security threats, much attention is paid to seeking solutions through designing economic and mathematical models that allow for a more integrated and full-fledged assessment of managerial decisions on environmental and economic security.

Skiter et al. [5] regards the static and dynamic approaches to assessing environmental and economic safety and suggests exploring the issue using a dynamic approach based on environmental and economic needs and interests.

In the course of our research, we have concluded that the threats to environmental and economic safety are mostly addressed in the part of assessing air and water pollution despite a significant number of works on this topic. Therefore, indicators for assessing environmental safety threats are designed mainly to estimate the level of air and water pollution.

Senchagov [6] in assessing the threats to the environmental safety of the regions considered only the level of pollution and reforestation.

However, as we have already noted, this is not enough. There are other types of threats that affect the land as a natural resource important for human life to be conserved. The issue of depletion, decreasing area of the most productive arable land is topical both in Russia and other countries. It entails the threat of losing their productivity, cutting down the output of foodstuffs and ultimately have an impact on the nutrition of the planet's population.

In our opinion, these approaches do not determine the depth of the environmental and economic security issues, since they regard threats only as factors affecting the natural environment, i.e. on the depletion of reserves of all types of natural resources, freshwater shortages, and others.

We believe that when assessing threats to environmental and economic security, it is necessary to take into account regional features and their impact on the natural environment (ecosphere), as well as on the technosphere and sociosphere. In assessing the quality of the environment, one should take into consideration the living conditions and activities of a man, reasonable utilization of natural and industrial resources.

Therefore, in the course of the research, we have come to the conclusion that environmental security cannot be assessed without regard to economic and social consequences, but should be considered as environmental and economic security. And the variety of natural, socio-economic, cultural, religious and other features of the regions should be reflected in the scientific analysis of the environmental and economic security of the regions and the development of assessment indicators.

86.3 Methodology

The research methodology is a rather complicated issue for any branch of science. It is multidimensional for the developing theory of economic security since it intertwines various fields of scientific research (economics, sociology, ecology and other branches of knowledge). In methodological terms, interdisciplinary analysis as a way of arranging scientific activity facilitates the research thanks to the participation and interaction of scientists from various branches. In the course of the research, the authors have employed methods of economic, environmental, and sociological research, mainly: factor, indicative, grouping, graphic, expert, etc. Coefficient and economic-mathematical methods have been applied in the development of indicators of environmental and economic safety threats.

86.4 Findings

Research of threats to environmental and economic security has been carried out by the authors for a long time in most of the arid territory of southern Russia. Studies were conducted both concerning the efficient use of land and the sustainable development of agricultural production.

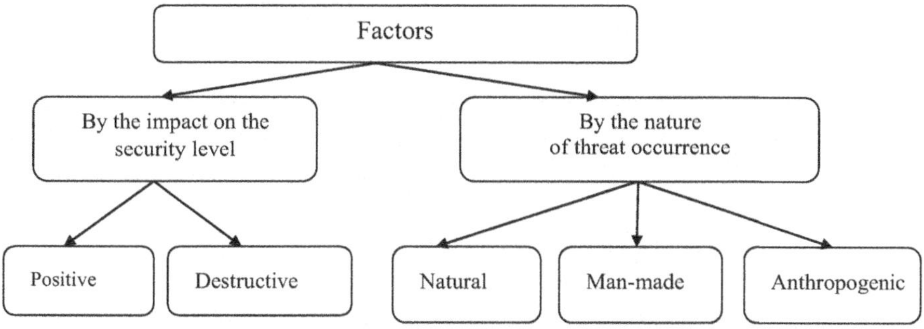

Fig. 86.1 Classification of environmental and economic safety factors

Exploring objects of environmental and economic safety made it possible to study the extent of influence of both positive and destructive factors and make their classification by the nature of threat occurrence and threatened objects. (Fig. 86.1)

Some of the factors are classified under various criteria. Thus, natural objects can be a threat to environmental and economic security, as they can lead to earthquakes, floods, forest fires, etc.

And at the same time, they can be an object exposed to unreasonable anthropogenic impact and, as a result, the occurrence of the greenhouse effect, land degradation, loss of biodiversity, floods, etc.

Recently, the emergence of local threats, floods, and forest fires has become threatening due to economic activity and other factors of human activity, which result in a loss of natural resources and poses sanitary-epidemiological threats. The above circumstances entail the challenges of maintaining the health of the population living in these areas.

Unreasonable utilization of production resources leads to man-made accidents, disasters and, as a result, to threats of rising morbidity and mortality.

Air pollution in cities with a developed heavy and non-ferrous metal industry exceeds MPC by two or more times. The morbidity rate per 1000 people in the cities of the Ural and Siberian federal districts increased from 786 in 2005 to 829.5 in 2016 and from 813.3 to 970.4, which is significantly higher than the Russian average.

Undoubtedly, air and water pollution are a very serious threat to environmental and economic security and therefore this issue is addressed in the studies of most authors (Table 86.1).

The study of approaches to the determination of environmental risks allowed us to conclude that the authors primarily highlight air and water pollution. Vechkanov [7] revealed to the fullest extent the existing threats to environmental security, including their impact on all types of natural resources: land, water, forest, animals. G. S. Vechkanov found out that the environmental situation is determined by the state of atmospheric air, waters, land, and soil, but he does not regard indicators for the assessment of threats to environmental and economic security.

Table 86.1 Threats to environmental security

Author	Environmental threats
V. K. Senchagov	Emissions of harmful substances into the atmosphere
	Discharge of untreated waters
	Toxic industrial waste storage with violation of safety requirements
	Man-made disasters
G. S. Vechkanov	Air pollution
	Soil erosion and pollution
	Deforestation
	The destruction of the earth's ozone layer
	The death and extinction of many species of animals
	Disruption of the World ocean's environment
	The death of large water reservoirs and other disasters
V. A. Bogomolov	Air pollutant emissions
	Discharge of pollutants into water reservoirs
	Subsoil and soil pollution
	Disposal of production and consumption waste
	Environmental pollution by noise, heat, various types of radiation

The more exhaustive list of environmental security indicators is given in the works by Bobylev and Khadzhaev [8]. However, he also did not consider land resources as an object for the assessment of threats and didn't determine indicators for this object.

Now there is no single approach to determining the set of indicators describing the extent of threats to environmental and economic security. Some authors regard them from the standpoint of assessing the impact of anthropogenic load, others—from man-made ones.

We have analyzed the available criteria for the assessment of environmental threats, different approaches, and attitudes to this issue.

As the analysis showed, almost all authors refer the standards of maximum permissible concentrations of harmful substances in air and water to the main indicators of environmental and economic security, which is undoubtedly a topical but unexhaustive range of threat assessment.

It should be noted that the criteria for assessing the environmental and economic threats typical for the region under study (the Republic of Kalmykia) are not also touched upon.

The same situation we see in the monograph by Senchagov [6], who considered only the level of pollution and reforestation.

He considered the volumes of discharged wastewater into surface water bodies per unit area of the territory occupied by the region as an indicator of assessment of water bodies' pollution and estimated an indicator of air pollution the same way.

The indicator of reforestation was defined as the proportion of regenerated forests of the territory's total area.

Table 86.2 Assessing the environmental state of the regions of the arid zone of the South-European part of Russia (according to [6])

Constituent entities	Indicator	Risk zone
Astrakhan region	Discharge of wastewater, th. m^3/km^2	Critical
	Air pollutant emissions, t/km^2	Critical
	Reforestation (share of regenerated forests, percent)	Critical
Volgograd region	Discharge of wastewater, th.m^3/km^2	Critical
	Air pollutant emissions, t/km^2	Substantial
	Reforestation (share of regenerated forests, percent)	Stability
Republic of Kalmykia	Wastewater discharge, th.m^3/km^2	Stability
	Air pollutant emissions, t/km^2	Stability
	Reforestation (share of regenerated forests, percent)	Stability

The results of the assessment (Table 86.2) showed controversial results. The territory of Kalmykia was found stable in terms of environmental security.

Thus, according to V. K. Senchagov, the environmental state of the Republic of Kalmykia is assessed as stable, which undoubtedly does not display the actual state of environmental and economic security of the republic, since indicators do not assess other risks and threats, namely large-scale land degradation and soil depletion. Certainly, it is impossible to take into account all types of threats and their significance for the territory in all regions, but it is unreasonable to conclude on the environmental stability of the territory from the perspective of the above indicators.

The regions of the arid zone of southern Russia are exposed to land degradation to one or another extent. According to the report "On the State and Use of Land in the Republic of Kalmykia in 2018 [9], over 4,000 ha are exposed to deflation and over 2,000 ha—to salinization. Large areas are vulnerable to desertification. Arable land in Kalmykia fall under threats of dehumidification and deflation (wind and water erosion).

Economic issues do not let to solve environmental ones in a prompt manner. Emerging vicious circle arises ultimately leads to global environmental challenges. The situation requires urgent measures to prevent and conserve the land, our national wealth.

Other threats to environmental and economic security are slipping out of attention, one of them is the lack of high-quality drinking water. According to the State Report [10], over 30 percent of drinking water samples from the distribution network of the Republic of Kalmykia do not meet sanitary and epidemiological requirements for sanitary and chemical indicators.

The threats to environmental and economic security should be surely assessed with regard to a wider range of indicators. For assessment of threats to the environmental and economic security of the region, we suggest the introduction of an integrated indicator of the level of environmental and economic threats, which takes into account all negative impacts on the threatened objects, i.e. environment, process environment, and man.

86.5 Recommendations

The research revealed that the methodology for the assessment of threats to environmental and economic security should be improved with account for regional features, threatened factors and objects. There are many indicators introduced for standardization of water and atmosphere quality, but as for land resources—restrictions are none. It is the loss of soil fertility and arable land that has recently become threatening.

We believe that available approaches to determining threat assessment indicators for environmental and economic security do not take into account the extent of threats and offer a wider range of them. All types of environmental and economic threats should be comprehensively investigated and an integrated environmental and economic indicator that could be used to estimate the territory's stability should be developed.

The improvement of methodological approaches to the assessment of threats to environmental and economic security is caused by the necessity of developing an effective mechanism for early prevention of negative processes in the nature management, which affect both the state of the environment and the population living in this territory.

Acknowledgements This article had been written under the support of the Russian Foundation for Basic Research within the implementation of the project No. 19-010-00796 "Improving Methodology for Assessment of Economic Security of a Multiethnic Region: An Interdisciplinary Approach".

References

1. Bogomolov VA, Eriashvili ND, Barikaev EN (2012) Economic security. Uniti-Dana, Moscow
2. Oleinikov EA (2005) Economic and national security. Publishing House "Examination", Moscow
3. Senchagov VK, et al. (2005) Economic security of russia. Delo, Moscow
4. Mamaev LN (2014) Risk management. Dashkov & Co. Publishing, Moscow
5. Skiter N, Rogachev A, Mazaeva T (2015) Modeling ecological security of a state. Mediterrean J Soc Sci 6(3):185–192
6. Senchagov VK, et al (2012) Economic security of russian regions. Nizhny Novgorod
7. Vechkanov GS (2007) Economic Security. Piter, SPb.
8. Bobylev SN, Khadzhaev ASh (2007) Environmental economics. INFRA-M, Moscow
9. Report (2019) On the State and Utilization of Land in the Republic of Kalmykia in 2018
10. State Report (2019) On the state of the sanitary-epidemiological well-being of the population in the Russian federation in 2018. Federal Service for Supervision of Consumer Rights Protection and Human Well-being, Moscow
11. Decree of the President of the Russian Federation (2017) Strategy of economic security of the Russian federation until 2030, No 208 as of May 13, 2017. https://rg.ru/2017/05/15/putin-podpisal-strategiiu-ekonomicheskoj-bezopasnosti-do-2030-goda.html
12. Ignatova TV (2014) Issues of managing regional social and labor and demographic security in the post-crisis period. Sci Educ: Econ Econ Entrep Law Manag 8(51). elibrary_21787966_83292214.pdf

13. Kurepina NL, Bogzykov YuS et al (2009) Sustainable rural development as a factor of ensuring economic security. APP Dzhangar, Elista
14. Tokarev KE (2015) Development of a decision-support system in the area of environmental and economic security. Mediterrean J Soc Sci 6(6)

Chapter 87
Eco-concept "Clean Region": Socio-Ecological and Economic Aspects of Innovative Design (Empirical Experience of Research)

Tatyana N. Ivanova

Abstract The paper examines the social, economic and environmental foundations of innovative design of the region. The paper includes informative empirical experience of large-scale prolonged sociological research devoted to the issues of spatial development of the region. The author focuses on the need to develop and implement an innovative target scenario Eco-concept "Clean Region". The paper presents an innovative concept of ecologization of the regional economy, taking into account changes in targeted foresight technologies in the socio-ecological-economic activity of the region.

Keywords Ecology · Concept · Society · Region · Economy · Design · Innovations · Foresight technologies · Spatial development · Ecologization of the economy · Scenario development · Culture · Potential · Eco-concept

In the framework of the implementation of the national Project "Ecology" of the Russian Federation, the study of social and economic problems of the region, assessment of efficiency of methods for solving them became the subject of a long-term sociological research of the Department of Sociology of the Togliatti State University from 2018 to 2019. A large-scale sociological research included a questionnaire survey of the population of the region and the formation of three focus groups with the use of foresight technology (design approach). The number of focus group members was 45 respondents. Focus group members live in cities of the Samara region, while a small proportion of members live in a non-urban area. Two fifths of respondents are married, one third of respondents are single. 43% of respondents graduated from higher education institutions, 21% of respondents currently go into higher education. Consequently, 64% of survey respondents obtained higher education. 22% of respondents graduated from colleges and training schools. 11% of respondents obtained secondary education, and 3% of respondents obtained School Certificate.

One half of respondents (52%) claim that the financial situation of their families is moderate. 34% of respondents are of the opinion that the financial situation of their

T. N. Ivanova (✉)
Togliatti State University, Togliatti, Russia
e-mail: Ivanovat2005@tltsu.ru

T. Kolmykova and E. V. Kharchenko (eds.), *Digital Future Economic Growth, Social Adaptation, and Technological Perspectives*, Lecture Notes in Networks and Systems 111, https://doi.org/10.1007/978-3-030-39797-5_87

families is "above average". As little as 3% of respondents claim that the financial situation of their families is "excellent". 11% of survey respondents claim that the level of their material well-being is "below average", and 4% of respondents are in severe difficulties.

The vision of the future by respondents—both their own future and the future of the region is primarily determined by how people live today, by the level of their material well-being, the state of the regional and urban environment and a range of problems the population is faced with. Based on the assessment of their daily lives today, residents of the Samara region form their expectations of the future and design a vision of the future of their region as well as their own future.

The study of social, environmental and economic problems of the region, assessment of the efficiency of methods for solving them became the subject of a sociological research of the Department of Sociology of the Togliatti State University. 754 people living in 35 settlements of the Samara region participated in the survey. One half of respondents live in cities of the Samara region, and the other half of respondents live in a non-urban area. Two fifths of respondents are married, one third of respondents are single. 42% of respondents graduated from higher education institutions, 15% of respondents currently go into higher education. Consequently, 57% of survey respondents obtained higher education. 28% of respondents graduated from colleges and training schools. 9% of respondents obtained secondary education, and 5% of respondents obtained School Certificate.

One half of respondents (54%) claim that the financial situation of their families is moderate. One fifth of respondents are of the opinion that the financial situation of their families is "above average". As little as 5% of respondents claim that the financial situation of their families is "excellent". 18% of survey respondents claim that the level of their material well-being is "below average", and 4% of respondents are in severe difficulties.

In the ranking of global environmental problems that is formed on the basis of the analysis of results of the survey, air contamination ranks first. This answer option was chosen by 48% of survey respondents. Slightly fewer number of respondents place emphasis on contamination of drinking water sources (45%), this answer option occupies the second position in the ranking. Garbage dumps and the lack of efficient waste utilization technologies as a topical global problem were emphasized by almost two fifths of survey respondents, which corresponds to the third position in the ranking. 35% of respondents place emphasis on chemical and radiation poisoning of the territory. This answer option occupies the fourth position in the ranking. Consequently, the top four positions in the ranking of global environmental problems pointed out by survey respondents are occupied by problems associated with the impact of modern civilization on environment.

22% of survey respondents point out the lack of efficient environmental legislation as a global environmental problem. This problem occupies the fifth position in the ranking. Positions 6 and 7 in the ranking are occupied by problems associated with consequences of environmental contamination: climatic change, deforestation.

19% of respondents claim that extinction of rare species is a global environmental problem. 18% of survey respondents point out ecological illiteracy of the population. These types of problems respectively occupy positions 8 and 9 in the ranking.

12% of survey respondents worry about the increase in the number of homeless animals, which corresponds to the tenth position in the ranking. Forest and peat fires were emphasized by 11% of respondents. This answer option occupies the eleventh position in the ranking. The bottom of the ranking is occupied by the problems associated with drainage of water reservoirs and poaching. These positions are noted by less than one tenth of respondents; they occupy 12th and 13 positions in the ranking respectively.

It may be concluded that approximately one half of survey respondents among global environmental problems place emphasis on the man-made pollution of air, water, and soil. These answer options occupy the top four positions in the ranking. Approximately one fifth of respondents place emphasis on climatic change, deforestation, extinction of rare species.

Slightly fewer number of respondents place emphasis on the shortcomings in environmental legislation, ecological illiteracy of the population. As a result, problems associated with environmental legislation and environmental education, are pointed out by survey respondents approximately two times less frequently than problems associated with direct infliction of environmental damage.

Survey respondents were invited to express their opinion on environmental problems in settlements of the Samara region and their suburbs. Formulated answers were put forward in the form of a nominal scale of 19 positions. As a result, the following distributions of answers were obtained. Two fifths of respondents pointed out gas-polluted air as a topical environmental problem. One third of respondents place emphasis on pollution of urban leisure areas (beaches, park, forest etc.) with waste and garbage; approximately the same number of respondents point out poor quality of drinking-water. These answer options occupy the top three positions in the ranking of the most severe environmental problems of the Samara region.

The following four positions (from position 4 to position 7) are occupied by answer options chosen by approximately by one fourth of respondents. These include: contamination of rivers and lakes with industrial waste; harmful substances in foods; the use of hazardous substances in construction materials; wide occurrence of animal vectors (rodents, insects, and birds).

14% of respondents place emphasis on the great number of poplars, abandoned fruit trees (fluff, litter, insects, etc.); degradation of soil, contamination with industrial waste. These answer options correspond to positions 8 and 9 in the ranking.

One tenth of respondents place emphasis on problems associated with the reduction of woodlands (deforestation, fires), increased radiation, radiation contamination of the territory, ground-water contamination, their poor quality. Approximately the same number of respondents are concerned about environmental problems associated with municipal improvement of urban environment: stray animals within the city, dust on the streets, increased noise. These answer options occupy positions from 10 to 15 in the ranking.

The bottom of the ranking is occupied by answer options that were chosen by 5% of respondents on each position: death of plants and trees from drought, extinction of certain species of birds, fish, and animals, climatic change, poaching. These answer options occupy positions from 16 to 19 in the ranking.

Respondents were invited to assess the state of the environmental situation in the world, in the country, in the region and in the city on a five-grade scale. Approximately two fifths of respondents are of the opinion that the situation is favorable (at a level of 3 points). 5% of survey respondents are of the opinion that the environmental situation is "excellent", while 13% of respondents are of the opinion that the environmental situation is "good". Consequently, one fifth of survey respondents take a favorable view of the environmental situation. One fourth of respondents are of the opinion that the environmental situation is "unsatisfactory", and 12% of respondents are of the opinion that the environmental situation is "very poor". Approximately 5% of respondents were at a loss to answer. Consequently, the overall assessment of the environmental situation by respondents was approximately equal to 2.5 points. This indicator remains almost the same depending on the range of the territory: on a world-wide scale, at the national level, in the Samara region, in the city.

The opinions of respondents with regard to changes in the environmental situation over the past ten years are reflected in the following distributions of answers. 6% of survey respondents are of the opinion that the environmental situation has significantly improved, 14% of respondents are of the opinion that it has slightly improved, and 22% of respondents are of the opinion that the environmental situation has remained on the same level which is quite good. Consequently, 42% of survey respondents are optimistic in their assessment of changes in the environmental situation, since they believe that it is consistently favorable or changes to the better. Consequently, 58% of survey respondents hold opposite opinions. One fourth of respondents are of the opinion that the environmental situation is in the same unfavorable state. 18% of survey respondents are of the opinion that the environmental situation slightly worsened, and 15% of respondents claim that it significantly worsened. Opinions of survey respondents with regard the assessment of changes in the environmental situation in the world, in Russia, in the Samara region, in the city are almost identical.

Survey respondents were asked the following question: "How do you assess the environmental situation in the Samara region as compared to other regions of Russia?". The following answer options were obtained. More than half of survey respondents (53%) are of the opinion that the environmental situation in the Samara region is no different from the environmental situation in other regions. Approximately one fourth of survey respondents are of the opinion that the situation is worse than in other regions. Only one tenth of respondents are of the opinion that the situation is better than in other regions. 13% of respondents were at a loss to answer.

Respondents were asked the following question: "What types of activity cause the most severe environmental damage in your opinion?". The following distributions were obtained in the analysis of answers. Approximately one half of respondents are of the opinion that the most severe environmental damage is caused due to contamination of rivers and lakes with industrial waste. One third of survey respondents place emphasis on the following problems: degradation of soil, contamination with industrial waste, deforestation, contamination of drinking water sources. These answer options occupy the top four positions in the ranking.

Positions 5 to 7 in the ranking are occupied by answer options chosen by one fourth of respondents. These include: garbage dumps, contamination of rivers, increased radiation, radiation contamination of the territory.

19% of respondents are of the opinion that the use of chemical fertilizers is a source of significant environmental damage. Forest fires were emphasized by 16% of respondents, and poaching was emphasized by 14% of respondents. These answer options occupy positions from 8 to 10 in the ranking of factors. 13% of survey respondents place emphasis on irresponsibility of enterprises as a source of environmental problems. 12% of respondents place emphasis on extermination of certain species of birds, fish, and animals. These positions are at the bottom of the ranking; they occupy positions 11 and 12 respectively.

Survey respondents were invited to express their opinion on what branches of industry cause the most severe damage to natural environment. Three fourths of survey respondents indicated chemical industry as an adverse factor. Two fifths of survey respondents are of the opinion that environmental situation is heavily affected by the mining industry.

One third of respondents point out the negative impact of power generating industry and fuel industry. Approximately one fourth of respondents are of the opinion that transportation industry, machine-building industry and food processing industry have a negative impact on environment. 18% of respondents refer to motor industry as a non-eco-friendly production. 14% of respondents point out the negative impact of consumer goods industry on natural environment, and 12% of respondents point out the negative impact of agricultural industry on natural environment.

5% of survey respondents are of the opinion that information technologies can act as a factor that reduces the ecological well-being of the territory.

Survey respondents were asked the following question: "What is the source of environmental problems in our region in your opinion?".

Approximately one half of respondents are of the opinion that environmental problems of the Samara region are primarily associated with a poor utilization of state-of-the-art waste management technologies, as well as with the activity of industrial enterprises (chemical plants). Two fifths of survey respondents are of the opinion that environmental problems are caused by the lack of efficient control over natural resources utilization, as well as outdated waste treatment facilities.

29% of survey respondents are of the opinion that the reduction of woodlands aggravates environmental problems in our region. Approximately one fifth of survey respondents place emphasis on global climatic change as a factor that impacts the environmental situation in the Samara region. The change of natural landscape due to the construction of hydroelectric station was noted by one fifth of survey respondents. 5% of survey respondents were at a loss to answer.

Comprehensive environmental education allows eliminating inconsistencies between the knowledge, emotional attitude and practical deeds, creating opportunities for the responsible behavior of a person.

Table 87.1 Opinions of respondents on the efficiency of legal regulation methods in the solution of environmental problems

Legal regulation methods	B%
More severe punishment for violations of environmental legislation	56
Strengthening personal responsibility for violation of environmental legislation	44
Increasing the fines for violation of environmental standards	42
Improvement of environmental legislation	38
Expansion of authorities of law enforcement agencies in the field of environmental protection	33
Raising of the efficiency of environmental compliance monitoring	31
Average value of efficiency	41

Respondents were asked a question about the most efficient and effective methods for solving environmental problems. These methods were grouped according to five grounds: legal regulation methods, socioeconomic methods, technical and technological methods, environmental education methods, and methods used to increase the degree of social engagement of the population.

We turn our attention to the description of opinions of survey respondents on the efficiency of these methods. Survey respondents were asked the following question: "What environmental compliance monitoring measures are the most efficient in your opinion?".

More than half of respondents (56%) point out more severe punishment for violations of environmental legislation as an efficient control measure.

Two fifths of respondents point out strengthening personal responsibility for violation of environmental legislation. 38% of respondents claim that it is necessary to improve environmental legislation. One third of respondents believe it necessary to expand authorities of law enforcement agencies in the field of environmental protection, as well as to increase the efficiency of environmental compliance monitoring (Table 87.1).

More than half of respondents are of the opinion that the introduction of sanctions against enterprises polluting the environment is an efficient measure for the improvement of the environmental situation. Approximately one half of respondents are of the opinion that sanctions should be introduced against the population for violation of environmental legislation. Two fifths of survey respondents believe that it is necessary to exercise strict control over waste treatment facilities of plants. Approximately one third of respondents are of the opinion that increasing the fines for environmental damage, as well as assignment of important environmentally protected sites to large enterprises are effective measures of environmental safety (Table 87.2).

One fourth of respondents favor the introduction of amendments to environmental legislation. As little as 7% of survey respondents are of the opinion that control should be introduced over vehicle exhaust gases. An average of 35% of survey respondents are of the opinion that socioeconomic methods are less efficient in changing the environmental situation (Table 87.3).

Table 87.2 Opinions of survey respondents concerning the efficiency of socioeconomic methods for solving environmental problems

Socioeconomic methods	B%
The introduction of sanctions against enterprises polluting the environment	59
The introduction of sanctions against the population for violation of environmental legislation	46
Exercising strict control over waste treatment facilities of plants	41
Increasing the fines for environmental damage	37
Assignment of important environmentally protected sites to large enterprises	33
Control over vehicle exhaust gases	7
Average value of efficiency	37

Table 87.3 Opinions of survey respondents with regard to efficiency of technical and technological methods for solving environmental problems

Technical and technological methods	B%
Implementation of a river purification program	33
Introduction of new waste management technologies	30
Introduction of resource-saving technologies	29
Construction of additional waste treatment facilities	29
Doing away with plastic packaging	22
Landscaping of cities	21
Strengthening incentives for organic production	17
Average value of efficiency	26

Among technical and technological methods of improvement of the environmental situation, respondents place emphasis on the implementation of a river purification program, introduction of new waste management technologies. One third of respondents place emphasis on these answer options. 29% of respondents are of the opinion that the environmental situation can be improved through the introduction of resource-saving technologies, as well as construction of additional waste treatment facilities. Approximately one fourth of respondents point out the need of doing away with plastic packaging, which can be achieved through the use of innovation technologies. 21% of survey respondents place emphasis on landscaping of cities as a measure for improving the ecological well-being of the territory. Strengthening incentives for organic production was pointed out by 17% of respondents. According to respondents, the average value of efficiency of technical and technological methods is 26%.

Many researchers make the solution of environmental problems under current conditions conditional on the formation of the value system which includes environmental imperatives. In this regard, environmental education based on the knowledge of humanities and natural sciences is taking on enormous significance. Environmental education integrates the knowledge about nature as a human environment, about the impact of industrial activity on natural environment. In addition, it forms the

Table 87.4 Opinions of survey respondents with regard to efficiency of educational methods for solving environmental problems

Environmental education methods	B%
Promotion of environmental culture	26
Promotion of healthy lifestyle	25
Introduction of special training courses on environmental protection in educational institutions	25
Social advertising of environmental protection	19
Average value of efficiency	24

Table 87.5 Opinions of survey respondents concerning the efficiency of methods used to increase the degree of social engagement of the population for solving environmental problems

Methods used to increase the degree of social engagement	B%
Creation of governmental organizations engaged in the solution of environmental problems	28
Establishing environmental volunteer movements	21
Increasing the degree of environmental engagement of local self-governing authorities	20
v6.1 Holding special campaigns	16
v7.10 Development of environmental movement	15
Average value of efficiency	20

body of expertise, knowledge, and skills of environmental protection. Environmental education includes the basic components of education in the sciences and the humanities. Environmental education is aimed not at learning scientific knowledge about the interaction of nature and society, but at forming a system of opinions and beliefs focused on conservation of natural environment and conservation of resources (Table 87.4).

26% of survey respondents are of the opinion that there is a need to promote environmental culture (See Table 87.5). One fourth of respondents are of the opinion that promotion of healthy lifestyle would be effective; in addition, they believe it necessary to introduce special training courses on environmental protection in educational institutions. One fifth of respondents place emphasis on the efficiency of social advertising of environmental protection. The average value of efficiency educational methods is 24%.

Among methods used to increase the degree of social engagement for solving environmental problems, 28% of survey respondents place emphasis on the creation of governmental organizations (See Table 87.6). Approximately one fifth of respondents believe it necessary to establish environmental volunteer movements, the same number of respondents are of the opinion that it is necessary to increase the degree of environmental engagement of local self-governing authorities. 16% of survey respondents are of the opinion that it is necessary to hold special environmental campaigns, and 15% of respondents are of the opinion that it is necessary to

Table 87.6 Opinions of survey respondents with regard to efficiency of various methods for solving environmental problems

Methods	Average values (in %)
Legal regulation methods	41
Socioeconomic methods	35
Technical and technological methods	26
Environmental education methods	24
Methods used to increase the degree of social engagement	20

develop environmental movements. The average value of efficiency of methods used to increase the degree of social engagement for solving environmental problems is 20%.

The table presents average distributions of answers by groups of methods that are used for solving environmental problems. Two fifths of respondents point out the efficiency of legal regulation methods in the solution of environmental problems. Approximately one third of respondents place emphasis on the efficiency of socioeconomic methods for solving environmental problems. This group of methods is the most popular among respondents; moreover, economic sanctions in the form of fines for environmental contamination prevail among these methods. Respondents place emphasis on other groups of methods approximately by 10% less frequently. 26% of respondents place emphasis on technical and technological methods associated with the introduction of new river purification technologies, waste management technologies, manufacture of products that reduce environmental damage. One fourth of respondents place emphasis on environmental education methods. They are aimed at promotion of environmental culture of the population, increasing the degree of environmental awareness, the formation of environmental values. One fifth of respondents are of the opinion that it is necessary to increase the degree of social engagement of the population in terms of its involvement in environmental campaigns.

Research results provide clear evidence of the fact that ecologization of the economy is a prerequisite and at the same time an essential constituent of environmentally sustainable regional development. The ecologization of the regional economy implies the change in priorities from intermediate results to long-run objectives of economic activity of the Samara region. Identification of these objectives is associated with the formation of environmental culture of the population in the region, which includes sustainable development values.

Environmental imperatives of regional development put forward the need to increase the importance of economic and social forecasts. The practical importance of ecology primarily consists in the implementation of scientifically grounded control over natural resources utilization. This control lacks efficiency under current conditions. As recommended by the OECD, sustainable development of the region is considered from the perspective of quantitative and qualitative changes. They are focused on the dynamic development of areas of economic activity and improvement of the quality of life the population of the region. The level of sustainable development is stipulated by the interaction of a number of groups of natural and environmental, political, public, economic and social factors.

High rates of economic growth do not provide the optimal balance of interaction of economic activity, solving social and environmental problems. Within the framework of the sustainable development concept, activity of management entities takes into account the interaction of social, environmental and economic factors. Management objectives include conservation of natural environment, maintaining the integrity of the natural complex, as well as improvement of the quality of life.

Mechanisms that can enhance the effectiveness of environmental control must be based on the improvement of environmental legislation. It should serve as a reference point in the regulation of relationship between various economic entities and structures exercising environmental control.

In the present context, there are deep contradictions between environmental standards and ideas of economic efficiency. Environmental standards are focused on efficient natural resources utilization, reduction of environmental damage in the process of transformational activity of society. A conceptual social, environmental and economic model of the Samara region must be supported by the following target scenarios of development and innovative designing of the eco-concept "Clean Region":

1. In the coming decades, it is necessary to introduce and to develop innovative technologies in the regional economy, which will allow reducing natural resources utilization and strengthening incentives for organic production. Innovation technologies must be designed not only to improve economic indicators, but also to observe environmental standards. Development of the economy must include economic incentives for the reduction of resources utilization and environmental contamination.
2. At present, enterprises of oil and gas, energy, and machine-building industries located in the Samara region generate effluents that run into water reservoirs and rivers in this territory. The increase in production in these industries results in increased environmental impact. The assessment of economic activity of enterprises must take into account the possibility of natural resources recovery, quantitative characteristics of the negative impact on the environment. This approach will make it possible to provide an idea of the criteria of ecological balance, as well as to change the content of the criteria for economic growth.
3. Natural resources utilization during production must comply with environmental standards. The location of enterprises in certain territories should take into account resource utilization. To all appearance, the introduction of progressive economic sanctions as compensation for environmental damage can become an efficient measure for solving environmental problems.
4. In order to implement the principles of sustainable development of the region, it is necessary to raise the level of environmental culture of the population, which largely depends on the level and the quality of environmental education in the region.

Acknowledgements The paper has been written with financial support from the Russian Foundation for Basic Research. RFBR grant project No. 17-46-630560 (2018-2019).

Chapter 88
Intelligent Transport System as a Concept of Environmental Innovation in the Transport System

Olga A. Serova and Aleksandr T. Naniev

Abstract The article examines the issues of legal regulation of reducing the negative impact of transport infrastructure on the environment. An important role in solving this problem is the construction of an intelligent transport system. The construction of a single digital platform, which includes ethereal (autonomous) transport, smart roads, Maas, as well as many other elements, is a prerequisite for building eco-mobility. The aim of the study is to identify the main problems of a legal nature in the construction of environmentally friendly transport infrastructure. The scope of the study is that during its conduct, an analysis of Russian and international legal acts governing the legal regulation of transport infrastructure was carried out. The problem of the lack of a legislative framework for the digitalization of roads, the imperfection of the terminological apparatus is being raised. The problem of determining the legal status of unmanned (autonomous) transport is investigated. The methodological basis of the study was a combination of general scientific and private scientific methods, such as: methods of legal modeling and forecasting, analysis, structural-functional, dialectical-materialistic, comparative legal, regulatory and logical, systemic methods. The results of the study are the identification and formulation of certain problems of the development of environmentally friendly transport infrastructure, which are associated with the lack of obligation to conduct an environmental review of design decisions on reconstruction, construction of transport infrastructure facilities; the possibility of recycling materials in the construction, reconstruction of transport infrastructure. Separate proposals are formulated to overcome legal barriers in the development of the intellectual transport system in the Russian Federation.

Keywords Ecology · Transport infrastructure · Intelligent transport system · Unmanned vehicles

---

O. A. Serova (✉)
Pskov State University, Pskov, Russia
e-mail: Olgaserova1974@mail.ru

A. T. Naniev
Immanuel Kant Baltic Federal University, Kaliningrad, Russia
e-mail: aleksandrnaniev@gmail.com

T. Kolmykova and E. V. Kharchenko (eds.), *Digital Future Economic Growth, Social Adaptation, and Technological Perspectives*, Lecture Notes in Networks and Systems 111, https://doi.org/10.1007/978-3-030-39797-5_88

JEL Code K190

88.1 Introduction

The road transport industry is rapidly moving into the field of digital technology. Digitalization of transport is becoming a key factor in achieving a balance between economic development and environmental conservation.

The impact of roads on the environment has not yet been adequately studied. The share of automobile transport in environmental pollution throughout the world varies up to 85%. Vehicles emit more than 280 toxic compounds into the atmosphere, which subsequently accumulate on the pavement and enter the river network through the drainage system, causing environmental damage. (https://www.bbc.com/russian/0901_3.shtml, [1]).

Intensive freight and passenger traffic today has a significant negative impact on the environment, which requires the state and society to immediately implement effective and efficient measures aimed at ensuring a favorable environment.

Automobile and road infrastructure is currently becoming the subject of active technological changes. And it is precisely new technologies that can become one of the means to increase their environmental friendliness and reduce the negative impact on the environment.

88.2 Methodology

During the study, a combination of general scientific and private scientific methods was used, such as: methods of legal modeling and forecasting, analysis, structural-functional, dialectical-materialistic, systemic comparative legal, regulatory and logical methods.

88.3 Results

The negative impact of road transport on the environment by many experts is assessed as significant. This includes emissions of various substances, pollution of soil, water, air. On the other hand, it is impossible to imagine a world without road transport. To reduce the negative impact of transport on the environment, it seems necessary at the regulatory level to consolidate the obligation to make design decisions on the construction and reconstruction of transport infrastructure facilities only after a positive conclusion of environmental review, on compliance with certain criteria: the degree of pollution of atmospheric air, sewage, soil and others.

At present, a mandatory environmental impact assessment is not provided for by applicable law. At the same time, it is roads that are the subject of active development of a new technological reality. The spread of unmanned vehicles, the transition to electric vehicles, etc. requires a radical restructuring of the entire road infrastructure. New construction should have a significant difference. Strict requirements for the environmental safety of the facility must be observed.

A prerequisite for solving environmental problems is to change the requirements of state standards for construction work. During the construction of new roads, the reconstruction of old objects of transport infrastructure, technology for the secondary use of resources should be provided. According to experts in the field of the road sector, the use of used asphalt in an amount of up to 20% has identical qualitative characteristics with new material. The use of such techniques in the construction industry provides significant economic benefits. It also solves the problem of recycling. This reduces the burden on the environment. Extremely relevant is the issue of reducing emissions of harmful chemicals by vehicles into the atmosphere. Currently, 44% of harmful substances are vehicles. To address the identified problems of the state resort to various methods. The German Federal Administrative Court allowed Stuttgart and Düsseldorf to prohibit the movement of cars with diesel engines. The introduction of such environmental requirements is associated with the need to reduce the level of harmful emissions into the atmosphere. (https://www.bbc.com/russian/news-43200971, [2]).

State participation in solving the problems of reducing the negative impact of transport on the environment is a comprehensive plan for the modernization of transport infrastructure. Requires the construction of a single digital platform. Such a platform would be an impetus for the development of eco-mobility.

Intelligent transport system integrates modern digital technologies, carries out automated search and decision-making in the field of transport system and vehicle management. An intelligent transport system is based on the integration of modern information and communication technologies and automation with transport infrastructure, vehicles and users. The purpose of such integration is to ensure mobility of the population, increase indicators of the use of the road network, increase the safety and efficiency of the transport process, increase comfort for drivers and users of vehicles.

Of particular importance for the successful development of an intelligent transport system is the improvement of the existing and the creation of a new road and information and telecommunication infrastructure. Such infrastructure provides unmanned (autonomous) vehicles with the necessary services and information. The environmental friendliness of roads is also increasing. The strategic objective is to radically reduce the level of air pollution in large industrial centers, including reducing by at least 20% the total volume of pollutant emissions into the air in the most polluted cities.

Active participation in the development of the intellectual transport system is taken by the Avtodor State Company, which ensures the competitive advantages of the Russian Federation in creating a digital transport infrastructure. A modern infrastructure for new types of vehicles is being created (http://russianhighways.ru/about/mission/, [3]).

To create an intelligent transport system (a meaningful and complex ecosystem) and ensure its high efficiency, a fundamental study of the basic steps and approaches to the digital transformation of the road industry is necessary. It is necessary to introduce mandatory interim measures, including conceptual and regulatory and technical support; regulatory support; development of platforms and technologies; application of public-private partnership mechanisms. Solving these tasks requires a systematic approach, combining the knowledge, experience and resources of state authorities, business, research and public organizations. An important component of the active development of the intellectual transport system is the careful elaboration of the legal framework and decisions, the need to identify legal restrictions that impede the development of the digital economy and formulate proposals to address them. The development of the digital economy at the state level is associated with solving a number of fundamental problems with the goal of creating a new regulatory environment that fully provides a favorable legal regime for the emergence and development of digital technologies, as well as for the implementation of economic activities related to their use [4].

To date, legislative digitization of roads in the Russian Federation has been developed by only 5% (http://transportrussia.ru/item/4406-tsifrovizatsiya-dorog-eto-tsifrovizatsiya-strany.html, [5]). The need to develop a legislative framework governing the creation of a digital economy in general and the transport industry, in particular, is discussed at all levels. The creation of an advanced and flexible legislative base for the development and widespread use of robotics, artificial intelligence, unmanned vehicles is required in the shortest possible time. Currently, there is an active discussion of the adoption of a federal law on the use of highly automated (unmanned) vehicles. This law would become the basis for legal regulation of relations in the field of unmanned (autonomous) automobile transport, would establish tax and other benefits in order to develop this direction [6].

The analysis of the current legislation on roads, as an element of the transport infrastructure, shows the following. The legal system is not ready to fully assume the function of creating an innovation infrastructure. The current legislation equates the infrastructure of unmanned (autonomous) transport with any other transport infrastructure. The problem of legal regulation of transport infrastructure facilities is not new to legal science. The law in Russia does not connect the signs of the highway with the characteristics of civil rights. This is the result of numerous scientific disputes. Similar problems arise in law enforcement practice. For example, solving the problem of classifying roads as real estate, registration of ownership of these objects.

Smart roads can calculate and analyze road traffic, monitor weather conditions by using an automated weather system, and use autonomous traffic lights. But now there are very few roads that are suitable for driving unmanned (autonomous) vehicles. The composition of such roads includes many objects of both movable and immovable property. The lack of legal norms gives rise to problems of interpretation and application of legislation.

Imperfection of the terminological apparatus is one of the main problems in the regulation of any new technology [7]. Currently, the Russian legal field lacks the concept of "roads designed for unmanned (autonomous) transport" or its equivalent. This impedes the development of an intelligent transport system in Russia. That is why it is necessary to create a new regulatory environment that provides a favorable legal regime for the emergence and development of modern technologies of the intelligent transport system [8]. The introduction at the normative level requires the concept of "roads designed for unmanned (autonomous) transport traffic", a clear legal description of these objects as real estate.

As noted, the lack of mandatory technical requirements for design technologies, road construction and operation for transport infrastructure facilities of unmanned (autonomous) transport is one of the factors that slows down the process of developing an intelligent transport system in Russia. Linking vehicles with transport infrastructure has been identified in the European Union as a priority area for the development and application of specifications and standards (https://eur-lex.europa.eu/legal-content/EN/ALL/?uri=CELEX%3A32010L0040, [9]). The current regulatory legal acts in the field of design, construction and operation of motor roads do not contain special requirements for motor roads intended for the movement of (unmanned) autonomous transport. Taking into account that smart roads, in technical terms, are more complex structures compared to roads, the need to approve the list of mandatory technologies for road design, construction, as well as operation in connection with the introduction of unmanned (autonomous) is obvious cars.

The central element of the intelligent transport system is the development of unmanned (autonomous) transport. Modern developments in the field of unmanned vehicles already allow the independent movement of a car on highways both for commercial and non-commercial purposes (https://nplus1.ru/news/2016/10/26/otto, [10]). In the Russian Federation, the Kazan–Naberezhnye Chelny highway was identified as a pilot project. At this site, preparations are being made for the movement of unmanned vehicles. The project was called "Caravan". It is considered as one of the conditions for obtaining the leading positions of the Russian automotive industry in the world. The state should solve the problems of determining the legal status of unmanned (autonomous) vehicles, autonomous vehicle control systems, ensure the conditions for cybersecurity, protection against hacker and terrorist attacks, and prevent possible hacking of intelligent transport systems. Modern developments in the field of unmanned vehicles allow the independent movement of a car on highways. Therefore, the advent of unmanned (autonomous) transport raises the question of the need to adjust legal regulation. Current legislation does not regulate the integration of artificial intelligence in driving a vehicle. The possibility of participating in the traffic of vehicles under the control of automatic control systems is not provided. At the same time, in March 2016, amendments to the Convention on Road Traffic entered into force (concluded in Vienna on 11/08/1968). Russia is a party to this Convention. The Convention provides for the traffic of autonomous (unmanned) vehicles. But the condition for such an admission is the conformity of the manufacture, installation and the automated system with international legal documents (Article 5, paragraph

5 of the Convention), or the driver can disable the automated system at any time (paragraph 1 of Article 13 of the Convention) [11].

88.4 Conclusions/Recommendations

Unmanned road transport is the future. This thesis is beyond doubt. But in conditions of ecological disaster, it is necessary to achieve a significant reduction in harm to the environment. This can be done in several ways.

First, to tighten the requirements for standards for the construction and reconstruction of road infrastructure. These standards should ensure compliance with environmental safety requirements.

Secondly, it is necessary to guarantee and encourage the application of new technologies to the automotive industry. The introduction of new technologies will minimize environmental damage. The construction of new roads can be a source of testing the most modern technologies, including the use of artificial intelligence.

Thirdly, the development of legislation on autonomous (unmanned) vehicles is required. The new norms should orient all participants in the turnover towards compliance with environmental safety standards.

References

1. Exhaust fumes kill (2019) https://www.bbc.com/russian/0901_3.shtml. Accessed 25 Sept 2019
2. Cities in Germany to deny entry for diesel cars (2018) https://www.bbc.com/russian/news-43200971. Accessed 25 Sept 2019
3. Mission, goals, and functions of the *Avtodor* (2019) http://russianhighways.ru/about/mission/. Accessed 18 Sept 2019
4. Egorova MA (2018) Features of regulatory regulation of the digital economy and the problems of antitrust regulation in digital markets as a means of protecting national interests. Lawyer, no 11, p 8
5. Digitization of roads is digitization of the country (2018) http://transportrussia.ru/item/4406-tsifrovizatsiya-dorog-eto-tsifrovizatsiya-strany.html. Accessed 8 Sept 2019
6. Vaipan VA, Egorova MA (2019) Legal regulation of economic relations in modern conditions of development of the digital economy, vol 2019. Justicinform, Moscow, p 103
7. Neznamov AV, Naumov VB (2018) Strategy for the regulation of robotics and cyberphysical systems. Law, no 2, p 72
8. Vaipan VA (2018) Legal regulation of the digital economy, business law. Appendix "law and business", no 1, p 14
9. Directive 2010/40/EU of the European Parliament and of the Council of 7 July 2010 on the framework for the deployment of Intelligent Transport Systems in the field of road transport and for interfaces with other modes of transport Text with EEA relevance (2010) https://eurlex.europa.eu/legalcontent/EN/ALL/?uri=CELEX%3A32010L0040. Accessed 27 Sept 2019
10. Unmanned truck made commercial flight for the first time (2016) https://nplus1.ru/news/2016/10/26/otto. Accessed 27 Sept 2019
11. Neznamov AV (2018) Unmanned driving rules: on amendments to the Vienna convention on road traffic. Law, no 1, p 177

Chapter 89
Transnational Legal System as a Response to Climate Change

Georgij Pisarev

Abstract The article is concerned with matter of the introduction of a new concept of transnational legal system. The purpose of the study is to study the transformation of the legal system under the influence of global climate change. Environmental threats are global in nature. Legal provisions are based on a national basis. It is necessary to compare these features and form a new paradigm of legal regulation. The methodological basis of the study is determined by a combination of general scientific and private scientific methods: methods of legal forecasting, analysis, structural-functional and dialectical-materialistic methods, normative-logical and systemic methods. The results of the study are related to the substantiation of the following provisions. Law takes on the features of a global system. It should make fundamental changes in the behavior of the world's population, have an impact on the implementation of political and legal reforms. A new concept of transnational law in the field of environmental protection should be developed. The basis of the concept will be the expansion of the rulemaking entities, to ensure the balance of preserving the environment and human economic activity.

Keywords Climate change · Natural resources · Legal regulation · Environmental standards · Transnational law

JEL Classification Codes A12 · K10 · K32

89.1 Introduction

The issue of environmental safety is now becoming a global one. While the main issue before was the protection of the environment, then today we are talking about recognition of a real disaster. The need for urgent measures is determined not by its prevention, but by minimization of the possible consequences.

G. Pisarev (✉)
Russian Presidential Academy of National Economy and Public Administration, Moscow, Russia
e-mail: pisarev.georgi@gmail.com

T. Kolmykova and E. V. Kharchenko (eds.), *Digital Future Economic Growth, Social Adaptation, and Technological Perspectives*, Lecture Notes in Networks and Systems 111, https://doi.org/10.1007/978-3-030-39797-5_89

Climate is the main point of recognition of global ecological changes. Most experts believe that climate change is inevitable and will have significant negative consequences. These consequences will affect all realms of the economy, the level and quality of living standards, the main economic management ways. One question must be answered. Can the law in such a situation retain its regulatory function? The analysis shows that the answer can be positive. However, a major transformation of the existing legal model is required. This transformation will affect not only national legal systems, but also the global structure. This decision is dictated by the global nature of the impending climatological changes.

89.2 Methodology

The study used the traditional general scientific and private scientific methods. The author has studied the scientific concepts of Russian and foreign authors related to the subject of the study. The system analysis method, comparative legal and normative logical method were used in this case. When summing up the research performed, the method of legal modeling was used. This allowed to identify possible options for the transformation of the legal model of regulation of relations in the field of the environment and natural resources management.

89.3 Results

Climate change is seen as a global problem of mankind. The reason is called unsustainable use of natural resources, causing harm to the environment. This topic is becoming the subject of scientific research in many branches of knowledge.

Humanity must strike a balance between increasing the volume of natural resources consumed and preserving its future in the face of the threat of a substantial environmental disaster. These targets were identified in 2015 by the Member States of the United Nations organization as 17 sustainable development goals for the entire world community. Two objectives that are excellent in objectives should be addressed. On the one hand, considerable efforts are required in the field of agriculture, providing humanity with water resources. An increase in the intensity of use of these resources is impossible. All countries are faced with the consequences of such activities. Climate change is associated with the effects of human activity. This defines the second task—to ensure the rational use of natural resources, to restore the ecosystem both on land and in the oceans.

It is possible to solve the indicated problems only by combining all available resources, including through the use of legal mechanisms of influence on the behavior of economic entities.

The law is increasingly taking on the features of a global regulator, since the problems themselves, which require changes in legal approaches, are global in nature and scope. The similar conclusion is drawn concerning both climatology and the science of sustainable development, which have become more international in scope and transdisciplinary by nature. The goals of both new scientific industries and the law coincide in this case—to influence collective actions, the behavior of a significant number of people in public settings. This should contribute to fundamental changes in society. The common features of both the law and the science of climate change coincide in the presence of "a common spatial frames (both global and regional) on the one hand, and the need to combine politico-relevant results and locally specific results on the other" [1].

Ensuring ecological balance is necessary for modern society, as it is associated with improving the quality of life. But no less important is the solution of this problem for the preservation of the future of mankind. Experts All, ordinary citizens are aware of the danger of postponing the solution of the problem of environmental disaster for future generations. But the legal mechanisms in these cases are not sufficiently effective. In Preventive functions, rights are limited to opportunities. Therefore, today it is necessary to strengthen legal regulation in this area. The law also performs other functions. It has an informational and organizational impact on society.

Significant political and legal reform is required to introduce new legal principles. Previously, for Western countries, the tasks of legislation were reduced to ensuring the inviolability of property, the unhindered exercise of civil rights. Currently, the importance of the quality of the natural environment as a condition for ensuring the safety of citizens and preserving their property is becoming increasingly important. This requires new approaches and significant outreach, because the negative impact of climate change in different countries is manifested with different intensities.

These changes pose the task of developing a new concept of transnational law in the field of environmental protection, as a response to protection of interests of the whole society. It is impossible to ensure prosperity in this area only in individual countries. The world has become dependent on the decisions and actions of each of the participants in the global world.

The population of most countries is characterized by a lack of knowledge about environmental management. Also a common sign of many laws and orders is the dispersion in numerous acts of norms related to responsible regulation of natural resources.

The concept of transnational law has an important difference from the traditional model of legal regulation. Previously, all regulatory functions came from authorized state bodies or interstate associations specifically charged with such functions. Now it is proposed to consider the possibility of extending the rules emanating from other entities that are not included in the list of those who form national or international sources of law [2].

For Russian law, the emergence of new participants in the process of regulating relations is also becoming a subject of discussion. Earlier in Russia, there was a rather strict delineation of regulation by branches of law. It has a negative impact on the behavior of participants in civil circulation. It is increasingly proposed to recognize the existence of the social function of civil legal regulation, when the "social "effect" is achieved due to the perception by the law of a changed social and economic reality" [3]. It must be admitted that, reality has really changed. Participants in any economic activity should be aware of the negative consequences that have human impacts on nature.

Entrepreneurial activities in the field of environmental management should be based on the principle of rational use of natural resources. This is "the scientifically based exploitation of natural resources, their involvement in economic turnover without environmental disruption, payment for all types of negative impact on them, as well as compliance with environmental requirements in the process of business operations and other activities" [4]. But such requirements should be addressed in the modern world to all entities whose activities may have negative implications for natural ecosystems or individual resources. For example, private individuals are practically not regulated and do not control over the process of groundwater extraction, there is no systematic control over the discharge of used water from the private sector into small rivers, streams, etc. Uncontrolled garbage dumps can harm both water bodies and animals, as well as people. In the conditions of sharp climatic changes, intensification of negative natural phenomena, this harm can be multiplied. Accordingly, each entity involved in economic relations on the use of natural resources should be recognized as the obligator in relation to any number of unspecified persons, and it should be responsible in natural resources exploitation.

In Western countries, the balance between economic development and environmental protection is significantly achieved through regional legislation. At the same time, experts note that regional development programs should include goals of three main types. This is a rational geographical distribution of industry, resource-ecological sustainable development as well as ensuring normal environmental and economic conditions and the equal living standards [5]. Such a practice should be implemented in our country. In this case, it is required to resolve all conflicts of regional norms with the general provisions of federal legislation.

The State should reserve the exercise of control and oversight functions. All other issues may be left to the discretion of private individuals, including the so-called non-state economic entities. Researchers note that their actions depend on the nature of economic and legal ties. For example, cross-border transactions can be concluded both between independent organizations and between affiliated companies. Solving of the climate change problems and other negative manifestations of environmental disasters is determined by the establishment of environmental standards or codes of conduct. These standards govern the behavior of both corporations and third parties related to them by contractual and organizational links:

- Companies may require compliance with high environmental standards in all their facilities, including those in third world countries, "exporting" the environmental legal requirements of their state;
- Companies can purchase goods and services only from certain companies that comply with established environmental and/or social standards;
- Process of trade or exchange within the company or between trading partners can be carried out taking into account certain regulatory (internal) requirements [2].

Dilling O. and Markus T. highlight as well the transnational industry manufacturing standards, which may include detailed manufacturing requirements. Experts and scientists who are also members of transboundary networks and communities are actively involved in the development of such standards. Researchers have identified three types of transnational scientific networks:

- Public sector-initiated networks (intergovernmental group of experts on climate change, intergovernmental platform on biological diversity and ecosystem services, etc.);
- Mixed organizations, which include both governments and groups of scientists and experts (International Union for Conservation of Nature and Natural Resources);
- Autonomous networks of scientists, engineers and people of other professions (International Council for Science, within the framework of which the Interdisciplinary Committee on Problems of the Environment is convened) [2].

These communities help experts to form a common understanding and common interpretation of environmental problems, to develop criteria and methodological standards for analysis.

In addition to changing the types of legal regulators, enhancement the importance of corporate standards and expert position, there is another possible option for legislation to go beyond the national framework. For example, in Russia, as in other countries, the search for new legal mechanisms to overcome the environmental crisis has led to the identification of a special category of business activity—environmental entrepreneurship [6]. Due to the global scale of environmental problems, the basic principle of conducting a business should be established at the international level—the principle of priority of the environmental arrangement of economic activity.

The introduction of this principle means that society is not able at this stage to stop the deterioration of the environmental situation as a result of economic activity. We presume the fact of causing environmental damage by business entities. It is necessary to balance the interests of business entities and other activities related to the impact on the environment and the interests of society as a whole.

Some types of business activity were initially recognized as environmentally hazardous. They harm the environment and at the legislative level only the goal is indicated—to minimize this damage [7]. The use of water resources may be called as an example [8]. Therefore, it is necessary to create comprehensive programs to achieve a balance between the implementation of business activities of entities (for example, the electric power complex) and measures aimed at minimization of the damage caused. The State can provide support (tax benefits, subsidies, etc.) to those business entities who invest significantly in scientific research on the conservation and recovery of water resources.

Many governments are aware that a change in public policy is needed. The rapid change of new technologies and environmental degradation have an impact. Time for manoeuvres—legislative or economic-is very short. Not so long ago, water was not considered a key economic and political resource. The lack of drinking water in the world is becoming more and more considerable, and it is impossible to create this resource artificially. There is no complex legal regulation of relations on water supply of the population, use of water resources for the industry, for life support systems of the state. Only some authors point to the need for a systematic approach to ensuring the rational use of water resources by all subjects on the basis of a single water management complex [9]. Before that, such an approach was lost in a market economy. And this conclusion applies to almost all areas of economic activity. The practical solution of a particular problem can be carried out only within the integrated, multidisciplinary approach.

Moreover, climatology, being an interdisciplinary field, is closely related to politics. Especially in the face of an uncertain future and the threat of climate change, which affects the decisions of States and State Unions [10]. Accordingly, many decisions in this area will also be authorized by the central authority.

In Russia, new requirements to economic activity through the prism of environmental responsibility of each subject can be introduced only at the level of Federal legislation. But branches of law of the Russian legislation have rather rigid borders, including in connection with various purposes. Civil law does not contain legal mechanisms to protect the environment. Environmental law, on the contrary, is aimed at compliance with the rules for the rational use of natural resources, but does not focus on modern requirements for economic activity. The market economy establishes the priority of private law and legal mechanisms of civil legislation. A new paradigm in the development of civil law is needed. Compliance with environmental standards of economic activity should receive the status of a principle of law. This condition is required for any branch of law. When deciding on the opening of new industries, the interaction of the state with citizens, it is impossible to conceal the fact that it is impossible to completely exclude the negative impact on the environment in business. Sometimes the interests of entrepreneurs and the population may coincide, as they are associated with the use of the same natural resources. For example, in fishing. Extraction of biological water resources is carried out by both entrepreneurs and citizens. The differences are related only to the purpose of using these resources. It is necessary to develop the promotion of ecological awareness, to discuss actively in various expert groups the problems of aquatic biological resources production. We should actively support the transition to innovative ways for restoration of these resources.

Scientists exactly point to the multiple meaning of the interpretation of the "rational use of natural resources" concept. Its use as a legal category cannot be based solely on doctrinal understanding. The law should provide general criteria for the "rational or irrational use of natural resources" (Petrova 2016). But separating the principles of conducting a business from the environmental conservation objectives seems to be an irrational task. Therefore, there are a lot of offers to adopt general laws, for example, the Environmental Code of the Russian Federation [11], where such issues were resolved at the intersectoral level. We believe that this solution is non-effective. The expansion of regional regulation is more rational, as it takes into account the specifics of the natural zone and the geographical location of objects.

89.4 Conclusions/Recommendations

The study allowed us to draw the following conclusions:

One of the most important conditions for the development of modern society is the problem of maintenance of favorable environment for humans. Environmental problems have become global, which leads to the need to change the prevailing approaches to their resolution in terms of legal regulation.

The global environmental crisis has highlighted the need to develop new economic concepts, scientific approaches to the organization of modern production and industry. It is impossible to completely exclude the negative impact of business and economic activities on nature and the environment. The company must establish strict environmental requirements for production, economic activity of the population, not related to profit. Citizens tend to indirectly harm the environment in the course of consuming natural resources. The exception is poaching. But they should also be subject to stricter rules regarding the environmental nature of their activities.

Previously, there was an installation on the individual nature of resource consumption and the actions of market participants in their own interests. Today we need active interaction of the subjects. Individual actions of each should be correlated with General requirements. Possible threats, including those of a climate nature, are linked to the pooling of efforts.

Ensuring ecological balance, rational use of natural resources, the development of new technologies to prevent climate threats, minimize the negative effects of climate change is the basis of modern political and socio-economic concepts of development of the state and society. It is impossible to achieve this goal only through the efforts of one State. The climate change impacts on the planet affect everyone, which predetermines the emergence of a new format of legal regulation, called transnationalization of law. Such regulation includes both the development of national legislation, and the active modification of international law. But it also implies significant activity of other legal entities in the development of environmental standards of activity, alternative mechanisms of conflicts and disputes resolution.

References

1. Meehan K, Klenk NL, Memdez F (2018) The geopolitics of climate knowledge mobilization: transdisciplinary research at the science-policy interface(s) in the Americas. Sci Technol Hum Values, 1–26
2. Dilling O, Markus T (2018) The transnationalisation of environmental law. J Environ Law (Oxford) 30(2):185–189
3. Serova OA (2013) The social function of civil law. Auth Law 3:32–33
4. Budnikova Yu E (2014) The concept and signs of entrepreneurship associated with the use of natural resources. Publ Inst State Law Russ Acad Sci 2:99–100
5. Bezgubova AA (2015) Foreign experience in natural resources management. Bulletin of Tula State University. Econ Legal Sci 4(1):505
6. Serova OA (2018) Environmental entrepreneurship subjects as participants of innovation activity. Legal Bull Samara Univ 4(3):64–69

7. Pisarev GA (2019) Principles of entrepreneurial law as a basis for the development of environmental-oriented business. Law Econ 5:33–38
8. Ismailov Sh M (2011) Rational use of water and energy resources of the region: legal aspects. Energy Law 1:30–34. Petrova TV (2016) Rational use of natural resources: concept, principle, authority, direction of state environmental policy. Environ Law 3:29–34
9. Romanova OA (2014) Water management activities as an object of scientific legal research. J Russ Law 5:158–164
10. Baker Z (2017) Climate science: science-state struggles and the formation of climate science in the US from the 1930s to 1960s. Soc Stud Sci 47(6):861–887
11. Volkov GA (2018) On the methodology of natural resources law branches. Environ Law 1:11–12

Chapter 90
Institutionalization of Hybrid Manufacturing as the Basis of New Digital Industrialization of the Russian Federation: Prospects and Contradictions

Anna V. Shkalenko, Daniil P. Frolov and Ekaterina A. Fadeeva

Abstract This article analyses production hybridization in the leading countries following the policy of neo-industrialization. The author allocates the features of state regulation system from the standpoint of institutional economy. Transformation models of production are described taking into account transition to digital economy. The main trends of industry digitalization are highlighted. The article provides a description of key market competences to form factories of the future. It also highlights the main barriers for Russia to shift to "Industry 4.0". Institutional contradictions within the new (digital) industrialization are outlined for Russia. Thus, the specific measures are proposed to eliminate institutional contradictions in the period of digital transformation of society.

Keywords Digitalization · Digital industrialization · Digital twin · Digital double · Failure prediction · Modelling

JEL Code O330

90.1 Introduction

Ensuring competitive, sustainable and safe development of the Russian Federation emphasizes the implementation of neo-industrial transformations in the Russian economy through the widespread use of hybrid manufacturing as the basis of new industrialization, which causes fundamental changes in the structure of factors in economic growth, state institutions, industrial structure and economic organization. Progress in the field of information and communication technologies is a driving

A. V. Shkalenko (✉) · D. P. Frolov
Volgograd State University, Volgograd, Russia
e-mail: oponir@volsu.ru; lavra.ne@mail.ru

D. P. Frolov · E. A. Fadeeva
Volgograd State Technical University, Volgograd, Russia

T. Kolmykova and E. V. Kharchenko (eds.), *Digital Future Economic Growth, Social Adaptation, and Technological Perspectives*, Lecture Notes in Networks and Systems 111, https://doi.org/10.1007/978-3-030-39797-5_90

force in the economic development of the 21st century in an ever-expanding world. The speed of innovation is growing exponentially, and new technologies, such as the Internet of things, artificial intelligence and robotics, can automate routine tasks and optimize the costs of production processes, reducing and taking it to a whole new level. Infocommunication technologies (ICT) are increasingly used in the industry. Indeed, in recent years, initiatives such as Industry 4.0, Factory of the Future, and Industrial Internet of Things, put forward by governments and industry leaders, allude to ICT as the future of manufacturing and management. To comply with this trend, the institutional environment for the formation of digital economy must consider all these changes in the formation of the country's new industrial policy. The vector of innovative development of the Russian economy, along with the government rethinking industrial policy, creates new impulses for the formation of complex systems of consolidation relations between diverse economic entities in the development of the domestic industry. Creation of digital twins and hybrid manufacturing reduces the total and individual transaction costs for participants in production chains in which they exchange information and make timely management decisions improving organization and creating a consolidated way for the most efficient production process.

90.2 Methodology

This study uses various methods of factorial, subject-object analysis as a part of a systematic approach to forming the entrepreneurial ecosystem of the university, including the innovative methodology of post-institutional analysis based on multi-paradigmatic and interdisciplinary synthesis.

90.3 Research and Development

Global market changes in the field of technology and high-tech innovation, increased competition and the emergence of highly complex scientific and technical dilemmas at the junction of several industries require the state to ensure the formation of the effective institutional environment that meets the new requirements for the development of hybrid manufacturing in the course of industry digitalization at a much faster pace development of production technologies, extremely short development cycles, low prices and high quality production. The demand for multi-functional engineering products stimulates the process of permanent modifications of the products themselves (products, constructions, various technical systems) and production technologies—tools to optimize the processes of development, production and management [1].

Among the national initiatives for the development of advanced production technologies, the most famous was the German Industrie 4.0, which was first discussed in 2011. Industrie 4.0 is one of the 10 Projects of the Future envisioned in the German High-Tech Strategy 2020 Action Plan, adopted in 2012.[1]

The keynote of Industrie 4.0 is the transition from the so-called embedded systems to cyber-physical systems. Embedded systems are central control units embedded in various objects that they control. Cyber-physical systems include a set of new enabling technologies connecting the virtual and physical worlds and interacting smart objects with each other through the Internet/networks and data. Industrie 4.0 documents also refer to this transition from embedded systems to cyber-physical systems as the fourth industrial revolution.

From an economic and political standpoint, the main goal of the Industrie 4.0 project is to increase the competitiveness of German industry in conditions when: (1) Germany cannot compete in labour costs with developing countries and even with some developed countries, for example, the USA; (2) Germany is strong in the development of industrial equipment, embedded control systems, machine-to-machine interaction systems, sensors but lags in the development of software, where American companies are the absolute leader, but simultaneously it is software development and software platforms that are the driver of digitalization.

The main means of achieving the goal is to unite the efforts of the state: the Ministry of Education and Research and the Ministry of Economics and Technology of the Federal Republic of Germany, and the scientific community: Fraunhofer, National Academy of Technical Sciences, and Private Business: Industrie 4.0 platform, selected companies, including Bosch and SAP, as well as the BITKOM IT Company Association, the VDMA Association of German Engineering and the ZVEI Electronics Manufacturers Association.

China is also developing programs to increase the competitiveness of its industry. In the spring of 2015, the Chinese government announced the "Made in China [3]" program, under which the priorities are[2]:

1. stimulation of innovative activity in industry;
2. harmonization of technology and industry;
3. the industrial base strengthening;
4. development of Chinese brands;
5. introduction of green technologies into production;
6. breakthrough in 10 key industry sectors;
7. industrial sector restructuring;
8. development of service-oriented industry and production services;
9. Internationalization of production.

The program includes a clear list of planned indicators to be achieved in the short and medium-term, measures to protect intellectual property, encourage companies to form their technical standards and encourage innovation in traditional industries reflected in previously adopted plans for the development of technologies and promising industrial sectors: The 15-year plan for the development of indigenous

[1]The New High-Tech Strategy [2].

[2]Made in China [3].

innovations, research and technological development in 2006, fully fulfilled and implemented in China itself and the 12th Five Year Plan on the Development of the Strategic Emerging Industries in 2010. But unlike the previous focused programs in the "[3]" strategy, the list of breakthrough sectors of the industry has been expanded to 10 both promising and traditional ones:

- information technology, to increase the share of Chinese servers in the telecommunications and financial sectors to 75% by 2020 and up to 90% by 2025;
- CNC and robotics, to increase the share of Chinese robots in the market to 50% by 2020 and up to 70% or more by 2025;
- aerospace technologies, to increase the turnover of the aerospace industry to 100 billion yuan ($15.6 billion) by 2020 and up to 2 trillion yuan by 2025;
- construction of high-tech ships and marine equipment, to occupy 40% of the global market by 2020 and 50% by 2025;
- railway machinery and equipment;
- energy saving and alternative fuel vehicles, to increase the share of Chinese cars on the market to 70% by 2020 (1 million cars) and 80% by 2025 (3 million cars);
- power engineering;
- materials science;
- medicine and medical devices;
- Agricultural machinery.

The Japanese concept of Society 5.0, initiated by the Keidanren Japanese Federation of Large Businesses, is aimed at solving Japan's problems such as population decline, ageing society, and childless working citizens. Society 5.0 is essentially a continuation of Industrie 4.0 and involves not only digitalization of industry and the economy but also the digital transformation of society as a whole.

The problems of advanced production technologies are actively discussed and considered as a priority area of industrial development in Russia at the level of federal authorities [4]. Today, this translates into the three main instruments of state support for such technologies: the National Technological Initiative, the Strategy for Scientific and Technological Development of the Russian Federation, and the Digital Economy program. All these programs are implemented in the IV industrial revolution and similar directions, Advanced Manufacturing Partnership in the USA, Industry 4.0 in Germany and others.

The transition to a digital economy requires transformation of the current traditional model of organizing production and development of the related infrastructure. Introduction of higher-level automation and control systems with elements of artificial intelligence requires appropriate changes in the current business models, organizational and managerial methods. Introduction of new digital systems needs an accurate picture of the resources available for setting goals, developing a strategy for further development and determining quantitative indicators. Digital production involves a continuous exchange of information between its components, the state of which is monitored by collecting data through various sensors. In the digital economy, the enterprise infrastructure builds on the industrial Internet of things for the interaction between the individual elements of the manufacturing system. The active use of cloud technologies makes possible storing and processing large amounts of information. Currently, so-called digital twins simulate various production processes, check their operability and effectiveness in a virtual environment, identify and eliminate errors with minimal financial and time resources. Digital twins combine modelling technology and the industrial Internet of things and are essential elements in organizing innovative production. Their application ensures the functioning of production processes at all stages of the product life cycle, from development to operation. For example, selecting the most appropriate equipment scenarios can reduce the likelihood of failures. When deciding to start manufacturing new products, the use of digital twins considers all the factors affecting quality and cost characteristics. The ability of technical systems to independently make decisions about optimal actions and control their behavior is ensured through artificial intelligence technologies. In real time, autonomous robots collect and process information to plan their next actions. Thus, an important feature of digital production is the introduction of automation technologies and industrial robotic systems. The main trend in automation remains the expansion of the functionality of robotic systems, improving their ability to perform tasks both autonomously and in the course of interaction with employees of the enterprise. New approaches to the organization of production make it possible to recoup investments and achieve high indicators of production efficiency, labour safety and the speed of developing the new production. However, the introduction of robotic systems in production includes significant risks that are characteristic of the innovation sphere as a whole, primarily in terms of information security. These risks can be reduced by applying modelling technologies that optimize all the most important functional characteristics of such systems and ensure their reliability.

Today, there is a struggle around the world to compete for efficient production and industry related to an initiative such as Industry 4.0 and the introduction of the latest developments in ICT, such as cyber-physical systems, Internet of things and cloud computing. To solve this problem, scientists and specialists in the field of institutional modelling of the production environment and effective management of the enterprise and the economic development of the country as a whole suggest using hybrid manufacturing mechanisms and digital twins.

Digital Twin is a technology to simplify and improve the work of physical prototypes of products or production equipment, entire systems and individual processes [5].[3] This is a physical and digital hybrid model that is created for specific purposes of the manufacturing company, for example, to indicate failures, reduce maintenance costs and prevent unplanned outages. The term Digital Twins is currently used primarily in industrial production. The concept of Digital Twins was introduced in 2003 by Professor Michael Grieves in its article Digital Twins: Manufacturing. Excellence through Virtual. Factory Replication (Michael, 2–14).

Another key principle of Industry 4.0 is to increase knowledge about the process gained through monitoring [6]. This knowledge can be applied, for example, to process control [7] optimize maintenance activities [8], and even to create digital twins. The combination of models and process monitoring will be useful not only at

[3]Komrakov and Sukhorukov [5].

the development and design stages but mainly at the production stage to check the continuity of production, to identify wear and tear without stopping production, or to predict component failure and other failures.

The combination of these new features will increase overall production efficiency [9] by impacting the various stages of the product creation life cycle: (1) accelerate the growth process over time, reducing the operation of equipment and/or adjusting the control process; (2) optimize performance, for example, forecasting and avoiding quality problems; (3) reduce failures due to faster customization of the new process and proactive service strategy based on root cause analysis and forecasting and preventing policies; (4) extend the life of the production system by reusing and restoring equipment and components.

Modelling tools are currently a key complement to European experience in improving production efficiency to increase competitiveness. In fact, according to Industry 4.0 [10], modelling plays a key role in managing the growing complexity of technology systems. A holistic engineering approach allows covering various technical disciplines and providing end-to-end design throughout the value chain. Also, more and more important life-cycle concepts, such as energy consumption, component life and wear are not always present in the calculations of machine and component manufacturers.

The concept of the digital twin embraces this holistic approach for modelling the production and treatment processes. In addition, digital twin provides interaction with the real performance of the machine [10]. The opportunity to combine real data with simulation models provides a new range of applications with clear benefits:

Digital twins can be evaluated with real data by supplying controlled input data, as in a real representation of a machine. Thus, models can serve to improve their performance [11].

Digital models are suitable for the so-called virtual commissioning [6, 12], which consists in using the digital twins of the physical system to configure the controller even before the physical system is ready for this. The inputs and outputs of the controller are connected to a digital twin as if they were connected to a real twin. This feature shortens the overall manufacturing phase of the product.

Digital twins can also improve health-monitoring capabilities. Digital models can determine the nominal conditions of the systems under study and, therefore, to improve the detection of abnormal characteristics [13].

Factory of the Future[4] is the type of business process system, the method of combining business processes, which has the following characteristics:

- creation of digital platforms, peculiar ecosystems of advanced digital technologies. Based on predictive analytics and big data, the platform approach brings together geographically distributed participants in the design and production processes, increases the level of flexibility and customization, considering the requirements of consumers;
- development of a system of digital models of both newly designed products and production processes. Digital models must have a high level of adequacy to real

[4]Mnogourovnevaya struktura fabrik budushchego [14].

objects and real processes, the convergence of the material and digital worlds that generate synergistic effects;
- digitalization of the entire product life cycle, from the concept idea, design, production, operation, maintenance and up to disposal. The later the changes are made, the greater their cost is. Therefore the centre of gravity shifts towards the design processes along with the characteristics of global competitiveness or high consumer requirements.

At the formative stage of Factory of the Future, new key competencies appear, for example:

- quick customization of the response to market or customer requests;
- system engineering, which allows at every moment to keep in sight the entire system, all its interacting components;
- the formation of a multi-level matrix of targets and limitations as the basis of a new design that significantly reduces risks, the volume of full-scale tests and the volume of work associated with the refinement of products based on tests;
- development and validation of mathematical models with a high level of adequacy to real objects and real processes, the so-called smart models;
- change management throughout the life cycle;
- digital certification based on thousands of virtual tests of both individual components and the entire system.

Under the program of the National Technological Initiative, the government of the Russian Federation is considering the transformation of the modern production cycle through Digital Factory, Smart Factory, and Virtual Factory. Digital factory is system of complex technological solutions that provide for the shortest time to design and manufacture globally competitive products of a new generation from the research and planning stage, when the basic principles of the product are laid down, and ending with the creation of Digital Mock-Up, Smart Digital Twin, prototype or small series: paperless production, digital turnover. A digital factory implies the presence of smart models of products, machines, structures, assemblies, devices, installations, etc. based on the new digital design and modelling paradigm Smart Digital Twin—((Simulation and Optimization) Smart Big Data)—Driven Advanced (Design and Manufacturing).

- Smart Factory is a system of integrated technological solutions that ensure the shortest possible time the production of globally competitive products of a new generation from procurement to finished products, the hallmarks of which are a high level of automation and robotics, eliminating the human factor and the associated errors leading to loss of quality. It is unmanned production. As an input product of Smart Factory, as a rule, is the results of the work of Digital Factory. Smart Factory usually implies the equipment for production: numerically controlled machines, industrial robots, etc., as well as automated process control systems, Industrial Control System and operational process control systems at the workshop level, Manufacturing Execution System.

- Virtual Factory is a system of integrated technological solutions that ensure the shortest time possible designing and manufacturing of globally competitive new generation products by combining Digital and/or Smart factory into a distributed network. A virtual factory implies the Enterprise Application Systems developing and using as a single object a virtual model of all organizational, technological, logistic and other processes at the global supply chain level: supply => production => distribution and logistics => sales => after-sales service, and/or at the level of distributed production assets.

Already at the turn of 2025–2035, it is expected a number of technological breakthroughs reflecting in-depth technological changes that will lead to the transformation of traditional industrial production and the transition to a new model of socio-economic development based on advanced production technologies and an exponential increase in their use: rapid growth in capacity and performance factors at an equally rapid drop in value, including:

- sensory revolution, the mass transition to the use of digital sensors, detectors, actuators and control systems;
- management based on mathematical models and digital data of all technological objects and processes;
- registration of legally significant actions and settlements on the Internet, based on distributed registries, including using blockchain technology;
- distribution of metamaterials and supermaterials with programmable and variable functional properties;
- transition to clean, low and post-carbon energy, which involves the formation of a new large energy storage market, the use of flexible intelligent networks, as well as distributed generation;
- distribution of non-traditional methods of processing materials: additive production, atomic-exact production, etc., of robotic and automated systems;
- scaling application for analytics and process control of augmented, virtual reality, artificial intelligence;
- connection of technological objects and a person to the Internet of the whole/industrial Internet, to neural networks;
- widespread use of nature-friendly technologies: bionics, bioengineering, synthetic biology, biotechnological production, etc.

90.4 Conclusion

One of the key barriers to Russia's transition to Industry 4.0 is the low level of digitization of the economy and the insufficient cost of enterprises to innovate. The main efforts of the state, business, scientific and academic sectors should be directed towards filling these gaps. The digital development model assumes the total digitalization of the economy and industry smart models and digital twins of products, equipment, and production.

It is important to note that, unlike the USA and the EU, whose industry development programs are aimed at supporting interdisciplinary, cross-industry technologies that have a significant impact on the transformation of various manufacturing industries; Chinese program highlights both the technological and sectoral areas of development. It is in these areas that China has the potential to become one of the leaders in global production. Given the accelerated pace of the emergence and development of new technologies, even countries that are indisputable leaders in terms of scientific and innovative potential are faced with increasingly active competition in the international arena. The task of productively incorporating into a new technological revolution to implement structural manoeuvre in the economy and social system is Russia's priority until 2035.

The technological revolution will have a profound and strong impact on the sectors of the Russian economy forming its traditional basic sector, which accounts for the bulk of GDP: oil, gas and other mineral extraction, chemical industry, metallurgy, mechanical engineering and other manufacturing industries, trade, financial sector, agriculture, etc.

The particularities of the current economic situation, the ongoing innovation and technology policy, as well as the presented indicators, reveal that Russia was delayed with the active inclusion in the new technological revolution. The country needs to make additional efforts not only to compensate for the gap but also to be proactive in many priority areas possessing the potential and formed backlogs.

For Russia, the technological revolution means not only the formation of new sectors but also profound technological and organizational changes in traditional industries. In the next 10–15 years, traditional sectors of the Russian economy will have to resolve the following institutional contradictions in the development of new digital industrialization in Russia:

1. an increase in structural imbalances in the national economy, while simultaneously exhausting the potential of the export-raw model of economic development;
2. maintenance of the high level of import dependence of certain sectors of the Russian economy;
3. lack of own resources and technologies necessary for the implementation of transitions in the framework of the new technological revolution.
4. low ability to convert high levels of human capital and research results into technological innovations.
5. ineffective system of management of scientific and technological development.
6. regulatory lag of changes related to technological modernization, digitalization and platformization of key industries and critical infrastructures;
7. maintenance of the competitive position in the national and global markets;
8. increase of productivity and accelerated modernization, including renewal of fixed assets;
9. adaptation to the rapid spread of breakthrough manufacturing technologies;

10. deep transformation of business models of economic agents that should become suppliers of not only physical goods and works but also digital products; in their business models, work with information and data begins to take a special place;
11. low level of activity of industrial companies in the implementation of innovative-activities, the orientation of these activities mainly on the domestic market.

90.5 Results

Thus, given the current structure of the Russian economy, the current level of development of the institutional environment of the national innovation system, the set of identified institutional contradictions, it can be concluded that the country's transition to a qualitatively new economic, industrial and technological paradigm will depend on the coordinated implementation of measures in a number of areas. These include technological modernization of industries traditional for the Russian economy, development of new high-tech sectors and ensuring the country's entry into new markets, restarting the R&D management system, reorganizing development institutions, measures to achieve a new quality of human capital and adequate labour market organization, as well as total/end-to-end digitalization of the real economy and the services sector.

Acknowledgements The work was supported by Russian Science Foundation (project №18-78-10075).

References

1. Borovkov AI, Ryabov YuA (2017) Prospective directions for the development of advanced manufacturing technologies in Russia. In: Information XVII April international scientific conference on economic and social development. Publishing House of the Higher School of Economics Moscow, p 381
2. The New High-Tech Strategy (2014) Innovations for Germany. Federal Ministry of Education and Research of Germany (BMBF)—2014. https://rio.jrc.ec.europa.eu/en/library/new-high-tech-strategy-innovations-germany
3. Made in China 2025 (2015) The state council of China—2015. https://www.mta.org.uk/system/files/resource/downloads/Made%20in%20China%202025%20Booklet%20One.pdf
4. Holicza P, Baimakova K, Lukina E (2016) Russian socio-economic development: the present situation and future directives 27:68. https://doi.org/10.1134/S1075700716010044
5. Komrakov AV, Sukhorukov AI (2017) The concept of a digital twin in the management of the life cycle of industrial facilities. Netw. Sci J Sci Idea, 3(3) (Electronic resource). www.nauch-idea.ru
6. Lee CG, Park SC (2014) Survey on the virtual commissioning of manufacturing systems. J Comput Des Eng 1(3):213–222
7. Denkena B, Fischer R, Euhus D, Neff T (2014) Simulation-based process monitoring for single item production without machine external sensors. In: 2nd international conference on

system-integrated intelligence: challenges for product and production engineering, procedia technology, vol 15, pp 341–348

8. Mahmood K, Otto T, Shevtshenko E, Karaulova T (2016) Performance evaluation by using overall equipment effectiveness (OEE) an analyzing tool. In: International conference on innovative technologies 2016, Prague
9. Alzaga A, Konde E, Bravo I, Arana R, Prado A, Yurre C, Monnin M, Medina-Oliva G (2014) New technologies to optimize reliability, operation and maintenance in the use of machine tools. In: Euro maintenance conference, Helsinki, Finland, 5–8 May 2014
10. Kagermann H, Helbig J, Hellinger A, Wahlster W (2013) Recommendations for implementing the strategic initiative INDUSTRIE 4.0: securing the future of German manufacturing industry. Final Report of the Industrie 4.0 Working Group. Forschungsunion
11. Atluru S, Huang S, Snyder J (2011) A smart machine supervisory system framework. Int J Adv Manuf Technol. 58. https://doi.org/10.1007/s00170-011-3405-4
12. Hoffmann P, Schumann R, Maksoud TMA, Premier GC Virtual commissioning of manufacturing systems—a review and new approaches for simplification. In: Proceedings of the 24th European conference on modelling and simulation (ECMS 2010), Kuala Lumpur, Malaysia, pp 175–181
13. Magargle R, Johnson L, Mandloi P, Davoudabadi P, Kesarkar O, Krishnaswamy S, Batteh J, Pitchaikani A (2017). A simulation-based digital twin for model-driven health monitoring and predictive maintenance of an automotive braking system. In: Proceedings of the 12th international Modelica conference, Prague, Czech Republic, 15–17 May 2017, no 132, pp 35–46, 4 July 2017
14. Mnogourovnevaya struktura fabrik budushchego (2019). https://technet-nti.ru/article/fabriki-buducshego

Conclusion: Digital Future—Consequences for Economic Growth and for Social and Technological Progress

A systemic view at the digital future of humanity showed that has a multi-aspect character. The obvious advantages of the leading technologies oppose their imperfections and risks. The conditions of aggressive competition in the global digital environment generate new markets, which are dominated by AI and robots. From companies and region to whole countries and unions of countries—all of them use the strategies of development, but face—during their implementation—support and opposition, possibilities and threats.

In the future 5–10 years, digital technologies will enter a new level of development, with a peak of economic growth that will be based on them. A new generation of digital technologies—robototronics and AI—could start a second wave of economic growth or increase the possible economic crisis in the case of realization of the security risks. Though most economic systems in the modern global economy have programs of digital development, the imbalance will be preserved—and perhaps increased—in the future.

Disproportions will be a usual phenomenon at the level of the regional and global economies. At the same time, their treatment could be various, depending on the level of strategy of development of the economic system taking into account its specifics. Economies should not necessarily orient at the external environment and strive to conform to international criteria of evaluation—in particular, their competitiveness. It is necessary to refuse from the idea of leveling of disproportions and to observe the internal landmarks of economic growth and development.

In the social aspect, it is necessary to expect contradictions—though they could be expressed differently and could have different scale. The best option of events is limited difficulties of digital adaptation of certain social categories. A less favorable option is also possible—unemployment and crisis of the labor market with mass inability of the population to adapt to the digital reality. That's why it is necessary to pay attention to the issues of education and development of the educational market.

In the technological aspect, it is possible to predict transition to the next technological mode, at which robots and AI will be replaced by their more perfect analogs,

T. Kolmykova and E. V. Kharchenko (eds.), *Digital Future Economic Growth, Social Adaptation, and Technological Perspectives*, Lecture Notes in Networks and Systems 111, https://doi.org/10.1007/978-3-030-39797-5

and "smart" companies will receive the network organization. However, the consequences of further digital modernization directly depend on attention to technological risks and successfulness of their management. The cookbook approach to risk management is incorrect, as high flexibility of economic subjects and systems is required.

In the ecological aspect, the situation is even tenser. Despite the actively developed and implemented "green" innovations, the problem of climate change won't probably be fully solved. The perspectives of overcoming the global ecological crisis lie in the legal field and envisage creation of new, more effective international institutes for stimulating the formation of the circular and "green" economy.

Thus, with joint efforts of the representatives of various sciences of the socio-humanitarian block, a multi-disciplinary view of the digital future is formed in this volume. The offered scientific and practical recommendations allow all interested parties to adapt to the future changes and to find their own niche in the digital world. At the same time, these inventions led to new issues—overcoming low desire of economic agents and systems to accept their position in the global economy and to adapt to new realia. This leaves a field for further research of the digital future.

The manufacturer's authorised representative in the EU is Springer Nature Customer Service Centre GmbH, Europaplatz 3, 69115 Heidelberg, Germany. If you have any concerns regarding our products, please contact ProductSafety@springernature.com

Printed and bound by CPI Group (UK) Ltd, Croydon, CR0 4YY
22/09/2026
02221053-0001